The Electronic Communications Code and Property Law

Life now without access to electronic telecommunications would be regarded as highly unsatisfactory by most of the UK population. Such ready access would not have been achieved without methodical and ultimately enforceable means of access to the land on which to install the infrastructure necessary to support the development of an electronic communications network. Successive governments have made such access a priority, regarding it as a principle that no person should unreasonably be denied access to an electronic communications network or electronic communications services. The enactment of the Telecommunications Act 1984 and its revision by the Communications Act in 2003 have played their role in the provision of an extensive electronic infrastructure in the UK, while their reshaping by means of the Digital Economy Act 2017 will continue that process. Throughout that process, a little publicised series of struggles has taken place between telecommunications operators and landowners, as they seek to interpret the Electronic Communications Code by which their rights and obligations have been regulated.

This book describes the problems that accompanied the Old Code (which will continue to regulate existing installations and agreements); and the intended solutions under the New Code. The eminent team of authors explain the background, provisions and operation of the old code and the new one, providing practical and jargon-free guidance throughout. It is sure to become *the* reference on this topic and is intended as a guide for telecommunications operators, landowners, and of course for their advisers in the legal and surveying professions.

All members of Falcon Chambers, comprising nine Queen's Counsel and 30 junior barristers, specialise in property law and allied topics, including the various incarnations of the Electronic Communications Code. Members of Falcon Chambers, including all the authors of this new work, have for many years lectured and written widely on the code, and have appeared (acting for both operators and landowners) in many of the few reported cases on the subject of the interface between property law and the code, including for example: *Geo Networks Ltd v The Bridgewater Canal Co. Ltd* (2010); *Geo Networks Ltd v The Bridgewater Canal Co. Ltd* (2011); *Crest Nicholson (Operations) Ltd v Arqiva Services Ltd* (2015); *Brophy v Vodafone Ltd* (2017).

The Electronic Communications Code and Property Law

Practice and Procedure

Falcon Chambers

LONDON AND NEW YORK

First published 2019
by Routledge
2 Park Square, Milton Park, Abingdon, Oxon OX14 4RN

and by Routledge
711 Third Avenue, New York, NY 10017

Routledge is an imprint of the Taylor & Francis Group, an informa business

British Library Cataloguing-in-Publication Data
A catalogue record for this book is available from the British Library

Library of Congress Cataloging-in-Publication Data
Names: Falcon Chambers, author.
Title: The Electronic Communications Code and property law : practice and procedure / Falcon Chambers.
Description: Abingdon, Oxon ; New York, NY : Routledge, 2018.
Identifiers: LCCN 2018014013 | ISBN 9781138543126 (hardback : alk. paper) | ISBN 9781351007283 (ebook)
Subjects: LCSH: Telecommunication–Law and legislation–Great Britain. | Right of way–Great Britain. | Great Britain. Office of Communications.
Classification: LCC KD2880 .F35 2018 | DDC 343.4109/94–dc23
LC record available at https://lccn.loc.gov/2018014013

ISBN: 978-1-138-54312-6 (hbk)
ISBN: 978-1-351-00728-3 (ebk)

Typeset in Times New Roman
by Out of House Publishing
Printed and bound by CPI Group (UK) Ltd, Croydon, CR0 4YY

Contents

Preface

At the Memorial Service for his friend Douglas Adams, author of *The Hitchhiker's Guide to the Galaxy*, our great friend and sadly missed colleague Jonathan Brock QC, late of these Chambers, gave a eulogy in which he recalled an incident from the early 1980s. Douglas had just invested in a mobile telephone – so called because it could be transported with some difficulty in a large rucksack. Newly arrived from the USA, it could neither transmit nor receive calls in London. The solution, Douglas decided, was to go as far west as possible, in order better to be able to pick up mobile phone signals from across the Atlantic. And so it was that Jonny accompanied Douglas in his MG to the furthest tip of Pembrokeshire that they could find. The rucksack was placed on the beach in an auspicious position; buttons were pressed; expectations were high. And expectations were dashed. Jonny recounted how Douglas, all patience exhausted, jumped up and down on the useless equipment. Then the two of them sat on a sand dune and watched the incoming tide cover the remains, while the sun set.

We have come a long way in a third of a century. OFCOM calculates that, by 2014, 93 per cent of adults personally owned or used a mobile phone in the UK, while the number of UK mobile subscriptions reached 88.4 million by the end of 2013. Life now without access to electronic telecommunications would be regarded as highly unsatisfactory by most of the population.

Little of this would have been achieved without methodical and ultimately enforceable means of access to the land on which to install the infrastructure necessary to support the development of an electronic communications network. Successive governments have made such access a priority, regarding it as a principle that no person should unreasonably be denied access to an electronic communications network or electronic communications services.

At the latest count, there were more than 23,000 telecommunications sites around the UK, occupied by a variety of infrastructure providers and mobile operators, with masts, equipment cabinets and other telecommunications paraphernalia.

Those sites are occupied under many different types of arrangement: licences, wayleaves, leases and easements. Some such arrangements enjoy protection under Part II of the Landlord and Tenant Act 1954. They may also enjoy protection under the Electronic Communications Code devised under the Telecommunications Act 1984, and revised by the Communications Act 2003. This mix of contractual and statutory rights has bedevilled the telecommunications market for as long as it has been in existence. It has not helped that the drafting of the 2003 Code is, according to the author of our Foreword, Lewison J (writing in his judgment in *Geo Networks Ltd v The Bridgewater Canal Co. Ltd* (2010)) 'one of the least coherent and thought-through pieces of legislation on the statute book'.

Following a long series of consultations, the Digital Economy Act 2017 has now introduced an amended Code, designed to deal with the shortcomings of the previous version of the Code. The New Code has its admirers and its detractors. Many will say that it is rather better than the Old Code. Others will say that an imperfect job has been done, and will prefer to stick with what they know. What is clear is that both Codes will coexist for years to come, complicating an already complex mix of contractual, property and statutory concepts.

As with so much of our domestic legislation in recent years, the European Union has played a large part in formulating our approach to electronic communications. As we write, that process continues, with a new EU code that is expected to be introduced before the end of 2018, with the result that it may become part of our national law (requiring changes to the New Code) before Brexit.

This book is primarily designed for property and telecommunications lawyers, surveyors, regulators and other professionals, but is also aimed at the landowner who has to grapple with the two Code regimes. As far as we are aware, this is the first proper textbook to deal with the interface between property law and telecommunications tenure.

Falcon Chambers has a long history of advising all parties in the telecommunications industry: landowners, operators, infrastructure providers and utility undertakers. This book does not set out to take sides: it attempts to provide an informed but neutral view of the issues that arise. Inevitably, given the number of authors and the dearth of authority, views may be expressed in the text with which not every author agrees.

We express our thanks to the assistance we have received from our external authors:

- Luke Maidens of Shulmans LLP, Solicitors, who has contributed a chapter on drafting;
- Gareth Hale and Kenny McLaren, formerly of Maclay Murray & Spens LLP, and now of Dentons UKMEA LLP, who have contributed our chapter on practice and procedure in Scotland.

And finally, we are grateful to Sir Kim Lewison for his Foreword – and for contributing so much to the authors' understanding, particularly of the Old Code.

Contributors Falcon Chambers

Guy Fetherstonhaugh QC

Jonathan Karas QC

Wayne Clark

Barry Denyer-Green

Stephanie Tozer

Oliver Radley-Gardner

Toby Boncey

Consultant Author on Scottish Law and Procedure: Gareth Hale, Partner, Dentons UKMEA LLP

Consultant Author on Drafting: Luke Maidens, Solicitor, Shulmans LLP

Foreword

By the Right Honourable Lord Justice Lewison

'Mr Watson – come here – I want to see you.' These were the first words spoken on the telephone in a conversation between Alexander Graham Bell and his assistant in March 1876. Now we carry in our pockets more computing power than it took to put the first man on the moon, complete with GPS tracking. The demand for data delivery at speed has grown exponentially in the last decade alone. The statutory provisions relating to the installation, operation and removal of the equipment required was, until recently, based on precedents first enacted in the days of Queen Victoria. By the early part of this century it was clear that they were no longer fit for purpose. After two years' work the Law Commission recommended wholesale reform, which has now found its way into the new Electronic Communications Code contained in the Digital Economy Act 2017.

This comprehensive and wide-ranging book, written by a team of experts, covers much more than a detailed explanation of both the old code and the new code. It traces matters historical, regulatory and European. It covers planning, rating, compulsory acquisition, competition and land registration as well as dealing with knotty points of interpretation of the code itself. It explains key concepts in property law, and considers how they fit into the structure of the new code. Long-standing problems under the old code, such as its interface with statutory security of tenure given to business tenants, have been addressed by the new code; and the authors consider whether those problems have truly been solved.

For anyone involved with the telecoms sector, whether as lawyer, valuer, operator or landowner, I have no doubt that this book will be an extremely valuable source of reference: perhaps the only one they will need.

Kim Lewison
Royal Courts of Justice
Strand
London WC2A 2LL

Table of Cases

Table of Statutes

Table of Statutory Instruments

Glossary

The purpose of this Glossary is to define the main terms used in this Book. It is not intended to replicate the annotated Code, which may be found in Part VI.

2003 Act: the Communications Act 2003.

2017 Act (or 'DEA 2017'): the Digital Economy Act 2017.

Code agreement: In its non-technical sense, an agreement, whether a lease or a licence, and whether consensual or imposed by the court, conferring one or more of the code rights provided for by paragraph 3 of the New Code. Technically a 'code agreement' is one to which Part 5 of the New Code applies. The meaning should be clear from the context or is made clear in the text.

Code Right: one or more of the rights provided for by paragraph 3 of the New Code.

Law Com No 336 (or simply Law Com 336): the 2013 report on the Electronic Communications Code by the Law Commission.

OFCOM: the Office of Communications.

Old Code: the Telecommunications Code as contained within Schedule 2 to the Telecommunications Act 1984, as amended by the 2003 Act, also known as the Electronic Communications Code 2003 or simply as "the Telecommunications Code".

New Code: the Electronic Communications Code 2017 as enacted by the 2017 Act and contained in Schedule 3A to the 2003 Act.

Paragraph: references to a paragraph number are either to the correspondingly numbered paragraph of the Old Code or the New Code; or to a paragraph of the text of this book. The meaning should be clear from the context.

Section: references to a section number are either to the correspondingly numbered section of a statute (but usually shortened to s.); or to a collection of paragraphs of the text of this book. The meaning should again be clear from the context.

Part I

Introduction

1 Introduction

1.1 The importance of electronic communications

1.1.1

In the introduction to the Electronic Communications Code in their 2013 report on that subject (Law Com No. 336), the Law Commission stated that 'It would be difficult to overstate the importance of electronic communications, both for business and individuals', and cite in support the following statistics from OFCOM *Communications Market Report* (July 2012):

> In 2011, there were 9.4 million business fixed lines in the UK, and businesses accounted for 1.7 million broadband subscriptions and 10.4 million mobile connections, including 1.4 million for mobile broadband. Of UK adults as a whole, 92% own a mobile phone and 39% own a smartphone, and it is reported that the number of households where at least one person uses a mobile phone to access data services is increasing. 76% of UK households have fixed or mobile broadband; overall, the number of fixed broadband connections in the UK is now over 20 million, and to that can be added 5.1 million active mobile broadband subscribers.

1.1.2

Only five years later, the OFCOM *Communications Market Report* (3 August 2017[1]) is testament to the huge growth in the market. As the authors say in their sector overview:

> By June 2016, 44% of all fixed broadband connections were able to receive actual download speeds of 30Mbit/s or more, up from 38% a year previously. Nearly two-thirds of mobile subscriptions were enabled for 4G, up from 46% in 2015. Consumers are also using these networks more – average data use per fixed line residential broadband connection increased by 36% year on year to 132GB in June 2016, and average data use per mobile connection increased by 44% to 1.3GB.
>
> Most households have both fixed broadband and a smartphone, and consumers are moving seamlessly between fixed and mobile connections. Our mobile-app-based research shows that around two-thirds of data connections made by our

1 www.ofcom.org.uk/__data/assets/pdf_file/0015/105441/uk-telecoms-networks.pdf

panel of Android smartphone users are via a Wi-Fi network, with the remaining third via a mobile network.

UK telecoms revenues grew by 0.4% in real terms (i.e. adjusted for inflation) in 2016 to £35.6bn. This has been driven by the growing take-up of superfast broadband services.

1.1.3

Speed of transmission is critical to enhanced use of digital technology, and this is something to which the Government has reiterated its commitment. In summer 2017, it launched a consultation on the design of a Universal Service Obligation (USO) to deliver high speed broadband across the UK. On 28 March 2018, the Government laid before Parliament the Electronic Communications (Universal Service) (Broadband) Order 2018 (2018 SI No. 445), which came into force on 23 April 2018. Subject to certain exceptions, this provides for universal high speed broadband, giving everyone in the UK access to speeds of at least 10 Mbps by 2020. This is the speed that OFCOM says is needed to meet the requirements of an average family. In a press release on 20 December 2017, the Culture Secretary stated that a regulatory approach was necessary in order to ensure that high speed broadband became a reality for everyone in the UK, regardless of where they live or work.

1.1.4

OFCOM's implementation of this USO is expected to take two years from the date when the secondary legislation was laid before Parliament, meeting the Government's commitment of giving everyone access to high speed broadband by 2020.

1.1.5

In order to meet this ambitious target, the Government and the operators upon which it relies will have recourse to the New Code. Whether the New Code will be fit for purpose is a matter of intense industry interest. As section 1.2 below explains, the Old Code had been the subject of withering criticism since its introduction, primarily because the mechanism it provides had been designed with fixed line technology in mind, and the attempt to update it in 2003 to accommodate mobile technology was largely unworkable. However, the New Code introduced by the Digital Economy Act 2017 (the 2017 Act)[2] is modelled a little too closely upon the provisions of the Old Code, as section 1.3 below shows, and the opportunity provided by the all-party support in Parliament for a radically different and effective new mechanism was passed up.

1.2 Shortcomings of the Old Code

1.2.1

For the purposes of this book, the chief shortcomings of the Old Code stemmed from the unsatisfactory interface between three types of law applicable to the siting of electronic communications apparatus: real property law; the statutory protection afforded

2 Please see Appendix B for the text of the relevant parts of the 2017 Act, and please see the Explanatory Notes for a helpful commentary upon it.

to operators by the Old Code itself; and the protection afforded to business tenants (as operators commonly are) by Part II of the Landlord and Tenant Act 1954. Chapter 14 examines the conflict between these various regimes in more detail. In practice, because with rare exceptions nobody was sure how the Old Code would be interpreted if tested in court, landowners and operators maintained an uneasy modus operandi, with neither being willing to push disagreements to the door of the court.

1.2.2

A particular point of tension concerned arrangements into which operators were keen to enter as the market consolidated this century, with operators willing to share sites and apparatus. This desired flexibility did not sit well with traditional landlord and tenant relationships. Landowners were not prejudiced by sharing – but wished to take their own slice of the savings which operators achieved.

1.2.3

From the point of view of operators, the Old Code was also seen to be too generous in its operation to landowners, whose otherwise worthless and miscellaneous bits of property (rooftops, woods and corners of fields) were turned to productive account by operators.

1.3 The New Code: an opportunity missed?

1.3.1

The problems briefly aired in section 1.2 were all too apparent when first the Law Commission and then the Government came to consult in the 2010s on the shape and content of a new Electronic Communications Code. The problems were rendered more difficult to resolve by the diametrically opposed views of the two main industry groupings – the landowners and the operators.

1.3.2

Ultimately, the New Code achieves much that is good, but at the expense of a series of unsatisfactory compromises that were perhaps inevitable. The chief shortcomings are examined in detail in Part III of this book, and are briefly summarised in this section, along with the main benefits. It should be borne in mind, however, that a shortcoming to a landowner is quite often perceived by an operator to be a benefit – and vice versa.

1.3.3

First and foremost, the confusing blend of property and statutory regimes that characterised the Old Code has been – albeit with some wrinkles – resolved in the New Code, by the device of compartmentalising agreements between landowners and operators either as business tenancies or as code agreements – but not both. As against that, there remains some doubt as to the meaning of the test of 'primary purpose'; and how termination under the 1954 Act in the case of a non-primary purpose agreement fits with the Part 6 requirements for removal, all of which is discussed in detail in Chapter 14.

1.3.4

Secondly, operators are provided with the rights both to share and to upgrade their apparatus, within certain constraints. This will be regretted by landowners, because of the loss of their ability to extract increased consideration, but welcomed by operators. However, operators who share apparatus with those who have code rights will not themselves acquire code rights. It will be interesting to see to what extent this has a dampening effect upon the new-found sharing freedom.

1.3.5

Thirdly, the financial provisions underpinning the calculation of code consideration have changed from their generous Old Code basis to one based upon the market value of the relevant person's agreement to confer or be bound by the code right, disregarding the fact that the site has a telecommunications use (see Chapter 30). Current industry commentary suggests that this new basis of assessing consideration is likely to result in reduction, possibly a substantial reduction, in the fee or rent required to be paid by operators for the code rights conferred.

1.3.6

Fourthly, the New Code has introduced a minimum 18-month notice requirement for termination under Part 5. Developers may well consider that this is too long to contemplate acquiring land subject to code rights, given that there remains the need to effect removal in accordance with Part 6 thereafter.

1.3.7

Fifthly, wholesale infrastructure providers can now number themselves amongst the ranks of operators who can benefit from code rights. There is however some lack of clarity associated with the division between land, on the one hand, and apparatus, on the other, which may take some time (and the expertise of the Lands Chamber) to work through.

1.3.8

Sixthly, the Lands Chamber, and cognate tribunals in Scotland and Northern Ireland, with their unique blend of concentrated specialist legal knowledge and surveying expertise, have now become the tribunals to which New Code disputes must primarily be referred, replacing the more diffuse, overworked, unspecialised and unfavoured county courts, which should result in problems arising being worked out and adjudicated upon rather than endured.

1.3.9

Overall, operators may feel that they have achieved much with their role in the emergence of the New Code; while landowners correspondingly are likely to be reluctant to enter voluntarily into New Code agreements.

1.3.10

The Old Code had reigned for 15 years at the point that the New Code was brought into force (and rather longer, if its pre-2003 incarnation under the 1984 Act is taken into account). That is a very long time in the digital world. The advent of the fifth generation of mobile technology in the 2020s may well require an entirely different regulatory landscape, rather than a bodged version of the New Code – and it may therefore be that this Code has a relatively limited shelf life, and that the opportunity for legislative reform to improve the product may not therefore arise. Landowners and operators are likely to have to live with the New Code in its current form for the time being. This book suggests how that may best be done.

1.4 Layout of this book

1.4.1

This book is laid out in a number of parts, dealing with the topics about which a practitioner, operator or landowner needs or may like to be informed when dealing with the interface between property law and telecommunications.

1.4.2

Part I deals with the history of the telecommunications legislation, before providing an outline of the Old and New Codes, and adding some reflections on the European dimension.

1.4.3

Part II, 'Telecommunications Code 2003', describes the Electronic Communications Code, devised in 1984 and revised in 2003, which will continue to apply to existing arrangements. It lists the deficiencies with the Old Code, and sets out the case for reform which has led to the enactment of the New Code.

1.4.4

Part III, 'Electronic Communications Code 2017', works through the New Code, dealing with the respects in which it represents an improvement over the Old Code.

1.4.5

Part IV, 'Matters common to both codes', considers such matters as notices, dispute resolution procedure and code avoidance; it also examines the interaction between the code and other legislative regimes – planning, compulsory purchase, rating, land registration and competition law; it adds a chapter on the position in Scotland; and it scrutinises the role of OFCOM.

1.4.6

Part V, 'Drafting', looks at the main drafting points that those concerned with the creation of code rights will wish to consider.

1.4.7

The Appendices contain the following material that will be useful to practitioners:

- Appendix A sets out the Old Code.
- Appendix B sets out the New Code.
- Appendix C contains the principal Regulations relevant to the New Code.
- Appendix D sets out the OFCOM Code of Practice.
- Appendix E contains the OFCOM Template Notices.
- Appendix F contains the OFCOM Template Terms.

2 Legislative history

2.1 Introduction

2.1.1

Telecommunications legislation in the UK has a long pedigree, with antecedents in the 19th century. The history with which this book is concerned is however confined to the legislative attempts which the Government has made in order to facilitate the provision of a telecommunications network, ensuring in particular that telecommunications operators can acquire rights in respect of land belonging to or occupied by others.

2.2 Early history

2.2.1

The British Government has long been keen to control the means by which its subjects communicate. In 1657, the Second Protectorate Parliament passed legislation that created one monopoly Post Office for the whole territory of the Commonwealth, under which the first Postmaster General was appointed in 1661. At the restoration of the monarchy in 1660, the General Post Office was officially established by Charles II.

2.2.2

When telegraphy rose to the fore as the first means of instant long-distance communication in the early 19th century, the Government lost little time in seizing control of this new medium, passing the first Telegraph Act in 1863.[1] This allowed it to take over telegraph companies and their operations, with the GPO, now a government department, having the monopoly over telegraph traffic.

2.2.3

In 1876, Alexander Graham Bell patented a device for converting air vibrations caused by sound into electrical current passed through wires to a receiver, at which point the process would be reversed, allowing the originating sound to be heard. Queen Victoria was an early enthusiast, following a demonstration at Osborne House in 1878. As telephony

1 www.legislation.gov.uk/ukpga/1863/112/pdfs/ukpga_18630112_en.pdf

caught on, the Government became involved. In *Attorney General v Edison Telephone Co of London Ltd*,[2] Stephen J and Pollock B held that the Telegraph Act 1869 covered 'communications by any wire and apparatus connected therewith used for telegraphic communication, or by any other apparatus for transmitting messages or other communications by means of electric signals'. The result was that the early telephone companies were subject to the licensing and monopoly provisions of the Telegraph Acts, which effectively allowed the GPO to take over their businesses. Thus, in 1901, began telecommunications regulation in the UK, with the licensing of private telephone companies. The GPO acquired the telephone trunk lines and most main routes, and by 1913 it owned the whole private sector telephone service in the UK save one municipal service in Hull.

2.2.4

Under the Post Office Act 1969, the GPO ceased to be a government department and became established as a public corporation – the Post Office Corporation – answerable to the Secretary of State for Industry, with two divisions: Post and Telecommunications. The new corporation had the exclusive right to run telecommunications systems, and to authorise others so to do.

2.3 The Telecommunications Act 1981

2.3.1

Following the election of a Conservative Government in 1979, Keith Joseph (Secretary of State for Industry between 1979 and 1981, subsequently Lord Joseph) announced to the House of Commons on 21 July 1980 the Government's intention of liberalising the telecommunications market. With a view to the eventual privatisation of the network, the Telecommunications Act 1981 split the Post Office Corporation into two separate corporations: the Post Office, and Telecommunications (which had been renamed British Telecommunications, trading as British Telecom, the previous year).

2.3.2

The 1981 Act also gave the Secretary of State for Industry and British Telecom power to license alternative telephone network operators, in order to introduce competition into the UK telecommunications industry, and thus break the monopoly over the telecommunications network. In 1982, a licence was granted to Cable & Wireless to run a public telecommunications network through its subsidiary, Mercury Communications Ltd as a prospective competitor to British Telecom. Monopoly thenceforth became duopoly.

2.4 The Telecommunications Act 1984

2.4.1

The Telecommunications Act 1984 was enacted in the wake of the privatisation of the public corporation British Telecom and as part of the move to a more competitive

2 [1880–81] LR 6 QBD 244.

market. It provided for the transfer of the business of British Telecom to British Telecommunications plc, a public corporation in which initially more than 50 per cent of the shares were sold to the public. The Government's remaining shares were sold successively in 1991 and 1993, while its Special Share, which allowed it to block a takeover of the company, was relinquished in 1997. The Act established Oftel, the Office of Telecommunications, a non-ministerial government department with close historic ties to the Department of Trade and Industry, as the regulator of the industry, charged with promoting competition in the telecommunications industry (with an eye on advances in digital technology) and protecting the rights of consumers.

2.4.2

The 1984 Act also abolished British Telecom's exclusive right to run telecommunications systems, and established a framework to safeguard the workings of competition. This meant that British Telecom finally lost its monopoly right to run telecommunications systems, which it had effectively retained under the 1981 Act, despite the Secretary of State's licensing powers. It now required a licence in the same way as any other telecommunications operator. The principal licence granted to British Telecom laid down strict and extensive conditions affecting the range of its activities. Operating a telecommunications system in the UK without a licence from Oftel was a criminal offence.

2.4.3

Section 10 of the 1984 Act provided for the telecommunications code set out in Schedule 2 to have effect, essentially to regulate landline telephone provision.

2.5 Developments in the 1980s and 1990s

2.5.1

In the two decades that followed the Telecommunications Act 1984, there were rapid advances in all communications technologies, particularly internet and mobile telephony, the demand for which increased exponentially.

2.5.2

On 5 March 1991, the Government issued the White Paper *Competition and Choice: Telecommunications Policy for the 1990s*. This effectively prepared the ground for the ending of the duopoly involving British Telecommunications plc and Mercury that had existed since privatisation. It also brought in a more open and fairer policy which allowed customers to acquire telecommunications services from competing providers using a variety of technologies. Independent retail companies thenceforth were permitted to bulk-buy telecommunications capacity, and sell it in packages to business and domestic users.

2.5.3

The international duopoly was ended in 1996, when other operators became free to offer international services within the UK.

2.5.4

In 1998, the Wireless Telegraphy Act allowed radio spectrum to be sold, and paved the way for subsequent auctions of spectrum for 3G mobile licences.

2.5.5

In response to the rapidly developing market, the European Council of Ministers issued five framework directives to encourage the market in competitive telecommunications services across Europe.[3] These effectively ended monopoly provision in all EU countries, requiring member states to adopt a common system of regulation for the telecommunications industries, the relaxation of restrictions on the supply of services, and common rules on interconnection, universal service and consumer protection.

2.5.6

In 2003, new European directives removed the need for telecommunication service providers to be licensed by national regulators, and generally simplified regulation.

2.6 The Communications Act 2003

2.6.1

This Act, which came into force on 25 July 2003, was prompted by developments in technology and in the regulatory regime, in the light of European intervention by way of the series of Directives referred to in paragraph 2.5.5 above. The Act abolished the licensing regime previously operated by Oftel, and replaced it with a scheme by which anyone could supply voice and data services, subject to compliance with the General Conditions of Entitlement for the time being. It introduced a new industry regulator, the Office of Communications (OFCOM), to replace Oftel.

2.6.2

Section 106 of the Act, together with Schedule 3, amended the Telecommunications Code (which it renamed the Electronic Communications Code) in Schedule 2 of the Telecommunications Act 1984 to reflect changes in technology (in particular, to support the infrastructure networks which supply and enable broadband, mobile internet, landlines and cable television). The revised Code is examined in Part II of this book.

3 Directive 2002/21/EC on a Common Regulatory Framework for Electronic Communications Networks and Services; Directive 2002/20/EC on the Authorisation of Electronic Communications Networks and Services; Directive 2002/19/EC on Access to, and Interconnection of, Electronic Communications Networks and Associated Facilities; Directive 2002/22/EC on Universal Service and Users' Rights Relating to Electronic Communications Networks and Services; and Directive 2002/58/EC of 12 July 2002 on Privacy and Electronic Communications.

2.7 The Law Commission Report 2013

2.7.1

In a letter dated 16 May 2011, the Secretary of State for Culture, Olympics, Media and Sport said:

> The Government is embarking on a wide-ranging review of the regulatory regime for the UK communications sector, to ensure the regulatory framework in place is fit for the digital age. The [Government's] ambition is to establish UK communications and media markets as amongst the most dynamic and successful in the world, with the review process culminating in a new communications framework by 2015, to support the sector for the next 10 years and beyond.

2.7.2

As part of its review, the Department for Culture, Media and Sport asked the Law Commission to conduct an independent review of the 2003 Code. This project was included in the Law Commission's Eleventh Programme of Law Reform, and began in September 2011.

2.7.3

On 28 June 2012, the Law Commission published a consultation paper (Law Commission Consultation Paper No. 205), seeking views on the reform of the Electronic Communications Code, with the consultation period running until 28 October 2012. The Commission published its report, No. 336, on 28 February 2013.[4] This endorsed the need for reform for three main reasons.

2.7.4

First, the Code was described as 'confusing and unduly complicated', with difficulty discerning the relationship of the Code with other elements of the law, such as the Land Registration Act 2002 and Part II of the Landlord and Tenant Act 1954.

2.7.5

Secondly, the approach of the Code was said to be outdated. The original draft was based on several 19th- and early 20th-century statutes (see section 2.2 above) dealing with telephone wayleaves. It was clear that the drafters of the Code did not contemplate that code operators might require a leasehold estate in land in order to site their apparatus on it. As a result, although attempts had been made to update the legislation for modern technology, important points had been left unclear.

4 The Law Commission, The Electronic Communications Code, 27 February 2013: Law Com Report No. 336. www.lawcom.gov.uk/wp-content/uploads/2015/03/lc336_electronic_communications_code.pdf.

2.7.6

Thirdly, there was concern that the 2003 Code was making the extension of electronic communications more difficult. The Code seeks to regulate the effects of agreements to confer specified rights, and to back this up with a system for compulsion where agreements cannot be reached. Yet it lacks clarity on several important matters, such as who is bound by rights conferred on code operators, how to assess the level of payments for the grant of rights, and how the termination of those rights is to be enforced. This, together with the absence of efficient dispute resolution, considerably hampers its usefulness to both Code Operators and landowners. In addition, it was not clear to the Law Commission that it struck the right balance between those parties.

2.7.7

Law Com Report No. 336 put forward 15 pages of recommendations to reform the 2003 Code. These included:

- providing a clearer definition of the market value that landowners can charge for the use of their land, giving them greater confidence in negotiating agreements and giving providers a better idea of what their network is likely to cost;
- clarifying the conditions under which a landowner can be ordered to give a network provider access to his or her land, bringing more certainty to both landowners and providers and helping them to reach agreement more easily;
- resolving the inconsistencies between the current Code and other legislation;
- clarifying the rights of landowners to remove network equipment from land;
- specifying limited rights for operators to upgrade and share their network equipment; and
- improving the procedure for resolving disputes under the Code.

2.7.8

As agreed with the Department for Culture, Media and Sport at the inception of the project, the Law Commission's report did not include a draft Bill; it would be for Government to draft and implement a revised Code. The Commission recommended that any attempt at reforming the Code should start again with 'a clean sheet of paper', rather than by way of amendment of the 2003 Code.

2.8 The Growth and Infrastructure Act 2013

2.8.1

Temporary amendments were made to the Code in 2013 as part of the Growth and Infrastructure Act. These amendments were not primarily intended to reform the Code, but were designed to promote economic growth by speeding up the deployment of broadband infrastructure.

2.8.2

These amendments were promulgated by secondary legislation, as follows:

(a) Regulations[5] to amend the Code to allow 'a more permissive regime' for installation of above ground fixed line broadband electronic communications apparatus. This secondary legislation also removed the requirement for prior approval by planning authorities for broadband cabinets and poles in protected areas.
(b) In order for these changes to take effect, complementary secondary legislation[6] was also required to amend the Town and Country Planning (General Permitted Development) Order 1995.

2.8.3

These amendments to the Code were given a sunset clause of five years, and expired in April 2018.

2.9 Legislative reform

2.9.1

In early 2015, the Department for Culture, Media and Sport indicated that they were intending to reform the Code in line with the Law Commission's recommendations. On 6 January 2015 the Minister of State for Digital Industries, Ed Vaizey MP, told the House:[7]

> DCMS plans to reform the Electronic Communications Code in line with the recommendations set out by the Law Commission in its report of February 2013. In this report, the Law Commission concluded that the existing market based approach to valuation was on the whole, functioning, but recommended some small modifications to improve the valuation regime. Government accepts this recommendation and will implement this reform. However, Government will also take a Power in the legislation, to allow the Secretary of State to make further changes to the wayleave valuation regime, if necessary following consultation, at a later date following full consultation. Government intends to lay amendments to the Infrastructure Bill.

2.9.2

On 13 January 2015, the Coalition Government put forward amendments to the Infrastructure Bill then going through Parliament (subsequently the Infrastructure Act

5 The Electronic Communications Code (Conditions and Restrictions) (Amendment) Regulations 2013 (SI 1403).

6 Town and Country Planning (General Permitted Development) (Amendment) (England) Order 2013 (SI 1101).

7 Telecommunications: Written question – 218396.

2015) to include a new Electronics Communications Code to replace the existing Code, following many of the recommendations of the Law Commission's review.

2.9.3

The Minister of State at the Department for Transport told the Public Bill Committee on 15 January 2015[8] that the amendments were designed to reform and update the Code by implementing the recommendations of the Law Commission:

> The amendments we are making to the code are the reasonable balance sought by the Opposition between the interests of mobile phone operators and the landowners on whose land they must put their masts to get the better coverage that I believe is necessary. …
>
> The current lack of clarity, given that the code has not been substantively amended since 1984, has led to countless complaints from all kinds of sources and a demand for action. To that end, the Law Commission looked at matters, and its February 2013 report suggested a wholesale rewrite of the provisions designed to improve procedures and the clarity of the drafting.

2.9.4

The Opposition was in agreement that reform of the Code was necessary. However, they opposed the Government's amendment because they felt it was an 'ineffective and rushed reform that will increase ambiguity, confusion and litigation'.[9] They argued that the Government's amendment failed to provide clarity on the issue of facilitating upgrades and site sharing – 'one of the key reasons for reform in the first place'.[10] Reference was also made in Committee to an email received from the Wireless Infrastructure Group,[11] which argued that the drafting of the revised Code left unnecessary uncertainty, particularly given:

> that an unintended consequence of these complex amendments on a complex sector could be to inadvertently interfere in the vital commercial relationships between Code Operators. The purpose of the Code is to govern the rights of Code Operators in their dealings with landowners yet the lack of clarity in the text could lead to confusion in the relationship between Code Operators including between wholesalers and network operators. This could discourage wholesale infrastructure providers from bringing further investment to the sector.

2.9.5

Representatives of landowners also voiced concerns regarding the valuation provisions in the draft code. The Central Association of Agricultural Valuers and the National Farmers' Union of England and Wales both submitted written evidence to the Public

8 PBC 15 January 2015, c.347.
9 PBC 15 January 2015 c.362.
10 PBC 15 January 2015, c.360.
11 PBC 15 January 2015 c.366.

Bill Committee in response to the tabling of the Government's amendment. They expressed concern about paragraph 23(4)(b) of the draft new code, which required the parties (or their valuers) to disregard the statutory limitations which the revised code would apply to agreements in permitting assignment or the sharing or upgrading of equipment. According to the CAAV:[12]

> The effect of this is that valuers will be asked to assess the consideration payable for a site on terms which cannot exist in practice because they are not permitted under the Code. This is akin to asking for a semidetached house to be valued as if it were a detached house, but in a world where no detached houses exist. The valuer would have no more direct evidence of such new agreements than he would if asked to value a horse by reference to the sale price of unicorns.

The NFU argued that the implication of paragraph 23 was that no rent would have to be paid for assignment or site sharing.[13] Both the NFU and CAAV were also concerned by paragraph 24, which would have given the Secretary of State the power to amend paragraph 23. The NFU felt that this paragraph should not be included and were 'strongly against the Secretary of State being given this power'. The CAAV also stated in their written evidence that this paragraph provided 'a reserve power to change the basic approach to valuation removing reference to the use for which the land is wanted, as prejudicial to the basic mechanism of the Code'. They therefore recommended that the paragraph should not be enacted. The Country Landowners Association was also highly critical of the lack of industry consultation and the attempt to introduce such detailed legislation at such a late stage. They also exposed 'significant flaws in the way the code would work' and expressed 'strong opposition to the concept of Ministers intervening to ensure mobile phone companies could impose terms, including rents and wayleave payments, on landowners for access to their land'.[14]

2.9.6

Following these and further representations from stakeholders voicing concerns as to inadequate time for proper scrutiny, the Government announced on 22 January 2015 that it was withdrawing the new draft code pending further consultation.

2.9.7

On 17 February 2015, the House of Lords Select Committee on Digital Skills, which had been appointed by the House of Lords on 12 June 2014 'to consider and report on information and communications technology, competitiveness and skills in the United Kingdom', produced a report named: *Make or Break: The UK's Digital Future*. This suggested that the UK was being held back as a digital economy by a number of factors including slow broadband speeds. It listed as a primary objective that 'the population as a whole has unimpeded access to digital technology [including]: (a) facilitation of universal internet access: the internet is viewed as a utility; and (b) removal of "not-spots" in urban areas'.

12 PBC, Infrastructure Bill Written Evidence, p.96.
13 PBC, Infrastructure Bill Written Evidence, p.104.
14 CLA, 'Landowners welcome withdrawal of flawed electronic communications code', January 2015.

2.9.8

On 26 February 2015, the Government issued its own consultation on a new draft Electronic Communications Code. The Bill accompanying the consultation was in similar form to the withdrawn code. The consultation ran for nine weeks, and closed on 30 April 2015. Submissions were invited on all areas of the Code. Following this, DCMS undertook further consultation with all stakeholders, and commissioned independent economic research into the impact of a range of reform options in the market.

2.9.9

In July 2015, the new Conservative Government published *Fixing the Foundations: Creating a More Prosperous Nation*, described as 'a comprehensive plan … to reverse the UK's long-term productivity problem and secure rising living standards and a better quality of life for our citizens'. Chapter 7 of this plan set out how the Government planned to deliver 'world-class digital infrastructure in every part of the UK' including introducing legislation in the first session of the (2015–20) Parliament to reform the Electronic Communications Code.[15]

2.9.10

On 17 May 2016, the Government published its response[16] to the consultation issued under the previous Government, setting out the proposals for reform of the Code it intended to take forward. The Queen's Speech on 18 May 2016 indicated that the Government would bring forward a Digital Economy Bill that would contain reform of the Electronic Communications Code, informed by consideration of the consultation responses. The proposals provided for the replacement of the Electronic Communications Code by a new code that was broadly aligned with the Law Commission's recommendations, with some key exceptions. In particular the Government proposed to adopt a different basis for the valuation of code rights, and to confer automatic rights to upgrade and share apparatus.

2.9.11

Ed Vaizey MP, now the Minister of State for Culture and the Digital Economy, stated in the response that the New Code would:

> vastly improve on the existing Code. It will make major reforms to the rights that communications providers have to access land – moving to a 'no scheme' basis of valuation regime. This will ensure property owners will be fairly compensated for use of their land, but also explicitly acknowledge the economic value for all of society created from investment in digital infrastructure. In this respect, it will put digital communications infrastructure on a similar regime to utilities like electricity and water. This will help deliver the coverage that is needed, even in hard to reach areas.

15 HM Treasury, *Fixing the Foundations: Creating a More Prosperous Nation*.
16 DCMS, A New Electronic Communications Code, 17 May 2016.

2.9.12

The Government's support for the Bill, and its reworking of the compensation provisions following industry pressure, were conditioned largely, it would appear, by its desire to ensure that the UK's digital economy was able to flourish. In this vein, Lord Adonis, the head of the National Infrastructure Commission, said on 26 May 2016 that a 5G network of ultrafast connections (to be launched in Britain in or after 2020) would need a huge increase in mobile phone transmitters, sweeping changes to planning laws and a big reduction to the rents that providers pay to site transmitters. The faster 5G networks would be needed to ensure that driverless cars are constantly connected, and to foster the current trend of automation and data exchange in manufacturing technologies (the 'internet of things'). The Government's objective was 'to provide a modern and robust legal framework for the rollout of electronic communications apparatus'.

2.9.13

The proposed reforms were summarised and explained in a House of Commons Library Briefing Paper *Reforming the Electronic Communications Code*, published on 1 June 2016.[17] The proposed version of the new Bill differed from that published as part of the Information and Infrastructure Bill in a number of substantive respects.

2.10 The Digital Economy Bill: enactment

2.10.1

A Digital Economy Bill enshrining the proposed reforms was introduced to the House of Commons on 5 July 2016. The sponsors of the Bill were Karen Bradley (DCMS) and Lord Ashton of Hyde.

2.10.2

The Bill was given its First Reading on 5 July 2016. The original Bill (Bill 45), as introduced, may be found at https://publications.parliament.uk/pa/bills/cbill/2016-2017/0045/cbill_2016-20170045_en_1.htm.

2.10.3

The Second Reading of the Bill took place in the Commons on 13 September 2016. The Public Bills Committee then held 11 sittings, hearing oral evidence from a number of industry, landowning and other representatives at the first three sittings on 11 and 13 October 2016, and considered other written submissions in October and on 2 November. There were no significant changes to the operative clauses of the Bill, but many changes to the Code in Schedule 1, as follows (new material in grey; deletions of old text shown by horizontal crossing out):[18]

17 http://researchbriefings.files.parliament.uk/documents/CBP-7203/CBP-7203.pdf

18 For the full text of the proposed amendments, see www.parliament.uk/documents/commons-public-bill-office/2016–17/compared-bills/Digital-Economy-Bill-161102.pdf.

1 The following proposed changes to para 3 (The code rights)

For the purposes of this code a 'code right', in relation to an operator and any land, is a right for the statutory purposes –

(a) to install ~~and keep~~ electronic communications apparatus on, under or over the land,

(b) to ~~inspect, maintain, adjust, alter, repair, upgrade or operate~~ keep installed electronic communications apparatus which is on, under or over the land,

(c) to inspect, maintain, adjust, alter, repair, upgrade or operate electronic communications apparatus which is on, under or over the land,

(d) to carry out any works on the land for or in connection with the ~~installation, maintenance, adjustment, alteration, repair, upgrading or operation~~ installation of electronic communications ~~apparatus~~ apparatus on, under or over the land or elsewhere,

(e) to carry out any works on the land for or in connection with the ~~installation~~, maintenance, adjustment, alteration, repair, upgrading or operation of electronic communications ~~apparatus~~ apparatus which is on, under or over the land or elsewhere,

[other paragraphs were unchanged].

2 The following proposed change to para 13 (Access to land)

(3) ~~The~~ A reference in ~~sub-paragraph (2)~~ this code to a means of access to or from land includes a means of access to or from land that is provided for use in emergencies.

3 The following proposed change to para 26 (Temporary code rights)

(1) (c): the person has the right to require the removal of the apparatus in accordance with paragraph 36 or as mentioned in paragraph 40(1) but the operator is not for the time being required to remove the apparatus.

...

(3) That objective is that, until the proceedings under paragraph 19 ~~or~~ and any proceedings under paragraph 39 are determined, the service provided by the operator's network is maintained and the apparatus is properly adjusted and kept in repair.

4 The following proposed change to para 35 (Introductory section on rights to require removal)

This Part of this code makes provision about –

a) the cases in which a person ~~with an interest in land~~ has the right to require the removal of electronic communications ~~apparatus~~ apparatus or the restoration of land, (b) the means by which a person can discover whether apparatus is on land pursuant to a code right, and (c) the means by which a right to require removal of apparatus or restoration of land can be enforced.

5 The following proposed change to para 36 (When does a person have the right to require removal of electronic communications apparatus?)

Heading: When does a ~~person~~ landowner have the right to require removal of electronic communications apparatus?

Sub-para (2): The first condition is that the landowner has never since the coming into force of this code been bound by a code right entitling an operator to keep the apparatus on, under or over the land.

This is subject to sub-paragraph (4).

Sub-para (3): The second condition is that a code right entitling an operator to keep the apparatus on, under or over the land has come to an end or has ceased to bind the landowner –

(a) as mentioned in paragraph 25(7) and (8)~~;~~,
(b) as the result of paragraph 31(1), or
(c) as the result of an order under paragraph 31(4) or 33(4) or (6)~~.~~, or
(d) where the right was granted by a lease to which Part 5 of this code does not apply.

This is subject to sub-paragraph (4).

New sub-para (9): This paragraph does not affect rights to require the removal of apparatus under another enactment (see paragraph 40).

6 Entirely new para 37: when does a landowner or occupier of neighbouring land have the right to require removal of electronic communications apparatus?

37 (1) A landowner or occupier of any land ('neighbouring land') has the right to require the removal of electronic communications apparatus on, under or over other land if both of the following conditions are met.

(2) The first condition is that the exercise by an operator in relation to the apparatus of a right mentioned in paragraph 13(1) interferes with or obstructs a means of access to or from the neighbouring land.

(3) The second condition is that the landowner or occupier of the neighbouring land is not bound by a code right within paragraph 3(h) entitling an operator to cause the interference or obstruction.

(4) A landowner of neighbouring land who is not the occupier of the land does not meet the second condition if –

(a) the land is occupied by a person who –
   (i) conferred a code right (which is in force) entitling an operator to cause the interference or obstruction, or
   (ii) is otherwise bound by such a right, and
(b) that code right was not conferred in breach of a covenant enforceable by the landowner.

(5) In the application of sub-paragraph (4)(b) to Scotland the reference to a covenant enforceable by the landowner is to be read as a reference to a contractual term which is so enforceable.

7 The following proposed change to para 38 (How does a person find out whether apparatus is on land pursuant to a code right?)

Heading change: How does a ~~person~~ landowner or occupier find out whether apparatus is on land pursuant to a code right?

New sub-para (1): A landowner may by notice require an operator to disclose whether –

(a) the operator owns electronic communications apparatus on, under or over land in which the landowner has an interest or uses such apparatus for the purposes of the operator's network, or

(b) the operator has the benefit of a code right entitling the operator to keep electronic communications apparatus on, under or over land in which the landowner has an interest.

Amended sub-para (2): A landowner or occupier of neighbouring land may by notice require an operator to disclose whether –

(a) the operator owns electronic communications apparatus on, under or over land ~~in which~~ that forms (or, but for the apparatus, would form) a means of access to the ~~landowner has an interest~~ neighbouring land, or uses such apparatus for the purposes of the operator's network, or

(b) the operator has the benefit of a code right entitling the operator to keep electronic communications apparatus on, under or over land ~~in which~~ that forms (or, but for the apparatus, would form) a means of access to the ~~landowner has an interest~~ neighbouring land.

Amended sub-para (4): Sub-paragraph (5) applies if –

(a) the operator does not, before the end of the period of three months beginning with the date on which the notice under sub-paragraph (1) or (2) was given, give a notice to the landowner or occupier that –
   (i) complies with paragraph 88 (notices given by operators), and
   (ii) discloses the information sought by the ~~landowner~~ landowner or occupier,

(b) the landowner or occupier takes action under paragraph 39 to enforce the removal of the apparatus, and [no further change]

Amended sub-para (5): The operator must nevertheless bear the costs of any action taken by the landowner or occupier under paragraph 39 to enforce the removal of the apparatus.

8 The following proposed change to para 39 (How does a person enforce removal of apparatus?)

Heading change: How does a ~~person~~ landowner or occupier enforce removal of apparatus?

Changes to each sub-para as follows:

(1) ~~A~~ The right of a landowner ~~who has the right~~ or occupier to require the removal of electronic communications apparatus on, under or over ~~land may~~land, under paragraph 36 or 37, is exercisable only in accordance with this paragraph, ~~require the operator whose apparatus it is –~~.

(2) The landowner or occupier may give a notice to the operator whose apparatus it is requiring the operator –

(a) **to remove the apparatus, and**
(b) to restore the land to its condition before the apparatus was placed on, under or over the land.

(3) The notice must –
(a) comply with paragraph 89 (notices given by persons other than operators), and
(b) specify the period within which the operator must complete the works.

(4) The period specified under sub-paragraph (3) must be a reasonable one.

(5) Sub-paragraph (6) applies if, within the period of 28 days beginning with the day on which the notice was given, the landowner or occupier and the operator do not reach agreement on any of the following matters –
(a) that the operator will remove the apparatus;
(b) ~~**to**~~ that the operator will **restore the land to its condition before the apparatus was placed on, under or over the land**~~.~~;
(c) the time at which or period within which the apparatus will be removed;
(d) the time at which or period within which the land will be restored.

(6) The landowner or occupier may make an application to the court for –
(a) an order under paragraph 43(1) (order requiring operator to remove apparatus etc), or
(b) an order under paragraph 43(3) (order enabling landowner to sell apparatus etc).

(7) If the court makes an order under paragraph 43(1), but the operator does not comply with the agreement imposed on the operator and the landowner or occupier by virtue of paragraph 43(7), the landowner or occupier may make an application to the court for an order under paragraph 43(3).

(8) On an application under sub-paragraph (6) or (7) the court may not make an order in relation to apparatus if an application under paragraph 19(3) has been made in relation to the apparatus and has not been determined.

9 The following proposed wholesale changes to para 40 (How are other rights to require removal of apparatus enforced?)

New heading: How are other rights to require removal of apparatus enforced?

Virtually entirely new sub-paras as follows:

(1) The right of a person (a 'third party') under an enactment other than this code, or otherwise than under an enactment, to require the removal of electronic communications apparatus on, under or over land is exercisable only in accordance with this paragraph.

(2) **The** ~~**landowner**~~ third party **may give a notice to the operator** whose apparatus it is, **requiring the operator** –
(a) **to remove the apparatus, and**
(b) to restore the land to its condition before the apparatus was placed on, under or over the land.

(3) The notice must –
(a) comply with paragraph 89 (notices given by persons other than operators), and
(b) specify the period within which the operator must complete the works.

(4) The period specified under sub-paragraph (3) must be a reasonable one.

(5) Within the period of 28 days beginning with the day on which notice under sub-paragraph (2) is given, the operator may give the third party notice ('counter-notice') –

(a) stating that the third party is not entitled to require the removal of the apparatus, or

(b) specifying the steps which the operator proposes to take for the purpose of securing a right as against the third party to keep the apparatus on the land.

(6) If the operator does not give counter-notice within that period, the third party is entitled to enforce the removal of the apparatus.

(7) If the operator gives the third party counter-notice within that period, the third party may enforce the removal of the apparatus only in pursuance of an order of the court that the third party is entitled to enforce the removal of the apparatus.

(8) If the counter-notice specifies steps under paragraph (5)(b), the court may make an order under sub-paragraph (7) only if it is satisfied –

(a) that the operator is not intending to take those steps or is being unreasonably dilatory in taking them; or

(b) that taking those steps has not secured, or will not secure, for the operator as against the third party any right to keep the apparatus installed on, under or over the land or to reinstall it if it is removed.

(9) Where the third party is entitled to enforce the removal of the apparatus, under sub-paragraph (6) or under an order under sub-paragraph (7), the third party may make an application to the court for –

(a) an order under paragraph 43(1) (order requiring operator to remove apparatus etc), or

(b) an order under paragraph 43(3) (order enabling third party to sell apparatus etc).

(10) If the court makes an order under paragraph 43(1), but the operator does not comply with the agreement imposed on the operator and the third party by virtue of paragraph 43(7), the third party may make an application to the court for an order under paragraph 43(3).

(11) An order made on an application under this paragraph need not include provision within paragraph 43(1)(b) or (3)(d) unless the court thinks it appropriate.

(12) Sub-paragraph (9) is without prejudice to any other method available to the third party for enforcing the removal of the apparatus.

10 Entirely new para 41 (How does paragraph 40 apply if a person is entitled to require apparatus to be altered in consequence of street works?)

(1) This paragraph applies where the third party's right in relation to which paragraph 40 applies is a right to require the alteration of the apparatus in consequence of the stopping up, closure, change or diversion of a street or road or the extinguishment or alteration of a public right of way.

(2) The removal of the apparatus in pursuance of paragraph 40 constitutes compliance with a requirement to make any other alteration.

(3) A counter-notice under paragraph 40(5) may state (in addition to, or instead of, any of the matters mentioned in paragraph 40(5)(b)) that the operator requires the third party to reimburse the operator in respect of any expenses incurred by the operator in or in connection with the making of any alteration in compliance with the requirements of the third party.

(4) An order made under paragraph 40 on an application by the third party in respect of a counter-notice containing a statement under sub-paragraph (3) must, unless the court otherwise thinks fit, require the third party to reimburse the operator in respect of the expenses referred to in the statement.

(5) Paragraph 43(3)(b) to (e) do not apply.

(6) In this paragraph –

'road' means a road in Scotland;

'street' means a street in England and Wales or Northern Ireland.

11 Largely new para 42 (When can a separate application for restoration of land be made?)

(1) This paragraph applies if –

(a) the condition of the land has been affected by the exercise of a code right, and

(b) **~~to restore~~** restoration of **the land to its condition before the ~~apparatus~~** code right **was ~~placed on, under or over~~** exercised does not involve **the** removal of electronic communications apparatus from any **land.**

(2) The occupier of the land, the owner of the freehold estate in the land or the lessee of the land ('the relevant person') has the right to require the operator to restore the land if the relevant person is not for the time being bound by the code right.

This is subject to sub-paragraph (3).

(3) The relevant person does not have that right if –

(a) the land is occupied by a person who –

(i) conferred a code right (which is in force) entitling the operator to affect the condition of the land in the same way as the right mentioned in sub-paragraph (1), or

(ii) is otherwise bound by such a right, and

(b) that code right was not conferred in breach of a covenant enforceable by the relevant person.

(4) In the application of sub-paragraph (3)(b) to Scotland the reference to a covenant enforceable by the relevant person is to be read as a reference to a contractual term which is so enforceable.

(5) A person who has the right conferred by this paragraph may give a notice to the operator requiring the operator to restore the land to its condition before the code right was exercised.

(6) **The notice must –**

(a) **comply with paragraph ~~85~~** 89 **(notices given by persons other than operators), and**

(b) **specify the period within which the operator must complete the works.**

(7) **The period specified under sub-paragraph ~~(3)~~**(6) **must be a reasonable one.**

(8) Sub-paragraph ~~(6)~~(9) applies if, within the period of 28 days beginning with the day on which the notice was given, the landowner and the operator do not reach agreement on any of the following matters –
   (a) ~~that the operator will remove the apparatus;~~
   (b) that the operator will restore the land to its condition before the ~~apparatus~~ code right was ~~placed on, under or over the land~~ exercised;
   (c) ~~the time at which or period within which the apparatus will be removed;~~
   (d) the time at which or period within which the land will be restored.

(9) The landowner may make an application to the court for –
   (a) an order under paragraph ~~39(1)~~ 43(2) (order requiring operator to ~~remove apparatus etc~~ restore land), or
   (b) an order under paragraph ~~39(2)~~ 43(4) (order enabling landowner to ~~sell apparatus etc~~ recover cost of restoring land).

(10) If the court makes an order under paragraph ~~39(1)~~43(2), but the operator does not comply with the agreement imposed on the operator and the landowner by virtue of paragraph ~~39(5)~~43(7), the landowner may make an application to the court for an order under paragraph ~~39(2)~~43(4).

(11) In the application of sub-paragraph (2) to Scotland the reference to a person who is the owner of the freehold estate in the land is to be read as a reference to a person who is the owner of the land.

12 Proposed changes to para 43 (What orders may the court make on an application under paragraph 38?)

New sub-para (2): An order under this sub-paragraph is an order that the operator must, within the period specified in the order, restore the land to its condition before the code right was exercised.

Changes to sub-para (3): An order under this sub-paragraph is an order that the ~~landowner~~ landowner, occupier or third party may do any of the following –

(a) remove or arrange the removal of the electronic communications apparatus;
(b) sell any apparatus so removed;
(c) recover the costs of any action under paragraph (a) or (b) from the operator;
(d) recover from the operator the costs of restoring the land to its condition before the apparatus was placed on, under or over the land;
(e) retain the proceeds of sale of the apparatus to the extent that these do not exceed the costs incurred by the ~~landowner~~ landowner, occupier or third party as mentioned in paragraph (c) or (d).

New sub-para (4): An order under this sub-paragraph is an order that the landowner may recover from the operator the costs of restoring the land to its condition before the code right was exercised.

Change to sub-para (5): An order under this paragraph on an application under paragraph 39 may require the operator to pay compensation to the landowner for any loss or damage suffered by the landowner as a result of the presence of the apparatus on the land during the period when the landowner had the right to require the removal of the apparatus from the land but was not able to exercise that right.

Change to sub-para (7): An order under sub-paragraph (1) or (2) takes effect as an agreement between the operator and the ~~landowner~~ landowner, occupier or third party that –

(a) requires the operator to take the steps specified in the order, and
(b) otherwise contains such terms as the court may so specify.

2.10.4

There were also proposed changes to the Transitional Provisions in Schedule 2, as follows (in addition to de-capitalising 'Code'):

1 Para 1(4) (Interpretation)

(4) A 'subsisting agreement' means –
  (a) an agreement ~~under~~ for the purposes of paragraph ~~2(1)~~ 2 or 3 of the existing ~~Code~~ code, or
  (b) an order under paragraph 5 of the existing code, which is in force, as between an operator and any person, at the time the new ~~Code~~ code comes into force (and whose terms do not provide for it to cease to have effect at that time).

2 New para 2(2) (Effect of subsisting agreement)

A person who is bound by a right by virtue of paragraph 2(4) of the existing code in consequence of a subsisting agreement is, after the new code comes into force, treated as bound pursuant to Part 2 of the new code.

3 Changes to para 3 (Limitation of code rights)

In relation to a subsisting agreement, references in the New Code to a code right are –
  (a) ~~In relation to a subsisting agreement, references in the new code to a code right are,~~ in relation to the operator and the land to which an agreement for the purposes of paragraph 2 of the ~~agreement~~ existing code relates, references to a right for the statutory purposes to do the things listed in paragraph 2(1) (a) to (c) of the existing code~~.~~;
  (b) in relation to land to which an agreement for the purposes of paragraph 3 of the existing code relates, a right to do the things mentioned in that paragraph.

4 New para 6 (Termination and modification of agreements)

(1) This paragraph applies in relation to a subsisting agreement, in place of paragraph 28(2) to (4) of the new code.
(2) Part 5 of the new code (termination and modification of agreements) does not apply to a subsisting agreement that is a lease of land in England and Wales, if –
  (a) it is a lease to which Part 2 of the Landlord and Tenant Act 1954 applies, and
  (b) there is no agreement under section 38A of that Act (agreements to exclude provisions of Part 2) in relation the tenancy.

(3) Part 5 of the new code does not apply to a subsisting agreement that is a lease of land in England and Wales, if –

(a) the primary purpose of the lease is not to grant code rights (the rights referred to in paragraph 3 of this Schedule), and

(b) there is an agreement under section 38A of the 1954 Act in relation the tenancy.

(4) Part 5 of the new code does not apply to a subsisting agreement that is a lease of land in Northern Ireland, if it is a lease to which the Business Tenancies (Northern Ireland) Order 1996 (SI 1996/725 (NI 5)) applies.

5 Modifications to para 7(1)

(1) Subject to paragraph 6, Part 5 of the new code ~~(termination and modification of agreements)~~ applies ~~in relation~~ to a subsisting agreement ~~subject to~~ with the following modifications.

6 Substantial modifications to para 12

(1) This paragraph applies where before the time when the new code comes into force –

(a) a notice has been given under paragraph 5(1) of the existing code, and

(b) an application has been made to the court in relation to the notice.

(2) Subject to ~~the following provisions of this paragraph~~ sub-paragraph (3), the existing code continues to apply in relation to the application.

(3) ~~An order made under the existing code by virtue of sub-paragraph (2) has effect as an order under paragraph 19 of the new code.~~

(4) ~~If the operator gives a notice in accordance with sub-paragraph (5), from the time when the notice takes effect –~~

(a) ~~sub-paragraph (2) does not apply, and~~

(5) An order made under the ~~application mentioned in~~ existing code by virtue of sub-paragraph ~~(1)(b)~~(2) has effect as ~~if made~~ an order under paragraph ~~19(3)~~ 19 of the new code; ~~but this is subject to sub-paragraph (6).~~

(6) ~~A notice under sub-paragraph (4) –~~

(a) ~~must be given to the person, or each person, on whom the notice mentioned in sub-paragraph (1)(a) was served;~~

(b) ~~must be given not later than the end of the period of 28 days beginning with the day on which the new code comes into force.~~

(7) ~~A notice under sub-paragraph (4) –~~

(a) ~~takes effect at the end of the period of 28 days beginning with the day on which the notice is given; but~~

(b) ~~does not have effect if before the end of that period the court or any person to whom the notice was given gives a notice under this subparagraph.~~

(8) ~~A notice under sub-paragraph (6) may be given by the court only if it appears to the court on its own motion that it would be unreasonable in all the circumstances for the application to have effect as if made under paragraph 19(3) of the new code.~~

(9) ~~For the purposes of sub-paragraph (7), any difference between the amount of any payment that would fall be to made under an order under paragraph 5 of~~

~~the existing code and under an order under paragraph 19 of the new code is to be disregarded.~~

(10) ~~Nothing in this paragraph prevents the operator from—~~

(a) ~~withdrawing an application falling within sub-paragraph (1)(b), or~~

(b) ~~giving a notice or making an application to the court under paragraph 19 of the new code in respect of the same right.~~

(11) ~~The operator must bear any costs arising from the service of a notice under sub-paragraph (4) or any action taken by the operator within sub-paragraph (9).~~

7 *Substantial modifications to paras 20 to 24* (Right to require removal of apparatus)

(20) ~~The repeal of the existing code does not affect the operation of paragraph 21 of that code in relation to an entitlement under sub-paragraph (3) of that paragraph arising before the repeal comes into force.~~

(21) (1) ~~This paragraph applies if—~~

(a) ~~a person has given notice under sub-paragraph (2) of paragraph 21 of the existing code before the time when the repeal of that code comes into force,~~

(b) ~~the 28 day period mentioned in sub-paragraph (3) of that paragraph ends after that time, and~~

(c) ~~no counter-notice is given under that paragraph within that period.~~

(2) ~~Paragraphs 38(6) and (7) and 39 of the new code apply in relation to that person as they apply in relation to the landowner mentioned in paragraph 38(6).~~

(22) (1) ~~This paragraph applies if—~~

(23) (1) This paragraph applies where before the repeal of the existing code comes into force a person has given notice under ~~sub-~~paragraph 21(2) of ~~paragraph 21 of the existing~~ that code ~~before~~ requiring ~~the time when the repeal~~ removal of ~~that code comes into force,~~ apparatus.

(a) ~~a counter-notice has been given under sub-paragraph (3) of that paragraph before or after that time, and~~

(b) ~~no application has been made to the court under that paragraph by the operator.~~

(2) ~~The counter-notice has effect as a notice under paragraph 19(2) of the new code.~~

(3) ~~On an application under paragraph 19 of the new code as it applies by virtue of sub-paragraph (2), the court may not make an order under that paragraph (in addition to the circumstances provided for in sub-paragraph (5) of that paragraph) if it appears to the court, in relation to any relevant agreement—~~

(a) ~~that there were substantial breaches by the operator of its obligations under the agreement, or~~

(b) ~~that there were persistent delays by the operator in making payments to the site provider under the agreement.~~

(4) ~~In sub-paragraph (3) 'relevant agreement' means any agreement between the operator and the landowner that was in force before the right to require removal arose under paragraph 21 of the existing code.~~

24 (1) ~~This paragraph applies if, before the repeal of the existing code comes into force –~~

(a) ~~an application has been made to the court under paragraph 21 of that code, and~~

(b) ~~the matter has not been determined by the court.~~

(2) The repeal ~~of the existing code~~ does not affect the operation of paragraph 21 ~~of that code~~ in relation to anything done or that may be done under that paragraph following the ~~application~~ giving of the notice.

(3) ~~But any party to the proceedings may apply to the court for an order that the application be treated as an application to the court under paragraph 19 of the new code.~~

(4) ~~The court must grant an application under sub-paragraph (3) unless it thinks it would be unreasonable in all the circumstances to do so.~~

(5) For the purposes of applying that paragraph after the repeal comes into force, steps specified in a counter-notice under sub-paragraph (4)(b) of that paragraph as steps which the operator proposes to take under the existing code are to be read as including any corresponding steps that the operator could take under the new code or by virtue of this Schedule.

2.10.5

The following changes were also proposed to the Consequential Amendments in Schedule 3, (in addition to all the amendments to specific Acts):

GENERAL PROVISION

Interpretation

1 In this Part –

'the commencement date' means the day on which Schedule 3A to the Communications Act 2003 comes into force;

'enactment' includes –

(a) an enactment comprised in subordinate legislation within the meaning of the Interpretation Act 1978,

(b) an enactment comprised in, or in an instrument made under, a Measure or Act of the National Assembly for Wales,

(c) an enactment comprised in, or in an instrument made under, an Act of the Scottish Parliament, and (d) an enactment comprised in, or in an instrument made under, Northern Ireland legislation;

'the existing code' means Schedule 2 to the Telecommunications Act 1984;

'the new code' means Schedule 3A to the Communications Act 2003.

References to the code or provisions of the code

2 (1) In any enactment passed or made before the commencement date, unless the context requires otherwise –

(a) a reference to the existing code is to be read as a reference to the new code;

(b) a reference to a provision of the existing code listed in column 1 of the table is to be read as a reference to the provision of the new code in the corresponding entry in column 2.

(2) This paragraph does not affect the amendments made by Part 2 of this Schedule or the power to make amendments by regulations under section 6.

(3) This paragraph does not affect section 17(2) of the Interpretation Act 1978 (effect of repeal and re-enactment) in relation to any reference to a provision of the existing code not listed in the table.

Existing code	*New code*
Paragraph 9	Part 8
Paragraph 21	Part 6
Paragraph 23	Part 10
Paragraph 29	Paragraph 17

References to a conduit system

3 In any enactment passed or made before the commencement date, unless the context requires otherwise –

(a) a reference to a conduit system, where it is defined by reference to the existing code, is to be read as a reference to an infrastructure system as defined by paragraph 7(1) of the new code, and;

(b) a reference to provision of such a system is to be read in accordance with paragraph 7(2) of the new code (reference to provision includes establishing or maintaining).

2.10.6

The Report and Third Reading took place on 28 November, and the Bill completed its final stages on 29 November 2016.[19]

2.10.7

The Bill was then introduced to the House of Lords for its Second Reading on 13 December 2016. It was amended by the House of Lords in Committee on 8 February 2017, and on Report on 29 March 2017.[20] Again, the amendments made in Committee, Grand Committee or in Special Public Bill Committee in the House of Lords did not entail significant changes to the operative clauses of the Bill, but there were many changes to the Code in Schedule 1, as follows (new text in bold; deletions in red).

1 Para 13 (Access to land)*: small changes to the following sub-paras*

(1) to change as follows: This paragraph applies to an operator by whom any of the following rights is exercisable in relation to land –

19 See www.parliament.uk/documents/commons-public-bill-office/2016–17/compared-bills/Digital-Economy-Bill-161102.pdf.

20 For the full text of the amendments, see www.parliament.uk/documents/commons-public-bill-office/2016–17/compared-bills/Digital-Economy-AAC-tracked-changes.pdf.

(a) a code right within paragraph (a) to (~~e~~g) or (~~g~~i) of paragraph 3;
(b) a right under Part 8 (street works rights);
(c) a right under Part 9 (tidal water rights);
(d) a right under paragraph 73 (power to fly lines).

2 to change as follows

The operator may not exercise the right so as to interfere with or obstruct any means of access to or from any other land unless, in accordance with this code, the occupier of the other land has conferred or is otherwise bound by a code right within paragraph (~~f~~h) of paragraph 3.

*3 Wholesale change to para 15 (*assignment of code rights*) as follows*

15 ~~(1) Any agreement under Part 2 of this code is void to the extent that –~~
~~(a) it prevents or limits assignment of the agreement to another operator, or~~
~~(b) it makes assignment of the agreement subject to conditions to be met by the operator (including a condition requiring the payment of money).~~
~~(2) In its application to England and Wales or Northern Ireland subparagraph (1) does not apply to the following terms of an agreement under Part 2 of this code –~~
~~(a) terms in a lease which require the operator to enter into an authorised guarantee agreement within the meaning of the Landlord and Tenant (Covenants) Act 1995 (see sections 16 and 28 of that Act) or (in Northern Ireland) a similar agreement;~~
~~(b) terms in an agreement other than a lease which have a similar effect to terms within paragraph (a).~~
~~(3) If an operator ('the assignor') assigns an agreement under Part 2 of this code to another operator ('the assignee'), the assignee is from the date of the assignment bound by the terms of the agreement.~~
~~(4) The assignor is not liable for any breach of a term of the agreement that occurs after the assignment if (and only if), before the breach took place, the assignor or the assignee gave a notice in writing to the other party to the agreement which –~~
~~(a) identified the assignee, and~~
~~(b) provided a contact address for the assignee.~~
~~(5) Sub-paragraph (4) is subject to the terms of any authorised guarantee agreement or similar agreement entered into by the assignor as mentioned in sub-paragraph (2).~~
(6) Any agreement under Part 2 of this code is void to the extent that –
(a) it prevents or limits assignment of the agreement to another operator, or
(b) it makes assignment of the agreement to another operator subject to conditions (including a condition requiring the payment of money).
(7) Sub-paragraph (1) does not apply to a term that requires the assignor to enter into a guarantee agreement (see sub-paragraph (7)).
(8) In this paragraph references to 'the assignor' or 'the assignee' are to the operator by whom or to whom an agreement under Part 2 of this code is assigned or proposed to be assigned.

(9) From the time when the assignment of an agreement under Part 2 of this code takes effect, the assignee is bound by the terms of the agreement.

(10) The assignor is not liable for any breach of a term of the agreement that occurs after the assignment if (and only if), before the breach took place, the assignor or the assignee gave a notice in writing to the other party to the agreement which –

(a) identified the assignee, and

(b) provided an address for service (for the purposes of paragraph 90(2)(b)) for the assignee.

(11) Sub-paragraph (5) is subject to the terms of any guarantee agreement.

(12) A 'guarantee agreement' is an agreement, in connection with the assignment of an agreement under Part 2 of this code, under which the assignor guarantees to any extent the performance by the assignee of the obligations that become binding on the assignee under sub-paragraph (4) (the 'relevant obligations').

(13) An agreement is not a guarantee agreement to the extent that it purports –

(a) to impose on the assignor a requirement to guarantee in any way the performance of the relevant obligations by a person other than the assignee, or

(b) to impose on the assignor any liability, restriction or other requirement of any kind in relation to a time after the relevant obligations cease to be binding on the assignee.

(14) Subject to sub-paragraph (8), a guarantee agreement may –

(a) impose on the assignor any liability as sole or principal debtor in respect of the relevant obligations;

(b) impose on the assignor liabilities as guarantor in respect of the assignee's performance of the relevant obligations which are no more onerous than those to which the assignor would be subject in the event of the assignor being liable as sole or principal debtor in respect of any relevant obligation;

(c) make provision incidental or supplementary to any provision within paragraph (a) or (b).

(15) In the application of this paragraph to Scotland references to assignment of an agreement are to be read as references to assignation of an agreement.

(16) Nothing in the Landlord and Tenant Amendment (Ireland) Act 1860 applies in relation to an agreement under Part 2 of this code so as to –

(a) prevent or limit assignment of the agreement to another operator, or

(b) relieve the assignor from liability for any breach of a term of the agreement that occurs after the assignment.

4 Minor change to para 36(8)(b)

that right has ceased to be exercisable in relation to the land by virtue of paragraph 53(9) ~~or 59(8)~~, and

5 The whole of para 59 is deleted

~~What happens to the street work rights if land ceases to be a street or road?~~

~~59 (1) This paragraph applies if an operator is exercising a street work right in relation to land immediately before a time when the land ceases to be a street or road.~~

~~(2) After that time, this Part of this code continues to apply to the land as if it were still a street or road (and, accordingly, the operator may continue to exercise any street work right in relation to the land as if it were still a street or road).~~

~~(3) But sub-paragraph (2) is subject to sub-paragraphs (4) to (8).~~

~~(4) The application of this Part of this code to land in accordance with sub-paragraph (2) does not authorise the operator to install or keep any electronic communications apparatus that is not in place at the time when the land ceases to be a street or road.~~

~~(5) But sub-paragraph (4) does not affect the power of the operator to replace existing apparatus (whether in place at the time when the land ceased to be a street or road or a replacement itself authorised by this sub-paragraph) with new apparatus which –~~

~~(a) is not substantially different from the existing apparatus, and~~

~~(b) is not in a significantly different position.~~

~~(6) The occupier of land may, at any time after the land ceases to be a street or road, give the operator notice specifying a date on which this Part of this code is to cease to apply to the land in accordance with this paragraph ('notice of termination').~~

~~(7) That date specified in the notice of termination must fall after the end of the period of 12 months beginning with the day on which the notice of termination is given.~~

~~(8) On the date specified in notice of termination in accordance with sub-paragraph (7), the street work rights cease to be exercisable in relation to the land in accordance with this paragraph.~~

2.10.8

The Bill returned to the Commons for consideration of the Lords' amendments on 6 April 2017. It went back to the House of Lords for reconsideration on 26 April 2017, and received Royal Assent on the same day.

2.10.9

The dates for all stages of the passage of the Bill, including links to the debates may be found at http://services.parliament.uk/bills/2016–17/digitaleconomy/stages.html. The full list of those stages is as follows:

Stage		*Date*
1	1st reading: House of Commons 5 July, 2016	05.07.2016
2	2nd reading: House of Commons 13 September, 2016	13.09.2016
	Money resolution: House of Commons 13 September, 2016	13.09.2016
	Programme motion: House of Commons 13 September, 2016	13.09.2016

Stage		*Date*
●	Ways and Means resolution: House of Commons 13 September, 201	13.09.2016
C	Committee Debate: 1st sitting: House of Commons 11 October, 2016 (1) (2)	11.10.2016
C	Committee Debate: 2nd sitting: House of Commons 11 October, 2016 (1) (2)	11.10.2016
C	Committee Debate: 3rd sitting: House of Commons 13 October, 2016 (1) (2)	13.10.2016
C	Committee Debate: 4th sitting: House of Commons 18 October, 2016 (1) (2)	18.10.2016
●	Programme motion No.2: House of Commons 18 October, 2016	18.10.2016
C	Committee Debate: 5th sitting: House of Commons 20 October, 2016 (1) (2)	20.10.2016
C	Committee Debate: 6th sitting: House of Commons 20 October, 2016 (1) (2)	20.10.2016
C	Committee Debate: 7th sitting: House of Commons 25 October, 2016 (1) (2)	25.10.2016
C	Committee Debate: 8th sitting: House of Commons 25 October, 2016 (1) (2)	25.10.2016
C	Committee Debate: 9th sitting: House of Commons 27 October, 2016 (1) (2)	27.10.2016
C	Committee Debate: 10th sitting: House of Commons 27 October, 2016 (1) (2)	27.10.2016
C	Committee Debates: compilation of sittings so far: House of Commons 1 November, 2016	01.11.2016
C	Committee Debate: 11th sitting: House of Commons 1 November, 2016 (1) (2)	01.11.2016
●	Programme motion No. 3: House of Commons 28 November, 2016	28.11.2016
R	Report stage: House of Commons 28 November, 2016	28.11.2016
●	Legislative Grand Committee: House of Commons 28 November, 2016	28.11.2016
3	3rd reading: House of Commons 28 November, 2016	28.11.2016
1	1st reading (Hansard): House of Lords 29 November, 2016	29.11.2016
1	1st reading (Minutes of Proceedings): House of Lords 29 November, 2016	29.11.2016
2	2nd reading (Hansard): House of Lords 13 December, 2016	13.12.2016
2	2nd reading (Minutes of Proceedings): House of Lords 13 December, 2016	13.12.2016
C	Committee: 1st sitting (Hansard): House of Lords 31 January, 2017	31.01.2017

	Stage	*Date*
C	Committee: 1st sitting (Hansard – continued): House of Lords 31 January, 2017	31.01.2017
C	Committee: 1st sitting (Minutes of Proceedings): House of Lords 31 January, 2017	31.01.2017
C	Committee: 2nd sitting (Hansard): House of Lords 2 February, 2017	02.02.2017
C	Committee: 2nd sitting (Hansard – continued): House of Lords 2 February, 2017	02.02.2017
C	Committee: 2nd sitting: House of Lords 2 February, 2017	02.02.2017
C	Committee: 3rd sitting (Hansard): House of Lords 6 February, 2017	06.02.2017
C	Committee: 3rd sitting Hansard – continued): House of Lords 6 February, 2017	06.02.2017
C	Committee: 3rd sitting (Hansard – continued): House of Lords 6 February, 2017	06.02.2017
C	Committee: 3rd sitting (Minutes of Proceedings): House of Lords 6 February, 2017	06.02.2017
C	Committee: 4th sitting (Hansard): House of Lords 8 February, 2017	08.02.2017
C	Committee: 4th sitting (Hansard – continued): House of Lords 8 February, 2017	08.02.2017
C	Committee: 4th sitting (Minutes of Proceedings): House of Lords 8 February, 2017	08.02.2017
R	Report: 1st sitting: House of Lords 22 February, 2017	22.02.2017
R	Report stage (Minutes of Proceedings): House of Lords 22 February, 2017	22.02.2017
R	Report: 2nd sitting (Hansard): House of Lords 20 March, 2017	20.03.2017
R	Report: 2nd sitting (Hansard – continued): House of Lords 20 March, 2017	20.03.2017
R	Report: 2nd sitting (Minutes of Proceedings): House of Lords 20 March, 2017	20.03.2017
R	Report: 3rd sitting (Hansard): House of Lords 29 March, 2017	29.03.2017
R	Report: 3rd sitting (Hansard – continued): House of Lords 29 March, 2017	29.03.2017
R	Report: 3rd sitting (Minutes of Proceedings): House of Lords 29 March, 2017	29.03.2017
3	3rd reading (Hansard): House of Lords 5 April, 2017	05.04.2017
3	3rd reading (Minutes of Proceedings): House of Lords 5 April, 2017	05.04.2017
C	Ways and Means resolution: House of Commons	26.04.2017
C	Ping Pong	26.04.2017
	Ping Pong (Hansard): House of Lords 27 April, 2017	27.04.2017

Stage		Date
●	Ping Pong (Minutes of Proceedings): House of Lords 27 April, 2017	27.04.2017
RA	Royal Assent (Hansard) 27 April, 2017	27.04.2017
RA	Royal Assent (Minutes of Proceedings) 27 April, 2017	27.04.2017
RA	Royal Assent (Hansard) 27 April, 2017	27.04.2017
RA	Royal Assent (Minutes of Proceedings) 27 April, 2017	

2.10.10

The New Code as enacted is considered in detail in Part III of this book.

2.11 Commencement

2.11.1

Section 118 of the 2017 Act brought certain specified provisions of the Act into force either on the same date as Royal Assent, or two months later, or on other specified dates. With the exception of s.7 (which deals with the protection of the environment under National Parks and other legislation in connection with the application of the Code), none of these provisions concerns either the Old Code or the New Code.

2.11.2

Section 118(6) provides:

> The other provisions of this Act come into force on whatever day the Secretary of State appoints by regulations made by statutory instrument.

2.11.3

On 12 July 2017, the Secretary of State made The Digital Economy Act 2017 (Commencement No. 1) Regulations 2017,[21] in exercise of the powers conferred by s.118(4), (6) and (7) of the 2017 Act. As far as concerns the New Code, paragraph 2 of these Regulations brought into force:

- (a) s.5 (power to make transitional provision in connection with the code);
- (b) s.6 (power to make consequential provision etc in connection with the code);
  - (ii) Schedule 1 (the electronic communications code), but only for the purpose of making regulations under paragraph 95 (power to confer jurisdiction on other tribunals) of Schedule 3A to the Communications Act 2003, and s.4 (the electronic communications code) so far as is necessary for that purpose;
  - (jj) paragraph 47 of Schedule 3 (electronic communications code: consequential amendments), and s.4 (the electronic communications code) so far as it relates to that paragraph.

21 www.legislation.gov.uk/uksi/2017/765/pdfs/uksi_20170765_en.pdf

2.11.4

On 21 November 2017, the Secretary of State made the Digital Economy Act 2017 (Commencement No. 2) Regulations 2017[22] in exercise of the powers conferred by s.118(6) and (7) of the 2017 Act. These brought into force on 22 November 2017 Schedule 1 to the 2017 Act, but only in relation to paragraph 106 (Lands Tribunal for Scotland procedure rules) of Schedule 3A to the 2003 Act, and s.4 so far as it relates to that paragraph.

2.11.5

On 14 December 2017, the Secretary of State made The Digital Economy Act 2017 (Commencement No. 2) Regulations 2017,[23] bringing into force with effect from 28 December 2017 the following provisions of the 2017 Act (in so far as not already in force):

(a) s. 4 (the electronic communications code);
(b) Schedule 1 (the electronic communications code);
(c) Schedule 2 (the electronic communications code: transitional provision);
(d) Schedule 3 (electronic communications code: consequential amendments).

Accordingly, the New Code came into full operation with effect from 28 December 2017.

2.12 The future

The sources for this section on the future are given in the note.[24]

2.12.1

The Universal Services Directive ('USD'), implemented in 2002,[25] and revised in 2009,[26] provides the framework within which the new UK broadband Universal Service Obligation ('USO') must operate. The USD requires Member States to provide users with a connection to the public communications network sufficient to permit 'functional internet access', on request, at an affordable price.[27]

2.12.2

'Functional internet access' is not defined in the USD, and Member States have the flexibility to define a broadband USO according to their own national circumstances. The telephony USO was introduced in 2003 to meet the requirements of the USD. It

22 (SI 2017 No. 1136) – www.legislation.gov.uk/uksi/2017/1136/pdfs/uksi_20171136_en.pdf.

23 (SI 2017 No. 1286) – www.legislation.gov.uk/uksi/2017/1286/made.

24 Source: the Government's Library Briefing Paper on Superfast Broadband Access in the UK (March 2017); the Library Briefing Paper on the USO, No. CBP 8146, 17 November 2017 – http://researchbriefings.files.parliament.uk/documents/CBP-8146/CBP-8146.pdf.

25 Directive 2002/22/EC on universal service and users' rights relating to electronic communications networks and services.

26 Directive 2009/136/Existing Code.

27 European Parliament Briefing, Broadband as a universal service, April 2016.

provides a right to request a dial-up internet connection (28 kbps, 0.028 Mbps), up to a cost threshold of £3,400.

2.12.3

In 2017, the Government introduced a new broadband USO. This upgrades the existing telephony USO, although it is not a mandatory requirement under the Directive. The new USO is also proposed to have a £3,400 cost threshold.

2.12.4

The broadband USO complements the Government's existing and past programmes to deliver broadband connections across the UK. This includes, first, the provision of subsidised broadband connections to premises unable to access download speeds of at least 2 Mbps, pursuant to the Government's Better Broadband Scheme (referred to as the 'Universal Service Commitment'), which began in December 2015, and ended at the end of 2017. Secondly, the government's Superfast Broadband Program supports the Government's target that 95 per cent of UK premises will have access to superfast broadband connections (download speeds greater than 24 Mbps) by the end of 2017. The Government has confirmed that it is on track to meet this target, and expects that this coverage should reach 97 per cent by 2020.

2.12.5

The USO is intended to fill the gaps left by these two programmes and provide connections to those premises that do not have access to download speeds of 10 Mbps. The Government has implemented the USO via secondary legislation under the 2017 Act,[28] and intends the USO to be in place by 2020 at the latest.

2.12.6

The DCMS has also conducted a Future Telecoms Infrastructure Review, to assess, among other things, the barriers to investment in digital infrastructure and next-generation digital connectivity. This led to a report published on 23 July 2018 by DDCMS.

2.12.7

The advent of fifth generation wireless systems (5G[29]) will also necessitate enhanced digital architecture and network infrastructure, to enable the huge increase in mobile traffic.[30]

28 See The Electronic Communications (Universal Service) (Broadband) Order 2018 (2018 SI No. 445), which came into force on 23 April 2018.

29 1G was introduced in 1982; 2G in 1992; 3G in 2001; and 4G in 2012.

30 Estimated at more than 15 additional ExaBytes (10^{18} bytes) per year in Europe by 2020 – https://5g-ppp.eu/wp-content/uploads/2016/02/BROCHURE_5PPP_BAT2_PL.pdf; while Ericsson and Machina estimate that there will be 25 billion connected devices by 2020.

3 The Electronic Communications Code 2003

An overview

3.1 Introduction

3.1.1

The Telecommunications Code was initially introduced by the Telecommunications Act 1984, and was revised by the Communications Act 2003. The code, as so revised, is referred to in this chapter and in Part III as 'the Old Code', to distinguish it from the Electronic Communications Code introduced by the Digital Economy Act 2017 ('the New Code'), which is dealt with in Part IV of this book.

3.1.2

Since the New Code is not retrospective, it will be necessary for some time to come to have regard to the provisions of the Old Code.

3.2 The genesis of the Old Code

3.2.1

The Telecommunications Act 1984 was enacted, in part, to open up the telecommunications sector to competition by privatising British Telecom and breaking its monopoly over the telecommunications network in the UK.

3.2.2

Section 10 of the 1984 Act provided for the telecommunications code set out in the Second Schedule to have effect. The intention was that this code would help other operators to develop networks to compete with and supplement the network which had been developed by the public sector, predominantly through British Telecom, over the preceding years.

3.2.3

Primarily, this first version of the Old Code was designed to regulate the relationship between electronic communications network operators and those landowners who provided the sites upon which the infrastructure necessary for the operation of the networks was to be sited. The Old Code therefore provided the legal framework for the installation and maintenance of the physical networks of apparatus that support the provision of electronic communications services throughout the UK.

3.3 The 2003 amendments to the Old Code

3.3.1

The 1984 original version of the Old Code was amended by the Communications Act 2003, prompted by the adoption by the UK in 2002 of a series of five EU Directives. The Old Code was also rebranded as 'the electronic communications code' (see s.106[1]).

3.3.2

The purpose of the amendments was to: '[translate] the telecommunications code into a code applicable in the context of the new regulatory regime established by the [Communications Act 2003]' (see s.106(2)). In effect, the aim was to extend the powers of the Old Code, from its original scope relating only to telephone services, to providers of electronic communications in the broad sense.

3.3.3

Many of the amendments sought to reflect a change in focus – made necessary by European Directives and significant developments in technology – from 'telecommunications' to the broader concept of 'electronic communications'. However, much of the Old Code remained as drafted in 1984.

3.4 The operation of the Old Code

3.4.1

The Old Code provides for network operators to obtain rights to install and maintain their apparatus on public and private land. Only those operators that apply for and receive approval from the Office of Communications ('OFCOM', the independent regulator and competition authority for the United Kingdom's communications industries) under s.106(3)(a) of the Communications Act 2003 are able to benefit from, and be subject to, the Old Code. In practice, all network operators have such approval, without which they would not be able to operate.

3.4.2

As part of its responsibility for applying the code to code operators, OFCOM must secure 'the optimal use for wireless telegraphy of the electro-magnetic spectrum' and

1 Although as Lewison J remarked in in *Geo Networks Ltd v The Bridgewater Canal Co. Ltd* [2010] 1 WLR 2576, 'the amendments made by the 2003 Act did not include changing the title to Schedule 2, so that in Schedule 2 itself it is still called "The Telecommunications Code"'.

'the availability throughout the United Kingdom of a wide range of electronic communications services' (s.3(2) of the Communications Act 2003).

3.4.3

At the heart of the Old Code, in paragraph 2(1) of Schedule 2, is the following set of the rights that an operator can acquire for the statutory purposes, with the agreement in writing of the affected landowner:

(a) to execute any works on that land for or in connection with the installation, maintenance, adjustment, repair or alteration of electronic communications apparatus; or
(b) to keep electronic communications apparatus installed on, under or over that land; or
(c) to enter that land to inspect any apparatus kept installed (whether on, under or over that land or elsewhere) for the purposes of the operator's network.

3.4.4

The expression 'electronic communications apparatus' is defined by paragraph 1(1) of the code to mean:

(a) any apparatus [which includes any equipment, machinery or device and any wire or cable and the casing or coating for any wire or cable] which is designed or adapted for use in connection with the provision of an electronic communications network; or
(b) any apparatus that is designed or adapted for a use which consists of or includes the sending or receiving of communications or other signals that are transmitted by means of an electronic communications network;
(c) any line;
(d) any conduit, structure, pole or other thing in, on, by or from which any electronic communications apparatus is or may be installed, supported, carried or suspended.

3.4.5

It is one of the more unsatisfactory features of the Old Code that agreements giving rise to such code rights can take a variety of different forms (wayleave or other simple licence; easement; tenancy), and may attract security of tenure under Part II of the Landlord and Tenant Act 1954 – quite apart from that bestowed by the Old Code itself.

3.4.6

Such agreements must be made with the 'occupier' of the land concerned. This is defined by paragraph 2(8) of the code to mean the physical occupier; or, where the land is unoccupied, 'the person … who for the time being exercises powers of management or control over the land or, if there is no such person, to every person whose interest in the land would be prejudicially affected by the exercise of the right in question'. This could be the freehold owner, a long leaseholder or even a weekly tenant. The occupier's agreement may bind others with an interest in the land. Even where it does not, an owner of the land with a superior interest to the occupier may find itself left with telecommunications equipment on its land that it will find difficult to remove.

3.4.7

Where agreement cannot be reached, the operator may use the notice procedure set out in paragraph 5 of the code, and apply to (typically the County) Court for an order conferring the proposed right. Paragraph 5(3) then provides:

> The court shall make an order if, but only if, it is satisfied that any prejudice caused by the order:
> (a) is capable of being adequately compensated for by money; or
> (b) is outweighed by the benefit accruing from the order to the persons whose access to an electronic communications network or to electronic communications services will be secured by the order;
>
> and in determining the extent of the prejudice, and the weight of that benefit, the court shall have regard to all the circumstances and to the principle that no person should unreasonably be denied access to an electronic communications network or to electronic communications services.

3.4.8

The 'principle' at the end of paragraph 5(3) was described by the Chancellor of the High Court in *Geo Networks Ltd v The Bridgewater Canal Co. Ltd*[2] as the 'overriding principle'. This is not, however, how the legislation is expressed. As matters stand, although this drafting has given rise to a great deal of controversy, there is no other case that provides an authoritative decision on its meaning, and on the weight to be given to the respective criteria to which the court is to have regard.

3.4.9

Once an operator has obtained, by agreement or by court order, the right to install electronic communications apparatus, the Old Code provides for a number of ancillary rights:

- the ability in certain circumstances to obstruct access to other land (paragraph 3);
- the power to fly overhead lines (paragraph 10);
- provision for cutting back trees where a tree overhangs a street (paragraph 19).

3.4.10

The Old Code does not in itself entitle the operator to assign its Code rights; to upgrade its equipment (or carry out other alterations); or to share its electronic communications apparatus with third parties. If such rights are needed, then they should be included in the agreement with the landowner.

2 [2011] 1 WLR 1487 at [25].

3.4.11

However, express (but opaque) provisions apply where an operator wishes to alter electronic communications apparatus, or where a landowner wishes to have the apparatus removed, in circumstances where the agreement between the parties does not provide for either circumstance. Paragraph 20 of the code gives an opportunity for any person with an interest in land to require the alteration of apparatus where 'the alteration is necessary to enable that person to carry out a proposed improvement of land in which he has an interest'. 'Alteration' is defined to include 'the moving, removal or replacement of the apparatus'. 'Improvement' is defined to include development and change of use. The right to require alteration can be exercised 'notwithstanding the terms of any agreement binding that person'.

3.4.12

The mechanism for requiring alteration follows a notice and counter-notice procedure. Where the court has to decide the question, it must make an order for an alteration only if, having regard to all the circumstances and to the principle set out in paragraph 5, it is satisfied that: (1) the alteration is necessary to enable the person requiring it to carry out a proposed improvement of his or her land; and (2) the alteration will not substantially interfere with any service which is or is likely to be provided using the operator's network. Moreover, the court must not make the order unless it is satisfied either: (1) that the operator has the rights needed for the purpose of making the alteration; or (2) that the operator could obtain those rights under the code.

3.4.13

Separately, paragraph 21 of the Old Code states that anyone who is entitled to require the removal of electronic communications apparatus from their land cannot do so except in accordance with the provisions of that paragraph. It then creates a procedure for serving notice requiring removal and enables the court to make an order requiring removal; but that order cannot be made unless a test (set out in paragraph 21(6)) is satisfied. If the order is made, the paragraph sets out a procedure for removal and for the landowner to recover expenses where necessary.

3.4.14

Paragraph 21 is intended to govern the position at the end of a code agreement. The drafting is particularly deficient, given the situation that arises where the code agreement falls within the security of tenure of the 1954 Act. In such a situation (by no means unusual), the landowner cannot give a notice under paragraph 21 unless and until it has observed the termination procedures under the 1954 Act; and it cannot go through those procedures unless and until it has done likewise in relation to the code.

3.4.15

The Old Code does not make clear how code rights are to be enforced. In the absence of any special provision, where a landowner interferes with the exercise of a right, or an operator fails to comply with the obligations to which the exercise is subject, then the remedies available would be those available under the general law, of which an injunction or damages are the most obvious remedies.

3.4.16

Finally, the code contains a number of provisions about payment in certain circumstances, described in a number of different ways, principal among which are (1) compensation; and (2) consideration. Again, the way in which these provisions are drafted has given rise to considerable debate.[3]

3.5 The case for reform of the Old Code

3.5.1

When the Old Code was devised, the telecommunications network in the UK relied primarily upon landlines, with little communication by either internet or mobile telephony. The situation has changed enormously since the closing years of the 19th century and the opening years of the 20th. The telecommunications sector presided over a huge investment in network expansion and upgrades; while correspondingly demand for the enhanced services it was able to offer had increased exponentially. The value of the telecommunications industry is enormous: according to a paper from the Department for Business, Innovation and Skills, the electronic communications market in the UK was valued at about £35 billion in 2010.[4]

3.5.2

The Government had come to acknowledge that there had been a profound shift in the way its citizens approached and accessed digital communications. What was once seen as a luxury had become a basic need, as people expected to have access to fast broadband at home, irrespective of where they live, and to use their mobile devices wherever they might be. To take the example of broadband, the average broadband speed is now (2017) eight times faster than in 2008, while superfast coverage has grown from 45 per cent in 2010 to 90 per cent today.[5]

3.5.3

A similar transformation was under way in mobile telephony, with an increase in UK data subscriptions from 33 million in 2011 to nearly 84 million in 2015, and 4G coverage

3 *Mercury Communications Ltd v London and India Dock Investments Ltd* (1995) 69 P & CR 135; *Cabletel Surrey and Hampshire Ltd v Brookwood Cemetery Ltd* [2002] EWCA Civ 720; *Geo Networks Ltd v The Bridgewater Canal Co. Ltd* [2011] 1 WLR 1487.

4 Implementing the Revised EU Electronic Communications Framework – Overall approach and consultation on specific issues (September 2010).

5 Section 2.12 of Chapter 2 of this book attempts to peer into the future, as the Government's Digital Strategy seeks to anticipate the exponential growth in wireless services that will accelerate with the introduction in 2020 and beyond of the fifth generation of wireless services (5G). While nothing is certain, we can be reasonably sure that there will be a greater demand in the future for land suitable for the siting of equipment, with which the New Code will have to cope.

now reaching 90 per cent of all UK premises since it was launched towards the end of 2012.

3.5.4

In order to accommodate this explosion in digital demand, the New Code was devised to make major reforms to the rights that communications providers have to access land, to deploy and maintain their infrastructure. New rights to upgrade and share would allow future generations of technology to be quickly rolled out as they become commercially viable. There were also to be administrative changes to court processes to allow for improved dispute resolution, ensuring that disagreements between communications providers and landowners do not hold up investment and create uncertainty.

3.5.5

Moreover, the basis of compensation for affected landowners was to change to a 'no scheme' basis of valuation regime – a basis that would ensure that landowners would be fairly compensated for use of their land, while acknowledging the economic value for all society created from investment in digital infrastructure.

3.5.6

Quite apart from these policy considerations, practitioners had long been concerned about the many drafting deficiencies in the Old Code. These had not gone unremarked in the Courts. In his judgment in *Geo Networks Ltd v The Bridgewater Canal Co. Ltd*[6], Lewison J remarked:

> The Code is not one of Parliament's better drafting efforts. In my view it must rank as one of the least coherent and thought-through pieces of legislation on the statute book.

3.5.7

In all those circumstances, the stage was clearly set for code reform.

6 [2010] 1 WLR 2576 at [7].

4 The European dimension

4.1 Introduction

4.1.1

This chapter gives an overview of the European background to the regulation of electronic communications insofar as they relate to property rights.

4.1.2

As seen in Chapter 2, the law relating to the use of land by electronic communications providers is, in origin, based on home-grown, domestic legislation in the form of the Telegraph Acts. The Telecommunications Code, as originally enacted, was a deliberate replication of the Telegraph Acts with, in some cases, rather minimal adaptation.[1] Certainly the Old Code in its unamended form did not bear any obvious traces of EU influence.

4.1.3

It is worth noting that the New Code in some instances, most notably in relation to the special regimes, replicates and builds on the Old Code (which in turn largely replicated the law that came before it). Even the new general regime is a hybrid of the old general regime, albeit heavily adapted to meet changing market conditions, and Part II of the Landlord and Tenant Act 1954.

4.1.4 EU regulation

The origins of European legislative intervention in the telecommunications sector can be traced to the Directive on the establishment of the internal market for telecommunications services through the implementation of open network provision, and the consultation that preceded it.[2] EU policy has been moving steadily towards much greater market integration and liberalisation through various phases of legislative intervention.[3]

1 *Access to Telephone Services: A Government Consultative Document on the Reform of the Telegraph Acts*, July 1982.
2 Council Directive 90/387/EEC of 28 June 1990.
3 For a detailed account, see Walden, *Telecommunications Law and Regulation* (4th ed., 2012), Chapter 3.

4.1.5

The pace quickened when the European Commission, in 1999, presented a communication to the European Parliament.[4] In that communication, the Commission reviewed the existing regulatory framework for telecommunications, in accordance with its obligation under Article 8 of Council Directive 90/387/EEC of 28 June 1990 on the establishment of the internal market for telecommunications services through the implementation of open network provision. It also presented a series of policy proposals for a new regulatory framework for electronic communications infrastructure and associated services for public consultation.

4.1.6

For present purposes, the most significant developments occurred in the early 2000s when the New Regulatory Framework was introduced, comprising four separate directives aimed at various aspects of the telecommunications, which became 'electronic communications' in 2003.

4.1.7

The UK Communications Act 2003 introduced a wider, European, dimension to electronic communications generally, but also specifically in relation to matters falling under the Code. Indeed, the adoption in 2002 of the five Directives referred to in the next section was a significant factor in the implementation of the 2003 Act.[5]

4.2 The European Framework

4.2.1

By a series of directives and regulations, the European Union has regulated a number of matters relation to the electronic communications sector, with a view to encouraging competition, improving the functioning of the digital single market and guaranteeing basic user rights.

4.2.2

The so-called 'framework' is a package of five directives[6] (which have since been amended and unofficially consolidated) and two regulations, as follows:

(a) The 'Framework Directive', which is based on the Framework Directive 2002/21/EC and the Better Regulation Directive 2009/140/EC;

4 On 10 November 1999, 'Towards a New Framework for Electronic Communications Infrastructure and Associated Services – the 1999 Communications Review'.

5 See the Explanatory Notes to the 2003 Act.

6 In a European Union context, Directives are binding on the member states 'as to the result to be achieved', but member states can take different approaches to their implementation provided that the result is achieved within the timeframe specified in the Directive. For the background to the Directives, see European

(b) The 'Access Directive', which is based on the Access Directive 2002/19/EC and the Better Regulation Directive 2009/140/EC;
(c) The 'Authorisation Directive', which is based on the Authorisation Directive 2002/20/EC and the Better Regulation Directive 2009/140/EC;
(d) The 'Universal Service Directive', which is based on the Universal Service Directive 2002/22/EC and the Citizens' Rights Directive 2009/136/EC;
(e) The Directive on Privacy and Electronic Communications is based on the Directive on Privacy and Electronic Communications 2002/58/EC, the Amending Directive 2006/24/EC and the Citizens' Rights Directive 2009/136/EC.[7]

4.2.3

The most relevant of the above for present purposes are the Framework and the Authorisation directives. It will be seen that these directives place the sharing of apparatus at the heart of the scheme for a single digital market.

4.2.4

However, it is worth drawing attention to the effect of the Better Regulation Directive 2009/140/EC, which amended three of the earlier directives. It recited, at (42), that

> Permits issued to undertakings providing electronic communications networks and services allowing them to gain access to public or private property are essential factors for the establishment of electronic communications networks or new network elements. Unnecessary complexity and delay in the procedures for granting rights of way may therefore represent important obstacles to the development of competition. Consequently, the acquisition of rights of way by authorised undertakings should be simplified. National regulatory authorities should be able to coordinate the acquisition of rights of way, making relevant information accessible on their websites.

4.2.5

It will therefore be noted that the Better Regulation Directive has, as one of its objectives, that access to private land should be more easily achievable across the EU. That directive further recites at (43) that

> It is necessary to strengthen the powers of the Member States as regards holders of rights of way to ensure the entry or roll-out of a new network in a fair, efficient and environmentally responsible way and independently of any obligation on an operator with significant market power to grant access to its electronic communications network. Improving facility sharing can significantly improve competition and lower

Commission, 'Towards a New Framework for Electronic Communications infrastructure and associated services – the 1999 Communications Review' COM (1999) 539.

7 The official website of the European Union gives a useful summary of the electronic communications framework, including all relevant legislation and amendments: see http://europa.eu/legislation_summaries/information_society/legislative_framework/124216 a_en.htm.

the overall financial and environmental cost of deploying electronic communications infrastructure for undertakings, particularly of new access networks. National regulatory authorities should be empowered to require that the holders of the rights to install facilities on, over or under public or private property share such facilities or property (including physical co-location) in order to encourage efficient investment in infrastructure and the promotion of innovation, after an appropriate period of public consultation, during which all interested parties should be given the opportunity to state their views. Such sharing or coordination arrangements may include rules for apportioning the costs of the facility or property sharing and should ensure that there is an appropriate reward of risk for the undertakings concerned. National regulatory authorities should in particular be able to impose the sharing of network elements and associated facilities, such as ducts, conduits, masts, manholes, cabinets, antennae, towers and other supporting constructions, buildings or entries into buildings, and a better coordination of civil works. The competent authorities, particularly local authorities, should also establish appropriate coordination procedures, in cooperation with national regulatory authorities, with respect to public works and other appropriate public facilities or property which may include procedures that ensure that interested parties have information concerning appropriate public facilities or property and on-going and planned public works, that they are notified in a timely manner of such works, and that sharing is facilitated to the maximum extent possible.

4.2.6

This accordingly enshrines the principle that infrastructure ought to be shared, and that Member States ought to be free to impose sharing and co-location conditions in relation to 'rights of way'. It is plain that 'rights of way' ought to be interpreted as meaning 'access to land' as opposed to a mere right to pass over land.

4.3 The Framework Directive following its 2009 Amendment

4.3.1

So far as relevant to this book, the objectives of the Framework Directive are as set out in Article 1(1):

> This Directive establishes a harmonised framework for the regulation of electronic communications services, electronic communications networks, associated facilities (19) OJ L 184, 17.7.1999, p. 23. (20) OJ L 336, 30.12.2000, p. 4. 45 and associated services, and certain aspects of terminal equipment to facilitate access for disabled users. It lays down tasks of national regulatory authorities and establishes a set of procedures to ensure the harmonised application of the regulatory framework throughout the Community.

4.3.2

The recitals to the Framework Directive further explain:

> (22) It should be ensured that procedures exist for the granting of rights to install facilities that are timely, non-discriminatory and transparent, in order to guarantee the conditions for fair and effective competition. This Directive is

without prejudice to national provisions governing the expropriation or use of property, the normal exercise of property rights, the normal use of the public domain, or to the principle of neutrality with regard to the rules in Member States governing the system of property ownership.

(23) Facility sharing can be of benefit for town planning, public health or environmental reasons, and should be encouraged by national regulatory authorities on the basis of voluntary agreements. In cases where undertakings are deprived of access to viable alternatives, compulsory facility or property sharing may be appropriate. It covers inter alia: physical co-location and duct, building, mast, antenna or antenna system sharing. Compulsory facility or property sharing should be imposed on undertakings only after full public consultation.

(24) Where mobile operators are required to share towers or masts for environmental reasons, such mandated sharing may lead to a reduction in the maximum transmitted power levels allowed for each operator for reasons of public health, and this in turn may require operators to install more transmission sites to ensure national coverage.

4.3.3

Whilst, therefore, the directive acknowledges that property law systems are a subsidiary matter for the relevant member state, it is clear that it is intended that rights to allow installation of apparatus to facilitate network roll-out are swiftly granted.

4.3.4

The Framework Directive contains definitions of various terms which now form part of the Old Code and the New Code, though the definitions in the directive are significantly different than those deployed in domestic legislation. The relevant definitions under Article 2 are:

(a) 'electronic communications network' means transmission systems and, where applicable, switching or routing equipment and other resources, including network elements which are not active, which permit the conveyance of signals by wire, radio, optical or other electromagnetic means, including satellite networks, fixed (circuit and packet-switched, including Internet) and mobile terrestrial networks, electricity cable systems, to the extent that they are used for the purpose of transmitting signals, networks used for radio and television broadcasting, and cable television networks, irrespective of the type of information conveyed

(b) 'electronic communications service' means a service normally provided for remuneration which consists wholly or mainly in the conveyance of signals on electronic communications networks, including telecommunications services and transmission services in networks used for broadcasting, but exclude services providing, or exercising editorial control over, content transmitted using electronic communications networks and services; it does not include information society services, as defined in Article 1 of Directive 98/34/EC, which do not consist wholly or mainly in the conveyance of signals on electronic communications networks;

[…]

(m) 'provision of an electronic communications network' means the establishment, operation, control or making available of such a network;

4.3.5

It is clear from the legislative background to s.32 of the Communications Act 2003 that that section was intended to implement those definitions, and transpose them into English domestic law. However, there is arguably a mismatch between the definitions in s.32 and the definitions in Article 2 of the Framework Directive, and as a result that directive has arguably been under-implemented, as this section explains.

4.3.6

Article 8 sets out the objectives which the relevant national regulatory authority is obliged to achieve. These include the following requirements:

3. The national regulatory authorities shall contribute to the development of the internal market by inter alia: (a) removing remaining obstacles to the provision of electronic communications networks, associated facilities and services and electronic communications services at European level; (b) encouraging the establishment and development of trans-European networks and the interoperability of pan-European services, and end-to-end connectivity;
4. The national regulatory authorities shall promote the interests of the citizens of the European Union by inter alia: (a) ensuring all citizens have access to a universal service specified in Directive 2002/22/EC (Universal Service Directive);

4.3.7

Specific regulation in relation to land use is then made by Articles 11 and 12.

4.3.8

Article 11 provides that:

1. Member States shall ensure that when a competent authority considers:
   - an application for the granting of rights to install facilities on, over or under public or private property to an undertaking authorised to provide public communications networks, or
   - an application for the granting of rights to install facilities on, over or under public property to an undertaking authorised to provide electronic communications networks other than to the public,
   - the competent authority:
   - acts on the basis of simple, efficient, transparent and publicly available procedures, applied without discrimination and without delay, and in any event makes its decision within six months of the application, except in cases of expropriation, and
   - follows the principles of transparency and non-discrimination in attaching conditions to any such rights. The abovementioned procedures can differ depending on whether the applicant is providing public communications networks or not.

2. Member States shall ensure that where public or local authorities retain ownership or control of undertakings operating public electronic communications networks and/or publicly available electronic communications services, there is an effective structural separation of the function responsible for granting the rights referred to in paragraph 1 from the activities associated with ownership or control.
3. Member States shall ensure that effective mechanisms exist to allow undertakings to appeal against decisions on the granting of rights to install facilities to a body that is independent of the parties involved.

4.3.9

Quite what Article 11(1) requires is a little obscure when applied to UK market practices. In relation to public land, it would appear to impose an obligation that consent for installations be given within a six-month period from any application. However, in relation to private land, those rights are, in the UK, a matter of private negotiation unless dispute resolution procedures are invoked on the Old or New Code to acquire rights compulsorily or have their terms settled by a Court, Tribunal or arbitrator (as the case may be).[8]

4.3.10

This article has been implemented in the United Kingdom by regulation 3 of the Electronic Communications and Wireless Telegraphy Regulations 2011. The regulation applies where:

(a) a person authorised to provide public electronic communications networks applies to a competent authority for the granting of rights to install facilities on, over or under public or private property for the purposes of such a network, [or]

(b) a person authorised to provide electronic communications networks other than to the public applies to a competent authority for the granting of rights to install facilities on, over or under public authority for the purposes of such a network ...

And it requires that:

(2) except in cases of expropriation, the competent authority must make its decision within 6 months of receiving the completed application.

8 This is not a universal approach to implementation. In Germany, for instance, under the Telekommunikationsgesetz (TKG), the acquisition of rights is not on a consensual basis. Although (as in the UK) operators are entitled to use public land as of right pursuant to §68(1) and 69 TKG. Public bodies owning infrastructure suitable for next generation electronic communications are required to share the same with operators on request (§77b). Operators have the right to request the use of federal long-distance roads (§77c), waterways (§77d) and railways (§77e). In relation to private land, private landowners may not refuse an operator access to install, operate or renew apparatus on such land insofar as either this installation, etc., makes use of an existing line on the land, or does not have a disproportionate adverse effect on the land in question. There are compensation provisions built into §§76 and 77.

4.3.11

The Department for Business, Innovation and Skills has described six months as a 'challenging timescale' where the competent authority is the county court,[9] but undertook to 'work with the Ministry of Justice and the courts' to meet it.[10]

4.3.12

Article 12 deals with co-location and sharing, and embeds the principle of apparatus sharing in EU law:

1. Where an undertaking providing electronic communications networks has the right under national legislation to install facilities on, over or under public or private property, or may take advantage of a procedure for the expropriation or use of property, national regulatory authorities shall, taking full account of the principle of proportionality, be able to impose the sharing of such facilities or property, including buildings, entries to buildings, building wiring, masts, antennae, towers and other supporting constructions, ducts, conduits, manholes, cabinets.
2. Member States may require holders of the rights referred to in paragraph 1 to share facilities or property (including physical co-location) or take measures to facilitate the coordination of public works in order to protect the environment, public health, public security or to meet town and country planning objectives and only after an appropriate period of public consultation, during which all interested parties shall be given an opportunity to express their views. Such sharing or coordination arrangements may include rules for apportioning the costs of facility or property sharing.
3. Member States shall ensure that national authorities, after an appropriate period of public consultation during which all interested parties are given the opportunity to state their views, also have the power to impose obligations in relation to the sharing of wiring inside buildings or up to the first concentration or distribution point where this is located outside the building, on the holders of the rights referred to in paragraph 1 and/or on the owner of such wiring, where this is justified on the grounds that duplication of such infrastructure would be economically inefficient or physically impracticable. Such sharing or coordination arrangements may include rules for apportioning the costs of facility or property sharing adjusted for risk where appropriate.
4. Member States shall ensure that competent national authorities may require undertakings to provide the necessary information, if requested by the competent authorities, in order for these authorities, in conjunction with national regulatory authorities, to be able to establish a detailed inventory of the nature, availability and geographical location of the facilities referred to in paragraph 1 and make it available to interested parties.

9 The County Court is the main forum for adjudication of issues arising under the Old Code in England and Wales.

10 Department for Business, Innovation and Skills, *Implementing the Revised EU Electronic Communications Framework: Overall Approach and Consultation on Specific Issues* (Sept. 2011), p.22. Policy responsibility for these issues has now passed to DCMS.

5. Measures taken by a national regulatory authority in accordance with this Article shall be objective, transparent, non-discriminatory, and proportionate. Where relevant, these measures shall be carried out in coordination with local authorities.

4.4 The Access Directive

4.4.1

Article 1 of the Access Directive sets out its scope and aims as follows:

1. Within the framework set out in Directive 2002/21/EC (Framework Directive), this Directive harmonises the way in which Member States regulate access to, and interconnection of, electronic communications networks and associated facilities. The aim is to establish a regulatory framework, in accordance with internal market principles, for the relationships between suppliers of networks and services that will result in sustainable competition, interoperability of electronic communications services and consumer benefits.
2. This Directive establishes rights and obligations for operators and for undertakings seeking interconnection and/or access to their networks or associated facilities. It sets out objectives for national regulatory authorities with regard to access and interconnection, and lays down procedures to ensure that obligations imposed by national regulatory authorities are reviewed and, where appropriate, withdrawn once the desired objectives have been achieved. Access in this Directive does not refer to access by end-users.

4.4.2

Further, it recites at (24) that:

> The development of the electronic communications market, with its associated infrastructure, could have adverse effects on the environment and the landscape. Member States should therefore monitor this process and, if necessary, take action to minimise any such effects by means of appropriate agreements and other arrangements with the relevant authorities.

4.4.3

Article 3 provides that:

> Member States shall ensure that there are no restrictions which prevent undertakings in the same Member State or in different Member States from negotiating between themselves agreements on technical and commercial arrangements for access and/or interconnection …

4.4.4

At the heart of these provisions lies infrastructure sharing. Article 12.1 provides that:

A national regulatory authority may, in accordance with the provisions of Article 8, impose obligations on operators to meet reasonable requests for access to, and use of, specific network elements and associated facilities, inter alia in situations where the national regulatory authority considers that denial of access or unreasonable terms and conditions having a similar effect would hinder the emergence of a sustainable competitive market at the retail level, or would not be in the end-user's interest. Operators may be required inter alia: (a) to give third parties access to specified network elements and/or facilities, including access to network elements which are not active and/or unbundled access to the local loop, to inter alia allow carrier selection and/or pre-selection and/or subscriber line resale offers; (b) to negotiate in good faith with undertakings requesting access; (c) not to withdraw access to facilities already granted; (d) to provide specified services on a wholesale basis for resale by third parties; (e) to grant open access to technical interfaces, protocols or other key technologies that are indispensable for the interoperability of services or virtual network services; (f) to provide co-location or other forms of associated facilities sharing; (g) to provide specified services needed to ensure interoperability of end-to-end services to users, including facilities for intelligent network services or roaming on mobile networks; (h) to provide access to operational support systems or similar software systems necessary to ensure fair competition in the provision of services; (i) to interconnect networks or network facilities. (j) to provide access to associated services such as identity, location and presence service. National regulatory authorities may attach to those obligations conditions covering fairness, reasonableness and timeliness.

4.5 The Digital Single Market

4.5.1

The Digital Single Market proposal was announced by the Juncker Commission in June 2015.[11] As explained in the Commission Staff Working Document of June 2015:[12]

The European Commission's Digital Agenda for Europe (DAE) formed one of the seven pillars of the Europe 2020 Strategy, which set objectives for the growth of the European Union (EU) by 2020. It defined in particular a strategy to take advantage of the potential offered by the rapid progress of digital technologies, in order to generate smart, sustainable and inclusive growth in Europe. The Digital Agenda's main objective, which is also one of the ten priorities of the new Commission, is to develop a Digital Single Market. In order to achieve this objective, on 6 May 2015 the Commission adopted a Digital Single Market Strategy. The strategy, which has a multiannual scope, focuses on key interdependent actions to be taken at EU level. The Strategy is built on three pillars, one of which is the creation of the right conditions for digital networks and services to flourish. This requires well-functioning markets that can deliver access to high-performance fixed and wireless broadband infrastructure at affordable prices. In this regard, the EU's telecoms rules aim to

11 http://eur-lex.europa.eu/legal-content/EN/TXT/HTML/?uri=CELEX:52015DC0192
12 SWD (2015) 126 (final), p.4: https://ec.europa.eu/transparency/regdoc/rep/10102/2015/EN/10102-2015-126-EN-F1-1.PDF.

ensure that markets operate more competitively and bring lower prices and better quality of service to consumers and businesses, while ensuring the right regulatory conditions for innovation, investment, fair competition and a level playing field.

4.5.2

Commenting on the fragmentary and disparate nature of the market within EU member states (albeit at a time when the Old Code was in force), the Working Document notes that:

> Complex, cumbersome and fragmented procedures on this issue were reported in Bulgaria, the Czech Republic and Luxembourg, although measures to address inefficiencies are in the pipeline. In Ireland and the United Kingdom, providers face burdensome negotiations with private landlords, while in Poland, the time it takes to grant permits is being drawn out by an increasing number of court cases. It usually takes more time to grant a permit for the deployment of mobile networks than for fixed ones, although standardised small antennae are sometimes exempt from the permit granting procedure. Tacit approval is applied in some Member States countries mostly with regard to permits for the deployment of fixed networks and, in Portugal and Romania, for rights of way.

4.5.3

Commenting on the position in the UK, the paper noted that:[13]

> Under the UK's Electronic Communications Code (ECC), Ofcom has powers to grant general entitlements to rights of way over public or private land to operators, subject to consultation; the average time is six months. In February 2013, the Law Commission made recommendations to the UK Government on reforming the ECC. These focus on private property rights between landowners and operators and do not consider planning issues. Many of them were included in a new ECC, which on 13 January 2015 the Government included in an amendment of the Infrastructure Bill then going through Parliament. However, on 22 January the Government withdrew the draft ECC from the legislative procedure. On 26 February it launched a consultation on a new draft and an accompanying draft bill that is very similar to the withdrawn amendments. The deadline for the consultation was 30 April 2015.

4.5.4

In 2016, the Commission proposed the roll-out of a European Electronic Communications Code, as part of the Digital Single Market project which the Commission is treating as a priority project.[14] As is plain from what has already been said, 'it is a key aspect of the Juncker Commission's political commitment to deliver the digital single market'.[15]

13 At p.321.
14 http://eur-lex.europa.eu/legal-content/EN/ALL/?uri=comnat:COM_2016_0590_FIN
15 SWD(2016) 305 final Commission Staff Working Document, p.2: https://ec.europa.eu/digital-single-market/en/news/proposed-directive-establishing-european-electronic-communications-code.

4.5.5

The draft directive consolidates the earlier directives, but also adds new provisions, for instance (see below) in relation to sharing and co-location. As matters stand, it remains to be seen whether the New Code will be superseded by more detailed directives promulgated under the auspices of the Digital Single Market plan, which the United Kingdom may be obligated to transpose into national law insofar as the obligation to do so arises prior to Brexit, or insofar as the further integration of the Digital Single Market forms part of the negotiated package of rights and obligations post-Brexit.

4.5.6

The proposed directive includes two broad provisions in relation to land use by operators. The first is Article 43, which is derived from the 2002 Framework Directive. This provides:

> Rights of way
>
> 1. Member States shall ensure that when a competent authority considers:
>   - an application for the granting of rights to install facilities on, over or under public or private property to an undertaking authorised to provide public communications networks, or
>   - an application for the granting of rights to install facilities on, over or under public property to an undertaking authorised to provide electronic communications networks other than to the public, the competent authority:
>   - acts on the basis of simple, efficient, transparent and publicly available procedures, applied without discrimination and without delay, and in any event makes its decision within six months of the application, except in cases of expropriation, and
>   - follows the principles of transparency and non-discrimination in attaching conditions to any such rights.
>
> The abovementioned procedures can differ depending on whether the applicant is providing public communications networks or not.
>
> 2. Member States shall ensure that where public or local authorities retain ownership or control of undertakings operating public electronic communications networks and/or publicly available electronic communications services, there is an effective structural separation of the function responsible for granting the rights referred to in paragraph 1 from the activities associated with ownership or control.

4.5.7

It also includes a wholly new co-location and sharing provision in the form of proposed Article 44:

> Co-location and sharing of network elements and associated facilities for providers of electronic communications networks

1. Where an operator has exercised the right under national legislation to install facilities on, over or under public or private property, or has taken advantage of a procedure for the expropriation or use of property, competent authorities shall, be able to impose co-location and sharing of the network elements and associated facilities installed, in order to protect the environment, public health, public security or to meet town and country planning objectives. Co-location or sharing of networks elements and facilities installed and sharing of property may only be imposed after an appropriate period of public consultation, during which all interested parties shall be given an opportunity to express their views and only in the specific areas where such sharing is deemed necessary in view of pursuing the objectives provided in this Article. Competent authorities shall be able to impose the sharing of such facilities or property, including land, buildings, entries to buildings, building wiring, masts, antennae, towers and other supporting constructions, ducts, conduits, manholes, cabinets or measures facilitating the coordination of public works. Where necessary, national regulatory authorities shall provide rules for apportioning the costs of facility or property sharing and of civil works coordination.
2. Measures taken by a competent authority in accordance with this Article shall be objective, transparent, non-discriminatory, and proportionate. Where relevant, these measures shall be carried out in coordination with the national regulatory authorities.

4.5.8

It will be interesting to see to what extent the Commission's proposals will ultimately require interference with private property rights in order to create a friction-free single digital market. The directives have, to date, been clear that member states' property laws will not be interfered with, and interference in internal property law would seem to be precluded by Article 345 TFEU,[16] which provides that: 'The Treaties shall in no way prejudice the rules in Member States governing the system of property ownership.'

4.5.9

However, that said, there is also an argument that Article 345 is subsidiary to the four fundamental freedoms guaranteed by the TFEU,[17] and must give way where local property laws are an impediment to securing those freedoms. As already indicated by the Commission (albeit under the Old Code), negotiations for the use of private land in the UK remain too cumbersome and slow from its perspective. It may yet be, therefore, that the opening up of private land is a matter that needs to be considered by the Commission in future. In practice, much of the UK's electronic communications network and service provision depends upon the availability of private land to host infrastructure – indeed it would be inconceivable to have such systems without such land being used.

16 The accepted abbreviation for the 'Treaty on the Functioning of the European Union', the governing consolidated treaty: http://eur-lex.europa.eu/legal-content/EN/TXT/?uri=celex%3A12012E%2FTXT.

17 See the discussion in van Erp and Akkermans, *Property Law* (2012), at pp. 1034–5.

4.5.10

Whether the New Code sufficiently addresses the needs of the Commission is an open decision, and whether the United Kingdom, either before Brexit or as part of the negotiated package of rights and obligations post-Brexit, will need to go further in order to give effect to the Digital Single Market proposals, is a matter that will need to be kept under review. It is clear that the future of the UK in the Digital Single Market is yet to be resolved.[18]

4.5.11

On 28th June 2018, EU ambassadors approved the rules which will be comprised in the new Electronic Communications Code.[19] The anticipated legislative timeline is that the new rules will be adopted and published in the Official Journal before the end of 2018[20]. This means that the new European Code could become law (requiring transposition into national law within two years) before Brexit.

18 https://publications.parliament.uk/pa/cm201617/cmselect/cmeuleg/71-xxxiv/7104.htm
19 http://www.consilium.europa.eu/en/policies/electronic-communications-code/
20 http://www.consilium.europa.eu/en/press/press-releases/2018/06/29/telecoms-reform-to-bolster-better-and-faster-connectivity-across-eu-approved-by-member-states

Part II

Electronic Communications Code 2003 (the Old Code)

5 Old Code

General and special regime overview

5.1 The structure of the Old Code

5.1.1

The Old Code draws a distinction between the 'general regime' (paragraphs 2–7) which is in principle applicable to all land, and the special regimes (paragraphs 9–15) which govern particular kinds of land, or particular land uses. In very general terms, the special regimes appear to be either cases in which the land in question is public or quasi-public in character, is land which in by its nature likely to be long and therefore require a number of crossings by the cables and wires making up a communications network, and where the cost to the occupier, landowner or undertaker whose land is crossed is likely to be slight. That risks oversimplification, however, and a number of key differences between the regimes are picked up in this chapter and further examined in the following chapters in this part of this book.

5.1.2

The distinction between the regimes was first made by Lewison J (as he then was) in *Bridgewater Canal Co. Ltd v Geo Networks Ltd:*[1]

> The general regime
>
> 8. In the ordinary case an operator of a network requires the agreement of an occupier of land in order to exercise rights under the Code. These rights include the right to execute works connected with the installation maintenance adjustment repair or alteration of apparatus; the keeping of apparatus on under or over the land; or the entry on to the land to inspect apparatus (§ 2(1)). The occupier's consent is needed before any of these rights can be exercised; and is needed for all of them. Normally a consent given by the occupier of land who is neither the freeholder nor a lessee will only bind the freeholder or lessee if they agree to be bound. If an operator requires an occupier to agree that any of these rights should be conferred on an operator, he may give notice to that effect (§ 5 (1)). Such a notice may also be given if the operator wants consent given by an occupier to bind holders of other interests in the

1 [2010] EWHC 548 (Ch); this part of the judgment is unaffected by the Court of Appeal's reversal of the decision.

land. If no agreement has been given after 28 days, the operator can apply to the court for an order conferring the proposed right (§ 5 (2)). On such an application the court must make an order in the operator's favour; but only if one of two conditions is satisfied. The first is that any prejudice caused by the order is capable of being compensated by money. The second is that any such prejudice is outweighed by the benefit accruing from the order (§ 5 (3)). In exercising this power the court must have regard to the principle that no person should unreasonably be denied access to an electronic communications network (§ 5 (3)). This principle recurs in the Code and is clearly one to which Parliament attached considerable importance. The terms of an order must include terms and conditions to ensure that the least possible loss and damage is caused by the exercise of the right (§ 5 (5)). In addition the order must include financial terms. The financial terms that the order must include are:

i) Terms for the payment of such consideration as appear to the court would have been fair and reasonable if the agreement had been given willingly; and
ii) Terms ensuring that persons bound by the right are adequately compensated for loss and damage (§ 7 (1)).

9. The payments may be periodical and the court may require their amount to be determined by arbitration (§ 7 (4)). The court may also determine the persons to whom payments must be made.
10. Where an operator exercises a right under the Code he may be liable to pay compensation for injurious affection to neighbouring land. This is assessed in the same way as compensation for injurious affection under section 10 of the Compulsory Purchase Act 1965 (§ 16). I note, in passing, that it is not easy to see how 'neighbouring land' is defined in a case where all that the operator wants to do is to run a cable through an existing duct.
11. The Code also contains restrictions on a person's ability to require the removal of apparatus. Where apparatus is kept installed in over or under land, a person with an interest in the land may give the operator notice requiring the alteration of the apparatus. He may do this 'notwithstanding the terms of any agreement binding' him (§ 20 (1)). Alteration of the apparatus includes its removal (§ 1 (2)). However, notice may only be given on the ground that the alteration is necessary to enable the carrying out of an improvement (which includes redevelopment and change of use) (§ 20 (1), (9)). The operator may give a counter-notice within 28 days. If he does, then the alteration will only be made if the court orders it to be made (§ 20 (2)). The Code circumscribes the court's power to make such an order. First the court must be satisfied that the alteration is necessary for the carrying out of the improvement. Second, it must be satisfied that the alteration will not substantially interfere with any service which is or is likely to be provided by the operator's network (§ 20 (4)). Third, the court must be satisfied that the operator already has all the rights needed to make the alteration, or could obtain them on an application made to the court (§ 20 (5)). If the court makes an order, the operator is entitled to reimbursement of his expenses of compliance with the order (§ 20 (8)); but is not otherwise entitled to compensation.
12. Where a person is entitled to require the removal of apparatus he must first serve notice on the operator. The operator may then serve a counter-notice

within 28 days. The counter-notice may either contest the alleged right to require removal or may specify steps which the operator will take in order to acquire a right to keep the apparatus on the land (§ 21). In other words, the operator may apply to the court for an order conferring the right to keep the apparatus. Once a counter-notice has been served, the apparatus can only be removed under a court order (§ 21 (5)).

13. I shall call these provisions (and in particular the provisions of paragraphs 2 to 7) 'the general regime'.

Special regimes

14. Running alongside the general regime are a number of special regimes. They are dealt with by paragraphs 9 to 12, which are expressly excluded from the general regime (§ 2 (9)). Thus, for example, an operator has the right to install apparatus over, under, along or across a publicly maintainable highway without going through the process of seeking agreement or going to court (§ 9). The highway authority has no right under the Code to object to the works. If an operator exercises this right, he is not required to pay compensation to the highway authority; although a charge may be payable under the New Roads and Street Works Act 1991. By contrast in the case of a private street he must go through the general regime (§ 9 (2)).
15. An operator may fly lines over land if the lines connect apparatus on adjacent or neighbouring land (§ 10). He may do this without giving notice to anybody; but the owner or occupier of the land over which the line flies or land in the vicinity which is prejudiced by the installation may object within three months (§ 17). The court will then decide the merits of any objection.
16. An operator also has the right to install apparatus on or under tidal waters or land (§ 11). If the land is one in which there is a Crown interest, he must obtain the consent of the Crown (§§ 11 (2), 26). The court does not have power to dispense with the Crown's consent. Land covered by tidal waters will normally be owned by the Crown, but this is not invariable. There may have been an ancient grant; or title to the bed of a tidal river may have been acquired by adverse possession. If the land is not owned by the Crown then it seems that no consent is needed. However, before executing any works, the operator must submit a plan of the proposed works to the Secretary of State who may approve it with or without modifications. The operator need not notify anyone else. The Secretary of State must consult authorities exercising functions in relation to the tidal water or lands in question (§ 11 (5)). He has no obligation to consult (or even notify) riparian owners. But he must not approve a plan unless he is satisfied that adequate arrangements have been made for compensating owners of interests in the land in question for loss or damage sustained in consequence of the execution of the works (§ 11 (6)). These provisions will be replaced in due course by the regime to be established by Part 4 of the Marine and Coastal Access Act 2009 under which the functions of the Secretary of State will be taken over by licensing authorities.
17. Paragraphs 12 to 14 deal with linear obstacles. They are at the heart of this appeal and I will need to consider them in detail. For the moment I pass over them.

18. Special provisions apply to undertakers' works. An undertaker is a person authorised by Act of Parliament to carry on any railway, tramway, road transport, water transport, canal, inland navigation, dock, harbour, pier or lighthouse undertaking (§ 23 (10)). If an undertaker wants to execute works which are likely to involve a temporary or permanent alteration to an operator's apparatus it must give notice to the operator (§ 23 (2)). The operator may give counter-notice either agreeing to carry out the work itself, or requiring the undertaker to carry out the work under the supervision of and to the satisfaction of the undertaker (§ 23 (4)). In either case the undertaker must pay the operator compensation for loss and damage suffered in consequence of the alteration (§ 23 (5), (6)). The operator has no right to object to the works; and there is no provision for arbitration or application to the court for the resolution of any dispute about the nature of the works. However, the compensation is recoverable by action.

5.1.3

This judicial classification into the two categories of general and special regimes (expressions that the Old Code does not itself use) are now considered in more detail.

5.2 The general regime

5.2.1

The most commonly encountered cases arise under the general regime, applying, in principle, to all land.

5.2.2

The Court of Appeal in *GEO Networks Ltd v The Bridgewater Canal Co. Ltd*[2] described the general regime in the following terms:

> Paragraphs 2 to 7 contain detailed provisions in relation to what the judge called 'the general regime'. The basic structure of the general regime is to require the operator, such as Geo, to acquire the rights it needs to install and keep electronic communications apparatus on land in the occupation and/or ownership of another by agreement. If agreement cannot be reached then Paragraph 5 provides a mechanism by which the necessary rights are obtained through the giving of a notice specifying the rights required and an order of the County Court if it is satisfied as to the matters set out in paragraph 5(3). The order of the County Court is required by paragraph 7(1) to make provision for the payment of consideration and compensation. The provisions which are particularly relevant are those contained in paragraphs 2(1), 5(2), (3) and (4), and 7(1). The overall effect of the general regime is that all those who have an interest in the relevant land, if bound by the order, are to be entitled to be paid for the right to install and keep the electronic apparatus on their land and to be compensated for any loss or damage sustained in consequence of its installation and maintenance thereon.

2 [2010] EWCA Civ 1348.

5.2.3

As we shall see below, but in more detail in Chapter 9, there are special regimes which apply to particular kinds of land, or particular kinds of activity on land. Both the general and the special regimes are all governed by the same 'security of tenure' provisions under paragraphs 20, 21 and 22, which are considered in Chapters 7, 10 and 11. As paragraphs 20, 21 and 22 span all regimes, it is inaccurate to describe them as part of the 'general regime', although it is the case that they are most often encountered in relation to apparatus installed under those regimes.

5.2.4

The general regime applies to all land (or use of land). Certain types of land, or certain activities on land, are specially regulated by the special regimes contained in other parts of the Old Code. Most uses by operators of private land encountered in practice arise under the general regime.

5.3 The special regimes

5.3.1

The special regimes that apply to particular kinds of land are:

- street works (paragraph 9);
- tidal waters (paragraph 11); and
- 'linear obstacles', meaning a railway, canal or tramway (paragraphs 12–14).

There is also a special regime applicable to any land where it is intended to fly lines over it (paragraph 10). The special regimes are separately considered in Chapter 9. As in each case, the special regimes regulate the activities of 'operators' under the Old Code, the definition of 'operator', considered in Chapter 6, will be equally relevant.

5.3.2

It is considered that, when one of the special regimes is engaged, then the special regime ought to be followed, particularly if it imposes stricter requirements than the general regime. In many cases it will in any event be more advantageous to the operator to do so, as in many cases the provisions for determining consideration payable under the special regimes (if any) favour the operator. That said, there is nothing explicit in the Old Code which prevents an operator from relying on the general regime (assuming the relevant conditions are met) if it chooses to do so.

5.4 General and special regimes: overlaps, similarities and differences

5.4.1

It is to be noted that, although the general and special regimes are distinct, there is some overlap between them.

5.4.2

First, the need for an agreement under paragraph 2 of the general regime is imported into paragraphs 9(2) (street works to a street which is not a maintainable highway) and 11(2) (tidal waters; agreement of the appropriate authority is required under paragraph 26(3) where there is a Crown interest).[3]

5.4.3

Secondly, an agreement to obstruct under paragraph 3 applies not just to the exercise of rights under the general regime, but also to the exercise of rights under paragraph 9 (street works), 10 (flying lines) and 11 (tidal waters) of the special regimes.

5.4.4

Thirdly, certain conduits are made the subject of special consent requirements under paragraph 15.

5.4.5

Fourthly, compensation for injurious affection is payable in relation to any right conferred or acquired by the Old Code (paragraph 16): see Chapter 8.

5.4.6

Fifthly, any apparatus installed, if 'overhead apparatus' under paragraph 17, can be objected to in accordance with that paragraph, see Chapter 9. Paragraph 18 (notices in relation to overhead apparatus) must also be complied with. See Chapter 9.

5.4.7

Sixthly, the 'sundry provisions' in paragraphs 22 and following are in principle applicable to the general and the special regimes; see Chapter 13.

5.4.8

Finally, as noted above, paragraphs 20 (alteration), 21 (removal) and 22 (abandonment) are available in relation to electronic communications apparatus whether installed under the general or the special regimes. These critically important provisions are considered in Chapters 10 to 12.

5.4.9

In addition to the different ways in which the various regimes operate procedurally there are fundamental differences between the general and the special regimes which must be appreciated.

3 See paragraph 2(9).

5.4.10

First, whereas under the general regime, rights must be bargained for under paragraph 2, in default of which they can be compulsorily acquired paragraphs 5–7, under some of the special regimes those rights are automatically conferred on operators (e.g. street works (paragraph 9(1), but only where the road is a highway maintainable at public expense; otherwise paragraphs 2–5 apply (see paragraph 9(2)), flying lines (paragraph 10(1)), tidal waters (paragraph 11(1), though with special consent requirements applicable to the Crown under paragraph 26 – see Chapter 9), 'linear obstacles' (paragraph 12(1)). The justification for the automatic conferral of rights would seem to be either (i) that the land is public or quasi-public (in the case of street works, tidal waters and linear obstacles), and that the land in question tends to be long, narrow tracts of land which any cable or wire would inevitably need to traverse more than once, or (ii) that the interference with use of the land crossed is likely to be trivial, and no more than a nominal trespass, in most cases (flying lines).

5.4.11

Secondly, whereas under the general regime, in default of agreement a 'fair and reasonable' amount of 'consideration', under most (but not all) special regimes, no consideration is payable.[4] This is the case with street works, flying lines and tidal waters. It is not, however, the case with linear obstacles, which tend to be private undertakings. It must be noted, however, that the manner of computing consideration under the linear obstacle special regime differs markedly from that under the general regime, as we shall see below and in Chapter 8.

5.4.12

Thirdly, in relation to tidal waters, the approval of the Secretary of State to the proposed works must be sought (see paragraphs 11(3) to (10)). In relation to linear obstacles (paragraph 12(4)–(6)), there is a right to object to the manner in which the rights conferred are to be implemented, as opposed to the conferral of the right itself (which is automatically granted in each case). There is also a special right to have the apparatus altered under paragraph 14, which seems to operate as an additional power to that contained in paragraph 20.

5.4.13

Fourthly, the dispute resolution machinery differs. The general regime places dispute resolution in the hands of the Court:[5] see paragraphs 5–7, governing whether a right should be granted and on what (financial and non-financial) terms. In relation to linear obstacles, there is a right to arbitration but limited to the mode of implementation of the right under paragraph 13.

4 Though compensation is payable by virtue of paragraph 16.
5 Meaning the County Court – see paragraph 1(1) of the Old Code and Chapter 32.

5.4.14

These differences were considered by the Court of Appeal in *GEO Networks Ltd v The Bridgewater Canal Co. Ltd.*[6] The Court explained the different policy choices made by the legislator as follows:

> [25] The overriding principle proclaimed in paragraphs 5(4), 13(5) and elsewhere is that 'no person should unreasonably be denied access to an electronics communications network or to electronic communications services.'
>
> [26] In furtherance of that principle the Code sought the agreement of the relevant occupiers/owners affected by the need to install and keep electronic communications apparatus. The general regime concentrated at the outset on the occupier no doubt because he was primarily affected but also most easily identified. If and so far as it was necessary for the agreement to bind others with an interest in the land detailed provision was also made in paragraph 2. But the procedure was intended to be swift; thus only 28 days had to elapse from the service of the relevant notice to the commencement of proceedings in the County Court. The order in those proceedings might bind any person having any interest in the land. And in assessing the consideration to be paid the County Court is specifically required to ensure that all persons bound by the order are adequately compensated for the grant of those rights. If necessary the order for compensation may take the form of an order for periodical payments. In that context it is plain that the compensation or consideration must extend to each of the rights specified in paragraph 2(1). Thus the consideration will extend to payment for the right to keep the apparatus on the land.
>
> [27] The special regimes are quite different. No consideration is payable in relation the installation of electronic apparatus in, on or under street works or in tidal waters. Nor is consideration or compensation payable for 'overflying'. Thus it is to be expected that the special regime entitled 'linear obstacles' will differ from the general regime. It also differs from the other special regimes in that some consideration or compensation is payable.

5.5 Relevance of the Old Code to the New Code transitional provisions

5.5.1

As Chapter 30 explains, it will remain necessary to consult the Old Code for some time to come given the operation of the New Code's transitional provisions, which preserve paragraphs 20 and 21 to the extent there provided for. It will therefore still be necessary to identify whether or not rights were validly conferred under the Old Code in the first place. In that regard, it will remain relevant to consider whether rights have been validly conferred or exercised under the general regime or one of the special regimes.

6 [2010] EWCA Civ 1348.

5.6 Pre-Old Code systems

5.6.1

As has been explained in Chapter 2, certain forms of telecommunication were already protected by statute prior to the coming into force of the Old Code, for instance pursuant to consents given under the Telegraph Acts of 1863 to 1916. These Acts were considered in *British Telecommunications Plc v Humber Bridge Board.*[7]

5.6.2

Such rights as may have been conferred before the Old Code came into force are expressly preserved by paragraph 28; however it is to be noted that the 'coming into force' of the Old Code is not the date on which it commenced, but rather the date on which 'the code comes into force in relation to the operator'. This means that a person with the benefit of a consent under the Telegraph Acts did not automatically have that consent converted on 5 August 1984, the Commencement Date of the Old Code,[8] but only when it became an 'operator' under the provisions of the Telecommunications Act 1984, as then in force.

5.6.3

Pursuant to paragraph 28(1), 'electronic communications apparatus' (as defined by the Old Code, see below) includes such apparatus as is installed on, under or over any land before the Old Code comes into force. Further, any apparatus which could lawfully have been installed under paragraph 12 (the right to cross 'linear obstacles', meaning a railway, canal or tramway, paragraph 12(10)) had it been in force is deemed to have been installed under paragraph 12. This is because the right to cross linear obstacles was a well-established right in relation to telegraphic apparatus.

5.6.4

Any consents given under the Telegraph Acts 1863 to 1916 are treated as if they were agreements under paragraph 2 of the Old Code, and are deemed to bind the persons who gave (or were deemed to have given) that consent: paragraph 28(3).[9] Consents converted into agreements by the Old Code continued on the same terms and conditions as previously applied. Insofar as compensation was payable under any previous enactment, or any matter fell to be determined by the Court, then those matters were to be determined as if that previous enactment remained in force: paragraph 28(4). If an entitlement to compensation, preserved by paragraph 28(4), existed under a previous enactment, then there was no right to seek that compensation under any applicable provision of the Old Code: paragraph 28(5).

7 [2000] NPC 144.

8 SI 1984/876, Art. 4.

9 See *British Telecommunications Plc v Humber Bridge Board* [2000] NPC 144, at paragraph [45] for a consideration of these provisions.

5.6.5

The Old Code expressly provides that nothing prejudiced any rights of liabilities under any pre-Old Code agreement relating to the installation, maintenance, adjustment, repair, alteration or inspection of any electronic communications apparatus, or the keeping of such apparatus on land. Paragraphs 28(7)–(11) makes provision for notices given and applications made under the Telegraph Acts, and other matters, which remained unresolved before the Old Code came into force.

6 Operators under the Old Code

6.1 Introduction

6.1.1

This chapter is concerned with the definition of 'operators' as it applied under the Electronic Communications Act 2003 in relation to the Old Code. It is important to note that the definition of 'operator' under the New Code is expanded, specifically to accommodate an innovation in the electronic communications industry, which is the wholesale infrastructure provider ('WIPs'). This is an entity that provides passive infrastructure for others who actually provide mobile networks and services. As set out below, there was some doubt as to whether WIPs could attain operator status under the Old Code. To the extent that there is genuine doubt about this, the Digital Economy Act 2017 has resolved that doubt into the future by altering the definition of 'operator' (see Chapter 17). What follows is accordingly *only* relevant when considering the position prior to those reforms.

6.1.2

As explained in Chapter 5, the Old Code only protects 'operators'. The register of operators with Old Code powers is provided by OFCOM on their website.[1]

6.1.3

Only an 'operator' within the meaning of the Old Code and the 2003 Act is able to acquire code rights, and only an operator will be subject to various general and special regimes under the Old Code.

6.1.4

Any entity that is not an 'operator', or any operator acting outside the parameters of the Old Code, will only have such rights as are conferred by the terms of the contractual arrangement in place, and the general law. However, an operator acting within the Old

1 www.ofcom.org.uk/phones-telecoms-and-internet/information-for-industry/policy/electronic-comm-code

Code is protected from civil liability save as is expressly preserved by paragraph 27 of the Old Code.

6.2 Attainment of operator status during currency of an agreement

6.2.1

The difficulty which is sometimes encountered is where an entity which is not, at that time, an operator, subsequently becomes one. Does this convert the agreement into a full code agreement? It may be thought unjust for an occupier of land to grant rights under a contract, only for those rights to acquire statutory protection by reason of a unilateral change in status of the conferee. Unfortunately, the answer to this apparently simple question turns out to be complex:

(a) On one construction of paragraph 2 of the Old Code (see section 7.2), it would appear that, to acquire code rights, the conferee of such rights must be an operator at the time of the agreement.
(b) It would appear that the provisions of paragraph 4 of the Old Code (see section 7.5) would only apply to rights which are conferred as code rights, that is, in conformity with paragraphs 2 and 3, which requires that they are granted to an operator.
(c) In relation to paragraphs 20 and 21, there is an argument on the literal wording of those paragraphs that there is no need for apparatus to have belonged to an operator from the beginning. In the Consultation Paper, the Law Commission certainly took the view that all that mattered was the status of the entity using the apparatus at the date on which those paragraphs were operated.[2] The Law Commission in its Report No.336 suggested that even trespassing apparatus might be protected under paragraph 21 as long as it belonged to an operator.[3] Both of these points suggest that the availability of paragraphs 20 and 21 is not contingent on apparatus being originally installed in conformity with paragraphs 2, 3 or 5 of the Old Code, or one of the special regimes. The correctness of this approach, at least insofar as it applies to trespassing apparatus, may be doubted: see paragraph 7.5.6 below.
(d) It is to be noted that paragraph 28(12) of the Old Code (relating to the conversion of consents previously given under the Telegraph Acts) has an express transformative effect, converting rights granted under predecessor legislation into code rights under the Old Code.

6.3 Definition of 'operator'

6.3.1

'Operator' is defined by paragraph 1 of the Old Code as follows:

'the operator' means –

2 Law Com 205, paragraphs 5.52–5.56.
3 Law Com 336 at 6.15, 6.16 and 6.123. This topic is discussed further in section 7.5.

(a) where the code is applied in any person's case by a direction under s.106 of Communications Act 2003, that person; and
(b) where it applies by virtue of s.106(3)(b) of that Act, the Secretary of State or (as the case may be) the Northern Ireland department in question.

6.3.2

It is therefore necessary to consider whether a particular entity falls within either limb of s.106(3) of the 2003 Act. This provides that:

> The electronic communications code shall have effect –
> (a) in the case of a person to whom it is applied by a direction given by OFCOM; and
> (b) in the case of the Secretary of State or any Northern Ireland department where the Secretary of State or that department is providing or proposing to provide an electronic communications network.

6.3.3

Accordingly, an 'operator' is either a private entity with the benefit of a s.106(3)(a) direction by OFCOM, or the Secretary of State or any Northern Ireland Department providing or proposing to provide a network. The former operator is by far the most common in practice.

6.3.4

These definitions of 'operator' are imported into the Old Code by paragraph 1, which provides:

> 'the operator' means –
> (a) where the code is applied in any person's case by a direction under s.106 of the Communications Act 2003, that person; and
> (b) where it applies by virtue of s.106(3)(b) of that Act, the Secretary of State or (as the case may be) the Northern Ireland department in question.

6.4 An operator by direction under s.106(3)(a)

6.4.1

In order to qualify for a direction under s.106(3)(a), the operator must demonstrate that it is engaging in an activity falling within that section, and must then go through a statutory procedure to secure the direction.

6.4.2 Operator activities within s.106(3)(a)

The activities which entitle a person to apply for operator status are those set out in s.106(4) of the 2003 Act. This provides that the 'only purposes' for which the Old Code can be applied in a person's case by direction under s.106(4) are either

(a) the purposes of the provision by him of an electronic communications network; or
(b) the purposes of the provision by him of a system of conduits which he is making available, or proposing to make available, for use by providers of electronic communications networks for the purposes of the provision by them of their networks.

6.4.3 Meaning of 'Electronic Communications Network'

Section 106(4)(a) therefore requires the person to show that he is 'providing' an 'electronic communications network'. These terms are defined in s.32 of the 2003 Act. Subsection (1) states that:

> In this Act 'electronic communications network' means –
> (a) a transmission system for the conveyance, by the use of electrical, magnetic or electro-magnetic energy, of signals of any description; and
> (b) such of the following as are used, by the person providing the system and in association with it, for the conveyance of the signals –
> (i) apparatus comprised in the system;
> (ii) apparatus used for the switching or routing of the signals;
> (iii) software and stored data; and
> (iv) (except for the purposes of sections 125 to 127) other resources, including network elements which are not active.

6.4.4

'Apparatus' is further defined in section 405 of the 2003 Act as including 'any equipment, machinery or device and any wire or cable and the casing or coating for any wire or cable'.

6.4.5

Under s.32(4), it is further stated that:

> In this Act –
> (a) references to the provision of an electronic communications network include references to its establishment, maintenance or operation;
> (b) references, where one or more persons are employed or engaged to provide the network or service under the direction or control of another person, to the person by whom an electronic communications network or electronic communications service is provided are confined to references to that other person; and
> (c) references, where one or more persons are employed or engaged to make facilities available under the direction or control of another person, to the person by whom any associated facilities are made available are confined to references to that other person.

6.4.6

Under s.32(6), it is stated that 'transmission system' includes, under s.32(6), 'a transmission system consisting of no more than a transmitter used for the conveyance of signals'. The definition provisions of s.32(4)(a) and (b) are applied to a transmission system by s.32(5).

6.4.7

An 'electronic communications network' is to be contrasted with an electronic communications service, which is defined in s.32(2) as meaning 'a service consisting in, or having as its principal feature, the conveyance by means of an electronic communications network of signals, except in so far as it is a content service'.[4]

6.4.8

'Signals' is defined in s.32(10) as meaning:

(a) anything comprising speech, music, sounds, visual images or communications or data of any description; and
(b) signals serving for the impartation of anything between persons, between a person and a thing or between things, or for the actuation or control of apparatus.

6.4.9

From the above, it is clear that an operator must provide the electronic communications network, but need not actually be providing an electronic communications service, that is, conveying signal. It would therefore seem that, for a person to satisfy s.106(4)(a), he needs to show that he is 'providing' (which bears an extended definition including 'establishing' or 'maintaining', s.32(4)(a)) an electronic communications system. He need not demonstrate that he is also using that network for the provision of a service, though if he does, that is also included under the statutory definition of 'providing' (as he would then be 'operating' the network as well). It is considered that, properly construed, the provision of an inactive systems capable of use for providing an electronic communications service is captured by the statutory definitions.

6.4.10

Paragraph 1 of the Old Code provides that the definitions of 'electronic communications network' and 'electronic communications service' mirror those under the 2003 Act:

> 'electronic communications network' has the same meaning as in the Communications Act 2003, and references to the provision of such a network are to be construed in accordance with the provisions of that Act;
>
> 'electronic communications service' has the same meaning as in the Communications Act 2003, and references to the provision of such a service are to be construed in accordance with the provisions of that Act.

6.4.11

Under s.106(5), OFCOM is entitled to restrict its direction to only part of the operator's network, or to a geographical area, and the statutory term 'operator's network' is then

4 A content service is defined in s.32(7).

limited to that part. The definition of 'operator's network' is at paragraph 1 of the Old Code and provides:

> 'the operator's network' means –
> (a) in relation to an operator falling within paragraph (a) of the definition of `operator', so much of any electronic communications network or conduit system provided by that operator as is not excluded from the application of the code under s.106(5) of the Communications Act 2003; and
> (b) in relation to an operator falling within paragraph (b) of that definition, the electronic communications network which the Secretary of State or the Northern Ireland department is providing or proposing to provide.

6.4.12 Meaning of 'system of conduits'

If a person cannot demonstrate that he is within s.106(4)(a), he may be able to show that he is within s.106(4)(b), namely 'the provision [...] of a system of conduits which he is making available, or proposing to make available, for use by providers of electronic communications networks for the purposes of the provision by them of their networks'.

6.4.13

Section 106(4)(b) therefore explicitly recognises as operators persons who are involved in providing conduits for the purposes of that other person's network, though without providing a complete network themselves.[5] 'Conduits' are defined under s.106(7), and include 'a tunnel, subway, tube or pipe'.

6.4.14

Although an inclusive definition, the scope of s.106(4)(a) is not clear. The specific inclusion of providers of a system of conduits was not a requirement of the Framework Directive. it would seem clear that, construed *eiusdem generis*, that definition would not extent to a mast, or switching gear, unless it could be said to be ancillary to a conduit. It would therefore seem that s.106(4)(b) is concerned with a narrow category of activity.

6.4.15 Is the provision of non-conduit apparatus to third parties outside s.106(4)?

The contrast between the definition in s.106(4)(a) and (b) has been thought to give rise to potential practical difficulties meaning that the protective scope of the Old Code was too narrow, and protects too limited a class of activity. This potential danger was recognised by the Law Commission.[6] It has been addressed by an amendment to s.106, meaning that in future infrastructure providers are expressly covered by the New Code.[7]

5 See explanatory notes to the 2003 Act, at paragraph 255.
6 *Reforming the Electronic Communications Code,* paragraph 2.23–2.28.
7 The amendment in introduced by s.4(6) of the Digital Economy Act, which provides in relation to s.106 that 'In subsection (4)(b) for "conduits" substitute "infrastructure"'.

6.4.16

It could be argued that a person can only be within s.106(4)(a) if he is providing a network for his own purposes. If 'all' the person does is make electronic communications apparatus available to third parties, then that might not suffice. Nor would such a person then be safe to assume that he could bring himself within the restrictive definition of persons who clearly do engage in the activity of providing apparatus to third parties, as they will only fall within s.106(4)(b) if they provide a system of conduits.

6.4.17

The practical difficulty that this gives rise to is that a great deal of electronic communications apparatus in the current market is provided by so-called 'wholesale infrastructure providers', that is, companies that are in the business of making land, masts and other apparatus available to those who provide electronic communications services. If the construction set out in this paragraph were correct, then such entities could not be operators or could not be acting for 'statutory purposes' under the Old Code, unless they incidentally could show that they fell within such a definition.

6.4.18

It is considered, however, that there is a strong argument that, while not perfect, the definition in particular of s.106(4)(a) is wider than some might consider. As set out above, the provision of an electronic communications network, as defined by the 2003 Act, does not in fact require that the provider of the network should use it to convey signal (though that is also covered). It is enough for the provider simply to establish or maintain it. That would seem to cover the activities of companies engaged in the 'mere' provision of infrastructure.[8] It is considered that the better reading of s.106(4) is not that it is contrasting operators who provide a network for their own use (under (a)) with those who are providing conduits for third parties (under (b)), but rather those who provide 'networks' (as defined, and bearing in mind also s.32(1)(b)(iv) and (6) of the 2003 Act, which define what is included in and what constitutes a 'transmission system') with those who are merely providing a component part of it (namely the conduit into which network cabling can be inserted).

6.4.19

The above position seems to be in line with Directive 2002/21/EC on a common regulatory framework for electronic communications networks and services ('the Framework Directive'), and with Directive 2002/19/EC on access to, and interconnection of, electronic communications networks and associated facilities (Access Directive), which the 2003 Act seeks to implement.

8 Though bare land may still be problematic, even in light of the definition of 'apparatus' under section 405 of the 2003 Act.

6.4.20

Significantly, the Framework Directive provides a wider definition for 'electronic communications network' than the 2003 Act (and the amended Old Code) does. This defines an 'electronic communications network' under Article 2(a) as meaning:

> 'electronic communications network' means transmission systems and, where applicable, switching or routing equipment and other resources which permit the conveyance of signals by wire, by radio, by optical or by other electromagnetic means, including satellite networks, fixed (circuit- and packet-switched, including Internet) and mobile terrestrial networks, electricity cable systems, to the extent that they are used for the purpose of transmitting signals, networks used for radio and television broadcasting, and cable television networks, irrespective of the type of information conveyed.

6.4.21

It will be noted that this definition includes not just 'systems' but also 'equipment and other resources' permitting the conveyance of signals. The Framework Directive further makes clear on its face that part of its policy is to facilitate sharing.[9] It may therefore be that, if scrutinised, the definitions adopted in s.32 need to be read more expansively against the background of the Directive to which they purport to give effect.[10] This is particularly so given that it is clear that s.32 was intended fully to implement Article 2(a), (c), (e) and (m) of the Framework Directive.[11]

6.4.22 Statutory procedure for a direction; acquisition and loss of Old Code application

A direction under s.106 may only be made on the application by a person seeking operator status (s.107(1)). OFCOM has the power to publish and review the formal and substantive requirements for such an application (s.107(2) and (3)). In considering whether to make a direction, OFCOM must have regard to the statutory criteria set out in s.107(4). If OFCOM propose to apply to the Existing Code in a particular case, then that proposal must be published and representations must be considered (s.107(6); representations must be made within a month (s.107(9)). The notification must contain the information set out in s.107(7), and set out any conditions attached to the proposed direction.

6.4.23

In making a direction, OFCOM may prescribe the places and localities to which the direction relates, and the purposes of the network or conduit system (s.106(5)). The

9 Recitals 23 and 24, and Article 12.

10 *Marleasing SA v La Comercial Internacional de Alimentacion SA* (1990) C-106/89; See e.g. *A v National Blood Authority* [2001] 3 All ER 289.

11 See Explanatory Notes to the Communications Act 2003, paragraphs 86–7, www.legislation.gov.uk/ukpga/2003/21/pdfs/ukpgaen_20030021_en.pdf.

Electronic Communications and Wireless Telegraphy Regulation 2011[12] places a six-month time limit on decisions.

6.4.24

A direction may be subject to conditions. The power to impose conditions and restrictions arises under s.109. Those conditions and restrictions are contained in regulations which the Secretary of State has the power to make.[13] The conditions and restrictions are policed by OFCOM.[14] OFCOM may suspend the application of the Old Code in the circumstances set out in s.113. The terms of any application of the Old Code to an operator may be modified, or a direction may be modified, this being the combined effect of sections 106 and 115.

6.4.25

Once a direction is made, OFCOM, pursuant to s.108, keeps a register of operators to whom the Old Code has been applied. This can be inspected on the OFCOM website.

6.4.26

If an operator ceases to operate an electronic communications network or, as the case may be, a system of conduits, then he must notify OFCOM (s.116). There is power to put in place transitional provisions (s.117). Further, OFCOM has the power to suspend, modify and revoke the application of the Old Code under s.113.

6.5 An operator under s.106(3)(b)

6.5.1

Where the Secretary of State or any Northern Ireland department is providing or proposing to provide an electronic communications network, it is an 'operator' for the purposes of the Old Code. The main practical effect of being an operator under s.106(3)(b) is that certain parts of the special regimes do not apply to them.[15]

12 SI 1210/2011.
13 See the Electronic Communications Code (Conditions and Restrictions) Regulations 2003, SI 2003/2553.
14 Sections 110–12.
15 See paragraph 26(4): 'Paragraphs 12(9) and 18(3) above shall not apply where this code [applies in the case of the Secretary of State or a Northern Ireland department by virtue of section 106(3)(b) of the Communications Act 2003'.

7 Old Code general regime

7.1 Introduction

7.1.1

This chapter considers the operation of the general regime, which applies to rights exercisable over 'any land' (paragraph 2(1) of the Old Code). In practice, most issues under the Old Code are general regime issues; however, as set out in Chapter 5, the general regime also makes an appearance within some of the special regimes, for instance in relation to 'streets' that are not maintainable highways (paragraph 9(2)), to which the general regime is applied, or in relation to tidal waters with a Crown interest (see paragraphs 11(2) and 26), in relation to which an agreement is required, which agreement must be one under paragraph 2.[1] Further, agreements to obstruct under paragraph 3 of the general regime are required in relation to street works (paragraph 9(2)), flying lines (paragraph 10(1)) and tidal waters (paragraph 11(1)).

7.1.2

Where the facts fall under the general regime, the 'Code Rights' individually enumerated in paragraph 2 can be acquired in accordance with the processes set out in paragraph 2 (that is, by agreement in conformity with the requirements of paragraph 2) or under paragraphs 5–7 (that is, by compulsion). Similarly, it is possible for rights of obstruction to allow works to be carried out to be acquired by agreement or compulsion under either paragraph 3 or the aforementioned paragraphs 5–7.

7.1.3

This chapter will consider the constituent components for the acquisition of code rights, whether by agreement or through the Court. It will also consider the scope of the Court's powers when granting code rights and dispensing with the need for agreement under its jurisdiction to do so under paragraph 5.

7.1.4

As an overarching point governing everything that follows, it will be remembered that only an 'operator' within the statutory definitions of the 2003 Act and the Old Code can

1 See generally paragraph 2(9).

take the benefit of code rights: see Chapter 6. If one is not dealing with an operator, or if what has been conferred is not a code right granted in conformity with the provisions of the Old Code, then the grantee of such rights has no Old Code protection, and, significantly, will not enjoy the protections afforded under paragraphs 4 and 27.

7.2 Voluntary conferral of code rights: paragraph 2 agreement

7.2.1 The difficulty in classifying code rights under the Old Code

The draftsman of the Old Code provided for only three rights to be 'Code Rights'. Paragraph 2(1) of the Old Code lists them as follows:

> a right for the statutory purposes –
> (a) to execute any works on that land for or in connection with the installation, maintenance, adjustment, repair or alteration of electronic communications apparatus; or
> (b) to keep electronic communications apparatus installed on, under or over that land; or
> (c) to enter that land to inspect any apparatus kept installed (whether on, under or over that land or elsewhere) for the purposes of the operator's network.

7.2.2

Nothing in paragraph 2(1) explains what in fact is the legal nature of these code rights. It seems clear that the parties to an agreement granting rights are free to 'customise' the rights, as paragraph 2(5) provides that rights conferred are exercisable in accordance with the terms 'whether as to payment or otherwise' subject to which they were granted. The Old Code is silent on assignment of code rights. Its primary concern is the ability for code rights to burden third parties (paragraphs 2(2)–(4)) and for successors in title to the grantor to take the benefit of the terms of any agreement conferring rights: see paragraph 2(5). The ability to customise is limited only by the contracting out provisions in paragraph 27, as to which see Chapter 34.

7.2.3

Commonly, (a), (b) and (c) above are called 'Code Rights', and are individual and separate rights that must be individually conferred by the agreement (which interpretation is consistent with paragraphs 2(3), (5) and (6) which speak of 'a right falling within' paragraph 2(1)), though in practice one is more likely to encounter a grant that does not specifically reference any of the three code rights. Instead, those code rights are generally accepted to be conferred by the grant or a right to occupy or a right to possess, and a bespoke right to install specific apparatus, and by the usual repairing obligations one finds in standard leases or licences.

7.2.4

It is extremely common to find code rights conferred by agreement under the general regime. There is no prescribed form of agreement, and there are no model terms to be followed. In practice, such agreement takes many different forms, from the barest agreement to allow works to be undertaken and then kept on site, to full commercial leases of mast sites with repairing covenants, rent reviews and all the other familiar features of a fully negotiated, commercial lease.

7.2.5

One also finds a corresponding range for formality, from unsigned single-page documents, through to contracts signed by both parties capable of conforming with s.2 of the Law of Property (Miscellaneous Provisions) Act 1989 (and therefore potentially creating an equitable property interest), through to a full deed of grant of legal rights in conformity with s.1 of the 1989 Act and ss.52 and 54 of the Law of Property Act 1925.

7.2.6

The reason for this great variety of agreements is that the exact nature of code rights has never been properly resolved. In the House of Lords debates, Lord Glenarthur explained that the Old Code was 'essentially concerned with land law'.[2] However, as we shall see, the Old Code effectively creates a parallel system of formalities, priorities and security of tenure without directly addressing whether this is in addition to, rather than instead of, the general law of property and landlord and tenant. As is shown in Chapter 11 in relation to removal of apparatus under paragraph 21, that creates logical and practical conundrums which the courts have never resolved.

7.2.7

One response to this position may have been to treat the Old Code as *lex specialis,*[3] that is, a special closed set of regulations which are outside the general law, and to treat code rights as *sui generis*. This would at least have avoided any inconsistency with the general law by avoiding it.

7.2.8

However, and understandably, this is not how the practical application of the code has developed in reality. Instead of treating code rights as a separate and self-contained area of law, drafters of agreements have sought to find ways of making them fit with conveyancing precedents, and courts and litigators have argued about them in terms of leases and licences. As there has never been a definitive review of the operation of the Old Code by any court, these practices have never been checked and have therefore continued to be applied for over 30 years.

7.2.9

Doubtless the shoehorning of code rights into tried and tested lease precedents is a response to the uncertainty of their nature. In practice, however, this has created

2 Hansard, 20 March 1984, col. 1169.

3 *EE Ltd v Office of Communications* [2017] 1 CMLR 23 applies this principle to certain duties of OFCOM in relation to the electromagnetic spectrum.

difficulties of its own, in that it leads to conflict between the requirements of the code on the one hand and the general law on the other. So for instance there is doubt as to the effect of packaging code rights as leases: are the code rights still exempt from the general law of real property (and in particular registration requirements) because they are code rights, or, given that a lease going beyond a bare code right has been granted, does this now fall outside the Old Code and is governed by the general law of real property?

7.2.10

Although the New Code addresses these questions (though still gives rise to considerable difficulty on the question of registration, see Chapter 38), these conundrums at the heart of the Old Code will continue to arise until all of the agreements entered into under it cease to be in force.

7.2.11 The general requirements of a code agreement

In order for there to be a valid agreement under the Old Code's general regime, the following elements must be present:

(1) What is conferred has to be a 'code right'.
(2) The code right has to be conferred in relation to 'any land'.
(3) It has to be conferred by an 'occupier'.
(4) It has to be conferred upon an 'operator'.
(5) It has to be conferred for 'statutory purposes'.
(6) It has to relate to 'electronic communications apparatus'.
(7) It has to comply with the minimal formality requirements of paragraph 2.

7.2.12

In addition to those basic components, the issue already touched on above, the consent and nature of code rights, will require further consideration below, along with registration requirements and the effect of code rights on non-privies, such as superior interest holders or other third parties.

7.2.13 'Land'

'Land' is not defined in the Old Code. Under the Interpretation Act 1978, it is provided that:[4]

> 'Land' includes building and other structures, land covered with water, and any estate, interest, easement, servitude or right in or over land.

7.2.14

There is a further, special rule relating to land where there is a Crown interest. This is to be found in paragraph 26(1), which states that:

4 Schedule 1.

> This code shall apply in relation to land in which there subsists, or at any material time subsisted, a Crown interest as it applies in relation to land in which no such interest subsists.

The persons responsible for giving their agreement in relation to land subject to a Crown interest are further set out in paragraph 26, which provision is considered in more detail in Chapter 9.

7.2.15 'Occupier'

An 'occupier', but no one else, is permitted to enter into an agreement under paragraph 2. The term 'occupier' is not a term of art and has a flexible meaning, and its meaning is further dependent upon the statutory context in which it is used.[5] The use of the term 'occupier' in the Old Code context would seem to fix on the person with physical control over a site. The Old Code does not further require the occupier to have any form of specific property right, though in some contexts a specific minimum property right is required for certain paragraphs to have effect (such as paragraph 2(3)).

7.2.16

This would seem to follow from paragraph 2(8)(a)(iii), which, in the context of the extended statutory definition of 'occupier', deals with vacant land as follows:

> in relation to any land (not being a street or, in Scotland, road which is unoccupied, as references to the person (if any) who for the time being exercises powers of management or control over the land or, if there is no such person, to every person whose interest in the land would be prejudicially affected by the exercise of the right in question

7.2.17

The use of the term 'occupier' as a concept in this sort of statute, though deliberate,[6] is still rather unfortunate and is apt to lead to uncertainty, particularly in cases where land is in multiple use, with various layers of ownership. In such cases (which are common), it could readily be said that more than one person can be an occupier for different purposes.[7] There may accordingly be questions over whether a landlord who has retained roofspace but is physically absent from the site is an 'occupier', where, for instance, a third party or tenant-owned management company has been put in place controlling access and egress to the common parts as part of its day-to-day management functions.

7.2.18

Further, whether or not the 'occupier' holds an interest in land which is bound by code rights may be relevant to the question of whether or not the operator is capable of

5 See e.g. *Wheat v E. Lacon & Co.* [1966] AC 552; *Linden v DSS* [1986] 1 WLR 164; *Wandsworth LBC v Singh* (1991) 62 P & CR 219.

6 Hansard, House of Lords, 20 March 1984, cols 1173–74.

7 *Wheat v E Lacon & Co.,* above, is an example of that.

protecting its interest (if it chooses to) under the Land Registration Act 2002. It is also relevant to the question whether paragraph 2(3) and (4) apply, which deal with third parties. They only apply where the grantor of a code right has a particular interest in land, and would not therefore apply to a licensee (see below).

7.2.19

Subject to the above, it is considered that the following classes of person can potentially be said to be in 'occupation':

(a) A freeholder;
(b) A leaseholder, including a tenant at will or a tenant at sufferance;
(c) A beneficiary under a trust of land exercising his statutory right of occupation;[8]
(d) A mortgagee who has gone into possession;
(e) A local authority which has taken possession of land, for instance following a notice of entry;
(f) A licensee whose rights are extensive enough to amount to occupation;
(g) A squatter who has gone into adverse possession.

7.2.20

It is not clear whether a person who merely has the benefit of an easement over land is an 'occupier' for these purposes. On the one hand, the Court of Appeal has held that the mere exercise of an easement in the form of a right of way is not 'actual occupation' under the Land Registration Act 2002.[9] On the other, in the context of the Landlord and Tenant Act 1954, which may be a closer analogy, it has been held that a person can occupy an incorporeal hereditament for the purposes of a business.[10]

7.2.21

Further, the owner of a reversionary interest (such as a freeholder who has granted an intermediate leasehold interest) would not appear to be an 'occupier' in relation to any land which he has demised away, as he will no longer be able to control it in any relevant way. The same is not true, however, in relation to undemised common parts retained by the reversioner, for the reasons which have been noted above, where the question of who is in sufficient factual control to be the occupier appears to be more debatable.

7.2.22

In this context, it is important to note that there is an express expanded definition of 'occupier' in paragraph 2(8)(a):

8 Under ss.12 and 13 of the Trusts of Land and Appointment of Trustees Act 1996.
9 *Chaudhary v Yavuz* [2011] EWCA Civ 1314. Nor was parking a car in *Epps v Esso Petroleum Ltd* [1973] 1 WLR 1071. All of these cases turned on their own facts. It is notable that the modern cases appear to accept that extensive use (e.g. for parking) by a dominant owner can be justified by reference to an easement: *Virdi v Chana* [2008] EWHC 2901 (Ch).
10 *Pointon York Group plc v Poulton* [2007] 1 P & CR 115.

In this paragraph and paragraphs 3 and 4 below –

(a) references to the occupier of any land shall have effect –

(i) in relation to any footpath, bridleway or restricted byway that crosses and forms part of any agricultural land or any land which is being brought into use for agriculture, as references to the occupier of that land;

(ii) in relation to any street or, in Scotland, road (not being such a footpath, bridleway or restricted byway), as references –

in England and Wales or Northern Ireland, to the street managers within the meaning of Part III of the New Roads and Street Works Act 1991 (which for this purpose shall be deemed to extend to Northern Ireland), and

in Scotland, to the road managers within the meaning of Part IV of that Act; and

(iii) in relation to any land (not being a street or, in Scotland, road) which is unoccupied, as references to the person (if any) who for the time being exercises powers of management or control over the land or, if there is no such person, to every person whose interest in the land would be prejudicially affected by the exercise of the right in question.

7.2.23

Paragraphs (i) and (ii) deal with streets that are not maintainable highways and revert to the general regime. Paragraph (iii) deals with vacant land, and states that the occupier is to be the person who manages or controls that land, or, in the absence of such a person, 'every' person whose interest would be prejudicially affected by that right.

7.2.24

The extended definition in paragraph (iii) gives rise to two principal difficulties:

(1) As to the first limb, 'management and control', it can be difficult to ascertain who is the person in control of a roofspace. This was recently illustrated in *Francia Properties Ltd v Aristos Aristou,*[11] a case concerning a proposal by a landlord to develop a flat roof above a residential block in relation to which the right to manage had been exercised under Part II of the Commonhold and Leasehold Reform Act 2002. On the one hand, the management company had become vested under the 2002 Act with the powers to manage the block, and the development plans of the landlord would conflict with those management powers. On the other hand the 2002 Act did not operate so as to remove the landlord's property interest in the roof, and its ability to redevelop. The case was settled before the appeal to the Court of Appeal was heard, but it can readily be seen that there might be more than one 'occupier' of the flat roof if the 'management and control' test under (iii) is applied.

(2) It is not clear how the second limb works if the 'occupier' is 'every person whose interest in the land would be prejudicially affected by the exercise of the right in question'. Taken literally, and on the basis that (iii) presupposes that the land is not physically occupied, this would require an operator to identify each and every

11 [2017] L & TR 5.

person with an interest in the land in question and ensure that they are joint grantors. That would include, presumably, mortgagees and other encumbrancers. If that is the correct interpretation, and assuming that the 'management and control' test is not met, that places a serious burden on operators to ensure that there has been a competent grant by a competent (collective) grantor.

7.2.25 'Operator'

The definition of operator has been considered in Chapter 6.

7.2.26 'Statutory purposes'

The code right must also be conferred for 'statutory purposes'. 'Statutory purposes' is defined in paragraph 1 as meaning 'the purposes of the provision of the operator's network'. The meaning of this is also discussed in Chapter 6.

7.2.27

In practice, an agreement occasionally (but infrequently) states that it is made for the statutory purposes, but it is not a requirement of paragraph 2 that this is stated in express terms. The question of whether an agreement has been granted for the statutory purposes or not will usually have an obvious answer from the contents of the rights granted, the identity of the parties and the nature of the land. It is considered that the question is one of substance and not form, so that the parties cannot 'contract in' to the Old Code by agreeing that it is for statutory purposes when in substance it is not.

7.2.28 'Electronic communications apparatus'

All three Code Rights are confined to rights to do things with 'electronic communications apparatus'. This is defined in paragraph 1 of the Old Code as follows:

> 'electronic communications apparatus' means –
> (a) any apparatus (within the meaning of the Communications Act 2003) which is designed or adapted for use in connection with the provision of an electronic communications network;
> (b) any apparatus (within the meaning of that Act) that is designed or adapted for a use which consists of or includes the sending or receiving of communications or other signals[12] that are transmitted by means of an electronic communications network;
> (c) any line;
> (d) any conduit, structure, pole or other thing in, on, by or from which any electronic communications apparatus is or may be installed, supported, carried or suspended;
>
> and references to the installation of electronic communications apparatus are to be construed accordingly'.

12 Under paragraph 1, 'signal' has the same meaning as in s.32 of the Communications Act 2003.

7.2.29

Section 405 of the 2003 Act further defines apparatus as including 'any equipment, machinery or device and any wire or cable and the casing or coating for any wire or cable'.

7.2.30

It is to be noted that there is no ownership requirement in relation to apparatus. It will be protected if used as described above. This means that, for instance, a transformer belonging to an electricity company but used to power a mast site will be 'apparatus' for these purposes, and infrastructure provided by a third party infrastructure provider will be 'electronic communications apparatus' if used in conjunction with an operator's network.

7.2.31

Under paragraph 1, 'structure' does not include a building. This means that a building can never constitute 'electronic communications apparatus' under the Old Code.

7.2.32

It is to be further noted that the Old Code provides at paragraph 27(4) that 'the ownership of any property shall not be affected by the fact that it is installed on or under, or affixed to any land by any person in exercise of a right conferred by or in accordance with this code'. It is considered that this provision displaces the ordinary presumption in the law of landlord and tenant, that anything that is affixed to the land becomes part of the land.[13]

7.2.33 Formalities

There are two separate facets to formalities in relation to code rights: what degree of formality is required of the instrument creating them; and what registration (if any) is applicable should the agreement give rise to a category of recognised interest in land.

7.2.34

The 'agreement' must simply be in writing, in accordance with paragraph 2. Under the Interpretation Act 1978, Schedule 1, '"writing" includes typing, printing, lithography, photography and other modes of representing or reproducing words in a visible form, and expressions referring to writing are construed accordingly'. It is considered that, for example, an emailed agreement could constitute 'writing' for those purposes. There is no obligation that an agreement be signed.

7.2.35

A question that is often encountered in such circumstances (although more usually in relation to formalities associated with notices) is whether an exchange of documents

13 See generally Woodfall, *Landlord and Tenant*, vol. 1, chapter 13, section 5, for a discussion of the law relating to fixtures.

transmitted by email, received therefore only in digital form, constitute an agreement in writing. The argument against an electronic document constituting written notice is that an electronic document in its digital form does not qualify as writing.[14] This argument has however been rejected by the Law Commission for England and Wales.[15] The Commission was of the view that while an electronic document may not be in writing, the screen display will satisfy the definition of writing.[16] In that respect, an electronic message is no different from a message contained in a document which could easily be delivered but not read. The fact that it remains unread does not affect its validity.[17] In the courts in the USA, analogies have been drawn between sending a facsimile and sending an email.[18] In Canada, it has been held[19] that an enforceable contract can be created by email. The court concluded that the electronic contract 'must be afforded the sanctity that must be given to any agreement in writing'.

7.2.36

Code rights, viewed purely within the four corners of the Old Code, are not required to be granted in conformity with the ordinary rules of formality applicable to the creation or transfer of legal or equitable interests in land. There is no need for writing to be in the form of a deed, for example, under ss.52 and 54 of the Law of Property Act 1925, or for a written agreement to comply with s.2 of the Law of Property (Miscellaneous Provisions) Act 1989, in order for code rights to be validly conferred.

7.2.37

However, in practice code rights are often (but not invariably) conferred in the form of a lease, that is, by granting to the operator exclusive possession for a term, usually with a full set of repairing obligations and other standard features of an ordinary commercial landlord and tenant relationship.[20] In order to take actual effect as a lease (rather than 'merely' as an agreement conferring code rights), such an arrangement would need to comply with the additional formality requirements applicable to the creation of a particular interest in land. If a lease has been created orally under ss.52 and 54 of the

14 In reality an electronic document is a series of numbers stored in the computer's memory. The binary code which represents the electronic information is not stored on the computer as one document but in a series of numbers which is only understandable to a person once the appropriate software has read and translated the numbers into words: C. Reed, *Digital Information Law: Electronic Documents and Requirements of Form* (London: Centre for Commercial Law Studies, 1996).

15 In their paper, 'Electronic Commerce: Formal Requirements in Commercial Transactions – Advice from the Law Commission' (2001): https://s3-eu-west-2.amazonaws.com/lawcom-prod-storage-11jsxou24uy7q/uploads/2015/09/electronic_commerce_advice.pdf.

16 For the purposes of the meaning of 'writing' in the Interpretation Act 1978.

17 See the Commission Advice at paragraphs 3.5–3.10.

18 *Shattuck v Klotzbach*, 2001 Mass. Super. LEXIS 642, 14 Mass. L. Rep. 360 (Mass. Super. Ct. 2001) where the parties created an agreement of sale by email. The court had no difficulty in concluding, and in fact the court accepted without argument, that the emails were 'writing' for the purposes of the Statute of Frauds. The more difficult question for the court was whether the emails were signed in compliance with the Statute of Frauds.

19 *Rudder v Microsoft Corp* [1999] 47 CCLT (2d) 168; 2. CPR (4th) 474 (Ont. SC); 40 CPC (4th) 394 (Ont. SCJ).

20 *Keith Prowse & Co. v National Telephone Co.* [1894] 2 Ch 147; *National Telephone Co. v Griffen* [1906] 2 IR 115.

Law of Property Act 1925,[21] then the arrangement might take effect as a proper lease between the occupier and the operator, but in such a case it would not confer code rights for want of writing satisfying paragraph 2. If the agreement does have a proprietary effect, then this gives rise to difficult questions as to the effect of registration requirements. Registration is considered in Chapter 38. It is further important to note that the Old Code does not exclude the Landlord and Tenant Act 1987. The grant of a telecommunications site on a residential building qualifying under that Act may therefore be a relevant disposal triggering the tenants' rights of first refusal.[22]

7.2.38

It is to be noted, however, that no registration requirements are applicable to code rights as such, and they will bind in accordance with the provisions of paragraph 2 (see paragraph 2(7); as to the binding nature of code rights as such, see the discussion later in this section. The likely issue is, therefore, whether the terms of a lease which go beyond the bare code right require separate protection under registration provisions in order to bind third parties, or whether they are automatically swept up on paragraph 2 as being the 'terms' on which the code rights are enjoyed pursuant to paragraph 2(5) (or, if imposed, under paragraph 5(4)). If the latter is the correct approach, then it would seem that an agreement would be exempted from registration requirements even if, apart from the Old Code, it would have been a registrable disposition.[23] In most cases, the registration question would seem to be academic, however, as it seems perfectly well arguable that keeping apparatus on land on a permanent basis would amount to actual occupation in many cases.[24]

7.2.39

It is possible to read the code as creating a *sui generis* right for operators, which operates outside and apart from general property law. Thus understood, code rights are not leases or licences at all, but purely statutory agreements entered into under the Old Code and otherwise unregulated by the common law or any other statutory security of tenure regimes and formality requirements. The significant attraction of this is that it renders the Old Code rather easier to use and eliminates the inconsistencies that arise when an agreement has a 'dual personality' as, for instance, a code right on the one hand and a lease with security of tenure under the Landlord and Tenant Act 1954 on the other.[25] However, there is nothing in the Telecommunications Act 1984 which makes

21 Under s.54(2), it is provided that 'Nothing in the foregoing provisions of this Part of this Act shall affect the creation by parol of leases taking effect in possession for a term not exceeding three years (whether or not the lessee is given power to extend the term) at the best rent which can be reasonably obtained without taking a fine'.

22 The New Code expressly disapplies the 1987 Act to prevent this from happening: see section 14.7 of this book.

23 It is not obvious why code rights should be exempt, as they must be expressly conferred. It therefore seems that the rational for making dispositions registrable identified in Law Commission 271, *Land Registration for the 21st Century*, ought to be equally applicable to them.

24 Though compare the 'parking' and 'easement' cases referred to above.

25 These inconsistencies are considered in Chapter 14. Other inconsistencies include the different rules of land registration applicable to code rights on the one hand, and leases on the other, and the differential bases and inconsistent procedures for renewal applications.

this clear or exempts code rights from, for instance, the 1954 Act. Further, the Law Commission's reform proposals in the form of the New Code assume that such 'dual personality' is possible, and have devised a scheme to eliminate it. As desirable as it may seem to confine code rights to regulation by the code, therefore, it may be difficult to argue that this is the case.

7.2.40 Substance

The Law Commission have expressed the view, correctly it is considered, that paragraph 2(1) contains not one composite, but three separate, rights, each of which must be individually conferred by the agreement in order for the right to be a code right, with all the protection that entails for the operator.[26] If the agreement does not confer one or more of those rights, then the operator does not enjoy them pursuant to the Old Code, but only contractually (if at all) and must either secure a further agreement in conformity with the Old Code or obtain a dispensation from the agreement under paragraphs 5 to 7 of the code. It is to be noted that, if not granted as a code right, the operator will not be protected by paragraph 27.

7.2.41

In practice, it is rare to find an agreement that is framed simply in terms of conferring the individual code rights. Agreements ordinarily take the form of site licences or leases, perhaps separately preceded by an agreement to allow works of installation to take place. It is comparatively rare to find an agreement framed in terms of the specific three code rights. Instead, agreements will ordinarily confer all three code rights in substance, in the form of a demise or occupation right, and easements for access, and express regulation of repair and maintenance, and similar provisions. Those provisions will rarely be limited to the four corners of the code rights under paragraph 2, and will often go further, permitting alteration and upgrading of antennae, for example, to keep the site up to date, and including sharing and so-called 'payaway' provisions (whereby a percentage of the subletting value is paid upwards to the reversioner by the operator in occupation).

7.2.42

The reason for these contractually enhanced rights is that the code rights themselves do not adequately cover all the requirements of either occupiers or operators in the modern electronic communications market. As a default position, the code right under paragraph 2(1)(a) would allow the upgrading of apparatus to meet new technical requirements, or the addition of further apparatus (such as further dishes) or the sharing of (for instance) a mast with another operator. All of these matters are of some practical importance to operators, and are accordingly usually separately negotiated.

7.2.43

The New Code addresses these limitations by expanding the range of code rights that can be conferred (see Chapter 15). Under the Old Code, provided that the anti-avoidance

26 Law Com 336, at paragraph 2.15.

provisions under paragraph 27 are not infringed, it appears that the parties to a paragraph 2 agreement are free to enter into bespoke agreements to meet their needs to the extent that the bare code rights in paragraph 2(1) do not: see paragraphs 2(5), (6) and 27(2).

7.2.44

Paragraph 2(5) is concerned with ensuring that every person from time to time bound by a code right has the ability to enforce the benefit of the terms on which it is exercised:

> A right falling within sub-paragraph (1) above shall not be exercisable except in accordance with the terms (whether as to payment of otherwise) subject to which it is conferred; and, accordingly, every person for the time being bound by such a right shall have the benefit of those terms.

7.2.45

Paragraph 2(5) is rather clumsily expressed. Although the focus of the second sentence of paragraph 2(5) is solely on ensuring that the person bound by the code right from time to time automatically has the ability to enforce the benefit of those terms (i.e. that the grantor's successor can take the benefit of terms), it is considered that the first sentence operates to ensure that the code rights are imprinted with the contractual terms on which they were granted. This reading also appears to be supported by (or at least consistent with) sub-paragraph (6), which is concerned with the variation of rights, or the terms on which they are exercised, and the effect of such a variation on successors in title. It appears to consider that both successors to the granting occupier and to the original grantee operator are bound by and can enforce such variations.

7.2.46

The Old Code does restrict the content of a paragraph 2 agreements negatively. Paragraph 27(2) states that 'the provisions of this code, except paragraphs 8(5) and 21 and sub-paragraph (1) above, shall be without prejudice to any rights or liabilities arising under any agreement to which the operator is party'. The reference to paragraph 8(5) is to prohibit clauses preventing potential subscribers from taking any step under paragraph 8. The reference to paragraph 21 is to the removal provisions under the Old Code (see Chapter 11), which cannot be contracted out of. The reference to sub-paragraph (1) is a reference to paragraph 27(1), which is the prohibition on operators acting in breach of statute save as provided for in Schedule 4. It is considered that these limited restrictions on the contents of a paragraph 2 agreement further support the suggestion that the parties are given freedom otherwise to formulate the content and terms and conditions upon which code rights can be exercised.

7.2.47

The most significant part of paragraph 27(2) is the prohibition on contracting out of paragraph 21. This means that clauses that are sometimes encountered in agreements under which operators agree not to give a counter-notice under paragraph 21(3), or state in general terms that no reliance will be placed on rights under the Old Code, are unenforceable (to the extent that they create enforceable obligations in the first place). However, a literal reading of paragraph 27(2) would not preclude a clause under which an operator agrees not to give a notice under paragraph 5 seeking to dispense with an occupier's agreement,[27] or agrees that no compensation will be payable in the event that a notice under paragraph 20 is given.[28] The anti-avoidance provisions are considered more fully in Chapter 34.

7.2.48

Additionally, the Old Code made provisions for sharing agreements in certain circumstances. This is dealt with by the overly complicated provisions of paragraph 29, which is difficult to understand.[29]

7.2.49

Section 134 of the 2003 Act introduced controls on sharing restrictions. In practice, it has been seldom used, and the Law Commission were unable to arrive at any certain conclusions in relation to its continuing effect, if any.[30]

7.2.50 ***Rights of sub-operators***

If an occupier grants a code right, for instance in the form of a lease, to an operator, and the operator then parts with or shares possession, or sublets, what is the position as between the occupier and the 'sub-operator'?

7.2.51

It is considered that, if an operator assigns an agreement into its own and another operator's name, then in such a case it is considered likely that the occupier will be faced with two operators with code rights from the date of assignment.

7.2.52

If, on the other hand, the operator does not assign its interest but creates instead a sub-interest, then the position if more complex:

(1) If there is no agreement between the occupier and the sub-operator, for instance because there is no restriction on alienation or the particular form of alienation is not caught, then it cannot be said that there is an agreement in writing for paragraph 2 purposes, and there will be no code rights conferred. Vis-à-vis the occupier, the sub-operator will therefore have no code rights at all. It is not considered that a

27 For paragraph 5, see section 7.6 of this chapter. It would appear most likely, however, that a Court would consider such a statement as part of its discretion when considering whether to dispense with agreement under paragraph 5.
28 For paragraph 20, see Chapter 10.
29 Reference is made to the discussion at 13.6
30 See paragraphs 3.55 to 3.59 of Law Com 336.

right under a lease or licence for the tenant or licensee to agree to share a site with a third party amounts to an agreement with the sub-operators for the purposes of paragraph 2. In those circumstances, regard is to be had to paragraph 4 (see below).

(2) If there has been an alienation pursuant to a licence to underlet or share possession under a qualified alienation covenant, then it might be said that the licence could amount to an agreement for paragraph 2 purposes to as to amount to a grant of Code Rights directly to the sub-operator. However, it is suggested that it would be rare for a licence to underlet or share possession to have this effect. The usual effect of such a licence is to permit the tenant or licensee to deal with its own interest, and it not usual for such a licence to intend to create fresh rights binding on the landlord/licensor.

7.3 Third parties and code rights under paragraph 2

7.3.1 Genesis of the special rule of priorities applicable to code rights

If registration requirements are disapplied, how, then, do third parties come to be bound by code rights conferred by an agreement under the Old Code? The matter is regulated by the complex provisions of paragraph 2. Evidently, the occupier who has granted the right is bound by it. Special rules then govern how third parties are affected.

7.3.2

These unnecessarily complex provisions were introduced by a House of Lords amendment in response to criticisms levelled against the draft code, which originally contemplated that an occupier should be free to grant code rights without regard to the interests of any other person with a right to the land over which they were granted. This principle of so-called 'occupier only consent' was subsequently watered down to take account of the rights of third parties. It would appear, however, that the provisions were not fully considered and Parliament was ultimately content to leave it to operators to consider whose consent to be sought, and to form their own view of the risks of not obtaining all necessary consents.[31]

7.3.3

The difficulty with this approach is that it creates overlapping and complex provisions which do not sit well with the settled principles of priorities of property rights familiar to real property lawyers and enshrined in the Land Registration Acts of 1925 and 2002, and in the principles of unregistered conveyancing (both of which regimes are expressly disapplied from code rights under paragraph 2(7)). Nor are the principles set out in the Old Code entirely coherent, which would appear to be explained by the superficial consideration given to the amendments, and by the fact that the provisions amounted to an imperfect compromise between those who wished for operators to be free to deal with occupiers only, and those who were concerned to protect the property interests of third parties.

31 Hansard, House of Lords, 20 March 1984, cols 1171–4.

7.3.4 General rule: freeholders and leaseholders

Paragraph 2(2) makes special provision for any owner of an estate in land, that is, a free-holder or a lessee.[32] The basic position is as follows:

> A person who is the owner of the freehold estate in any land or is a lessee of any land shall not be bound by a right conferred in accordance with sub-paragraph (1) above by the occupier of that land unless –
> (a) he conferred the right himself as occupier of the land; or
> (b) he has agreed in writing to be bound by the right; or
> (c) he is for the time being treated by virtue of sub-paragraph (3) below as having so agreed; or
> (d) he is bound by the right by virtue of sub-paragraph (4) below.

7.3.5

Evidently, if the freeholder or lessee is the conferring 'occupier', then he is bound, as recognised by paragraph 2(2)(a). Agreeing to be bound by the right is equivalent to conferring it under paragraph 2(2)(b). Whether an alienation by a tenant or licensee made pursuant to an unqualified or qualified alienation provision might be sufficient to trigger paragraph 2(2)(b) is discussed above.

7.3.6 Special sub-rule: freeholders or lessees granting code rights to secure own electronic communications services access

Paragraph 2(3) is a limited provision, dealing with rights conferred on an operator by a person with a relevant interest in land in order that the grantor receives electronic communications services. The policy underlying this provision is that, if an occupier with a sufficient interest in the land needs to grant code rights to access such services, it should be free to do so, irrespective of whether or not third parties have consented.[33]

7.3.7

Paragraph 2(3) states as follows:

> If a right falling within sub-paragraph (1) above has been conferred by the occupier of any land for purposes connected with the provision, to the occupier from time to time of that land, of any electronic communications services and –
> (a) the person conferring the right is also the owner of the freehold estate in that land or is a lessee of the land under a lease for a term of a year or more, or

32 Paragraph 2(8) contains definitions of lease and lessee: (b) 'lease' includes any leasehold tenancy (whether in the nature of a head lease, sub-lease or underlease) and any agreement to grant such a tenancy but not a mortgage by demise or sub-demise and 'lessee' shall be construed accordingly.

33 Hansard, House of Lords, 20 March 1984, cols 1173–4.

(b) in a case not falling within paragraph (a) above, a person owning the freehold estate in the land or a lessee of the land under a lease for a term of a year or more has agreed in writing that his interest in the land should be bound by the right,

then, subject to paragraph 4 below, that right shall (as well as binding the person who conferred it) have effect, at any time when the person who conferred it or a person bound by it under sub-paragraph (2)(b) or (4) of this paragraph is the occupier of the land, as if every person for the time being owning an interest in that land had agreed in writing to the right being conferred for the said purposes and, subject to its being exercised solely for those purposes, to be bound by it.

7.3.8

What this provision therefore does is preclude any person with an interest in land from disputing that they are bound by the right during the currency of the occupation of (i) the grantor of the right or (ii) any person who is bound by it by reason of paragraphs 2(2)(b) or (2(4). For the duration of that period, the code right is incapable of being challenged by any person with an interest, notwithstanding that they have not agreed to it, and are not treated as if they have agreed to it.

7.3.9

This paragraph would cover, for instance, a right granted by an office tenant for the receipt of fibre-optic broadband in his offices. If the occupier is freeholder, or tenant with a lease of more than one year, or is a person with such an interest who has agreed in writing to be bound by such a right conferred by the occupier, then any occupier is bound for the duration of the code right, provided that it is exercised for the purposes it was granted.

7.3.10

The result of this is that an office tenant can, for the duration of the term of his lease, and for so long as he (or someone else bound) remains occupier and receives the services bind each and every other person with an interest in the land. The evident policy behind the provision is to ensure that all occupiers are able to access electronic communications services for the duration of their occupation.

7.3.11

The Old Code does not address whether a grant of such a right in breach of alienation provisions, and in breach of covenants against alteration to install apparatus to allow services into the demised premises, perhaps also necessitating what would otherwise be a trespass to the common parts (such as the use of ducts or the cutting into the fabric of the building) gives a reversioner a remedy against the offending occupier. The difficulty that the aggrieved reversioner faces, however, is that it is treated as having agreed to the intrusion by reason of paragraph 2(3), so that it ought not to be possible to dispute the lawfulness of what has happened.

7.3.12

Paragraph 4 of the Old Code, which contains a number of important provisions, is considered in section 7.5 below. It contains a specific compensation provision applicable to paragraph 2(3) which should be noted, which are considered in more detail below. It would appear that the sole remedy of a superior interest holder whose interest is caught by the deemed agreement provision under paragraph 2(3) is to statutory compensation under paragraph 4.

7.3.13

Paragraph 2(3) does not have effect if the occupier's interest is not one falling within paragraph 2(3)(a) and (b). Therefore, it will be open to a reversioner to object to the grant of a code right by a licensee, or a tenant under a periodic tenancy or with a fixed term of less than a year and seek removal of the apparatus under paragraph 4(2).

7.3.14 General sub-rule: effect on third parties generally

In addition to sub-paragraph 2(3), which is limited to rights granted by an occupier for the provision of services to that occupier, priorities are further regulated by sub-paragraph 2(4).

7.3.15

It is to be noted that paragraph 2(4) is in addition to paragraph 2(3), if applicable. It provides as follows:

> In any case where a person owning an interest in land agrees in writing (whether when agreeing to the right as occupier or for the purposes of sub-paragraph (3)(b) above or otherwise) that his interest should be bound by a right falling within sub-paragraph (1) above, that right shall (except in so far as the contrary intention appears) bind the owner from time to time of that interest and also –
>
> (a) the owner from time to time of any other interest in the land, being an interest created after the right is conferred and not having priority over the interest to which the agreement relates; and
>
> (b) any other person who is at any time in occupation of the land and whose right to occupation of the land derives (by contract or otherwise) from a person who at the time the right to occupation was granted was bound by virtue of this sub-paragraph.'

7.3.16

Accordingly, if a right is granted by the owner of an interest in land, then that code right will bind:

(1) his successors in title,
(2) any person with an interest in the land post-dating the grant of Code-Rights and not having 'priority' over the interest to which the agreement relates;
(3) any occupier who right to occupy was conferred by a person already bound by a code right.

7.3.17

It will be noted that, if the grantor of the code right has no interest in land at all, then neither paragraph 2(3) nor (4) will apply. However, if a person with an interest in land purports to grant a mere licence, then (subject to the next paragraph) that licence could, it seems, bind successors in title by reason of paragraph 2(4) as the only relevant question for this paragraph to operate against third parties is whether the grantor had an interest in land, and not whether he granted one.

7.3.18

Unlike paragraph 2(3), paragraph 2(4) is subject to a contrary intention, so that the owner who grants the code right seemingly has the option of deciding whether or not it is to bind successors in title.

7.3.19

It is not clear what would constitute a sufficient contrary intention. If, for instance, what is granted is expressly and in substance a licence, then as a matter of common law that would expire on transmission of the licensed land by the licensor. A clause that the agreement is not to constitute a landlord and tenant relationship, or is not intended to be assignable, are also indications that an agreement is to be 'personal'. However, it is not clear whether this is sufficient to displace the presumption that successors will be bound in paragraph 2(4). In *Wood v Waddington,*[34] the Court of Appeal (in the context of an easement to be implied under s.62 of the Law of Property Act 1925) held that the fact that the parties to a conveyance had expressly intended to grant a lesser right of access did not constitute a 'contrary intention' ousting an implied, more expansive right under that section.

7.3.20

As with the grant of rights, it is also clear that variations of such rights will bind successors by reason of paragraph 2(6). If the parties subsequently vary the code right, or the terms on which the code right can be exercised, then that variation will also bind them on the basis of the rules in sub-paragraphs 2(2) – (4) governing who is bound by the grant of the code right. This means that a superior interest holder, or owner of an interest in land with priority over the code right, will need to consent to any variation in order to be bound by it. This again raises the difficult question of the effect of a variation undertaken between occupier and operator without the involvement of a superior interest holder. If that variation cannot be shown to bind the superior interest, then, to the extent that it is varied, the agreement may be vulnerable to a claim under paragraph 4(2) for removal, if applicable.

7.4 Voluntary agreement to obstruct access: a paragraph 3 agreement

7.4.1

A code right under paragraph 2(1) does not permit the operator to obstruct access (including emergency access: paragraph 3(3)) to other land, unless the occupier

34 [2015] 2 P & CR 11.

of the other land conferred or was bound by a right to allow this to happen: see paragraph 3(1).

7.4.2

Paragraph 3 provides that

(1) A right conferred in accordance with paragraph 2 above or by paragraph 9, 10 or 11 below to execute any works on any land, to keep electronic communications apparatus installed on, under or over any land or to enter any land shall not be exercisable so as to interfere with or obstruct any means of entering or leaving any other land unless the occupier for the time being of the other land conferred, or is otherwise bound by, a right to interfere with or obstruct that means of entering or leaving the other land.
(2) The agreement in writing of the occupier for the time being of the other land shall be required for conferring any right for the purposes of sub-paragraph (1) above on the operator.
(3) The references in sub-paragraph (1) above to a means of entering or leaving any land include references to any means of entering or leaving the land provided for use in emergencies.
(4) Sub-paragraphs (2) to (7) of paragraph 2 above except sub-paragraph (3) shall apply (subject to the following provisions of this code) in relation to a right falling within sub-paragraph (1) above as they apply in relation to a right falling within paragraph 2(1) above.
(5) Nothing in this paragraph shall require the person who is the occupier of, or owns any interest in, any land which is a street or to which paragraph 11 below applies to agree to the exercise of any right on any other land.

7.4.3

It follows that, if the works under paragraph 2, 9, 10 or 11 would obstruct access to other land, then the agreement of the other land's occupier is required (paragraph 3(2)). The agreement under paragraph 3(2) is subject to the same rules as an agreement under paragraph 2(1), save that paragraph 2(3) does not apply. It again has to be in writing (as to the meaning of which, see above, paragraph 7.4.3). Paragraph 3(5)'s apparently intended effect is to spell out that the occupier of or owner of an interest in a street or tidal waters is not required to agree that rights be exercised over other land.

7.5 The exercise of rights: paragraph 4

7.5.1

Unless 'operators' can justify their actions by reference to a right conferred under the Old Code, they are subject to the general law and do not enjoy any special status.[35]

35 This is subject to the argument, explored in this section, that the language of paragraph 21 of the Old Code might be broad enough to protect even a trespassing operator who has *ex hypothesi* not obtained the consent of the relevant occupier or land, nor had that consent requirement dispensed with.

7.5.2

If operators are validly exercising a right under the Old Code, then they are not liable to compensate any person for any loss or damage, or subject to any other liability, caused by the exercise of any such right.[36] Except so far as authorised by Schedule 4 to the Telecommunications Act 1984, operators under the Old Code are not authorised to act in contravention of any statute.[37] Operators exercising code rights are deemed to be acting in pursuance of a statutory power.[38]

7.5.3

Paragraph 4 contains important general provisions about the effect of code rights, and compensation for their exercise.

7.5.4 Availability of defence of exercise of a statutory power

Paragraph 4(1) of the Old Code states that

(1) Anything done by the operator in exercise of a right conferred in relation to any land in accordance with paragraph 2 or 3 above shall be deemed to be done in exercise of a statutory power except as against –

(a) a person who, being the owner of the freehold estate in that land or a lessee of the land, is not for the time being bound by the right; or

(b) a person having the benefit of any covenant or agreement which has been entered into as respects the land under any enactment and which, by virtue of that enactment, binds or will bind persons deriving title or otherwise claiming under the covenantor or, as the case may be, a person who was a party to the agreement.

7.5.5

Accordingly, anyone bound by a code right under the provisions of paragraphs 2 and 3 who seeks to bring a claim in nuisance will be met by the defence that the acts complained of were done in pursuit of a statutory power. In general terms, however:[39]

(1) The statutory defence is available only in relation to damage resulting from the *intra vires* exercise of the statutory power.[40]

(2) The statutory defence is not available where damage is caused by some act collateral to the power (such as an accident in the course of the installation of the mast).

(3) Substantial interference by the exercise of the statutory power will only be allowed if all reasonable care is taken when exercising the statutory power.[41]

36 Paragraph 27(3).

37 Paragraph 27(1).

38 Paragraph 4(1), except as there provided. See further below.

39 See generally, *Clerk and Lindsell's Law of Torts* (21st ed., 2014), paragraphs 20–87 and following

40 *Home Office v Dorset Yacht Co. Ltd* [1970] AC 1004, 1064–1071; *Wildtree Hotels v Harrow London Borough Council* [2001] 2 AC 1.

41 *Manchester Corp v Farnworth* [1930] AC 171; *Allen v Gulf Oil Refining Ltd* [1981] AC 1001.

7.5.6

Paragraph 4 extends the defence of exercise of statutory power only to operators who have been conferred a code right in conformity with paragraphs 2 or 3.[42] Against that background, the suggestion (made for instance by the Law Commission[43]) that trespassing apparatus must still be removed using the machinery of paragraphs 20 and 21 must be open to doubt. Paragraph 4 does not protect the operator from a trespass claim in such a case, as the operator cannot point to a right conferred under the Old Code to justify its presence on land.

7.5.7

It would therefore seem that this defence of exercise of statutory power is not available against a freeholder or lessee[44] not bound by the code right (which would logically follow from the opening words of paragraph 4(1) in any event). Nor will it operate as a defence against the beneficiary of a covenant or agreement in relation to the land entered into under an enactment and which, because of the enactment, binds successors in title to that person.[45] It is considered that a person is 'bound' for the purposes of this provision if he is treated as bound pursuant to paragraph 2(3).

7.5.8 Right of removal for grants by inferior interest holder

A superior interest holder, who is not bound by a code right that has been granted by an occupier under his inferior interest, has the right to require the removal of the apparatus from the land in question under paragraph 4(2), using the procedure under paragraph 4(3), that is, the procedure under paragraph 21 (as to the latter, see Chapter 11).

7.5.9

In effect, the superior interest holder is entitled to require the reinstatement of any works that have been undertaken by the operator to give effect to that right. The only response to such an application can be an application by the operator under paragraph 5 of the Old Code, as allowed for in the scheme under paragraph 21.

7.5.10

It is to be noted that 'all' that paragraph 4(2) entitles the freeholder or lessee to do is to have the works undone. That provision says nothing at all about what the status of the conferred code right is as a result. Although the right cannot then be implemented by physical works (or such work must be undone), can the operator sue the occupier for derogation from grant or breach of contract? Although it would seem possible that such a claim could be maintained, an occupier would be well advised to record that the

42 It would presumably also extend to the situation where the need for agreement has been dispensed with under paragraph 5, though it does not say so in terms.

43 Law Com 336, paragraphs 6.15, 6.16 and 6.123.

44 As defined in paragraph 2(8)(b).

45 It is unclear what this relates to.

onus of ensuring that all parties who need to be bound, are bound, is on the operator (as Parliament apparently envisaged).[46]

7.5.11

It is easy to understand how this provision operates in a case where a subsidiary interest holder lets an operator on site under paragraph 2(4). In such a case, unless superior interest holders can be shown by the operator to be bound, or to be treated as bound, by paragraph 2's provisions, the operator is vulnerable to a removal order claim by such a superior interest holder. It is, however, at first glance difficult to see how this provision operates in a case governed by paragraph 2(3). In such a case, provided that a person with a relevant interest had granted the right in question, all persons with an interest in land are deemed to be bound by the code right for its duration. It is suggested that in a case where paragraph 2(3) applies, paragraph 4(2) cannot be invoked by a freeholder to require the undoing of works by an operator, because all interest holders are deemed bound as provided for in paragraph 2(3). It is considered that the 'subject to paragraph 4' qualification to paragraph 2(3) is a reference to the compensation provisions in paragraphs 4(4) to (10A).

7.5.12 Compensation entitlement

Paragraph 4(4) creates a special entitlement to compensation for 'relevant interest' owner, as defined in paragraph 4(5). This term is defined as follows:

> In sub-paragraph (4) above 'relevant interest', in relation to land subject to a right conferred or varied in accordance with paragraph 2 above, means any interest in respect of which the following two conditions are satisfied at the time the right is conferred or varied, namely –
>
> (a) the owner of the interest is not the occupier of the land but may become the occupier of the land by virtue of that interest; and
>
> (b) the owner of the interest becomes bound by the right or variation by virtue only of paragraph 2(3) above.

7.5.13

Again, these provisions are difficult to follow. The first limb appears to confine compensation to those persons who have an 'interest' (not, note, necessarily an interest in land) which may entitle them to occupy (though they are not in fact in occupation). The second limb appears to be a reference to the fact that, when paragraph 2(3) applies, every person with an interest in the land is treated as bound by that right as long as either the person granting the code right under paragraph 2(2)(b) or the person treated as agreeing to be bound under paragraph 2(4) is in occupation. This would appear to mean that any interest holder's interest will be similarly bound by the code right for that period, whether they have agreed to or not. It is these interest holders, whose agreement is deemed by paragraph 2(3), that seem to be the subject matter of this compensation measure.

46 *Hansard*, 20 March 1984, cols 1173–4.

7.5.14

Paragraph 4(4) provides that

Where –

(a) on a right in relation to any land being conferred or varied in accordance with paragraph 2 above, there is a depreciation in the value of any relevant interest in the land, and

(b) that depreciation is attributable to the fact that paragraph 21 below will apply to the removal from the land, when the owner for the time being of that interest becomes the occupier of the land, of any [electronic communications apparatus] installed in pursuance of that right,

the operator shall pay compensation to the person who, at the time the right is conferred or, as the case may be, varied, is the owner of that relevant interest; and the amount of that compensation shall be equal (subject to sub-paragraph (9) below) to the amount of the depreciation.

7.5.15

It therefore appears that a claim in compensation lies where a person with an interest in reversion to that of the grantor of the code right under paragraph 2(3) can show that he is likely to have to deal with the removal of the apparatus under paragraph 21 when the inferior interest of the grantor of the code right comes to an end, and the reversioner's interest becomes an interest in possession.

7.5.16

The question of compensation if not agreed is to be dealt with by the Upper Tribunal (Lands Chamber),[47] paragraph 4(6), and s.4 of the Land Compensation Act 1961(costs) shall apply, with the necessary modifications, in relation to any such determination. There is a statutory claim mechanism set out in paragraph 4(7):

(7) A claim to compensation under sub-paragraph (4) above shall be made by giving the operator notice of the claim and specifying in that notice particulars of –

(a) the land in respect of which the claim is made;

(b) the claimant's interest in the land and, so far as known to the claimant, any other interests in the land;

(c) the right or variation in respect of which the claim is made; and

(d) the amount of the compensation claimed;

and such a claim shall be capable of being made at any time before the claimant becomes the occupier of the land in question, or at any time in the period of three years beginning with that time.

47 Paragraph 4(10A) In this paragraph 'the appropriate tribunal' means –:

(a) in the application of this Act to England and Wales, the Upper Tribunal;

(b) in the application of this Act to Scotland, the Lands Tribunal for Scotland;

(c) in the application of this Act to Northern Ireland, the Lands Tribunal for Northern Ireland.

(8) For the purposes of assessing any compensation under sub-paragraph (4) above, rules (2) to (4) set out in s. 5 of the Land Compensation Act 1961 shall, subject to any necessary modifications, have effect as they have effect for the purposes of assessing compensation for the compulsory acquisition of any interest in land.

(9) Without prejudice to the powers of the [appropriate tribunal][5] in respect of the costs of any proceedings before the Tribunal by virtue of this paragraph, where compensation is payable under sub-paragraph (4) above there shall also be payable, by the operator to the claimant, any reasonable valuation or legal expenses incurred by the claimant for the purposes of the preparation and prosecution of his claim for that compensation.

(10) Subs.s (1) to (3) of s. 10 of the Land Compensation Act 1973 (compensation in respect of mortgages, [trusts of land][6] and settled land) shall apply in relation to compensation under sub-paragraph (4) above as they apply in relation to compensation under Part I of that Act.

7.5.17

For compensation generally, see Chapter 8.

7.6 Compulsory (non-consensual) code rights: dispensing with agreement

7.6.1

If an occupier is unwilling to agree to confer (or, upon giving notice under paragraph 21, agree to renew) a code right, or if an interest holder in the land not automatically bound under paragraph 2 declines to be bound by a code right, then the operator is able to invoke the paragraph 5 machinery under the code to dispense with the need for the agreement. Paragraph 5 can also be used to dispense with the need for an obstruction agreement under paragraph 3.

7.6.2 Transitional provisions

For the transitional provisions applicable to paragraph 5, see Chapter 31.

7.6.3 Paragraph 5

Paragraph 5(1) and (2) state as follows:

(1) Where the operator requires any person to agree for the purposes of paragraph 2 or 3 above that any right should be conferred on the operator, or that any right should bind that person or any interest in land, the operator may give a notice to that person of the right and of the agreement that he requires.

(2) Where the period of 28 days beginning with the giving of a notice under sub-paragraph (1) above has expired without the giving of the required agreement, the operator may apply to the court for an order conferring the proposed right or providing for it to bind any person or any interest in land, and (in either case) dispensing with the need for the agreement of the person to whom the notice was given.

7.6.4

OFCOM has prescribed a form of notice, and the provisions of paragraph 24 of the Old Code are applicable to such notices (see Chapter 32 for consideration of service provisions). There are two separate notices, one in relation to required rights and agreement (i.e. dispensing with agreement under paragraph 2), and the other relating to obstruction (i.e. in relation to paragraph 3).[48]

7.6.5

If a notice is given on either basis under paragraph 5(1), and is not responded to, that amounts to a refusal to agree and triggers the right to invoke the court[49] jurisdiction under paragraph 5(2).[50] The Court must approach such application on the basis of paragraph 5(3):

> The court shall make an order under this paragraph if, but only if, it is satisfied that any prejudice caused by the order –
> (a) is capable of being adequately compensated for by money; or
> (b) is outweighed by the benefit accruing from the order to the persons whose access to an electronic communications network or to electronic communications services will be secured by the order;
>
> and in determining the extent of the prejudice, and the weight of that benefit, the court shall have regard to all the circumstances and to the principle that no person should unreasonably be denied access to an electronic communications network or to electronic communications services.

7.6.6

Under section 405(1) of the 2003 Act, '"*access*" is to be construed in accordance with subsection (4)'. Sub-section (4) states that in the 2003 Act, access means:

(a) in relation to an electronic communications network or electronic communications service, are references to the opportunity of making use of the network or service; and

48 www.ofcom.org.uk/phones-telecoms-and-internet/information-for-industry/policy/electronic-comm-code/notices

49 Court means County Court under paragraph 1. Whether the County Court is the best forum for this litigation is discussed in *Mercury Communications Ltd v London and India Dock Investments Ltd* [1993] 69 P & CR 135, at 142.

50 Paragraph 5(6): 'For the purposes of proceedings under this paragraph in a county court in England and Wales or Northern Ireland, s.63(1) of the County Courts Act 1984 and Article 33(1) of the County Courts (Northern Ireland) Order 1980 (assessors) shall have effect as if the words 'on the application of any party' were omitted; and where an assessor is summoned, or, in Northern Ireland, appointed, by virtue of this sub-paragraph –:

(a) he may, if so directed by the judge, inspect the land to which the proceedings relate without the judge and report on the land to the judge in writing; and

(b) the judge may take the report into account in determining whether to make an order under this paragraph and what order to make.

In relation to any time before 1st August 1984, the reference in this sub-paragraph to s. 63(1) of the County Courts Act 1984 shall have effect as a reference to s. 91(1) of the County Courts Act 1959.'

(b) in relation to a programme service, are references to the opportunity of viewing in an intelligible form the programmes included in the service or (as the case may be) of listening to them in such a form.

7.6.7

Accordingly, if the prejudice caused by the person whose agreement is required can be compensated, then his agreement will be dispensed with on the basis that such compensation will be paid. What the 'compensation' covers is considered below. If it is not compensatable, the Court must balance that prejudice against the benefit to those who will benefit from the electronic communications network or to electronic communications services secured by the order. The Court is given no assistance as to how to balance what seem to be incommensurable matters, save that the Court is to have regard to 'all the circumstances' and to the fact that no person should 'unreasonably be denied' access to a network or services. The latter provision seems intended to weight the Court's discretion in favour of access for operators, which would, again, seem consistent with the underlying Framework and Access Directives. It is to be noted that there is no right to access a specific network or services – only access to a network or services.

7.6.8

It is clear that the right to access a network or services is of the first importance. In *Bridgewater Canal Co. v GEO Networks,*[51] the Court of Appeal explained that:

25. […] it is necessary first to consider the ambit of the Code in general and of paragraphs 12 and 13 in particular before, not after, subjecting the words used in specific sub-paragraphs to a textual analysis. The Code was originally enacted as part of the privatisation of British Telecommunications and consequential expansion of the telecommunications market. It was substantially amended by the Communications Act 2003 which was enacted for the like purpose but in the light of all the technological developments in the intervening period. Thus the overriding principle proclaimed in paragraphs 5(4), 13(5) and elsewhere is that 'no person should unreasonably be denied access to an electronics communications network or to electronic communications services'.
26. In furtherance of that principle the Code sought the agreement of the relevant occupiers/owners affected by the need to install and keep electronic communications apparatus. The general regime concentrated at the outset on the occupier no doubt because he was primarily affected but also most easily identified. If and so far as it was necessary for the agreement to bind others with an interest in the land detailed provision was also made in paragraph 2. But the procedure was intended to be swift; thus only 28 days had to elapse from the service of the relevant notice to the commencement of proceedings in the county court. The order in those proceedings might bind any person having any interest in the land. And in assessing the consideration to be paid the county court is specifically required to ensure that all persons bound by the order are adequately compensated for the grant of those rights. If necessary the order for

51 [2010] EWCA Civ 1348.

> compensation may take the form of an order for periodical payments. In that context it is plain that the compensation or consideration must extend to each of the rights specified in paragraph 2(1). Thus the consideration will extend to payment for the right to keep the apparatus on the land.

7.6.9

Whether or not to dispense in a particular case was considered by the Court of Appeal in *St Leger-Davey & Anor v First Secretary of State.*[52] In that case, the Court of Appeal had to consider the terms of paragraph 5 of the code. The consideration of paragraph 5 arose in the context of an appeal against a decision of an inspector, following an inquiry, to grant planning permission for the erection of a telecommunications mast and equipment cabin to an operator. The inspector, in considering the question for planning purposes whether the proposed site was the best location, had considered paragraph 5 of the code with respect to alternative sites which had been suggested at the inquiry. The inspector had concluded in relation to these alternative sites that because of the objections by the landowners to the use of their land for the installation of the mast there was no achievable alternative site available. Thus, he had concluded that a county court would have refused an order under paragraph 5 on the ground that the appeal site was more suitable. The appellants argued that this was to misunderstand the nature of the paragraph 5 right. Although the appeal did not succeed for reasons which are not material for present purposes, the Court nevertheless endorsed the criticism which was levelled at the inspector's approach to paragraph 5.

7.6.10

At paragraph [24], it explained that:

> the limitations on the power are narrowly confined and do not permit the court to make an environmental appraisal as between sites. The 'circumstances' to which the court, under the closing part of the paragraph, shall have regard, are the circumstances relevant to determining the extent of the prejudice and the weight of the benefit of the site to users, as specified in the sub-paragraph. It does not permit the court to conduct that overall assessment of the benefits and disbenefits of land use which is appropriate to the decision of a planning authority. The narrowness of the court's power under paragraph 3(b) of Schedule 2 is also illustrated [...] by the use, on two occasions in the sub-paragraph, of the word 'access' to a system which shall not unreasonably be denied, and not to improved access or reception.

7.6.11

The Court of Appeal were clear that the exercise required under paragraph 5 involved an evaluation of the site in question, and not a comparative exercise between the site in question and other potential sites. In other words the Court cannot refuse to make an order under para 5 because of the existence of a better and more feasible location.

52 [2004] EWCA Civ 1612.

This narrow construction of course favours the operator rather than the landowner. Accordingly, it would seem that the fact that the landowner may be able to point to an alternative site which is more suitable and confers upon users of the network a potentially greater benefit than the proposed site is irrelevant to the Court's determination under paragraph 5(3).

7.6.12

It is suggested that, in most cases, an objector's prejudice will be capable of monetary compensation, assessed in accordance with paragraph 7. It is only in cases where, for instance, apparatus is to be installed on or near a building where the objector has aesthetic or sentimental objections that the issue might become more difficult. In those circumstances, it will be necessary for the Court to have regard to the effect on the specific operator's customers under paragraph 5(3)(b), and to the wider available systems. Undoubtedly, operators will call evidence in such a case concerning the lack of availability of other sites.

7.6.13 Paragraph 5 notices in relation to apparatus already on site

If apparatus is already installed, and new code rights are sought (for instance, in response to a paragraph 21 notice), then paragraph 6 applies. This provides that:

(1) The following provisions of this paragraph apply where the operator gives notice under paragraph 5(1) above to any person and –
   (a) that notice requires that person's agreement in respect of a right which is to be exercisable (in whole or in part) in relation to electronic communications apparatus already kept installed on, under or over the land in question, and
   (b) that person is entitled to require the removal of that apparatus but, by virtue of paragraph 21 below, is not entitled to enforce its removal.

(2) The court may, on the application of the operator, confer on the operator such temporary rights as appear to the court reasonably necessary for securing that, pending the determination of any proceedings under paragraph 5 above or paragraph 21 below, the service provided by the operator's network is maintained and the apparatus properly adjusted and kept in repair.

(3) In any case where it is shown that a person with an interest in the land was entitled to require the removal of the apparatus immediately after it was installed, the court shall, in determining for the purposes of paragraph 5 above whether the apparatus should continue to be kept installed on, under or over the land, disregard the fact that the apparatus has already been installed there.

7.6.14

The onus of the application for temporary rights, and the only person with standing to make such an application, is the operator. In practice, the operator never takes any such steps at all, and this paragraph is little used. It suggests, however, that where the right is not sought and obtained under paragraph 6(2) to secure 'the service provided by the operator's network is maintained and the apparatus properly adjusted and kept in repair', that the operator has no further entitlement to exercise his code rights. His apparatus will be on the land lawfully under the deemed lawfulness provision under paragraph 21, but he will not be able to undertake any works under code rights. This provision is rather overlooked, but supports the argument that operators do not have full rights in the nature of, for instance, a tenant holding over under the statutory continuation of his tenancy under the Landlord and Tenant Act 1954, Part II.

7.6.15 Terms (non-financial)

The Court has the ability to impose terms on the dispensation under paragraphs 5(4) and (5):

> (4) An order under this paragraph made in respect of a proposed right may, in conferring that right or providing for it to bind any person or any interest in land and in dispensing with the need for any person's agreement, direct that the right shall have effect with such modifications, be exercisable on such terms and be subject to such conditions as may be specified in the order.
>
> (5) The terms and conditions specified by virtue of sub-paragraph (4) above in an order under this paragraph, shall include such terms and conditions as appear to the court appropriate for ensuring that the least possible loss and damage is caused by the exercise of the right in respect of which the order is made to persons who occupy, own interests in or are from time to time on the land in question.

7.6.16

On one reading of paragraph 5(4), it would appear that the Court is limited to qualifying the code right rather than expanding it. However, there is limited evidence in the form of decided cases that suggest that the Court has jurisdiction to impose a wide range of terms going beyond mere limitations on the exercise of code rights, including the imposition of a fixed term or an indefinite term terminable on 12 months' notice for the agreement: see *Brophy v Vodafone Ltd.*[53]

7.6.17

Whilst the point cannot be said to be free from doubt, and whilst in *Brophy* the issue appears to have been common ground between the parties, it is considered more sensible that the Court has a wide range of powers in relation to the terms on which any agreement can be conferred. It is not considered that, in dealing with paragraph 5, the Court is limited to conferring code rights without any modification, or simply to imposing conditions or qualifications upon such rights.

53 [2017] EWHC B9 (TCC).

8 Old Code general regime
Financial provisions

8.1 Introduction

8.1.1

This chapter considers the financial provisions in the Old Code. Chapter 30 analyses the substantially different financial regime under the New Code.

8.1.2

In essence, three distinct types of claim for financial remedies are enabled by the Old Code:

(a) Claims for consideration under paragraph 7;
(b) Claims for compensation, also under paragraph 7; and
(c) Claims for Injurious Affection to Neighbouring Land under paragraph 16 – see section 8.7 below.

8.2 Paragraph 7 of the Old Code

8.2.1

The question of financial terms is expressly regulated by paragraph 7 of the Old Code.[1] Paragraph 7(1) provides that:

> The terms and conditions specified by virtue of sub-paragraph (4) of paragraph 5 above in an order under that paragraph dispensing with the need for a person's agreement, shall include –
>
> (a) such terms with respect to the payment of consideration in respect of the giving of the agreement, or the exercise of the rights to which the order relates, as it appears to the court would have been fair and reasonable if the

1 These provisions are usefully considered at first instance by Lewison J in *Bridgewater Canal Co. Ltd v Geo Networks Ltd* [2010] EWHC 548 (Ch). That part of the judgment has since been applied in *Brophy v Vodafone Ltd* [2017] EWHC B9 (TCC) and is unaffected by the Court of Appeal's reversal of Lewison J's decision.

agreement had been given willingly and subject to the other provisions of the order; and

(b) such terms as appear to the court appropriate for ensuring that that person and persons from time to time bound by virtue of paragraph 2(4) above by the rights to which the order relates are adequately compensated (whether by the payment of such consideration or otherwise) for any loss or damage sustained by them in consequence of the exercise of those rights.

8.2.2

It is therefore expressly contemplated that the person whose agreement is dispensed with receives a payment for the agreement given or for the fact that he is bound by the right, and compensation for losses sustained. It seems that the payment, though expressed to be for the agreement being given, reflects the fair value of the code right the operator secures.

8.2.3

Sub-paragraph 7(2) states that

> In determining what terms should be specified in an order under paragraph 5 above for requiring an amount to be paid to any person in respect of –
> (a) the provisions of that order conferring any right or providing for any right to bind any person or any interest in land, or
> (b) the exercise of any right to which the order relates,
>
> the court shall take into account the prejudicial effect (if any) of the order or, as the case may be, of the exercise of the right on that person's enjoyment of, or on any interest of his in, land other than the land in relation to which the right is conferred.'

8.2.4

It would therefore appear that the combined effect of paragraphs 7(1) and (2) is that:

(1) The person whose agreement is dispensed with is entitled to 'consideration' for the right or exercise of them.

(2) Under paragraph 7(2), that payment can be adjusted to reflect the prejudicial effect of the order or the detrimental effect of it on other land. On the face of it, if the paragraph 5 dispensation renders an otherwise feasible planning project impossible, then the order should reflect that loss in the consideration.

(3) There is a further entitlement for a term in the order to allow the person to claim future compensation (a form of indemnity). In that regard, paragraph 7(4) expressly states 'for questions arising in consequence of those terms (whether as to the amount of any loss or damage caused by the exercise of a right or otherwise) to be referred to arbitration or to be determined in such other manner as may be specified in the order.'

8.3 Judicial consideration of paragraph 7

8.3.1

The operation of paragraph 7 has been considered on a number of occasions. The first decision is that of HHJ Hague QC in *Mercury Communications Ltd v London and India Dock Investments Ltd*:[2]

(A) The words 'willingly given' clearly refer to the grantor, who must therefore be a 'willing' grantor. However, in my judgment, a 'willing' grantee must also be assumed, that is to say a grantee who is willing to take on 'fair and reasonable' terms. After all, it is the grantee who is applying to the court for a grant on those terms, and it necessarily follows that he is a 'willing grantee' in that sense. Statute frequently refers only to the 'willing' vendor; lessor or grantor, without mention of the purchaser, lessee or grantee, but it is always assumed that the latter too will be 'willing': see, *e.g.* rule 2 of section 5 of the Land Compensation Act 1961; section 25(1) of the Finance (1909–10) Act 1910, mentioned below, section 34 of the Landlord and Tenant Act 1954 and section 9(1) of the Leasehold Reform Act 1967.

(B) In my judgment, I have to consider what would be 'fair and reasonable' terms as between these parties, *i.e.* LIDI and Mercury as opposed to a hypothetical grantor and a hypothetical grantee. That appears to me to be the consequence of the wording of paragraph 7(1)(a), particularly in the light of the earlier parts of the Code which assume that there have been prior negotiations between the parties. Mr Aikens argued to the contrary, saying that paragraph 7(1)(a) requires the court to consider a hypothetical situation. That is to a certain extent true, but I see no reason to depart from reality more than is necessary or is required by the relevant wording of the legislation. The wording of paragraph 7(1)(a) is significantly different in this respect from other legislation, *e.g.* section 8(2) of the Mines (Working Facilities and Support) Act 1966, which refers to 'a willing grantor' and 'a willing grantee'.

(C) What I have to decide is what would have been 'fair and reasonable' if the grant had been willingly given. That, in my view, necessarily involves an element of subjective judicial opinion, for there can be no proof or objective determination of what is fair and reasonable. To a certain extent, the answer must depend on the judge's own perception of what is 'fair and reasonable'. There is thus a distinction between this case and the more common category of case where the legislation or the agreement in question requires a determination of 'market value' or 'market rent' or the like; such a determination involves an objective assessment of a factual matter not involving any discretion or subjective opinion. This distinction may perhaps account for the fact that the Code directs the consideration to be determined by a judge rather than the Lands Tribunal.

(D) Following on from that, it is in my judgment clear that what I have to determine is not the same as what the result in the market would have if the grant had been given willingly. That is, however, far from saying that the market result is irrelevant or can afford no guidance. Indeed, in my view the market result is

2 [1993] 69 P & CR 135 at 144–5.

the obvious starting point; and in most cases it will come to the same thing as what is 'fair and reasonable', because prima facie it would be neither fair nor reasonable for the grantor to receive less than he would in the market or for the grantee to have to pay more than he would in the market. But there may be circumstances, of which the absence of any real market may be one, in which a judge could properly conclude that what the evidence may point to as being the likely market result is not a result which is 'fair and reasonable'.

(E) In my view the word 'willingly' in paragraph 7(1)(a) cannot be taken in isolation. For it is a meaningless concept if considered apart from the financial terms of the grant. Whether or not a vendor or grantor is 'willing' must necessarily depend on the price or consideration he is to receive. Thus a vendor cannot sensibly be stigmatised as being unwilling because he refuses to accept an offer at an inadequate price. 'Willing' must mean 'willing to sell at a fair price'. Likewise 'given willingly' in paragraph 7(1)(a) must in my judgment mean 'given willingly for a consideration which is fair and reasonable'. In his submissions, Mr Aikens argued that *ex hypothesi* LIDI was not willing because it would not give consent to the proposed grant except on what he described as its own prohibitive terms. In my judgment, that does not advance the argument and is logically fallacious, because it begs the real question and indeed really assumes what it is setting out to demonstrate.'

8.3.2

The effect of *Mercury* is that the occupier under paragraph 7 is entitled to a fair and reasonable consideration for the right – going beyond mere diminution in value, but not going so far as a ransom price. Regard can be had to comparable transactions: the market will show what is fair and reasonable.[3]

8.3.3

The Law Commission have expressed the view that the approach in *Mercury* may have been called into question by the decision in *Bocardo SA v Star Energy UK Onshore Ltd*:[4]

> In the Consultation Paper, we also noted the decision of the Supreme Court in *Bocardo SA v Star Energy UK Onshore Ltd* on similar wording in the statutory framework for rights to extract oil. In that case payment for the right to bore pipelines beneath land was assessed on compulsory purchase principles, even though those principles were not expressly incorporated into the legislation. The value of the rights was therefore based on the value to the owner of the land at the time of the acquisition, rather than to the potential buyer; the fact that the rights were being used to extract oil from underneath the land was disregarded. This may cast doubt upon the approach taken by His Honour Judge Hague QC in *Mercury*, though – as many consultees reminded us – the statutory wording was not identical, and the factual situation differed from that of a Code Operator seeking Code Rights from a Site Provider.

3 See at 163, and 168–9.
4 [2011] 1 AC 380. Law Com 336, at 5.19.

8.3.4

The provisions were further considered in *Cabletel Surrey & Hampshire Ltd v Brookwood Cemetery Ltd.*[5] Mance LJ explained the effect of *Mercury* as follows:

7. First, the exercise required by paragraph 7 is not one of ascertaining market terms or value, although any market terms or value are a relevant consideration to take into account. The test, when fixing terms with respect to either the payment of consideration or the exercise of the rights to which the order relates, is what 'it appears to the court would have been fair and reasonable if the agreement had been given willingly'. This formulation was no doubt chosen because of the public interest in enabling ordinary members of the public to be offered and to obtain new telecommunications services without individual landowners being able to insist on perhaps excessive sums, for example because of the need to use what might in some cases amount to no more than ransom strips.
8. However, as His Honour Judge Hague remarked at page 144G, this formulation does introduce an element of subjective judgment into the process of fixing of terms. His Honour Judge Hague found that assistance was to be obtained when making such a judgment from examining comparables, and so did the experts called in the present case. When considering comparables allowance should, however, be made if it could be shown that the paying party had, for whatever reason, been ready to concede a high value for pragmatic reasons: for example time constraints, the expense or uncertainty of litigation, or (I might add) the small size of the works and of any payment: see page 168.
9. His Honour Judge Hague further identified as relevant considerations 'the importance and value of the proposed right to the grantee' and the parties' relative bargaining positions, although the last factor may, as already indicated, cut both ways: see pages 159 and 169.

8.3.5

Lewison J in *Bridgewater Canal Co. Ltd v Geo Networks Ltd*,[6] further explained the effect of paragraph 7 as follows:

35. I start with the general regime. A court order under paragraph 7 of the Code must include financial terms. Those terms are:
    i) such terms with respect to the payment of consideration in respect of the giving of the agreement, or the exercise of the rights to which the order relates, as it appears to the court would have been fair and reasonable if the agreement had been given willingly and subject to the other provisions of the order; and
    ii) such terms as appear to the court appropriate for ensuring that that person and persons from time to time bound by the rights to which the order relates are adequately compensated (whether by the payment of such consideration or otherwise) for any loss or damage sustained by them in consequence of the exercise of those rights.
36. The first component of the money payment contains a number of instructions. First, what is to be paid is consideration. This word is used in contra-distinction

5 [2002] EWCA Civ 720.
6 [2010] EWHC 548 (Ch).

to 'compensation' in the second component and simply means 'price'. Second, what the consideration is paid for is 'the giving of the agreement or the exercise of the rights'. This is very strange drafting. In the first place it assumes that agreement is *in fact* being given, when the whole point of the application resulting in the making of an order is to *dispense* with agreement, as the opening words of paragraph 7 itself make clear. In the second place the order will itself confer rights on the operator (which he can choose to exercise or not) so it is difficult to see why he should pay consideration for the *exercise* of the rights, rather than the rights themselves (or, to be pedantic, the right to exercise the rights). It may be that what the draftsman had in mind was, for example, a term that requires the operator to make a payment each time he comes onto the land, rather like a wayleave rent. In the third place the subsequent reference to agreement having been willingly given does not in terms apply to the exercise of rights (in so far as that is meant to be different from the giving of agreement). So it is difficult to avoid the conclusion that the draftsman has not been precise with his choice of words.

37. Third, the payment terms must be 'fair and reasonable'. Fourth, they must be assessed 'as if' the agreement had been willingly given. These last two instructions mean that the price will not necessarily be the same as the market value of the rights, although the market value may be a good starting point (*Mercury Communications Ltd v London and India Dock Investments Ltd*).[7] A person may not be willing to agree unless proper (but not undue) account is taken of the potentiality of his land in unlocking development or other value (*Mercury Communications Ltd* at 159–60). But this still leaves open the question what is 'fair and reasonable' which may involve a measure of subjective opinion (*Mercury Communications Ltd* at 144 and 159). In *Cabletel Surrey and Hampshire Ltd v Brookwood Cemetery Ltd*,[8] Mance LJ said, without expressing any concluded view, (§ 7) that this formulation:

> '... was no doubt chosen because of the public interest in enabling ordinary members of the public to be offered and to obtain new telecommunications services without individual landowners being able to insist on perhaps excessive sums, for example because of the need to use what might in some cases amount to no more than ransom strips.'

38. Later in his judgment Mance LJ seems to have approved the rejection in *Mercury Communications Ltd* of the argument that a ransom value may be set by the owner (§ 25). It seems to me, therefore, that even in the case of rights acquired over land of a private owner under the general regime, ransom payments are excluded by the statutory formula.
39. I should mention here that the general regime incorporates the landowner's power to require alteration or removal of apparatus where that is necessary for development (§ 20). Since this power applies irrespective of the terms of any agreement, it is plainly something that must be taken into account in assessing the amount of the consideration.
40. The second component of the money payment is called compensation. It is tied to loss or damage sustained by persons bound by the rights 'in consequence of the exercise of those rights'. This component seems to be tied to the actual (and future) exercise of rights, rather than the initial grant by court order. If

7 [1993] 69 P & CR 135, 144.
8 [2002] EWCA Civ 720.

this is right, then it seems that what the draftsman had in mind was some form of contractual indemnity rather than the fixing of an actual money payment at the date of the court order.'

8.3.6

In *Brophy v Vodafone Ltd*,[9] paragraph 64, HHJ Stephen Davies explained that:

> I should also refer to the question of compensation under paragraph 5(1)(b). I respectfully agree with the view expressed by Lewison J in *Geo Networks* that this requires not payment of compensation but the inclusion of terms appropriate to ensure that adequate compensation is paid for loss and damage sustained by them in consequence of the exercise of the rights conferred by the order. I am satisfied that these are included here.

8.3.7

The principles emerging from this extensive judicial treatment may be summarised as follows:

(1) Paragraph 7 entitles the occupier of land to receive both consideration, the price of the right, and compensation, that is, financial recognition of any losses suffered. This latter element is in the form of an ongoing indemnity for losses sustained in exercising the right in question.
(2) The 'price' is to be 'fair', which is not market value but is informed by it. What is fair introduces a subjective element which means that the market value, though a starting point, may be departed from.
(3) There is no entitlement to a ransom value.

8.4 The special rule where the paragraph 5 notice is in response to a paragraph 21 notice

8.4.1

Invoking the paragraph 5 process is one possible response by an operator to a notice given under paragraph 21, and can amount to a relevant step being taken to secure further rights for the purposes of paragraph 21(4)(b).

8.4.2

Where the Court is asked to determine a paragraph 5 application in that context, sub-paragraph 7(3) provides that:

> In determining what terms should be specified in an order under paragraph 5 above for requiring an amount to be paid to any person, the court shall, in a case where the order is made in consequence of an application made in connection with proceedings under paragraph 21 below, take into account, to such extent as it thinks fit, any period during which that person –

9 [2017] EWHC B9 (TCC).

(a) was entitled to require the removal of any electronic communications apparatus from the land in question, but
(b) by virtue of paragraph 21 below, was not entitled to enforce its removal;

but where the court takes any such period into account, it may also take into account any compensation paid under paragraph 4(4) above.

8.4.3

This provision therefore entitles the person who gave a paragraph 21 notice to retrospective occupation payments from the operator. This raises the difficult question of whether or not an operator, holding over beyond the term of any code right, is liable to make any payments. Under the Landlord and Tenant Act 1954, the lease (and the obligations under it, including as to rent) are continued. Under the Old Code, this is less clear and it is to be noted that paragraph 21(9), which deems that occupation after expiry of the code right is lawful, makes this subject only to paragraphs 6(3) and 7(3). The deeming of lawfulness appears to call into question whether the operator is liable for damages for use and occupation, or for mesne profits. On one reading, the operator's only liability to pay for the holding over period is therefore if paragraph 5 has been invoked, which is entirely up to the operator, or if paragraph 6 has been operated. This seems deeply unsatisfactory, if it is the position.

8.5 Form of payments to be made

8.5.1

The Court (unlike under, for example, the linear obstacle special regime in paragraphs 12–14) can structure the payments to be made by the operator, as follows under paragraph 7(4):

> The terms specified by virtue of sub-paragraph (1) above in an order under paragraph 5 above may provide –
> (a) for the making of payments from time to time to such persons as may be determined under those terms; and
> (b) for questions arising in consequence of those terms (whether as to the amount of any loss or damage caused by the exercise of a right or otherwise) to be referred to arbitration or to be determined in such other manner as may be specified in the order.

8.5.2

In *Brophy v Vodafone Ltd*,[10] HHJ Stephen Davies considered his jurisdiction to award either a lump sum of a periodic payment. He said as follows:

> The court clearly has jurisdiction under the Code to make either order. Paragraph 7(4)(a) permits the court to make an order for payments to be made from time to time to persons to be determined, so that the court clearly has jurisdiction to make an order as sought by Vodafone, and it is not suggested that the absence of specific

10 [2017] EWHC B9 (TCC).

provision for a lump sum payment prevents the court from so doing. I note that in *Cabletel* Mance LJ firmly rejected at [43–4] an argument that the judge below ought not to have ordered a lump sum as opposed to annualised payments. However whilst that confirms that the court has jurisdiction to make a lump sum order, it does not provide any guidance as to whether or not that is appropriate in this case, since it is clear that the decision in that case turned on the particular circumstances of, and evidence adduced in, the case rather than on any considerations of wider application. Finally, and insofar as relevant, I note that paragraph 23(5)(a) of the draft Code accompanying the Digital Economy Bill makes clear that the court has the option of providing either for a lump sum or for periodic payments.

8.6 Payment into court

8.6.1

Paragraph 7 gives the Court powers to order payments into Court, as follows:

(5) The court may, if it thinks fit –
  (a) where the amount of any sum required to be paid by virtue of terms specified in an order under paragraph 5 above has been determined, require the whole or any part of any such sum to be paid into court;
  (b) pending the determination of the amount of any such sum, order the payment into court of such amount on account as the court thinks fit.

(6) Where terms specified in an order under paragraph 5 above require the payment of any sum to a person who cannot be found or ascertained, that sum shall be paid into court.

8.7 Injurious affection to neighbouring land

8.7.1

We have already seen under the general regime paragraph 4,[11] that persons with an interest in land subject to the code right but who are not bound by a code right have a claim for injurious affection. Paragraph 16 of the Old Code gives such a right to neighbouring land.

8.7.2

Paragraph 16 provides:

(1) Where a right conferred by or in accordance with any of the preceding provisions of this code is exercised, compensation shall be payable by the operator under section 10 of the Compulsory Purchase Act 1965 (compensation for injurious affection to neighbouring land etc.) as if that section had effect in relation to injury caused by the exercise of such a right as it has effect in relation to injury caused by the execution of works on land that has been compulsorily purchased.

11 See Chapter 7 of this book.

(2) Sub-paragraph (1) above shall not confer any entitlement to compensation on any person in respect of the exercise of a right conferred in accordance with paragraph 2 or 3 above, if that person conferred the right or is bound by it by virtue of paragraph 2(2)(b) or (d) above, but, save as aforesaid, the entitlement of any person to compensation under this paragraph shall be determined irrespective of his ownership of any interest in the land where the right is exercised.

(3) Compensation shall not be payable on any claim for compensation under this paragraph unless the amount of the compensation exceeds £50.

8.7.3

This right is accordingly not available to anyone agreeing to be bound or treated as bound by the code right under paragraphs 2 or 3. Special provision is made for Scotland and Northern Ireland.

8.7.4

Section 10 of the Compulsory Purchase Act 1965 has been the subject of interpretation by the House of Lords in *Wildtree Hotels Ltd v Harrow London Borough Council*,[12] in which Lord Hoffmann restated a number of principles, originally known as the *McCarthy* rules.[13] The first rule is that the injury must have been done by reason of what is statutorily authorised. For example, if in the course of erecting a mast in the corner of a field, the contractor is careless or negligent and damage is caused in consequence of that, such negligently caused damage would not normally be statutorily authorised and the landowner would have to pursue a claim under the civil law against the contractor rather than a claim under the New Code against the operator. Second, there must be an injury which, but for the statutory authorisation, would have been actionable at law. Third, the damage must arise from a physical interference with some public or legal right, which the claimant has as owner of an interest in property is, by law, entitled to make use of. This usually means interference or an obstruction to a right of way, a right to light or a right of access onto a public highway. Fourth, the damage must arise from the execution of the authorised works, and not from their use. That means that it is only the consequence of the erection of apparatus, which would give rise to a claim for injurious affection compensation. The use of that apparatus, in terms of any adverse effects from electrical fields or otherwise, would not be compensatable under s.10.

8.7.5

The measure of compensation for injurious affection is the diminution in value of the affected interest in land caused by the interference with or obstruction of the legal right affected. If there is a permanent diminution in value, such values are used. If the interference is only temporary, then the measure of compensation is diminution in the rental value of the affected property for the relevant period.[14]

12 [2001] 2 AC 1.

13 So-called, after *Metropolitan Bd of Works v McCarthy* [1874] LR 7 HL 243.

14 *Wildtree Hotels Ltd v Harrow LBC* [2001] 2 AC 1; *Ocean Leisure Ltd v Westminster City Council* [2004] 3 EGLR 9.

8.7.6

The effect of Old Code paragraph 16(2) is to preclude any entitlement to compensation under s.10 by a person who conferred a code right or would be bound by it, but it does not preclude any such entitlement if a person holds an interest in the land, which is not bound by a code right. Thus, where a code right has been granted by the person having an interest in the roof and airspace above a building, the owner of a lease of a flat or a floor in such a building, whose interest is not bound by the code right, may have a claim under s.10, if that person can satisfy the rules set out above.

9 Old Code special regimes

9.1 Introduction

9.1.1

As explained in Chapter 5, Parliament distinguished between the general regime on the one hand, and separate special regimes on the other. Different policy decisions find expression in the different ways the regimes are formulated.

9.1.2

Chapter 11 explained that the Old Code operates a General Regime for the voluntary or compulsory acquisition of code rights in relation to 'any land' which can be used by operators. However, operators[1] have the ability to bring themselves within a number of 'special regimes' if they apply. These are:

(a) Street works (paragraph 9)
(b) Power to fly lines (paragraph 10)
(c) Tidal waters (paragraph 11)
(d) Linear obstacles (meaning a railway, canal or tramway) (paragraphs 12–14)
(e) Tree lopping (paragraph 19).

9.1.3

As also described above, even within the special regimes, it is possible to revert to the general regime:

(1) 'streets' that are not maintainable highways (paragraph 9(2)), to which the general regime is applied,
(2) tidal waters with a Crown interest (see paragraphs 11(2) and 26), in relation to which an agreement is required, which agreement must be one under paragraph 2;[2]
(3) agreements to obstruct under paragraph 3 of the general regime are required in relation to street works (paragraph 9(2)), flying lines (paragraph 10(1)) and tidal waters (paragraph 11(1)).

1 This is the same definition as discussed in relation to the general regime, in Chapter 6.
2 See generally paragraph 2(9).

9.1.4

It is considered that it was open to parties exercising a special regime right to do so on a time-limited basis in the same way that it was possible to limit the term length of a right conferred under the general regime. In that case, in the event that a special regime right was conferred and then expired, it is considered that the provisions of paragraph 21 of the Old Code applied. It is also considered that paragraph 20 applied generally to the special regimes.

9.2 Overview of the special regimes

9.2.1

The Court of Appeal in *GEO Networks Ltd v The Bridgewater Canal Co. Ltd*[3] described the various special regimes as follows:

> 6. Paragraphs 9 to 13 provide for what the judge called special regimes in respect of street works, a power to fly lines, tidal waters and linear obstacles. Each of the special regimes differs from the general regime in a number of different, but important, respects. In the case of street works paragraph 9 confers on the operator […] the right to install, keep, inspect and maintain electronic apparatus under, in, across and along a street maintainable by the public and any incidental works. In the case of street works no provision is made for notices, court orders or payment of consideration or compensation.
> 7. Paragraph 10 applies in a case where some electronic communications apparatus has been installed and the operator wishes to attach a line to it, overfly neighbouring land and attach the other end to similar apparatus there. In such a case the paragraph itself confers the requisite power on the operator. No provision is made for notices, court orders or the payment of consideration or compensation.
> 8. Paragraph 11 deals with tidal waters. It confers on the operator the right to install and keep electronic apparatus on or over tidal waters and land. But where the tidal waters or land are part of the Crown Estate such right is only exercisable with the agreement of the particular emanation of the Crown concerned. Further, in all cases, the operator is required, prior to seeking to exercise the right, to submit plans of the proposed works to the Secretary of State for his approval. No provision is made for any court order in default of agreement. Nor is consideration or compensation payable to anyone.
> 9. Paragraphs 12, 13 and 14 deal with what are called ‘linear obstacles’. Not only is that term not defined but it only appears in the headings to these three paragraphs. This omission may be important in other cases because it may provide a limit to the application of the general regime the terms of which are materially different to those of this special regime in a number of important respects. Paragraph 12 confers the relevant right but subjects its exercise to certain conditions.

3 [2010] EWCA Civ 1348.

9.3 Special regime: compendious nature of the right granted

9.3.1

Paragraphs 9 to 14 and 19 of the Old Code will now be considered in turn. It will be noted from what follows that there is significant similarity between the three separate code rights in paragraph 2(1), and the three limbs of the single code right conferred under paragraphs 9, 10, 11 and 12.

9.3.2

The sole but significant difference is that paragraph 2(1) treats the three elements as separate rights to be individually conferred, whereas the special regimes are framed in terms of a single right with three limbs.

9.3.3

The construction issues as to the scope of the rights granted under paragraph 2(1) limbs (a)–(c) of paragraph 2(1) are therefore, subject to that difference, equally applicable to the special regime rights, subject to that qualification.

9.4 Street works: paragraph 9

9.4.1

Unlike the general regime, the operator has, under paragraph 9, the right for the statutory purposes,[4] to do *any* of the things listed in paragraph 9. This amounts to the automatic statutory conferral of a single, compendious right to:[5]

(a) install electronic communications apparatus, or keep electronic communications apparatus installed, under, over, in, on, along or across a street or, in Scotland, a road;
(b) inspect, maintain, adjust, repair or alter any electronic communications apparatus so installed; and
(c) execute any works requisite for or incidental to the purposes of any works falling within paragraph (a) or (b) above, including for those purposes the following kinds of works, that is to say –
  (i) breaking up or opening a street;
  (ii) tunnelling or boring under a street; and
  (iii) breaking up or opening a sewer, drain or tunnel.

9.4.2

The similarity between the rights enumerated in paragraph 9(1) to those in paragraph 2(1) is to be noted, but the 'and' between rights (b) and (c) makes clear that they are

4 See the definition in paragraph 6.3.2 of Chapter 6.
5 There is a separate provision dealing with Scotland.

part of the single compendious right granted by paragraph 9, and are not individual and separate rights as under paragraph 2(1) (where (b) and (c) are separated by an 'or').

9.4.3

This provision was amended by the New Road and Street Works Act 1991, Schedule 8. 'Street' is defined in s.48(1) of the 1991 Act as follows:

(1) In this Part a 'street' means the whole or any part of any of the following, irrespective of whether it is a thoroughfare –
  (a) any highway, road, lane, footway, alley or passage,
  (b) any square or court, and
  (c) any land laid out as a way whether it is for the time being formed as a way or not.

Where a street passes over a bridge or through a tunnel, references in this Part to the street include that bridge or tunnel.

(2) The provisions of this Part apply to a street which is not a maintainable highway subject to such exceptions and adaptations as may be prescribed.

(3) In this Part 'street works' means works of any of the following kinds (other than works for road purposes) executed in a street in pursuance of a statutory right or a street works licence –
  (a) placing apparatus, or
  (b) inspecting, maintaining, adjusting, repairing, altering or renewing apparatus, changing the position of apparatus or removing it,

  or works required for or incidental to any such works (including, in particular, breaking up or opening the street, or any sewer, drain or tunnel under it, or tunnelling or boring under the street).

(3A) For the purposes of subsection (3), the works that are street works by virtue of being works required for or incidental to street works of any particular kind include –
  (a) reinstatement of the street, and
  (b) where an undertaker has failed to comply with his duties under this Part with respect to reinstatement of the street, any remedial works.

9.4.4

It will be noted that, at first sight, the right applies to any 'street',[6] irrespective of whether or not it is 'maintainable at public expense'.[7] An 'operator' is an 'undertaker' for the purposes of Part III of the 1991 Act and is subject to the provisions of that Part governing the execution of 'street works'. Where Part III of the 1991 Act applies (as it will in the case of all operator works on a 'street'), it prescribes detailed procedures for the carrying out of works.

6 Presumably in *Mercury Communications Ltd v London and India Dock Investments Ltd* [1993] 69 P & CR 135, which involved works under a street, the street was not a maintainable highway so that the valuation procedures reverted to the general regime.

7 For a definition of 'highway maintainable at public expense', see ss.36 and 329(1) of the Highways Act 1980.

9.4.5

That right conferred by paragraph 9(1) is however specifically qualified by paragraph 9(2):

> This paragraph has effect subject to paragraph 3 above and the following provisions of this code, and the rights conferred by this paragraph shall not be exercisable in a street which is not a maintainable highway or, in Scotland, a road which is not a public road without either the agreement required by paragraph 2 above or an order of the court under paragraph 5 above dispensing with the need for that agreement.

9.4.6

It therefore follows from the above that:

(1) In the case of any street, an obstruction agreement under paragraph 3 is required where an obstruction will be caused by the works.
(2) Where the street is not a maintainable highway, a paragraph 2 agreement or paragraph 5 Court dispensation is required. A 'maintainable highway' is defined in paragraph 1 of the Old Code as follows:

> 'maintainable highway' –
> (a) in England and Wales, means a maintainable highway within the meaning of Part III of the New Roads and Street Works Act 1991 other than one which is a footpath bridleway or restricted byway that crosses, and forms part of, any agricultural land or any land which is being brought into use for agriculture; and
> (b) in Northern Ireland, means a highway maintainable by the Department of the Environment for Northern Ireland.

9.4.7

In relation to streets that are not maintainable highways, where the position reverts to the general regime, it will be recalled that 'occupier' is given an extended definition in paragraph 2(8) to mean:

> In this paragraph and paragraphs 3 and 4 below –
> (a) references to the occupier of any land shall have effect –
> (i) in relation to any footpath or bridleway that crosses and forms part of any agricultural land or any land which is being brought into use for agriculture, as references to the occupier of that land;
> (ii) in relation to any street or, in Scotland, road (not being such a footpath or bridleway), as references – in England and Wales or Northern Ireland, to the street managers within the meaning of Part III of the New Roads and Street Works Act 1991 (which for this purpose shall be deemed to extend to Northern Ireland), and in Scotland, to the road managers within the meaning of Part IV of that Act.

9.5 Power to fly lines (paragraph 10)

9.5.1

Paragraph 10 is another special regime which confers an automatic right for an operator – in this case, it gives the operator the automatic right to fly lines to electronic communications apparatus over adjacent land:

(1) Subject to paragraph 3 above and the following provisions of this code, where any electronic communications apparatus is kept installed on or over any land for the purposes of the operator's network, the operator shall, for the statutory purposes, have the right to install and keep installed lines which –
   (a) pass over other land adjacent to or in the vicinity of the land on or over which that apparatus is so kept;
   (b) are connected to that apparatus; and
   (c) are not at any point in the course of passing over the other land less than 3 metres above the ground or within 2 metres of any building over which they pass.

(2) Nothing in sub-paragraph (1) above shall authorise the installation or keeping on or over any land of –
   (a) any electronic communications apparatus used to support, carry or suspend a line installed in pursuance of that sub-paragraph: or
   (b) any line which by reason of its position interferes with the carrying on of any business carried on on that land.

(3) In this paragraph *'business'* includes a trade, profession or employment and includes any activity carried on by a body of persons (whether corporate or unincorporate).

9.5.2

This is again a single, compendious right comprising the matters in paragraph 10(1)(a)–(c).

9.5.3

'Line' is defined by paragraph 1 of the Old Code as meaning:

> any wire, cable, tube, pipe or similar thing (including its casing or coating) which is designed or adapted for use in connection with the provision of any electronic communications network or electronic communications service

9.5.4

It will be noted that the right conferred is limited in a number of respects. First, under paragraph 10(1)(c), the line must be no less than 3 metres about ground, and must not be within 2 metres of any building on the land over which the lines pass. The right under paragraph 10(1) is expressly defined by sub-paragraph (2) to exclude supports, such as masts or poles. It is also expressly provided (under 2(b)) that such a line cannot interfere with a business being carried on, on the land. Paragraph 10(3) defines business inclusively as a 'trade, profession and employment', but also 'any' activity[8] being carried on by a body of persons. That latter part of the definition would seem to simply require that activity by a body of persons whether through the medium of an unincorporated body (not limited to an unincorporated association) or the vehicle of a single corporate entity is protected.

9.5.5

The definition in paragraph 10 is the same as the definition of 'business' in s.23(2) of the Landlord and Tenant Act 1954, and the above construction is supported by the interpretation that has been placed on that definition by commentators.[9] Applying the case law under the 1954 Act, it would seem that the term 'business' connotes an activity undertaken with a view to making a profit, even if none is made: see *Rolls v Miller.*[10]

9.5.6

An activity carried out in the spirit of benevolence without a view to making a profit is not captured by the definition, even if it is 'any activity' by a 'body of persons': *Secretary of State for Transport v Jenkins.*[11] However, an activity undertaken with a view to generating profit without distributing the same by dividend to shareholders is still a business: see *Hawkesbrook Leisure Ltd v The Reece-Jones Partnership.*[12] A member's club has been found to be a business.[13] There is no direct equivalent to s.56 of the 1954 Act by which the activities of a government department are a 'business' for these purposes.

9.5.7

Paragraph 17 provides a time-limited right for the occupier or owner of the affected land to object in certain circumstances. This right is available only 'before the expiration of the period of 3 months beginning with the completion of the installation of the apparatus'. In *Jones v T Mobile (UK) Ltd,*[14] the Court of Appeal considered what was meant by 'completion of the installation'. Holman J indicated (at [40]) that he agreed with the trial judge that the 'installation was complete when physical installation was complete, and not when the apparatus was operational' (see paragraphs 11 and 16 of the code). It was also held (at [13]) that the three months begins whether or not the code operator has attached a notice pursuant to the code under paragraph 18.

9.5.8

In the event that such apparatus requires alteration or removal, the provisions of paragraphs 20 and 21 must be employed.

8 See *Willis v British Commonwealth Universities Association* [1965] 1 QB 140; *Hillil Property & Investment Co Ltd v Naraine Pharmacy Ltd* [1979] 2 EGLR 65; *Parkes v Westminster Roman Catholic Diocese Trustee* [1978] 2 EGLR 50; *Groveside Properties Ltd v Westminster Medical School* [1983] 2 EGLR 68.

9 Reynolds and Clark, *The Renewal of Business Tenancies* (5th ed. 2017), at 1–111.

10 [1884] LR 25 Ch D 206.

11 [2000] 79 P & CR 118.

12 [2003] EWHC 3333 (Ch).

13 *Addiscombe Garden Estates Ltd v Crabbe* [1958] 1 QB 513.

14 [2003] 3 EGLR 55.

9.6 Tidal waters (paragraph 11)

9.6.1

'Tidal waters' are defined by paragraph 11(11) as including 'any estuary or branch of the sea, the shore below mean high water springs and the bed of any tidal water'. Although there was a complex set of limitations attaching to agreements for the use of rights over land in which there was a 'Crown interest' (see below), many of those provisions were repealed by the Marine and Coastal Access Act 2009.[15]

9.6.2

Under paragraph 11, it is provided (again subject to paragraph 3 of the general regime, relating to agreements to obstruct) that:

> the operator shall have the right for the statutory purposes –
> (a) to execute any works (including placing any buoy or sea-mark) on any tidal water or lands for or in connection with the installation, maintenance, adjustment, repair or alteration of electronic communications apparatus;
> (b) to keep electronic communications apparatus installed on, under or over tidal water or lands; and
> (c) to enter any tidal water or lands to inspect any electronic communications apparatus so installed.

9.6.3

It will be seen that the right referred to in paragraph 11(1) is again a single, compendious right as with the prior two special regimes considered.

9.6.4

There is a limitation to land in which the Crown has an interest. Where a Crown interest subsists in tidal waters, then paragraph 11(2) states that:

> A right conferred by this paragraph shall not be exercised in relation to any land in which a Crown interest, within the meaning of paragraph 26 below, subsists unless agreement to the exercise of the right in relation to that land has been given, in accordance with sub-paragraph (3) of that paragraph, in respect of that interest.

9.6.5

Paragraphs 26(2) and (3) of the Old Code sets out where that is the case:[16]

15 Transitional provisions appear at Schedule 9, part IV, paragraph 11 of that Act.

16 A modified consent rule under sub-paragraph (3) applies in Scotland: '(3A) In sub-paragraph (3), "*relevant person*", in relation to any land to which section 90B(5) of the Scotland Act 1998 applies, means the person who manages that land.'

(2) In this paragraph *'Crown interest'* means an interest which belongs to Her Majesty in right of the Crown or of the Duchy of Lancaster or to the Duchy of Cornwall or to a Government department or which is held in trust for Her Majesty for the purposes of a Government department and, without prejudice to the foregoing, includes any interest which belongs to Her Majesty in right of Her Majesty's Government in Northern Ireland or to a Northern Ireland department or which is held in trust for Her Majesty for the purposes of a Northern Ireland department.

(3) An agreement required by this code to be given in respect of any Crown interest subsisting in any land shall be given by the appropriate authority, that is to say –
   (a) in the case of land belonging to Her Majesty in right of the Crown, the Crown Estate Commissioners or, as the case may require, the government department having the management of the land in question or the relevant person;[17]
   (b) in the case of land belonging to Her Majesty in right of the Duchy of Lancaster, the Chancellor of that Duchy;
   (c) in the case of land belonging to the Duchy of Cornwall, such person as the Duke of Cornwall, or the possessor for the time being of the Duchy of Cornwall, appoints;
   (d) in the case of land belonging to Her Majesty in right of Her Majesty's Government in Northern Ireland, the Northern Ireland department having the management of the land in question;
   (e) in the case of land belonging to a government department or a Northern Ireland department or held in trust for Her Majesty for the purposes of a government department or a Northern Ireland department, that department;

   and if any question arises as to what authority is the appropriate authority in relation to any land that question shall be referred to the Treasury, whose decision shall be final.

9.6.6

It seems that the form of 'agreement' must correspond with paragraph 2 of the general regime, as paragraph 2(9) provides that 'Subject to paragraphs 9(2) and 11(2) below, this paragraph shall not require any person to give his agreement to the exercise of any right conferred by any of paragraphs 9 to 12 below'. It therefore follows that paragraph 11(2) does require that paragraph 2 be complied with.

9.7 Linear obstacles (paragraph 12)

9.7.1

Linear obstacles are not directly defined, but must be either 'a railway, canal or tramway' (paragraph 12(10)). The Court of Appeal in *Bridgewater Canal Co. v GEO Networks* explained that:

17 [2010] EWCA Civ 1348, at paragraph [10].

This definition operates to confine the linear obstacle regime to a railway, canal or tramway and associated land. Further the person to whom a notice must be given and by whom a counter-notice may be given is the person carrying on the undertaking of a railway, canal or tramway whether or not he owns the land on or over which it runs.

9.7.2

'Railway' includes a light railway (see paragraph 1). Paragraph 12 provides that:

> Subject to the following provisions of this code, the operator shall, for the statutory purposes, have the right in order to cross any relevant land with a line, to install and keep the line and other electronic communications apparatus on, under or over that land and –
>
> (a) to execute any works on that land for or in connection with the installation, maintenance, adjustment, repair or alteration of that line or the other [electronic communications apparatus; and
>
> (b) to enter on that land to inspect the line or the other apparatus.

9.7.3

It is to be noted that the right relates to both a 'line' and 'electronic communications apparatus'. Given that all linear obstacles have in common that they are long, narrow transportation networks, it is clear that an electronic communications network will, from time to time, have to pass on, under or over 'relevant land'. 'Relevant land' means:

> land which is used wholly or mainly either as a railway, canal or tramway or in connection with a railway, canal or tramway on that land, and a reference to the person with control of any such land is a reference to the person carrying on the railway, canal or tramway undertaking in question.

9.7.4

As the Court of Appeal observed in *Bridgewater Canal Co. v GEO Networks*[18] at paragraph [9],

> Paragraphs 12, 13 and 14 deal with what are called 'linear obstacles'. Not only is that term not defined but it only appears in the headings to these three paragraphs. This omission may be important in other cases because it may provide a limit to the application of the general regime the terms of which are materially different to those of this special regime in a number of important respects.

9.7.5

It described the mechanism for the exercise of such rights as (*ibid.*):

18 [2010] EWCA Civ 1348.

Paragraph 12 confers the relevant right but subjects its exercise to certain conditions. They include the giving of notice by the operator not to the owner of any interest in the land but to the person with control of it. That person may object to the proposed works and/or require compensation. In those events an arbitrator is appointed in accordance with the provisions of paragraph 13(1) with the powers specified in paragraph 13(2). The compensation and consideration referred to in paragraph 13(2)(e) is assessed in accordance with paragraph 13(6) and paid, not to the owner of any interest in the land but to the person in control of it. Thus the linear obstacle regime differs from the general regime in at least the following respects. (1) In the former the notice under paragraph 12(4) is given to the person in control of the land, in the latter paragraph 5(1) requires the notice to be given to the person whose agreement is required to bind his or any relevant interest in the land. (2) The linear obstacle regime provides for the person to whom the notice was given to serve a counter-notice of objection and requirement for payment of compensation under paragraph 12(6)(8) but the general regime does not. (3) In the case of the linear obstacle regime differences as to what may be done and what should be paid are referred to arbitration but in the case of the general regime they are determined by the county court. The issue on this appeal is whether the matters for which consideration or compensation should be paid are also different.

9.7.6

The right is further circumscribed as follows:

(1) Under sub-paragraph 12(2), it is provide that '[a] line installed in pursuance of any right conferred by this paragraph need not cross the relevant land in question by a direct route or by the shortest route from the point at which the line enters that land, but it shall not cross that land by any route which, in the horizontal plane, exceeds the said shortest route by more than 400 metres'.
(2) Under sub-paragraph 12(3), it is provided that 'electronic communications apparatus shall not be installed in pursuance of any right conferred by this paragraph in any position on the relevant land in which it interferes with traffic on the railway, canal or tramway on that land.'

9.7.7 Non-emergency works: notice procedure

If the works are not in the nature of an emergency, then a notice procedure has to be followed by the operator wishing to exercise the right to cross land which is wholly or mainly used as a railway, canal or tramway.

9.7.8

Sub-paragraph (4) states that the operator shall not execute any works on any land in pursuance of any right conferred by this paragraph unless he has either given the person with control of the land 28 days' notice[19] of his intention to do so, or the works are emergency works. Emergency works, and their definition, are considered below at paragraph 9.7.11.

19 As to Old Code notices, formality and service, see Chapter 32.

9.7.9

The notice under sub-paragraph (4) must 'contain a plan and section of the proposed works or (in lieu of a plan and section) any description of the proposed works (whether or not in the form of a diagram) which the person with control of the land has agreed to accept for the purposes of this sub-paragraph'. This means that a plan and section must be provided, unless the person with control of the land has agreed to make do with a description.

9.7.10

Under paragraph 12(6), if an objection is received within the 28-day notice period, then the operator can only execute the works if either:

(1) within the period of 28 days beginning with the giving of the notice of objection, neither the operator not that person has given notice to the other requiring him to agree to an arbitrator to whom the objection may be referred under paragraph 13 below; or
(2) in accordance with an award made on such a reference; or
(3) to the extent that the works have at any time become emergency works.

9.7.11 Emergency works

'Emergency works' are defined in paragraph 1 of the Old Code. This provides that:

> '*emergency works*', in relation to the operator […] means works the execution of which at the time it is proposed to execute them is requisite in order to put an end to, or prevent, the arising of circumstances then existing or imminent which are likely to cause –
> (a) danger to persons or property,
> (b) the interruption of any service provided by the operator's network […]; or
> (c) substantial loss to the operator […]
>
> and such other works as in all the circumstances it is reasonable to execute with those works.

9.7.12

Accordingly, the operator may bypass the need for notice in relation to linear obstacles in limited circumstances. It is not clear from the drafting whether works of installation to provide an alternative network route, where it is anticipated that another part of the operator's network will be interfered with, is an 'emergency'.

9.7.13

The notice procedure does not apply at all where the works are a recognised emergency from the outset, however paragraph 12(6)'s wording would also allow the operator to step outside the notice procedure if, once a notice is given, the works are first understood to be, or have become, emergency works. Once the works are commenced, the operator must give notice to the person with control of the land notice identifying the works, explaining the reason why there was an emergency, and the information to be contained in a paragraph 21(4)-compliant notice. That notice triggers a right for the person to whom notice has been given to give notice of the compensation by way of loss and damage suffered in consequence of the works being carried out: paragraph 12(8). If that amount is not agreed, the matter can be referred to arbitration under paragraph 13.

9.7.14 Works in breach of paragraph 12

If works are carried out otherwise than under paragraph 12, that constitutes a criminal offence under paragraph 12(9):

> If the operator commences the execution of any works in contravention of any provision of this paragraph, he shall be guilty of an offence and liable on summary conviction to a fine not exceeding level 3 on the standard scale.

9.7.15

Presumably an operator acting outside the provisions of paragraph 12 will be treated as an ordinary person at common law, liable in tort in the usual way.

9.7.16 The paragraph 13 arbitration

A statutory arbitration procedure is provided for in paragraph 13,[20] for issues arising out of paragraph 12 to be referred to arbitration. Those are to resolve objections by persons with control of the land under the notice procedure in paragraphs 12(3)–(6), or in relation to compensation payable for emergency works under paragraph 12(8).

9.7.17

The objection or question to be referred to the arbitrator is referred to a single arbitrator appointed by agreement between the parties or, in default, by the President of the Institution of Civil Engineers (paragraph 13(1)). This latter provision can be problematic, as often, linear obstacle arbitrations are not matters of engineering, but law and valuation.

9.7.18

The scope of what the arbitrator can award is set out in paragraph 13(2). This provides that:

> Where an objection under paragraph 12 above is referred to arbitration under this paragraph the arbitrator shall have the power –
> (a) to require the operator to submit to the arbitrator a plan and section in such form as the arbitrator may think requisite for the purposes of the arbitration;

20 For Scotland, see paragraph 13(7).

(b) to require the observations on any such plan or section of the person who objects to the works to be submitted to the arbitrator in such form as the arbitrator may think requisite for those purposes;
(c) to direct the operator or that person to furnish him with such information and to comply with such other requirements as the arbitrator may think requisite for those purposes;
(d) to make an award requiring modifications to the proposed works and specifying the terms on which and the conditions subject to which the works may be executed; and
(e) to award such sum as the arbitrator may determine in respect of one or both of the following matters, that is to say –
   (i) compensation to the person who objects to the works in respect of loss or damage sustained by that person in consequence of the carrying out of the works, and
   (ii) consideration payable to that person for the right to carry out the works.

9.7.19

In relation to an arbitration concerning emergency works compensation under paragraph 12(8), paragraph 13(3) provides that:

> Where a question as to compensation in respect of emergency works is referred to arbitration under this paragraph, the arbitrator –
> (a) shall have the power to direct the operator or the person who requires the payment of compensation to furnish him with such information and to comply with such other requirements as the arbitrator may think requisite for the purposes of the arbitration; and
> (b) shall award to the person requiring the payment of compensation such sum (if any) as the arbitrator may determine in respect of the loss or damage sustained by that person in consequence of the carrying out of the emergency works in question.

9.7.20

It is provided under paragraph 13(4) that the payment of sums under 13(2)(e) or 13(3)(b) may be made a condition for the award.

9.7.21

As to the computation of the amounts under paragraph 13(2) and (3), paragraph 13(5) provides that:

> For the purposes of the making of an award under this paragraph –
> (a) the references in sub-paragraphs (2)(e) and (3)(b) above to loss shall, in relation to a person carrying on a railway, canal or tramway undertaking, include references to any increase in the expenses of carrying on that undertaking; and
> (b) the consideration mentioned in sub-paragraph (2)(e) above shall be determined on the basis of what would have been fair and reasonable if the person who

objects to the works had given his authority willingly for the works to be executed on the same terms and subject to the same conditions (if any) as are contained in the award.

9.7.22

The most problematic aspect of paragraph 13 is the definition of 'consideration'. On the face of it, paragraph 12 makes available the land of a private undertaking to a private communications operator in order to use it for the purposes of the latter's business. The person with control of the land is not entitled to object to that use (provided that the substantive and procedural requirements applicable under paragraph 13 are met). This is because paragraph 12 proceeds from the position that the operator does not have to secure the right, but from the position that he already has it.

9.7.23

In *Bridgewater Canal Co. Ltd v Geo Networks Ltd*[21] at first instance, Lewison J held that the operator seeking to use a paragraph 12 right was required to make a payment for the enjoyment of the right into the future. He stated:

49. [Paragraph 13(2)(e)(ii)] is concerned with 'consideration' (i.e. price). Since loss and damage to the undertaker has been dealt with in paragraph 13(2)(e)(i), the natural reading of paragraph 13(2)(e)(ii) is that it is dealing with something else. What is that something else, if not the value to the operator of acquiring the right? If, as I think, part of the notion of carrying out works is the permanent (or indefinite) consequence of carrying them out, then it seems to me that paragraph 13(2)(e)(ii) envisages that the consideration will take into account the fact that the right to carry out the work will carry with it the right to retain on or under the land whatever apparatus has been installed as a result of those works. The notion that the consideration must be assessed on the assumption that the operator has no right to retain or use the apparatus after it has been installed seems to be to be both counter-factual and far divorced from the real world. If a counter-factual assumption is to be made, then it must be spelled out clearly. There is only one counter-factual assumption referred to in this part of the Code, and that is the counter-factual assumption that authority has been willingly given. So far as the real world is concerned, if an operator and an undertaker were discussing the consideration that the operator would be required to pay in order to acquire the right to carry out works to install apparatus, the first question that the operator would ask would surely be: can I keep it there once I have installed it? If the answer is no, it is difficult to see why the operator should agree to pay anything, particularly since he will have to compensate the undertaker for any loss or damage suffered in consequence of the works. Yet paragraph 13(2)(e)(ii) must have contemplated that consideration would be payable, at least in some circumstances. [Counsel for Geo] was not able to explain satisfactorily in what circumstances consideration would be payable.
50. In addition, the consideration must be 'fair and reasonable'. It does not seem to me to be fair that an operator should have something for nothing. If he acquires something of value to him, fairness seems to me to require that he pays for it.

21 [2010] EWHC 548 (Ch).

> As I have said, [Counsel for Geo] warned against allowing persons in control of linear obstacles extracting ransom payments from operators. However, it seems to me that the use of the phrase 'fair and reasonable' precludes the extraction of a ransom payment, as Mance LJ observed in Cabletel. Once that objection has been cleared out of the way, I do not consider that there is a compelling argument against the payment of consideration by an operator.

9.7.24

On appeal, in *GEO Networks Ltd v The Bridgewater Canal Co. Ltd*,[22] the Court of Appeal, overruling the decision of the High Court on this point, did not entitle the person with control of relevant land under paragraph 12 for consideration payable in respect of the operator's enjoyment of rights under paragraph 12. Sir Andrew Morritt contrasted the general and the special regimes. In relation to the general regime, he explained at paragraph [26] that:

> The general regime concentrated at the outset on the occupier no doubt because he was primarily affected but also most easily identified. If and so far as it was necessary for the agreement to bind others with an interest in the land detailed provision was also made in paragraph 2. But the procedure was intended to be swift; thus only 28 days had to elapse from the service of the relevant notice to the commencement of proceedings in the County Court. The order in those proceedings might bind any person having any interest in the land. And in assessing the consideration to be paid the County Court is specifically required to ensure that all persons bound by the order are adequately compensated for the grant of those rights. If necessary the order for compensation may take the form of an order for periodical payments. In that context it is plain that the compensation or consideration must extend to each of the rights specified in paragraph 2(1). Thus the consideration will extend to payment for the right to keep the apparatus on the land.

9.7.25

In contrast, the 'special regimes' did not so provide.

> 27. The special regimes are quite different. No consideration is payable in relation the installation of electronic apparatus in, on or under streetworks or in tidal waters. Nor is consideration or compensation payable for 'overflying'. Thus it is to be expected that the special regime entitled 'linear obstacles' will differ from the general regime. It also differs from the other special regimes in that some consideration or compensation is payable. The question is 'for what?'. There are a number of pointers to the answer to that question.
> 28. The first is that paragraph 12 is concerned to enable the operator to cross a railway, canal or tramway. Although the crossing does not have to be by the most direct or shortest route the excess must not exceed 400 metres. Thus the interference or intrusion on the land of others is minimal. Second, the land in question must be used by or incidental to a railway, canal or tramway. Land used for any other purpose does not fall within the linear obstacle regime. An undertaker of those activities is not necessarily a public body, but is likely to

22 [2010] EWCA Civ 1348.

be providing a public facility. There is no principle of which I am aware which requires the provider of one public facility, a railway, to be paid by another, a provider of electronic communications networks or services for such a minimal intrusion as crossing the railway with a line and, if necessary, a pylon. Third, it is noteworthy that under paragraph 13(2)(e) the compensation or consideration is payable to the person in control of the land, that is the person carrying on the railway, canal or tramway undertaking. It is payable to him either in respect of loss sustained by him or as consideration 'for' the right. There are no provisions applicable to this special regime comparable to those of the general regime for binding persons with an interest in the land in question or assessing compensation in respect of those interests. Indeed the payments under paragraph 13 are made to the person in control of the land, namely the person carrying on the railway, canal or tramway as the case may be, not, as in the general regime, to the occupier.

29. Given these features I do not approach the question [...] on the basis that the code should be interpreted so as not to burden a person's property without some compensatory payment. Moreover in my view paragraphs 12 and 13 do not require it. It is clear from the opening words of paragraph 12 that subparagraphs (a) and (b) are included for implementing the grant of the right to install and keep the apparatus crossing the railway not for conferring it. The compensation and consideration payable under paragraph 13(2)(e) is in respect of the right to carry out of those works and the loss sustained by reason of doing so in implementation of the right to install and keep. Given the structure of paragraphs 12(1) and 13(2)(e) I see no reason to interpolate into the words 'the right to carry out the works' in paragraph 13(2)(e) the additional words 'and to keep the same ...'.
30. This interpretation avoids the windfall and also the anomaly referred to in para 22 above which might arise on the judge's construction. I would add that I derive no support from the fact that the appointor is the President of the Institution of Chartered Engineers. The powers of the arbitrator conferred by paragraph 13(2) are wide. The issues which may arise are not confined to any one professional discipline. Further there are other professional bodies whose members may be equally or more concerned with particular issues capable of arising in such arbitrations: see *Whitaker's Almanack* (2001), pp 637–640.

9.7.26

Consequently, it appears that 'consideration' is confined to the conferral of the right to carry out the works (i.e. for some form of 'works access licence'), but not, then, any right to make use of the works, once in place. This must follow from the Court of Appeal's acceptance of the submissions of the Appellant, GEO Networks, as recorded at paragraph [21] of the judgment. This, it was said, was to avoid a windfall arising in favour of a person with control of relevant land, as such a person might not have a proprietary interest out of which to grant a right in the nature of an easement or lease in favour of an operator.

9.7.27

In determining a paragraph 13 arbitration, paragraph 13(5) requires that

> In determining what award to make on a reference under this paragraph, the arbitrator shall have regard to all the circumstances and to the principle that no person should unreasonably be denied access to [an electronic communications network or to electronic communications services.

9.7.28

As to the 'unreasonably denied access to an electronic communications network or to electronic communications services test, see section 7.6 of Chapter 7.

9.7.29 Additional right to alter

A special alteration right exercisable on notice arises in relation to electronic communications apparatus. This is in addition to the rights under paragraphs 20 and 21, which apply generally (and also apply to rights under paragraph 12).

9.7.30

Paragraph 14(1) operates 'without prejudice to the following provisions of this code', and states that

> the person with control of any relevant land[23] may, on the ground that any electronic communications apparatus kept installed on, under or over that land for the purposes of the operator's [network] [2]interferes, or is likely to interfere, with –
> (a) the carrying on of the railway, canal or tramway undertaking carried on by that person, or
> (b) anything done or to be done for the purposes of that undertaking,
> give notice to the operator requiring him to alter that apparatus.

9.7.31

The notice provisions under paragraph 24 of the Old Code apply (see Chapter 32 of this book). As can be seen from grounds 14(1)(a) and (b), this special right to alter arises where there is interference with the undertaking. The extended meaning of 'alter' as including 'removal' applies to the above provisions (see paragraph 1(1) and (2) of the Old Code).

9.7.32

Under paragraph 14(2), it is then stated that

23 Paragraph 14(6): 'In this paragraph references to relevant land and to the person with control of such land have the same meaning as in paragraph 12 above'.

(2) The operator shall within a reasonable time and to the reasonable satisfaction of the person giving the notice comply with a notice under sub-paragraph (1) above unless before the expiration of the period of 28 days beginning with the giving of the notice he gives a counter-notice to the person with control of the land in question specifying the respects in which he is not prepared to comply with the original notice.

(3) Where a counter-notice has been given under sub-paragraph (2) above the operator shall not be required to comply with the original notice but the person with control of the relevant land may apply to the court for an order requiring the alteration of any electronic communications apparatus to which the notice relates.

9.7.33

Under paragraph 14(4), the Court may then make an order for alteration in the following circumstances:

> The court shall not make an order under this paragraph unless it is satisfied that the order is necessary on one of the grounds mentioned in sub-paragraph (1) above and in determining whether to make such an order the court shall also have regard to all the circumstances and to the principle that no person should unreasonably be denied access to [an electronic communications network or to electronic communications services].

9.7.34

Paragraph 14(5) adds:

> An order under this paragraph may take such form and be on such terms as the court thinks fit and may impose such conditions and may contain such directions to the operator or the person with control of the land in question as the court thinks necessary for resolving any difference between the operator and that person and for protecting their respective interests.

9.7.35

As to the 'unreasonably denied access to an electronic communications network or to electronic communications services' test, see section 7.6 of Chapter 7 of this book.

9.8 Tree lopping

9.8.1

Paragraph 19(1) gives a right to lop trees where any tree overhangs any street and, in doing so, either –

(a) obstructs or interferes with the working of any electronic communications apparatus used for the purposes of the operator's network, or

(b) will obstruct or interfere with the working of any electronic communications apparatus which is about to be installed for those purposes.

9.8.2

It will be noted that this provision concerns trees overhanging streets.[24] There is again a notice procedure, by which the operator can give notice to the 'occupier' of the land on which the tree is located for it to be lopped to prevent the obstruction or interference.[25] If within the period of 28 days beginning with the giving of the notice by the operator, the occupier of the land on which the tree is growing gives the operator a counter-notice objecting to the lopping of the tree, the notice shall have effect only if confirmed by an order of the court (paragraph 19(2)).

9.8.3

If, however, no notice is given or a Court order confirming the notice has been obtained, then the operator is entitled to lop the tree himself (paragraph 19(3)). That must be done in a husband-like manner and in such a way as to cause the minimum damage to the tree (paragraph 19(4)).

9.8.4

Paragraph 19(5) states that where:

(a) a notice under sub-paragraph (1) above is complied with either without a counter-notice having been given or after the notice has been confirmed, or

(b) the operator exercises the power conferred by sub-paragraph (3) above,

the court shall, on an application made by a person who has sustained loss or damage in consequence of the lopping of the tree or who has incurred expenses in complying with the notice, order the operator to pay that person such compensation in respect of the loss, damage or expenses as it thinks fit.

24 See also the discussion in Chapter 29 in respect of the New Code provisions, which are very similar to those of the Old Code save that they extend to vegetation as well as trees.

25 As to notices, see paragraph 24 and Chapter 32.

10 Alteration and removal of apparatus under paragraph 20

10.1 Introduction

10.1.1

This chapter and Chapter 11 consider the rights which the Old Code contains which entitle the occupier/owner to seek alteration or removal of the apparatus. They are contained in paragraphs 20 and 21 respectively. Both of these Old Code provisions will have continuing relevance, notwithstanding the coming into force of the New Code, having regard to the transitional provisions contained in Schedule 2 to the Digital Economy Act 2017.[1] Under those transitional provision, paragraph 20 may continue to be relied upon with respect to an agreement which is a subsisting agreement at the date of the coming into force of the New Code. Paragraph 21 may be relied upon with respect to a subsisting agreement where notice was served under that paragraph prior to the coming into force of the New Code.

10.1.2

Both provisions provide for a similar procedure, namely:

(1) Service of a notice by the party seeking the alteration or removal;
(2) Service of a counter-notice by the operator in response within a period of 28 days beginning with the giving of the notice;
(3) A requirement to comply with the notice upon failure of the operator to serve a counter-notice;
(4) The institution of proceedings by the party seeking the alteration or removal where a counter-notice has been served;
(5) An appropriate order of the court providing for the alteration or removal.

10.1.3

The provisions of paragraphs 20 and 21 are often combined by a party seeking to remove electronic communications apparatus and operators often respond to any notice served upon them, e.g. a notice pursuant to s.25 of Part II of the Landlord and Tenant

1 As to the transitional provisions, see Chapter 31 generally and section 10.14 of this chapter.

Act 1954 ('the 1954 Act') as if the same were a notice requiring alteration or removal pursuant to paragraphs 20 and/or 21 of the Old Code.

10.1.4

This chapter is arranged in the following way:

(a) Section 10.2 deals with an overview of the paragraph.
(b) Section 10.3 deals with who may serve a paragraph 20 notice.
(c) Section 10.4 deals with when notice pursuant to paragraph 20 may be served.
(d) Section 10.5 considers what is meant by 'alteration of electronic communications equipment'.
(e) Section 10.6 considers the meaning of the phrase 'necessary to enable ... a proposed improvement of the land' to be carried out.
(f) Section 10.7 considers the procedure for obtaining an order for alteration.
(g) Section 10.8 looks at the statutory criteria to which the court will have regard in determining whether or not to make an order for alteration.
(h) Section 10.9 considers the orders which may be made by the court.
(i) Section 10.10 considers the impact of paragraph 5 of the Old Code.
(j) Section 10.11 considers the question of whether an order may be enforced before the expiry of a fixed term agreement.
(k) Section 10.12 deals with who has to bear the costs of effecting removal.
(l) Section 10.13 considers the interrelation of paragraph 20 with the 1954 Act.
(m) Section 10.14 considers the transitional provisions of the New Code.

10.1.5

The Old Code applies whether the agreement grants a lease, whether or not contracted out from the 1954 Act, or a licence. In this chapter, any reference to a lease (which is the most common form of arrangement encountered in practice) should also be taken to include a contractual arrangement between the landowner and the operator.

10.2 Alteration of apparatus: overview

10.2.1

The heading to paragraph 20 is 'Power to require alteration of apparatus'. That description of what paragraph 20 is dealing with is both accurate and misleading. It is accurate in that, unlike paragraph 21, it is not a restriction but an enabling provision. As one commentator has noted,[2] it amplifies rather than restricts the rights of the landowner.[3] It is misleading in that 'alteration' is not confined to what one would ordinarily consider is an alteration;[4] it includes removal or replacement.[5]

2 Malcolm Dowden, solicitor.

3 Here used in a neutral sense to refer to the party seeking the alteration. Who may seek to invoke paragraph 20 is considered in section 10.3.

4 See e.g. some form of reconfiguration. In practice it is invariably the case that a paragraph 20 notice is served in circumstances where complete removal is required by the landowner.

5 Paragraph 1(2) of the Old Code. It has been held that the detachment of any part of the apparatus or its supporting structure during the process of alteration pursuant to paragraph 20 does not engage paragraph 21 of the Old Code: *PG Lewins Ltd v Hutchison 3G UK Ltd* 9 March 2018, Bristol CC.

10.2.2

In summary, paragraph 20 enables any person with an interest in the land on, under or over which the apparatus is installed to serve notice on the operator that the owner requires alteration of the electronic communications apparatus on the ground that it is necessary to enable the landowner to carry out a proposed improvement of the land. There are two preliminary points to note with respect to the paragraph 20 power, namely (1) the power is not exhaustive as to the circumstances in which an alteration may be sought and (2) it may be invoked irrespective of the terms of the agreement with the operator.

10.2.3

As to the first point, paragraph 20 is not intended to be exhaustive of the circumstances in which an alteration to the electronic communications apparatus of the operator may be required. It is clear from the terms of paragraph 27(2) of the Old Code that the provisions of the Old Code, except paragraphs 8(5) and 21, are without prejudice to any rights arising under the code agreement. Thus, the power need not be invoked if there exists a contractual entitlement enabling the landowner to require the operator to effect an alteration which is an alteration falling short of full removal.[6] It is considered that if there is a contractual provision providing for a unilateral right enabling the landowner to require the operator to remove the electronic communications apparatus, this will not be enforceable because of the restriction contained in paragraph 21 as to the circumstances in which complete removal of electronic communications apparatus may be effected and paragraph 21 is not capable of being excluded by the express terms of the agreement: paragraph 27(2).[7] Complete removal of the apparatus other than in reliance on paragraph 21 can be required only by relying on paragraph 20 and satisfying the various criteria under it. A failure to comply with the contractual entitlement to require an alteration to be effected will give rise to a claim in damages.[8]

10.2.4

Paragraph 20 is not affected by the contractual arrangements that exist with the operator. It is considered that it may be invoked irrespective of what is said in the terms of the agreement. Clearly it may be invoked notwithstanding the absence of a contractual entitlement to require an alteration to be effected by the operator. It would also

6 See *PG Lewins Ltd v Hutchison 3G UK Ltd* [2018] Bristol CC at [32]–[35].

7 As was said in *PG Lewins Ltd v Hutchison 3G UK Ltd* at [49]:

> The scheme of the Code, which provides the context and the purpose for the provisions in para 21 is, in my judgment, that para 21 is intended to provide general protection to an operator against the removal from a landowner's land of telecommunications equipment, while para 20 enables the landowner to seek to redevelop his land and, subject to a balancing of the owner's and the operator's and the public interests, and subject to maintaining the integrity of the network and services, to require the alteration of the location of that equipment. This is illustrated in the instant case, in that PGL did not want the apparatus removed from the land (defined in the R/A as Greyfriars generally) but simply its relocation on the roof.

8 *PG Lewins Ltd v Hutchison 3G UK Ltd, ibid.*

seem that paragraph 20 would override a contractual prohibition restricting the landowner from carrying out alterations. This, it is considered, is made clear by paragraph 20(1) which refers to the entitlement to serve a notice 'notwithstanding the terms of any agreement binding that person', i.e. the landowner. Paragraph 27(3) of the Old Code furthermore provides that no compensation is payable to the operator for exercising a code right other than that provided for by the code. Thus, if and in so far as the exercise of the code right is contrary to the terms of the agreement, there would appear to be no contractual entitlement to damages for breach of contract.[9]

10.3 Who may invoke paragraph 20?

10.3.1

The paragraph may be invoked by a 'person with an interest in that land or adjacent land'. There is no definition within the Old Code of what is an 'interest in land'. Ordinarily the reference to an 'interest in land' would exclude a person who has a mere contractual interest; some form of proprietary interest in land is required. An 'interest in land' would clearly include a freehold or leasehold interest. A specifically enforceable agreement for lease giving rise to a lease in equity would probably also, in the view of the Authors, come within the definition. It would also cover an interest in land such as an easement.

10.3.2

The party that may serve the paragraph 20 notice is one who has either (1) an interest in the land upon which the electronic communications apparatus is kept installed on, under or over for the purposes of the operator's network or (2) a person with an interest in land which is adjacent to such land.

10.3.3

The category of persons entitled to serve a paragraph 20 notice is, for convenience, referred to in this chapter as 'the landowner'.

10.4 When may notice first be served?

10.4.1

There is no restriction contained within paragraph 20 as to the date before which a paragraph 20 notice may be served. It is, therefore, possible for a paragraph 20 notice to

9 But conversely, para 27(3) does not protect the operator where it is in breach of a contractual obligation which is not struck down by 27(2). As was said in *PG Lewins Ltd v Hutchison 3G UK Ltd* [2018] Bristol CC, at [44]:

> So far as para 27(3) is concerned this protects the operator from liability to compensate any person in respect of any loss or damages caused by the lawful exercise of any right conferred by or in accordance with the Code. However, there are two elements to that provision which deprive it of application here. One is that it is not the exercise of a Code right of which the Claimant complain, but a failure to comply with a contractual obligation (to move the apparatus). The second is that, if one characterised the failure to move the equipment as the continued exercise of the previously existing right to locate the

be served before the expiry of a fixed term agreement. As to whether an order pursuant to paragraph 20 can be enforced prior to the expiry of any fixed term agreement, see section 10.11 below.

10.4.2

It matters not that the operator's agreement, if a lease, is one protected by the 1954 Act. As the paragraph 20 notice is not one seeking to terminate the lease it need not comply with the terms of the 1954 Act and may be served at any time.

10.4.3

It is not considered that service of a paragraph 20 notice can be said fall foul of s.38 of the 1954 Act. Section 38 strikes down 'any agreement relating to a tenancy' to which the 1954 Act applies, in so far as it 'purports to preclude the tenant from making an application or request under' the 1954 Act. However, the service of paragraph 20 notice is a statutory right and does not form the subject matter of any agreement between the parties.

10.5 Alteration of electronic communications apparatus

10.5.1

As previously noted, 'alteration' includes removal.[10] The alteration is of 'electronic communications apparatus'. Electronic communications apparatus is given a wide definition in paragraph 1 of the Old Code as meaning:

(a) any apparatus (within the meaning of the Communications Act 2003) which is designed or adapted for use in connection with the provision of an electronic communications network;
(b) any apparatus (within the meaning of that Act) that is designed or adapted for a use which consists of or includes the sending or receiving of communications or other signals that are transmitted by means of an electronic communications network;
(c) any line;
(d) any conduit, structure, pole or other thing in, on, by or from which any electronic communications apparatus is or may be installed, supported, carried or suspended;

and references to the installation of electronic communications apparatus are to be construed accordingly.

10.5.2

This definition would clearly include e.g. a mast from which apparatus is carried or suspended.

equipment in its old location, then that was not a lawful exercise of the right because it was in breach of the contractual provision which para 2(5) [of the Old Code] provides shall only be exercisable in accordance with the terms subject to which it was conferred.

10 See paragraph 10.2.1.

10.5.3

It is important to note that the paragraph applies only 'where any electronic communications apparatus is kept installed on, under or over any land *for the purposes of the operator's network*'. The 'operator's network' is defined in paragraph 1 of the Old Code as meaning 'so much of any electronic communications network or conduit system provided by [the] operator as is not excluded from the application of the code under section 106(5) of the Communications Act 2003'.

10.5.4

Thus, a piece of apparatus such as a mast erected under a tenancy agreement, but which is utilised for apparatus serving networks of third party operators but not any network of the tenant, is not, vis-à-vis the tenant, apparatus falling within paragraph 20. In those circumstances the landowner would have to serve notice upon the third party operator even if the apparatus was owned by the tenant. There is nothing within paragraph 20 which requires the apparatus required to be altered to be owned by the operator; it is sufficient that the operator against whom alteration is sought is utilising the apparatus for the purposes of the operator's network.

10.6 Necessary to enable a proposed improvement of the land

10.6.1

A notice under paragraph 20 may be served only if the removal of the apparatus is 'necessary' to enable the person serving the notice to carry out 'a proposed improvement of the land in which he has an interest'.

10.6.2

There is no definition of whether the proposed alteration can be said to be 'necessary'. One can think of obvious examples e.g. where electronic communications apparatus is situated on a roof of a building and the proposed improvement consists of its demolition and rebuilding. The land which the apparatus is installed on, under or over will simply disappear and clearly the presence of the apparatus renders it necessary for it to be removed in order to enable the demolition to be effected. Another common example which arises in practice is that of a residential building which is proposed to be improved by providing additional floors for flats.

10.6.3

Alternatively, where the improvement is relatively minor in terms of its impact upon the apparatus, it may be difficult to suggest that the removal is 'necessary' for the improvement to be effected. Thus, for instance, 'improvement' is defined in paragraph 20(9) as including 'development and change of use'. If the improvement involved simply a change of e.g. premises from office use to residential use, although the developer may consider it desirable that the electronic communications apparatus situated on the roof of the building should be removed to facilitate the sale of the proposed flats, it is unlikely that it could be said that the alteration is necessary to enable the improvement of the land to be undertaken.

10.6.4

A number of matters arise:

(1) First, the definition of 'improvement' in paragraph 20(9) is inclusive and is not intended to be exhaustive. In the context of Part 1 of the Landlord and Tenant Act 1927 it has been held that the complete demolition of the premises comprised in the holding and the erection of buildings of an entirely different nature is an 'improvement': see *National Electric Theatres Ltd v Hudgell.*[11] In *Price v Esso*[12] the phrase 'improvements, alterations or additions' was held to include the total demolition of the existing buildings comprised in the holding and the erection of buildings of a different layout and design.
(2) Secondly, it is considered that the removal of the apparatus of itself cannot be said to be an 'improvement' of the land so as to fall within paragraph 20(9). The alteration is one which is required to be 'necessary' to effect the development of the land. In other words it is the existence of the apparatus which would otherwise prevent a development of the land from being undertaken. The apparatus does not by its presence prevent a development of the land if all that is proposed is work to the apparatus itself.
(3) Thirdly, there is no indication within the Old Code as to the date on which the improvement is required to be shown to be intended to be effected. How imminent the proposed improvement is required to be in order for a notice to be served and/or an order made is unclear. Like many such matters it is probably a question of degree. If there is an immediate requirement to undertake the work this will obviously satisfy the statutory wording. If an improvement is proposed but will not be undertaken for, say, two years the landowner may have greater difficulty.
(4) Fourthly, there is, it is to be noted, no reference to the landowner needing to establish an 'intention' to effect the alteration. However, it is considered that the phrase 'the alteration is necessary to enable that person [i.e. the landowner] to carry out a proposed improvement of the land' necessarily implies that the landowner will be able to establish that he is in a position such that he can carry out the proposed improvement. This may import a number of elements of the familiar concept of 'intention' derived from the case law under s.30(1)(f) of the 1954 Act. It is a 'proposed improvement' for which the alteration has to be shown to be necessary.[13] Thus there must at the very least be some evidence from the landowner of preparatory steps having been undertaken to be able to enable it to effect the improvement.

11 [1939] Ch 553. The cases under s.19(2) of the 1927 Act (dealing with covenants in leases restricting the right of the tenant to effect improvements) were not considered by the judge to be of assistance in determining the meaning of 'improvement' under Part 1 of the 1927 Act.
12 [1980] 2 EGLR 58 CA.
13 It may be that 'proposed' signifies an intention. The *OED* defines 'propose' as meaning, inter alia, intend, resolve (on), propose (to do). The word 'intends' is, of course, used in both grounds (f) and (g) in s.30(1) of the 1954 Act.

(5) Fifthly, it is unclear whether planning permission is needed or its prospects of being obtained are required to be shown. Under the terms of section 30(1)(f) of the 1954 Act, it is sufficient that one can show that there is a reasonable prospect of obtaining planning permission.[14] Similarly, if one does not have planning permission nevertheless it may be said, in the context of the Old Code, that the improvement is 'proposed' in the sense that it reflects the desire of the landowner. However, it must be implicit that the proposal be a realistic one. If there is no prospect of planning permission then it is difficult to see how a court could find that the improvement is one which is proposed to be carried out.

10.7 Procedure

10.7.1

The procedure is fairly straightforward:

(1) The landowner is required to serve notice on the operator requiring the alteration to be effected (paragraph 20(1)). There is no prescribed form for such a notice.
(2) The notice is to be served by registered post or recorded delivery: paragraph 24(2). The address to which the notice is to be sent is prescribed by paragraph 24(2A). Where it is not practicable to serve at the prescribed address in the circumstances provided for by paragraph 20(5), the notice may be served as provided for by that sub-paragraph.
(3) The operator is required to comply with the notice served unless within 28 days beginning with the day of the giving of the notice the operator serves a counter-notice: paragraph 20(2).
(4) Any form of counter-notice to be served by the operator is required to be in a form approved by OFCOM 'as adequate for indicating to that person the effect of the notice and so much of this code as is relevant to the notice and to the steps that may be taken by that person under this code in respect of that notice': paragraph 24(1).

14 In a situation where under the 1954 Act there is any uncertainty as to whether the landlord's plans require planning permission in order to be carried out, the court need not and normally should not try to resolve that question. an objective test is applied, namely:

> An enquiry whether the landlords on the evidence have established a reasonable prospect either that planning permission is not required or, if it is, that they would obtain it. That does not necessitate the determination by the court of any of the questions which may one day be submitted to the planning authority or to the Minister; it is the practical appraisal upon the evidence before the Court as to whether the landlords upon whom … the onus lies, have established a reasonable prospect of success.
> (Upjohn LJ in *Gregson v Cyril Lord* [1963] 1 WLR 41 at 48)

In the later decision of *Cadogan v McCarthy and Stone (Developments) Ltd* [2000] L & TR 249 CA at 254 (followed in *The Gulf Agencies Ltd v Adbul Salam Seid Ahmed* [2016] EWCA Civ 44, CA, a case under s.30(1)(g)) Saville LJ said:

> A reasonable prospect in this context accordingly means a real chance, a prospect that is strong enough to be acted on by a reasonable landlord minded to go ahead with plans which require permission, as opposed to a prospect that should be treated as merely fanciful or as one that should sensibly be ignored by a reasonable landlord. A reasonable prospect does not entail that it is more likely than not that permission will be obtained.

The OFCOM website provides a form of Model Counter Notice (as well as other notices).[15]

(5) Where a counter-notice is given, the operator shall make the alteration only if the court on an application by the landowner makes an order requiring the alteration to be made.

(6) The application to the court is to be made by the landowner. The court is the county court: paragraph 1(1) of the Old Code.

10.8 Court criteria for making order

10.8.1

The terms of paragraph 20 itself circumscribe the court's power to make an order. It must be shown that:

(1) The alteration is necessary to carry out the improvement to the land (paragraph 20(4)(a)).

(2) The alteration will not substantially interfere with any service which is or is likely to be provided using the operator's network (paragraph 20(4)(b)).

(3) The operator has sufficient rights for making the alteration or would have such rights if an application were made under paragraph 5 of the Old Code and the court were to dispense with the need for the landowner's agreement (paragraph 20(5)).

10.8.2

The first of these points does not call for comment. As for the second, it would ordinarily be expected that the operator would wish to call evidence as to the impact that the loss of the site will have on the operator's network for, in most cases, full removal is required by the landowner. Looking at the matter simplistically, any loss of a site will have some impact on the services provided, but such an impact may not amount to 'substantial interference'. This might be so where the operator's network has inbuilt resilience, and where there may be little difficulty in transferring customers and services hosted by the relevant apparatus at the site to other networks operated by the operator or to other apparatus hosting similar services for other customers.

10.8.3

The third matter is a rather curious one. One would have thought that there would be little difficulty in the operator being able to remove that which it has installed. However, it may be that the installation is such that in order to be able to remove it the operator needs rights over the landowner's adjoining property i.e. access to land adjacent to the demised premises, which access is not conferred by the terms of the existing arrangement with the landowner. It is expressly provided in paragraph 20(5) that in considering

15 www.ofcom.org.uk/phones-telecoms-and-internet/information-for-industry/policy/electronic-comm-code/notices. See also paragraph 32.4.

whether any rights need to be conferred upon the operator in order to effect the alteration, the court can essentially confer such rights as if an application had been made under paragraph 5 of the Old Code and 'the court shall have the same powers as it would have if an application had been duly made under paragraph 5 above for an order dispensing with the need for that person's agreement'.[16]

10.8.4

These criteria have to be considered against the backdrop of two overriding considerations, namely: that the court will (1) have regard to all the circumstances, and (2) the principle that no person should unreasonably be denied access to an electronic communications network or to electronic communications services. As to the latter principle of access to an electronic communications network/service Lewison J put it in *Geo Networks Ltd v Bridgewater Canal Co. Ltd*[17] at first instance:

> The formulation of the principle ... is not that no person should be denied access to a network. It is that no person shall *unreasonably* be denied access. Necessarily, as it seems to me, formulating the principle in this way entails the conclusion that there may be circumstances in which it is reasonable to deny such access.

10.9 Orders of the court

10.9.1

The order to be made is one requiring the operator to effect the alteration of the apparatus. It is clear from the terms of paragraph 20(2) that the operator is the party required to comply with the notice. If a counter-notice is served by the operator it is provided that 'the operator shall make the required alteration only if the court on an application' by the landowner 'makes an order requiring the alteration to be made'.

10.9.2

An order by the court pursuant to paragraph 20(7) may:

> provide for the alteration to be carried out with such modifications, on such terms and subject to such conditions as the court thinks fit, but the court shall not include any such modifications, terms or conditions in its order without the consent of the applicant, and if such consent is not given may refuse to make an order under this paragraph.

10.9.3

Thus any form of qualification to the order which the landowner seeks requires the court first to seek the consent of the landowner applicant. If consent is not forthcoming

16 The court may provide that for the purposes of conferring appropriate rights upon the operator, the applicant landowner is required to bring the application to the notice of such other interested persons as it thinks fit: paragraph 20(6).

17 [2010] 1 WLR 2576.

the court may either (1) make the order subject to the appropriate condition or modification which the court has proposed or (2) refuse to make the order at all. The fact that the court has power to refuse to make the order where the landowner applicant objects to any terms, conditions or modifications to the order sought, invariably leads to the landowner agreeing to the same.

10.10 Paragraph 5 of the Old Code

10.10.1

There is no reason why an operator cannot seek to invoke paragraphs 2 and 5 of the Old Code in response to either a paragraph 20 or paragraph 21 notice. Often operators respond to such notices with a claim by the operator for an agreement for installation elsewhere on the land of the person serving the notice. Paragraph 5 of the Old Code itself requires service of a notice claiming the rights required and an application to the court for the dispensation of the landowner's consent. If the alteration is necessary for the purposes of carrying out the development, it might be difficult to satisfy the court that any prejudice caused by the order would be compensated by money, in which case one would not expect paragraph 5 rights to override the paragraph 20 rights which the landowner is invoking.

10.11 Can an order be enforced before the expiry of a fixed term agreement?

10.11.1

Given the terms of the order which the court may make (as to which see paragraph 10.9 above) there seems to be no reason why the operator should not be required to effect the alteration prior to the expiry of any fixed term agreement. It may be considered that there is a practical benefit to the landowner in ensuring that the alteration is effected prior to the end of the fixed term. However, given that many developers are under tight time constraints, the developer may much prefer having a vacant site at the end of the term rather than having to take appropriate steps to ensure that the operator complies with the terms of the order post the expiry of the term.

10.11.2

Although it is not explicitly so stated in the Old Code, it may be thought obvious that the landowner cannot enter upon the land demised to the operator until the lease has been terminated. The service of the paragraph 20 notice and the application to the court for an order for the operator to effect the alteration says nothing about the continuing existence of the underlying lease between the parties. Neither the notice nor the application terminates the lease. Accordingly, from a practical point of view there often appears to be little point in serving a paragraph 20 notice until shortly prior to the expiry of any fixed term agreement. Experience suggests that ordinarily landowners serve paragraph 20 notices 12 to 18 months before the expiry of a fixed term contract. Furthermore, there is the overriding requirement that the alteration is 'necessary' to enable the carrying out of a 'proposed improvement of the land' in which the landowner has an interest.

Removal would have to be shown to be necessary as at the date of the making of the order and, in a case where the fixed term agreement will not expire for several years, it is difficult to see how this requirement could be satisfied. It is for this reason that where the tenancy is one protected by the provisions of 1954 Act the earliest time at which one normally encounters paragraph 20 notice is in the 12–18 months preceding the contractual expiry date of the lease.

10.11.3

Paragraph 20 does not contain any express provision similar to that which one sees in paragraph 21(7) enabling the court to confer upon the landowner authority to remove the apparatus himself. However, given the broad terms of paragraph 20(7) it may be thought that such authority may be conferred by the court by way of the imposition of a term or condition as to the carrying out of the alteration, providing for the landowner to enter upon the land himself to remove the apparatus in default of the operator effecting the alteration.

10.12 Costs of effecting the alteration

10.12.1

One of the principal deterrents to serving a paragraph 20 notice is the requirement in paragraph 20(8) that the person who serves the notice should reimburse the operator 'in respect of any expenses which the operator incurs in or in connection with the execution of any works in compliance with the order'. Those expenses are to be paid 'unless the court otherwise thinks fit'. The obligation is to pay 'any expenses', and this is ordinarily interpreted as meaning a full indemnity of all of the costs which the operator will incur.

10.12.2

There is no express qualification to the effect that such costs must be reasonable. The expenditure sought to be recovered must be incurred 'in or in connection with the execution of any works'. It seems, therefore, that the common case of requiring complete removal will only cover the costs of the removal of the apparatus not its installation elsewhere. It would be sensible for any order to provide that upon default not only may the landowner effect the work, but, furthermore, recover the costs of so doing. It seems reasonably arguable that such an order may be made in light of the breadth of paragraph 20(7) of the Old Code.

10.12.3

It is not uncommon to find in leases (and licences) a covenant by the tenant (or licensee) to deliver up vacant possession and to reinstate the premises. This may give rise to a potential argument as the operator`s entitlement to the costs of effecting the alteration. The landowner may argue that it can seek to set off the costs of effecting the alteration because he has a cause of action against the operator in respect of the operator`s breach of covenant. The operator ought to have effected removal itself in accordance with the covenant, and its failure to do so has led to a loss, namely the costs which the landowner has to pay the operator. On the other hand the operator may argue that any such provisions are contrary to the terms of paragraph 27(2) of the Old Code and thus should not be enforced.[18]

10.13 Interrelation with the 1954 Act

10.13.1

Paragraph 20 of the code is concerned with carrying out development. So is s.30(1)(f) of the 1954 Act. That section entitles the landlord to oppose renewal if on the termination of the current tenancy he intends to carry out work of demolition reconstruction or construction. Although the statutory criteria are different, there seems little difficulty in combining a paragraph 20 notice with a notice under s.25 of the 1954 Act.[19] A s.25 notice opposing renewal under s.30(1)(f) is often regarded in practice as a paragraph 20 notice and invariably one sees counter notices under the Old Code being served in response.

10.13.2

There is no reason why a paragraph 20 notice cannot be served in advance of the date upon which the current tenancy would otherwise come to an end in accordance with the provisions of the 1954 Act. Ordinarily, one would seek to combine service of a paragraph 20 notice with the service of the s.25 notice. There are, as is well known, restrictions on the service of s.25 notices, namely, that they are to be served not less than six months and not more than 12 months prior to the date of termination specified in the notice. There is nothing within paragraph 20 restricting the date at which a paragraph 20 notice may be served other than, of course, the practical fact that the server may be unable to enforce the alteration (by reason of the inability to bring to an end the termination of the operator's interest in the land).[20] However, if it so happens that the paragraph 20 proceedings are heard prior to the 1954 Act proceedings, there is no reason why the court cannot seek to provide, as a term of any order under the Old Code, that the alteration is to be effected upon, and subject to the determination of, the operator's tenancy under the provisions of the 1954 Act. Such an order appears to be possible in reliance upon paragraph 20(7).

10.14 Transitional provisions of the New Code

10.14.1

The New Code contains transitional provisions in Schedule 2 to the 2017 Act. Paragraph 16(1) to Schedule 2 provides that 'the repeal of the existing code[21] does not

18 See Chapter 34, sections 34.2 et seq.

19 Or a counter-notice in response to a s.26 request served by a tenant under the 1954 Act. What is said in the text applies equally to a counter-notice to a s.26 request.

20 It may also give rise to a difficulty in establishing that the alteration 'is necessary' to carry out the proposed development.

21 i.e. the Old Code.

affect paragraph 20 of that code as it applies in relation to anything whose installation was completed before the repeal comes into force'.

10.14.2

It is expressly provided that paragraph 20 is not 'exercisable in relation to any apparatus by a person who is a party to, or is bound by, an agreement under the new code in relation to the apparatus', but a 'subsisting agreement is not an agreement under the new code' for the purposes of these provisions: paragraphs 16(2) and (3) of Schedule 2 to the 2017 Act. Paragraph 1(4) of Schedule 2 to the 2017 Act defines 'subsisting agreement' as follows:

> A 'subsisting agreement' means –
> (a) an agreement for the purposes of paragraph 2 or 3 of the existing code, or
> (b) an order under paragraph 5 of the existing code,
>
> which is in force, as between an operator and any person, at the time the new code comes into force (and whose terms do not provide for it to cease to have effect at that time).

10.14.3

The upshot of these provisions appears to be that the rights under paragraph 20 continue to apply in relation to a 'subsisting agreement' (i.e. an agreement made under the Old Code). There is no need in order for the paragraph 20 rights to continue to apply in relation to the subsisting agreement that a paragraph 20 notice was served before the coming into force of the New Code (in contrast to paragraph 21, where such a requirement is imposed[22]).

10.14.4

One matter which is unclear is whether, in relying upon paragraph 20 of the Old Code, proceedings are instituted in the County Court (as was the position before the New Code came into force) or whether proceedings are to be in the Tribunals as directed by the Electronic Communications Code (Jurisdiction) Regulations 2017 (SI/1284). One possibility is that as a 'subsisting agreement' is treated as a Part 2 New Code Agreement (see paragraph 2 of Schedule 2 to the 2017 Act), the new Regulations apply so as to ensure that Old Code procedures come within the remit of the Tribunals. Regulation 3 provides that 'the functions conferred by the code on the court are also exercisable by the following tribunals'. However, the terms of Regulations 2 and 4 of the Electronic Communications Code (Jurisdiction) Regs 2017 (No1284), appear to enable Old Code paragraph 20 claims to be issued in the County Court. A paragraph 20 Old Code claim is not part of "the electronic communications code set out in Schedule 3A to the Communications Act 2003" within Regulation 2 nor is such a claim a claim for termination within Part 5 or removal within Part 6 of the New Code and thus not "relevant proceedings" which "must" be issued in the Upper Tribunal.[23]

22 As to which, see Chapter 31 and section 10.9 of Chapter 11.
23 See also paragraph 33.7.

11 Removal of apparatus under paragraph 21

11.1 Introduction

11.1.1

This chapter, and Chapter 10, consider the rights under the Old Code which entitle the occupier/owner to seek alteration or removal of the apparatus. They are contained in paragraphs 20 and 21 respectively.[1] This chapter deals with the right of removal.

11.1.2

This chapter is arranged in the following way:

(a) Section 11.2 deals with an overview of paragraph 21.
(b) Section 11.3 deals with who may serve a paragraph 21 notice.
(c) Section 11.4 deals with procedure.
(d) Section 11.5 deals with the orders which the court may make.
(e) Section 11.6 considers who has to bear the costs of effecting removal.
(f) Section 11.7 considers the impact of paragraph 5 of the Old Code.
(g) Section 11.8 considers the interrelation with the 1954 Act.
(h) Section 11.9 considers the transitional provisions of the New Code.

11.1.3

The Old Code applies whether the agreement grants a lease, whether or not contracted out from the 1954 Act, or a licence. In this chapter any reference to a lease (which is the most common form of arrangement encountered in practice) should also be taken to include a contractual arrangement between the landowner and the operator.

11.2 Removal of apparatus: overview

11.2.1

The heading to paragraph 21 is 'restriction on right to require the removal of apparatus'. Unlike paragraph 20, which is an enabling provision amplifying rather than

1 As to the special right of removal in respect of railways, canals or tramways, see Chapter 12, paragraph 12.7.29.

restricting the rights of the landowner, the terms of paragraph 21 is a restriction on the rights of the landowner.

11.2.2

Paragraph 21(1) provides:

> Where any person is for the time being entitled to require the removal of any of the operator's electronic communications apparatus from any land (whether under any enactment or because that apparatus is kept on, under or over that land otherwise than in pursuance of a right binding that person or for any other reason) that person shall not be entitled to enforce the removal of the apparatus except, subject to sub-paragraph (12) below, in accordance with the following provisions of this paragraph.

11.2.3

There are three preliminary points to note with respect to the paragraph 21 restriction, namely:

(1) where the entitlement to removal falls within sub-paragraph (1) of paragraph 21, the right of removal can be effected only in accordance with the terms of paragraph 21;
(2) furthermore paragraph 21 overrides the terms of the agreement with the operator;
(3) the restriction applies to prevent removal of 'any of the operator's electronic apparatus'.

11.2.4

As to the first point, analytically it is to be noted that paragraph 21 in fact entitles a notice to be served pursuant to its terms in three sets of circumstances:

(1) Where any person is for the time being entitled to require the removal of any of the operator's electronic communications apparatus from any land under any enactment (Category 1); or
(2) Where any person is for the time being entitled to require the removal of any of the operator's electronic communications apparatus from any land because that apparatus is kept on, under or over that land otherwise than in pursuance of a right binding that person (Category 2); or
(3) Where any person is for the time being entitled to require the removal of any of the operator's electronic communications apparatus from any land for any other reason (Category 3).

11.2.5

Thus the restriction on removal provided for by paragraph 21 would arise where:

(1) The contractual term of the agreement (whether a lease or a licence) has come to an end.

(2) The agreement constitutes a sub-lease and the intermediate lease is forfeited or terminated, e.g. by a break clause.[2]
(3) The agreement constitutes a licence and the interest out of which the licence was created is assigned. As a licence does not create a proprietary interest in land it is not binding on successors of the licensor.[3]
(4) Removal is sought by a person with title paramount, e.g. where a lessee has conferred the right on the operator but the freehold owner has not agreed to be bound pursuant to paragraph 2(2)(b) of the Old Code.
(5) Removal is required by a statutory body, e.g. the apparatus has been installed in breach of planning control and the planning authority is entitled to take enforcement action for its removal.[4]

11.2.6

The three categories referred to above will be considered in section 11.3. In that section, the question as to whether or not notice may be served within Category 2 prior to the expiry of a fixed term agreement is also considered. The current state of first instance authority (at County Court level both in England and Scotland) is that it cannot.

11.2.7

As to the second matter, namely whether the terms of the agreement can disapply paragraph 21, it is provided that the terms of the Old Code are without prejudice 'to any rights or liabilities arising under any agreement to which the operator is a party': paragraph 27(2). However, this exception does not apply to paragraph 21: paragraph 27(2). Thus, it is considered that any contractual entitlement to enforce removal which would have the effect of circumventing paragraph 21 is simply ineffective.

11.2.8

It is quite common in pre-New Code agreements to find a provision pursuant to which the operator agrees not to serve a counter-notice in response to a paragraph 21 notice served by the landowner.[5] It is considered that such a provision which subsists in an agreement with the operator does not prevent an operator from serving a counter-notice

2 The break clause served by the superior landlord upon the intermediate lessee will bring to an end all derivative interests: *Pennell v Payne* [1995] QB 192 CA.

3 *Ashburn Anstalt v WJ & Co Arnold* [1989] Ch 1 CA.

4 See the terms of the Electronic Communications Code (Conditions and Restrictions) Regulations 2003 (SI 2003/2553) Reg 5, relating to the requirement to serve notice on the planning authority of an intended installation of electronic communications apparatus.as (amended by the Electronic Communications Code (Conditions and Restrictions) Amendment Regulations 2009 (SI 2009/584), Electronic Communications Code (Conditions and Restrictions) Amendment Regulations 2013 (SI 2013/1403), Electronic Communications Code (Conditions and Restrictions) Amendment Regulations 2016 (SI 2016/1049) and the Electronic Communications Code (Conditions and Restrictions) Amendment Regulations 2017 (SI 2017/753)).

5 Here used in a neutral sense to refer to the party seeking the removal. Who may seek to invoke paragraph 21 is considered in scetion 11.3.

pursuant to the terms of paragraph 21. Whether the operator by serving a counter-notice in defiance of such a provision in the terms of the agreement with the landowner is in breach of contract giving rise to a claim in damages is a matter which has yet to be the subject of judicial consideration. Given the fact that the New Code will essentially render paragraph 21 redundant even in the case of agreements subsisting at the date that it came into force,[6] it is probably the case that this issue is now simply of historic academic interest.

11.2.9

The third point relates to the apparatus which is protected by the restriction. The apparatus which is protected does not have to be vested in the user. It is sufficient that the operator is using electronic communications apparatus, albeit that apparatus may belong to others, so long as it is being used for the purposes of the operator's network: paragraph 21(11). The important consideration is not who owns the apparatus but whether the apparatus is being used for the purposes of the operator's network. Thus, if one has a mast vested in one operator who provides facilities to others and both are using the mast for apparatus in connection with their own networks each will have protection. If the mast owner is simply providing a structure for the carrying or suspension of the electronic communications apparatus of others and is not utilising the mast in connection with his own network, the mast agreement is not caught by the restriction.[7]

11.3 Who may serve a paragraph 21 notice?

11.3.1 'Entitled to require removal'

Unlike paragraph 20, which refers to the entitlement to serve notice being conferred upon the person with an interest in the land on which the electronic communications apparatus is kept on, under or over or land adjacent to such land, paragraph 21 identifies, as noted above in section 11.2, three categories of entitlement. In all three cases the entitlement to serve notice vests in 'any person' who 'is for the time being entitled to require removal' by reason of the matters specified.

6 Paragraph 1(4) of Schedule 2 to the 2017 Act provides:
A 'subsisting agreement' means –:

(a) an agreement for the purposes of paragraph 2 or 3 of the existing code, or
(b) an order under paragraph 5 of the existing code, which is in force, as between an operator and any person, at the time the new code comes into force (and whose terms do not provide for it to cease to have effect at that time).

By the terms of paragraph 20 of Schedule 2 to the 2017 Act, the rights of a landowner in reliance on paragraph 21 apply only where, in the case of a subsisting agreement, the landowner has given notice under paragraph 21(2) requiring the removal of the apparatus. See further section 11.9.

7 Although of course those operators using the mast in connection with their own networks will have the benefit of the restriction. It is to be noted by way of contrast that under the New Code the 'statutory purposes' in respect of which the code rights are being conferred upon the operator by the code agreement are defined so as to include 'providing an infrastructure system': paragraph 4 of the New Code. 'Infrastructure system' is elaborated upon in paragraph 7 of the New Code. See section 15.4 of Chapter 15 of this book.

11.3.2

Two issues arise on this wording:

(1) May a paragraph 21 notice be served before the expiry of the contractual term of a lease/licence?
(2) Is it necessary, at the date of service of the paragraph 21 notice, to have a present entitlement, capable of immediate enforcement, to require removal of the electronic communications apparatus?

The two issues are related but legally and logically separate.

11.3.3 Issue 1

As to the first issue, it has been held in the Cambridge County Court, on an application to strike out, that the words 'is for the time being entitled to require the removal' meant that a paragraph 21 notice could be served only where there was an immediate right at the date of service to require removal. Thus, if the term of the code agreement had not been terminated as at the date of the service of the paragraph 21 notice, the paragraph 21 notice was void and of no effect: *Crest Nicholson (Operations) Ltd v Arqiva Services Ltd*[8] and more recently in Scotland, *Tait v Vodafone*.[9]

11.3.4

There is no doubt that a paragraph 21 notice can be served in numerous circumstances prior to the expiry of any contractual term of a lease or licence. This is apparent from the terms of Category 1 and Category 3. Category 1 enables a third party, e.g. a planning authority, to seek removal and clearly a third party to the agreement would have no entitlement to bring it to an end. Thus the right 'to require removal' arises pursuant to the statute even though the contractual term of the operator with the landowner continues.

11.3.5

Equally Category 3 is clearly not confined to circumstances where the contractual term of the lease or licence has expired. One may think of a number of circumstances where the person calling for removal wishes to do so e.g. apparatus is installed contrary to the terms of a covenant as between him and the lessee and he seeks a mandatory injunction for its removal.

11.3.6

Some support for the view that a paragraph 21 notice may be served during the currency of the contractual term of the operator is derived from paragraph 21(12). That sub-paragraph is important in two respects:

8 28 April 2015. Transcript available on Egi.
9 24 July 2017. Unreported, Sheriff Court at Edinburgh.

(1) By its terms it precludes service of a paragraph 21 notice simply because a person would be entitled to serve a paragraph 20 notice. It was accepted in *Crest Nicholson v Arqiva* that a paragraph 20 notice may be served during the term of a lease or licence. Thus, but for the prohibition in paragraph 21(12), a paragraph 21 could be invoked simply because a paragraph 20 could be served. Accordingly, it is necessarily implicit that the draftsman was contemplating that a paragraph 21 notice could be served during the term and thus needed to protect the operator against the use of paragraph 21 in circumstances where a paragraph 20 notice could be served by the landowner.

(2) The terms of paragraph 21(12) also provide that 'this paragraph [i.e. paragraph 21] is without prejudice to paragraph 23 below and the power to enforce an order of the court under the said paragraph 11, 14, 17 or 20.' An *order* under paragraph 20 may be obtained prior to the contractual expiry date of the operator's lease or licence. If an *order* is obtained under paragraph 20 during the term its enforcement will not be precluded by paragraph 21. Accordingly the qualification in paragraph 21, with respect to the enforcement of an order under paragraph 20, recognises that paragraph 21 could operate with respect to an entitlement arising during the continuation of the operator's lease or licence. The draftsman thus needed to exclude the restrictive effect of its operation where e.g. the owner had obtained an order under paragraph 20 requiring removal.

11.3.7

Thus it is considered that to the extent that the two authorities referred to above are said to be authority for the proposition that a paragraph 21 notice cannot be served prior to the expiry of the contractual term of the lease or licence, it is considered that the authorities are incorrect. However, in light of the implementation of the New Code and the fact that the transitional provisions contained in paragraph 20 Schedule 2 to the 2017 Act[10] enable reliance, with respect to a subsisting agreement, on a paragraph 21 notice only where it was served prior to the New Code coming into force, the debate is now perhaps rather academic. That said, a number of landowners have served paragraph 21 notices prior to the New Code coming into force.

11.3.8 Issue 2

It may be said that all of the categories identified above, when properly considered, are, ultimately, based upon an immediate right to seek removal at the time the notice is served. In relation to Category 1, the right of removal arises by reason of the enactment; under Category 2 the right of retention on the land has expired and under Category 3 the breach of the covenant or the nuisance, by way of example, confers upon the server the entitlement, immediately, to seek removal (subject, of course, to establishing the same at trial).

11.3.9

There is nothing within paragraph 21 that indicates that the entitlement to require removal must be one which can be enforced immediately by the person serving the

10 As to which see Chapter 31 generally and section 11.9 of this chapter.

notice although it may be suggested that it is necessarily implicit in the words 'is for the time being entitled'; one cannot, under Category 2, be entitled until the right to possession has accrued and that is only upon expiry of the term.

11.3.10

The argument is that the verb 'to require' means, in this context, to have the right to demand removal not as an absolute right as at the date of service but demanding removal by reference to some legal right which is sought to be invoked. All that paragraph 21 is seeking to do is to enable the person who asserts an entitlement to seek removal, to initiate the procedure for so doing. He does not have to show at the time of service of the paragraph 21 notice an unassailable right to require removal.

11.3.11

At present the authorities which have been reported which have given consideration to this issue require an immediate right to seek removal equating, in the case of Category 2, to a right to immediate possession. This conclusion causes considerable difficulty with respect to a lease which is protected by the 1954 Act. This is discussed in section 11.8.

11.3.12 Category 1

As to Category 1, the entitlement to require removal arises where the person seeking removal does so in reliance 'under any enactment'. If there is an enactment entitling removal, the term may not have expired, e.g. a planning authority may seek removal pursuant to some form of enforcement action or apparatus may be required to be removed following the stopping up or diversion of a highway, under the Town and Country Planning Act 1990, s.256(2).

11.3.13 Category 2

Category 2 arises where the apparatus which is 'is kept on, under or over' the land 'is otherwise than in pursuance of a right binding that person'. The simple scenario is the termination of a fixed term agreement.[11]

11.3.14

Category 2 extends not only to where the right has expired (e.g. by effluxion of time or by forfeiture) but also to a number of situations where the operator's lease or licence has not come to an end. For example:

11 It is to be noted that, notwithstanding the owner's entitlement to seek removal of the apparatus, it would appear that no claim for damages for trespass can be made. So long as the paragraph 21 restriction applies the apparatus is deemed to have been lawfully kept on the land: paragraph 21(9). Thus, for instance, in the case of a fixed term agreement which is not protected by the 1954 Act, service of a counter-notice may arguably defer any entitlement to damages to the date of the court order providing for removal.

(1) The lease or licence remains extant but the relevant right is not binding on the person serving the notice, e.g. say the operator has a lease and grants, contrary to the terms of the lease, a sub-right which is not binding on the superior lessor. Thus, rights by title paramount may be exercised by a paragraph 21 notice albeit the sub-lease continues as between the lessee and sub-lessee;
(2) The operator does not own the equipment and the owner of the equipment seeks to recover it. (The installation of the equipment does not affect ownership: paragraph 27(4).)

11.3.15 Category 3

The 'other reason' may relate, as noted above, to a breach e.g. of an alterations covenant and the landowner seeks a mandatory injunction for restoration. Further examples are where the apparatus is said to cause a nuisance, or the apparatus has been abandoned: paragraph 22.[12]

11.4 Procedure

11.4.1

The procedure is as follows:

(1) The person entitled to require removal (for convenience referred to as 'the landowner') is required to serve notice on the operator requiring the removal to be effected (paragraph 21(2)). There is no prescribed form for such a notice.
(2) The notice is to be served by registered post or recorded delivery: paragraph 24(2). The address to which the notice is to be sent is prescribed by paragraph 24(2A). Where it is not practicable to serve at the prescribed address in the circumstances provided for by paragraph 24(5), the notice may be served as provided for by that sub-paragraph.
(3) The operator is required to comply with the notice served unless within 28 days beginning with the day of the giving of the notice the operator serves a counter-notice upon the landowner: paragraph 21(3).
(4) Any form of counter-notice to be served by the operator is required to be in a form approved by OFCOM 'as adequate for indicating to that person the effect of the notice and so much of this code as is relevant to the notice and to the steps that may be taken by that person under this code in respect of that notice': paragraph 24(1). The OFCOM website provides a model form of counter-notice (as well as other notices).[13]
(5) Failure to serve a counter-notice results in the person serving it becoming 'entitled to enforce the removal of the apparatus': paragraph 21(3). The landowner may, upon default of the service of a counter-notice, obtain court authority to remove it himself (paragraph 21(7)) and recover the costs of so doing (paragraph 21(8)).

12 As to abandoned apparatus, see Chapter 12.
13 www.ofcom.org.uk/phones-telecoms-and-internet/information-for-industry/policy/electronic-comm-code/notices. See section 32.4.

(6) Where a counter-notice is given, the operator shall be required to effect removal only if the court on an application by the landowner makes an order requiring him to do so (paragraph 21(6)).
(7) The counter-notice served in response to a paragraph 21 notice may:
  (a) state that the person giving the notice is not entitled to require the removal of the apparatus; and/or
  (b) specify the steps which the operator proposes to take for the purpose of securing a right as against the person giving the notice to keep the apparatus on the land. The steps which the operator proposes can include steps for retention pursuant to a right to be acquired pursuant to paragraph 5 (paragraph 21(4)).
(8) If the counter-notice specifies steps that the operator proposes to take to keep the apparatus on the land, the court shall not make an order unless it is satisfied either:
  (a) that the operator is not intending to take those steps or is being unreasonably dilatory in the taking of those steps; or
  (b) that the taking of those steps will not secure for the operator as against the giver of the notice, any right to keep the apparatus installed on, under or over the land or as the case may be, to reinstall it if it is removed (paragraph 21(6)).
(9) The application to the court is to be made by the landowner. The court is the county court: paragraph 1(1) of the Old Code.[14]

11.5 Orders of the court

11.5.1

Where the court is satisfied as to the entitlement of the landowner to require removal the court also has to be satisfied, as noted above, that where the counter-notice specifies steps which the operator is proposing to take to secure a right to keep the apparatus on the land, that:

(1) the operator is not intending to take those steps or is being unreasonably dilatory in the taking of those steps; or
(2) the taking of those steps has not secured, or will not secure, for the operator as against the landowner any right to keep the apparatus installed on, under or over the land or, as the case may be, to reinstall it if it is removed.

11.5.2

It is often the case that operators indicate that they intend to serve notices pursuant to paragraph 5 of the Old Code in response to a paragraph 21 notice seeking to assert that they will ask the court to impose an agreement upon the landowner to secure the right to keep the apparatus on the land.[15] Invariably no paragraph 5 notice will be served. It is rare to encounter a contested paragraph 21 claim for removal where the operator has, in response to the paragraph 21 notice, in fact served a paragraph 5 notice and seeks to sustain a case that an agreement should be imposed upon the landowner with a dispensation of the landowner's consent.

14 And see further Chapter 33.
15 See section 11.7.

11.5.3

The court may:

(1) Order the operator to remove the electronic communications apparatus; and
(2) Confer authority upon the landowner to remove the apparatus: paragraph 21(7).

The wording of sub-paragraph (7) may suggest that authority will be conferred only where there has been default on the part of the operator to remove the apparatus in accordance with the order of the court made. The wording provides for the landowner to 'apply to the court for authority to remove it himself'. However, there ought to be no reason why the court cannot at the same time as making the order confer the relevant authority for the operator to remove the apparatus, with it being provided that the removal by the landowner is in default of the operator in compliance with the order made.

11.6 Cost of removal

11.6.1

As noted in Chapter 10, under paragraph 20 there is no entitlement for the landowner to recover the costs of the alteration. On the contrary, the landowner will be required to pay the costs of the operator in effecting the alteration. In contrast, paragraph 21 entitles the landowner to seek to recover costs.

11.6.2

Expenses incurred by the landowner 'in or in connection with the removal of the apparatus shall be recoverable by him from the operator' (paragraph 21(8)). However, it is to be noted that this entitlement to recover costs arises where the apparatus 'is removed by any person under any authority given by the court under sub-paragraph (7) above'. The effect of this provision is that expenditure incurred by the landowner is recoverable only where:

(1) there is a default in serving a counter-notice, and the giver of the notice seeks and obtains the court's consent for authority to go in and do the work, or
(2) a counter-notice is served and an order of the court is made and the order of the court confers authority to go in and remove the apparatus.

11.6.3

The court may order the landowner to effect a sale of the apparatus and for him to recoup the expenses incurred from the proceeds of sale: paragraph 21(8).

11.7 Paragraph 5 of the Old Code: orders of the court

11.7.1

There is no reason why an operator cannot seek to invoke paragraphs 2 and 5 of the Old Code in response to a paragraph 21 notice. Often operators respond to such notices with a claim by the operator for an agreement for installation elsewhere on the land of the person serving the notice. Paragraph 5 of the Old Code itself requires service of a notice claiming the rights required and an application to the court for the dispensation of the landowner's consent. As previously noted, paragraph 21 provides for service of a counter-notice and the counter-notice can include steps which the operator proposes to take for the purposes of securing a right to keep the apparatus on, under or over the land. This will reflect the paragraph 5 rights that the operator will seek to acquire. Paragraph 21 specifically provides that that claim to such rights in the counter-notice will not prevent a paragraph 21 order being made, if the paragraph 5 rights will 'not secure for the operator against the landowner any right to keep the apparatus installed on, under or over the land or, as the case may be, to reinstall it if it is removed'. The paragraph 5 claim will not secure such a right if the prejudice cannot be compensated for in money or the benefit to the network or the service does not otherwise override the prejudice.

11.8 Interrelation with the 1954 Act

11.8.1

There is much debate by commentators upon the interrelationship between the Old Code and the 1954 Act and in particular the difficulty that the landlord faces in seeking to oppose renewal under s.30(1)(f) (or (g)) of the 1954 Act because of the stipulation in paragraph 21 that no notice for removal can be served until the person giving the notice is 'entitled' to do so. The argument which has held sway to date is that the landowner will not be entitled to serve a paragraph 21 notice until the operator's legal rights have come to an end. However, if the tenant makes an application to the court for renewal in accordance with the 1954 Act, there remains a legally binding right imposed upon the landowner which prevents him from serving a paragraph 21 notice, namely the continuation tenancy under the 1954 Act.

11.8.2

Thus, the landlord is caught either way. He cannot serve a paragraph 21 notice because the tenant remains entitled to keep the apparatus on the land by reason of the continuation of the tenancy under the 1954 Act. Equally, he cannot bring the tenancy to an end because he cannot serve a paragraph 21 notice so is unable to show, for the purposes of the 1954 Act, an intention capable of being implemented within a reasonable time of termination of the current tenancy.

11.8.3

It has been suggested that one possible way of overcoming this absurdity would be to provide for a broad construction of the reference to 'entitled' in paragraph 21; according to this suggestion, the owner is 'entitled' to require removal if he is entitled to serve a s.25 notice under the 1954 Act which will have the effect, eventually, of requiring the operator to remove the apparatus. However, as those commentators note, the s.25 notice is not per se a right to require removal, but a right to have the question of whether the tenant's 1954 Act continuation rights should be brought to an end determined. However one seeks to accommodate paragraph 21 with the 1954 Act one will be required to strain the language of the Old Code to achieve a 'fit'.

11.9 Transitional provisions of the New Code

11.9.1

By paragraph 20 of Schedule 2 to the 2017 Act it is provided that the repeal of the Old Code does not affect the service of a notice pursuant to paragraph 21(2) served prior to the coming into force of the New Code. Accordingly, where a paragraph 21 notice has been served by the landowner requiring removal prior to the coming into force of the New Code, the landowner can continue to rely upon the paragraph 21 notice to seek removal.

11.9.2

One matter which is unclear is whether in relying upon paragraph 21 of the Old Code, proceedings are to be instituted in the County Court (as was the position before the New Code came into force) or whether proceedings are to be in the Tribunals as directed by the Electronic Communications Code (Jurisdiction) Regulations 2017 (SI/1284). One possibility is that, as a 'subsisting agreement' is treated as a paragraph 2 New Code Agreement (see paragraph 2 of Schedule 2 to the 2017 Act), the new Regulations apply so as to ensure that Old Code procedures come within the remit of the Tribunals. Regulation 3 provides that 'the functions conferred by the code on the court are also exercisable by the following tribunals …' However, the terms of Regulations 2 and 4 of the Electronic Communications Code (Jurisdiction) Regs 2017 (No1284), appear to enable Old Cod paragraph 20 claims to be issued in the County Court. A paragraph 20 Old Code claim is not part of "the electronic communications code set out in Schedule 3A to the Communications Act 2003" within Regulation 2 nor is such a claim a claim for termination within Part 5 or removal within Part 6 of the New Code and thus not "relevant proceedings" which "must" be issued in the Upper Tribunal.[16]

16 See also paragraph 33.7.

12 Abandonment of apparatus

12.1 Introduction

12.1.1

This chapter considers what rights the landowner has with respect to abandoned apparatus.

12.2 Old Code: paragraph 22

12.2.1

Paragraph 22 of the Old Code provides:

> Without prejudice to the preceding provisions of this code, where the occupier has a right conferred by or in accordance with this code for the statutory purposes to keep electronic communications apparatus installed on, under or over any land, he is not entitled to keep the apparatus so installed if, at a time when the apparatus is not, or is no longer, used for the purposes of the operator's network, there is no reasonable likelihood that it will be so used.

12.2.2 Abandonment

It is to be noted that, although the terms of paragraph 22 are headed 'abandonment of apparatus', the word 'abandonment' is not referred to in the substantive provision itself. 'Abandonment' is here referring to the fact that the apparatus is no longer being used for the operator's network and of there being no reasonable likelihood that it will be so used.

12.2.3 Entitlement to keep apparatus installed

Where the terms of paragraph 22 are satisfied, the operator 'is not entitled to keep that apparatus so installed'. It is to be noted that what paragraph 22 provides for is a cessation of the entitlement to keep the apparatus installed on, under or over any land. It makes no provision for the landowner to enter and remove the apparatus.

12.2.4 Paragraph 21

Where apparatus has been 'abandoned' within the terms of paragraph 22, the landowner, in order to exercise a right of removal, is, nevertheless, still required to serve a notice pursuant to paragraph 21 of the Old Code. Paragraph 21 provides that the landowner may serve a notice enforcing the removal of the apparatus where, inter alia, the landowner 'is for the time being entitled to require the removal … of the operator's electronic communications apparatus from any land … because that apparatus is kept on, under or over that land otherwise than in pursuance of a right binding that person'. Accordingly, the operator's right to keep the apparatus on, under or over the land has ceased to bind the landowner by reason of the fact that the terms of paragraph 22 are satisfied.

12.2.5

What is interesting about the right to seek removal of the apparatus in reliance on paragraph 22, is that the code agreement does not have to be determined in order to exercise the right of removal. The landowner 'is for the time being entitled to require the removal', within the terms of paragraph 21, because the apparatus is kept on, under or over land 'otherwise than in pursuance of a right binding' the landowner, because the operator's right has ceased by reason of the terms of paragraph 22.

12.3 Conditions to which paragraph 22 are subject

12.3.1

Paragraph 22 applies where at any time:

(1) the apparatus is not, or is no longer, used for the purposes of the operator's network; and
(2) there is no reasonable likelihood that it will be so used.

12.3.2

It is difficult to discern what, if any, difference is intended by the statutory draughtsman between the expressions 'not used' and 'no longer used', but, presumably, the draughtsman is drawing a distinction between apparatus which was never used for the operator's network and where it was so used but is now no longer used.

12.3.3

The right to keep the apparatus installed ceases where both conditions are satisfied. It would appear that the entitlement to keep the apparatus installed terminates immediately if, objectively, it may be said that the two conditions are satisfied. Of course, as a matter of practicality, the two conditions will only become contentious, and need to be established by the landowner by appropriate evidence, where:

(1) the landowner becomes aware that the apparatus is no longer being utilised by the operator for the operator's network; and

(2) the landowner is able to establish that there is no reasonable likelihood that the apparatus will be utilised for the operator's network.

12.3.4

There is nothing within the terms of the Old Code to enable a landowner to determine either of the matters referred to in paragraph 12.3.3 above. How a landowner is expected to become aware of the fact that apparatus is no longer being utilised by the operator for the operator's network and of the fact that there is little likelihood of it being so used in the future, are matters which are not dealt with by the terms of the Old Code.

12.3.5

Most Old Code agreements would ordinarily contain provision for the landowner to enter for the purpose of ensuring compliance with the relevant obligations on the part of the operator under the agreement. It may be possible, therefore, for a landowner to have become aware of the non-use of the apparatus. However, even in the case of entry, a landowner ordinarily is only able to undertake a visual inspection of the apparatus which is unlikely to reveal anything about its functionality. The Authors are unaware of Old Code agreements containing provision for the operator to notify the landowner when apparatus ceases to be utilised by the operator for the operator's network.

12.4 Recommencing use of apparatus

12.4.1

An interesting matter which arises under paragraph 22 is whether the operator can avoid service of a paragraph 21 notice by the simple expedient of recommencing use of the apparatus. Paragraph 22 refers to the fact that the operator 'is not entitled to keep that apparatus so installed if, at a time when' the conditions are fulfilled. Thus it would appear that if the conditions are, subsequently, fulfilled there is no longer 'a time when' the apparatus satisfies the two conditions to which paragraph 22 is subject. Accordingly, it would seem that the right to keep the apparatus installed may be restored upon recommencing use of the apparatus in connection with the operator's network.

12.4.2

What happens if the apparatus is used again for the operator's network after service of a paragraph 21 notice, and which notice was served at the time when the terms of paragraph 22 were satisfied? Paragraph 21 enables notice to be served where the landowner 'is for the time being entitled to require the removal' because the apparatus has been abandoned. Thus service of the paragraph 21 notice was valid, but by the time any proceedings pursuant to paragraph 21 are heard, the apparatus is no longer 'abandoned'.

12.4.3

In those circumstances it is probably the case that, assuming the operator were to serve a counter-notice, the operator can establish that there are 'steps which the operator proposes to take for the purposes of securing a right as against the person to keep the apparatus on the land', namely, the step of reusing it. Taking such a step would appear to satisfy the terms of paragraph 21(6), namely, that he has in fact taken the intended step and the taking of that step will secure the right to keep the apparatus installed on, under or over the land, because the apparatus is no longer abandoned and, therefore, the right to keep the apparatus installed appears to be 'resuscitated'.

12.5 New Code: paragraph 37(6)

12.5.1

The New Code replicates, to an extent, the provisions of paragraph 22 in Part 6 of the New Code, dealing with the rights to require removal of electronic communications apparatus, and in particular, paragraph 37(6).

12.5.2

By paragraph 37(1) it is provided that 'a person with an interest in land (a "landowner") has the right to require the removal of electronic communications apparatus on, under or over the land if (and only if) one or more of the following conditions are met'. The various sub-paragraphs of paragraph 37 which follow then identify five conditions which a landowner may satisfy in order to require removal of the apparatus.

12.5.3

By paragraph 37(6) it is provided that:

> The third condition is that –
> (a) an operator has the benefit of a code right entitling the operator to keep the apparatus on, under or over the land, but
> (b) the apparatus is not, or is no longer, used for the purposes of the operator's network, and
> (c) there is no reasonable likelihood that the apparatus will be used for that purpose.

12.5.4

It will be seen that this provision all but mirrors that of paragraph 22 of the Old Code. However, the New Code does make provision (see section 12.6 below) enabling the landowner to determine whether or not apparatus is being used for the purposes of the operator's network. It would appear that, like paragraph 22 of the Old Code, the operator's code agreement does not need to be terminated before reliance can be placed upon the right of removal to be found in paragraph 37(6) (see section 12.7 below).

12.6 Conditions to which paragraph 37(6) are subject

12.6.1

Paragraph 37(6) applies where, as with paragraph 22:

(1) the apparatus is not, or is no longer, used for the purposes of the operator's network; and
(2) there is no reasonable likelihood that it will be so used.

12.6.2

Unlike the terms of paragraph 22 of the Old Code, the New Code does make provision to enable the landowner to find out whether apparatus is being used by the operator for the operator's network. By paragraph 39(1) of the New Code notice may be served by the landowner on an operator to disclose whether, inter alia, 'the operator owns electronic communications apparatus on, under or over land in which the landowner has an interest or uses such apparatus for the purposes of the operator's network'.

12.6.3

It would appear from the terms of paragraph 39(1) of the New Code that the notice may be served by the landowner at any time.

12.7 Termination of New Code agreement?

12.7.1

The terms of paragraph 22 of the Old Code made provision for the right to keep the apparatus installed to cease where the apparatus had been abandoned. The operator was 'not entitled to keep the apparatus so installed' where the conditions of paragraph 22 were satisfied. There was no requirement to terminate the code agreement in order to invoke paragraph 22 so as to seek to require removal pursuant to paragraph 21 of the Old Code (see section 11.2).

12.7.2

The terms of paragraph 37(6) of the New Code provide simply for those conditions as to abandonment now to be conditions for enforcing a right to require removal. The terms of paragraph 37(6) say nothing about the continuance of the code right and the cessation of such a right as against the landowner. However, there would appear to be no requirement to terminate the code agreement in order to enforce the right of removal pursuant to paragraph 37(6). This appears to be deliberate. The Law Commission in paragraph 6.126 of their Report,[1] expressed the view that 'where apparatus has been

1 *The Electronic Communications Code*, Law Com No. 336.

abandoned the landowner should be in the same position as if he was not bound by Code Rights'. When one has regard to paragraph 37(1), it is clear that paragraph 37(6) is simply one of five conditions pursuant to which the landowner has the right to require the removal of the apparatus. The second condition, contained in paragraph 37(3), makes provision for removal where the code right entitling the operator to keep the apparatus on, under or over land has come to an end because of a termination order pursuant to paragraph 32 or 34 of the New Code. Thus, termination of the code agreement is unnecessary where reliance is placed on paragraph 37(6); it is a self-contained provision entitling the landowner to require removal so long as the conditions referred to are satisfied.

12.8 The code agreement

12.8.1

The terms of paragraph 22 of the Old Code require the operator to have 'a right conferred by or in accordance with this code for the statutory purposes to keep electronic communications apparatus installed on, under or over any land' and a provision for the cessation of that right if the conditions to which paragraph 22 was subject were satisfied.

12.8.2

The first condition contained in paragraph 37(6)(a) of the New Code makes reference to the fact that the 'operator has the benefit of a code right entitling the operator to keep the apparatus on, under or over the land'. Thus, it is clear that paragraph 37(6) cannot be invoked simply because the code agreement has come to an end. One cannot seek to argue that apparatus is no longer used for the operator's network because of the expiry of the code agreement. In fact given the terms of paragraphs 30(1) and (2) of the New Code, the mere expiry by effluxion of time does not bring about a cessation of the code right. Nor does paragraph 37(6) bring about a cessation of the code right. Unlike paragraph 22 of Old Code which, if satisfied, provided that the entitlement to keep the apparatus ceased, paragraph 37(6) has no such effect. There is no need under the New Code to make provision for the code right to keep the apparatus on, under or over the land to cease; the terms of Part 6 make provision simply for removal without the necessity for termination either of the code right or of the code agreement itself.

12.8.3

Paragraph 37(6)(a) it is to be noted does not refer to the fact that the operator who has the benefit of a code right does so vis-à-vis the landowner who is seeking removal in accordance with Part 6. The landowner who is seeking removal has to seek removal against an operator who has the benefit of a code right to keep the apparatus installed. It does not provide that the landowner has himself to be bound by the code right of the operator. Accordingly, paragraph 37(6) may be utilised by e.g. a superior landowner against the operator where albeit the code agreement is not binding on the landowner, the landowner wishes to seek removal in reliance upon paragraph 37(6) because, although he could also seek to rely upon the first condition justifying removal contained in paragraph 37(2),[2] he may e.g. have difficulty in satisfying paragraph 37(4)(a)[3] of the New Code.

2 Which enables the landowner to seek removal under the New Code where 'the landowner has never since the coming into force of this code been bound by a code right entitling an operator to keep the apparatus on, under or over the land' see Chapter 24, section 24.3.

3 Which provides that the first condition is not satisfied if the party who conferred the code right remains in occupation.

13 Old Code sundry matters

13.1 Introduction

13.1.1

In addition to the special and general regimes, the Old Code contains a number of other provisions which are of practical significance. This chapter will discuss the disparate provisions contained in paragraphs 15–19:

(a) Paragraph 15: Use of Certain Conduits;
(b) Paragraphs 17 and 18: Objections to and Notices Affixed to Overhead Apparatus;
(c) Paragraph 23: Undertaker's Works;
(d) Paragraph 26: Crown Land;
(e) Paragraph 29: Sharing Agreements.

13.1.2

Other issues of importance are dealt with in separate chapters:

(1) Injurious Affection to Neighbouring Land (paragraph 16) is dealt with in Chapter 8.
(2) Notices (paragraph 24) are dealt with in Chapter 32.
(3) 'The Court' is dealt with in Chapter 22.
(4) Contracting out (paragraph 27) is dealt with in Chapter 34.

13.2 Use of certain conduits (paragraph 15)

13.2.1

Paragraph 15 provides that nothing in the preceding provision of the code authorise 'the doing of anything inside a relevant conduit without the agreement of the authority with control of that conduit'. These are specific classes of conduit identified within s.98 of the Telecommunications Act 1984, and the following consent requirements appear required instead of (though, perhaps, in some cases, in addition to) any right conferred by the general or special regimes.

13.2.2

A 'relevant conduit' is defined in s.98(6) of the 1984 Act as follows:[1]

(a) any conduit which, whether or not it is itself an electric line, is maintained by an electricity authority for the purpose of enclosing, surrounding or supporting such a line, including where such a conduit is connected to any box, chamber or other structure (including a building) maintained by an electricity authority for purposes connected with the conveyance, transmission or distribution of electricity, that box, chamber or structure; or
(b) a water main or any other conduit maintained by a water authority for the purpose of conveying water from one place to another; or
(c) a public sewer;[2] or
(d) a culvert which is a designated watercourse within the meaning of the Drainage (Northern Ireland) Order 1973.

13.2.3

Regrettably, s.98 contains a slightly different definition for conduit under s.98(9), where it is defined to include 'a tunnel or subway'.

13.2.4

An 'authority with control of a relevant conduit' is defined by reference to s.98(7) and (8):

(a) in relation to a conduit or structure falling within paragraph (a) or (b) of subsection (6) above, shall be construed as a reference to the authority by whom the conduit or structure is maintained;
(b) in relation to a public sewer, shall be construed, subject to subsection (8) below, as a reference to the person] in whom the sewer is vested; and
(c) in relation to a culvert falling within paragraph (d) of subsection (6) above, shall be construed as a reference to the Department of Agriculture for Northern Ireland.

(8) Where –
(a) the functions of an authority with control of a public sewer are, in pursuance of any enactment, discharged on its behalf by another person, and
(b) the other person is authorised by the authority with control of the sewer to act on its behalf for the purposes of the matters referred to in subsection (1) above, this section shall have effect in relation to that sewer as if any reference to the authority with control of the sewer included, to such extent as may be necessary for the other person so to act, a reference to the other person.

1 The definition in paragraph 1 of 'conduit' must be displaced by the special definition in s.98 for the purposes of paragraph 15. The definition of conduit in s.98 is narrower than that in paragraph 1.
2 Defined in s.98(9).

13.2.5

As to public sewers, it is provided that (paragraph 15(2)):

> The agreement of the authority with control of a public sewer shall be sufficient in all cases to confer a right falling within any of the preceding provisions of this code where the right is to be exercised wholly inside that sewer.

13.2.6

Where a relevant conduit is concerned, s.98(1) then provides that:

> The functions of an authority with control of a relevant conduit shall include the power –
> (a) to carry out, or to authorise another person to carry out, any works in relation to that conduit for or in connection with the installation, maintenance, adjustment, repair or alteration of electronic communications apparatus;
> (b) to keep electronic communications apparatus installed in that conduit or to authorise any other person to keep electronic communications apparatus so installed;
> (c) to authorise any person to enter that conduit to inspect electronic communications apparatus kept installed there;
> (d) to enter into agreements, on such terms (including terms as to the payments to be made to the authority) as it thinks fit, in connection with the doing of anything authorised by or under this section; and
> (e) to carry on an ancillary business consisting in the making and carrying out of such agreements.

13.2.7

Section 98(2)–(5) then further provide:

(2) Where any enactment or subordinate legislation expressly or impliedly imposes any limitation on the use to which a relevant conduit may be put, that limitation shall not have effect so as to prohibit the doing of anything authorised by or under this section.

(3) Where the doing by an authority with control of a public sewer of anything authorised by this section would, apart from this subsection, constitute a contravention of any obligation imposed (whether by virtue of any conveyance or agreement or otherwise) on the authority, the doing of that thing shall not constitute such a contravention to the extent that it consists in, or in authorising, the carrying out of works or inspections, or keeping of apparatus, wholly inside a public sewer.

(4) Subject to subsections (2) and (3) above, subsection (1) above is without prejudice to the rights of any person with an interest in land on, under or over which a relevant conduit is situated.

(5) Without prejudice to subsections (1) to (4) above, the Secretary of State may by order provide for any local Act under or in accordance with which any

conduits (whether or not relevant conduits) are kept installed in [roads] 2 to be amended in such manner as appears to him requisite or expedient for securing –

(a) that there is power for those conduits to be used for [the purposes of any electronic communications network or of any electronic communications service];
(b) that the terms (including terms as to payment) on which those conduits are used for those purposes are reasonable; and
(c) that the use of those conduits for those purposes is not unreasonably inhibited (whether directly or indirectly) by reason of the terms of any consent, licence or agreement which has been given, granted or made in relation to any of those conduits for the purposes of that Act.

13.3 Overhead apparatus

13.3.1 Objecting to overhead apparatus: rights to object, standing and procedure

Under paragraphs 17 and 18, special provisions govern overhead apparatus.

13.3.2

Paragraph 17 permits objections to be made to overhead apparatus, a term which covers electronic communications apparatus the whole or part of which is over 3 metres or more above the ground (paragraph 17(1)). The objection rights and process are set out by paragraph 17(2):

(2) At any time before the expiration of the period of 3 months beginning with the completion of the installation of the apparatus a person who is the occupier of or owns an interest in –
  (a) any land over or on which the apparatus has been installed, or
  (b) any land the enjoyment of which, or any interest in which, is, because of the nearness of the land to the land on or over which the apparatus has been installed, capable of being prejudiced by the apparatus, may give the operator notice of objection in respect of that apparatus.

(3) No notice of objection may be given in respect of any apparatus if the apparatus –
  (a) replaces any electronic communications apparatus which is not substantially different from the new apparatus; and
  (b) is not in a significantly different position.

13.3.3

The objection provisions are therefore only concerned with ‘new’ or ‘different’ apparatus. The right to object within the initial three-month period from the completion of installation, both (a) occupiers or owners of an interest in the land containing the apparatus, and (b) occupiers or owners of an interest in land which is near, and which is prejudiced by apparatus, can object within the timeframes and using the procedure under paragraph 17.

13.3.4

Under paragraph 18, electronic communications apparatus used for the purposes of an operator's network must have affixed to it a notice before the expiration of the period of three days from completion of the installation. Paragraph 18(1) requires the notice to be affixed to (a) to every major item of apparatus installed; or (b) if no major item of apparatus is installed, to the nearest major item of electronic communications apparatus to which the apparatus that is installed is directly or indirectly connected. If no notice is affixed, this is a criminal office and liable on summary conviction to a fine not exceeding level 2 on the standard scale (paragraph 18(3)). In any proceedings for an offence under paragraph 18 it is a defence for the person charged to prove that he took all reasonable steps and exercised all due diligence to avoid committing.

13.3.5

Under paragraphs 17(4) and (5), the following procedural rules then apply following the giving of a notice:

(4) Where a person has both given a notice under this paragraph and applied for compensation under any of the preceding provisions of this code, the court –
  (a) may give such directions as it thinks fit for ensuring that no compensation is paid until any proceedings under this paragraph have been disposed of, and
  (b) if the court makes an order under this paragraph, may provide in that order for some or all of the compensation otherwise payable under this code to that person not to be so payable, or, if the case so requires, for some or all of any compensation paid under this code to that person to be repaid to the operator.

(5) At any time after the expiration of the period of 2 months beginning with the giving of a notice of objection but before the expiration of the period of 4 months beginning with the giving of that notice, the person who gave the notice may apply to the court to have the objection upheld.

13.3.6

Under paragraph 17(6), the Court is able to determine the dispute as follows:

> Subject to sub-paragraph (7) below, the court shall uphold the objection if the apparatus appears materially to prejudice the applicant's enjoyment of, or interest in, the land in right of which the objection is made and the court is not satisfied that the only possible alterations of the apparatus will –
> (a) substantially increase the cost or diminish the quality of the service provided by the operator's network to persons who have, or may in future have, access to it, or
> (b) involve the operator in substantial additional expenditure (disregarding any expenditure occasioned solely by the fact that any proposed alteration was not adopted originally or, as the case may be, that the apparatus has been unnecessarily installed), or
> (c) give to any person a case at least as good as the applicant has to have an objection under this paragraph upheld.

13.3.7

If there is material prejudice caused to the applicant's enjoyment of land or an interest in it, *prima facie* the Court will uphold the objection. However, a person who has sold their interest in the land no longer has standing to object under paragraph 17(6): see *Petursson & Anor v Hutchison 3G UK Ltd.*[3] Health consequences of proximity to a mast could be 'material prejudice' for the purposes of paragraph 17(6), but must be objectively established: *ibid.*, paragraph [53]. In *Petursson*, the applicants held a sincere subjective belief of a link with their own ill-health, but failed to prove this was in fact the case on medical evidence.

13.3.8

If the Court believes that one of conditions (a) to (c) is however met, then the Court cannot uphold the objection. It will be noticed that the three conditions impose a lower threshold on operators than, say, the provisions of paragraph 5 discussed in Chapter 7 at section 7.6. Paragraph 17(8) further provides that

> In considering the matters specified in sub-paragraph (6) above the court shall have regard to all the circumstances and to the principle that no person should unreasonably be denied access to an electronic communications network or to electronic communications services.

13.3.9

This term has already been explained in the context of paragraph 5 of the special regime.

13.3.10

Paragraph 17(7) gives the Court the ability to refuse to uphold objections made by persons who conferred or are bound by code rights, if the giving of a notice of objection by such a person is 'unreasonable':

> The court shall not uphold the objection if the applicant is bound by a right of the operator falling within paragraph 2 or 3(1) above to install the apparatus and it appears to the court unreasonable, having regard to the fact that the applicant is so bound and the circumstances in which he became so bound, for the applicant to have given notice of objection.

13.3.11

In relation to objections by persons with an interest in land, or occupiers of the land, on which the apparatus is located, although paragraph 17 gives the right to object to apparatus shortly after its installation, it sets up an extra hurdle for such persons (otherwise they could simply go back on their agreement as they wished within the statutory timeframe). It seems plain that, in a case covered by paragraph 17(7), an objection cannot be made by an occupier simply for tactical reasons, or because of a simple 'change

3 [2005] EWHC 920 (TCC) at paragraph [49].

of heart'. However, such an objection may be reasonable should it be established that the burden imposed by the apparatus is greater than anticipated, or if the installation of the apparatus has had unforeseen consequences. The same is true of persons who agreed to be bound or are treated as bound by paragraph 2. Those who are not bound have remedies under paragraphs 4 and 21: see Chapters 7 and 11.

13.3.12

If the Court is persuaded to make an order, then it must be done under paragraphs 17(9) and (10). This significantly mirrors paragraph 20 (see Chapter 10) and is as follows:

> (9) If it upholds an objection under this paragraph the court may by order –
> (a) direct the alteration of the apparatus to which the objection relates;
> (b) authorise the installation (instead of the apparatus to which the objection relates), in a manner and position specified in the order, of any apparatus so specified;
> (c) direct that no objection may be made under this paragraph in respect of any apparatus the installation of which is authorised by the court.
>
> (10) The court shall not make any order under this paragraph directing the alteration of any apparatus or authorising the installation of any apparatus unless it is satisfied either –
> (a) that the operator has all such rights as it appears to the court appropriate that he should have for the purpose of making the alteration or, as the case may be, installing the apparatus, or
> (b) that –
> (i) he would have all those rights if the court, on an application under paragraph 5 above, dispensed with the need for the agreement of any person, and
> (ii) it would be appropriate for the court, on such an application, to dispense with the need for that agreement;
>
> and, accordingly, for the purposes of dispensing with the need for the agreement of any person to the alteration or installation of any apparatus, the court shall have the same powers as it would have if an application had been duly made under paragraph 5 above for an order dispensing with the need for that person's agreement.

13.3.13

As is the case elsewhere in the Old Code, 'alter' includes 'remove': see paragraphs 1(1) and (2).

13.3.14

Under paragraph 17(11), the Court has power to give directions to bring the application under 17(6)(c) (i.e. where it is considered that, if an alteration is ordered, it will simply enable another person to make an objection, to bring the application to that person's attention) and 17(10) (in relation to the alteration).

13.4 Undertaker's works (paragraph 23)

13.4.1

Under paragraph 23, it is provided for where a 'relevant undertaker' wishes to carry out undertaker's works involving the temporary or permanent alteration of electronic communications apparatus on land for the purposes of the operator's network. A relevant undertaker is defined in paragraph 23(10):

(a) any person (including a local authority) authorised by any Act (whether public general or local) or by any order or scheme made under or confirmed by any Act to carry on –
  (i) any railway, tramway, road transport, water transport, canal, inland navigation, dock, harbour, pier or lighthouse undertaking;

13.4.2

'Undertaker's works' are defined as follows:

(a) in relation to a relevant undertaker falling within paragraph (a) of the preceding definition, any works which that undertaker is authorised to execute for the purposes of, or in connection with, the carrying on by him of the undertaking mentioned in that paragraph;
(b) in relation to a relevant undertaker falling within paragraph (b) of that definition, any works which that undertaker is authorised to execute by or in accordance with any provision of this code; and
(c) in relation to a relevant undertaker falling within paragraph (c) of that definition, the works for the purposes of which this paragraph is applied to that undertaker.

13.4.3

In relation to element (c) of each definition above, it is provided by sub-paragraph (11) that

> The application of this paragraph by virtue of paragraph (c) of each of the definitions in sub-paragraph (10) above to any person for the purposes of any works shall be without prejudice to its application by virtue of paragraph (a) of each of those definitions to that person for the purposes of any other works.

13.4.4

There is a notice procedure for such works, set out in sub-paragraph (2):

(2) The relevant undertaker shall, not less than 10 days before the works are commenced, give the operator a notice specifying the nature of the undertaker's works, the alteration or likely alteration involved and the time and place at which the works will be commenced.
(3) Sub-paragraph (2) above shall not apply in relation to any emergency works of which the relevant undertaker gives the operator notice as soon as practicable after commencing the works.

(4) Where a notice has been given under sub-paragraph (2) above by a relevant undertaker to the operator, the operator may within the period of 10 days beginning with the giving of the notice give the relevant undertaker a counter-notice which may state either –
(a) that the operator intends himself to make any alteration made necessary or expedient by the proposed undertaker's works; or
(b) that he requires the undertaker in making any such alteration to do so under the supervision and to the satisfaction of the operator.

(5) Where a counter-notice given under sub-paragraph (4) above states that the operator intends himself to make any alteration –
(a) the operator shall (subject to sub-paragraph (7) below) have the right, instead of the relevant undertaker, to execute any works for the purpose of making that alteration; and
(b) any expenses incurred by the operator in or in connection with the execution of those works and the amount of any loss or damage sustained by the operator in consequence of the alteration shall be recoverable by the operator from the undertaker in any court of competent jurisdiction.

(6) Where a counter-notice given under sub-paragraph (4) above states that any alteration is to be made under the supervision and to the satisfaction of the operator –
(a) the relevant undertaker shall not make the alteration except as required by the notice or under sub-paragraph (7) below; and
(b) any expenses incurred by the operator in or in connection with the provision of that supervision and the amount of any loss or damage sustained by the operator in consequence of the alteration shall be recoverable by the operator from the undertaker in any court of competent jurisdiction.

(7) Where –
(a) no counter-notice is given under sub-paragraph (4) above, or
(b) the operator, having given a counter-notice falling within that sub-paragraph, fails within a reasonable time to make any alteration made necessary or expedient by the proposed undertaker's works or, as the case may be, unreasonably fails to provide the required supervision, the relevant undertaker may himself execute works for the purpose of making the alteration or, as the case may be, may execute such works without the supervision of the operator; but in either case the undertaker shall execute the works to the satisfaction of the operator.

(8) If the relevant undertaker or any of his agents –
(a) executes any works without the notice required by sub-paragraph (2) above having been given, or
(b) unreasonably fails to comply with any reasonable requirement of the operator under this paragraph, he shall, subject to sub-paragraph (9) below, be guilty of an offence and liable on summary conviction to a fine which –
(i) if the service provided by the operator's [network] is interrupted by the works or failure, shall not exceed level 4 on the standard scale; and
(ii) if that service is not so interrupted, shall not exceed level 3 on the standard scale.

(9) Sub-paragraph (8) above does not apply to a Northern Ireland department.

13.4.5

Lewison J explained these provisions as follows in *The Bridgewater Canal Co. Ltd v Geo Networks Ltd*:[4]

> Special provisions apply to undertakers' works. An undertaker is a person authorised by Act of Parliament to carry on any railway, tramway, road transport, water transport, canal, inland navigation, dock, harbour, pier or lighthouse undertaking: paragraph 23(10). If an undertaker wants to execute works which are likely to involve a temporary or permanent alteration to an operator's apparatus it must give notice to the operator: paragraph 23(2). The operator may give counter-notice either agreeing to carry out the work itself, or requiring the undertaker to carry out the work under the supervision of and to the satisfaction of the operator: paragraph 23(4). In either case the undertaker must pay the operator compensation for loss and damage suffered in consequence of the alteration: paragraph 23(5), (6). The operator has no right to object to the works; and there is no provision for arbitration or application to the court for the resolution of any dispute about the nature of the works. However, the compensation is recoverable by action.

13.5 Crown land (paragraph 26)

13.5.1

Special rules apply to land in which the Crown has an interest. Paragraph 26 provides:

(1) This code shall apply in relation to land in which there subsists, or at any material time subsisted, a Crown interest as it applies in relation to land in which no such interest subsists.

(2) In this paragraph 'Crown interest' means an interest which belongs to Her Majesty in right of the Crown or of the Duchy of Lancaster or to the Duchy of Cornwall or to a Government department or which is held in trust for Her Majesty for the purposes of a Government department and, without prejudice to the foregoing, includes any interest which belongs to Her Majesty in right of Her Majesty's Government in Northern Ireland or to a Northern Ireland department or which is held in trust for Her Majesty for the purposes of a Northern Ireland department.

(3) An agreement required by this code to be given in respect of any Crown interest subsisting in any land shall be given by the appropriate authority, that is to say –

  (a) in the case of land belonging to Her Majesty in right of the Crown, the Crown Estate Commissioners or, as the case may require, the government department having the management of the land in question or the relevant person;

  (b) in the case of land belonging to Her Majesty in right of the Duchy of Lancaster, the Chancellor of that Duchy;

  (c) in the case of land belonging to the Duchy of Cornwall, such person as the Duke of Cornwall, or the possessor for the time being of the Duchy of Cornwall, appoints;

  (d) in the case of land belonging to Her Majesty in right of Her Majesty's Government in Northern Ireland, the Northern Ireland department having the management of the land in question;

4 [2010] EWHC 548 (Ch), paragraph [18].

(e) in the case of land belonging to a government department or a Northern Ireland department or held in trust for Her Majesty for the purposes of a government department or a Northern Ireland department, that department;

and if any question arises as to what authority is the appropriate authority in relation to any land that question shall be referred to the Treasury, whose decision shall be final.

(3A) In sub-paragraph (3), 'relevant person', in relation to any land to which section 90B(5) of the Scotland Act 1998 applies, means the person who manages that land.

(4) Paragraphs 12(9) and 18(3) above shall not apply where this code applies in the case of the Secretary of State or a Northern Ireland department by virtue of section 106(3)(b) of the Communications Act 2003.

13.6 Sharing apparatus (paragraph 29)

13.6.1

A paragraph purporting to deal with sharing was introduced by the 2003 Act, being paragraph 29. It is almost completely incomprehensible.

13.6.2

Paragraph 29(1) states that

> This paragraph applies where –
> (a) this code has been applied by a direction under section 106 of the Communications Act 2003 in a person's case;
> (b) this code expressly or impliedly imposes a limitation on the use to which electronic communications apparatus installed by that person may be put or on the purposes for which it may be used; and
> (c) that person is a party to a relevant agreement or becomes a party to an agreement which (after he has become a party to it) is a relevant agreement.

13.6.3

It is unclear what the 'limitation' is that is referred to in paragraph 29(1)(b), though it could be a reference to the fact that the use of apparatus has to be in connection with the operator's network under paragraph 2, which may suggest that it cannot be used in connection with a different operator's network.

13.6.4

Paragraph 29 considers what happens when a limitation of that kind conflicts with the terms of a 'relevant agreement'. As to paragraph 29(1)(c), and the meaning of a 'relevant agreement', further definitions are given in paragraphs 29(4)–(6). In particular:

> 'relevant agreement' means an agreement in relation to electronic communications apparatus which –

(a) relates to the sharing by different parties to the agreement of the use of that apparatus; and
(b) is an agreement that satisfies the requirements of sub-paragraph (5).

13.6.5

Paragraph 29(5) to (7) then state:

(5) An agreement satisfies the requirements of this sub-paragraph if –
   (a) every party to the agreement is a person in whose case this code applies by virtue of a direction under section 106 of the Communications Act 2003; or
   (b) one or more of the parties to the agreement is a person in whose case this code so applies and every other party to the agreement is a qualifying person.

(6) A person is a qualifying person for the purposes of sub-paragraph (5) if he is either –
   (a) a person who provides an electronic communications network without being a person in whose case this code applies; or
   (b) a designated provider of an electronic communications service consisting in the distribution of a programme service by means of an electronic communications network.

(7) In sub-paragraph (6) –
'designated' means designated by an order made by the Secretary of State;
'programme service' has the same meaning as in the Broadcasting Act 1990.

13.6.6

The above would suggest that paragraph 29 only captures sharing agreements where every party is, or could be, an operator. If those conditions are met, then the terms of that sharing agreement take primacy over any implied limitation in the Old Code, so that, at least as far as the agreement between operators and/or potential operators is concerned in relation to sharing, nothing in the Old Code precludes such an agreement. Paragraph 29(2) accordingly provides:

(2) The limitation is not to preclude –
   (a) the doing of anything in relation to that apparatus, or
   (b) its use for particular purposes,

to the extent that the doing of that thing, or the use of the apparatus for those purposes, is in pursuance of the agreement.

13.6.7

Paragraph 29(3) expressly states that, although an arrangement between operators (or potential operators) to share is not to be treated as inconsistent with the Old Code, it does not override any obligations on the operator seeking to share to obtain consent under a contractual obligation to do so (for instance, under its own lease) or under any statutory provision:

> This paragraph is not to be construed, in relation to a person who is entitled or authorised by or under a relevant agreement to share the use of apparatus installed

> by another party to the agreement, as affecting any consent requirement imposed (whether by a statutory provision or otherwise) on that person.

13.6.8

Accordingly, if a covenant is a qualified alienation covenant, the consent of the landlord or licensor is still required. This is further amplified by paragraph 29(4) which explains that

> In this paragraph –
>
> 'consent requirement', in relation to a person, means a requirement for him to obtain consent or permission to or in connection with –
> (a) the installation by him of apparatus; or
> (b) the doing by him of any other thing in relation to apparatus the use of which he is entitled or authorised to share.

13.6.9

In the Law Commission's Consultation Paper, these provisions were explained as follows (at 3.81–3.82):

> The Code contains very little on sharing, although paragraph 29 is relevant; it applies where: (1) the Code has been applied to an operator ('Operator A'); (2) the Code expressly or impliedly limits the use to which apparatus installed under the Code can be put; and (3) Operator A is (or becomes) party to an agreement to share apparatus installed under the Code. Where this is the case, the limitation referred to in point (2) above is deemed not to preclude doing anything, or using any apparatus, in pursuance of that sharing agreement. Although convoluted, paragraph 29 may have the effect of permitting a Code Operator to share its apparatus with another operator.
>
> However, paragraph 29 does not override the constraints of agreements with landowners and occupiers. So where a landowner includes a term in an agreement with a Code Operator that the rights conferred benefit that Code Operator only, paragraph 29 will not operate. For landowners and occupiers this is beneficial – if a Code Operator wants to share its apparatus, then another agreement (with attendant delays in concluding negotiations or seeking an order) must be sought, and a second payment of consideration will be due.

13.6.10

The simplest reading of paragraph 29 is that the Old Code did not preclude operators or potential operators from entering into agreements between themselves to share. However, if the operator seeking to share was subject to an external consent requirement (such as under a qualified alienation covenant), that consent still had to be sought and obtained.

Part III

Electronic Communications Code 2017 (the New Code)

14 The Electronic Communications Code and property law

Key concepts

14.1 Introduction

14.1.1

Whereas the Old Code was bolted on to existing property law concepts, adding a layer of statutory protection to operators, and a corresponding layer of difficulty for landowners when they came to seek to terminate code agreements, the New Code has sought (but not altogether successfully) to avoid some of the difficulties experienced with the Old, by creating a statutory creature – the agreement conferring code rights – which can, but need not, depend upon any existing property law concepts for its continued existence.

14.1.2

Having said that, the New Code does not prevent parties using any wrapper they like for their code agreement, be it a lease, licence or wayleave (or none of these), provided always that the agreement complies with the criteria spelt out in Part 2 of the New Code (discussed in Chapter 19).

14.1.3

For that reason, it remains important for those drafting and entering into code agreements to have an appreciation of the property law consequences of using different wrappers, since these will affect how apparatus may be accessed, used, shared and improved; and how rights may be terminated.

14.1.4

Section 14.2 below deals with the different types of property wrapper which may be used for a code agreement. Section 14.3 analyses the courts' approach to the interpretation of property agreements. Section 14.4 examines the extent to which property agreements allow for those aspects of site use which will be of importance for operators and landowners, such as sharing, upgrading and termination. Section 14.5 looks at the statutory protection available to those in occupation of business premises. Section 14.6 then provides an overview of the working of the Old Code; while section 14.7 finally does the same in relation to the New Code.

14.2 Property law concepts: leases, licences, easements and wayleaves

14.2.1

It is our experience that the vast majority of telecommunications equipment under the Old Code will have been installed pursuant to some form of written agreement. The one thing that can uniformly be said of all such agreements is that they display no uniformity. Thus:

- Some are described as leases, and demise portions of airspace, with associated rights of access, sometimes for fixed terms of 10 or 20 years, at a reviewable rent; sometimes on an annual or other periodic basis.
- Some are described as licences, easements or wayleaves, with widely varying provisions.
- Some – perhaps the majority – avoid labels altogether, and use neutral language to describe the principal terms of the occupational arrangements.

14.2.2

A *lease* (also referred to less formally as a tenancy agreement) creates an interest in land which may (and in some circumstances must) be registered at HM Land Registry in order to bind successors in title to the landlord's interest.[1] The characteristics of a lease are that it must grant exclusive possession (that is, a right to keep out all, including the landlord, unless the landlord reserved a right to enter), of a fixed or ascertainable interest in land – see the speech of Lord Templeman in *Street v Mountford*[2] (in which he added that the reservation of a rent was another characteristic), as supplemented by the decision of the Court of Appeal in *Ashburn Anstalt v WJ Arnold & Co*[3] (reservation of a rent not a critical ingredient). A lease must also satisfy certain formalities contained in ss.52 and 54 of the Law of Property Act 1925. The effect of these is outside the scope of this book, and reference should be made as necessary to Woodfall on *Landlord and Tenant*. Leases can only be terminated in accordance with their own contractual terms, as supplemented by a dense layer of common law and statute.

14.2.3

A *licence*, by contrast, may not (and usually does not) grant exclusive possession; and may not (and sometimes does not) specify the period for which it is to last. Licences are usually simple, informal and flexible arrangements, in contrast to leases. Because they do not create interests in land, they are not registrable.

14.2.4

An *easement*, by further contrast, is a property right (and in the circumstances considered in this book, invariably a right of way) granted by a landowner enabling an adjoining property owner to cross a specified part of its land, either with equipment (for example,

1 See Chapter 38, which discusses land registration generally.
2 [1985] AC 809.
3 [1989] Ch 1.

an easement allowing electricity to flow through a conduit), or in person (for example, a right of access). Being an interest in land, an easement is registrable, and there are complex rules regarding their grant and termination, which must be strictly observed.

14.2.5

Wayleaves, lastly, are simply a species of licence (although some may in substance be easements), much used by utility companies, that allow services to cross land without creating proprietary interests in land. They accordingly are not registrable interests.

14.2.6

Why does it matter whether the agreement takes effect as a lease or a licence? There are two principal reasons:

(a) In general terms, a lease of premises occupied for business purposes will attract statutory security of tenure under Part II of the Landlord and Tenant Act 1954, which will affect the landlord's freedom of manoeuvre. We deal with this in Chapters 11 and 14 (Old Code) and 25 (New Code).
(b) A lease grants an interest in the land occupied, which the telecommunications operator is free to exploit (subject to any provisions to the contrary in the lease). It is rare for a licence to do so. This consideration particularly impacts on network sharing (see paragraph 14.4.9 below, and Chapter 20).

14.3 The courts' approach to the interpretation of property agreements

14.3.1

Because of the differences between leases and licences highlighted in section 14.2 above, it is usually critical for parties to such agreements to know to which they have subscribed. In very general terms, landowners wish to create licences, in order to retain greater control of their land, while operators will wish for leases, in order to have greater security of tenure. It is common to find that parties have entered into agreements drafted by landowners which use expressions which are characteristic of licences, only for the operator to contend that the labelling is not representative of the reality of the agreement.

14.3.2

In *Street v Mountford,*[4] the House of Lords held that the proper resolution to the question whether a property agreement took effect as a lease or a licence depended upon the substance of the agreement, and not upon the label which the parties had attached to it. A document which grants the right to exclusive possession for a term, at a rent, creates a lease, no matter what the parties have chosen to call it. As Lord Templeman observed (at p.819):

> the consequences in law of the agreement, once concluded, can only be determined by consideration of the effect of the agreement. If the agreement satisfied all the requirements of a tenancy, then the agreement produced a tenancy and the parties

4 [1985] AC 809.

> cannot alter the effect of the agreement by insisting that they only created a licence. The manufacture of a five-pronged implement for manual digging results in a fork even if the manufacturer, unfamiliar with the English language, insists that he intended to make and has made a spade.

14.3.3

Their Lordships in *Street* were willing to contemplate that there may be cases where the finding of the three hallmarks does not inexorably lead to a lease, such as where the relationship was not intended to give rise to a legal relationship at all. Further, it is now accepted that there are only two hallmarks, and not three (rent being inessential – see s.205(1)(xxvii) of the Law of Property Act 1925, and the decision of the Court of Appeal in *Ashburn Anstalt v WJ Arnold & Co*).[5] Save for those caveats, the ruling in *Street* represents the law: substance prevails over form.[6]

14.3.4

There are, however, limitations to the extent to which the courts are prepared to accede to argument that the parties should be governed by the labels they have chosen, at any rate where the parties are commercial organisations that were professionally represented when they entered into their agreement. In *Clear Channel UK Ltd v Manchester City Council*,[7] the Court of Appeal held that a contract for the claimant to erect and operate advertising stations created a licence rather than a tenancy. Jonathan Parker LJ was particularly influenced by the fact that the contract contained a statement that it should not create a tenancy, and added this trenchant comment to the end of his judgment:

> I find it surprising and (if I may say so) unedifying that a substantial and reputable commercial organisation like Clear Channel, having (no doubt with full legal assistance) negotiated a contract with the intention expressed in the contract … that the contract should not create a tenancy, should then invite the Court to conclude that it did.
>
> In making that comment … [I do not] intend to cast any doubt whatever upon the principles established in *Street v Mountford*. On the other hand, the fact remains that this was a contract negotiated between two substantial parties of equal bargaining power and with the benefit of full legal advice. Where the contract so negotiated contains not merely a label, but a clause that sets out in unequivocal terms

5 [1989] Ch 1.

6 Lord Hoffman said in *Bruton v London & Quadrant Housing Trust* [2000] 1 AC 406 at 413E, that *Street v Mountford* [1985] AC 809 'is authority for the proposition that a "lease" or "tenancy" is a contractually binding agreement, not referable to any other relationship between the parties, by which one person gives another the right to exclusive occupation of land for a fixed or renewable period or periods of time, usually in return for a periodic payment in money. An agreement having these characteristics creates a relationship of landlord and tenant to which the common law or statute may then attach various incidents.' In *Mexfield Housing Co-operative Ltd v Berrisford* [2012] 1 AC 955 at 101, Lord Mance said 'The three characteristic hallmarks of a contractual tenancy, as distinct from a contractual licence, are: (a) exclusive occupation; (b) rent; and (c) a term which the law regards as certain.'

7 [2006] 1 EGLR 27.

the parties' intention as to its legal effect, I would in any event have taken some persuading that its true effect was directly contrary to that expressed intention.[8]

14.3.5

In the same vein, in *National Car Parks Ltd v Trinity Development Co. (Banbury) Ltd*,[9] the Court of Appeal upheld as a licence a car park management agreement containing a declaration by the parties that it created a licence and not a tenancy. The licensee asserted that the agreement was in substance a tenancy (and which therefore qualified for statutory protection). Arden LJ said, in the course of her judgment (with which the Vice-Chancellor and Buxton LJ agreed):

> The court must, of course, look at the substance but as I see it it does not follow from that that what the parties have said is totally irrelevant and to be disregarded. For my part, I would agree with the judge that some attention must be given to the terms which the parties have agreed. On the other hand it must be approached with healthy scepticism, particularly, for instance, if the parties; bargaining positions are asymmetrical. ... It would in my judgment be a strong thing for the law to disregard totally the parties' choice of wording and to do so would be inconsistent with the general principle of freedom of contract and the principle that documents should be interpreted as a whole. On the other hand, I agree ... that it does not give rise to any presumption. At most it is relevant as a pointer.

14.3.6

In reaching its judgment, the Court of Appeal combed through the terms of the agreement, commenting in relation to each whether they were more indicative of a licence or a tenancy. In those circumstances, the only safe course, it would seem, for the party attempting to find out whether its agreement takes effect as a licence or a tenancy, is similarly to comb through its provisions to see which have the better 'fit'. Pointers which may assist with that exercise include:

- If the agreement is described as a lease or a tenancy, it would be very difficult to argue with the conclusion that that is exactly what it is, unless the terms of the agreement are uncertain in some important respect (for example, if they lack any reference to the term of the agreement).
- If the agreement contains a statement that it is not intended to create a tenancy, then that is likely to carry great weight (see *Clear Channel*).
- If the landowner provides attendance or services which require unrestricted access to and use of the premises the agreement is more likely to be construed as one negating the grant of exclusive possession.
- If the agreement contains a unilateral right on the part of the landowner to move the operator to other premises this will negate the grant of exclusive possession: *Dresden Estates Ltd v Collinson*[10] (relating to commercial premises).

8 This passage was followed in *Scottish Widows plc v Stewart* [2006] EWCA Civ 999, CA (and recently in Ireland in *Car Park Services Ltd v Bywater Capital (Wine Tavern) Ltd* (2016) 16 September 2016, Lands Tribunal, Northern Ireland, Horner J).

9 [2002] 2 P & CR 18.

10 [1987] 1 EGLR 45.

- If the agreement contains a right of re-entry this is more consistent with the grant of a lease rather than a licence, for a licensor is entitled to enter the premises by reason of his retention of possession subject only not to derogate from the occupiers right to use and occupy the premises: *Essex Plan v Broadminster*;[11] *Car Park Services Ltd v Bywater Capital (Wine Tavern) Ltd* (2016)[12] (Lands Tribunal, Northern Ireland).
- The nature of the right is such as to negate the grant. Thus, if an operator is simply granted rights in connection with a pre-existing mast but there is no grant of the mast as a piece of land per se it may be difficult to suggest that the operator has exclusive possession of any identifiable piece of land.
- The nature of the relationship is such as to negate the grant of exclusive possession e.g. it may be explicable other than on the basis of the existence of a landlord and tenant relationship. Thus, for instance it has been held that where occupation is referable to the relationship of vendor and purchaser under an option exclusive possession has not been granted: *Essex Plan v Broadminster,* CA[13]. And more recently, exclusive occupation granted by a charitable trust in breach of the terms of the trust: *Watts v Stewart*[14] (discussed in detail below).
- If the agreement is wholly unclear as to where the telecommunications equipment in question is to be stationed, then it may be arguable that no tenancy has been created, since tenancies require a specific area to be demised.
- If the parties agreed that the operator was to have the benefit of an agreement for an uncertain term, then, on the face of it, that cannot be a tenancy.[15] It is unsafe to assume that this automatically creates a licence, however. It can be 'cured' by implying a periodic tenancy from the regular payment of 'rent': *Prudential Insurance Co v London Residuary Body*;[16] such periodicity to be inferred from the payments made.
- Finally, subject to the normal caveats that attend generalisations, it will be difficult to argue in most cases that an agreement for the installation and use of telecommunications equipment, *however expressed,* takes effect as a licence rather than a tenancy. The reason is that the telecommunications equipment requires a defined location for

11 [1988] 56 P & CR 353 Ch D.
12 16 September 2016.
13 [1988] 56 P & CR 353 Ch D.
14 [2017] 2 WLR 1107.
15 In *Mexfield Housing Co-operative Ltd v Berrisford* [2012] 1 AC 955, the Supreme Court considered the effect of an 'occupancy agreement' between the claimant fully mutual housing association (FMHA), set up as part of a mortgage rescue scheme, and an 'occupier'. The claimant agreed to 'let', and the defendant 'occupier' to take, a property from month to month for a weekly rent. Clause 5 enabled the defendant to end the agreement by giving a month's notice. Clause 6 enable the claimant to end the agreement only if (among other things) the rent was in arrears or the defendant ceased to be a member of the FMHA. The agreement contained a forfeiture provision. The rent was not in arrears, but the claimant served a notice to quit on the defendant. The claimant did not rely on cl.6. The Court of Appeal held by a majority that as a lease could not be created for an uncertain term, the agreement did not create a tenancy on the terms set out in the agreement. The effect of the entry into possession and the payment of rent was that there was a monthly periodic tenancy.

The Supreme Court allowed the appeal. While, on their own, the words 'from month to month' were apt to create a monthly periodic tenancy, which could be ended by either side's giving the other a month's notice, the agreement provided that it could only be ended by the means provided by cl.5 (to the defendant) and by cl.6 (to the claimant). Such an agreement was an agreement for an uncertain term and could not take effect as a tenancy in accordance with its terms. It would have taken effect as a tenancy for life before the coming into force of the Law of Property Act 1925 (the LPA). The effect of s.149(6) of the LPA was that the tenancy was to be treated as a tenancy for a term of 90 years, determinable on the defendant's death, or in accordance with cls 5 and 6. The defendant, accordingly, was not entitled to possession.
16 [1992] 2 AC 288.

its installation, which, for obvious reasons, will be exclusive to the telecommunications operator in question. Those ingredients favour a tenancy rather than a licence.

14.3.7

The Court of Appeal have had to consider a similar question in the residential context, but with a twist. In *Watts v Stewart*,[17] Mrs Watts, following a letter of appointment by her 'landlord' – a charitable trust – was given a 'tenancy' of an almshouse. 'Rent' was paid. A right of entry was reserved to the 'landlord'. She appeared to enjoy exclusivity. There was a term. So far, so *Street v Mountford*. The charitable trust wanted to evict her and brought possession proceedings, arguing she was a licensee.

14.3.8

In favour of the trust was a case called *Gray v Taylor*,[18] where the Court of Appeal had previously held that the beneficiary of such a charitable trust was given an occupation right as licensee, and not as tenant. That case was based on *Errington v Errington and Woods*,[19] where Denning LJ said: 'Parties cannot turn a tenancy into a licence merely by calling it one. But if the circumstances and the conduct of the parties show that all that was intended was that the occupier should be granted a personal privilege, with no interest in the land, he will be held to be a licensee only.' That passage was approved in *Street v Mountford*. The document in *Gray v Taylor* was expressly not a tenancy, but a licence. The position in *Watts* was the opposite. The question was not whether the label 'licence' could be upheld, but whether the trust could escape from the misuse of the label 'lease'. Surely the latter case is much less promising for the trust, and it should be held to what it had promised?

14.3.9

As in *Gray v Taylor*, the constitutional instrument of the charity in *Watts* forbade the trustees from parting with possession, so that a lease would have been *ultra vires* their powers as trustees. However, it is to be noted that an *ultra vires* disposal by a trustee is not null and void – it is legally effective, subject to personal liability on the part of the trustee for breach of trust. Despite the terminology of Mrs Watts' agreement, the Court of Appeal found that she was a mere licensee. The reasoning was as follows. First, adopting *Street v Mountford* (though this time to find that a document labelled lease was a licence, and not the other way round), what parties chose to call a document was immaterial. Secondly, a distinction had to be drawn between what the Court called 'legal exclusive possession' on the one hand, and 'exclusive occupation' on the other. A tenant has both – that is, the legal right to control possession which, when exercised, amounts to factual exclusive occupation. A licensee may have the latter if the terms of the licence are expansive enough to bring a possession claim under CPR r.55.3: *Manchester Airport Plc v Dutton*;[20] whether the correct level of control is vested in the licensee is a matter

17 [2016] EWCA Civ 1247, CA.
18 [1998] 1 WLR 1093.
19 [1952] 1 KB 290, 298.
20 [2000] QB 133.

of construction of the licence: *Countryside Residential (North Thames) Ltd v Tugwell.*[21] We have moved on from the forms of action, and from requiring that a person seeking possession must be in possession of an estate in land: *Mayor of London v Hall.*[22]

14.3.10

The question that *Watts v Stewart* raises is, when is one dealing with a clause conferring a right to land, how does one tell whether the clause merely confers exclusive occupation as opposed to exclusive possession? The answer appears to be different from the commercial context. The terms of an agreement insofar as they are labels are to be ignored. Agreement terms which describe the rights of the parties are relevant to identify what 'package' of rights have been conferred, and to see whether, cumulatively, they amount to exclusive possession. Interestingly, in *Watts*, it does not appear that any of the clauses which pointed against a lease were particularly strong, the best being a provision that an almshouse occupier could be relocated. What appears to have carried the day was that fact that the trustees cannot possibly have intended to grant a lease as that was a breach of their trust instrument. It is therefore possible that, after *Watts*, the Courts might be receptive to an argument that, when construing a clause which might confer mere occupation but equally might confer exclusive possession, the background to the transaction is relevant. Although the legal theoretician may well appreciate the difference between exclusive possession on the one hand, and exclusive occupation on the other, the legal practitioner is also likely to find it quite difficult to distinguish between the two in practice, especially when looking at bare words on a page. It may be that, after all, category one of the *Street v Mountford* exceptions is the true difference. It may simply be that, on some cases, and having regard to who the parties are, it is simply unfair to find that a lease has been created.

14.3.11

Against this background, how does the clear dictum in *Street* that 'the only intention which is relevant is the intention demonstrated by the agreement to grant exclusive possession for a term at a rent. Sometimes it may be difficult to discover whether, on the true construction of an agreement, exclusive possession is conferred' stand up? Absent a small and diminishing number of residual tenancies, the Rent Acts no longer hold sway. The policies that formed part of the background to *Street v Mountford* are no longer as strong as they were. It is suggested by the Authors that a divergence of approach can be discerned between commercial and residential cases.[23] In commercial cases, arguments based on *Street v Mountford* formalism are likely to be met with some judicial resistance – at least where there is no question of exploitation of a dominant position. The Court of Appeal has stated in all manner of contexts, most recently forfeiture, that it is not attracted to arguments where litigants seek to get a windfall – and why should a well-resourced commercial tenant be able to turn around at the end of a long contractual relationship and claim the benefits of a tenancy even if it never bargained for them at the outset?

21 [2001] 81 P & CR 2.
22 [2011] 1 WLR 504.
23 As Glidewell LJ observed in *Dresden Estates Ltd v Collinson* [1987] 1 EGLR 45 at p.47: 'the attributes of residential premises and business premises are often quite different'.

14.3.12

In the residential setting, *Watts* nudges the door opened by the secure tenancy cases ending in *Westminster CC v Clarke*[24] wider still, and opens the prospect that arguments based on *Street v Mountford* formalism can still be headed off by adducing evidence of circumstances and background to show that, whatever the parties may have written down, they cannot possibly have intended there to be a lease. Often, it seems that the crucial factor is the identity and status of the alleged landlord. Sometimes, the fork can be a spade after all.

14.4 Rights under property agreements

14.4.1

Leases and licences work in very different ways. In essence, a lease grants the tenant an interest in the landlord's land for the duration of the tenancy agreement: in effect, the land belongs exclusively to the tenant for that limited period, and the land is therefore the tenant's, to do with as it wishes, save to the extent that there is any provision curtailing the tenant's freedom of manoeuvre.

14.4.2

A licence, by contrast, is a mere permission; the land remains the landowner's own property throughout the terms of the licence, and the licensee may not do anything on the land unless it is expressly or impliedly sanctioned by the terms of the licence.

14.4.3

To put the matter in a nutshell, leases allow the tenant to do anything it likes, unless the agreement says otherwise; whereas licences do not allow the licensee to do anything, unless the agreement says otherwise.

14.4.4

The most important features of any property agreement are those concerning duration, payment, termination, use, alienation and works. These features are now examined in turn.

14.4.5 Duration

In the case of a lease, the duration must be defined (failing which the agreement will not create a lease). The lease will either be for a fixed term (e.g. 'ten years from 1 January 2017'), or it will be periodic (e.g. 'yearly until terminated from 1 January 2017'). Although a licence will usually be drafted in much the same way, there is much more flexibility: a licence may simply provide that it should continue 'until terminated by the

24 [1992] 2 AC 288.

licensor by 28 days notice'. Such an agreement could not be a lease, because it would be void for uncertainty.

14.4.6 Payment

Both leases and licences commonly provide for consideration to be provided by the operator. This may be in the form of a fixed sum (a 'premium'), although more commonly it will be a periodic payment, referred to as a 'rent', in the case of a lease; and a 'licence fee', in the case of a licence.

14.4.7 Termination

A fixed term lease will simply expire by effluxion of time on the specified contractual term date, although there will usually be provision for the lease to expire on an earlier date (e.g. by forfeiture, in the event of default by the tenant; or upon exercise of a break clause by either party). Licences will contain similar provisions, albeit with different labels. Over and above that, there are other, common law, means of termination available in either case, such as surrender.

14.4.8 Use

In the context of an agreement with an operator, the use will usually be specified by reference to the operator's proposed plans, whether the agreement be a lease or a licence.

14.4.9 Alienation

Unless a lease dictates otherwise (which it usually will), the tenant's leasehold interest is freely assignable to whomsoever it pleases. Moreover, the tenant is treated in law as an indivisible entity, irrespective of the number of individuals constituting the tenant, so that, for example, a tenancy vested in A and B must be assigned by both A and B, and not by either severally. By contrast, a licence is usually personal to the licensee, and may not therefore be assigned. Moreover, in the case of a licence, for the same reason, the licensee will not ordinarily be able to share its site with another. In the case of a lease, by contrast, the tenant will be free to share, save to the extent (again) that its lease dictates otherwise.

14.4.10 Works

A lease will ordinarily provide for the tenant to keep the demised premises in repair, and will impose controls upon the extent to which the tenant may alter the premises. The Landlord and Tenant Acts 1927 and 1954 impose an important statutory overlay upon such controls. Moreover, there are complex rules concerning the ownership of fixtures: structures installed in the premises by the tenant. The drafting of licences is far more variable, although as a generalisation it is less often the case that a licensee will be subject to an onerous repairing obligation. Licences are not subject to the same statutory overlay, and do not raise problems of fixtures.

14.5 Statutory protection for business tenants: the position under the Old Code

14.5.1

Section 23(1) of the Landlord and Tenant Act 1954 provides:

> this Part of this Act applies to any tenancy where the property comprised in the tenancy is or includes premises which are occupied by the tenant and are so occupied for the purposes of a business carried on by him or for those and other purposes.

14.5.2

In those cases where the view is taken that the agreement with the telecommunications operator created a tenancy rather than a licence, the next question is whether the operator can be said to *occupy* the area in question.

14.5.3

It may well be the case that the operator's only contact with the premises in question will have been to install the necessary machinery and associated cabling at the beginning of the agreement, and to undertake very occasional servicing visits, perhaps at annual periods, thereafter. The question that arises in those circumstances is whether this level of use and activity can be said to constitute 'occupation'.

14.5.4

Part II of the 1954 Act does not require personal occupation by the tenant. Occupation by its chattels will suffice, provided that that occupation is for the purposes of the tenant's business. It is well settled, for example, that long-term storage will suffice for these purposes. In *Northern Electric plc v Addison*,[25] the proposition that an electricity substation came within the Act appears to have been common ground between the parties, and was not questioned either by the Judge at first instance, or by the Court of Appeal. That said, an argument may still exist in the Landlord and Tenant Act 1954 context as to whether the use of land merely to contain machinery amounts to occupation: see *Commissioners for Customs and Excise v Sinclair Collis Ltd*.[26]

14.5.5

However, in two rating cases, the Court of Appeal has treated telecommunications operators as being in rateable occupation. First, in *Orange PCF v Bradford (Valuation Officer)*,[27] the Court proceeded on the basis that a telecommunications operator with a licence under the Telecommunications Act 1984, and therefore subject to the Electronic Communications Code, was occupying the land on which telecommunications masts had been erected, but otherwise upon which no activity occurred.

25 [1997] 2 EGLR 111.
26 [2001] UKHL 30.
27 [2004] EWCA Civ 155.

14.5.6

Secondly in *Vtesse Networks Ltd v Bradford*,[28] the Court held that a person with exclusive use of fibre optic cables could be said to be in 'actual occupation'. In that case, Vtesse had a network of its own, and third party owned, fibre optic cables which formed part of its network. The Court of Appeal decided that it did not matter that some of those cables were owned by third parties, as in fact Vtesse had sole use over all of the cables, and was in occupation of them for rating purposes. The 147 km network of cables therefore counted as a single hereditament for ratings purposes.

14.5.7

In those circumstances, it would be difficult to argue that there is insufficient occupation of the premises by a telecommunications operator for the purposes of the 1954 Act. Accordingly, if the operator's agreement in question takes effect as a tenancy rather than a licence, then that tenancy is likely to be protected by the 1954 Act.

14.5.8

This conclusion had significant ramifications for the operation of the Old Code, which is summarised in the next section.

14.6 The Old Code and the 1954 Act: overview

14.6.1

The Old Code is dealt with in detail in Part 2 of this book. This section summarises the way in which the Old Code interacts with the 1954 Act (a matter which is covered in greater depth in sections 11.8 and 14.6 of Chapters 11 and 14). In 1984, s.10 of the Telecommunications Act of that year introduced a code known as the Telecommunications Code ('the Code'), granting extensive powers over land to telecommunications operators. Although the operative section has now been replaced by the Communications Act 2003, the Code itself (now relabelled 'the Electronic Communications Code') was retained, with the modifications introduced by Schedule 3 to that Act. The powers granted to electronic communications operators were extended to include, for example, rights for the providers of electronic communications networks compulsorily to acquire land (see s.118).

14.6.2

Moreover, the Office of Communications ('OFCOM') was given various duties, including a duty (see s.10(1)):

> to encourage others to secure:
> (a) that domestic electronic communications apparatus is developed which is capable of being used with ease, and without modification, by the widest possible range of individuals … ; and

28 [2006] EWCA Civ 1339.

(b) that domestic electronic communications apparatus which is capable of being so used is as widely available as possible for acquisition for those wishing to use it.

The Government's commitment to electronic communications was difficult to overstate.

14.6.3

The Telecommunications Act 1984 and the Communications Act 2003, which provide for the application of the Old Code in its current form, have the following effect (in summary):

(a) They apply the code to licensed operators, relating to the equipment used by the operators and forming part of their network.
(b) The code confers upon the operator the right to keep apparatus installed on, under or over land or buildings, until that right is terminated under either paragraph 20 or 21 of the code, or following abandonment and other specified eventualities.
(c) There is no provision allowing either party to contract out of paragraph 21 of the code. Except for paragraphs 8(5) and 21 of the code, the provisions of the code were expressed to be 'without prejudice to any rights or liabilities arising under any agreement to which the operator is a party': paragraph 27(2) of the code.
(d) An operator can apply to court under paragraph 5 of the code to force an owner of land to enter into a contract to confer the rights permitted under the code. On such an application, the court must have regard to the overriding principle that 'no person should unreasonably be denied access to an electronic communications network'.

14.6.4

Paragraph 20[29] gives a landowner power to require 'alteration' of electronic telecommunications apparatus installed on the owner's land, 'notwithstanding the terms of any agreement binding' the owner. Its purpose is therefore to override contractual stipulations so as to allow redevelopment or other improvements of land in certain circumstances. If the operator refuses to comply with the requisite notice from the landowner, the matter must then be referred to court.

14.6.5

Paragraph 21(1) provides:

> Where any person is for the time being entitled to require the removal of any of the operator's electronic communications apparatus from any land whether under any enactment or because that apparatus is kept on, under or over that land otherwise than in pursuance of a right binding that person or for any other reason) that person shall not be entitled to enforce the removal of the apparatus except … in accordance with the following provisions of this paragraph …

29 Considered in detail in Chapter 10.

14.6.6

Each of these paragraphs contains a detailed notice and counter-notice procedure that must be followed to the letter – and that must be considered in conjunction with the statutory regime under the Landlord and Tenant Act 1954.

14.6.7

Paragraph 20 of the code permits a landowner to carry out an 'improvement' of its property notwithstanding the presence of telecommunications equipment, and in so doing to require the 'alteration' of that equipment. The word 'improvement' is defined in the same paragraph to include redevelopment, while the word 'alteration' is defined (misleadingly, in a completely different paragraph) to include removal.

14.6.8

Accordingly, supposing the court agrees, paragraph 20 may be used by a landowner to dislodge a sitting operator, in cases where the landowner is seeking to redevelop its property. In cases where the tenancy is one to which Part II of the Landlord and Tenant Act 1954 applies, it is arguable that the landowner will also have to ensure that the procedure for terminating the tenancy that that Act lays down is followed through. Here, however, there is a bizarre divergence between the workings of the two statutes.

14.6.9

The 1954 Act, first, requires the landlord to establish that, upon termination of the tenancy, it intends to demolish the premises or carry out substantial works of construction (etc.) to them. In the ordinary case, the operator will not have been demised anything other than an airspace in which to secure its equipment, and it is therefore difficult to see how the landlord could satisfy this requirement.

14.6.10

By contrast, under paragraph 20 of the code, the court must not make an order for the alteration of the telecommunications equipment unless satisfied about certain requirements.

14.6.11

The position under paragraph 21 of the code is even more obscure.[30] This provision presupposes that, in a case of an agreement providing both contractual security of tenure, and statutory protection under the 1954 Act, both contractual and statutory rights must first be terminated before paragraph 21 can apply. Prior to that termination, the opening words of paragraph 21(1) cannot be satisfied, because the landowner will not, by definition, be 'entitled to require the removal' of the apparatus. The landowner

30 Paragraph 21 is considered in detail in Chapter 11.

would, of course, be in that position only once it has received a determination from the court in its favour on a preliminary application under s.30(1) of the Landlord and Tenant Act 1954 (aside from s.30(1)(f), with which paragraph 20 of the code presumably deals).

14.6.12

If the literal interpretation is correct, then it will do severe damage to plans for a landowner to occupy the premises in question for its own purposes, because it will mean that a notice requiring removal can only be served once the whole of the 1954 Act procedure has been exhausted. At that stage, another round of litigation may well then ensue, if any of the operators gives a counter-notice under paragraph 21(3). The landowner will then have to go to court to secure an order for the removal of the apparatus under paragraph 21(5).

14.6.13

The court will then have to weigh up the ingredients set out in paragraph 21(6). Paraphrasing those various provisions, the court must approach the matter as if the operators were applying for an order under paragraph 5, which imports a wide measure of discretion, in which the maintenance of the electronic communications network and the need for the redevelopment will be paramount.

14.6.14

Although this position may appear draconian in the extreme, it is fair to say that paragraph 21 is not quite so extreme in the operators' favour as is paragraph 20. Paragraph 20(4) introduces three ingredients that must be taken into account by the court in considering whether an alteration to telecommunications apparatus should be made at the suit of the owner of the land on which it is installed.

14.6.15

First, the court must have regard 'to all the circumstances and the principle that no person should unreasonably be denied access to an electronic communications network or to electronic communications services'.

14.6.16

Secondly, the court must be satisfied that the alteration is necessary to enable the owner to carry out the proposed development.

14.6.17

Thirdly, the court must be satisfied that the alteration 'will not substantially interfere with any service which is or is likely to be provided using the operator's network'.

14.6.18

That list of ingredients is surprisingly stark, and appears to weight the scales heavily in favour of the operator and against the owner, even in the case of properties that are ripe for redevelopment.

14.6.19

In the case of paragraph 21, too, the availability of alternative sites within the area will be a matter of the first importance. There may well be considerable difficulty in finding alternative facilities in the area.

14.6.20

It is tempting to speculate that the draftsmen of the code were in ignorance of the provisions of Part II of the Landlord and Tenant Act 1954. With very limited exceptions, none of the provisions of the code referred to above has been analysed in court, and therefore the many problems arising from the interaction between the two statutory regimes has not been resolved in practice.

14.7 The Old Code and the New Code and the Landlord and Tenant Act 1987 Part 1

14.7.1 The 1987 Act

Part 1 of the Landlord and Tenant Act 1987 provides for a right of first refusal to tenants of qualifying premises where the landlord[31] of those qualifying tenants[32] proposes to make a relevant disposal (as defined by the terms of s.4 of the 1987 Act). The qualifying premises are premises which consist of the whole or part of a building, contain two or more flats held by qualifying tenants, and the number of flats held by such tenants exceed 50 per cent of the total number of flats contained in the premises: s.1(2) of the 1987 Act.

14.7.2

The definition of 'relevant disposal'[33] includes a disposal of any estate or interest 'in any common parts' of the qualifying premises.[34] 'Common parts' are defined 'in relation

31 The 'landlord' who qualifies for the purposes of the 1987 Act is set out in s.2. The 'landlord' for the purposes of the 1987 Act is either the immediate landlord of the qualifying tenants (as to which see further below) of the flats contained in the qualifying premises or if those tenants are statutory tenants the person who, apart from the statutory tenancy, would be entitled to possession of the flat in question: s.2(1).

32 In order for the provisions of the 1987 Act to bite there must be at least two or more qualifying tenants: s.1(2)(b). The terms of s.3 of the 1987 Act excludes a tenant who holds under 'an assured tenancy' from being a qualifying tenant. This includes an assured shorthold tenancy which is a form of assured tenancy with a specific statutory right to obtain possession: see Radevsky and Clark, 3rd ed. (2017).

33 In s.4. The section also contains detailed provisions as to what is not a relevant disposal in s.4(2).

34 Section 4(1).

to any building or part of a building' as including 'the structure and exterior of that building or part and any common facilities within it'.[35] The airspace is itself probably to be viewed as part of the 'common parts' of the building (as defined in the 1987 Act): *Dartmouth Court Blackheath Ltd v Berisworth*.[36]

14.7.3

Where the landlord of the qualifying tenants proposes to make a relevant disposal he must first serve notice upon the tenants in accordance with s.5 of the 1987 Act.[37] If he fails to do so the qualifying tenants have the right to serve a purchase notice upon the purchaser of the landlord`s interest so as to acquire the interest disposed of.[38]

14.7.4 The Old Code

The Old Code made no express provision for the terms of the 1987 Act. Accordingly, the grant e.g. of a lease of the roof space to an operator of qualifying premises would be a relevant disposal under the 1987 Act and would be subject to the requirement for the landlord to first serve notice upon the qualifying tenants.

14.7.5 The New Code

By the terms of paragraph 26 of Part 2 of Schedule 3 to the Digital Economy Act 2017, s.4 of the 1987 Act (which makes detailed provision as to what is or is not a relevant disposal) is amended[39] subject that the conferral of a code right under the New Code is not to be treated as a relevant disposal under the 1987 Act.[40]

14.8 The New Code: overview

14.8.1

The New Code is dealt with in detail in the remainder of this part of this book. This section summarises the way in which the New Code treats with property interests, and the way in which it interacts with the 1954 Act (a matter which is covered in greater depth in Chapter 25).

35 Section 60(1).

36 [2008] 2 P & CR 3. The tenants in that case had no rights to go on the roof and that suggested that the airspace was not 'appurtenant'. However, Warren J stated that the airspace, at least to the height of the chimneys 'is an essential part of the space over which any owner of the [block of flats] with repairing obligations would need to have adequate rights of access' and accordingly should be regarded as appurtenant to the building if not actually part of it. 'If that is wrong' he said, 'I would conclude that the airspace above the roof to that height is a "common part" being part of the exterior of the building'. In the earlier decision of *Savva v Galway Cooper* [2005] 3 EGLR 40, CA the grant of a lease of roof space was conceded to be a relevant disposal.

37 There are four forms of notice contained in ss.5A–5D.

38 In accordance with ss.12A–12C.

39 By inserting a new s.4(2)(db).

40 This is in accordance with the recommendation in paragraph 2.74 of Law Com Report No. 336, in which the Law Commission set out their view that the conferral of code rights should not be a relevant disposal for the purposes of Part I of the 1987 Act.

14.8.2

Unlike the Old Code, which bolted its bespoke system of statutory protection on to property agreements which, if leases, were already protected in many cases by the 1954 Act, the New Code applies to any agreement which satisfies the requirements laid down (as to which, see Chapter 19). Such an agreement need not be for a fixed term, provided that it states (i) for how long the code right is exercisable, and (ii) the period of notice (if any) required to terminate the agreement.[41] It is clearly therefore an agreement which need not be a conventional lease, and which enjoys its own statutory protection. The parties are free to make their agreement a lease if they wish, but they need not.

14.8.3

The important features of the New Code which distinguish it from the Old, for these purposes, are as follows.

14.8.4

First, as Chapter 25 explains, a lease which is granted after the coming into force of the New Code cannot be one to which the 1954 Act applies, if the primary purpose of it is to grant code rights within the meaning of the New Code: see s.43(4) of the 1954 Act, inserted by Schedule 3 paragraph 4 to the 2017 Act. The important consequence of this drafting is that the problem with the interaction between the Old Code and the 1954 Act highlighted in section 14.6 above should largely disappear.

14.8.5

Secondly, and as against that, although the problems associated with dual statutory protection may be over (save in transitional cases where the Old Code continue to apply), the termination procedures under the New Code are lengthened, with 18-month notice periods for termination.

14.8.6

Thirdly, a landowner seeking to have a code agreement terminated must use a procedure similar in many respects to the termination procedure for business tenancies under the 1954 Act, albeit via the Lands Chamber of the Upper Tribunal rather than the county court.

14.8.7

Fourthly, even where a code agreement has been terminated, a landowner seeking to remove the operator's apparatus may have to reapply to the tribunal for that purpose. The combined effect of these provisions will therefore be that lead-in times for developers will be substantial, even if the procedure has been greatly clarified.

41 Paragraph 11 of the New Code. See Chapter 19.

15 What are code rights?

15.1 Introduction

15.1.1

The New Code sets out a package of the rights that an occupier of land can agree to confer on an operator, or that in default of such agreement can be imposed by a court under Part 4 of the code.

15.1.2

These code rights are clearly defined in the New Code for the most part, although the drafting does prompt a number of questions, to which this chapter seeks to suggest answers.

15.2 'Code rights': the definition

15.2.1

Paragraph 3 of the New Code provides:

> For the purposes of this code a 'code right', in relation to an operator and any land, is a right for the statutory purposes –
> (a) to install electronic communications apparatus on, under or over the land,
> (b) to keep installed electronic communications apparatus which is on, under or over the land,
> (c) to inspect, maintain, adjust, alter, repair, upgrade or operate electronic communications apparatus which is on, under or over the land,
> (d) to carry out any works on the land for or in connection with the installation of electronic communications apparatus on, under or over the land or elsewhere,
> (e) to carry out any works on the land for or in connection with the maintenance, adjustment, alteration, repair, upgrading or operation of electronic communications apparatus which is on, under or over the land or elsewhere,
> (f) to enter the land to inspect, maintain, adjust, alter, repair, upgrade or operate any electronic communications apparatus which is on, under or over the land or elsewhere,
> (g) to connect to a power supply,

(h) to interfere with or obstruct a means of access to or from the land (whether or not any electronic communications apparatus is on, under or over the land), or
(i) to lop or cut back, or require another person to lop or cut back, any tree or other vegetation that interferes or will or may interfere with electronic communications apparatus.

15.2.2

A number of the terms used in this list of code rights are separately defined elsewhere, viz:

'*operator*' is defined in paragraph 2 of the New Code, and is discussed in Chapter 16 of this book;
'*land*' is negatively defined by paragraph 108(1) of the New Code to mean that it 'does not include electronic communications apparatus';
'*the statutory purposes*' are defined in paragraph 4 of the New Code, and are explained in section 15.4 below;
'*electronic communications apparatus*' is defined in paragraph 5 of the New Code, and is discussed in Chapter 18 of this book.

15.2.3

By contrast, the following terms are not defined, and were presumably considered by the draftsman to be so obvious as to need no explanation:

'works';
'power supply'.

One might add that the unhelpful definition of 'land' (see paragraph 15.2.2 above) leaves the reader with the task of deciding what the draftsman has meant by that expression. This topic is discussed in Chapter 18.

15.2.4

The next two sections in this chapter devote some time to examining the scope of and purpose of the code rights, while the following section elaborates on a particular problem that may arise in practice when considering that scope.

15.3 Scope of the code rights

15.3.1

Paragraph 3 of the New Code carefully, and in some detail, describes the code rights by reference to what they include. Like all very specific drafting, this approach makes it easier to see what has been omitted, whereas more general drafting would have provided room for manoeuvre. The following examples in this section make the point.

15.3.2

First, the various verbs used in paragraph 3 may be said not merely to be inclusive, but also (by their sheer variety) to be exhaustive. For example, those arguing in favour of the proposition that the verb 'upgrade' must include compete replacement (as to which, see paragraph 15.3.4 below) will no doubt be met with the observation that the draftsmen clearly thought carefully about what to include as legitimate works, and that there is therefore no room for any further implication. The Law Commission were alert to this drafting conundrum. Responding in para 2.39 of Law Com 336 to the suggestion that the words 'the installation, maintenance, adjustment, repair or alteration' be removed, on the basis that they generate 'sterile argument about definitions', they said:

> Arguably a right 'to execute works on land in connection with electronic communications apparatus' includes those additional verbs. We agree; but of course in this Report we are not drafting the revised Code. Our recommendation below sets out our policy; the drafter of the revised Code will wish to bear in mind our suggestion that the list be kept as simple and inclusive as possible.

The drafter of the code evidently chose to proceed down the all-inclusive route.

15.3.3

Secondly, it might be thought that code right (b) (to keep installed) is implicit in code right (a) (to install). In adding (b) to (a), the draftsmen have stated the obvious. In so doing, they have followed the approach of approach taken by the Law Commission in their 2013 Report on the Electronic Communications Code (Law Com 336). In para 2.33 of that report, the authors propose the addition of code right (b) 'for the avoidance of doubt'. The inclusive nature of this approach makes it difficult to argue in favour of further implicit rights (such as a right to carry out a survey).

15.3.4

Thirdly, code right (c), with its list of permitted works to electronic communications apparatus, may be said to be limited to works to existing apparatus – but not its replacement by new apparatus altogether. The argument relies upon the facts that (a) the works verbs all implicitly refer to something that is already there, without which the verbs would have no subject matter upon which to operate; and (b) works of installation of apparatus are dealt with in code right (a). This argument is premised upon the code rights being severally available, rather than available only as a package – see section 15.5 below. There is a counter-argument, which requires the verb 'upgrade' to be read in the sense of allowing replacement of the existing apparatus with new apparatus. However, although this approach is clearly right insofar as it would allow replacement of old with new in relation to *parts* of the apparatus (by analogy with the position in relation to repairing covenants, where patch repair is often within the covenant, whereas wholesale replacement often is not[1]), it may be said to be doubtful whether 'upgrading' extends to replacement of the whole apparatus with a new installation.

1 As to the concept of 'repair' comprising the replacement of subsidiary parts, see *Lurcott v Wakeley* [1911] 1 KB 905, where Buckley LJ said at 923:

15.3.5

Thirdly, whereas the drafting refers repeatedly to the code rights being exercisable 'on, under or over the land' code right (d) restricts itself to works 'on the land'. A literal reading of this right might be taken to suggest that an operator cannot carry out works *under* or *over* the land. This reading is almost surely wrong: code rights (a) and (c) quite clearly refer to works of installation and repair (and so forth) to apparatus 'on, under or over the land', from which it logically follows that works under or over the land may be necessary in order to exercise such rights. The omission of similar drafting in relation to code right (d) is curious, but the remainder of the drafting of the code right makes it clear that the work is in connection with installation of the electronic communications apparatus 'on, under or over the land', the facilitation of which would surely require works on, under or over the land. Perhaps the draftsman considered that, in this connection, the repetition of the qualifying words at the beginning of the right would be superfluous.

15.3.6

Fourthly, code right (g) restricts itself to connection to 'a power supply'. This expression is normally used in connection with supplies of electricity or gas, which may be used to provide or generate energy in connection with the use or installation or other works to the electronic communications apparatus. It would not, however, appear to include the supply of water, which might be necessary in order to perform a cooling function.

15.3.7

Fifthly, code right (h) is expressly concerned only with means of access. It is thought that this means the physical ability to gain access on foot or with vehicles or other means of transport. Code right (h) therefore allows the operator to interfere with or obstruct such access – for example by blocking a drive. Nothing is said about other rights the land might enjoy – such as a right of light. While it would be inconceivable for a mast to interfere with the access of light to such an extent as to constitute a nuisance, it is possible that a building housing electronic communications apparatus might do so. Code right (h) does not afford an operator a right to interfere with such a right.

15.3.8

Sixthly, the terms of paragraph 3(h) do not refer to interfering with electronically generated signals. But presumably it was thought that there was no need to do so given the decision in *Hunter v Canary Wharf Ltd*[2], where the House of Lords was concerned with a claim for interference with reception of television broadcasts in private homes caused by the construction of a building erected by the defendant in London Docklands

'Repair always involves renewal; renewal of a part; of a subordinate part … repair is restoration by renewal or replacement of subsidiary parts of a whole. Renewal, as distinguished from repair, is reconstruction of the entirety, meaning by the entirety not necessarily the whole but substantially the whole subject-matter under discussion.'

2 [1997] AC 655.

pursuant to a planning permission which had been granted. It was held that a right to receive a signal was not capable of existing at law as an easement. As Lord Lloyd said at 699:

> The annoyance caused by the erection of Canary Wharf and the consequential interference with television reception must have been very considerable. But unfortunately the law does not always afford a remedy for every annoyance, however great. The house-owner who has a fine view of the South Downs may find that his neighbour has built so as to obscure his view. But there is no redress, unless, perchance, the neighbour's land was subject to a restrictive covenant in the house-owner's favour. It would be a good example of what in law is called 'damnum absque injuria:' a loss which the house-owner has undoubtedly suffered, but which gives rise to no infringement of his legal rights. In the absence of a restrictive covenant, there is no legal right to a view. The analogy between a building which interferes with a view and a building which interferes with television reception seems to me, as it did to the Court of Appeal, to be very close.

15.3.9

Seventhly, code right (i) gives the operator rights to deal with vegetation that interferes or that may or will interfere with electronic communications apparatus.[3] Nothing is said about physical structures that may interfere, such as, for example, (flags on) flagpoles, sheds or cranes.

15.3.10

For further discussion of the nature and content of individual code rights, please see sections 42.7–42.17.

15.4 Purpose of the code rights

15.4.1

The New Code does not simply enable the conferment upon the operators of the unfettered code rights: paragraph 3 of the New Code stipulates that the code rights are 'for the statutory purposes'.

15.4.2

These 'statutory purposes' are defined by paragraph 4 of the New Code as follows:

> In this code 'the statutory purposes', in relation to an operator, means –
> (a) the purposes of providing the operator's network, or
> (b) the purposes of providing an infrastructure system.

3 Note this right is not limited to trees overhanging highways which is dealt with in Part 13 of the New Code and considered in detail in Chapter 29.

15.4.3

The expressions italicised above are defined further as follows. First, paragraph 6 provides:

> 'network' in relation to an operator means if the operator falls within paragraph 2(a), so much of any electronic communications network or infrastructure system provided by the operator as is not excluded from the application of the code under section 106(5).

15.4.4

'Network' has already been considered in Chapter 6 under the Old Code, but a critical difference has to be noted. Pursuant to the amendments made by the 2017 Act, the phrase 'system of conduits' in s.106(4) of the 2003 Act has been altered to mean 'infrastructure system'. The express intention behind this change was that it was considered that provision of infrastructure which was not conduits fell between the two stools of s.106 as originally drafted.[4]

15.4.5

To remove doubt, the Law Commission recommended that the definition should be expanded to cover infrastructure providers.[5] It therefore follows that the provision of masts, equipment cabins or other electronic communications apparatus which has no active use and which will only be activated by a third party operator will nevertheless be 'infrastructure' falling within the statutory purposes of an operator. It is to be noted that passive apparatus will also be 'electronic communications apparatus' within the definition of that term, and exempted from code rights as a result (see below section 15.7).

15.4.6

It is also to be noted that there is no separate definition of 'infrastructure' under the New Code. The Law Commission report[6] gives a clue as to what the legislator might have had in mind when considering 'infrastructure':

> [w]here a wholesale infrastructure provider holds a lease of land, installs infrastructure upon it, and then confers contractual rights upon mobile operators to install antennae on the mast. The terms of the provider's lease enable (explicitly or otherwise) the installation of electronic communications apparatus

15.4.7

The above passage suggests that the paradigmatic infrastructure provider is an entity which has constructed network-related equipment on land. But what of the land itself?

4 Though see the discussion in Chapter 6 as to the arguments to the contrary.
5 Law Com 336 at paragraph 2.28.
6 Law Com 336 at paragraph 2.26.

There is nothing that expressly excludes land from infrastructure, save that, as a matter of English language, a distinction is sometimes drawn between the two. That said, 'land banking' is an essential part of the activities of an infrastructure provider. It is arguable that this is catered for by a further definition, namely that contained in paragraph 7.

15.4.8

Paragraph 7 adds:

(1) 'infrastructure system' means a system of infrastructure provided so as to be available for use by providers of electronic communications networks for the purposes of the provision by them of their networks.
(2) References in this code to provision of an infrastructure system include references to establishing or maintaining such a system.

15.4.9

In relation to that paragraph:

a. Although the definition in 7(1) is circular, it does make the point that infrastructure is protected even if intended for third party use (and there is no requirement that those users be operators).
b. Paragraph 7(2) states that provision of an infrastructure system includes the establishment of such a system. It is arguable that this means that a bare site, not yet developed with apparatus, could still be covered by the statutory purposes as amended by the 2017 reforms, and that the retention of bare land for that intended purpose will be a protected operator function.

15.4.10

In light of the intentions behind the reforms, and the range of services provided by Wholesale Infrastructure Providers, for whom land banking can be an essential part of their business, it is considered that OFCOM or the Courts ought to take a pragmatic view of the above issues. There is nothing in the New Code that suggests that the retention of a bare site with a view to making it up into a mast site should not be a permissible operator function.

15.4.11

The above issues are simply about whether or not 'land banking' is within an operator's statutory purposes. There is a separate, and more important, question as to the point at which bare land is so adapted as to contain 'electronic communications apparatus' over which code rights cannot be acquired. This is considered below, at section 15.7.

15.5 Are the code rights a composite package?

15.5.1

Logic suggests that the code rights together form the body of rights that an operator needs in connection with the occupation of land for the statutory purposes, and that the grant of the rights should mean all of them, as a package. For example, the right to install (code right (a)) is likely to be ineffective unless it is accompanied by the rights to repair (code right (c)) and to supply with power (code right (g)).

15.5.2

A closer reading of the text of the New Code would suggest, however, that logic is not the correct guide in this instance. First, the drafting displays some singularity – see:

(a) The reference in paragraph 3 of the New Code to 'a' code right;
(b) The consistent references elsewhere in the text to '*a* code right' – see especially paragraph 9, which is headed 'Who may confer code rights?', but which then refers to 'A code right in respect of land …';
(c) The word 'or', rather than 'and', separating the penultimate and last rights, which suggests a shopping list of individual and several rights, as opposed to a compendious package;
(d) Paragraph 13(1), which refers to an operator having 'a code right within paragraph (a) to (g) or (i) of paragraph 3', which is consistent only with code rights being several;
(e) Paragraph 13(2), which refers to an occupier of other land having conferred or being bound by a code right within paragraph (h) of paragraph 3, which suggests that the rights are capable of being conferred individually;
(f) Paragraph 34(4), which commences, 'Where under the code agreement more than one code right is conferred by or otherwise binds the site provider';
(g) Paragraph 37, which is equally consistent with several rather than compendious rights.
(h) Furthermore, the termination and removal provisions are consistent only with the existence of several rights. For example, paragraph 32(3) refers to the operator wanting 'a new code right in place of the existing code right'.

15.5.3

Secondly, it is easy to see why an occupier would be prepared to agree only to some code rights but not others. For example:

(a) An occupier might be prepared only to grant a right to install specific electronic communications apparatus, and to maintain and upgrade it – but not the right to install entirely new apparatus; or
(b) An occupier might be unwilling to allow any interference with its means of access.

While operators might be supposed to want the full panoply of code rights, they might be prepared to settle for less, if by doing so they acquire in substance all that they reasonably need. Further, the consideration for a lesser package may also be reduced commensurately. It is also right to recall that operators may acquire less than the whole panoply of code rights, and then seek later to acquire more compulsorily.[7]

7 This is expressly catered for in Part 4 of the New Code.

15.5.4

Thirdly, the several interpretation is supported by the approach taken by the Law Commission in Law Com 336. First, albeit by reference to the Old Code, the authors commented in paragraph 2.15:

> the fact that an operator has one item on the list does not give that operator any of the others; a right to keep equipment on land does not give the operator the right to enter the land to inspect or repair it – although of course the agreement conferring the right to install equipment may, and generally will, address the issue of access.

15.5.5

Next, in para 4.44 of Law Com 336, the authors say:

> the tribunal's order may be for the grant of one Code Right or a number of Code Rights, depending upon the nature of the Code Operator's requirements.

15.5.6

The point may rightly be made that the draftsman of the New Code had the opportunity, against the background of these comments, to make it plain that the code rights are several, but did not take that opportunity. However, the point cuts both ways, and it would have been simple to provide that the code rights were to be conferred on an all or nothing basis.

15.5.7

The arguments in favour of code rights being severally available are therefore substantial. This may turn out to be a source of contention, both at the point of negotiation of rights, and subsequently. A well-advised operator may well wish to acquire the whole package of rights, whereas the owner may wish to confine the operator to those which are strictly needed at the time of the agreement.

15.5.8

A possible compromise for the parties will be for the owner to restrict the operator to a specified list of code rights which are essential for its operation, and to agree the grant of further rights (but not code rights) purely as matter of contract, and not for the statutory purposes. This expedient will mean at least that the operator will not gain the statutory protection afforded by the New Code, while enabling it to argue its case for more extensive code rights in the future.

15.6 Exercise of code rights

15.6.1

The code rights are drafted in clear language, and require little elucidation, beyond:

(a) installation, which is dealt with below;
(b) upgrading, which is dealt with in Chapter 20;
(c) tree lopping, which is dealt with in Chapter 29.

15.6.2

Paragraph 3(c) of the New Code refers to a general right for the statutory purposes 'to install electronic communications apparatus'. That right cannot sensibly be read as an open right to install any, and as much apparatus as the operator likes, subject to other public policy considerations. Any site provider would baulk at the notion that, having been granted that code right, the operator might then install anything from a small cabinet and antenna to a 30 metre mast. In practice, this code right is likely to be interpreted as a right to install *specified* apparatus, with the apparatus having to be specified with a reasonable amount of detail in the operator's approach to the site provider. The operator will then of course have the flexibility to seek to upgrade its apparatus pursuant to paragraph 17 of the code, or alternatively to seek more extended rights subsequently.

15.7 Code rights and apparatus only agreements

15.7.1

The New Code applies only to code rights (elaborated upon as discussed above in relation to paragraph 3 of the New Code). Those rights must be 'in relation to an operator and any land': paragraph 3. If the rights do not relate to land but only to electronic communications apparatus ('ECA'), one does not have a code agreement, because land is defined so as to exclude ECA: paragraph 108(1). ECA is elaborated upon in paragraph 5 of the New Code, which is discussed in detail in Chapter 18.

15.7.2

Accordingly, the mere right to use something which qualifies as ECA exempts it from code rights, whether or not it was installed or utilised by an operator. As there are no code rights conferred by the grant of a right simply to use ECA, any ECA-only agreement will either be governed by the general law of property or may be a protected 1954 Act tenancy of the ECA.[8]

15.7.3

It may that some of the rights conferred under the relevant agreement (e.g. a right to access the apparatus, a right to house an equipment cabin on bare land), if exercisable over *land*, are code rights. Other rights, such as the right to affix an antenna on a mast which has been installed by another operator (or indeed anyone else), is not a code

8 Although even here it must be questioned whether the operator can be said to have 'premises' within s.23(1) of the 1954 Act. The ECA may be a fixture which would be land as a matter of general property law, and if demised by the landowner would probably be sufficient to constitute 'premises' under the 1954 Act. The general rule is that which constitutes a fixture becomes part of the realty: *Melluish v BMI (No.3) Ltd* [1996] AC 454, HL.

right, as it is a right over ECA and not land, so long as it can be said that the mast is e.g. 'apparatus … designed or adapted' or 'a structure or thing designed or adapted' for use in the provision of an electronic communications network within paragraph 5 of the New Code. This result may seem odd but appears to us to be consistent with the policy of allowing wholesale infrastructure providers New Code protection, and preventing operators from obtaining code rights as against the infrastructure provider.

15.7.4

ECA is defined to include 'structures or things designed or adapted for use in connection with the provision of an electronic communications network': para 5(1)(d). Thus conferring a right to use an equipment cabin may simply be a right to use ECA not land. However, even if this is the case one would expect there to be ancillary rights of access e.g. a right to access the landowner`s land to e.g. inspect and maintain the cabin. Such a right is a code right: paragraph 3(c).

15.7.5

This dichotomy between 'land' and 'ECA' may lead to a market whereby landowners agree only to enter into two forms of agreement, First, one limited to the right to ECA which will not have code protection; and the second which relates to the ancillary rights with respect to land (e.g. the right to inspect the ECA) which will be code protected. The practical advantage of this double agreement arrangement is that the ECA agreement can be determined, free of code protection. The 'ancillary land rights' agreement is code protected but it matters not, if the landowner has the simple expedient of removing the ECA by terminating the unprotected ECA-only agreement. Once the ECA has been removed the 'ancillary land rights' agreement may itself lose code protection as the rights are, once the ECA only agreement has been terminated, no longer being used for the statutory purpose as required by paragraph 3 of the New Code.

16 Who may confer code rights?

16.1 Introduction

16.1.1

Paragraph 9 of the New Code is headed 'Who May Confer Code Rights?', and provides:

> A code right in respect of land may only be conferred on an operator by an agreement between the *occupier of the land* and the operator.

16.1.2

This drafting is interesting and important. It is critical to note (as this chapter makes clear) that the grantor of code rights:

(a) need have no legal or other interest in the land – and indeed may be a squatter;
(b) need not have the landowner's permission to grant code rights; and
(c) (arguably) need not even be in exclusive occupation of the land.

16.1.3

This chapter is concerned with the following issues that are prompted by that drafting, and in particular with:

(a) the identity of the *occupier of the land*;
(b) what happens where the occupier of the land is not the ultimate owner of the land;
(c) what happens where the occupier of the land has no property interest in the land;
(d) the categories of occupier;
(e) who may become bound by code rights;
(f) the practical effect of the occupier granting code rights when it has no right to do so.

16.2 Meaning of 'the occupier of the land'

16.2.1

Paragraph 105 of the New Code, which is headed 'Meaning of "occupier"' provides the following comprehensive detail in relation to the identity of the occupier of land:

> (1) References in this code to an occupier of land are to the occupier of the land for the time being.

(2) References in this code to an occupier of land, in relation to a footpath or bridleway that crosses and forms part of agricultural land, are to the occupier of that agricultural land.

(3) Sub-paragraph (4) applies in relation to land which is –
 (a) a street in England and Wales or Northern Ireland, other than a footpath or bridleway within sub-paragraph (2), or
 (b) a road in Scotland, other than such a footpath or bridleway.

(4) References in this code to an occupier of land –
 (a) in relation to such a street in England and Wales, are to the street managers within the meaning of Part 3 of the New Roads and Street Works Act 1991,
 (b) in relation to such a street in Northern Ireland, are to the street managers within the meaning of the Street Works (Northern Ireland) Order 1995 (SI 1995/3210 (NI 19)), and
 (c) in relation to such a road in Scotland, are to the road managers within the meaning of Part 4 of the New Roads and Street Works Act 1991.

(5) Sub-paragraph (6) applies in relation to land which –
 (a) is unoccupied, and
 (b) is not a street in England and Wales or Northern Ireland or a road in Scotland.

(6) References in this code to an occupier of land, in relation to land within sub-paragraph (5), are to –
 (a) the person (if any) who for the time being exercises powers of management or control over the land, or
 (b) if there is no person within paragraph (a), to every person whose interest in the land would be prejudicially affected by the exercise of a code right in relation to the land.

(7) In this paragraph –
 (a) 'agricultural land' includes land which is being brought into use for agriculture, and
 (b) references in relation to England and Wales to a footpath or bridleway include a restricted byway.

16.2.2

The term 'occupier' itself is not actually defined. In *Newnham College v Revenue and Customs Commissioners*[1], Lord Hoffmann noted in that the cognate expression 'occupation' is 'a word which can mean different things in different contexts', citing the remark of Viscount Cave in *Madrassa Anjuman Islamia v Johannesburg Municipal Council*[2] that it is 'a word of uncertain meaning'. Lord Neuberger added:

> Occupation can have a variety of different meanings, some of which can vary quite subtly from others, and its precise meaning in any case is thus particularly prone to be governed by its context.

16.2.3

We are therefore left with recourse to common sense and the general law. Common sense provides an uncertain guide, suggesting that the question who is the occupier of

1 [2008] 1 WLR 888.
2 [1922] 1 AC 500.

land is simply the person who is found on the land from time to time. Such a definition would however include the postman delivering letters, a builder carrying out repairs and other invitees. The Authors suggest that a surer guide is provided by analogy with the approach encountered in other areas of statutory control: land registration, taxation and rating.

16.2.4

In relation to land registration, it is possible for the interests of those in 'actual occupation' of registered land to bind others as overriding interests. In this context, the expression 'actual occupation' has a settled meaning. In *Baker v Craggs*[3], Newey J analysed the relevant authorities, concluding (a) that what was required to constitute actual occupation was physical presence, not merely some entitlement in law; (b) that occupation is to be distinguished from mere use; (c) that the nature of the relevant property can be significant; (d) that occupation was a concept that may have different connotations according to the nature and purpose of the property; and (e) that it was necessary that there be some degree of permanence and continuity and more than a fleeting presence. It should be noted, however, that before the coming into force of the Land Registration Act 2002, 'actual occupation' included the right to receive rents and profits. Since that Act, however, actual occupation no longer includes a right to receive rents and profits. In the context of the New Code, the latter is the more applicable definition.

16.2.5

Secondly, in the field of taxation, the term 'occupation' has been given a similar meaning. In *Newnham College v Revenue and Customs Comrs*[4], the House of Lords had to consider whether the College was in occupation of its college library for the purposes of the Value Added Tax Act 1994, by virtue of its control of the staff who worked there, as well as the access to it. The House held, by a majority, that the term 'occupation' bore what they called its 'ordinary and well understood meaning', which they expressed to be a sufficient degree of possession and control of the land that gave the occupier the right to occupy it as if he were the owner, and to exclude any other person (by which their Lordships meant any person other than the actual owner) from enjoyment of that right. In so expressing themselves, the whole House held that the words 'in occupation of the land' should be interpreted in accordance with the principles laid down by the European Court of Justice[5] and the majority of the House of Lords in *Customs and Excise Comrs v Sinclair Collis Ltd*[6]. In that case, Lord Scott said:

> So what are the characteristics that distinguish a licence to occupy from a mere licence to use? There are, in my opinion, two characteristics, one or other of which must, in some sufficient degree, be present. One is possession. The other is control. If neither is present, I find it difficult to understand how the licensee could be said to occupy.

3 [2017] 2 WLR 1483.
4 [2008] 1 WLR 888.
5 [2003] ECR I-5965.
6 [2001] STC 989.

16.2.6

In the *Newnham* case, Lord Hope, agreeing with Lord Hoffmann (and with whom Lord Mance agreed) said:

> I do not think that there is much doubt about what the word occupation means, although it may be more difficult to apply its ordinary meaning to the facts in some contexts than it is in others. In its ordinary meaning it requires more than just a right to use the land or to enjoy the facilities that are to be found there. Physical presence is an essential element. But there is more to it than that. It requires actual possession of the land, and the possession must have some degree of permanence.

Dissenting in the result, but agreeing with the test, Lord Walker, with whom Lord Neuberger agreed, said:

> [Occupation] is in general taken to import an element of physical presence and some element of control over the presence of others. But these generalities are strongly influenced by the statutory context and purpose. ... [Its] meaning is also strongly influenced by the nature of the premises in question. Acts which amount to occupation of a stretch of entirely undeveloped land ... are obviously of a different quality from those which amount to occupation of an inner-city flat. Moreover some premises are designed for particular specialised purposes (unmanned bank premises were mentioned in argument, to which one might add unmanned launderettes and indeed unmanned public lavatories). Such premises may be physically occupied, to all outward appearances, only by the customers or members of the public, and the element of control may be correspondingly tenuous.

Lord Neuberger added:

> occupation in the present context need not be exclusive. Indeed, that is implicit in what Lord Scott of Foscote said, in para 79, in the *Sinclair Collis* case: 'it seems to me unnatural to treat the room in which the vending machine is installed as being partly occupied by the owner and partly occupied by the company. In common sense and commercial terms the owner remains in occupation of the whole of the room.'

16.2.7

Thirdly, in the context of rating, rateable occupation is normally made up of four ingredients:

(a) there must be actual occupation;
(b) it must be exclusive for the particular purposes of the occupier;
(c) it must be of some value or benefit to the occupier; and
(d) it must not be for too transient a period.

For authorities supporting this approach, see the decisions of, respectively, the Court of Appeal and the House of Lords, in *John Laing & Sons Ltd v Kingswood Assessment Committee*[7]; *London County Council v Wilkins*[8]; and see more generally Chapter 37.

16.2.8

Where the same unit of property is occupied concurrently by more than one occupier, each occupying for a purpose of their own, the person in rateable occupation is identified by asking which of them is in 'general control' so that their position in relation to occupation may be seen as 'paramount' and that of the other occupier as 'subordinate'. The proper approach was described by Lord Russell in *Westminster Council v Southern Railway Company*[9]:

> The general principle applicable to the cases where persons occupy parts of a larger hereditament seems to be that if the owner of the hereditament (being also in occupation by himself or his servants) retains to himself general control over the occupied parts, the owner will be treated as being in rateable occupation; if he retains to himself no control, the occupiers of the various parts will be treated as in rateable occupation of those parts.

16.2.9

The statutory treatment summarised above, all of which is said to be based upon the 'ordinary meaning' of the term 'occupation', does not entirely coincide – for the reason that the term takes its colour from its statutory context and purpose. The Authors consider that the elements which can conveniently be derived from the other statutory treatment for the purposes of paragraph 105 of the New Code, are that the occupier is that person who has a physical presence on the land, whether in person, or through possessions, and whether exclusively, or in company with others; who exercises a measure of control; and who is there for a non-trivial period.

16.2.10

One possible difficulty may arise where the land is occupied by an employee of the landowner, on terms that the occupation is required for the better performance of the employment. Examples would include a school caretaker or a concierge. The law has generally regarded the occupation of such an employee as representative occupation on behalf of the employer, such that the employer is regarded as occupying the property vicariously through the employee (see *Commissioner of Valuation for Northern Ireland v Fermanagh Protestant Board of Education*[10]; *Street v Mountford*[11]). The New Code does not provide any guidance to resolve this difficulty.

7 [1949] 1 KB 344.
8 [1957] AC 362.
9 [1936] AC 511.
10 [1969] 1 WLR 1708.
11 [1985] AC 809.

16.3 The categories of land occupier

16.3.1

The drafting of paragraph 105 of the New Code effectively divides land for the purposes of considering its occupation into three categories:

(a) The first is land that is conventionally occupied in the sense considered in section 16.2 above – i.e. land where someone is physically present. The draftsman has been careful to include in this category agricultural land that is crossed by a footpath or bridleway (adopting for this purpose the same drafting device used in paragraph 2(8)(a)(i) of the Old Code). It is not clear why the draftsman has taken this trouble, because under the general law the user of any right of way is not considered to be in occupation of the servient tenement. Neither is it clear why the draftsman has only singled out agricultural land for this treatment.
(b) The second category is land which is a street in England and Wales or Northern Ireland (other than such a footpath or bridleway over agricultural land), or a road in Scotland. The occupier in these cases is taken to be the relevant street or road manager.
(c) The third category is land which is neither occupied, nor a street in England and Wales or Northern Ireland or a road in Scotland. In such cases, the 'occupier' for the purposes of paragraph 105 is taken to be either (i) the person (if any) who for the time being exercises powers of management or control over the land; or failing that, (ii) every person whose interest in the land would be prejudicially affected by the exercise of a code right in relation to the land.

16.3.2

These definitions are tolerably clear, although it will be noted that in a category (c) case (i.e. non-street or road, where nobody is in physical occupation, and nobody exercises management or control), it is possible that more than one person may qualify for consideration as the occupier.

16.4 What happens where the occupier of the land is not the ultimate owner of the land?

16.4.1

It follows from the drafting of the New Code, and in particular the consideration of who might be 'the occupier' examined in section 16.2 above, that the occupier could be any of the following:

(a) The freehold proprietor, with no encumbrances;
(b) A leasehold proprietor, with an interest in possession;
(c) The holder of a beneficial interest with occupation rights under the Trusts of Land and Appointment of Trustees Act 1996;
(d) A person with a licence to occupy the land; or
(e) A person who is trespassing upon the land, but in a non-transient way (i.e. a squatter).

16.4.2

Each of these categories of person is capable of being an occupier. It is important to note that persons in categories (c), (d) and (e) have no property interest in the land over which code rights are proposed to be conferred. This does not matter (although it is of course relevant – see section 16.5 below), because code rights are not property interests, but a *sui generis* statutory creation. The New Code neither requires nor enables code rights to be carved out of, or sustained by, a property interest.

16.5 What happens where the occupier of the land has no property interest in the land?

16.5.1

Put simply, where the occupier of land who confers code rights is not the ultimate owner of the land, the rights do not bind those who do own the land. To take an obvious example, if an occupational underlessee grants code rights, neither its landlord nor the freeholder will be bound by those rights.

16.5.2

This does not mean that the owners of superior interests *cannot* become bound. Paragraph 10(4) of the New Code provides:

> The code right also binds any other person with an interest in the land who has agreed to be bound by it.

16.5.3

Accordingly, in the example given in paragraph 16.5.1 above, if either the landlord or the freeholder (or both) agree to be bound by the code rights, then they will become bound as if they had been a party to the agreement conferring the rights.

16.5.4

Moreover, if any such person becomes bound, its successor in title and the holders of any derivative interest created after the code agreement will also be bound, by virtue of paragraph 10(5), which provides:

> If such a person ('P') agrees to be bound by the code right, the code right also binds –
> (a) the successors in title to P's interest,
> (b) a person with an interest in the land that is created after P agrees to be bound and is derived (directly or indirectly) out of –
> (i) P's interest, or
> (ii) the interest of a successor in title to P's interest, and
> (c) any other person at any time in occupation of the land whose right to occupation was granted by –

(i) P, at a time when P was bound by the code right, or
(ii) a person within paragraph (a) or (b).

16.5.5

The operation of this provision may be illustrated by the following example:

(a) A tenant, T, occupies land;
(b) T confers code rights on an operator;
(c) T's landlord, L, is *not* automatically bound by the code rights;
(d) However, L may agree to be bound, and if it does so, then not merely L, but also the following will become bound in turn:
  (i) L's assignee; and
  (ii) Any person, New T, to whom L relets the land following the expiry of T's tenancy, as well as New T's own subtenants; and
  (iii) New T's assignee (and so on).

16.5.6

This drafting does not, of course, catch owners of *inferior* interests *existing at the time of the code agreement* (but note section 16.6 below for the position regarding inferior interests that are *subsequently* created by the occupier). Suppose, to take the same example, the occupational underlessee has sublet part of the land to T. Although T is 'any other person with an interest in the land', it will occupy in its own right, and the underlessee is not therefore the occupier in relation to that part of the land, and will not therefore be entitled to grant code rights over it (although T itself of course may).

16.6 Who else is bound by code rights?

16.6.1

Paragraph 10 of the New Code is headed 'Who else is bound by code rights?', and provides (so far as relevant):

(1) This paragraph applies if, in accordance with this Part, a code right is conferred on an operator in respect of land by a person ('O') who is the occupier of the land when the code right is conferred.
(2) If O has an interest in the land when the code right is conferred, the code right also binds –
  (a) the successors in title to that interest,
  (b) a person with an interest in the land that is created after the right is conferred and is derived (directly or indirectly) out of –
    (i) O's interest, or
    (ii) the interest of a successor in title to O's interest, and
  (c) any other person at any time in occupation of the land whose right to occupation was granted by –
    (i) O, at a time when O was bound by the code right, or
    (ii) a person within paragraph (a) or (b).

(3) A successor in title who is bound by a code right by virtue of sub-paragraph (2)(a) is to be treated as a party to the agreement by which O conferred the right.

16.6.2

This drafting provides for the occupier who has granted code rights to bind others whose interest in, or right to occupy, the land is subsequently derived from that occupier. It is worth noting three limitations to this.

16.6.3

First, although paragraph 10(1) would appear to suggest that the occupier can bind all those to whom it assigns its occupational right, this ability to bind successors and others applies only where the occupier has an interest in land. So, if the occupier is a mere licensee or squatter, paragraph 10 will not apply. This is consistent with the general law, whereby the burdens taken on by a licensee cannot be passed to a successor in title in any event.

16.6.4

Secondly, this ability to bind applies only to third parties who have gained their interest *after* the code rights have been created. So, if O shares space with a third party, Q, and grants code rights to an operator, Q will not be bound by those rights.

16.6.5

Thirdly (and although this probably goes without saying, the draftsman has said it), the holder of the derivative interest will only be bound if, at the time the interest was granted, O was bound by a code right.

16.6.6

An example should suffice to make this clear:

(a) O is a tenant in occupation of land;
(b) O confers code rights upon an operator;
(c) For so long as they subsist, those code rights will bind:
 (i) O's assignee, A (and successive assignees);
 (ii) O's tenant, S;
 (iii) S's assignees and tenants.

16.7 The practical effect of the occupier granting code rights when it has no property right to do so

16.7.1

It follows from the analysis above that the occupier may have no right at all to grant code rights. If the occupier is a squatter, then this goes without saying: just as it will not have the permission of the landowner to occupy the land, so it will follow that it will have no permission to grant code rights either.

16.7.2

The same applies where the occupier is on the land with the permission of the landowner, but has no permission to grant code rights.

16.7.3

In either of the circumstances set out above, the landowner will be faced with a fait accompli in the event that the occupier creates code rights. The landowner will not be bound by the code rights (see section 16.5 above) – but that will not stop them existing. Moreover, the landowner may struggle to obtain an order for removal of the electronic communications apparatus – see Chapter 24.

16.7.4

However, this is not to say that the landowner is without remedy. We say this in light of the decision in *Arqiva Ltd v Everything Everywhere Ltd.*[12] The matter before the court arose out of the joint venture arrangement between T-Mobile (now Everything Everywhere) and Orange. Arqiva provided sites for use by Orange and T-Mobile. Those companies wished to roam[13] over each other's networks which involved the sending, receipt and automatic changeover of signals by Everything Everywhere and Orange on each other's frequencies and equipment throughout the duration of a call and for internet access. There was no sharing of frequencies; each network operator still operated its own frequencies within the radio spectrum.[14] In all cases it was the customer's handset that changed automatically the radio frequencies when switching between the networks. This arrangement was held to give rise to breaches of a whole host of covenants which had been entered into with Arqiva under various agreements. The judge held in particular that the roaming gave rise to a breach of a clause, cl.2.1.1(d), which permitted 'the Client to share use of such part or parts of the Station(s) and such of the BBC's accommodation and equipment therein in common with others including the BBC, as

12 [2011] EWHC 1411 (TCC).

13 Roaming does not involve any physical access to a site or structure, e.g. a mast being operated by the tenant. It essentially involves a reconfiguration of network software and equipment. Seamless roaming allows a call to be uninterrupted as a customer moves between cells of two networks of two different network operators. Roaming occurs in very familiar everyday situations, e.g. by using a mobile phone abroad, where the network operator will have entered into an agreement with the operator of another network to permit that to happen, and the making of emergency calls in areas where customers do not have access to their own operator's network. Thus, roaming involves the process of a customer of one network operator accessing the network and frequency of another operator of which it is not a customer.

14 For a recent decision explaining how and why radio spectrum is a critical raw material for the provision of mobile telephony services, see the decision of Green J in *The Queen on the application of Hutchison 3G UK Ltd v Office of Communications v Telefónica UK Ltd, EE Ltd, British Telecommunications Plc, Vodafone Ltd; The Queen on the application of British Telecommunications Plc v Office of Communications Telfó̲nica UK Ltd, Vodafone Ltd, Hutchison 3G UK Ltd* [2017] EWHC 3376 (Admin).

the BBC may from time to time approve (such approval not to be unreasonably withheld or delayed)'. The judge said:

> Clause 2.1.1(d) sets out the use which OPCS is permitted to make of Arqiva's Station, accommodation and equipment. It permits OPCS 'to share use … in common with others including the BBC, as the BBC may from time to time approve (such approval not to be unreasonably withheld or delayed).' By allowing roaming OPCS are allowing the voice and data traffic of EE's T-Mobile brand customers to use the OPCS equipment which is located at the Station. There is therefore shared use of the 'Client's Equipment' which is installed at the Station and is, in turn, making use of part of the Station, accommodation and equipment. This shared use of the 'Client's Equipment' is also shared use of the Station, accommodation and equipment. [At 143.]

16.7.5

Arqiva had sought an interim injunction to prevent such sharing. The report of the decision does not record the relief which was finally granted in light of the Judge`s findings, as submissions were invited as to the appropriate relief to be granted. However, it highlights that a breach of a contractual stipulation may enable a landowner to seek injunctive relief without fear of the code, the matter being one merely of property rights. Accordingly, it appears to the Authors that the grant of code rights without the permission of the landowner does not leave the landowner wholly without remedy.

17 To whom may code rights be granted?

17.1 Introduction

17.1.1

Paragraph 9 of the New Code is headed 'Who May Confer Code Rights?', and provides:

> A code right in respect of land *may only* be conferred on *an operator* by an agreement between the occupier of the land and the operator.

17.1.2

It is vital, therefore, for those who wish to install or operate electronic communications apparatus to become operators, without which designation they will not be able to take advantage of code rights.

17.1.3

Paragraph 2 of the New Code provides:

> In this code 'operator' means –
> (a) where this code is applied in any person's case by a direction under section 106, that person, and
> (b) where this code applies by virtue of section 106(3)(b), the Secretary of State or (as the case may be) the Northern Ireland department in question.

17.1.4

The reference to s.106 is a reference to s.106 of the Communications Act 2003, rather than the Digital Economy Act 2017, for the reason that the New Code is to be inserted before Schedule 4 to the 2003 Act, with the result that references to section numbers within it are to that Act. Accordingly, all references to sections in the New Code are to the 2003 Act (save where otherwise specified).

17.1.5

There is therefore no difference in the *mechanism* for the appointment of those entitled to be operators under the Old Code (therein called 'code operators'), and those entitled under the New Code (simply called 'operators').

17.1.6

However, the 2017 Act made a significant amendment to s.106 of the 2003 Act, which widened the category of operators to include wholesale infrastructure providers (WIPs). This is considered in section 17.2 below.

17.2 The procedure for designation as operators

17.2.1

Section 106 of the 2003 Act provided, as originally drafted:

> (3) The electronic communications code shall have effect –
> (a) in the case of a person to whom it is applied by a direction given by OFCOM; and
> (b) in the case of the Secretary of State or any Northern Ireland department where the Secretary of State or that department is providing or proposing to provide an electronic communications network.
> (4) The only purposes for which the electronic communications code may be applied in a person's case by a direction under this section are –
> (a) the purposes of the provision by him of an electronic communications network; or
> (b) the purposes of the provision by him of a system of conduits which he is making available, or proposing to make available, for use by providers of electronic communications networks for the purposes of the provision by them of their networks.
> (5) A direction applying the electronic communications code in any person's case may provide for that code to have effect in his case –
> (a) in relation only to such places or localities as may be specified or described in the direction;
> (b) for the purposes only of the provision of such electronic communications network, or part of an electronic communications network, as may be so specified or described; or
> (c) for the purposes only of the provision of such conduit system, or part of a conduit system, as may be so specified or described.
> (6) …
> (7) In this section 'conduit' includes a tunnel, subway, tube or pipe.

17.2.2

In its Report No. 336, the Law Commission recommended that this restriction, beyond network operators, to those providing conduits, should be released, so as to allow all infrastructure providers to apply to have the code applied to them. As it rightly observed in paragraph 2.28 of the report:

> It is perhaps strange that a provider of conduits for the use by others for electronic communications apparatus can be a Code Operator, but a provider of other infrastructure cannot.

17.2.3

Section 4 of the 2017 Act made a number of changes to the drafting of s.106(4)–(7) of the 2013 Act, so that the amended version now reads (so far as relevant – deleted words shown struck through):

> (4) The only purposes for which the electronic communications code may be applied in a person's case by a direction under this section are –
> (a) the purposes of the provision by him of an electronic communications network; or
> (b) the purposes of the provision by him of a system of ~~conduits~~ infrastructure which he is making available, or proposing to make available, for use by providers of electronic communications networks for the purposes of the provision by them of their networks.
> (5) A direction applying the electronic communications code in any person's case may provide for that code to have effect in his case –
> (a) in relation only to such places or localities as may be specified or described in the direction;
> (b) for the purposes only of the provision of such electronic communications network, or part of an electronic communications network, as may be so specified or described; or
> (c) for the purposes only of the provision of such ~~conduit~~ system of infrastructure, or part of a ~~conduit~~ system of infrastructure, as may be so specified or described.
> (6) …
> (7) In this section 'conduit' includes a tunnel, subway, tube or pipe.

17.2.4

This drafting marries up with two further provisions in the New Code. The first is the definition of 'system of infrastructure' in paragraph 7(1) of the New Code to mean:

> a system of infrastructure provided so as to be available for use by providers of electronic communications networks for the purposes of the provision by them of their networks.

17.2.5

The second is the definition of 'the statutory purposes' in paragraph 4 of the New Code, in relation to an operator, which provides that this includes not merely the purposes of providing the operator's network, but also 'the purposes of providing an *infrastructure system*'.

17.2.6

The effect of this amendment is to bring within the category of potential code operators not merely those who provide a *network* (including, as under the Old Code, infrastructure limited to conduits such as tunnels, subways, tubes or pipes), but also all those who provide the rest of the infrastructure necessary for the provision of that network – including thereby masts, equipment cabins, cabling and so forth.

17.2.7

That important distinction between the codes aside, in simple terms, an operator for the purposes of both the Old Code and the New Code is an electronic communications provider which has received a direction under s.106 of the Communications Act 2003.

17.2.8

The process by which a direction is made is set out in s.107 of the 2003 Act, headed 'Procedure for directions applying code'. Prior to the entry into force of the 2003 Act on 25 July 2003, the Telecommunications Act 1984 set out the rules governing the application of code powers. It enabled the Secretary of State to apply the code to a particular person by a licence granted under section 7 of the 1984 Act. As a result of the implementation of the 2003 Act on 25 July 2003, the licensing regime established under the 1984 Act was abolished (subject to certain transitional provisions). For those persons that were granted Code powers by the Secretary of State before 25 July 2003, paragraph 17(2) of Schedule 18 (Transitional Provisions) to the 2003 Act provides that those persons shall be treated as a person in whose case the code applies by virtue of a direction given by OFCOM.

17.2.9

The current process commences with an application made by a prospective operator – see s.107(1). Subsection (2) provides that if OFCOM (as to which, see Chapter 41 of this book) publishes a notification setting out its requirements with respect to the content and manner of such an application, then the application must be made in accordance with those requirements. Such a notification has indeed been published, and is updated from time to time in accordance with subsection (3).[1]

17.2.10

When an application is made by a prospective operator, OFCOM is required to take into consideration the following matters under s.107(4) (which are to be given equal priority to its general duties under ss.4, 24 and 25):

(a) the benefit to the public of the electronic communications network or conduit system by reference to which the code is to be applied to that person;
(b) the practicability of the provision of that network or system without the application of the code;

1 The details of the requirements are currently set out in Annex B of the document at www.ofcom.org.uk/__data/assets/pdf_file/0028/8578/ecc.pdf. OFCOM is in the process of redrafting the guidance, to simplify the information; the new version is not yet available. Although the requirements remain the same, the only change which has come about due to the 2017 Act is that the term 'conduits' is replaced by 'infrastructure'.

(c) the need to encourage the sharing of the use of electronic communications apparatus;
(d) whether the person in whose case it is proposed to apply the code will be able to meet liabilities arising as a consequence of –
  (i) the application of the code in his case; and
  (ii) any conduct of his in relation to the matters with which the code deals.

17.2.11

If having considered the application OFCOM is minded to make a direction, it must first publish (in such manner as it considers appropriate for bringing the notification to the attention of the persons who, in its opinion, are likely to be affected by it – see s.107(10)[2]) a notification containing the following matters prescribed by s.107(7):

(a) a statement of OFCOM's proposal, (i) containing a statement that it proposes to apply the code in the case of the person in question; and (ii) setting out any proposals to impose terms under s.106(5) (see s.107(8)). The publication of such a statement is subject to OFCOM's right at any time to suspend the application of the code to that operator under s.113(7), or to modify terms or revoke the direction applying the code under s.115(5);
(b) a statement of its reasons for that proposal;
(c) a statement of the period within which representations may be made to it about the proposal (which, under s.107(9), must end no less than one month after the day of the publication of the notification).

17.2.12

OFCOM must then consider any representations about the proposal that are made to it within the period specified in the notification (see s.107)(6)). No period was originally specified by the 2003 Act for OFCOM to carry out this task, but the Electronic Communications and Wireless Telegraphy Regulations 2011 provided for the insertion of a new s.107(1A), which requires the application to be determined within the six-month period specified in Regulation 3.

17.2.13

At the conclusion of this process, OFCOM then invariably makes a direction that the code is to apply in the applicant's case. There is no known case of OFCOM refusing to make such a direction following the consultation process referred to above.

17.2.14

OFCOM is required to keep a register of persons in respect of which a direction has been made – see section 17.4 below.

2 In practice, OFCOM publishes these notifications on its website – see www.ofcom.org.uk/phones-telecoms-and-internet/information-for-industry/policy/electronic-comm-code/notifications-communications-act2003.

17.3 OFCOM

17.3.1

OFCOM (referred to in the Act in the plural, but in this text in the singular) was formally established as a body corporate, the Office of Communications, by s.1 of the Office of Communications Act 2002, with effect from 29 December 2003.[3]

It is the regulator for the UK communications industries, having replaced five legacy regulator organisations: Oftel, the Independent Television Commission, the Radio Authority, the Radiocommunications Agency and the Broadcasting Standards Commission.

17.3.2

OFCOM regulates television, radio and video-on-demand sectors, fixed line telecoms, mobiles and postal services, together with the airwaves over which wireless devices operate. As such, it sets and enforces regulatory rules for those sectors. It also has powers to enforce competition law in those sectors, alongside the Competition and Markets Authority.

17.3.3

OFCOM operates under a number of Acts of Parliament, including in particular the Communications Act 2003, but also the Wireless Telegraphy Act 2006, the Broadcasting Acts 1990 and 1996, the Digital Economy Act 2010 and the Postal Services Act 2011. It acts independently from governments and commercial interests in performing its duties, but is accountable to Parliament. OFCOM is funded by fees from industry for regulating broadcasting and communications networks, and grant-in-aid from the Government.

17.3.4

OFCOM's primary duties are set out in s.3 of the Communications Act 2003. These include the requirement to ensure the availability throughout the United Kingdom of a wide range of electronic communications services.

17.4 Operators under the codes

17.4.1

Section 108(1) of the 2003 Act imposes a duty upon OFCOM to establish and maintain a register of persons in whose case the electronic communications code applies by virtue of a direction under s.106.

3 A fuller account of OFCOM's role is provided in Chapter 41.

17.4.2

The register of persons with powers under the code is on the OFCOM website.[4] It currently contains over 120 names, predominant among which are providers of actual electronic communications services (usually Mobile Network Operators), latterly supplemented by some Wholesale Infrastructure Providers, whose function it is to install and maintain apparatus in, over and under land (rather than operate electronic communications networks).

4 See www.ofcom.org.uk/phones-telecoms-and-internet/information-for-industry/policy/electronic-comm-code.

18 Over what may code rights be granted?

18.1 Introduction

18.1.1

Paragraph 3 of the New Code provides:

> For the purposes of this code a 'code right', in relation to an operator and any land, is a right for the statutory purposes –
> (a) to install electronic communications *apparatus* on, under or over the *land* …

18.1.2

The other code rights are similarly described in relation to 'apparatus' and 'land'.

18.1.3

It is of critical importance to an understanding of code rights to appreciate the scope of the two expressions 'apparatus' (considered in section 18.2 below) and 'land' (sections 18.3 and 18.4).

18.1.4

Section 18.5 below analyses whether the ownership of electronic communications apparatus may change as a matter of law.

18.1.5

Section 18.6 discusses the reasons for, and consequences of, the code distinction between apparatus and land.

18.2 The meaning of 'apparatus'

18.2.1

Paragraph 5(1) of the New Code explains:

In this code 'electronic communications apparatus' means –

(a) *apparatus designed or adapted for* use in connection with the provision of an *electronic communications network*,

(b) *apparatus designed or adapted for a use* which consists of or includes the sending or receiving of communications or other signals that are transmitted by means of an electronic communications network,

(c) *lines, and*

(d) *other structures or things designed or adapted for use in connection with the provision of an electronic communications network*.

The italicised expressions are discussed in the text below.

18.2.2

Two preliminary points may be made about this definition as a whole. First, the definition of electronic communications apparatus does not stipulate that it should actually belong to, or be used by, an 'operator' who has been made the subject of a direction under s.106 of the 2003 Act. Whether an item of equipment qualifies as electronic communications apparatus will simply depend upon whether it fulfils one of the four definitions in paragraph 5(1)(a)–(d) of the New Code. Thus, what constitutes electronic communications apparatus is not dependent on the status of the person providing it. It therefore follows that the operator need not own the apparatus it installs, and indeed may acquire code rights over that apparatus, irrespective of ownership.

18.2.3

It is worth noting in this context that, insofar as the main operator is using a code right for installation, then the right must be exercised 'for the statutory purposes'. That is defined by paragraph 4 of the New Code to mean the purposes of providing the operator's network, or an infrastructure system. The term 'network' is then defined by paragraph 6 to mean 'so much of any electronic communications network or infrastructure system provided by the operator'. 'Infrastructure system' is defined by paragraph 7 to mean 'a system of infrastructure provided so as to be available for use by providers of electronic communications networks'. In relation to the operator's network, it was the Law Commission's aim to remove the qualification 'provided by the operator', which was brought in by the Communications Act 2003. It said, in paragraph 2.3 of its Report No. 336:

> ... removing the restriction to 'the operator's network' would provide greater flexibility, and would be consistent with the inclusion of the provision of infrastructure among the purposes for which Code Rights can be conferred. This would deal at a stroke with the problem encountered under the old code whereby an infrastructure provider which is a Code Operator has the right to keep electronic communications apparatus – for example, a mast and a structure that supports it – on land but is not itself providing an electronic communications network. Under the Old Code that arrangement would have the consequence that the infrastructure provider's right would not be a Code Right, and the infrastructure provider's relationship with the landowner would not be a regulated relationship.

18.2.4

Ultimately, this recommendation was not carried forward into the drafting referred to above. However, none of the statutory language refers to 'ownership' of the apparatus, the drafting referring only to 'provision' of the network, which does not carry with it any necessary ingredient of ownership. So much would appear to be confirmed by the definition of the term 'provides' in s.405(1) of the 2003 Act by reference to s.32(4) of that Act, which states that 'references to the provision of an electronic communications network include references to its establishment, maintenance or operation'.

18.2.5

The second overall point to make in relation to paragraph 5 is that it should also be noted that the definition of electronic communications apparatus is not dependent on its actual *use*. It is apparatus if, within sub-paragraphs (a), (b) or (d) it is 'designed or adapted' for use, *irrespective of whether it is actually used or not*.

18.2.6

The remainder of this section focuses upon the individual components of the paragraph 5 definition of electronic communications apparatus.

18.2.7 'Apparatus'

The term 'apparatus' is not defined in the New Code itself, although the definition in s.405(1) of the 2003 Act applies, remaining unchanged from its Old Code use:

> 'apparatus' includes any equipment, machinery or device and any wire or cable and the casing or coating for any wire or cable.

18.2.8

Although this definition is inclusive rather than exhaustive, and may be suggestive of operational equipment rather than inoperative structures ('passive assets'), the intention of the Law Commission was plain:

> Overall, the apparatus that is protected by the Code should include not only the core equipment – masts, antennae, cables, and so on – but also the additional items and facilities needed – physically within a secure area – to make it workable.[1]

18.2.9

Construed with that assistance from the Law Com Report, the term 'apparatus' may therefore reasonably be considered to include every single part of the hardware in question. Some assistance in the definition of the term may also be derived from authority.

1 See paragraph 2.54 of Law Com 336.

In *Rudd v Secretary of State For Trade and Industry*[2], the House of Lords had to decide whether a forfeiture order had been properly made against a pirate radio station operator, who had been found guilty of using apparatus for wireless telegraphy without a licence under the Wireless Telegraphy Act 1949, as amended by the Telecommunications Act 1984. Section 14 of the 1949 Act provided:

> the court may, in addition to any other penalty, order all or any of the apparatus of the station, or (as the case may be) of the apparatus in connection with which the offence was committed, to be forfeited to the Secretary of State. The power conferred by virtue of paragraph (a) or (c) above does not apply to wireless telegraphy apparatus not designed or adapted for emission (as opposed to reception).

18.2.10

The forfeited items included a number of music record discs and cassettes. The Divisional Court had held that these did not constitute 'apparatus'. The House of Lords disagreed. In the words of Lord Goff (with whom all the Law Lords agreed):

> Apparatus is no more than equipment prepared for a purpose: in the Shorter Oxford English Dictionary, the second meaning given is 'The things collectively in which preparation consists, and by which its processes are maintained; equipments, material, machinery; material appendages or arrangements.' Consistently with that broad definition, I can see no reason why discs or cassettes should not be described as 'apparatus'; this indeed accords with my own understanding of the ordinary use of that word.

18.2.11

However, the House concluded that, although the items were apparatus, they did not constitute part of 'wireless telegraphy apparatus' within the meaning of the statute. The meaning of the term clearly therefore depends upon the context in which it is used.

18.2.12

It is clear from paragraph 5(1) itself that 'apparatus' may depart widely from its conventional meaning of machinery or equipment, and may include lines and structures. Paragraph 5(3) of the New Code defines these separately: see paragraphs 18.2.26–31 below.

18.2.13 'Designed or adapted for use'

This drafting makes it plain that, even if a structure was not originally conceived of and built as electronic communications apparatus, it can yet become electronic communications apparatus if it is subsequently adapted for use as such.

18.2.14

The New Code contains no definition of 'adapted'. As a matter of ordinary English 'to adapt' something is 'to adjust, to make suitable for' or to 'alter or modify to fit for a new

2 [1987] 1 WLR 786.

use or new conditions' or 'undergo modification' to fit a new use: *Shorter OED* 2007. In other contexts it has been said that the mere fact something is suitable for a use does not make it adapted for such a use; some form of process must be undertaken to the thing to make it suitable for the use.[3]

18.2.15

The definition of 'apparatus' will rarely be relevant to a consideration of active assets such as equipment cabins, transmission equipment, antennae and cabling, but it is of obvious application where a mast used for other purposes is converted for telecommunications purposes. Unlike the case of a building, considered below, there is no 'sole purpose' condition for the design/adaptation qualification in relation to a structure such as a mast.

18.2.16

It is therefore an open question whether a mast that is built to perform several different functions, including the installation of dishes or antennae, will become apparatus for the purposes of the New Code. The arguments for and against that proposition are as follows. In favour of the proposition that the mast may constitute apparatus is the fact that there is no sole purpose condition: that is to say, paragraph 5(1)(d) does not stipulate that the 'other structures or things' should be designed or adapted for use *only* in connection with the provision of an electronic communications network. That observation militates in favour of the whole mast being apparatus. If that is right, however, then the owner of the mast would not be entitled to grant code rights over it to another operator, for the reason that all the code rights apply in relation only to 'land', and the mast (now 'apparatus' for these purposes) would not be 'land'.

18.2.17

The argument against the proposition that a mast designed or adapted for a range of purposes, only one of which is the provision of a network, picks up on the point last

3 Thus (simply by way of example):

(1) In *French v Champkin* [1920] 1 KB 76 at 79 it was said that 'The justices seem to have treated the word "adapted" [in the Customs and Inland Revenue Act 1888] as if it were synonymous with "suitable" or "apt", whereas it must be construed as meaning altered so as to make the vehicles apt for the conveyance of goods. The words are intended to cover the case of a vehicle which was not constructed solely for the purpose of conveying goods in the course of trade, but after its construction has been made apt for that purpose.'

(2) In *Davison v Birmingham Industrial Cooperative Society* [1920] 90 LJKB 206 at 208 it was said that 'It is necessary that the building in question should be a "building constructed or adapted to be used for human habitation [within the relevant housing statute"] … I think "adapted" is used to mean changed or altered or transformed.'

(3) In *Grove v Lloyds British Testing Co* [1931] AC 450 it was said that 'adapting for sale' points clearly to something being done to the article in question which, in some way, makes it in itself different from what it was before.

(4) In *Backer v Secretary of State for the Environment* [1983] 2 All ER 1021 the mere provision of a bed in a caravan and living in it did not make a motor vehicle 'adapted for human habitation' within the Caravan Sites and Control of Development Act 1960.

made. In the Government's strong support for the proposition that operators should share apparatus, and that there should be a move away from the proliferation of masts, it is to be expected that the use of a multi-purpose mast should not be restricted by such an interpretation. The answer to that is not that the encouragement of sharing will deal with the problem, because it is likely that rival operators will be excluded rather than included.

18.2.18

This situation becomes more complex when tall and often substantial structures with an original non-telecommunications use (water towers, fire drill towers, floodlighting columns) are considered. The following examples, all of which start with the erection of a substantial lattice tower upon land, demonstrate the complexity:

(a) The tower was erected by the landowner, and is used for birdwatching. Subsequently, the landowner grants code rights to an operator to install a small equipment cabin at the foot of the tower, run cabling up one side of the structure, and bolt a dish to the top, leaving plenty of room for other equipment for code purposes, and on terms that the landowner may continue to watch birds.
(b) As above, but the operator's equipment leaves no room for any other equipment.
(c) The tower was erected by an operator, with the specific purpose of using it as part of its network.

18.2.19

In the case of the first two examples, there is no demonstrably conclusive answer to the question whether either necessarily has the result that the *whole tower becomes apparatus* for the purposes of the New Code. If either does, the consequence will be that the landowner will be unable to grant any further code rights over the tower, for the simple reason that the tower will then be apparatus rather than land – and code rights can only be granted over land.

18.2.20

In the case of the third example, where the operator has itself erected the tower as part of its electronic communications network, it will be very difficult to argue that the tower is anything other than apparatus, just as much as any other part of the operator's apparatus, such as its cabling and antennae. But if that is right, then how can the situation be any different in relation to the second example, where the factual circumstances concerning the actual kit on the tower may be identical?

18.2.21

This conundrum is thrown into sharper focus where the tower in question is vast, and would lend itself to a variety of telecommunications or non-telecommunications uses. A water tower provides a good example. At first sight, it may be obvious that an operator who negotiates for the code rights to install simple cabling with a small cabin and antenna would be in difficulties if it sought to argue that it had thereby adapted the whole tower for use as apparatus, and could therefore prevent code rights being granted to any other. After all, an operator who has been granted code rights to install antennae on the roof of a building can hardly be said to have adapted the whole building for use for its network.[4] But if so, then what is the difference in principle between this example and those considered above? If the answer is that the tower is not really being 'adapted', then at what point does adaptation occur?

18.2.22

There may be no ready answer to such questions. The courts are used to picking their way through legislation where definitional difficulties produce no obvious solution, leaving it to rough and ready 'fact and degree' tests to determine what may constitute the requisite degree of any adaptation in any given case.

18.2.23

The Authors however venture the following considerations as factors that may influence the tribunal in arriving at a decision as to whether a structure designed for a non-telecommunications purpose, but over which code rights are subsequently granted, may have become 'adapted for use' in connection with the provision of an electronic communications network:

(a) the size of the structure: the larger the structure, the less the potential for an operator to contend that the whole structure has been 'adapted for use' (as opposed to works simply having been carried out, leaving the structure in substantially the same condition);
(b) the nature of the works: the more substantial they are, the greater the potential for the operator to contend that the whole structure has been adapted;
(c) the capacity for the structure to be lent to other uses beyond the operator's own use: if, in other words, there is plenty of room for the owner of the structure to allow others to take space upon it, the argument that the whole structure has become 'apparatus' becomes harder to sustain;
(d) whether the owner of the structure is itself subject to statutory duties, which may conflict with those of the operator, if the whole structure is to become 'apparatus';[5]
(e) whether the structure is in the form of a building: if so, and if the operator's equipment is on, rather than within it, the building itself will not comprise a structure, for the reasons set out later in this section.

18.2.24

The parties may be safe in concluding that the Tribunal which has first to rule on this question will not accept that a structure erected originally for a different purpose (the

4 Of course, the definition of the term 'building', considered below, would preclude such an argument in any event. But the very existence of this drafting suggests that, in its absence, such an argument could have been maintained in appropriate circumstances.

5 The precise situation contemplated in paragraph 6.112 of Law Com 336, reflecting the possibility that owners of water towers might be subject to statutory duties.

redundant water tower or fire drill tower being cases in point), and which was clearly not 'designed' for use (etc.) will not become a structure merely because electronic communications apparatus is subsequently bolted on. Such work would be unlikely to count in normal parlance as 'adaptation': much of the structure in each case remains exactly the same. The structure is no more adapted for use (etc.) than is a building rooftop which has apparatus stationed upon it.

18.2.25 'The provision of an electronic communications network'

In order to qualify for apparatus status, the equipment in question must be 'designed or adapted for use in connection with *the provision of an electronic communications network*'. Such a network is defined by paragraph 6 of the New Code to mean:

> so much of any electronic communications network or infrastructure system *provided by the operator* as is not excluded from the application of the code under section 106(5) [of the 2003 Act].

Although the words '*provided by the operator*' are a critical component of the definition, feeding in turn into the definition of the statutory purposes for which code rights may be granted, the same qualification is not imported into the definition of 'apparatus'. That is to say, the infrastructure[6] forming the apparatus need not itself be provided by an operator.

18.2.26 'Line'

'Line' is defined to mean:

> any wire, cable, tube, pipe or similar thing (including its casing or coating) which is designed or adapted for use in connection with the provision of any electronic communications network or electronic communications service.

It is not thought that this definition will cause any difficulty.

18.2.27 'Structure'

'Structure' is defined to include:

> a building only if the sole purpose of that building is to enclose other electronic communications apparatus.

6 Unlike the definition of 'electronic communications apparatus' in paragraph 5, there is nothing in the New Code that excludes 'land' from infrastructure. Unless it falls outside the meaning of infrastructure as used within the New Code, land could conceivably therefore be infrastructure. However, 'infrastructure' is not a defined term in the New Code. The dictionary definition of infrastructure connotes installations forming the subordinate parts of a larger undertaking. A bare site is not obviously 'infrastructure' in this sense, so that on this basis the bare holding of such land one would expect would not of itself be an activity being carried out for statutory purposes within New Code, and ought not be a purpose for which a s.106 direction ought to be given.

Hence, a building will not constitute 'apparatus' to which code rights might be capable of applying, unless it is a room housing equipment or an equipment cabin, and used for no other purpose. Similarly, a field or a rooftop upon which a telecommunications mast is stationed will not be a 'structure', although the mast itself may be.

18.2.28

It must be acknowledged that this drafting too conceals possible complexity in its application. The following questions arise, and more are likely to be prompted in the early years of operation of the New Code:

(a) Suppose a building is designed to house a large piece of electronics equipment for the statutory purpose: it will clearly be 'apparatus'. However, if the operator's plans change, and it uses part of the building for another purpose, will the building lose its status as 'apparatus'?
(b) Conversely, if a building is used for other purposes, and is then converted to house electronic communications apparatus, will it then become 'apparatus'?

18.2.29

The definition of 'structure' in paragraph 5(3) of the New Code would suggest that the test is to be applied from time to time, rather than at the point the use commenced. It might however be thought incongruous that apparatus should change its special character (and the code rights attaching to it) in this way.

18.2.30

The wider expression 'other structures or things' which paragraph 5(1)(d) of the New Code uses may generate argument over the precise ambit of electronic communications apparatus in a given case, although the qualifying words 'designed or adapted for use in connection with the provision of an electronic communications network' should serve to curtail the potential for serious dispute.

18.2.31

In conclusion, it seems safe to say that the draftsman of paragraph 5 of the New Code wished the class of possible structure that might constitute apparatus to be as wide as possible, including every form of man-made erection, other than a non-qualifying building.

18.3 The meaning of 'land'

18.3.1

The New Code provides no definition of 'land' as such, although paragraph 108 states that 'land' does not include 'electronic communications apparatus'. The ramifications of this drafting are considered in section 18.6 below.

18.3.2

Section 36(7) of the Telecommunications Act 1984 (since repealed by the Communications Act 2003), which allowed for the compulsory purchase of land by a public telecommunications operator, provided:

> In this section 'land' has the meaning assigned to it by section 45(1)(a) of the Interpretation Act (Northern Ireland) 1954 c. 33 1954.[7]

18.3.3

The comparable provision for England and Wales in s.34 of the 1984 Act contained no such definition, and made no reference to the domestic Interpretation Act 1978. Section 5 of the 1978 Act provides that:

> In any Act, unless the contrary intention appears, words and expressions listed in Schedule 1 to this Act are to be construed according to that Schedule.

Schedule 1 in turn stipulated that:

> 'Land' includes buildings and other structures, land covered with water, and any estate, interest, easement, servitude or right in or over land'.

18.3.4

The question which therefore arises in relation to the 2017 Act (and which also arose in relation to the 1984 and 2003 Acts) is whether there is anything in the drafting of that legislation that expresses a contrary intention which ousts the Interpretation Act definition of 'land'.

18.3.5

A similar question, but with regard to other legislation, arose in *British Waterways Board v London Power Networks plc*.[8] In that case, the court was asked to decide whether the Secretary of State had power to grant a wayleave for the installation of four cables for the transmission of electricity and associated telephone and signalling cables through a services tunnel belonging to the Board, which in turn depended upon whether services tunnels fell within the meaning of 'land' for the purposes of the Electricity Act 1989. Paragraph 6(1) of Schedule 4 to that Act provided:

> This paragraph applies where – for any purpose connected with the carrying on of the activities which he is authorised by his licence to carry on, it is necessary or expedient for a licence holder to install and keep installed an electric line on, under or over any land.

7 '"land' shall include – (i) messuages, tenements and hereditaments of any tenure; (ii) land covered by water; (iii) any estate in land or water; and (iv) houses or other buildings or structures whatsoever'.

8 [2003] 1 All ER 187.

The Board argued that the tunnel was not 'land', and that the definition in the Interpretation Act 1978 did not apply. Its arguments, as summarised in the judgment, apparently depended upon the drafting of comparable provisions in the Telecommunications Act 1984. In particular, the Board contended that (1) in the context of the Electricity Act 1989, the word 'land' did not include installations or equipment attached to the land; and (2) the words 'on, under or over [any][the] land' where they appear in paragraph 6(1) and (2) of Schedule 4 to the Act did not include 'through installations or equipment' on that land.

18.3.6

In paragraph 13 of his judgment, the Vice-Chancellor said:

> The word 'land' is not defined in the Electricity Act 1989. Accordingly resort must be had to the Interpretation Act 1978 s.5 and Schedule 1 which provide that, 'unless the contrary intention appears', 'land' includes buildings and other structures, land covered with water, and any estate, interest, easement, servitude or right in or over land'. It is not disputed that the service tunnels are land within the principles of *Elitestone Ltd v Morris*[9] and, in the absence of a contrary intention, fall within this definition of land.

18.3.7

He therefore dismissed the claim, and granted a declaration to the effect that the power conferred on the Secretary of State by paragraph 6(3) of Schedule 4 to the Electricity Act 1989 enabled him to grant a wayleave entitling the defendant to install and maintain in the Board's services tunnel four cables for the transmission of electricity and associated telephone and signalling cables.

18.3.8

Whatever may have been the position in relation to the drafting of the Telecommunications Act 1984, it is strongly arguable, given the similarity in the drafting between the 2017 Act and the Electricity Act 1989, that the same approach should inform the interpretation of 'land' in the New Code. That is to say, that 'land' will include not merely the physical ground, but also the buildings and other structures over it, land covered with water, and any estate, interest, easement, servitude or right in or over the land.

18.3.9

There is, however, one important point of distinction in the New Code, which is analysed in section 18.6 below.

18.4 Is there any 'land' over which apparatus may not legitimately be installed?

18.4.1

Potentially expropriatory measures such as the telecommunications legislation considered in this book usually contain savings provisions designed either to prevent conflict

9 [1997] 1 WLR 687.

with other legislation, with which such expropriation would conflict, or to safeguard the rights of landowners such as the Crown, or to protect areas in the public interest, such as defence installations.

18.4.2

The New Code adopts the device of stipulating that the code does not authorise the contravention of any provision of an enactment passed or made before the coming into force of the code, unless and to the extent that an enactment makes provision to the contrary (see paragraph 99).

18.4.3

One example of a provision to the contrary (albeit enacted in the context of the Old Code, which contained provision similar to paragraph 99, in the shape of paragraph 27) is provided by s.194 of the Law of Property Act 1925, which prohibits the erection of any building or fence, or the construction of any other work, whereby access to commons and waste lands is prevented or impeded. As amended by the 2003 Act, however, that section does not apply to any electronic communications apparatus installed for the purposes of an electronic communications code network. Electronic communications apparatus may therefore be installed on common land.

18.4.4

Another substantial category consists of marine areas (i.e. the territorial sea, any area of land submerged at mean high water spring tide, and the waters of every estuary, river or channel, so far as the tide flows at mean high water spring tide), in respect of which there is general legislation controlling the right to carry out works.[10]

18.4.5

Unusually, there is no general exemption for land owned by the Crown. Paragraph 104 of the New Code provides that the code 'applies in relation to land in which there subsists, or at any material time subsisted, a Crown interest as it applies in relation to land in which no such interest subsists'. However, this does not apply to tidal waters. Under both the Old Code (paragraph 11(2)) and the New Code (paragraph 64(1)), an operator may not exercise a right to install electronic communications apparatus in tidal waters subject to a Crown interest, without the consent of the Crown.

18.4.6

Special provision is made for statutory undertakers. Under both codes, code operators have rights to install and keep apparatus in, over or under certain types of land, and to cross land: streets which are maintainable highways, tidal waters and lands, and railways, canals and tramways (paragraphs 9, 11 and 12 of the Old Code; Parts 7, 8 and 9 of the New Code). By contrast, the use of some utility conduits, such as water mains

10 See in particular the Marine and Coastal Access Act 2009.

and sewers, is limited under both codes so that the consent of the authority controlling the conduit is needed in order to place apparatus there (paragraph 15 of the Old Code; paragraph 102 of the New Code).

18.5 Ownership of electronic communications apparatus

18.5.1

It is of the first importance to note that the code rights that are granted over land are simply a right *to access and use the land for the statutory purposes*. Unlike other typical forms of land use agreements, agreements conferring code rights do not in themselves grant any interest in the land, which would otherwise gain statutory protection under Part II of the Landlord and Tenant Act 1954: they are a *sui generis* statutory right to access the land; to install apparatus; and to use and maintain that apparatus as necessary.

18.5.2

In such circumstances, it might though be obvious that the operator would retain the ownership of its equipment[11] as a matter of course, and that ownership could not pass to the landowner. In the consultation period prior to enactment of the 2017 Act, a number of respondents felt that it was important for the New Code to make clear that an operator's apparatus remained the property of its owner unless there was explicit agreement between both parties otherwise, in order to ensure that operators could protect their property and site providers could require removal at the end of an agreement.

18.5.3

The Law Commission did not feel that this concern was adequately addressed by paragraph 27(4) of the Old Code, which provided that 'the ownership of any property shall not be affected by the fact that it is installed on or under, or affixed to, any land'. Notwithstanding that uncertainty, very similar drafting appears in paragraph 101 of the New Code, which provides:

> The ownership of property does not change merely because the property is installed on or under, or affixed to, any land by any person in exercise of a right conferred by or in accordance with this code.

18.5.4

This drafting may be said to deal with the problem that can arise in relation to the law of fixtures (see Chapter 14), when equipment may be installed by a tenant in such a way as to make it irremovable without substantial damage to itself or the property to which it is affixed, in which case it becomes part of the landlord's property.[12]

11 If indeed it owns the equipment in the first place, which it need not – see section 18.2 above.
12 See the exposition of the law by the House of Lords in *Elitestone v Morris* [1997] 1 WLR 687.

18.5.5

Taken together with the point discussed in section 18.6 below, the result is that the conferment of the appropriate code rights upon an operator simply gives that operator the right (for example) to install apparatus upon land. It does not give the operator any proprietary interest in the land; it does not give the landowner any right in the apparatus; and it does not give any other operator any opportunity to gain code rights over that apparatus (although the landowner or occupier may of course confer other code rights over the same land, provided that that is physically possible).

18.5.6

The primary significance of this is that a rival operator cannot piggy-back upon another operator's apparatus (without that operator's consent – see Chapter 20 on sharing apparatus). This is so even if the structure is not actually being used by the operator: it suffices if it is 'designed or adapted for use in connection with the provision of an electronic communications network'.

18.6 Conclusion: reasons for and consequences of the apparatus/land distinction

18.6.1

Paragraph 108 of the New Code states that 'land' does not include 'electronic communications apparatus'. The draftsman clearly envisaged that in any given case, it ought to be possible to distinguish between (a) the land upon which electronic communications apparatus is installed, on the one hand, and (b) the apparatus itself, on the other.

18.6.2

In the paper issued by the DCMS in May 2016: 'A New Electronic Communications Code', the Government said:

> there has been considerable debate on the definition of land within the Code, and in particular whether 'apparatus' should be regulated under the new Code. The Government received a number of responses on this issue, and there were strongly opposing views on all sides, suggesting the legal position under the current Code to be ambiguous. However, the original purpose of the Code was to allow access to land so that communications infrastructure could be installed rather than to allow access to the infrastructure itself. That rationale has not changed, and Government does not want to increase regulation and risk disruption of market incentives for investment in passive infrastructure. There is an existing and well understood legal framework in place to provide for access to apparatus in cases where there is significant market power and/or anticompetitive behaviour. As the UK's independent regulator for telecommunications, Ofcom is responsible for ensuring effective competition in telecommunications markets. Given this, the Government will exclude apparatus from the scope of land within the Code and avoid 'gold-plated' regulation.

18.6.3

The intended purpose of the New Code drafting was therefore to enable access to land, rather than to allow operators access to another operator's apparatus, at least without that other's consent.

18.6.4

Of course, where the operator does consent, then the sharing provision in the New Code will help to ensure that the incoming operator may install its apparatus – even where the landowner is resistant.

18.6.5

Where the existing operator is unwilling to allow sharing, and is abusing a dominant position in so doing, OFCOM has a number of statutory powers which could enable it, in principle, to regulate the terms on which operators should nevertheless grant access to their infrastructure. This topic is considered further in section 41.12.

19 The agreement conferring code rights

19.1 Introduction

19.1.1

A number of provisions in the New Code refer to Part 2 of the New Code as being the means by which an agreement conferring code rights is made (see for example paragraph 15 in Part 3, which refers to 'agreements under Part 2').

19.1.2

There is a trap here for the unwary. Such an agreement is not necessarily a 'Code Agreement'. That expression is defined by paragraph 29(5) of the New Code to mean an agreement to which Part 5 applies. Paragraph 29(1) stipulates that that Part applies to an agreement under Part 2, but only where (a) the primary purpose[1] of the agreement is to grant code rights, and (b) the agreement is not a lease to which Part 2 of the Landlord and Tenant Act 1954 applies.[2] The result is that an agreement under Part 2 *is not necessarily a Code Agreement.*

19.1.3

For ease of expression, an agreement under Part 2, with which this chapter deals, is referred to as a 'conferring agreement', and it should be kept in mind that, when it comes to termination of the agreement, both parties may discover that their agreement is not a full code agreement, but requires an entirely different mode of termination to that laid down under Part 5 of the New Code. This important topic is dealt with in Chapter 21.

19.1.4

Paragraph 9 of the New Code provides, under the heading 'Who may confer code rights?':

1 As to the meaning of the expression 'primary purpose' see Chapter 22, paragraphs 22.2.5 et seq and Chapter 25, paragraphs 25.2.2 et seq.

2 Paragraph 29 in fact uses a double negative, by providing that an agreement under Part 2 is not a code agreement if (a) its primary purpose is not to grant code rights and (b) it is a lease to which Part 2 of the Landlord and Tenant Act 1954 – but the effect is the same.

> A code right in respect of land may only be conferred on an operator by an agreement between the occupier of the land and the operator.

What is therefore required in order for code rights to be conferred (in default of the compulsory process described in Chapter 20) is an agreement between the 'occupier of the land' (as to which, see Chapter 15) and 'the operator' (see Chapter 16). Because of the cautionary point made in section 19.1.3, such an agreement is referred to in this chapter as a 'conferring agreement' rather than a 'code agreement'.

19.1.5

Paragraph 11(1) of the New Code sets out formalities which must be complied with for a valid conferring agreement, under the heading 'Requirements for agreements', as follows:

> An agreement under this Part –
> (a) must be in writing,
> (b) must be signed by or on behalf of the parties to it,
> (c) must state for how long the code right is exercisable, and
> (d) must state the period of notice (if any) required to terminate the agreement.

19.1.6

This chapter deals with the detail of these requirements (section 19.2); with the consequences of inadvertent or deliberate failures to comply with these requirements (section 19.3); and with contracting out of the New Code (section 19.4).

19.2 The requirements for conferring agreements

19.2.1

The conferring agreement requirements are simply stated in paragraph 11(1) of the New Code (but always remembering that compliance with these requirements will create an agreement conferring code rights, but not necessarily a code agreement – see parargraph 19.1.3 above), and at first sight it seems that there will therefore be no room for any doubt as to whether what needs to be done in order to satisfy the requirements. Practice will determine whether this is so – and in the meantime this section sets out a number of considerations that may be material in ascertaining whether the particular ingredients have been satisfied.

19.2.2 Requirement (a): writing

The requirement that the conferring agreement be in writing obviously does not contemplate only handwriting for its text. Writing, in other words, encompasses typewritten and other text.[3] Quite how far the substitution of technology goes is debatable,

3 So much is confirmed by Schedule 1 to the Interpretation Act 1978, which applies here, and which provides: '"Writing" includes typing, printing, lithography, photography and other modes of representing or reproducing words in a visible form, and expressions referring to writing are construed accordingly'.

however: for example, does email text suffice for an agreement in writing? The problem is at its most acute when it comes to the assessment of typed or digital signatures (see paragraphs 19.2.3 et seq below).

19.2.3 Requirement (b): signed by or on behalf of the parties to it

At least two problems may arise in relation to the requirement that the conferring agreement be *signed by or on behalf of the parties to it*. The first is what qualifies for a signature. The following permutations may be encountered (although others will no doubt arise):

(a) Occupier (A) and Operator (B) both sign the same document with ink signatures.
(b) A and B apply inked stamps bearing their names above a signature block.
(c) A and B apply their digital signatures to the agreement above a signature block.
(d) A and B exchange emails referring to an attached draft agreement, and stating in the email text that the agreement is confirmed as final and that the email exchange is to amount to execution.

Permutation (a) is obviously effective. The other permutations are examined briefly in paragraph 19.2.4.

19.2.4

The New Code is silent on the question what constitutes a signature, and the Interpretation Act gives no assistance either. Judicial authority is also relatively sparse, with most of it to do with signature requirements in other statutes, where the policy considerations will or may have been quite different.[4] More generally, in *Golden Ocean Group Ltd v Salgaocar Mining Industries Pvt Ltd*,[5] the Court of Appeal held that a sequence of negotiating emails in which terms were agreed was capable of satisfying the requirements of the Statute of Frauds that a contract of guarantee (or some memorandum or note thereof) must be in writing and signed by or on behalf of the party to be charged. It held too that an electronic signature was likewise sufficient to satisfy the statute.

19.2.5

Although this authority is helpful, until the judicial approach to the signature requirement in the New Code becomes settled it will clearly be good practice to adopt conventional means of signing conferring agreements. In the case of a company signatory, it will of course be necessary in addition to comply with the requirements of s.44 of the Companies Act 2006.

4 See e.g. *43 Cowthorpe Road 1-1A Freehold Ltd v Wahedally* [2017] L & TR 4, in which HH Judge Dight said, in construing the requirement of s.99(5) of the Leasehold Reform, Housing and Urban Development Act 1993 that 'Any notice which is given … must be signed by or on behalf of each of the tenants': 'It is, in my judgment, not possible in the ordinary sense of the word to sign an electronic document with an original signature'.

5 [2012] 1 WLR 3674.

19.2.6

The second problem that may arise in relation to the requirement in paragraph 11(1)(b) of the New Code that the conferring agreement must be signed *by or on behalf of the parties to it* occurs if identical copies of a purported conferring agreement are exchanged, with each party having signed one copy. The effect will be that no one copy is signed by *both* parties to it. It is uncertain whether this will suffice for the purposes of paragraph 11(1).

19.2.7

A contrast is provided by the rather clearer drafting of s.2(3) of the Law of Property (Miscellaneous Provisions) Act 1989, which provides, in relation to contracts for the sale or other disposition of an interest in land:

> The document incorporating the terms or, where contracts are exchanged, one of the documents incorporating them (but not necessarily the same one) must be signed by or on behalf of each party to the contract.

Given that contrast, it may be said to be arguable that a purported code agreement which is signed by one party on each copy is not a valid code agreement.

19.2.8 Requirement (c): state for how long the code right is exercisable

The formulation of this requirement neatly avoids the problems that arise in relation to purported leasehold arrangements where certainty of term is a fundamental requirement. It is suggested that the following would be compliant conferring agreements:

(a) An agreement of unspecified length, to run until the landowner should require the site of the apparatus for redevelopment (compare *Prudential Assurance Co Ltd v London Residuary Body*,[6] in which the House of Lords held that a purported tenancy to continue until the land was required by the council for the purposes of the widening of a highway, was void for uncertainty);
(b) An agreement to endure until the landowner should die (again, a tenancy for such an uncertain term would be void);
(c) An agreement with no fixed term, to continue until terminated by three months' notice. Again, a tenancy to this effect would be void.

19.2.9 Requirement (d): state the period of notice (if any) required to terminate

Again, this drafting is flexible, allowing for a conferring agreement that provides for any length of termination notice – or even none at all. Where no notice is required (for example where the conferring agreement is for a fixed period), it is questionable where this requirement nevertheless means that the conferring agreement should stipulate words to the effect 'no notice is required to terminate this agreement'. The authors suggest that this would be a very technical reading of the New Code, although cautious draftsmen may like to include a statement to this effect until the law has been tested.

6 [1992] 2 AC 386.

19.3 The effect of failure to comply with the requirements for code agreements

19.3.1

The requirements set out in paragraph 11(1) of the New Code which must be complied with for a valid conferring agreement (see paragraph 19.1.2 above) are mandatory. Accordingly any failure to comply will have the consequence that, although the parties may have contracted validly, they will not have created a conferring agreement, and the operator will not therefore have attracted code rights.

19.3.2

Under the Old Code, there were fewer requirements: it sufficed for there to be an 'agreement in writing of the occupier for the time being of any land' (see paragraph 2(1)). Such an agreement might be (and in practice, many were) signed by only one party; or of indeterminate length; or lacking a termination provision. Such defects will be fatal to the validity of an agreement under the New Code.

19.3.3

While the conferring agreement requirements may be simply stated and easy to observe, it may not always be possible to be confident regarding compliance in any given case, as section 19.2 above shows. Non-compliance might appear to be attractive from the point of view of the occupier, and disadvantageous to the operator, but this will not necessarily be the case. The occupier may find, for example, that instead of facing code rights, it is confronted by a tenancy to which Part II of the Landlord and Tenant Act 1954 applies (see Chapter 25).

19.3.4

The effect of paragraph 11 is that informal agreements will not be conferring agreements. Thus in the not uncommon situation of a tenant holding over after the expiration of its contractual term, the continuing payment and acceptance of rent may be said to give rise to an implied periodic tenancy. However, given the terms of paragraph 11, such an implied periodic tenancy will not be a conferring agreement, and accordingly there will be no New Code protection, at least with respect to its termination under Part 5 of the New Code. (It may have protection under the Landlord and Tenant Act 1954.[7]) Albeit such an informal agreement may not be within the terms of Part 5,

7 It is in fact unclear whether an informal agreement which is not a Conferring Agreement because it does not comply with Part 2 may nevertheless fall within the protection of the 1954 Act. Para 4 of Schedule 3 of the DEA 2017 excludes from the 1954 Act, a tenancy 'the primary purpose of which is to grant code rights within the meaning of Schedule 3A to the Communications Code 2003 (the electronic communications code) …' An informal tenancy agreement may nevertheless be a primary purpose agreement in the sense that the purpose of the tenancy may be to grant rights which fall within the code rights provided for by paragraph 3. However, as it is not a Conferring Agreement it has no code protection nor, arguably, any 1954 Act protection. One reading of paragraph 4 of Schedule 3 to the DEA 2017 is that it is necessarily implicit that the tenancy granting code rights is one which would ordinarily satisfy the terms of Part 2 to

it would seem to be arguable from the terms of Part 6 that an informal agreement which is not a conferring agreement, is, nevertheless, within the ambit of Part 6, thus requiring a court order for the removal of any electronic communications apparatus. Paragraph 37(1) provides that 'a person with an interest in land … has the right to require the removal of electronic communications apparatus on under or over the land *if (and only if)* one or more of the following conditions are met' (emphasis added). There is no requirement in Part 6, unlike Part 5, for the agreement to be an 'an agreement under Part 2' (see paragraph 29).

19.4 Contracting out of the New Code

19.4.1

Section 19.3 above deals with *inadvertent* non-compliance with the requirements for a valid conferring agreement. It prompts the question whether the parties can *deliberately* avoid the New Code, and enter into an agreement which is not a code agreement. Common sense suggests that this must be so: if they can avoid the New Code inadvertently, then it must surely follow that they can do so deliberately.

19.4.2

Chapter 34 deals with the extent to which the New Code makes provision to prohibit contracting out, offering a contrast with many comparable pieces of property legislation (see e.g. s.25 of the Landlord and Tenant (Covenants) Act 1995). Further, the draftsman of the New Code has clearly found himself able, where necessary, to provide that certain provisions (for examples restrictions on assignment, upgrading or sharing – see Chapter 19) should be void, which suggests that the draftsman was fully alert to the task of prohibiting certain contractual avoidance, but chose not to exercise it in relation to complete code avoidance. On the face of it, therefore, there would appear to be nothing preventing the parties arranging their affairs so as to engineer an agreement for the installation of apparatus that is not a conferring agreement. It is to be noted that the anti-avoidance provisions contained in paragraph 100 apply only to Parts 3–6 of the New Code: paragraph 100(2).

19.4.3

That was not the original intention of the draftsman of the New Code. The paper issued by the Department for Culture, Media and Sport in May 2016, 'A New Electronic Communications Code',[8] stated:

> We have given careful consideration to all the views expressed by stakeholders on this issue. We recognize that there is a divergence of views as to whether there

the code, and the draftsman was not contemplating 'code rights' falling within paragraph 3 granted by an agreement which happened to fall outside Part 2 but was nevertheless one which could be described as 'primary purpose' agreement.

8 www.gov.uk/government/uploads/system/uploads/attachment_data/file/523788/Electronic_Communications_Code_160516_CLEAN_NO_WATERMARK.pdf

> should be measures in the reformed Code on the ability to contract out. We also recognize that the existing Code is unclear on the flexibility to contract out. The Government is bringing forward a new Code that makes significant policy changes in important areas in order to support investment in network growth and sustainability, and equip the UK with the best possible digital communications infrastructure. Given this, on balance, the Government considers that any attempts by one or more parties to gain advantage by circumventing the new Code's provisions must be prohibited if the Code is to be truly effective. We will therefore make provision in the revised Code to prohibit the ability to contract out and stop parties making private agreements capable of excluding Code provisions.

19.4.4

This legislative intention does not appear to have survived to its fullest extent into the final draft of the New Code. Accordingly, to the extent that they so wish, occupier and operator may indeed ensure that their agreement is not a conferring agreement.

19.4.5

There are various ways in which this might be accomplished, which are explored in more detail in Chapter 34. At its simplest, the agreement could be oral; or it could be signed by just one of the parties; and so on.

20 Assignment, upgrading and sharing apparatus

20.1 Introduction

20.1.1 General

Part 3 of the New Code brackets together a number of provisions that govern the assignment of code rights, and upgrading and sharing of apparatus. This chapter takes the same approach, with the sections that follow dealing separately with each of the three topics.

20.1.2

As a general comment, it may be said that these provisions in the New Code impose a liberal regime in the operators' favour. In the more recent days of the Old Code, disputes between landowners and operators concerning the ability of the latter to diversify by sharing their sites with other operators, or by installing more equipment, had become commonplace, particularly given that increasing cooperation between operators made it sensible to seek to share equipment on one site, rather than use equipment on two sites. Landowners naturally sought, and often gained, higher rents in return for such flexibility. The policy of these provisions in the New Code is to allow operators to do that which they had to secure by negotiation, and pay for, under the Old Code.

20.1.3

For ease of expression, an agreement conferring code rights is referred to in this chapter as a 'code agreement', although strictly speaking it may not be an agreement to which Part 2 of the New Code applies (see the explanation given in section 19.1 of this book).

20.1.4 Interpretation of provisions

On the face of the provisions, there appears to be a drafting ambiguity. It is unclear as to whether the paragraphs are power-conferring, i.e. providing for an entitlement which would not otherwise exist under the terms of the code agreement conferring the code right, unless expressly agreed; or whether they simply limit the ability of the parties to contract out what is otherwise conferred expressly or impliedly by the code right. This issue is relevant not only for the purposes of considering the true ambit of paragraphs 16 and 17, but also impacts upon the determination of consideration under paragraph

24, where paragraph 24(3)(b) provides expressly that 'subparagraphs 16 and 17 ... do not apply to the right or any apparatus to which it could apply' in determining the assessment of the consideration under Part 4.

20.1.5

Paragraph 15, by way of introduction to Part 3, provides that it 'makes provision for (a) operators to assign agreements under Part 2, (b) operators to upgrade the electronic communications apparatus to which such an agreement relates, and (c) operators to share the use of any such electronic medications apparatus'. Paragraph 16 appears by its terms to be simply a limitation on the ability of the parties to contract out of any right to assign the code agreement to another operator. The inference is that the code agreement is freely assignable unless restricted, and any restriction is prohibited to the extent provided for by paragraph 16. In contrast, paragraph 17 appears from its wording to be a power-conferring provision. It provides that the operator 'may, if the conditions' to which the paragraph refers are satisfied, 'upgrade the electronic medications apparatus to which the agreement relates' or 'share the use of such electronic communications apparatus with another operator'. Although not an aid to construction, the heading of paragraph 17 refers to 'Power for operator to upgrade or share apparatus', in contrast to paragraph 16, which refers to simply 'Assignment of code rights'.

20.1.6

In paragraph 3.42 of Law Com 336, the Law Commission refers to paragraph 17 as 'an exception' to the general rule that there can be no automatic right to share or upgrade apparatus unless negotiated or granted by the tribunal under Part 4, thus:

> So in general, it is not possible for Code Operators to have an automatic right to share or to upgrade equipment. Such rights must be negotiated for, or granted by the tribunal; it may be right for there to be additional consideration payable, depending upon the market itself. The same goes for rights to maintain and repair equipment, which cannot be conferred automatically; the range of technical implications, from access to safety to structural integrity, is such that automatic rights cannot be given and it is for the parties to negotiate them or for the tribunal to confer them.

The Commission added in paragraph 3.45 of the Report:

> We take the view that there may be an exception to the concerns expressed [above], in that there are clearly identifiable cases where upgrading and sharing have no physical implications at all and cannot be seen because they are physically confined to a space controlled by the Code Operator. The obvious example is the addition of fibre in a duct or sub-duct, which can be achieved by simply 'blowing' the fibre without impact on the land.

20.1.7

Although as a matter of language it is possible to read paragraph 17 as a limitation on the ability to contract out of an automatic right to upgrade or share apparatus, so that it operates in a manner similar to paragraph 16, (with the inference that the right to upgrade or share apparatus is unlimited and automatic unless otherwise restricted by the terms of the code agreement), it is considered that this is not the correct view have regard to the terms of the Law Commission Report and the contrasting wording between paragraphs 16 and 17. The Law Commission Report will be admissible as an aid to the true interpretation of the provisions and may be thought to be persuasive if not positively determinative.

20.2 Assignment of a code agreement

20.2.1

Assignment of tenancies is a topic that has attracted much legislative attention in England and Wales over the years (see in particular s.19 of the Landlord and Tenant Act 1927 and the Landlord and Tenant (Covenants) Act 1995), and disputes over process and substance in relation to proposed assignments frequently occur in leasehold.

20.2.2

The New Code aims to avoid these complications, first by defining code agreements in such a way as to sidestep leasehold tenure altogether (see Chapter 14), together with all its legislative baggage; and secondly by laying down, in paragraph 16 of the New Code, clear stipulations as to the rights and obligations of owners and operators in relation to proposed assignments of code agreements.

20.2.3

The text of paragraph 16 of the New Code, which is headed 'Assignment of code rights', is as follows:

(1) Any agreement under Part 2 of this code is void to the extent that –
  (a) it prevents or limits assignment of the agreement to another operator, or
  (b) it makes assignment of the agreement to another operator subject to conditions (including a condition requiring the payment of money).

(2) Sub-paragraph (1) does not apply to a term that requires the assignor to enter into a guarantee agreement (see sub-paragraph (7)).

(3) In this paragraph references to 'the assignor' or 'the assignee' are to the operator by whom or to whom an agreement under Part 2 of this code is assigned or proposed to be assigned.

(4) From the time when the assignment of an agreement under Part 2 of this code takes effect, the assignee is bound by the terms of the agreement.

(5) The assignor is not liable for any breach of a term of the agreement that occurs after the assignment if (and only if), before the breach took place, the assignor or the assignee gave a notice in writing to the other party to the agreement which –
  (a) identified the assignee, and
  (b) provided an address for service (for the purposes of paragraph 91(2)(a)) for the assignee.

(6) Sub-paragraph (5) is subject to the terms of any guarantee agreement.

(7) A 'guarantee agreement' is an agreement, in connection with the assignment of an agreement under Part 2 of this code, under which the assignor guarantees to any extent the performance by the assignee of the obligations that become binding on the assignee under sub-paragraph (4) (the 'relevant obligations').

(8) An agreement is not a guarantee agreement to the extent that it purports –
- (a) to impose on the assignor a requirement to guarantee in any way the performance of the relevant obligations by a person other than the assignee, or
- (b) to impose on the assignor any liability, restriction or other requirement of any kind in relation to a time after the relevant obligations cease to be binding on the assignee.

(9) Subject to sub-paragraph (8), a guarantee agreement may –
- (a) impose on the assignor any liability as sole or principal debtor in respect of the relevant obligations;
- (b) impose on the assignor liabilities as guarantor in respect of the assignee's performance of the relevant obligations which are no more onerous than those to which the assignor would be subject in the event of the assignor being liable as sole or principal debtor in respect of any relevant obligation;
- (c) make provision incidental or supplementary to any provision within paragraph (a) or (b).

(10) In the application of this paragraph to Scotland references to assignment of an agreement are to be read as references to assignation of an agreement.

(11) Nothing in the Landlord and Tenant Amendment (Ireland) Act 1860 applies in relation to an agreement under Part 2 of this code so as to –
- (a) prevent or limit assignment of the agreement to another operator, or
- (b) relieve the assignor from liability for any breach of a term of the agreement that occurs after the assignment.

20.2.4

A number of points deserve to be made about this drafting.

20.2.5

First, the restrictions upon alienation only apply where the proposed assignment is to *another operator*. There is therefore nothing preventing the parties agreeing that an assignment to an entity that is *not* an operator should be restricted. Moreover, there is of course nothing preventing the parties stipulating that the operator should *only* be permitted to assign *to another operator.*

20.2.6

Secondly, although paragraph 16 is headed 'assignment of code rights', it in fact deals with the assignment of code agreements. There is no process whereby code rights may be assigned individually. This in itself presents a novel point under English (but not Scottish) law, which has long held that a contract cannot usually be assigned, for the reason that, although the benefit of some contracts is usually freely assignable, the burden is not.[1] The code appears to present a statutory route around this problem. Upon assignment, therefore, not merely will the assignee (if an operator) enjoy code rights under the agreement, but it will also become subject to the obligations under the agreement, without any novation (see paragraph 16(4)).

20.2.7

Thirdly, paragraph 16 prohibits any clause which prevents, limits, or imposes conditions upon the assignment of a code right agreement, only 'to the extent' that it purports to have that effect. Accordingly, a compendious clause which offends paragraph 16, but also provides for other matters, will only have a blue pencil taken to the part that is offensive, leaving intact the remainder. There is much law concerning the severability of provisions in relation to the Landlord and Tenant (Covenants) Act 1995 (which uses the same drafting device), which may be useful by analogy on the question just how widely the blue pencil should be wielded.[2]

20.2.8

Fourthly, paragraph 16(1) has the consequence that a provision of the sort routinely encountered, and regarded as reasonable in leasehold tenure, to the effect that assignment may not be effected without consent, with consent not to be unreasonably withheld, will be void, as a term that prevents or limits assignment of the agreement to another operator. Occupiers may therefore reasonably apprehend that the covenant strength of their operator may quickly be lost, in return for a poorly performing or start-up operator. However, the guarantee provision (see paragraph 20.2.11 below) will be of some comfort to occupiers. There is nothing in paragraph 16 (although operators may insist upon this in the code agreement) that imposes any obligation upon the occupier to act reasonably in relation to the proposed guarantee, and occupiers would therefore be prudent to ensure that the value of their land is not diminished upon assignment, by seeking to ensure a suitable guarantee provision.

20.2.9

Fifthly, paragraph 16 is silent on the question of subletting or sub-licensing, from which it would appear to be safe to draw the conclusion that a code agreement may validly prohibit the operator allowing another operator to rent apparatus from it (although it may of course validly share instead – see section 20.4 below).

20.2.10

Sixthly, the provisions for release of the assignor upon assignment follow the approach taken by the Law Commission in their 2013 Report (Law Com 336). In paragraph 3.27, the authors recommended:

1 See the decision of the Court of Appeal in *Budana v The Leeds Teaching Hospitals NHS Trust* [2017] EWCA Civ 1980, and in particular the summary of the law in the judgment of Gloster LJ at paragraph 26.
2 See e.g. *Tindall Cobham 1 Ltd v Adda Hotels* [2015] 1 P & CR 5.

that the revised Code should provide that on the assignment of the benefit of an agreement or the lease pursuant to the recommendations made above: (1) either the assignor or the assignee shall give notice to the Site Provider of the identity, and address for service, of the assignee; and (2) the assignor shall not be released from its obligations under the agreement or lease until this notice has been given (notwithstanding the provisions of section 5 of the Landlord and Tenant (Covenants) Act 1995).

20.2.11

In this respect, there are three further points to make. First, if the assignor does not take the course provided by paragraph 16(5), then although the assignee will become liable from the point of assignment notwithstanding the assignor's failure, the assignor will remain liable to the occupier until the appropriate notice has been given. If through oversight no notice is ever given, the assignor will remain liable until the end of the code agreement. There is no reference in paragraph 16(5) to the terms of the Landlord and Tenant (Covenants) Act 1995, the effect of which is that upon assignment of a lease the tenant is released from the tenant covenants of the tenancy: s.5 of the 1995 Act. The release of the tenant does not affect any liability of his arising from a breach of the covenant occurring before the release: s.24(1) of the 1995 Act. However, there is no continuing liability of the tenant assignor by reason of the failure to serve any form of notice upon the landlord as would appear to be provided for by paragraph 16(5). How, in the case of a lease, the terms of paragraph 16(5) 'fit' with the 1995 Act is unclear.

20.2.12

The second point is that paragraph 16(2) makes it clear that the prohibition upon the owner imposing conditions on the occasion of an assignment does not prevent the occupier requiring the assignor to guarantee the assignee's code agreement liabilities for so long as the assignee remains liable (see paragraph 16(8)(b)). As Law Com 336 suggests, the drafting is reminiscent of the position under the 1995 Act, whereby, in an express exception to the principle that assignors escape further tenant liability upon assignment, the landlord can require the assignor to guarantee the assignee's liabilities under s.16. Paragraph 16(7) is, however, rather simpler in its requirements.

20.2.13

The third point is that, just as the assignor's liability may continue after the assignment if it fails to give the appropriate notice, so too may the guarantor's liability continue (assuming appropriate drafting).

20.3 Upgrading apparatus

20.3.1

The operator's ability to upgrade electronic communications apparatus is dealt with in paragraph 17 of the New Code, under the title 'Power for operator to upgrade or share apparatus'. Those parts of paragraph 17 which deal with upgrading provide:

(1) An operator ('the main operator') who has entered into an agreement under Part 2 of this code may, if the conditions in sub-paragraphs (2) and (3) are met –
(a) upgrade the electronic communications apparatus to which the agreement relates …
(2) The first condition is that any changes as a result of the upgrading … to the electronic communications apparatus to which the agreement relates have no adverse impact, or no more than a minimal adverse impact, on its appearance.
(3) The second condition is that the upgrading … imposes no additional burden on the other party to the agreement.
(4) For the purposes of sub-paragraph (3) an additional burden includes anything that –
(a) has an additional adverse effect on the other party's enjoyment of the land, or
(b) causes additional loss, damage or expense to that party.
(5) Any agreement under Part 2 of this code is void to the extent that –
(a) it prevents or limits the upgrading …, in a case where the conditions in sub-paragraphs (2) and (3) are met, of the electronic communications apparatus to which the agreement relates, or
(b) it makes upgrading … of such apparatus subject to conditions to be met by the operator (including a condition requiring the payment of money).
(6) …

20.3.2

This drafting is straightforward, although a small number of points deserve to be made.

20.3.3

First, as paragraph 15.3.2 notes, there is no definition of the term 'upgrading'. The Authors there suggest that it may be said to be doubtful whether 'upgrading' extends to replacement of the whole apparatus with a new installation. As against that, the Law Commission noted in paragraph 2.39 of their Report No. 336, a reference to the suggestion that the comparable Old Code words 'installation, maintenance, adjustment, repair or alteration' be removed, on the basis that they generated 'sterile argument about definitions'. The Law Commission said, in response to this:

> Arguably a right 'to execute works on land in connection with electronic communications apparatus' includes those additional verbs. We agree; but of course in this Report we are not drafting the revised Code. Our recommendation below sets out our policy; the drafter of the revised Code will wish to bear in mind our suggestion that the list be kept as simple and inclusive as possible.

20.3.4

The draftsman of the New Code did not embrace this suggestion, choosing instead to enlarge the number of verbs used to describe permissible works, and thereby, as always with more detailed drafting, omitting other works by default. The Law

Commission's view was clearly that operators should have a considerable degree of flexibility in relation to the works they are entitled to do as part of their code rights. Ultimately, however, Parliament has seen fit to confine, or at any rate describe, further works following the initial installation as 'upgrading'. It therefore remains arguable that complete replacement is not upgrading. Even if it is not, the replacement of subsidiary parts falling short of the whole would be accommodated within the term 'upgrade', such that the entire replacement of one component such as an antenna on a mast would be an upgrade, although the replacement of the mast itself might not be.

20.3.5

Secondly, this paragraph will be irrelevant if the operator has been granted, or had imposed, a full code right to upgrade (see paragraph 3(c) of the New Code, discussed in paragraph 15.2.1). If no such code right has been granted, then this provision affords some assistance to the operator, by allowing upgrading even in the absence of the relevant code right, provided that the two specified conditions are met.

20.3.6

Thirdly, although both the specified conditions involve nuanced value judgments, the application of which may be a source of contention, the drafting of them is markedly different. The first condition, which deals with external appearance, is subject to a de minimis exception ('no more than a minimal adverse impact'). By contrast, the second condition has no such exception, prompting the conclusion that any additional adverse effect or loss, damage or expense, no matter how small, will have the consequence that this condition will not be met, and thus that the upgrading may not be carried out. It remains to be seen whether the Tribunal will adopt that view. It is worth adding that the point deepens in complexity when the change is minor in itself, but when taken together with previous changes imposes a considerable burden.

20.3.7

The first condition is most obviously directed at the upgrading of apparatus in the shape of replacement of parts within an existing equipment or structure, or the placing of additional receivers or transmitters on an existing pylon.

20.3.8

The second condition would appear to be concerned with remedial measures that may be required if the upgrading is carried out, such as additional security or insurance costs that the occupier may have to meet as a result of more intensive access to the site.[3]

3 The need for such works may arise as a result of The Network and Information Systems Regulations 2018 (SI 2018/506) (in force from 10 May 2018), regulation 10 of which stipulates that an operator of essential services 'must take appropriate and proportionate technical and organisational measures to manage risks'.

20.3.9

Where these tests are unlikely to be met, the operator may choose either to proceed with the works anyway, on the footing that, if the site provider objects, it has its own steps open to it under the code; or to seek extended code rights to encompass the upgrading.

20.3.10

Lastly, paragraph 17(5) imposes an anti-avoidance provision, to which the comments made above in paragraph 20.2.7 in relation to anti-avoidance in the case of assignment also apply.

20.4 Sharing telecommunications apparatus

20.4.1

The term 'sharing' in relation to telecommunications covers a wide range of possibilities, some of which were highlighted by the Law Commission in paragraph 3.31 of its Report No. 336, as follows:

> Code Operator A, which is occupying a site with a mast or conduit, may allow Code Operator B to install physical apparatus there, or to have access to the existing apparatus to route its electronic communications network (for example, for its customers' mobile phone calls). Sharing in a wider sense – A's infrastructure being used for B's business – may effectively be achieved by a variety of methods, subject to the terms of the agreement with the Site Provider where relevant. If Code Operator X is purchased by Code Operator Y, yet retains its corporate identity, the contractual arrangement with a Site Provider may be unchanged yet it may seem that an assignment has taken place; if Y then uses X's mast, that may seem like sharing, rather than the expansion of the Code Operator which is on site.

20.4.2

Against that background, paragraph 3.32 of the same report posed the question:

> Whether the Code should make special provision to permit sharing of apparatus – say, a mast, or a conduit – irrespective of the terms of the agreement with the Site Provider.

20.4.3

The facilitation of sharing is something which the Old Code had promoted, but largely without effect. For example:

(a) s.107(4) of the 2003 Act provided:

> In considering whether to apply the electronic communications code in any person's case, OFCOM must have regard, in particular, to each of the following matters – …

(c) *the need to encourage the sharing* of the use of electronic communications apparatus ...

(b) to this, s.109(1) of the 2003 Act: (Restrictions and conditions subject to which code applies) added:

> Where the electronic communications code is applied in any person's case by a direction given by OFCOM, that code is to have effect in that person's case subject to such restrictions and conditions as may be contained in regulations made by the Secretary of State. (2) In exercising his power to make regulations under this section it shall be the duty of the Secretary of State to have regard to each of the following – ... (d) *the need to encourage the sharing* of the use of electronic communications apparatus.

(c) s.134 of the 2003 Act imposed further controls on practices restrictive of sharing, although these were rarely encountered in practice.[4]

20.4.4

Article 12 of the Framework Directive,[5] draws attention to the ability of national regulatory authorities to impose the sharing of electronic communications facilities or of property. As a result, the issue of site and infrastructure sharing has been expressly included in governmental planning policy guidance for several years.[6]

20.4.5

The applicable domestic regulations were to the same effect. Thus, paragraph 3(4) of the Electronic Communications Code (Conditions and Restrictions) Regulations 2003 (2003 No. 2553) provides:

> A code operator, where practicable, *shall share* the use of electronic communications apparatus.

20.4.6

Ultimately, these exhortations did not have the effect the operators desired, in that many arrangements under the Old Code did not permit sharing, at least without the payment of additional consideration to the occupier.

20.4.7

The Law Commission considered that there could be little objection to sharing and upgrading within a duct or a cabinet on land, without physical or visual impact on the site provider. As it said in paragraph 3.8 of Law Com 336:

4 See the commentary in paragraphs 3.55 to 3.59 of Law Com 336.

5 Directive 2002/21/EC of 7 March 2002 on a common regulatory framework for electronic communications networks and services.

6 See Department for Communities and Local Government, *National Planning Policy Framework* (March 2012), paragraphs 43 to 45 (which has superseded *Planning Policy Guidance 8: Telecommunications* (August 2001)), available at www.gov.uk/government/publications/national-planning-policy-framework – 2.

We think that the situation where sharing or upgrading takes place within the confines of a duct, or even of a cabinet on land, without physical or visual impact on the Site Provider, without requiring a power supply or the addition of an antenna for example, and without conferring Code Rights on additional Code Operators, ought to be permitted. These are cases where there is no possible additional burden on the Site Provider and no technical or safety issues. If, in the case of sharing, the additional Code Operator requires Code Rights, it can negotiate these independently with the Site Provider, or apply for them to be imposed by the Tribunal.

20.4.8

These sentiments have found their way into the New Code. The operator's ability to share electronic communications apparatus is dealt with in paragraph 17 of the New Code, under the title 'Power for operator to upgrade or share apparatus'. Those parts of paragraph 17 which deal with sharing provide:

> (1) An operator ('the main operator') who has entered into an agreement under Part 2 of this code may, if the conditions in sub-paragraphs (2) and (3) are met –
> (a) …
> (b) share the use of such electronic communications apparatus with another operator.
> (2) The first condition is that any changes as a result of the … sharing to the electronic communications apparatus to which the agreement relates have no adverse impact, or no more than a minimal adverse impact, on its appearance.
> (3) The second condition is that the … sharing imposes no additional burden on the other party to the agreement.
> (4) For the purposes of sub-paragraph (3) an additional burden includes anything that –
> (a) has an additional adverse effect on the other party's enjoyment of the land, or
> (b) causes additional loss, damage or expense to that party.
> (5) Any agreement under Part 2 of this code is void to the extent that –
> (a) it prevents or limits the … sharing, in a case where the conditions in sub-paragraphs (2) and (3) are met, of the electronic communications apparatus to which the agreement relates, or
> (b) it makes … sharing of such apparatus subject to conditions to be met by the operator (including a condition requiring the payment of money).
> (6) References in this paragraph to sharing electronic communications apparatus include carrying out works to the apparatus to enable such sharing to take place.

20.4.9

It is thought that this New Code facility will be of considerable benefit to operators. All the points made above in section 20.3 above apply with equal effect in the case of proposed sharing. There are however three points to add which are of special relevance to sharing.

20.4.10

First, if the act of sharing requires no facilitative works which would trigger a consideration of the pre-condition in paragraph 17(2), it may be said to be unlikely that the sharing would contravene the other pre-condition in paragraph 17(3) (unless the sharing would involve increased access leading to cost considerations). It is therefore only ever likely to be works of upgrading that will offend against these conditions, rather than the act of sharing itself.

20.4.11

Secondly, paragraph 18 of the New Code contains a set of provisions which concern the effect of agreements enabling sharing between operators and others. This is dealt with in section 20.5 below.

20.4.12

Thirdly, paragraph 17 deals with the sharing of *apparatus* by one operator (say 'Operator 1') with another ('Operator 2'). It is critical to note that Operator 2 *cannot acquire code rights* in such circumstances. That is because (a) paragraph 3 of the New Code provides that a code right is a right 'to install electronic communications apparatus on, under or over the *land*'; while (b) paragraph 107 of the New Code states that 'land' does not include 'electronic communications apparatus'.

20.4.13

Suppose therefore that an MNO wishes to install a piece of broadcasting equipment on another operator's mast, and that other operator and the occupier of the land are content to allow the installation. If the mast is *apparatus*, then the MNO will not acquire code rights, for the reasons just explored. If however the mast is part of the land and not apparatus (as to which, see Chapter 18 of this book), then the converse will follow.

20.4.14

This distinction was drawn by the Law Commission in paragraph 3.52 of its Report No. 336 as follows:

> Sharing, in this sense, of course generates no Code Rights because the additional Code Operator allowed onto the site because of this limited permission to share does not have any legal relationship with the Site Provider. The latter is bound by the same Code Rights as before, and once he or she is entitled to have apparatus removed he or she will encounter the provisions about removal which we discuss in Chapter 6 and which will replace the current paragraph 21. It is important to ensure that where sharing has been allowed automatically, by way of exception to the general rule and within the confines of another structure, the Site Provider is not required to deal with an additional Code Operator. So where Operator 1 has shared its ducts with Operator 2 under the provisions that we have recommended, the duct remains protected under the provisions we discuss in Chapter 6. Once the

Site Provider has the right, under those provisions, to have the duct removed, he will be entitled to have it removed as a whole, along with Operator 2's fibres. The same would apply where the structure in question is a cabinet containing equipment.

20.5 Effect of agreements enabling sharing of apparatus

20.5.1

Paragraph 18 of the New Code is headed 'Effect of agreements enabling sharing between operators and others', and provides:

> (1) This paragraph applies where –
> (a) this code has been applied by a direction under section 106 in a person's case,
> (b) this code expressly or impliedly imposes a limitation on the use to which electronic communications apparatus installed by that person may be put or on the purposes for which it may be used, and
> (c) that person is a party to a relevant agreement or becomes a party to an agreement which (after the person has become a party to it) is a relevant agreement.
> (2) The limitation does not preclude –
> (a) the doing of anything in relation to that apparatus, or
> (b) its use for particular purposes,
> to the extent that the doing of that thing, or the use of the apparatus for those purposes, is in pursuance of the relevant agreement.
> (3) This paragraph is not to be construed, in relation to a person who is entitled or authorised by or under a relevant agreement to share the use of apparatus installed by another party to the agreement, as affecting any consent requirement imposed (whether by an agreement, an enactment or otherwise) on that person.
> (4) In this paragraph – 'consent requirement', in relation to a person, means a requirement for the person to obtain consent or permission to or in connection with –
> (a) the installation by the person of apparatus, or
> (b) the doing by the person of any other thing in relation to apparatus the use of which the person is entitled or authorised to share;
> 'relevant agreement' means an agreement in relation to electronic communications apparatus which –
> (a) relates to the sharing by different parties to the agreement of the use of that apparatus, and
> (b) is an agreement that satisfies the requirements of sub-paragraph (5).
> (5) An agreement satisfies the requirements of this sub-paragraph if –
> (a) every party to the agreement is a person in whose case this code applies by virtue of a direction under section 106, or
> (b) one or more of the parties to the agreement is a person in whose case this code so applies and every other party to the agreement is a qualifying person.
> (6) A person is a qualifying person for the purposes of sub-paragraph (5) if the person is either –

(a) a person who provides an electronic communications network without being a person in whose case this code applies, or

(b) a designated provider of an electronic communications service consisting in the distribution of a programme service by means of an electronic communications network.

(7) In sub-paragraph (6) –

'designated' means designated by regulations made by the Secretary of State;

'programme service' has the same meaning as in the Broadcasting Act 1990.

20.5.2

The meaning of this obscure and elaborate drafting is hard to divine. Stripped of its complexity, it appears to permit *either* party to an apparatus sharing agreement to use the apparatus as it pleases, subject to any consent that the code agreement (or any other stipulation) requires. It may therefore have been intended to provide for the situation where a code agreement permits an operator to exercise code rights, but says nothing about any third party – which, in the absence of this paragraph of the New Code, would have no right to do anything at all.[7] If that was indeed its intended purpose, it would have been rather simpler to say 'An operator allowed by paragraph 17 of this Code to share apparatus shall have the following rights as against the occupier, namely ...'

20.5.3

There is one further point to note about paragraph 18. Suppose that an operator with the benefit of a code agreement enters into a sharing agreement, of whatever nature, with a third party, which is either an operator itself, or a qualifying person within the meaning of paragraph 18(6). Paragraph 18(2) allows 'the doing of anything in relation to that apparatus, or its use for particular purposes', to the extent permitted by the agreement between the main operator and the sharer. It is, however, relevant to note the limitations in the wording of paragraph 18.

20.5.4

First, notwithstanding the reference to 'doing anything in relation to that apparatus', which might be said to include the carrying out of alterations by the sharer, it may be said to be doubtful that that is the correct interpretation. That is because the 'relevant agreement' is defined by paragraph 18(4) as an agreement which 'relates to the sharing by different parties ... of the use of that apparatus'. There is no reference to the carrying

7 The Explanatory Notes to the Bill when first introduced in November 2016 supported this interpretation:

> 278 Paragraph 17 [now 18] has the same effect as paragraph 29 of the existing code. Its purpose is to facilitate sharing of apparatus between: (i) one or more operators to whom the code has been applied by Ofcom under section 106 of the 2003 Act; (ii) other providers of electronic communications networks to whom the code has not been applied by Ofcom; and (iii) qualifying persons whom the Secretary of State may designate by regulations.
>
> 279 At least one of the parties to such an agreement must be an operator to whom Ofcom has applied the code. The effect of paragraph 17 [now 18] is that the provisions of the code may not be interpreted so as to restrict the operators' agreement to share apparatus with each other. Paragraph 17 does not affect the interpretation of an agreement between an operator and a site provider.

out of works (although paragraph 17(6), dealt with below, should be noted). Given that the permission afforded by paragraph 18(2) is limited to the extent to which the relevant agreement so provides, the correct conclusion is arguably that paragraph 18 as a whole is dealing only with the legitimisation of the sharing of use by the sharer, and not the grant of any other rights *to the sharer*, such as the right to alter apparatus.

20.5.5

If that is right, then it follows that a 'sharer' would have no rights other than the right to use the apparatus. Even that formulation may be inaccurate, since technically the proper formulation is that the main operator has the right to share; the sharer simply shares by virtue of the main operator's right: it has no independent right of its own.

20.5.6

It would also follow that the sharer has no right to install its own equipment, since it gains no code rights. That this was intended is clear from the Law Commission's Report No. 336, paragraph 3.52 of which states:

> Sharing, in this sense, of course generates no Code Rights because the additional Code Operator allowed onto the site because of this limited permission to share does not have any legal relationship with the Site Provider. … It is important to ensure that where sharing has been allowed automatically, by way of exception to the general rule and within the confines of another structure, the Site Provider is not required to deal with an additional Code Operator. So where Operator 1 has shared its ducts with Operator 2 under the provisions that we have recommended, the duct remains protected under the provisions we discuss in Chapter 6. Once the Site Provider has the right, under those provisions, to have the duct removed, he will be entitled to have it removed as a whole, along with Operator 2's?providers?. The same would apply where the structure in question is a cabinet containing equipment.

20.5.7

Paragraph 3.8 of the same report adds:

> If, in the case of sharing, the additional Code Operator requires Code Rights, it can negotiate these independently with the Site Provider, or apply for them to be imposed by the Tribunal.

Of course, the additional code operator could acquire no such code rights if the structure which it wishes to attach its apparatus itself constitutes apparatus – see Chapter 18.

20.5.8

However, the right to share the use of the main operator's apparatus must implicitly carry with it the right to operate, transmit from and gain access to, the apparatus in question, without which use would not be possible (although there could be circumstances in which the sharer could access the site remotely, without having to access it physically). Equally, however, this is not a right the sharer has vis-à-vis the site provider: it is simply a statutory right the *main operator* has (to share with another), to which the site provider cannot object.

20.5.9

It could be argued that, if access for the purposes of use is implicit, why should works of alteration necessary in order to allow the use to be shared not be equally implicit? The answer to this point may be that the main operator may itself make the requisite alterations, or appoint the sharer its agent so to do (but see the discussion of paragraph 17(6) below). Although this point may well become litigious, the prudent course to adopt, at least until the point is tested, will be to assume that the ability to share under the New Code carries with it no right *for the sharer* to carry out any works at all.

20.5.10

In this context, paragraph 17(6) of the code provides:

> References in this paragraph to sharing electronic communications apparatus include carrying out works to the apparatus to enable such sharing to take place.

20.5.11

Two points arise in relation to this drafting. The first is that there is a marked contrast between this general reference to 'carrying out works', and the precise way in which the works under the code rights are defined. It may be said to be likely that 'carrying out works' in this context will encompass works of installation and repair, provided that in each case the purpose of those works is in order to enable the sharing to take place. Whether these works also include upgrading is more moot. First, initial upgrading, the purpose of which would be to allow the sharer to operate the apparatus, is likely to be part and parcel of the works referred to in paragraph 17(6). To be distinguished from that is works of upgrading which the sharer may wish to have done subsequently, which would improve the extent to which it is able to use the apparatus. This may be said to be more of a stretch for the language used in paragraph 17(6). Ultimately, however, it may be doubted that the Tribunal would be persuaded that there is any meaningful difference between initial upgrading and subsequent upgrading. The only substantive difference is that the degree of sharing may change – but sharing is still involved, and the works of upgrading are carried out to achieve that purpose.

20.5.12

The second point concerns the identity of the person carrying out the works. Paragraph 17(6) is silent on the point. The argument that this must encompass the sharer is founded upon the fact that the main operator would need no such right if it has a code right to upgrade in any event. Paragraph 17(6) must therefore be directed at, or at least include, the sharer. As against that, the point has already been made that the right to share is a right for the main operator to do so, with no right as such being extended to the sharer. If that follows in relation to sharing, it would be incongruous were the same approach not adopted in relation to the works to facilitate that sharing. This point gains in strength from the observation that the ability to upgrade is expressly granted to the main operator under paragraph 17(1), but the sharer is not expressly identified in paragraph 17(6) as being able to carry out the works to which that paragraph refers.

20.5.13

By parity of reasoning (but a fortiori in the case of other code rights), there is no support for the proposition that the sharer may enjoy a mirror version of the main operator's code rights. There is nothing in the New Code that either gives the main operator permission to grant code rights to its sharer, or which extends those rights directly to the sharer.

20.5.14

The question that may be said to arise in these circumstances is how the sharer may maintain any apparatus it may have installed as part of its sharing arrangement with the main operator. As indicated above, the New Code does not envisage anyone other than the main operator carrying out works to the electronic communications apparatus. In the absence of any such provision, but given that the code expressly envisages the use of apparatus, the only way in which any necessary works for that purpose could be carried out is by the main operator executing the works itself. In practice, it will be in the site provider's interests to have the work carried out by the sharer, because in that event (since the sharer will have no code rights itself in relation to its equipment) the site provider will be in a position easily to compel the removal of the apparatus on termination of the code agreement. In practice, therefore, the identity of the operator carrying out work should not matter greatly: what is important is that paragraph 17(6) caters for the works to be done. How, and by whom, they are done should not a matter of great concern to the site provider.

21 Imposition of code agreements

21.1 Introduction

21.1.1

Where the occupier and operator cannot agree whether code rights should be conferred, or what should be the extent of those code rights, the court has power to impose agreement (referred to in this chapter for ease of understanding, but not in the New Code itself, as an 'imposed agreement').

21.1.2

Part 4 of the New Code sets out the provisions that deal with that power, which is explored in this chapter.

21.1.3

The structure of this chapter follows the topics outlined in paragraph 19 of the New Code, namely:

(a) the circumstances in which the court can impose an agreement – see section 21.2;
(b) the test to be applied by the court in deciding whether to impose such an agreement – see section 21.3;
(c) the impact of human rights – see section 21.4;
(d) the effect of an imposed agreement – see section 21.5;
(e) the terms of an imposed agreement – see section 21.6;
(f) how consideration is determined for an imposed agreement – see section 21.7;
(g) rights to the payment of compensation – see section 21.8;
(h) the imposition of an agreement on an interim basis – see section 21.9;
(i) the imposition of an agreement on a temporary basis – see section 21.10.

21.2 The circumstances in which the court can impose an agreement

21.2.1

Paragraph 20 of the New Code is headed 'When can the court impose an agreement?', and provides:

(1) This paragraph applies where the operator requires a person (a 'relevant person') to agree –
   (a) to confer a code right on the operator, or
   (b) to be otherwise bound by a code right which is exercisable by the operator.

(2) The operator may give the relevant person a notice in writing –
   (a) setting out the code right, and all of the other terms of the agreement that the operator seeks, and
   (b) stating that the operator seeks the person's agreement to those terms.

(3) The operator may apply to the court for an order under this paragraph if –
   (a) the relevant person does not, before the end of 28 days beginning with the day on which the notice is given, agree to confer or be otherwise bound by the code right, or
   (b) at any time after the notice is given, the relevant person gives notice in writing to the operator that the person does not agree to confer or be otherwise bound by the code right.

(4) An order under this paragraph is one which imposes on the operator and the relevant person an agreement between them which –
   (a) confers the code right on the operator, or
   (b) provides for the code right to bind the relevant person.

21.2.2

Before turning to an examination of the sequence of events envisaged by this paragraph, it is worth making three preliminary points.

21.2.3

First, the concept of the 'relevant person' is used throughout this Part of the New Code, in contradistinction to the 'occupier', to which paragraph 9 of the New Code refers as the only person who may be entitled to grant a code rights. The reason for the different terminology is not explained, but is likely to reside in the fact that the operator will wish the code rights to be granted by that person who has the right to deal with the land over which the rights are proposed during the whole period for exercise of those rights. Take, for example, an occupier who has a licence to occupy the land for one month. Although that occupier has the right to grant code rights, that facility will not be of much use to an operator who wishes to be conferred code rights for a period of ten years.

21.2.4

Secondly, the paragraph 20 mechanism is applicable not merely where an operator wishes to obtain a code right which it does not have, but also where it already has a code right, but wishes to procure or ensure that the right binds the relevant person. The purpose of this drafting is presumably to pick up the point just made, for example where the operator has a code agreement with an occupier whose own tenure is on the point of expiry. In such circumstances, the operator will wish to ensure that the code rights existing under the code agreement come to bind the superior interest in the land.

21.2.5

Thirdly, the singularity of the drafting is notable: the reference is to *a* code right. This is consistent with the Authors' suggestion in section 15.5 to the effect that the drafting of the New Code overall envisages an à la carte approach to code rights, with the operator having the ability to select which code rights it wishes, rather than the whole array available under paragraph 3; while the relevant person may wish for its part to confine the operator to certain only of the rights.

21.2.6

The process outlined in paragraph 20 involves the following sequence of events:

(a) the giving by the operator to the relevant person of a notice;
(b) either (i) a positive response; or (ii) a negative response by that person;
(c) (as necessary) an application to court; and
(d) (if appropriate) the making of a court order.

These steps are examined in the remainder of this section.

21.2.7 (a) The notice

Paragraph 20(2) stipulates that the notice must specify three matters:

(i) the required code right;
(ii) all the other terms of the agreement that the operator seeks;
(iii) that the operator seeks the relevant person's agreement to those terms.

21.2.8

Paragraph 20 does not refer to a prescribed form for such a notice, but the reference to the need for a notice imports the requirements of Part 15 of the New Code, which lays down detailed further matters which must be observed by the operator. These are explained in Chapter 32, but in essence, the operator must in addition (see paragraph 88(1) of the New Code) explain:

(iv) the effect of the notice;
(v) which provisions of the code are relevant to the notice; and
(vi) the steps that may be taken by the recipient in respect of the notice.

21.2.9

Moreover, if OFCOM has prescribed the form of the notice (which paragraph 90(1) of the New Code requires it to do), then the notice given by the operator must be in that form (see paragraph 88(2)). If the notice is not in the prescribed form, it is invalid (paragraph 88(3)).[1]

1 See further Chapter 41 concerning OFCOM's duties to specify forms of notice.

21.2.10

Finally, it is worth noting (as Appendix E explains in greater depth) that paragraph 91 of the New Code also sets out detailed procedures for the giving of such notices.

21.2.11

The record of defective notice-giving in other statutory contexts is not a happy one, demonstrating both that the givers of notices frequently fail to observe the detailed stipulations as to the form, content and mode of delivery of the notices, and also that the recipients often find it in their commercial interest to challenge the validity of the notices. Operators would therefore do well to observe the detail of Part 15 of the New Code to the letter.

21.2.12

Appendix E of this Book contains examples of the notice required in this case.

21.2.13 (b) The response

Paragraph 20(3) of the New Code appears to suggest the following possible reactions to the operator's paragraph 20(2) notice proposing the conferment of code rights:

(a) the relevant person does not, before the end of 28 days beginning with the day on which the notice is given, agree to confer or be otherwise bound by the code right; or
(b) the relevant person gives notice in writing to the operator that the person does not agree to confer or be otherwise bound by the code right.

These two alternatives conceal other possibilities. For example, the relevant person may:

(c) agree to confer or be otherwise bound by the code right and the proposed terms; or
(d) agree to confer or be otherwise bound by the code right, but disagree some or all of the other terms.

21.2.14

If the relevant person gives notice not agreeing to confer or be otherwise bound by the code right (i.e. the possibility considered in paragraph 20(3)(b) of the New Code), such a notice may itself need to be in a prescribed form (depending upon whether OFCOM has by then prescribed such a form): see paragraph 89(5) of the New Code, which is discussed more fully in Chapter 32. A non-compliant notice is not invalid, but the relevant person will have to meet any costs incurred by the operator as a result of the notice not being in the right form (paragraph 89(6)).

21.2.15 (c) The court application

Paragraph 20(3) of the New Code gives the operator the right to apply to the court for an order regarding its requested code right, where either no positive response has been received within 28 days, or a negative response has been received before then. The application process is described in Chapter 33.

21.2.16 (d) The court order

Paragraph 20(4) of the New Code describes the order the court may make as one which either confers the code right on the operator, or provides for the code right to bind the relevant person. Had matters been left there, that would have left the court with a rather unsatisfactory binary choice, in that this paragraph does not cater for the owner agreeing the code right, but disagreeing some of the other proposed terms of the code agreement (see paragraph 21.2.13(d) above). However, paragraph 20(3) (see section 21.5 below) makes it clear that the court will in fact carry out a rather more nuanced exercise.

21.3 The test to be applied by the court in deciding whether to impose an agreement

21.3.1

Paragraph 21 of the New Code is headed 'What is the test to be applied by the court?', and provides:

> (1) Subject to sub-paragraph (5), the court may make an order under paragraph 20 if (and only if) the court thinks that both of the following conditions are met.
> (2) The first condition is that the prejudice caused to the relevant person by the order is capable of being adequately compensated by money.
> (3) The second condition is that the public benefit likely to result from the making of the order outweighs the prejudice to the relevant person.
> (4) In deciding whether the second condition is met, the court must have regard to the public interest in access to a choice of high quality electronic communications services.
> (5) The court may not make an order under paragraph 20 if it thinks that the relevant person intends to redevelop all or part of the land to which the code right would relate, or any neighbouring land, and could not reasonably do so if the order were made.

In relation to sub-paragraph (4), s.405(4) of the 2003 Act adds: 'References in this Act to *access* ... in relation to an electronic communications network or electronic communications service, are references to the opportunity of making use of the network or service'.

21.3.2

At first sight, the drafting of paragraph 21 is tolerably clear. However, a closer examination reveals a number of possible difficulties.

21.3.3

First, the condition that the prejudice caused to the relevant person by the order is capable of being adequately compensated by money, is almost certainly a condition to ensure that the imposition on a relevant person of code rights is human rights compliant.[2] Article 1, First Protocol, European Convention on Human Rights and Fundamental Freedoms, provides that no one shall be deprived of his possessions except in the public interest and subject to the conditions provided for by law and by the general principles of international law. A deprivation will include a taking of land or a right over land, and the validity of any such taking, in relation to Article 1, will depend on a requirement of proportionality between the public interest in the taking and the private interest in the possession or property affected.[3] The existence or adequacy of any compensation provided for under any relevant legislation may be relevant to the question as to whether a taking of, or an interference to, property is human rights compliant.[4] This matter was fully discussed in the Law Commission's Report, where the Commission recorded that it might be hard to justify a code where a landowner cannot be adequately compensated with money.[5] The Commission, in considering the appropriate test in relation to the Human Rights Convention, posed the question as follows:

> So what test should the revised Code prescribe for the imposition of Code Rights by the tribunal? That test will only be relevant if money will provide adequate compensation, because we have concluded that if monetary compensation is impossible, Code Rights should not be imposed.[6]

The Commission was certainly concerned with the adequacy of the consideration, and not simply the adequacy of the supplementary compensation provisions. It then recommended a test in the form of words in the first condition at New Code paragraph 21(2) above.[7] It is quite clear that the Law Commission was using the expression 'compensated in money' in a broad sense to cover all provisions for compensation, including those for the payment of a consideration under New Code paragraph 24, as well the compensation to which paragraph 25 refers. It is likely that the New Code will be interpreted similarly, and it is highly unlikely that the expression 'compensated in money' refers only to the compensation provisions in paragraph 25(6), and its reference to paragraph 84.

21.3.4

Secondly, the balancing exercise required by paragraph 21(3). The background to this requirement is almost certainly the balancing test that arises under Article 1, First

2 See Law Com 336, paragraphs 2.16–2.17 (the imposition of rights for code operators must be human rights compliant: Article 1, First Protocol, European Convention on Human Rights and Fundamental Freedoms), 4.3–4.4 (Article 1 of the Convention requires that a deprivation, or control of use, of possessions should be in accordance with the law).

3 As to the general principles for the engagement of Article 1, see *Sporrong & Lönnroth v Sweden* [1982] 5 EHRR 35 and *Wilson v First County Trust* [2004] 1 AC 816.

4 *James v United Kingdom* [1986] 8 EHRR 123.

5 Law Com 336, paragraph 4.22.

6 Law Com 336, paragraph 4.27.

7 Law Com 336, paragraph 4.43(1).

Protocol, European Convention on Human Rights and Fundamental Freedoms, as explained by the Law Commission.[8] The factors identified by the Commission were expressed as:

> A fair balance must be struck between the community interest and the individual's rights, and several factors are relevant. These include the provision of compensation, the conduct of the state, the conduct of the individual, the effect of the interference on the individual, the strength of the benefit to the wider community and whether the measure effecting the interference has retrospective effect.[9]

Whether compensation is payable and its measure will usually be the most important factor in deciding whether the imposition of an agreement was proportionate, if not decisive.[10] But the available procedure safeguards is another important factor.[11] It seems highly unlikely that the New Code, both as to the measure of compensation, as used to include both consideration and additional compensation, and the procedures available, would fail the balancing test under the Convention of Human Rights. This is further discussed below.

21.3.5

Thirdly, and leading on from the point just considered, what if the electronic communications apparatus is installed inside a private arena, to which the public has access at certain times for certain events, upon payment? Would that meet the public benefit test set out in paragraph 21? This is likely to be a matter of fact and degree. Paying members of the public are no less members of the public, with an interest in the provision of access to a choice of high quality electronic communications services, within the meaning of paragraph 21(4) of the code. It is suggested that the answer is likely to depend upon the number of benefited individuals in any given case.

21.3.6

Fourthly, the redevelopment ground for refusal set out in paragraph 21(5) is very short on detail, leaving unanswered the following questions:

(a) How proximate need the redevelopment plans be in order to engage this provision? The comparable provision in section 30(1)(f) of the Landlord and Tenant Act 1954 is usually understood to mean that the landlord will commence its works within a reasonable time after recovering possession of the holding, which itself will be postponed by more than three months as a result of the operation of s.64. Is a similar timeframe likely to be applied by the court? What if the relevant person's plan is to commence redevelopment within a year? The paragraph provides no answer.

8 Law Com 336, paragraph 4.4.

9 Relying on R. Clayton and H. Tomlinson, *The Law of Human Rights* (2nd ed. 2009) paragraph 18.114, citing *The Former King of Greece v Greece* [2000] 33 EHRR 21.

10 *The Former King of Greece v Greece* [2000] 33 EHRR 21.

11 *Jokela v Finland* [2003] 37 EHRR 26 at [45].

(b) How extensive should be the part of the land to which the code right would relate which the relevant person intends to redevelop before the court would exercise its power to refuse an order? 10 per cent? Half? Again, the paragraph provides no guidance.

21.4 The impact of human rights

21.4.1

The imposition of an agreement by the court necessarily involves a deprivation of the landowner's right. Article 1 of the First Protocol to the European Convention on Human Rights and Fundamental Freedoms (which forms part of the domestic law of the United Kingdom, by virtue of the Human Rights Act 1998) stipulates that no one shall be deprived of his possessions except in the public interest and subject to the conditions provided for by law and by the general principles of international law. It adds that this stipulation shall not, however, in any way impair the right of a State to enforce such laws as it deems necessary to control the use of property in accordance with the general interest.

21.4.2

It is well settled that any deprivation of such a right must have a basis in national law which is accessible, sufficiently certain and provides protection against arbitrary abuses: see *The Former King of Greece v Greece*.[12]

21.4.3

As explained above, paragraph 21 of the New Code has been drafted in a way that seeks to meet this test. It does so by balancing a number of competing interests in a way that is designed to be fair and proportionate. Those interests were described by the Law Commission in para 4.4 of their 2013 Report on the Electronic Communications Code (Law Com 336) as follows:

> These include the provision of compensation, the conduct of the state, the conduct of the individual, the effect of the interference on the individual, the strength of the benefit to the wider community and whether the measure effecting the interference has retrospective effect.

21.4.4

In formulating the test for imposition in paragraph 21 of the New Code, the Law Commission followed the view that it had put forward in its earlier consultation paper, to the effect that it was arguably inappropriate that it should be possible for a landowner's consent to be dispensed with on no other basis than that he or she should be compensated in money. Such an approach makes no attempt to balance the public benefit against the private prejudice.

12 [2000] 33 EHRR 21.

21.4.5

The test thus formulated would appear to be human rights-compliant, although some of those thus deprived will no doubt consider that they have been unfairly treated.

21.5 The effect of an imposed agreement

21.5.1

Paragraph 22 of the New Code is headed 'What is the effect of an agreement imposed under paragraph 20?', and provides:

> An agreement imposed by an order under paragraph 20 takes effect for all purposes of this code as an agreement under Part 2 of this code between the operator and the relevant person.

21.5.2

The effect of this is to make an imposed agreement exactly equivalent to a (voluntary) code agreement. The purpose of this drafting is not spelled out, but it is presumably to ensure that the imposed agreement takes effect as an agreement voluntarily entered into. One effect of this is to ensure that Part 5 (termination) applies to the imposed agreement. Part 5 provides that 'This Part of this code applies to an agreement under Part 2 of this code'. Thus the mere fact that the agreement is imposed by the court does not deprive the operator of Part 5 protection.

21.6 The terms of an imposed agreement

21.6.1

Paragraph 23 of the New Code is headed 'What are the terms of an agreement imposed under paragraph 20?', and provides:

> (1) An order under paragraph 20 may impose an agreement which gives effect to the code right sought by the operator with such modifications as the court thinks appropriate.
> (2) An order under paragraph 20 must require the agreement to contain such terms as the court thinks appropriate, subject to sub-paragraphs (3) to (8).
> (3) The terms of the agreement must include terms as to the payment of consideration by the operator to the relevant person for the relevant person's agreement to confer or be bound by the code right (as the case may be).
> (4) Paragraph 24 makes provision about the determination of consideration under sub-paragraph (3).
> (5) The terms of the agreement must include the terms the court thinks appropriate for ensuring that the least possible loss and damage is caused by the exercise of the code right to persons who –
> (a) occupy the land in question,
> (b) own interests in that land, or
> (c) are from time to time on that land.

(6) Sub-paragraph (5) applies in relation to a person regardless of whether the person is a party to the agreement.

(7) The terms of the agreement must include terms specifying for how long the code right conferred by the agreement is exercisable.

(8) The court must determine whether the terms of the agreement should include a term –

(a) permitting termination of the agreement (and, if so, in what circumstances);

(b) enabling the relevant person to require the operator to reposition or temporarily to remove the electronic communications equipment to which the agreement relates (and, if so, in what circumstances).

21.6.2

This paragraph deals with a number of discrete matters prompted by the drafting of paragraph 23, which are addressed in the following paragraphs of this section:

(a) what sort of code right may be included in the imposed agreement;
(b) what other terms should be included in the imposed agreement;
(c) the consideration payable under the imposed agreement;
(d) protective provisions for relevant persons and third parties;
(e) the length of the imposed agreement;
(f) provision for termination of the imposed agreement;
(g) provision for temporary or permanent repositioning of the electronic communications equipment.

21.6.3

The first point to note about paragraph 23(1) is that it provides that the Tribunal will seek to give 'effect to the code right sought by the operator with such modifications of the court thinks appropriate'. Thus, the operator may seek such terms, to give effect to the code right or rights sought in the agreement to be imposed by the Court, as it desires subject to review by the Court. From a practical point of view it would thus seem sensible for operators to seek terms which are as full as possible with respect to the code rights which are sought. The court's starting position is with the terms requested and it is those terms which are subject to the power for modification 'as the court thinks appropriate'. It is important for the operator to give due consideration to the terms it seeks, because the landowner may simply agree the terms which are requested so as to avoid any consideration by the Tribunal with respect to the terms which are to be imposed. There would appear to be no reason why the operator cannot, on a consideration by the court of the terms requested, seek appropriate amendment to accommodate concerns expressed by the court or the relevant person, and this may include the possibility of completely new terms not previously requested. It is unclear whether the operator can ask for completely new terms not contained in its notice, and not arising by reference to concerns expressed by the Court or the relevant person (as opposed to amendments to requested terms).

21.6.4

Paragraph 23(1) gives the court a discretion as to whether the imposed agreement should contain a modified version of the code right sought by the operator. This drafting does not appear to contemplate that the Court would simply refuse to include the code right at all. The reason for this is presumably that, in order for the process to have come this far, the Court must have decided already, as part of the exercise under paragraph 21, that a code right should in principle be included in the imposed agreement. Admittedly, this drafting does not expressly cater for the situation where the operator seeks two or more code rights, and the Court decides to allow one but not another. Again, the answer to this is probably that each code right is to be decided as a discrete exercise.

21.6.5

The sort of modification that the draftsman of paragraph 23(1) had in mind is not clear. Examples might include a different routing for any desired access; or less frequent inspections.

21.6.6

As to what terms should be included in the imposed agreement, paragraph 23 lists two types. First, there are the mandatory terms, with which sub-paragraphs (3) to (8) deal (and to which must be added the considerations affecting assignment, upgrading and sharing, discussed in Chapter 20). Secondly, there are all the other terms, the inclusion of which will depend upon the court's view of what is appropriate (sub-paragraph (2)). The court's attitude to the discretionary terms may be shaped both by those terms which have become standard in commercial leases, such as obligations to repair and insure, and also by those which OFCOM has included in its standard template agreement.[13] There is a distinction between 'modifications' (paragraph 23(1)) and 'terms' (paragraph 23(2)). The Court is, when imposing 'terms' accordingly not limited to terms which modify or restrict a code right, but appears able to impose terms which supplement them subject to paragraphs 23(3)–(8), whereas 'modification' appears to be limited to modification of the code right requested by the operator. It is considered that in principle the Court is able to impose any sort of further term it sees fit provided that the statutory test for so doing is met. Perhaps most importantly, the Law Commission in its Report clearly contemplated that this would extend to the imposition of a bespoke sharing and upgrading provision going further than the more limited powers conferred by paragraph 17 (see paragraph 20.1.6).

21.6.7

As paragraph 23(3) states, the imposed agreement must include terms as to the payment of consideration by the operator to the relevant person for the relevant person's agreement to confer or be bound by the code right (as the case may be). The determination of that consideration is dealt with by paragraph 24 of the New Code.

21.6.8

Paragraph 23(5), as supplemented by 23(6), contains provisions protective of the interests of the relevant person and other parties. Those other parties who are listed are persons who:

13 As to which, see Chapter 41 and Appendix F to this book.

(a) occupy the land in question,
(b) own interests in that land, or
(c) are from time to time on that land.

A familiar example might be a farm which (a) has a tenant; (b) also has a house occupied by a farm manager; (c) is burdened by restrictive covenants in favour of an adjoining landowner; and (d) is subject to a right of way to another adjoining owner's land. Under this example, the interests of the freeholder, tenant, house occupant, restrictive covenant beneficiary and easement owner will all potentially fall to be protected under this paragraph.

21.6.9

In deciding what term to impose, the Court is required to decide what is appropriate to ensure 'that the least possible loss and damage' is caused by the exercise of the code right. There is however no express provision for the Court to award compensation to third parties, that being restricted to the relevant person under paragraph 25. There is an argument that this omission may incline the Court towards a preference for a solution that minimises the loss and damage caused to those in this category who stand no chance of obtaining compensation.

21.6.10

Paragraph 23(7) stipulates that the terms of the imposed agreement must include terms specifying for how long the code right conferred by the agreement is exercisable. This is the exact counterpart of the requirement in paragraph 11(1) of the New Code for a valid code agreement. As paragraph 19.2.8 explains, this requirement is not to be equated to the leasehold principle that a term of years should be of ascertainable length at its outset. It should suffice for these purposes that the length of the code agreement is governed by a formula that may be uncertain until the happening of an event (e.g. 'until the land is required for redevelopment').

21.6.11

Paragraph 23(8) requires the court to determine whether the terms of the imposed agreement should include a term permitting termination of the agreement (and, if so, in what circumstances). This goes further than its counterpart in paragraph 11 of the New Code regarding 'voluntary' code agreements (see paragraph 19.2.9), which merely stipulates that the agreement must 'state the period of notice (if any) required to terminate the agreement'. The reason for the difference in language is unclear.

21.6.12

Given the language of paragraph 23(7), which requires the length of the code agreement to be specified, it is not considered that paragraph 23(8)(a) is dealing with the stated end of the term of the imposed agreement, because that would involve a redundancy. In other words, why should the draftsman of paragraph 23(8)(a) of the New Code have troubled to provide for the inclusion of a term permitting termination, when the length is catered for by the previous sub-paragraph, with the result that termination upon the happening of the given event will be automatic, needing no permission? Instead, the authors suggest that paragraph 23(8)(a) is dealing with termination of the imposed agreement other than by effluxion of time – in effect, to borrow from landlord and tenant language, a break clause provision for either party.

21.6.13

Paragraph 23(7) also deals with the quite different topic of provision for temporary relocation of the electronic communications equipment. This facility is a helpful addition to the protection provisions for the landowner, in contrast to the position under the Old Code, whereby operators' rights to retain their apparatus frequently frustrate or delay redevelopment plans. This provision will entail a degree of forward thinking, but the outcome should make for a more harmonious relationship between the relevant person and the operator.

21.7 Determination of the consideration for an imposed agreement

21.7.1

Paragraph 24 of the New Code is headed 'How is consideration to be determined under paragraph 23?', and provides:

> (1) The amount of consideration payable by an operator to a relevant person under an agreement imposed by an order under paragraph 20 must be an amount or amounts representing the market value of the relevant person's agreement to confer or be bound by the code right (as the case may be).
> (2) For this purpose the market value of a person's agreement to confer or be bound by a code right is, subject to sub-paragraph (3), the amount that, at the date the market value is assessed, a willing buyer would pay a willing seller for the agreement –
>   (a) in a transaction at arm's length,
>   (b) on the basis that the buyer and seller were acting prudently and with full knowledge of the transaction, and
>   (c) on the basis that the transaction was subject to the other provisions of the agreement imposed by the order under paragraph 20.
> (3) The market value must be assessed on these assumptions –
>   (a) that the right that the transaction relates to does not relate to the provision or use of an electronic communications network;
>   (b) that paragraphs 16 and 17 (assignment, and upgrading and sharing) do not apply to the right or any apparatus to which it could apply;
>   (c) that the right in all other respects corresponds to the code right;
>   (d) that there is more than one site which the buyer could use for the purpose for which the buyer seeks the right.
> (4) The terms of the agreement may provide for consideration to be payable –
>   (a) as a lump sum or periodically,
>   (b) on the occurrence of a specified event or events, or
>   (c) in such other form or at such other time or times as the court may direct.

21.7.2

This subject is dealt with in depth in Chapter 30.

21.8 The rights to the payment of compensation for an imposed agreement

21.8.1

Paragraph 25 of the New Code is headed: 'What rights to the payment of compensation are there?', and provides:

> (1) If the court makes an order under paragraph 20 the court may also order the operator to pay compensation to the relevant person for any loss or damage that has been sustained or will be sustained by that person as a result of the exercise of the code right to which the order relates.
> (2) An order under sub-paragraph (1) may be made –
> (a) at the time the court makes an order under paragraph 20, or
> (b) at any time afterwards, on the application of the relevant person.
> (3) An order under sub-paragraph (1) may –
> (a) specify the amount of compensation to be paid by the operator, or
> (b) give directions for the determination of any such amount.
> (4) Directions under sub-paragraph (3)(b) may provide –
> (a) for the amount of compensation to be agreed between the operator and the relevant person;
> (b) for any dispute about that amount to be determined by arbitration.
> (5) An order under this paragraph may provide for the operator –
> (a) to make a lump sum payment,
> (b) to make periodical payments,
> (c) to make a payment or payments on the occurrence of an event or events, or
> (d) to make a payment or payments in such other form or at such other time or times as the court may direct.
> (6) Paragraph 84 makes further provision about compensation in the case of an order under paragraph 20.

21.8.2

This subject too is dealt with in depth in Chapter 30.

21.9 The imposition of an agreement on an interim basis

21.9.1

Paragraph 26 of the New Code is headed 'Interim code rights', and provides:

> (1) An operator may apply to the court for an order which imposes on the operator and a person, on an interim basis, an agreement between them which –
> (a) confers a code right on the operator, or
> (b) provides for a code right to bind that person.

(2) An order under this paragraph imposes an agreement on the operator and a person on an interim basis if it provides for them to be bound by the agreement –
  (a) for the period specified in the order, or
  (b) until the occurrence of an event specified in the order.

(3) The court may make an order under this paragraph if (and only if) the operator has given the person mentioned in sub-paragraph (1) a notice which complies with paragraph 20(2) stating that an agreement is sought on an interim basis and –
  (a) the operator and that person have agreed to the making of the order and the terms of the agreement imposed by it, or
  (b) the court thinks that there is a good arguable case that the test in paragraph 21 for the making of an order under paragraph 20 is met.

(4) Subject to sub-paragraphs (5) and (6), the following provisions apply in relation to an order under this paragraph and an agreement imposed by it as they apply in relation to an order under paragraph 20 and an agreement imposed by it –
  (a) paragraph 20(3) (time at which operator may apply for agreement to be imposed);
  (b) paragraph 22 (effect of agreement imposed under paragraph 20);
  (c) paragraph 23 (terms of agreement imposed under paragraph 20);
  (d) paragraph 24 (payment of consideration);
  (e) paragraph 25 (payment of compensation);
  (f) paragraph 84 (compensation where agreement imposed).

(5) The court may make an order under this paragraph even though the period mentioned in paragraph 20(3)(a) has not elapsed (and paragraph 20(3)(b) does not apply) if the court thinks that the order should be made as a matter of urgency.

(6) Paragraphs 23, 24 and 25 apply by virtue of sub-paragraph (4) as if –
  (a) references to the relevant person were to the person mentioned in sub-paragraph (1) of this paragraph, and
  (b) the duty in paragraph 23 to include terms as to the payment of consideration to that person in an agreement were a power to do so.

(7) Sub-paragraph (8) applies if –
  (a) an order has been made under this paragraph imposing an agreement relating to a code right on an operator and a person in respect of any land, and
  (b) the period specified under sub-paragraph (2)(a) has expired or, as the case may be, the event specified under sub-paragraph (2)(b) has occurred without (in either case) an agreement relating to the code right having been imposed on the person by order under paragraph 20.

(8) From the time when the period expires or the event occurs, that person has the right, subject to and in accordance with Part 6 of this code, to require the operator to remove any electronic communications apparatus placed on the land under the agreement imposed under this paragraph.

21.9.2

Paragraph 26 operates as a short cut to the full-blown imposed agreement procedure, allowing the court to make an order in appropriate circumstances for such an agreement on an interim basis (in this section, for ease of reference, an 'interim imposed agreement'). For the most part, the steps to be taken and considerations to be born in mind replicate those applicable to the full procedure. It seems plain from paragraph 26(7) that the draftsman envisages that an interim imposed agreement order will be made as a preliminary to an eventual paragraph 20 order. However, it is difficult to see why this interim procedure should have any proper place in the New Code, unless delays in the courts are such that it will take a long time to obtain a date for a hearing for the full blown procedure.

21.9.3

This expedient appears to stem from the recommendation made by the Law Commission in Report No. 336. They say, under the heading 'Enabling Early Access':

> 9.50 … In many cases a landowner will not be wholly opposed to the installation of electronic communications apparatus. Installation in itself may be uncontentious, and may be welcomed as a source of revenue (and of a mobile phone or broadband service). The sticking point is more usually the terms on which Code Rights will be conferred, and most often the price. …
>
> 9.51 … Some landowners would be content to allow early access if they could do so safely, as it would allow them to secure the benefits of the installation (consideration and, in some cases, also enhanced communications service). This is, however, problematic under the 2003 Code, because once apparatus is installed under the 2003 Code it benefits from the protection of paragraph 21, leaving the landowner locked into the security provisions of the Code and the Code Operator with little incentive to make progress towards agreeing consideration. Accordingly, the landowner cannot safely allow early access until it is satisfied that all the terms and conditions of the parties' legal relationship are in place. …
>
> 9.61 We have concluded that the revised Code should allow early access for Code Operators, but only on an interim basis. We recommend that this should be achieved by enabling Code Operators, when making an application for Code Rights under the revised Code, to apply to the Lands Chamber for an interim order for access pending the resolution of disputes over payment. Such orders would only be granted on terms that give the landowner the right to vacant possession if a final order in favour of the Code Operator is not made before the expiry of the interim order.

21.9.4

The interim procedure was designed to achieve this aim. Particular points to note about the interim procedure are addressed in the remainder of this section.

21.9.5

First, it will be just as critical for the operator to comply, in relation to the form, contents and delivery of the notice required by paragraph 26(3) with the notice requirements in Part 15 of the New Code – see paragraph 21.2.8 above.

21.9.6

Secondly, paragraph 26(3) provides that an interim order will be made (all other things being equal) if either the parties agree (presumably after the onset of the litigation, for agreement beforehand would have avoided the need for litigation in the first place); or the court thinks that there is a 'good arguable case' that the test in paragraph 21 for the making of an order under paragraph 20 is met (as to which see section 21.3 above). The reference to a 'good arguable case' will be familiar to litigators, who encounter it routinely in relation to summary judgment cases and appeals in civil litigation. In essence, it does not mean that the operator has to show that it will persuade the court in due course to order an imposed agreement – merely that its prospects of doing so are reasonably good.

21.9.7

Thirdly, in cases of urgency, paragraph 26(5) entitles the court to make an order for an interim imposed agreement even within the 28-day period mentioned in paragraph 20(3)(a) for the relevant person to respond to the operator's notice.

21.9.8

Fourthly, an interim imposed agreement will only be ordered for a specified period (e.g. 'until 4 pm on the date four weeks after the making of this order'), or to take effect upon the happening of a specified event (e.g. 'until the later of the date four weeks after the making of this order or the completion of the building now in the course of construction at …'): see paragraph 26(2). Paragraph 26(7) takes this further forward by providing that if the period expires or the event occurs without a paragraph 20 order having been made in the meantime, the person against whom the order was made then has the right, under paragraph 26(8), subject to and in accordance with Part 6 of the New Code, to require the operator to remove any electronic communications apparatus placed on the land under the interim imposed agreement.

21.10 The imposition of an agreement on a temporary basis

21.10.1

Paragraph 27 of the New Code is headed 'Temporary code rights', and provides:

> (1) This paragraph applies where –
>   (a) an operator gives a notice under paragraph 20(2) to a person in respect of any land,
>   (b) the notice also requires that person's agreement on a temporary basis in respect of a right which is to be exercisable (in whole or in part) in relation to electronic communications apparatus which is already installed on, under or over the land, and
>   (c) the person has the right to require the removal of the apparatus in accordance with paragraph 37 or as mentioned in paragraph 40(1) but the operator is not for the time being required to remove the apparatus.

(2) The court may, on the application of the operator, impose on the operator and the person an agreement between them which confers on the operator, or provides for the person to be bound by, such temporary code rights as appear to the court reasonably necessary for securing the objective in sub-paragraph (3).

(3) That objective is that, until the proceedings under paragraph 20 and any proceedings under paragraph 40 are determined, the service provided by the operator's network is maintained and the apparatus is properly adjusted and kept in repair.

(4) Subject to sub-paragraphs (5) and (6), the following provisions apply in relation to an order under this paragraph and an agreement imposed by it as they apply in relation to an order under paragraph 20 and an agreement imposed by it –

- (a) paragraph 20(3) (time at which operator may apply for agreement to be imposed);
- (b) paragraph 22 (effect of agreement imposed under paragraph 20);
- (c) paragraph 23 (terms of agreement imposed under paragraph 20);
- (d) paragraph 24 (payment of consideration);
- (e) paragraph 25 (payment of compensation);
- (f) paragraph 84 (compensation where agreement imposed).

(5) The court may make an order under this paragraph even though the period mentioned in paragraph 20(3)(a) has not elapsed (and paragraph 20(3)(b) does not apply) if the court thinks that the order should be made as a matter of urgency.

(6) Paragraphs 23, 24 and 25 apply by virtue of sub-paragraph (4) as if –

- (a) references to the relevant person were to the person mentioned in sub-paragraph (1) of this paragraph, and
- (b) the duty in paragraph 23 to include terms as to the payment of consideration to that person in an agreement were a power to do so.

(7) Sub-paragraph (8) applies where, in the course of the proceedings under paragraph 20, it is shown that a person with an interest in the land was entitled to require the removal of the apparatus immediately after it was installed.

(8) The court must, in determining for the purposes of paragraph 20 whether the apparatus should continue to be kept on, under or over the land, disregard the fact that the apparatus has already been installed there.

21.10.2

In contrast to the position dealt with by paragraph 26, paragraph 27 provides a temporary solution to the operator who has already had electronic communications apparatus installed pursuant to a previous code agreement, but is now faced by a situation whereby it has no right as against the landowner to keep the apparatus installed (although the owner is not currently requiring its removal).

21.10.3

This temporary solution applies only in two situations. The first is where the operator is seeking an imposed agreement under paragraph 20, but has already installed its apparatus under a previous agreement (so that paragraph 26 will be of no application). The second is where there is a dispute between the owner and occupier concerning the removal of the operator's apparatus upon termination of the existing agreement (as to which, see Chapter 24), and the operator wishes to maintain the status quo pending resolution of that dispute.

21.10.4

Paragraph 27(3) governs the nature and duration of the temporary code rights that are to be granted upon an application for an order under paragraph 27. The factors the court is to take into account exclusively favour the operator: they are that the service provided by the operator's network is maintained, and the apparatus is properly adjusted and kept in repair. The fact that the landowner may have a pressing need to proceed with a much needed redevelopment is not taken into account.

22 Termination of code agreements

22.1 Introduction

22.1.1

Part 5 of the New Code deals with the procedure for the termination and modification of code agreements. Modification is dealt with by Chapter 23. This chapter deals with the termination of code agreements.

22.1.2

The termination procedures in the New Code are modelled closely upon those set out in Part 2 of the Landlord and Tenant Act 1954, employing the following identical or similar features:

(a) the termination procedure is applicable only to certain types of agreement (in this case, code agreements);
(b) a notice is required to bring a code agreement to an end;
(c) until such notice is served, and the procedure has run its course, the code agreement will continue;
(d) unless the site provider can establish a ground for termination, the operator will be entitled to a new code agreement, which may involve different or additional code rights, and different or additional terms;
(e) pending determination of such matters, the operator is to pay a sum determined by the court.

Those matters are analysed in detail in the following sections.

22.1.3

The New Code is dissimilar to the 1954 Act in other respects. First, if and when the site provider proves one of the statutory grounds for termination, that is not the end of the process: it has still to apply to court for removal of the apparatus from its land. This is dealt with in Chapter 24 of this book.

22.1.4

Secondly, there is no provision in the New Code for the site provider to initiate a procedure for the renewal of the code agreement: all that it is entitled to do is to seek to prevent continuation of an agreement by establishing a ground of termination (although it is of course always open to a site provider to reach agreement with the operator for a new code agreement).

22.1.5

In contrast to the position under the Old Code, both the New Code and the 1954 Act provide that an agreement cannot attract protection under both statutory codes.[1]

22.2 The termination provisions apply only to code agreements

22.2.1

Paragraph 29 of the New Code provides:

> (1) This Part of this code applies to an agreement under Part 2 of this code, subject to sub-paragraphs (2) to (4).
> (2) This Part of this code does not apply to a lease of land in England and Wales if –
>   (a) its primary purpose is not to grant code rights, and
>   (b) it is a lease to which Part 2 of the Landlord and Tenant Act 1954 (security of tenure for business, professional and other tenants) applies.
> (3) In determining whether a lease is one to which Part 2 of the Landlord and Tenant Act 1954 applies, any agreement under section 38A (agreements to exclude provisions of Part 2) of that Act is to be disregarded.
> (4) This Part of this code does not apply to a lease of land in Northern Ireland if –
>   (a) its primary purpose is not to grant code rights, and
>   (b) it is a lease to which the Business Tenancies (Northern Ireland) Order 1996 (SI 1996/725 (NI 5)) applies.
> (5) An agreement to which this Part of this code applies is referred to in this code as a 'code agreement'.

22.2.2

As section 19.1 cautions, this drafting creates a trap for the unwary. Although the operator and the owner may have entered into an arrangement conferring code rights which is fully compliant with Part 2 of the New Code, it may not in fact be a code agreement – and if it is not a code agreement, then the termination procedures cannot be used. It is possible, therefore, that parties will waste a great deal of time and money pursuing the wrong termination procedure.

1 See Chapter 25.

22.2.3

Using the negative drafting technique employed in paragraph 29 of the New Code, an agreement conferring code rights will *not* be a code agreement if:

(a) it is a lease; *and*
(b) its primary purpose is not to create code rights; *and*
(c) the lease is one to which Part 2 of the Landlord and Tenant Act 1954 applies.

These ingredients are now analysed separately.

22.2.4 (a) Lease

In summary (and reference should be made to one of the main landlord and tenant texts for the detail), a lease is an agreement the right for a person exclusively to possess land for a fixed or periodic term, and usually at a rent.[2] A licence (which may lack any or all of these features) falls outside this definition, as do other forms of limited tenure such as a tenancy at will, under which the right to occupy land is precarious.

22.2.5 (b) Primary purpose

This expression is not usually employed in property legislation, although it has received the attention of the courts in some 1954 Act cases.[3] In their 2013 Report on the Electronic Communications Code (Law Com 336), the Law Commission commented, in relation to the employment of this proposed expression:

> 6.86 There is of course room for doubt and for dispute as to the primary purpose of a lease. But we think that difficulties will arise in only a few cases; the lease of a mast site falls clearly on one side of the line, the lease to a Code Operator of a retail unit, where the lease incidentally permits the tenant to install a cell site on the roof, falls on the other.
>
> 6.87 It follows that in a mixed use lease where Code Rights are not the primary purpose of the letting, which is contracted out of the 1954 Act, the Code Operator will have no security. Where security is important, therefore, the Code Operator will want a separate lease for the apparatus.

22.2.6

It remains to be seen whether the confidence of the Law Commission will have been misplaced. Mixed use sites – particularly in inner city areas – are common, and there may well be considerable uncertainty as to whether the primary purpose test is met in any given case.

2 Although the reservation of a rent is not a prerequisite to the grant of a lease: *Ashburn Anstalt v WJ & Co Arnold* [1989] Ch.1, CA. See also Chapter 14.
3 See e.g. the decision of the Court of Appeal in *Fisher v Taylors Furnishing Stores Ltd* [1956] 2 QB 78.

22.2.7

Two more comments deserve to be made in connection with the expression 'primary purpose'. The first concerns the time at which the primary purpose test is to be applied. The authors consider that this is a once and for all test to be considered at the date of grant of the lease, because that would appear to be the natural and ordinary meaning of the reference in paragraph 29(2)(a) to the primary purpose of the lease being 'not to *grant* code rights'. Even though a lease is, for these purposes, a living, breathing document, with rights and obligations that may alter during its term, there was only ever one time at which the code rights under it were granted. Any other interpretation would have the consequence that a lease could move in and out of different forms of statutory protection, which would be unsatisfactory.

22.2.8

The second concerns the drafting of the agreement. In other areas of statutory control where leases commonly acquire protection but licences do not, it has become common for agreements to assert that they create licences rather than tenancies. Despite the deprecation of Lord Templeman in *Street v Mountford*[4] in the context of residential agreements of parties seeking to mislabel agreements in an attempt to avoid statutory protection, the Court of Appeal has taken the view that where professionally advised commercial parties willingly enter into documents described as licences, it is no part of the function of the court to arrive at a different interpretation. In *Clear Channel UK Ltd v Manchester City Council.*[5] Jonathan Parker LJ, with whom the other judges agreed, said:

> the fact remains that this was a contract negotiated between two substantial parties of equal bargaining power and with the benefit of full legal advice. Where the contract so negotiated contains not merely a label but a clause which sets out in unequivocal terms the parties' intention as to its legal effect, I would in any event have taken some persuading that its true effect was directly contrary to that expressed intention.

22.2.9

In the light of these comments, it is to be expected that draftsmen of agreements conferring code rights, which are intended to be code agreements, will include express statements of intention to that effect.

22.2.10 (c) 1954 Act protection

A lease to which Part 2 of the Landlord and Tenant Act 1954 applies[6] cannot be a code agreement, in the sense provided for by paragraph 29(5). Part 5 will not apply to a 1954

4 [1989] AC 805.
5 [2006] L & TR 7.
6 Section 23(1) of the Landlord and Tenant Act 1954 provides: 'Subject to the provisions of this Act, this Part of this Act applies to any tenancy where the property comprised in the tenancy is or includes premises which are occupied by the tenant and are so occupied for the purposes of a business carried on by him or for those and other purposes.'

Act agreement where the primary purpose is not to grant code rights and the agreement is one within the 1954 Act: paragraph 29(2). For these purposes, a lease which is contracted out of the security of tenure provisions of the 1954 Act is treated as a lease to which that Act applies (see paragraph 29(3) of the New Code).

22.2.11

The provision in paragraph 29(2)(b) of the New Code is complemented by a mirror amendment to the 1954 Act, contained in paragraph 4 of Part 2 of Schedule 3 to the 2017 Act, providing for a new subsection (4) to be inserted in s.43 of the 1954 Act in these terms:

> This Part does not apply to a tenancy –
> (a) the primary purpose of which is to grant code rights within the meaning of Schedule 3A to the Communications Act 2003 (the electronic communications code), and
> (b) which is granted after that Schedule comes into force.

22.2.12

The result is that a lease the primary purpose of which is to grant code rights cannot be one falling within the protection of the 1954 Act and falls within Part 5 and is thus a code agreement as defined by paragraph 29(5).[7] Where the primary purpose is *not* to grant code rights and it is a lease which would fall within the terms of Part II of the 1954 Act albeit there may be an exclusion order under s.38A of the 1954 Act, the lease will not be a code agreement and fall outside Part 5. Thus, a non-primary purpose agreement which is one to which Part II of the 1954 Act applies (irrespective of any exclusion order under s.38A) is (1) an agreement conferring code rights (as it will confer code rights albeit that is not its primary purpose); (2) Part 5 will not apply to it but (3) Part 6 will apply to the removal of the apparatus. However, a non-primary purpose agreement being a lease which falls outside Part II of the 1954 Act (other than simply by reason of there being in existence an exclusion agreement under s.38A) can be one caught by both Parts 5 and Part 6 of the New Code. All that para 29 is doing is making it clear that Part 5 does not apply to a tenancy which falls within the terms of Part 2 of the 1954 Act.

7 It is in unclear whether an informal agreement which is not a conferring agreement because it does not comply with Part 2 may nevertheless fall within the protection of the 1954 Act. Paragraph 4 of Schedule 3 of the DEA 2017 excludes from the 1954 Act, a tenancy 'the primary purpose of which is to grant code rights within the meaning of Schedule 3A to the Communications Code 2003 (the electronic communications code)'. An informal tenancy agreement may nevertheless be a primary purpose agreement in the sense that the purpose of the tenancy may be to grant rights which fall within the code rights provided for by paragraph 3. However, as it is not a conferring agreement it has no code protection nor, arguably, any 1954 Act protection. One reading of paragraph 4 of Schedule 3 to the DEA 2017 is that it is necessarily implicit that the tenancy granting code rights is one which would ordinarily satisfy the terms of Part 2 to the code, and the draftsman was not contemplating 'code rights' falling within paragraph 3 granted by an agreement which happened to fall outside Part 2 but was nevertheless one which could be described as 'primary purpose' agreement.

22.2.13 Summary

The possible modes of protection for agreements conferring code rights are therefore as follows:

(a) A licence, or other simple contract not creating an interest in land but conferring code rights in respect of land, will therefore be a code agreement i.e. an agreement to which Part 5 applies – irrespective of its primary purpose.
(b) A tenancy or lease (the expressions have the same effect) will be a code agreement (as defined in paragraph 29(5)) if the primary purpose of the letting is to grant code rights. It will not be the subject of 1954 Act protection.
(c) A tenancy or lease which has a different primary purpose cannot be a code agreement as defined by paragraph 29(5) in that Part 5 will not apply to it *unless* it is one to which the 1954 Act does not apply: paragraph 29(2) and (3).
(d) Part 6 will apply to both a primary purpose and non-primary purpose agreement.

22.3 Continuation of code agreements

22.3.1

As with tenancies to which Part 2 of the 1954 Act applies, paragraph 30 of the New Code provides that a code agreement will not end on what might colloquially be called its contractual expiry date, but will be continued under the code until terminated in accordance with Part 5.

22.3.2

The text of paragraph 30, which is headed 'Continuation of code rights', is as follows:

> (1) Sub-paragraph (2) applies if –
>   (a) a code right is conferred by, or is otherwise binding on, a person (the 'site provider') as the result of a code agreement, and
>   (b) under the terms of the agreement –
>   (i) the right ceases to be exercisable or the site provider ceases to be bound by it, or
>   (ii) the site provider may bring the code agreement to an end so far as it relates to that right.
>
> (2) Where this sub-paragraph applies the code agreement continues so that –
>   (a) the operator may continue to exercise that right, and
>   (b) the site provider continues to be bound by the right.
>
> (3) Sub-paragraph (2) does not apply to a code right which is conferred by, or is otherwise binding on, a person by virtue of an order under paragraph 26 (interim code rights) or 27 (temporary code rights).
>
> (4) Sub-paragraph (2) is subject to the following provisions of this Part of this code.

22.3.3

This drafting is short, clear and effective, although several comments should be made.

22.3.4

First, it is to be noted that the terms of paragraph 30 apply where the code right was binding on the site provider. The site provider need not be a party to the code agreement to be subject to the terms of paragraph 30. This is in contrast to the right of termination referred to in paragraph 31, which is conferred on a site provider who is a party to the code agreement.

22.3.5

Secondly, the drafting of paragraph 30(1)(b) refers to the circumstances where the code agreement is terminable, either because it was granted for a fixed period, or is for a series of periods, or because it contains within it the means of termination upon a given event. In so doing, it might appear to suggest that there are other circumstances in which a code agreement is not so terminable. That is not, of course, the case: it is of the essence of a code agreement that it must contain provision for termination in any event (see paragraph 11(1) of the New Code, considered in paragraph 18.1.5). So, the drafting of this paragraph merely sets the scene for the continuation.

22.3.6

Thirdly, paragraph 30(2) states baldly that in such circumstances, the code agreement continues. It expressly stipulates that the operator may continue to exercise the code rights, while the site provider continues to be bound by the code rights. Presumably this drafting is not intended to suggest that the parties are not bound and cannot benefit in any other respect. That is to say, the draftsman cannot have meant by the specific references to the code rights continuing to bind and benefit that the other elements of the code agreement should not also continue to bind and benefit.

22.3.7

Finally, paragraph 30(3) provides that sub-paragraph (2) does not apply in the case of an order for interim code rights or temporary code rights, because these have their own internal termination procedures (see Chapter 21).

22.4 The notice to terminate

22.4.1

Paragraph 31 of the New Code sets out the following notice procedure for termination of a code agreement, under the heading 'How may a person bring a code agreement to an end?':

> (1) A site provider who is a party to a code agreement may bring the agreement to an end by giving a notice in accordance with this paragraph to the operator who is a party to the agreement.
>
> (2) The notice must –

(a) comply with paragraph 89 (notices given by persons other than operators),
(b) specify the date on which the site provider proposes the code agreement should come to an end, and
(c) state the ground on which the site provider proposes to bring the code agreement to an end.

(3) The date specified under sub-paragraph (2)(b) must fall –
(a) after the end of the period of 18 months beginning with the day on which the notice is given, and
(b) after the time at which, apart from paragraph 30, the code right to which the agreement relates would have ceased to be exercisable or to bind the site provider or at a time when, apart from that paragraph, the code agreement could have been brought to an end by the site provider.

(4) The ground stated under sub-paragraph (2)(c) must be one of the following –
(a) that the code agreement ought to come to an end as a result of substantial breaches by the operator of its obligations under the agreement;
(b) that the code agreement ought to come to an end because of persistent delays by the operator in making payments to the site provider under the agreement;
(c) that the site provider intends to redevelop all or part of the land to which the code agreement relates, or any neighbouring land, and could not reasonably do so unless the code agreement comes to an end;
(d) that the operator is not entitled to the code agreement because the test under paragraph 21 for the imposition of the agreement on the site provider is not met.'

22.4.2

The following comments deserve to be made about this drafting.

22.4.3

First, the term 'site provider' is defined by paragraph 30(1)(a) to mean the person who conferred the code rights, or upon whom the code rights are binding as a result of a code agreement. This term was presumably used in preference to the 'occupier' with which paragraph 10 deals, on the footing that the occupier may have changed hands; or the code agreement came to bind others by virtue of paragraph 10(4) of the New Code (see paragraph 15.5.2).

22.4.4

Secondly, only a site provider who is a party to a code agreement may serve the termination notice provided for by paragraph 31. A person may be treated as a party to a code agreement where bound by it albeit he may not be a signatory to the code agreement: see paragraph 10(3)(a) (which concerns successors in title, and deems them to be bound by as if they were parties to the conferring agreement). Less clear is whether a person is 'party' to a code agreement simply because he is bound by it (by virtue of paragraph 10(2), or by reason of an agreement to be bound under paragraph 10(4), or by reason of being deemed to be bound under paragraph 10(6)). Although paragraph

10 does say that such persons are 'bound', they are not (unlike under paragraph 10(3)(a) expressly made or treated as a party to the conferring agreement creating the code right.[8] The distinction between being an actual or statutorily deemed party on the one hand, and simply being 'bound' on the other, is critical for termination purposes. This is because only a 'site provider who is a party to a code agreement' has standing to give a termination notice under paragraph 31(1).[9] Given the distinction made in paragraph 10, it will be critical in each case to understand whether the proposed giver of the notice can be described as a 'party' to the code agreement. It will, for instance, not be possible for a mortgagee seeking possession to give such a notice in its own name in relation to an agreement between the mortgagor and an operator, unless it is a real or deemed party.

22.4.5

Thirdly, although the notice periods are long, the site provider's notice can be served while the contractual period of the code agreement is still running, thus enabling the renewal or modification machinery to run its course in tandem, hopefully eliminating or minimising any delay.

22.4.6

Fourthly, the notice requirements are lengthy and detailed (although in practice the detail will be set out in a prescribed form, similar to a s.25 notice under the 1954 Act). The notice must:

(a) Comply with the requirements of paragraph 89 of the New Code. These requirements are discussed in Chapter 29. Essentially, if the notice is not in the prescribed form (that is to say, in accordance with the model forms issued from time to time by OFCOM – see paragraph 90 of the New Code, and Chapter 41 of this book), then it will be invalid.
(b) Specify the date on which the site provider proposes the code agreement should come to an end (see paragraph 22.4.7 below).
(c) State the ground on which the site provider proposes to bring the code agreement to an end (see paragraph 22.4.9 below).

22.4.7

The date on which the site provider proposes the code agreement should come to an end for the purposes of a valid termination notice must itself be in accordance with the requirements in paragraph 31(3). This provides that the termination date must fall:

(a) after the end of the period of 18 months beginning with the day on which the notice is given, and

8 So, paragraph 10(4) provides that the code right 'binds any other person with an interest in the land who has agreed to be bound by it', but does not provide that such a person is to be treated as a party to the code agreement. Paragraph 10(6) provides only that 'a successor in title who is bound by a code right by virtue of sub-paragraph (5)(a) [successors to the party who has agreed to be bound] is to be treated as a party to the agreement by which P agreed to be bound by the right'.

9 Or a variation notice under paragraph 33.

(b) after the time at which, apart from paragraph 30, the code right to which the agreement relates would have ceased to be exercisable or to bind the site provider or at a time when, apart from that paragraph, the code agreement could have been brought to an end by the site provider.

22.4.8

An example should serve to explain the effect of this drafting. Suppose that a site provider, A, grants a code agreement to an operator, B, providing for code rights to apply to a particular site for ten years, ending on 31 December 2020. Suppose that A then serves an 18-month notice on B on 31 December 2019. The earliest date A can insert in that notice (ignoring complications such as whether the date of service of the notice should be included for the purpose of calculation) is 30 June 2021,[10] since that is 18 months after the date of service of the notice, and clearly after the date on which the agreement would have ended had it not been for the continuation provisions in paragraph 30 (as to which, see section 22.3 above). Clearly, therefore, should the site provider wish to terminate the code agreement at the earliest possible opportunity, it should aim to serve its notice 18 months before the code agreement expiry date.

22.4.9

The grounds on which the site provider might propose to bring the code agreement to an end are set out in paragraph 30(4), and are four in number, in most cases reminiscent of, but not entirely duplicating, the six grounds set out in s.30(1) of the 1954 Act:

(a) substantial breaches of the code agreement;
(b) persistent delay in payment;
(c) redevelopment; and
(d) that the continuation or renewal of the code agreement would result in prejudice that could not be adequately compensated, and that the public benefit does not outweigh that prejudice.

22.4.10

Experience with the similar grounds in s.30(1) of the 1954 Act suggests that the first two grounds will be rarely used in practice, essentially because tenants usually improve their

10 For the purposes of the New Code, 'month' means a calendar month: Interpretation Act 1978 ss.5 and 22(1), Sch 1 and Sch.2 para.4(1). The 'corresponding date' rule will apply to the computation of time. In *Dodds v Walker* [1981] 1 WLR 1027, the House of Lords explained the 'corresponding date rule' as follows:

(1) Where there is a date in the month during which the notice is to expire which 'corresponds' to the date in the month in which the notice is served, then the notice should specify that 'corresponding date'. Accordingly, if one had to give two months' notice from October 31, the appropriate date to specify would be December 31, because both October and December are months which have 31 days.
(2) If there is no 'corresponding date' in the month in which the notice is to expire, because that month contains fewer days than the month in which the notice is served, then the notice should be expressed to expire on the last day of the month in which the notice expires. Thus, if one were giving one month's notice from August 31, it should be expressed to expire on September 30.

record as termination draws closer; while the court is unlikely to penalise a tenant who has shown improvement. The same is likely to apply to operators.

22.4.11

The redevelopment ground is a shortened version of that set out in paragraph (f) of s.30(1) of the 1954 Act, which applies where:

> on the termination of the current tenancy the landlord intends to demolish or reconstruct the premises comprised in the holding or a substantial part of those premises or to carry out substantial work of construction on the holding or part thereof and that he could not reasonably do so without obtaining possession of the holding.

A number of comments about this drafting are worth making.

22.4.12

First, the difference in the wording of paragraph 31(4)(c) ('that the site provider intends to redevelop all or part of the land to which the code agreement relates, or any neighbouring land, and could not reasonably do so unless the code agreement comes to an end') suggests that the draftsman wished to constrain the type of works that might constitute a redevelopment. This would probably be an unsafe conclusion to draw. Quite apart from the fact that statutes fall to be construed in accordance with their own terms, rather than as a comparative exercise, a court is perhaps unlikely to be persuaded that there is a meaningful distinction between redevelopment of part, and 'substantial works of construction … on part' (the latter being the wording to be found in paragraph (f) of s.30(1) of the 1954 Act).

22.4.13

Secondly, paragraph 31(4)(c) allows for termination on the ground of redevelopment to neighbouring land, whereas s.30(1)(f) is confined to works to 'the holding'.[11] Having said that, it may be said to be likely to be rare for a redevelopment of land to require the termination of the code agreement on the adjoining land.

22.4.14

Thirdly, paragraph 31(4)(c) does not state the date at which the intention must be shown to be intended to be implemented. Under the 1954 Act the landlord's intention must be one which he intends to implement 'on the termination of the current tenancy'. The reference to 'the termination of the current tenancy' under that Act includes the continuation of the tenancy pursuant to ss.24 and 64, since the landlord cannot possibly

11 The 'holding' is defined by s.23(3) as: 'the property comprised in the tenancy, there being excluded any part thereof which is occupied neither by the tenant nor by a person employed by the tenant and so employed for the purposes of a business by reason of which the tenancy is one to which this Part of this Act applies'.

implement his intention until the current tenancy has come to an end and he is entitled to vacant possession.

22.4.15

Fourthly, those differences aside, the language of the two redevelopment grounds is similar, and until a body of case law on the New Code has collected, decisions on s.30(1)(f) of the 1954 Act may provide some assistance.

22.4.16

Fifthly, each provision employs the concept of the site provider (or the landlord, under the 1954 Act) being unable reasonably to carry out the relevant works unless (in this case) the code agreement comes to an end, or (in the case of the 1954 Act) the landlord recovers possession. In the context of the 1954 Act, if the existing agreement allows the landlord to enter the premises and carry out the relevant works, then it will follow that it does not need possession in order to carry out the works, because it has a current legal right to do so in any event.[12] The same finding is likely to apply in the case of code agreements which reserve extensive rights for site operators to enter and carry out works.

22.4.17

The fourth termination ground engages the same factors as those which are discussed in section 21.3 of this book, to which reference should be made.

22.4.18

The last point to make about paragraph 31 is that its provisions override (but without making redundant) contractual means of termination such as the operation of a forfeiture proviso or break clause. Paragraph 34.4.5 of Chapter 34 expands on this point.

22.5 The effect of a notice under paragraph 31

22.5.1

Paragraph 32 of the New Code provides, under the heading 'What is the effect of a notice under paragraph 31?':

> (1) Where a site provider gives a notice under paragraph 31, the code agreement to which it relates comes to an end in accordance with the notice unless –
> (a) within the period of three months beginning with the day on which the notice is given, the operator gives the site provider a counter-notice in accordance with sub-paragraph (3), and
> (b) within the period of three months beginning with the day on which the counter-notice is given, the operator applies to the court for an order under paragraph 34.

12 See e.g. the decision of the Court of Appeal in *Price v Esso Petroleum Co* [1980] 2 EGLR 58.

(2) Sub-paragraph (1) does not apply if the operator and the site provider agree to the continuation of the code agreement.

(3) The counter-notice must state –

(a) that the operator does not want the existing code agreement to come to an end,

(b) that the operator wants the site provider to agree to confer or be otherwise bound by the existing code right on new terms, or

(c) that the operator wants the site provider to agree to confer or be otherwise bound by a new code right in place of the existing code right.

(4) If, on an application under sub-paragraph (1)(b), the court decides that the site provider has established any of the grounds stated in the site provider's notice under paragraph 31, the court must order that the code agreement comes to an end in accordance with the order.

(5) Otherwise the court must make one of the orders specified in paragraph 34.

22.5.2

It will be noted that paragraph 32 does not merely specify what is the effect of a site provider's paragraph 31 notice: it also goes on to deal with five other substantial matters, which are considered in this section.

22.5.3

First, only one effect is specified for a paragraph 31 notice: the code agreement will come to an end in accordance with the notice, in the absence of protective action on the part of the operator. While on the face of it this would appear to be good news for the site provider, the possibility remains that if the operator has failed to give counter-notice through inadvertence (or subsequently has a change of heart), there will be nothing stopping it seeking a fresh imposed agreement under Part 4 of the New Code (see Chapter 21).

22.5.4

Secondly, upon being served with a paragraph 31 notice, the operator has a choice available to it. If three months elapse with it having done nothing, then the code agreement will come to an end. If the operator wishes to negotiate renewal or extension of the code agreement with the site provider, then it may of course seek to do so – but if it fails to protect its position by serving a counter-notice (see paragraph 22.5.5 below), then it will lose its rights once the three month period has elapsed. If it manages to secure agreement with the owner during that period, however, then paragraph 32(1) no longer applies – see paragraph 32(2).

22.5.5

Interestingly, paragraph 32(2) refers to paragraph 32(1) not applying 'if the operator and the site provider agree to the continuation of the code agreement'. If the site provider and the operator agree a *new* agreement that would appear to take effect not as a 'continuation of the code agreement' but rather as the conferring of code rights under a new arrangement. Thus it is unclear what paragraph 32(2) is intending to provide for. Perhaps it simply means an agreement that the notice is to have no effect? See also paragraph 32(3)(a), where reference is made to the operator not wanting the 'existing code agreement to come to an end'. Further, it may be asked what does 'continuation' mean? Does it mean that it continues in accordance with paragraph 30? If this is right, is it subject to notice in accordance with paragraph 30(3)? If not it seems difficult to see on what basis the continuation agreement is to be brought to end. When the court makes an order for continuation it is provided by paragraph 34(2) that it will be 'for such period as may be specified in the order (so that the code agreement has effect accordingly)'.

22.5.6

Upon being served with a paragraph 31 notice, therefore, the prudent operator will serve a counter-notice under paragraph 32(3). This must state that the operator:

(a) does not want the existing code agreement to come to an end,
(b) wants the site provider to agree to confer or be otherwise bound by the existing code right on new terms, or
(c) wants the site provider to agree to confer or be otherwise bound by a new code right in place of the existing code right.

22.5.7

In addition to these requirements the operator must ensure that it uses a counter-notice in the prescribed form issued by OFCOM – see paragraph 90 of the New Code.[13]

22.5.8

The third matter with which paragraph 32 deals is the application to court by the operator for an order under paragraph 34, which must be made within three months beginning with the day on which the counter-notice was given. As with the case of failure to deliver a timeous counter-notice, if the operator fails to apply to the court in time, then the code agreement will come to an end – see paragraph 32(1) of the New Code.

22.5.9

Fourthly, paragraph 32 deals with the result of the site provider succeeding in establishing one of its grounds of opposition (see section 22.4 above). In those circumstances, the court must order that the code agreement comes to an end in accordance with the order – see sub-paragraph (4). Unhappily, this sub-paragraph does not state what the order should actually say – but the Authors suggest that it should be straightforward to persuade the court to make an order simply providing for the code agreement to come to an end on the date specified in the owner's paragraph 31 notice. Even if, by the time of the hearing, that date has passed, that should not affect the order, because the owner will still need to procure the removal of the operator's electronic communications apparatus under Part 6 of the New Code – see Chapter 24.

13 See Chapter 32 for notices.

22.5.10

Lastly, paragraph 32(4) provides that if no such order is made (i.e. if the owner fails to make out one of its grounds), then the court must instead make one of the orders specified in paragraph 34 – see the next section.

22.6 What orders a court may make on an application under paragraph 32

22.6.1

Paragraph 34 of the New Code sets out the orders that the court may make on an application under paragraph 32(1)(b) – i.e. an order protecting the operator's rights in the event of service by the owner of a paragraph 31 notice seeking to terminating the code agreement.

22.6.2

Paragraph 34 sets out the following possibilities:

(2) The court may order that the operator may continue to exercise the existing code right in accordance with the existing code agreement for such period as may be specified in the order (so that the code agreement has effect accordingly).

(3) The court may order the modification of the terms of the code agreement relating to the existing code right.

(4) Where under the code agreement more than one code right is conferred by or otherwise binds the site provider, the court may order the modification of the terms of the code agreement so that it no longer provides for an existing code right to be conferred by or otherwise bind the site provider.

(5) The court may order the terms of the code agreement relating to the existing code right to be modified so that –
   (a) it confers an additional code right on the operator, or
   (b) it provides that the site provider is otherwise bound by an additional code right.

(6) The court may order the termination of the code agreement relating to the existing code right and order the operator and the site provider to enter into a new agreement which –
   (a) confers a code right on the operator, or
   (b) provides for a code right to bind the site provider.

22.6.3

The menu of options for the court is therefore as might be expected: continuation of the existing code right for a specified period, with or without modification of other terms of the code agreement; modification of the existing code agreement to provide for the addition or deletion of a code right; or entry into an entirely new code agreement.

22.6.4

An order continuing the existing code right in accordance with the existing code agreement may, as noted above, be provided to be 'for such period as may be specified in the order' (paragraph 34(2)) and may be subject to 'the modification of terms of the code agreement relating to the existing code right' (paragraph 34(3)). Those terms can presumably provide for a variation of the consideration payable. It is clear from a consideration of paragraphs 23(2) and (3) that the draftsman of the New Code when referring to 'terms' was intending to include the consideration payable. Any continuation of the existing code agreement will post the order of the court be once again subject to the right of termination provided for by Part 5.

22.6.5

Paragraph 34 ends by dealing with a long list of ancillary matters to accompany the process of making any of the orders referred to.

22.6.6

First, paragraph 34(7) provides that the existing code agreement continues until the new agreement (meaning thereby whichever of the options the court orders, rather than necessarily a new code agreement) takes effect.

22.6.7

Secondly, paragraph 34(8) provides that the New Code applies to the new agreement as if it were an agreement under Part 2 of the code, with the effect that all the provisions that apply to Part 2 agreements will also apply to the new arrangements effected by an order under paragraph 34.

22.6.8

Thirdly, paragraph 34(9) provides that the terms conferring or providing for an additional code right under sub-paragraph (5), and the terms of a new agreement under sub-paragraph (6), are to be such as are agreed between the operator and the site provider, to be decided in default of agreement by the court under sub-paragraph (10).

22.6.9

Fourthly, paragraph 34(11) applies the provisions of paragraphs 23(2) to (8), 24, 25 and 84 (all of which deal with imposed agreements) to the new arrangements arising from orders made under paragraph 34.

22.6.10

Fifthly, paragraph 34(12) stipulates that in making an order under sub-paragraph (10), the court must also have regard to the terms of the existing code agreement; while paragraph 34(13) adds that the court must have regard to all the circumstances of the case, and in particular to –

(a) the operator's business and technical needs,
(b) the use that the site provider is making of the land to which the existing code agreement relates,
(c) any duties imposed on the site provider by an enactment, and
(d) the amount of consideration payable by the operator to the site provider under the existing code agreement.

22.6.11

Sixthly, paragraph 34(14) provides that where the court makes an order under paragraph 34:

> it may also order the operator to pay the site provider the amount (if any) by which A exceeds B, where –
> (a) A is the amount of consideration that would have been payable by the operator to the site provider for the relevant period if that amount had been assessed on the same basis as the consideration payable as the result of the order, and
> (b) B is the amount of consideration payable by the operator to the site provider for the relevant period.

In other words, if the amount of the consideration determined by the court exceeds the consideration payable by the operator for the relevant period, the court may order the excess to be paid by the operator to the site provider. Paragraph 34(15) defines the relevant period as:

> the period (if any) that –
> (a) begins on the date on which, apart from the operation of paragraph 30, the code right to which the existing code agreement relates would have ceased to be exercisable or to bind the site provider or from which, apart from that paragraph, the code agreement could have been brought to an end by the site provider, and
> (b) ends on the date on which the order is made.

22.6.12

There does not appear to be any provision whereby the court can provide for the consideration, assessed on the basis of the terms provided for by the court order, to be less and thus to provide for reimbursement to the operator. Paragraph 34(14) refers only to circumstances where the consideration exceeds that currently payable by the operator. However, it would seem that either party can apply for the current consideration to be reassessed on the basis of the terms of the current code rights under the current code agreement in accordance with paragraph 35(2) (see 22.7 below) for the period from the date of the making of the court application until the application is finally determined.

22.7 The arrangements for payment pending determination of an application

22.7.1

Under the heading 'What arrangements for payment can be made pending determination of the application?', paragraph 35 provides:

> (1) This paragraph applies where –
> (a) a code right continues to be exercisable under paragraph 30 after the time at which, apart from the operation of that paragraph, the code right would have ceased to be exercisable or to bind the site provider or from which, apart from that paragraph, the code agreement relating to the right could have been brought to an end by the site provider, and
> (b) the operator or the site provider has applied to the court for an order under paragraph 32(1)(b) …
> (2) The site provider may –
> (a) agree with the operator that, until the application has been finally determined, the site provider will continue to receive the payments of consideration from the operator to which the site provider is entitled under the agreement relating to the existing code right,
> (b) agree with the operator that, until that time, the site provider will receive different payments of consideration under that agreement, or
> (c) apply to the court for the court to determine the payments of consideration to be made by the operator to the site provider under that agreement until that time.
> (3) The court must determine the payments under sub-paragraph (2)(c) on the basis set out in paragraph 24 (calculation of consideration).

22.7.2

This flexible provision seeks to ensure that the site provider is not disadvantaged by the delay that may occur as a result of the operator using the provisions of Part 5 to continue operating from the provider's site. It is akin to the interim rent procedure available under s.24A of the 1954 Act, but shorn of much of the complexity of that procedure.

23 Modification of code agreements

23.1 Introduction

23.1.1

Part 5 of the New Code deals with the procedure for the termination and modification of code agreements. Termination is dealt with by Chapter 22. This chapter deals with the modification of code agreements.

23.1.2

The principles and procedure applicable to modification of a code agreement are the same as those which apply to termination, although modification is addressed in one dedicated paragraph of the New Code, paragraph 33, which is examined in this chapter.

23.2 How to modify a code agreement

23.2.1

Paragraph 33 of the New Code is headed 'How may a party to a code agreement require a change to the terms of an agreement which has expired?', and provides:

> (1) An operator or site provider who is a party to a code agreement by which a code right is conferred by or otherwise binds the site provider may, by notice in accordance with this paragraph, require the other party to the agreement to agree that –
> (a) the code agreement should have effect with modified terms,
> (b) where under the code agreement more than one code right is conferred by or otherwise binds the site provider, that the agreement should no longer provide for an existing code right to be conferred by or otherwise bind the site provider,
> (c) the code agreement should –
> (i) confer an additional code right on the operator, or
> (ii) provide that the site provider is otherwise bound by an additional code right, or
> (d) the existing code agreement should be terminated and a new agreement should have effect between the parties which –

(i) confers a code right on the operator, or
(ii) provides for a code right to bind the site provider.

(2) The notice must –
(a) comply with paragraph 88 or 89, according to whether the notice is given by an operator or a site provider,
(b) specify –
(i) the day from which it is proposed that the modified terms should have effect,
(ii) the day from which the agreement should no longer provide for the code right to be conferred by or otherwise bind the site provider,
(iii) the day from which it is proposed that the additional code right should be conferred by or otherwise bind the site provider, or
(iv) the day on which it is proposed the existing code agreement should be terminated and from which a new agreement should have effect, (as the case may be), and
(c) set out details of –
(i) the proposed modified terms,
(ii) the code right it is proposed should no longer be conferred by or otherwise bind the site provider,
(iii) the proposed additional code right, or
(iv) the proposed terms of the new agreement,
(as the case may be).

(3) The day specified under sub-paragraph (2)(b) must fall –
(a) after the end of the period of 6 months beginning with the day on which the notice is given, and
(b) after the time at which, apart from paragraph 30, the code right to which the existing code agreement relates would have ceased to be exercisable or to bind the site provider or at a time when, apart from that paragraph, the code agreement could have been brought to an end by the site provider.

(4) Sub-paragraph (5) applies if, after the end of the period of 6 months beginning with the day on which the notice is given, the operator and the site provider have not reached agreement on the proposals in the notice.

(5) Where this paragraph applies, the operator or the site provider may apply to the court for the court to make an order under paragraph 34.

23.2.2

The following points should be noted about this modification procedure in paragraph 33.

23.2.3

First and foremost, it applies only where, but for the New Code, the code agreement has expired, or is shortly to expire. In that respect, the heading of paragraph 33 is inappropriate, since code agreements do not expire as such: they continue by virtue of paragraph 30 until terminated by agreement or court order.

23.2.4

Secondly, and as a development of the first point, there is no modification procedure available which enables a change to the terms of a code agreement at any other time. Accordingly, a site provider who wishes to constrain a particular type of activity, or to alter an access to the site, must build a term to that effect into the agreement conferring code rights at the outset, for it will be too late to do so later, at least until the point of expiry of the agreement. Correspondingly, an operator who wishes to alter the terms of the agreement must ensure that the right to do so is included within the original agreement.

23.2.5

Thirdly, the modification procedure is a facility for *both* parties: see paragraph 33(1) – 'An operator or site provider who is a party to a code agreement'. This contrasts with termination under paragraph 31, which only the site provider is entitled to instigate. Note again, however, the need for the site provider to be a party to a code agreement, as to the significance of which see paragraph 22.4.4 above.

23.2.6

Fourthly, the modification procedure applies only to *code agreements*. As section 19.1 explains, not all agreements conferring code rights will be code agreements.

23.2.7

Fifthly, there is only one time limit that is relevant to modification, in contrast to the array of limits that apply to termination. Paragraph 33(3) stipulates that the notice which triggers the modification process must specify a date from which the proposed change is to take effect. This date must fall both after the end of the period of 6 months beginning with the day on which the notice is given, and after the time at which, but for the New Code itself, the code rights would have ceased to apply. In simple terms, if the code agreement provides for a code right to be exercisable until 31 December 2020, the date to be inserted in any notice given under paragraph 33 cannot be earlier than that date, and must be six months and a day after the notice is given. From a practical point of view it may, accordingly, be appropriate for the landowner to seek to incorporate a break clause to enable earlier termination of the contractual term either unconditionally or conditionally to enable it to trigger paragraph 33 simultaneously with service of the break notice (see paragraph 42.24.3)

23.2.8

Sixthly, the procedure under paragraph 33 is much simpler than the termination procedure under paragraph 31. Apart from the notice procedure and time limit dealt with in paragraph 33(3), no other notice is required, and there are no other time limits to be observed (other than the stipulation in sub-paragraphs (4) and (5) that the application to the court for modification may not be made until six months after the notice has been given). There is no procedure for a counter-notice to be served by the counter-party.

It would seem that where notice has been served under paragraph 33 but neither party applies to the court after the six-month period provided for by sub-paragraphs (4) and (5), the existing agreement simply continues unmodified.

23.2.9

Lastly, although paragraph 33 deals with termination of the existing code agreement and its replacement with another, there is no provision in paragraph 33 for out and out termination.

23.3 Modification: ancillary matters

23.3.1

Paragraphs 29, 34 and 35 all apply as much to modification as they do to termination of a code agreement. The matters with which those paragraphs deal are examined in sections 21.2, 21.6 and 21.7 of Chapter 22, and need not be repeated here.

23.4 Modification contrasted with termination

23.4.1

There is plainly a degree of cross-over between paragraphs 31 and 33. Each describes a process which may ultimately result in an order under paragraph 34, providing for continuation of the existing code right for a specified period, with or without modification of other terms of the code agreement; modification of the existing code agreement to provide for the addition or deletion of a code right; or entry into an entirely new code agreement. The summary headings for the paragraphs – respectively 'How may a person bring a code agreement to an end?' and 'How may a party to a code agreement require a change to the terms of an agreement which has expired?' do not therefore do full justice to their subject matter, and indeed are positively misleading.

23.4.2

In practice, therefore, the only outcome which the procedure under paragraph 33 does not allow is actual termination with no further agreement, for which resort must be had to paragraph 31, with its various grounds for termination.

23.4.3

The last point prompts the observation that in practice it is likely that paragraph 31 will be used by site providers only for termination, properly so called. It is difficult to think of a reason why a site provider should wish to use paragraph 31, with its 18-month time limit, for any other sort of change to a code agreement. For any other type of change other than termination, paragraph 33 provides a swifter and easier route – and of course the only route available to an operator.

24 Rights to require removal

24.1 Introduction

24.1.1

Under paragraph 21 of the Old Code, the procedure for removal of apparatus (see Chapter 11) involves the service of a notice, in the absence of which the site owner may take such steps as are open to it to remove the apparatus, without involving the court. If a counter-notice is served by the operator, then the owner must go to court to secure an order for removal and may apply to the court for authority to remove it himself. Although the procedure is far from simple, the intervention of the court is comparatively limited.

24.1.2

The position is different under the New Code: an order for termination of the code agreement under Part 5 will not give the landowner the right to have the electronic communications apparatus on the site removed. For that, the landowner will need to resort to the removal procedure in Part 6 of the New Code.

24.1.3

In nine long paragraphs, Part 6 of the New Code sets out the detail of the removal process. This chapter sets out and explains the relevant provisions.

24.1.4

This chapter is arranged in the following way:

(a) Section 24.2 introduces 'the landowner' – the person who has the right to require removal of apparatus.
(b) Section 24.3 considers the landowner's removal rights.
(c) Section 24.4 then turns to an analysis of the neighbour's removal rights.
(d) Section 24.5 considers how code rights may be detected.
(e) Section 24.6 discusses how the landowner can enforce the removal of apparatus.
(f) Section 24.7 considers the enforcement of removal rights under other powers.
(g) Section 24.8 looks at the alteration of apparatus in the case of street works.

(h) Section 24.9 analyses the separate application for restoration of land.
(i) Section 24.10 finally considers the orders that may be made under paragraphs 40 to 43.

24.2 'The landowner'

24.2.1

In contrast to Part 2 of the New Code, which designates the person with the right to create an agreement conferring code rights 'the occupier', who is someone who does not necessarily have any interest in the land over which the rights are to be conferred, such as a licensee or a trespasser; and Part 5, which refers to the person with the right to seek termination or modification of a code agreement as 'the site provider', Part 6 uses the concept of 'the landowner', and provides that the right to require removal with which it deals may only be exercised by that person.

24.2.2

'The landowner' is defined by paragraph 37(1) of the New Code as 'a person with an interest in land' – i.e. the land on, over or under which the electronic communications apparatus is situated. This will exclude the occupier who may have no interest in the land – even if it was the very same occupier who granted the agreement conferring code rights (in this chapter, as with Chapter 19, although this expression is not used in the New Code, the 'conferring agreement'), and who may still occupy the land.

24.2.3

As a result, although a licensee who has granted a code right is within the class of persons who can terminate a code agreement under Part 5 of the New Code, it cannot then enforce the removal of the apparatus under Part 6. This curious result is not explained or justified in either the New Code or the Law Commission Report that preceded it.

24.2.4

The precise interest in land with which this paragraph (and indeed the whole of Part 6 of the New Code) is concerned is nowhere spelled out. For property law purposes (to which the expression 'interest in land' is obviously directed), an interest in land may include not merely the obvious interests, such as a freehold or leasehold estate, but also other less obvious interests, such as an easement, or a right of re-entry.

24.2.5

It is to be doubted whether the owners of any of the lesser interests in land would interest themselves in Part 6 of the New Code. In practice, therefore, the landowner for the purposes of Part 6 is likely to be the owner of the freehold or leasehold estate in the land. Part 6 does not distinguish between the two, or provide any mechanism by which the interests of one may rank above the interests of the other. Again, the resolution of this conundrum is likely to be found in property law: a freeholder may have no right of access to its lessee's demise for the purposes of apparatus removal. In such circumstances, the landowner for all practical purposes will be the leaseholder.

24.2.6

Having said that, circumstances may well arise where this will not be the case, such as the following:

(a) if the leaseholder has only a short reversion remaining on its lease, with no renewal or extension rights, then it is likely to be the freeholder who will wish to be 'the landowner' for the purposes of Part 6;
(b) the same may well be the case if the owner of the derivative interest has a short periodic tenancy.

24.2.7

In this regard, note needs to be made of paragraph 37(4) of the New Code, which restricts the right of the landowner to seek removal of the electronic communications apparatus where the land is occupied by the person who granted the conferring agreement entitling the operator to keep the apparatus on, under or over the land or is occupied by a person who is bound by the operator's rights. In such a case it would seem that paragraph 37(4) requires the landowner to first remove the occupier or other party bound by the operator's rights before the landowner is able to seek removal as against the operator. This is discussed further below.

24.2.8

Finally, it is to be noted that the landowner may be required to seek removal under Part 6, albeit the conferring agreement is not a code agreement for the purpose of Part 5. In other words, Part 6 may be engaged albeit Part 5 may not. By way of example, the effect of paragraph 11 (mandatory requirements for a Part 2 agreement) is that an informal agreement will not be a conferring agreement. Thus in the not uncommon situation of a tenant holding over after the expiration of its contractual term, the continuing payment and acceptance of rent may be said to give rise to an implied periodic tenancy. However, given the terms of paragraph 11, such an implied periodic tenancy will not be a conferring agreement and accordingly there will be no New Code protection, at least with respect to its termination under Part 5 of the New Code. (It may of course have protection under the Landlord and Tenant Act 1954.) Albeit such an informal agreement may not be within the terms of Part 5, it would seem to be arguable from the terms of Part 6 that an informal agreement which is not a conferring agreement, is, nevertheless, within the ambit of Part 6 thus requiring a court order for the removal of any electronic communications apparatus. Paragraph 37(1) provides that 'a person with an interest in land ... has the right to require the removal of electronic communications apparatus on under or over the land *if (and only if)* one or more of the following conditions are met' (emphasis added). There is no requirement in Part 6, unlike Part 5, for the agreement to be an 'an agreement under Part 2' (paragraph 29).

24.3 The landowner's removal rights

24.3.1

Paragraph 37 of the New Code is headed 'When does a landowner have the right to require removal of electronic communications apparatus?', and contains a considerable amount of detail as to the grounds which must be established in order to put the landowner into a position whereby it may be entitled to seek removal of electronic communications apparatus.

24.3.2

Paragraph 37(1) of the New Code provides that the landowner has the right to require the removal of electronic communications apparatus on, under or over its land if (and only if) one or more of a number of conditions is met. The conditions are then listed in the remaining sub-paragraphs of paragraph 37, and are dealt with separately below.

24.3.3 First condition: landowner never bound

Paragraph 37(2) provides:

> The first condition is that the landowner has never since the coming into force of this code been bound by a code right entitling an operator to keep the apparatus on, under or over the land.

24.3.4

This condition (and the second condition, below), is subject to sub-paragraph (4), which provides:

> The landowner does not meet the first or second condition if –
> (a) the land is occupied by a person who –
> (i) conferred a code right (which is in force) entitling an operator to keep the apparatus on, under or over the land, or
> (ii) is otherwise bound by such a right, and
> (b) that code right was not conferred in breach of a covenant enforceable by the landowner.

24.3.5

Putting sub-paragraphs (2) and (4) together, this first condition will be met where:

(a) the conferring agreement was made with an occupier who was not the landowner, or its predecessor in title;
(b) the landowner did not agree to be bound by the relevant code right;

(c) the land continues to be occupied by the person who made the conferring agreement (or someone who is otherwise bound); and
(d) the conferring agreement was made in breach of a covenant, enforceable by the landowner, under which the occupier was entitled to occupy the land.

24.3.6

Two examples may serve to explain the purport of this drafting.

24.3.7

First, suppose that F owns the freehold of land; T is F's tenant; and P is a prospective operator. T and P enter into a conferring agreement. They do not secure F's agreement to be bound by the conferring agreement, with the result that neither F nor F's successor in title to the freehold, R, is bound by the conferring agreement. Suppose further that at the end of the conferring agreement, T remains in occupation of the land (as does P). Suppose too that, under this example, there is nothing in the lease between F and T which prohibits T entering into a conferring agreement. As a result, although R passes the test in paragraph 37(2), it does not after all meet the first condition, because of sub-paragraph (4), since the conferring agreement was not made in breach of the agreement under which the occupier was entitled to occupy the land.

24.3.8

Secondly, suppose that F has let the land on a long lease to L, with a prohibition upon the grant of conferring agreements; and L then sublets the land to T, with no such prohibition. T then enters into a conferring agreement with P. Neither F nor T agrees to be bound by it. At the end of the conferring agreement, F wishes to have P's apparatus removed. Although both F and L satisfy paragraph 37(2), L cannot satisfy paragraph 37(4). F does satisfy that sub-paragraph, and therefore meets the first condition as a result, but whether this will be of any practical use to it will depend upon the terms of its lease with L. It is possible, therefore, that although F may be 'the landowner' for the purposes of paragraph 37, and although it may satisfy the first condition, it will not in practical terms be able to procure removal of P's apparatus.

24.3.9

There a number of miscellaneous matters to note about this first condition.

24.3.10

First, paragraph 37(1) refers to the fact that 'the landowner has never since the coming into force of this code been bound by a code right entitling an operator to keep the apparatus on, under or over the land'. Accordingly, in order to determine whether the first condition has been satisfied, this would appear to require regard to be had to any earlier code agreements pursuant to which the apparatus had been permitted to be installed on, under or over the land. Thus, if the apparatus has previously been the subject matter of a conferring agreement binding the landowner, but the code agreement has been renewed without obtaining the landowner's consent, it would seem, from a literal interpretation of the words that the landowner fails to satisfy this first condition. The landowner has, since the coming into force of the New Code, been bound by a code right, entitling an operator with respect to the apparatus in question albeit under an earlier code agreement. Given the wording, it may be that in considering the first condition one must pay particular attention to the apparatus which is the subject matter of the code agreement, for it may be that if e.g. on renewal additional apparatus is permitted, any such additional apparatus cannot be said to have been the subject matter of a code agreement which has previously been one binding the landowner. Thus, the first condition may be satisfied in respect of some but not all of the apparatus.

24.3.11

Secondly, the first condition is not satisfied if, among other things, the land continues to be occupied by the person who has granted the conferring agreement. Accordingly, if the occupier who granted the conferring agreement is e.g. protected by Part II of the Landlord and Tenant Act 1954, removal cannot be sought pursuant to the first condition unless and until the tenant occupier has been removed pursuant to the 1954 Act. Accordingly, the landowner will have to have regard to the grounds of opposition it may have in order to resist renewal of any tenancy of the conferring party.

24.3.12

Thirdly, the first condition is not satisfied if 'the land is occupied' by the conferring party under the conferring agreement. It does not require the occupation to be lawful. Thus, taking the example of the 1954 Act tenant occupier referred to previously, a successful termination claim under the 1954 Act by the landowner resisting a tenant's claim to renewal merely brings the tenant occupier's tenancy to an end three months and 21 days after final disposal pursuant to s.64 of the 1954 Act. In order to obtain possession the landowner would have to institute separate possession proceedings after the s.64 date had expired. Until those proceedings have reached their successful conclusion it would appear arguable that the first condition remains unsatisfied.

24.3.13

Fourthly, the reference in paragraph 37(4)(a)(ii) to occupation by a person who 'is otherwise bound by such a right', requires the landowner to remove all intermediate rights or interests of third parties, whether or not they confer a proprietary interest in land unless the code right was conferred in breach of covenant enforceable by the landowner. Thus, if F grants a tenancy to T who sublets to ST, who grants the conferring agreement upon the operator, there being no restriction under any of the tenancies in granting conferring agreements, if F as landowner wishes to effect removal it would need to remove T and ST from occupation of the land before triggering the right of removal under Part 6 of the New Code.

24.3.14

Fifthly, the code right must not have been 'conferred in breach of covenant enforceable by the landowner': paragraph 37(4)(b). The 'covenant' may include a restrictive covenant binding on those in occupation of the land albeit it may not have been imposed directly upon the operator by the landowner. To take as an example, if F grants a lease to T and restricts the use of the land for electronic communications purposes, then, if T were to sublet to ST, which contained no such restriction on use, and ST were to grant a code agreement to O, O would be bound by the restriction contained in the lease as between F and T and thus the grant of the code agreement by ST to O would have been granted 'in breach of a covenant enforceable by the landowner'. Similarly, the words would extend to a covenant enforceable by the landowner in reliance on s.56 of the Law of Property Act 1925, s.15 of the Landlord and Tenant (Covenants) Act 1995 or the Contracts (Rights of Third Parties) Act 1999.

24.3.15

Sixthly, the date on which the landowner does ultimately remove the conferring party under the conferring agreement from the land determines the earliest date from which the relevant notice may be served on the operator in accordance with paragraph 40. Paragraph 37 (1) refers to 'the right to require the removal … if (and only if) one or more of the following conditions are met'. Thus, until the satisfaction of sub-paragraph (4) the first condition is not met and notice cannot be served pursuant to paragraph 40. A notice served e.g. in anticipation of say obtaining possession as against the occupational conferring party under the conferring agreement would appear to be premature.

24.3.16

Before leaving this first condition, it is important to note the way in which the code right is expressed. Paragraph 37(4) refers 'a code right entitling an operator to keep the apparatus on, under or over the land'. This is surely a reference just to code right (b) in paragraph 10(4) of the New Code (as to which, see section 15.4), namely a right 'to keep installed electronic communications apparatus which is on, under or over the land'. If that is right, then this condition does not apply if the landowner is bound by any other code right. That is a curious limitation for the draftsman to have intended – but it appears to be the literal meaning of the words used.

24.3.17 Second condition: landowner not bound

Paragraph 37(3) provides:

> The second condition is that a code right entitling an operator to keep the apparatus on, under or over the land has come to an end or has ceased to bind the landowner –
>
> (a) as mentioned in paragraph 26(7) and (8),
> (b) as the result of paragraph 32(1), or
> (c) as the result of an order under paragraph 32(4) or 34(4) or (6), or
> (d) where the right was granted by a lease to which Part 5 of this code does not apply.

24.3.18

This condition too is subject to sub-paragraph (4) – see paragraph 24.3.4 above.

24.3.19

Paragraph 37(3) scrambles together a number of different situations:

(a) Sub-paragraph (a) deals with the expiry of an interim code right without an imposed agreement having been ordered (see paragraph 20.8.7).
(b) Sub-paragraph (b) is concerned with the situation where a site provider has given a paragraph 31 notice seeking the termination of a code agreement, and the operator has failed to protect its rights by serving a counter-notice;
(c) Sub-paragraph (c) applies where the site provider has under Part 5 of the New Code established a ground of opposition to the renewal (etc.) of the code agreement under paragraph 31; and
(d) Sub-paragraph (d) applies where the right was granted by a lease to which Part 5 of the New Code does not apply – i.e. a lease to which Part II of the 1954 Act applies.

24.3.20

The first three situations are self-explanatory. The fourth calls for more comment. If the conferring agreement is not a code agreement because of the 1954 Act exemption, then it might be thought that the 1954 Act regime alone should apply, consistently with the Law Commission's sensible proposal that the two regimes should be kept separate (see paragraph 25.2.5). However, paragraph 37(3)(d) makes it clear that even where an operator has installed electronic communications apparatus under a non-code agreement, the landowner's right to remove the apparatus at the end of the term of the agreement may be constrained by this paragraph. This is consistent with the dichotomy introduced by the New Code of separating out the procedure for termination of the conferring agreement and the procedure for removal of the apparatus.

24.3.21

The fourth situation enables a landowner to seek removal in circumstances where the conferring agreement, being a lease, is not one to which Part 5 of the New Code applies, and where (1) the lease has come to an end or (2) has ceased to be binding on the landowner. To illustrate the distinction between these two circumstances:

(1) F has granted a tenancy to T and T has granted a non-primary purpose agreement conferring 1954 Act rights upon the operator, O. As landowner, F will be entitled to rely upon the second condition where O's sublease 'has come to an end', which would, it is considered, require T to have terminated O's sublease in accordance with the 1954 Act. (As noted above, the second condition is also subject to paragraph 37(4) and thus F as landowner would need to ensure, if it wished to exercise the right of removal as against O, that T's occupation of the land has also ceased). The mere determination of the contractual term of O's sublease is not, it is considered,

a situation where it could be said that the 'code right entitling an operator to keep the apparatus on, under or over the land has come to an end'. Insofar as O has 1954 Act rights, the right will not end unless terminated in accordance with the 1954 Act.

(2) F has granted a tenancy to T and T has granted a non-primary purpose agreement conferring 1954 Act rights upon the operator, O. F forfeits T's lease which would destroy O's sublease being a derivative interest of T. In those circumstances, subject to any right of relief from forfeiture, it may be said that O's right has 'ceased to bind the landowner'.

24.3.22

It is important to note that the terms of paragraph 37(3)(d) refer to a lease 'to which Part 5 of this code does not apply'. It must not be thought that in this context the New Code is dealing only with leases which are caught by the 1954 Act. Part 5 is disapplied not only to leases which are within the 1954 Act but also to leases which are excluded from the 1954 Act, e.g. where there is a non-primary purpose agreement and Part II of the 1954 Act has been excluded, so long as, notwithstanding the exclusion pursuant to section 38A of the 1954 Act, the lease would otherwise be one to which Part 2 of the 1954 Act would apply: paragraph 29(2) and (3).

24.3.23 The third condition: apparatus obsolete

Paragraph 37(6) provides:

> The third condition is that –
> (a) an operator has the benefit of a code right entitling the operator to keep the apparatus on, under or over the land, but
> (b) the apparatus is not, or is no longer, used for the purposes of the operator's network, and
> (c) there is no reasonable likelihood that the apparatus will be used for that purpose.

24.3.24

The drafting of this condition is straightforward, although it does prompt the curious notion that the landowner may procure removal even though the conferring agreement or code agreement is still running, with consideration presumably being paid under it. However, although in theory the landowner could require the removal of the apparatus, the land will remain bound by the code right, and the landowner's ability to put its land to better use will therefore be precarious.

24.3.25

Obsolescence is not a ground for termination of the code agreement under Part 5.[1] One way of dealing with this is for the conferring party to make provision by way of a break

1 Although query whether paragraph 31(4)(d) may be utilised to enable termination in these circumstances. This provides that the agreement may be terminated where 'the operator is not entitled to the code

clause (or where an intermediate interest is conferring the code right, for the superior landowner to insist upon the insertion of a break clause in the conferring agreement) enabling earlier termination of the operator's contractual term to overcome the otherwise practical difficulty that albeit the landowner may be able to remove the apparatus it may be unable to utilise the land due to the continuing existence of the code agreement.

24.3.26

It is considered that this provision would enable landowners to exert pressure upon sharers of electronic communications apparatus. A sharer of electronic communications apparatus is one that does not share 'land': paragraph 108. A sharer accordingly has no agreement which can constitute a code agreement falling within Part 2 of the New Code. If the operator upon whom code rights were conferred is no longer using the electronic communications apparatus for its own network but the sharer is utilising the apparatus for its network, on the face of it the landowner would appear to be able to invoke the third condition.

24.3.27 The fourth condition: code no longer applies

Paragraph 37(7) provides:

> The fourth condition is that –
> (a) this code has ceased to apply to a person so that the person is no longer entitled under this code to keep the apparatus on, under or over the land,
> (b) the retention of the apparatus on, under or over the land is not authorised by a scheme contained in an order under section 117, and
> (c) there is no other person with a right conferred by or under this code to keep the apparatus on, under or over the land.

24.3.28

This condition applies where the operator is no longer designated as such (see section 16.2), and neither is any other person who might e.g. share *the land* with the operator. The third criterion, in referring to s.117, is referring to s.117 of the 2003 Act, pursuant to which authorisation of such schemes may be effected.

24.3.29 The fifth condition: transport/street exemption

Paragraph 37(8) provides:

> The fifth condition is that –
> (a) the apparatus was kept on, under or over the land pursuant to –
> (i) a transport land right (see Part 7), or
> (ii) a street work right (see Part 8),

agreement because the test under paragraph 21 for the imposition of the agreement on the site provider is not met'. Given the terms of paragraph 21, it may be sought to be argued that obsolete apparatus may not satisfy the second condition in paragraph 21(3) and (4).

(b) that right has ceased to be exercisable in relation to the land by virtue of paragraph 54(9), and
(c) there is no other person with a right conferred by or under this code to keep the apparatus on, under or over the land.

24.3.30

This condition is straightforward, and needs no further explanation.

24.3.31

Finally, paragraph 37(9) stipulates:

> This paragraph does not affect rights to require the removal of apparatus under another enactment (see paragraph 41).

24.3.32

This sub-paragraph is important, as it enables persons to seek removal of apparatus under another enactment. Paragraph 41, which is dealt with under section 24.7 below, defines such a person as a 'third party'. This is to distinguish their entitlement to remove the apparatus from their right to do so as a 'landowner'. A third party as defined may, in some circumstances, be a landowner as a matter of fact. However, if the landowner is exercising rights under paragraph 41, he is doing so as a third party. The right is based it seems on the provisions of paragraph 21 of the Old Code.

24.4 The neighbour's removal rights

24.4.1

Paragraph 38 of the New Code is headed 'When does a landowner or occupier of neighbouring land have the right to require removal of electronic communications apparatus?', and details the grounds which must be established in order to put a neighbour into a position whereby it may be entitled to seek removal of electronic communications apparatus on the adjoining land.

24.4.2

As the title of this paragraph suggests, the right to require removal of electronic communications apparatus from neighbouring land applies not merely to the owner of that land (ie the holder of a property interest in it), but also to its 'occupier' – an expression which is no doubt intended to bear the same meaning as the term used in Part 2, which is itself defined in paragraph 105 of the New Code (see section 16.2). The draftsman has thus provided that the removal rights should benefit a wider audience in relation to the neighbouring land, but not the land upon which the electronic communications apparatus is actually situated. This is perhaps because (1) the neighbour will not ordinarily have been a party to the code agreement and thus ought not to be prejudiced; and (2) the neighbour should be entitled to a full right over the access in question subject to the operator seeking a code agreement against him.

24.4.3

For ease of exposition, the owner and occupier of the neighbouring land are jointly referred to as 'the neighbour' in this section.

24.4.4

The starting point for the consideration of the neighbour's right in paragraph 38 is that an operator should have code rights, street works rights under Part 8, tidal water rights under Part 9 or a power to fly lines in respect of the adjoining land under paragraph 74 of Part 11 (see paragraph 38(1)).

24.4.5

Like paragraph 37, this paragraph sets out the conditions – this time only two – which must be satisfied before the neighbour acquires the right to require removal of the electronic communications apparatus from the adjoining land. In this case, however, the conditions are cumulative rather than alternatives.

24.4.6

The first condition is that the apparatus interferes with or obstructs a means of access to or from the neighbouring land – see paragraph 38(2). This needs no elaboration: it is most obviously aimed at apparatus which traverses or impinges upon a right of way to the neighbouring land, even if the apparatus is located wholly on the adjoining land, and not the neighbouring land (as this paragraph posits).

24.4.7

The second condition, which must be read together with sub-paragraph (4), is that the landowner or occupier of the neighbouring land is not bound by a code right within paragraph 3(h) entitling an operator to cause the interference or obstruction. If the operator is bound by such a right in the first place, then it will have no right to seek removal of the apparatus. This may be said to be unfortunate for the neighbour, who may have been content to be bound by the relevant code right in the past, but whose use for the neighbouring land has changed.

24.4.8

Paragraph 38(4) adds to this second condition by providing that the neighbour will not meet the condition if the code right in question was conferred by an occupier of the neighbouring land other than in breach of covenant to the neighbouring land owner.

24.4.9

The second condition is relevant only to code rights conferred upon an operator by an occupier of land under Part 2. The rights conferred under Parts 8, 9 and 11 of the New Code are not the subject of agreement with an occupier of land.

24.4.10

It is worth observing that, if the neighbour satisfies both conditions, it will be entitled to seek an order for removal: there will be no balancing test to be carried out. An operator may, if met with a claim by a neighbouring land owner for removal, seek a code right under paragraph 3(h) by serving notice in accordance with paragraph 20. The right of removal by the neighbouring land owner is by notice (paragraph 40) and the court may not make an order for removal if an application under paragraph 20(3) has been made in relation to the apparatus and the application has not been determined: paragraph 40(8).

24.5 Detecting code rights

24.5.1

It may sometimes be difficult for a landowner or a neighbour, where the landowner's land contains electronic communications apparatus, or is otherwise subject to code rights, to ascertain the ownership of the apparatus or the rights, especially since there is no need for operators to register such rights at the Land Registry.[2]

24.5.2

No doubt with this in mind, paragraph 39 of the New Code, which is headed 'How does a landowner or occupier find out whether apparatus is on land pursuant to a code right?', provides a notice procedure containing incentives for the operator to disclose its role in relation to any apparatus and code rights in respect of the land.

24.5.3

The procedure is available both for the landowner and for the neighbour. Oddly, the procedure applies only to a landowner in respect of land in which a landowner has an interest (paragraph 39(1)). It does not enable an occupier of such land to serve notice. However, when dealing with *neighbouring* land, paragraph 39(2) provides that the notice procedure is available to both the landowner and occupier of the neighbouring land. Why there is this distinction is not clear.

24.5.4

In the case of the landowner, the procedure commences with a notice to the operator and provides in paragraph 39(1) for the disclosure of three matters, namely:

(1) whether the operator owns electronic communications apparatus on, under or over land in which the landowner has an interest;
(2) whether the operator uses, for the purposes of the operator's network, apparatus on, under or over land in which the landowner has an interest

2 See paragraph 14 of the New Code: 'Where an enactment requires interests, charges or other obligations affecting land to be registered, the provisions of this code about who is bound by a code right have effect whether or not that right is registered.' See further Chapter 38.

(3) whether the operator has the benefit of a code right entitling the operator to keep electronic communications apparatus on, under or over land in which the landowner has an interest.

24.5.5

It is to be noted that the terms of paragraph 39(1) thus entitle the landowner to seek disclosure of apparatus which is being *used* by the operator even if not *owned* by the operator. There are, however, it is considered, a number of deficiencies in the disclosure notice provisions, namely:

(1) They presuppose that the landowner knows who is 'the operator' upon whom notice is to be served. An operator under a code agreement is entitled to upgrade the electronic communications apparatus to which the agreement relates and to share the use of apparatus with another operator: paragraph 17 (see Chapter 20). It may be that the sharer provides upgraded apparatus which the operator then installs on the land to enable the sharing to take place (see paragraph 17(6)). There is nothing within paragraph 17 to require the operator who is the party to the code agreement to notify the landowner of the identity of any sharer or of the apparatus installed to facilitate sharing.
(2) They do not go on to provide that the operator is required to identify who is the owner of the apparatus used. Given that a notice of removal which is to be served pursuant to paragraph 40 has to be served upon owner of the apparatus (see paragraph 40(2)) one would have thought that it would have been sensible to require the operator to disclose the identity of the owner of the apparatus.
(3) The disclosure of the code right of which the operator has the benefit, is limited to 'a code right entitling the operator to keep electronic communications apparatus on, under or over land in which the landowner has an interest'. This code right is that contained within paragraph 3(b) of the New Code. The code rights provided for in paragraph 3 to the New Code are much more extensive than simply a right to keep electronic communications apparatus on, under or over land.

24.5.6

Paragraph 30(3) expressly requires that the notice must comply with paragraph 89; quite apart from that express requirement, the notice must also be in the form prescribed under paragraph 90; and delivered in accordance with paragraph 91 (as to which requirements, see Chapter 32).

24.5.7

In the case of the *neighbouring* land, the owner or occupier (and not just the landowner, as in the case of the adjoining land), is entitled by virtue of paragraph 39(2) to serve notice requesting the like details, in respect of apparatus or code rights concerning the land over which the neighbour has rights of access.

24.5.8

Paragraph 39(4) and (5) then detail the effect of a failure, or late failure, of the operator to respond to a notice from either the landowner or the owner or occupier of the neighbouring land, as follows:

(4) Sub-paragraph (5) applies if –
 (a) the operator does not, before the end of the period of three months beginning with the date on which the notice under sub-paragraph (1) or (2) was given, give a notice to the landowner or occupier that –
  (i) complies with paragraph 88 (notices given by operators), and
  (ii) discloses the information sought by the landowner or occupier,
 (b) the landowner or occupier takes action under paragraph 40 to enforce the removal of the apparatus, and (c) it is subsequently established that –
  (i) the operator owns the apparatus or uses it for the purposes of the operator's network, and
  (ii) the operator has the benefit of a code right entitling the operator to keep the apparatus on, under or over the land.
(5) The operator must nevertheless bear the costs of any action taken by the landowner or occupier under paragraph 40 to enforce the removal of the apparatus.

24.5.9

In summary, the effect of non-compliance by the operator is as follows:

(a) The operator has three months from receipt of a notice from either the landowner or the neighbour to provide the information requested.
(b) If the operator fails to provide the information, following which the landowner or occupier takes steps to enforce removal, but it is subsequently established that the (same) operator does in fact own, use, or have code rights in respect of the apparatus, then the landowner or occupier is entitled to be recouped its costs of the removal application.

24.5.10

Subject to any prescribed form,[3] there is nothing within paragraph 39 to indicate the extent of the particularity required in response to the disclosure notice with respect to the identification of the apparatus owned or used by the operator. Given that one would expect a paragraph 39 disclosure notice to be a prelude to removal pursuant to paragraph 40, it would be sensible, absent any prescribed form, for the landowner or occupier to seek details in the disclosure notice served of each and every piece of apparatus owned or used.

24.5.11

In referring to 'landowner or occupier', these sub-paragraphs do not make it clear whether they are concerned only with the position of the landowner or occupier of the land upon which the apparatus is situated, or the landowner or occupier of the neighbouring land as well. However, the wider context provides the clarification: paragraph 39(4) refers to sub-paragraph (2), which deals with the apparatus on land which is a

3 As to which, see section 41.6 of Chapter 41.

means of access to the neighbouring land and enables the owner or occupier of the neighbouring land to serve the notice.

24.5.12

In conclusion, paragraph 39 does not give either the landowner or the neighbour a right to remove apparatus: instead they provide assistance to those parties seeking to ascertain from the operator what rights it has, with protection to the parties if the operator does not cooperate.

24.6 Enforcing removal of apparatus

24.6.1

Paragraph 40 of the New Code is headed 'How does a landowner or occupier enforce removal of apparatus?' Paragraph 40(1) makes it clear that the landowner or occupier in question are those of not merely the land, but also the neighbouring land; and that removal is exercisable only in accordance with paragraph 40 itself.

24.6.2

In common with many of the mechanisms employed by the New Code, paragraph 40 sets out a procedure, in summary, for the landowner or occupier to serve a notice containing requirements as to removal and restoration of the land within a reasonable time; for the landowner or occupier to apply to court for appropriate orders in default of a satisfactory agreement.

24.6.3

The following points should be noted about this procedure.

24.6.4

First, it is clear from the terms of paragraph 40(1) that removal of electronic communications apparatus on, under or over land 'is exercisable only in accordance with this paragraph'. Accordingly, removal cannot be effected in exercise of some self-help remedy.[4] Any attempt by a landowner or occupier to effect removal other than in accordance with the terms of paragraph 40 would appear to be unlawful and thus capable of being restrained by an injunction.

24.6.5

Secondly, paragraph 40 has to be utilised where the right to require removal arises under paragraph 37 or 38. The terms of sub-paragraph (1) provide that 'the right of a landowner or occupier to require the removal of … apparatus… under paragraph 37 or 38, is exercisable only in accordance with this paragraph'.

4 Contrast the terms of paragraph 41(12) in respect of the rights of removal by a third party.

24.6.6

Thirdly, the notice must be given 'to the operator whose apparatus it is' (paragraph 40(2)). As noted above the New Code makes provision enabling the landowner to identify the ownership of apparatus: paragraph 39. It is quite often the case that operators when sharing apparatus or sites (by way of a sub-letting or licence) permit the sharer to install its own apparatus.

24.6.7

Fourthly, paragraph 40(2) provides that the notice may require the operator:

(a) to remove the apparatus, and
(b) to restore the land to its condition before the apparatus was placed on, under or over the land.

Requirement (a) is plainly an alternative to requirement (b): that is to say that the landowner or occupier may be content to require removal but not restoration (for example, if it intends to redevelop the land, which might make restoration nugatory). What is less clear is whether requirement (a) is itself an alternative – i.e. whether the landowner or occupier may simply require restoration (perhaps in a case where the operator has removed the apparatus, but not restored the land). Good sense would suggest that both requirements are optional.

24.6.8

Paragraph 40(2) does not make clear is the extent to which the notice must particularise what is to be removed or what is to be required by way of restoration once removal has been effected. Thus, it is unclear whether, for instance, a notice which simply requires the operator to 'remove all of its apparatus situated on the land' is sufficient or whether it is necessary for the land owner or operator to be more specific in terms of identifying the specific items apparatus to be removed. Similarly, is it sufficient simply to require the operator to 'restore the land to its former condition prior to the apparatus having been placed on, under or upon it' without identifying what is required by way of restoration or what is alleged to have been the former condition?

24.6.9

The terms of paragraph 40 thus make it clear that, from a practical point of view, any landowner or operator entering into a code agreement, would be well advised to prepare a schedule of condition of the land, which is to be attached to the agreement, so as to be able to establish, upon termination of the code agreement, the 'former condition' to which restoration is required upon removal of the apparatus.

24.6.10

Fifthly, as with all other notices required by the New Code, the notice must expressly comply with the requirements (in this case) of paragraph 89 – see paragraph 40(3)(a).

As in the case of the notice discussed in paragraph 24.5.4 above, the notice must also be in the form prescribed under paragraph 90; and delivered in accordance with paragraph 91 (as to which requirements, see Chapter 32).

24.6.11

Sixthly, the notice must 'specify the period within which the operator must complete the works' (paragraph 40(3)(b)), which must be 'a reasonable one' (paragraph 40(4)). This drafting may well cause significant problems for the landowner or occupier, who may have little idea of the time involved in such works. In such circumstances, the draftsman of the notice might like to adopt the expedient used for notices under s.146 of the Law of Property Act 1925, which frequently require remedy of a breach of covenant 'within a reasonable time'. The letter enclosing the paragraph 40 notice could go on to suggest a reasonable time, and ask for the operator's comments. Such a course would at least avoid the notice being held to be void from its outset for failure to specify a time which was reasonable, so long as this device of referring to, rather than specifying, a reasonable time, is valid (which may be said to be open to doubt).

24.6.12

The 'reasonable time' which is required to be given is a period for the operator to 'complete the works'. The 'works' must refer back to the two matters which may be contained within the notice, namely, works to effect removal of the apparatus and works of restoration. Accordingly, a landowner or occupier who serves notice which specifies both removal and restoration must take both matters into account in determining the 'reasonable period' to be specified. It is probably the case that a notice which specifies a period which, objectively, is reasonable for removal of the apparatus but is insufficient to accommodate any or all of the works of restoration (such restoration having been required by the notice) will be held to be unreasonable

24.6.13

It is considered that, as is the case with many issues concerning the validity of notices, what is a 'reasonable time' is to be determined objectively. Thus, the fact that the specific operator upon whom notice has been served may be able to effect 'the works' within the time specified matters not if, objectively, the period specified would not enable a reasonable person in the position of the operator to effect removal. Conversely, an operator who is unable to effect the works because of personal financial difficulty, whereas a reasonable person would, will be disadvantaged.

24.6.14

Seventhly, in addition to the point just made, the operator's failure to carry out the works required within a reasonable time does not trigger any right for the landowner or occupier to take any action; the trigger is instead the failure of the landowner or occupier and the operator to reach agreement on various matters within 28 days beginning with the day on which the notice was given – see paragraph 40(5).

_24.6.15_

In such an event (that is to say, the failure of the parties to reach agreement within 28 days), paragraph 40(6) entitles the landowner or occupier to apply to court for orders for removal or sale of the apparatus under paragraph 44. It is therefore immaterial whether the operator commences the works required (although obviously no further action will be required if the works are completed): it is the failure to agree which is determinative.

_24.6.16_

It is to be noted that it is only the landowner or occupier that can apply to the court for the removal or sale order. The operator need and can do nothing to resist the application for removal, other than to serve a notice pursuant to paragraph 20. Where the operator has served a paragraph 20 notice and made an application to the court for the appropriate code right to be conferred so as to be able to retain the electronic communication apparatus upon the land, the court may not make an order in relation to removal or sale of the operator's apparatus if the paragraph 20(3) application has not been determined: paragraph 40(8) (see further paragraph 24.6.17).

_24.6.17_

Eighthly, and rather oddly, paragraph 40(7) deals with a matter that is consequential upon paragraph 44, rather than paragraph 40: this is dealt with under section 24.10 below.

_24.6.18_

Ninthly, and importantly, paragraph 40(8) provides:

> On an application under sub-paragraph (6) or (7) the court may not make an order in relation to apparatus if an application under paragraph 20(3) has been made in relation to the apparatus and has not been determined.

_24.6.19_

This is likely to be more than just a sting in the tail for the landowner or occupier, who may find that they have gone through virtually the whole cumbersome apparatus provided by the New Code for the termination of code rights and the removal of apparatus with an uncooperative operator, only to find that the operator wishes after all to continue to station its apparatus on the land and claim code rights. There is no similar provision in relation to the termination procedure set out in Part 5, providing for example for that procedure to be put into abeyance if an application is made under paragraph 20(3). The reason for this may be that when the court comes to make an order on an application for termination, the extent of the orders that the court can make include modification of the existing code right or for the site provider to enter into a new agreement: paragraph 34(5) and (6). Thus there is no need to make provision similar to that contained within paragraph 40(8), because the court can in essence do this of its own motion.

24.6.20

Lastly, paragraph 40(9) makes it clear that paragraph 40 'does not affect rights to require the removal of apparatus under another enactment' in accordance with paragraph 41. Thus if there is an independent right entitling the landowner under an enactment[5] to seek removal, the landowner may opt to rely upon paragraph 41. Paragraph 41 is not subject to any of the conditions to be found in paragraph 37.

24.7 Enforcement of removal rights under other powers

24.7.1

Section 24.3 above has already made the point that even where an operator has installed electronic communications apparatus under a non-code agreement, the landowner's right to remove the apparatus at the end of the term of the agreement may be constrained by paragraph 37(3). This is reinforced by paragraph 41 of the New Code, which is headed 'How are other rights to require removal of apparatus enforced?'

24.7.2

Thus, paragraph 41(1) provides:

> The right of a person (a 'third party') under an enactment other than this code, or otherwise than under an enactment, to require the removal of electronic communications apparatus on, under or over land is exercisable only in accordance with this paragraph.

24.7.3

Paragraph 41 is dealing with the rights of a third party, other than the landowner, to require removal. It makes no reference to the prior termination of the agreement.

24.7.4

A 'third party' is an expression commonly used to refer either to a third party to an agreement, such as a guarantor, or to an interested person who is not party to the agreement, but affected by it. In this context, however, it is used to mean the right of any person to require the removal of electronic communications apparatus on, under or over land, other than a right under the New Code.

24.7.5

Paragraph 41(1) provides that the right in question may, but need not, stem from an enactment. No examples are given, but two obvious situations will be where:

(a) a planning authority seeks to remove apparatus installed in breach of planning legislation; or

5 Defined in paragraph 108(1).

(b) the conferring agreement is not a code agreement, perhaps because it is a lease to which the 1954 Act applies; a termination order under the 1954 Act is obtained; and the operator has to deliver up possession of the land. In such circumstances, if the operator does not remove its apparatus, the landowner will have a right, by virtue of its ownership of the land, to require removal by obtaining an injunction restraining trespass.

24.7.6

But there is one situation that is not obvious. Paragraph 41(1) refers to the right of a person to require the code operator to remove electronic communications apparatus. Where powers of compulsory purchase of land are conferred by an order, the acquiring authority is authorised to take possession of the affected land, including anything on it, such as electronic communications apparatus.[6] As the acquiring authority will have possession of the land and the apparatus, it would not seem that the authority needs to require the code operator to remove any apparatus; it can remove the apparatus itself.

24.7.7

There is not ordinarily any specific power in an order authorising compulsory purchase by which an acquiring authority can require the removal of electronic communications apparatus, and there is no enactment specifically conferring any such power. But, there are two situations of doubt raised by paragraph 41(1). First, an acquiring authority might contend that, where a code operator leaves apparatus on land being compulsorily acquired, the operator has failed to give up possession of the land or part of it, and possession should be enforced by the sheriff.[7] Would the sheriff be requiring the removal of the apparatus in terms of paragraph 41(1)? It is certainly so arguable. The preferable answer is that as a code right is not an interest in land, it cannot prevail against a compulsory purchase, and that no special exception has been made in the compulsory purchase legislation to preserve the application of the code; further, that the sheriff is enforcing statutory rights of possession. Secondly, where an order authorises the compulsory taking of a right in or over land, it is conceivable, depending on its terms, that the right might require, directly or indirectly, the removal of apparatus. Paragraph 41 might have application if the contention that compulsory purchase prevails against code rights is not accepted.

24.7.8

However, it may be thought that in the case of example (b) above, the example in fact falls within the terms of paragraph 37(3)(d) of the New Code. Paragraph 40(1) says that the right of a landowner to require removal of apparatus under paragraph 37 'is exercisable only in accordance with *this paragraph*'. Accordingly, it would appear that if a landowner has a right of removal in accordance with paragraph 37, that right is exercisable 'only in accordance with' para 40, and the landowner does not have the

6 Compulsory Purchase Act 1965, s.11.
7 Compulsory Purchase Act 1965, s.13.

option of utilising paragraph 41. If paragraph 41 could be invoked in the circumstances postulated, it would suggest that the landowner could simply sidestep the conditions to which paragraph 37 is subject, arguing simply that it has an entitlement to possession notwithstanding the terms of paragraph 37(3). It is the case that paragraphs 37 and 41 do not provide any form of 'priority provision' but it seems arguable that the reference to requiring removal 'otherwise than under an enactment' must necessarily be read as meaning 'otherwise than under an enactment in circumstances not otherwise provided for by this code'.

24.7.9

In such circumstances (that is to say, a right of removal arising other than under the New Code), the right is (only) exercisable using the notice procedure with which paragraph 41 deals in familiar detail. The procedure is identical in some respects to that stipulated in paragraph 40, but departs from it in others.

24.7.10

It would appear from the terms of paragraph 41 that it is providing for an independent right of removal without the necessity to comply with the conditions contained within paragraph 37. Thus there is no need, in order for the third party to rely upon the terms of paragraph 41, for the third party to bring about a termination of the code agreement in accordance with Part 5. In fact paragraph 41 may be relied upon by someone other than a party to the code agreement, e.g. a planning authority might seek removal because of a breach of the terms of planning legislation. The planning authority has no right to seek termination of the code agreement but may, notwithstanding the absence of any termination of the code agreement, seek to enforce removal in accordance with paragraph 41.

24.7.11

As in the case of paragraph 40, the notice under paragraph 41 must expressly comply with the requirements (in this case) of paragraph 89 – see paragraph 41(3)(a). As in the case of the notice discussed in paragraph 24.5.6 above, the notice must also be in the form prescribed under paragraph 90; and delivered in accordance with paragraph 91 (as to which requirements, see Chapter 32).

24.7.12

Next, the notice must 'specify the period within which the operator must complete the works' (paragraph 41(3)(b)), which must be 'a reasonable one' (paragraph 41(4)). As discussed above in section 24.6, this drafting may well cause significant problems for the third party, who may have little idea of the time involved in such works. In such circumstances, the draftsman of the notice might like to adopt the expedient used for notices under s.146 of the Law of Property Act 1925, which frequently require remedy of a breach of covenant 'within a reasonable time'. The letter enclosing the paragraph 41 notice could go on to suggest a reasonable time, and ask for the operator's comments. Such a course would at least avoid the notice being held to be void from its outset for failure to specify a time which was reasonable, provided that this device of referring to, rather than specifying, a reasonable time is valid (which may be open to doubt).

24.7.13

At this stage, the paragraph 41 procedure differs from that set out in paragraph 40. Paragraph 41(5) provides:

> Within the period of 28 days beginning with the day on which notice under sub-paragraph (2) is given, the operator may give the third party notice ('counter-notice') –
>
> (a) stating that the third party is not entitled to require the removal of the apparatus, or
>
> (b) specifying the steps which the operator proposes to take for the purpose of securing a right as against the third party to keep the apparatus on the land.

24.7.14

Failing such counter-notice, the third party is entitled to enforce the removal of the apparatus (paragraph 41(6)), whether via a court order under paragraph 41(9), or such other lawful means for enforcing the removal of the apparatus as are appropriate (see paragraph 41(12)).

24.7.15

If a timely counter-notice is given, however, then the third party must apply to court if it wishes to enforce removal of the apparatus. Paragraph 41(5) refers to two grounds which the operator may put in its counter-notice opposing removal (effectively, no entitlement to remove, or operator proposing to take steps to secure rights to retain). Paragraph 41(7) takes this forward, by providing:

> If the operator gives the third party counter-notice within that period, the third party may enforce the removal of the apparatus only in pursuance of an order of the court that the third party is entitled to enforce the removal of the apparatus.

24.7.16

Paragraph 41(8) adds, in the case of the second ground:

> If the counter-notice specifies steps under paragraph (5)(b), the court may make an order under sub-paragraph (7) only if it is satisfied –
>
> (a) that the operator is not intending to take those steps or is being unreasonably dilatory in taking them; or
>
> (b) that taking those steps has not secured, or will not secure, for the operator as against the third party any right to keep the apparatus installed on, under or over the land or to reinstall it if it is removed.

24.7.17

Paragraph 41(9) then provides:

> Where the third party is entitled to enforce the removal of the apparatus, under sub-paragraph (6) or under an order under subparagraph (7), the third party may make an application to the court for –
> (a) an order under paragraph 44(1) (order requiring operator to remove apparatus etc.), or
> (b) an order under paragraph 44(3) (order enabling third party to sell apparatus etc.).

24.7.18

By this stage, it will be observed that the third party may already have had to launch another set of proceedings, most obviously to secure the termination of the operator's rights, whether under the New Code, or the 1954 Act, or perhaps at common law.[8]. However, the draftsman's appetite for court proceedings to reinforce removal reaches a peak in paragraph 41(10), which provides:

> If the court makes an order under paragraph 44(1), but the operator does not comply with the agreement imposed on the operator and the third party by virtue of paragraph 44(7), the third party may make an application to the court for an order under paragraph 44(3).

It is not obvious why these repeated applications to court could not have been conflated in the New Code.

24.8 Alteration of apparatus in the case of street works

24.8.1

Paragraph 42 of the New Code is headed 'How does paragraph 40 apply if a person is entitled to require apparatus to be altered in consequence of street works?' There appears to be an error in the heading to this paragraph in the body of the New Code, as it is apparent from the terms of paragraph 42 that it is intending to refer to paragraph 41, not paragraph 40.

24.8.2

In essence, paragraph 42 applies the paragraph 41 removal procedure, but with some amendments noted in this section. It should be read together with Part 10 of the New Code, which applies to undertaker's works requiring alteration to apparatus. The

8 But as noted above in paragraph 24.7.6 the third party need not, in order to exercise its rights under paragraph 41, be required to terminate the code agreement. The third party may not in fact be a party to the code agreement.

drafting notably favours the position of the operator as opposed to the entity responsible for carrying out street works.

24.8.3

It should be noted that the term 'alteration' is defined in paragraph 108(2) to include 'moving, removal or replacement' of apparatus. Thus, when paragraph 42 refers to requiring alteration, it is in fact referring to removal as well as alteration properly so called. Paragraph 42 is not therefore dealing with alterations other than removal. This is reinforced by paragraph 42(2) which provides that 'removal of the apparatus in pursuance of paragraph 41 constitutes compliance with the requirement to make any other alteration.'

24.8.4

The first amendment to the paragraph 41 removal procedure is that paragraph 42 applies to a right to require the alteration of the operator's apparatus in consequence of the stopping up, closure, change or diversion of a street or road or the extinguishment or alteration of a public right of way, as opposed to the works of removal with which paragraph 41 is concerned. The meaning of this provision, which is carried over from the Old Code, is obscure, and the following example may serve to explain what might have been intended. Suppose that the third party is under a statutory duty to undertake some form of work to the apparatus – for example where the operator has installed apparatus and the planning authority serve an enforcement notice upon the landowner, contending that there is a breach of planning control because it affects the external appearance of the site. The landowner might argue that it can rely on paragraph 41 as amended by paragraph 42. Rather than undertaking some form of cosmetic work to satisfy the planners regarding the external appearance, the landowner simply seeks removal. Thus he discharges his requirement to comply with any enforcement notice by obtaining removal.

24.8.5

Secondly, paragraph 42(3) provides:

> A counter-notice under paragraph 41(5) may state (in addition to, or instead of, any of the matters mentioned in paragraph 41(5)(b)) that the operator requires the third party to reimburse the operator in respect of any expenses incurred by the operator in or in connection with the making of any alteration in compliance with the requirements of the third party.

24.8.6

Thirdly, paragraph 42(4) provides:

> An order made under paragraph 41 on an application by the third party in respect of a counter-notice containing a statement under sub-paragraph (3) must, unless

the court otherwise thinks fit, require the third party to reimburse the operator in respect of the expenses referred to in the statement.

24.8.7

Finally, paragraph 42(5) disapplies paragraph 44(3)(b) to (e) (which allow the third party to sell the apparatus, retaining the proceeds of sale).

24.9 The separate application for restoration of the land

24.9.1

Paragraph 43 of the New Code is headed 'When can a separate application for restoration of land be made?' Broadly, it comes to the assistance of a relevant party who wishes the land to be restored to its condition prior to the installation of the electronic communications apparatus, but without wishing to have the apparatus itself removed, and where that party is not bound by a code right. The remainder of this section examines this paragraph in more detail.

24.9.2

First, the restoration right is available only to another addition to the corpus of land-connected people, the 'relevant person', defined by paragraph 43(2) to mean the occupier of the land, the owner of the freehold estate in the land, or the lessee of the land. The definition is inclusive, intended to encompass everyone to do with the control of the land, from a trespasser all the way up to the full legal owner.

24.9.3

Secondly, the restoration must 'not involve the removal of electronic communications apparatus from any land'. There would appear to be no reason, therefore, why the relevant person cannot seek restoration in circumstances where the electronic communications apparatus has already been removed from the land. For instance, it may be that the landowner, being a lessee and the operator's immediate landlord, has terminated the operator's code agreement in accordance with Part 5 and obtained removal in accordance with paragraph 40 but did not, in its removal notice in accordance with paragraph 40, seek restoration of the land. The freehold owner would, in those circumstances, appear to be able to seek restoration. The freehold owner is not bound by the code right, because the code right has come to an end. This appears to be so even if during the currency of the code agreement the freehold owner was bound by the code right.

24.9.4

Thirdly, paragraph 43(2) gives the relevant person the right to the right to require the operator to restore the land, but only where the relevant person is not for the time being bound by the code right. To this, however, paragraph 43(3) adds that, even if that relevant person is not itself bound by a code right, it has no right to require restoration if:

(a) the land is occupied by a person who –
   (i) conferred a code right (which is in force) entitling the operator to affect the condition of the land in the same way as the right mentioned in subparagraph (1), or
   (ii) is otherwise bound by such a right, and
(b) that code right was not conferred in breach of a covenant enforceable by the relevant person.

The way in which this familiar New Code drafting technique works has been explained in paragraph 24.3.4 et seq above.

24.9.5

Fourthly, once it has established that it is entitled to require restoration, the mechanism for achieving that result is set out in the remainder of the paragraph, employing the usual process of notice affording a reasonable time, followed as necessary by a court order.

24.9.6

The notice requirements are set out in paragraph 43(6) and (7), which are drafted in identical terms to the removal and restoration procedure in paragraph 40 (see section 24.6 above). Once the relevant 28-day period has expired without agreement between the parties for the operator to carry out the requisite restoration, and the time to be taken in that process, the landowner (not, note, the relevant person) may make an application to court pursuant to paragraph 43(9) for either an order under paragraph 44(2) requiring the operator to carry out the restoration, or an order under paragraph 44(3), enabling the landowner to recover the cost of restoring the land itself. For the sake of completeness, if the landowner selects the former option, but the operator does not carry out the work, the landowner may select the latter option, always assuming that it has the appetite for all the litigation that this course will involve.

24.9.7

As noted in the immediately preceding paragraph, it is only the 'landowner', not the 'relevant person' that may make an application to the court. 'Landowner' is defined in paragraph 37(1) to be 'a person with an interest in land'. Thus the only material difference between 'relevant person' and 'landowner' is that the person entitled to seek restoration would exclude (1) a licensee or (2) a trespasser of the land. Neither of these exclusions is likely to have any practical consequence.

24.10 Paragraph 44: orders under paragraphs 40 to 43

24.10.1

The text above has made frequent reference to orders that the Court may be invited to make, usually at the suit of the landowner. The New Code brackets all the various order-making powers together in paragraph 44. The reason is that the draftsman deals compendiously with the effect of the orders, as this section explains, so that paragraph 44 serves as a useful point of reference for other paragraphs in Part 6.

24.10.2

Paragraph 44 does not deal with all the possible orders the Court may make under Part 6 of the New Code, but focuses upon those requiring various works to be carried out, to which paragraphs 40 (removal of apparatus under the code), 41(removal of apparatus pursuant to any other power), 42 (alteration as a result of street works) and 43 (restoration of land) refer.

24.10.3

Paragraph 44 introduces a number of ancillary aspects to the orders that the landowner is likely to find beneficial. Thus:

(a) An order for removal of apparatus may include an order for restoration of the land (sub-paragraph (1)).
(b) An order for removal or restoration may include an order for the relevant person to sell the apparatus so removed; to recover the costs of the removal or restoration and sale; and to retain the proceeds of sale to defray those costs (but apparently not the costs of the litigation) (sub-paragraph (3)).
(c) An order under paragraph 44 may also require the operator to pay compensation to the landowner for any loss or damage suffered by the landowner as a result of the presence of the apparatus on the land during the period when the landowner had the right to require the removal of the apparatus from the land but was not able to exercise that right (sub-paragraph (5)).[9] The question arises whether the operator is liable for mesne profits or some other similar form of damages when holding over post-termination under Part V, but before a removal order is made. It would appear that paragraph 44(5) is the primary avenue for recovering 'any loss or damage suffered by the landowner as a result of the presence of the apparatus on the land during the period when the landowner had the right to require the removal of the apparatus from the land but was not able to exercise that right'. There seems to be no reason why that wording should not cover some compensation for the operator's use and occupation for a period during which it had no right to be on the land at all. Such a claim would not be caught by the exclusion of compensation (except as provided for in the New Code) in paragraph 86: first of all, on the above reading of paragraph 44(5), the New Code does allow recovery of such losses. Secondly, if it did not, the paragraph 86 exclusion does not bite, because the operator is, following termination, not in occupation 'in the lawful exercise of any right'. The only argument to the contrary is that paragraph 44(6) states that 'Paragraph 84 makes further provision about compensation under sub-paragraph (5)'. Paragraph 84(2) then contains certain heads of loss, which are expenses, diminution in value or the land and reinstatement costs. It could be argued that paragraph 84(2) sets out a closed list of heads of loss recoverable under paragraph 44(6), and a charge for occupation is not one of them. However, (a) it would be remarkable if an operator with

9 Further provision with respect to compensation is made in paragraph 84. See further Chapter 30.

no rights would not have to pay for the period between termination and removal (when an operator seeking temporary rights would), (b) there is no strong statutory language pointing to this result, and (c) paragraph 84(2) can just as readily be understood as setting out three specific heads of loss that are recoverable under the more generous power to compensate for '*any*' loss under paragraph 44(5).

24.10.4

Finally, paragraph 44(7) adopts the following device:

> An order under sub-paragraph (1) or (2) takes effect as an agreement between the operator and the landowner, occupier or third party that –
> (a) requires the operator to take the steps specified in the order, and
> (b) otherwise contains such terms as the court may so specify.

24.10.5

This provision appears to provide for some form of statutory contract requiring the operator to take the relevant steps and otherwise comply with the terms of the court order. Given that it may be that an order for e.g. removal may be made without termination of the code agreement (e.g. pursuant to paragraph 41), thought may need to be given as to how any such 'agreement' is to be binding upon successors in title to the operator. Nothing is said in the terms of Part 6 or in the New Code as to whether any such agreement needs to be registered at HM Land Registry in order to be binding on successors in title to the operator. Although paragraph 14 disapplies the requirement of registration at the Land Registry, this is limited to 'the provisions of this code about who is bound by a code right'. Presumably, the limitation period for enforcement of the agreement commences with effect from the date of the making of the court order.

25 The New Code and the 1954 Act

25.1 Introduction

25.1.1

The New Code deals with a number of matters with respect to the application of Part II of the Landlord and Tenant Act 1954 ('the 1954 Act') to code agreements,[1] including agreements subsisting at the time the New Code came into force. In particular:

(1) a lease granted after the New Code comes into force the primary purpose of which is to grant code rights within the terms of New Code is not one to which the 1954 Act applies: Schedule 3, paragraph 4 to the 2017 Act;
(2) a lease which is granted after the New Code comes into force may be 1954 Act protected and also one to which the New Code applies, but the New Code disapplies the termination provisions contained within Part 5 of the New Code, leaving termination of the lease to be dealt with by the terms of the 1954 Act or the common law: paragraph 29; and
(3) there are detailed transitional provisions dealing with 'subsisting agreements' in existence at the date that the New Code came into force: Schedule 2, paragraph 6 to the 2017 Act.

25.1.2

The 1954 Act may also impact on the ability of the 'landowner' (defined in paragraph 37(1) as 'a person with an interest in land') to seek, in accordance with Part 6, to remove electronic communications apparatus from the land on, under or over which the apparatus is situated.

1 Under the New Code a 'code agreement' is expressly defined as one to which Part 5 of the New Code (dealing with termination) applies: paragraph 29(5). However, in this chapter it is used in a more general sense as referring to an agreement conferring code rights, whether consensual or imposed. On the distinction between a conferring agreement and code agreement see Chapter 19 generally. However, the distinction between a conferring agreement and a code agreement is an important one. By way of example, an informal agreement (e.g. say a periodic tenancy arising by the payment and acceptance of rent) would not comply with the terms of paragraph 11 and thus would not be a conferring agreement. Thus, such an agreement would not be within the ambit of Part 5. However, there is nothing preventing the application of Part 6 to such an agreement and the terms of paragraph 37(3)(d) clearly envisages its application in circumstances where Part 5 does not apply. It is unclear whether an informal agreement would in fact be protected by Part II of the Landlord and Tenant Act 1954: see paragraph 19.3.4.

25.1.3

This chapter is arranged in the following way:

(a) Section 25.2 deals with primary purpose New Code agreements.
(b) Section 25.3 deals with New Code agreements which may be caught by the 1954 Act.
(c) Section 25.4 deals with the transitional provisions.
(d) Section 25.5 considers the implication of the 1954 Act with respect to removal of electronic communications apparatus under Part 6.
(e) Section 25.6 discusses the issue of a renewal of a subsisting lease protected by the 1954 Act the primary purpose of which is to grant code rights.
(f) Section 25.7 considers the termination of a code agreement to which the 1954 Act applies in accordance with the 1954 Act, whether that is a New Code agreement or is a subsisting agreement to which the transitional provisions of the New Code apply.

25.2 New Code agreements

25.2.1

Under the Old Code regime, there was nothing preventing a lease which conferred rights under the Old Code from being protected both by the 1954 Act and the existing code. This had caused problems for landowners[2] wishing to seek removal of apparatus for the purposes of redevelopment in reliance upon paragraph 21 of the existing code by serving paragraph 21 notices prior to the termination of the tenancy in accordance with the 1954 Act (see Chapter 11).

25.2.2

The New Code avoids this difficulty, at least in relation to leases where the primary purpose is to grant code rights under the New Code. Paragraph 29 of the New Code provides:

> (1) This Part of this code applies to an agreement under Part 2 of this code, subject to sub-paragraphs (2) to (4).
> (2) This Part of this code does not apply to a lease of land in England and Wales if –
> (a) its primary purpose is not to grant code rights, and
> (b) it is a lease to which Part 2 of the Landlord and Tenant Act 1954 (security of tenure for business, professional and other tenants) applies.
> (3) In determining whether a lease is one to which Part 2 of the Landlord and Tenant Act 1954 applies, any agreement under section 38A (agreements to exclude provisions of Part 2) of that 40 Act is to be disregarded.
> (4) […]
> (5) An agreement to which this Part of this code applies is referred to in this code as a 'code agreement'.

2 Here used in its non-New Code sense as meaning simply the party who is seeking removal of the apparatus.

25.2.3

There is a mirroring amendment to Part II of the Landlord and Tenant Act 1954: a lease which is granted after the coming into force of the New Code cannot be one to which the 1954 Act applies if the *primary purpose* of it is to grant code rights within the meaning of the New Code: s.43(4) of the 1954 Act, inserted by Schedule 3 paragraph 4 to the 2017 Act.

25.2.4

It should be noted that the 'primary purpose' test is not one to determine whether or not the lease is a code agreement, but simply to determine whether the code agreement is or is not within the terms of the 1954 Act. Any agreement which confers code rights falling within paragraph 3 to the New Code is potentially one to which the New Code applies. It is unclear whether the 'primary purpose' is to be tested as at the date of grant of the lease[3] or from time to time.[4]

25.2.5

The New Code does not define what is a 'primary purpose' code agreement. No doubt the commonplace agreement for use of roof space for the installation of a mast[5] to which is attached electronic communications apparatus will be the paradigm example of a primary purpose agreement. As with the issue as to the distinction between a 'lease' and a 'licence' the matter is probably to be treated as one of substance not form, although if the parties were to declare in their agreement that it was intended to be a 'primary purpose' code agreement the court may, at least in the case of commercial parties of equal bargaining power, be prepared to give effect to it.[6]

3 As under the Rent Acts where the expression is whether the premises 'are let as a separate dwelling house': s.1 of the Rent Act 1977, Thus in that context it has been held that attention is directed to the object of the letting at its commencement: *Wolfe v Hogan* [1949] 2 KB 194, CA. Denning LJ said (204/205):

> In determining whether a house or part of a house is 'let as a dwelling' within the meaning of the Rent Restriction Acts, it is necessary to look at the purpose of the letting. If the lease contains an express provision as to the purpose of the letting, it is not necessary to look further. But, if there is no express provision, it is open to the court to look at the circumstances of the letting. If the house is constructed for use as a dwelling-house, it is reasonable to infer the purpose was to let it as a dwelling. But if, on the other hand, it is constructed for the purpose of being usedas a lock-up shop, the reasonable inference is that it was let for business purposes. If the position were neutral, then it would be proper to look at the actual user. It is not a question of implied terms. It is a question of the purpose for which the premises were let.

4 As under Part II of the Landlord and Tenant Act 1954, where the test is whether 'the property comprised in the tenancy is or includes premises which are occupied by the tenant and are so occupied for the purposes of a business carried on by him or for those and other purposes': s.23(1) of the 1954 Act. Under the 1954 Act attention is drawn to the purposes for which the premises are being occupied at the relevant time i.e. the time of determining whether there is security of tenure under that Act. Thus a tenancy may fall in and out of protection depending on the purposes for which the premises are occupied.

5 The mast is itself probably electronic communications apparatus, being a structure or thing designed for use in connection with the provision of an electronic communications network: paragraph 5(1)(d). See Chapter 18.

6 In the context of the lease/licence distinction, fairly recent Court of Appeal authority reflects a judicial approach to giving effect, where the contract is negotiated between two substantial parties of equal

25.2.6

It will be rare for a lease which confers code rights not have that as its primary purpose. As the Law Commission said in their Report No. 336:

> 6.86 There is of course room for doubt and for dispute as to the primary purpose of a lease. But we think that difficulties will arise in only a few cases; the lease of a mast site falls clearly on one side of the line, the lease to a Code Operator of a retail unit, where the lease incidentally permits the tenant to install a cell site on the roof, falls on the other.
>
> 6.87 It follows that in a mixed use lease where Code Rights are not the primary purpose of the letting, which is contracted out of the 1954 Act, the Code Operator will have no security. Where security is important, therefore, the Code Operator will want a separate lease for the apparatus.

25.2.7

Whilst there will be little room for doubt in the common mast agreement, if there are more complex agreements granting code rights as well as other rights (for instance, the grant of an office lease with permission to put a mast on top), then there will be more room for debate – and delay.

25.2.8

The 'primary purpose' of the lease must be to grant 'code rights'. Paragraph 3 of the New Code is headed 'the code rights'. Paragraph 3 then provides that 'for the purposes of this code a "code right," in relation to an operator and any land, is a right for the statutory purposes' to exercise the rights set out in sub-paragraph (a) to sub-paragraph (i) of paragraph 3. The 'statutory purposes' are defined in paragraph 4 and mean '(a) the purposes of providing the operator's network' or '(b) the purposes of providing an infrastructure system'.[7]

25.2.9

There is no composite definition of 'code rights'; each of the nine rights enumerated in paragraph 3 is referred to as a 'code right' in the singular. However, albeit the exclusion of a primary purpose agreement from the protection of the 1954 Act is expressed in terms of being one 'the primary purpose of which is to grant code rights', it is considered that an agreement which provides simply for a right to e.g. 'keep installed electronic communications apparatus which is on, under or over the land' within paragraph 3 (b) is nevertheless one which would be said to be granting 'code rights'.

bargaining power and with the benefit of full legal advice, to the description which the parties themselves confer upon the agreement: *Clear Channel UK Ltd v Manchester City Council* [2006] 1 EGLR 27, CA, per Parker LJ at paragraphs [28] and [29], followed in *Scottish Widows plc v Stewart* [2006] EWCA Civ 999, CA (and recently in Ireland in *Car Park Services Ltd v Bywater Capital (Wine Tavern) Ltd* (2016)).

7 As to the meaning of which, see section 15.4.

25.2.10

In conclusion, an agreement can therefore *only* be outside Part 5 of the New Code if it is a lease but is *not* for the primary purpose of conferring code rights, and has, or would have but for contracting out, protection under the 1954 Act. The point of this is to avoid the dual protection issue noted above, namely the fact that it is currently quite widely argued that a particular agreement under the Old Code enjoys duplicate protection under the 1954 Act and under that Code. This means that leases with such a primary purpose (likely to be most of them, according to the Law Commission – see above) and all licences will be governed by Part 5 of the New Code.

25.3 New Code agreements which may be caught by the 1954 Act

25.3.1 Primary purpose agreements

A code agreement under the New Code is required to be terminated in accordance with the New Code. Accordingly, a primary purpose code agreement can be terminated only in accordance with Part 5 of the New Code; the 1954 Act has no role to play. Part 5 applies only to 'an agreement under Part 2 of this code': paragraph 29(1). So what of informal tenancy agreements which are not conferring agreements[8] because they do not comply with para 11? Para 4 of Schedule 3 of the DEA 2017 excludes from the 1954 Act, a tenancy 'the primary purpose of which is to grant code rights within the meaning of Schedule 3A to the Communications Code 2003 (the electronic communications code)'. An informal tenancy agreement may nevertheless be a primary purpose agreement in the sense that the purpose of the tenancy may be to grant rights which fall within the code rights provided for by paragraph 3. However, as it is not a conferring agreement it has no code protection nor, arguably, any 1954 Act protection. One reading of paragraph 4 of Schedule 3 to the DEA 2017 is that it is necessarily implicit that the tenancy granting code rights is one which would ordinarily satisfy the terms of Part 2 to the code, and the draftsman was not contemplating 'code rights' falling within paragraph 3 granted by an agreement which happened to fall outside Part 2 but was nevertheless one which could be described as a 'primary purpose' agreement.

25.3.2 Non-primary purpose agreements

However, this is not necessarily the case with all New Code agreements. Although the 1954 Act does not apply to a lease the primary purpose of which is to grant code rights, a lease the primary purpose of which is not to grant code rights may, it seems, have New Code protection as well as 1954 Act protection.

25.3.3

Where one has a non-primary purpose agreement which is subject to the New Code, but which is one falling within the terms of the 1954 Act, the New Code provides for termination in accordance with the 1954 Act rather than in accordance with Part 5 of the

8 See Chapter 19.

New Code. Paragraphs 29(2) and (3), contained within Part 5 of the New Code dealing with termination of code agreements, provide:

> (2) This Part of this code does not apply to a lease of land in England and Wales if –
> (a) its primary purpose is not to grant code rights, and
> (b) it is a lease to which Part 2 of the Landlord and Tenant Act 1954 (security of tenure for business, professional and other tenants) applies.
>
> (3) In determining whether a lease is one to which Part 2 of the Landlord and Tenant Act 1954 applies, any agreement under section 38A (agreements to exclude provisions of Part 2) of that Act is to be disregarded.

25.3.4

Thus, a lease the primary purpose of which is not to grant code rights is not excluded from the 1954 Act by reason of section 43(4) of the 1954 Act, However, the termination provisions of Part 5 of the New Code will not apply if the non-primary purpose lease is one falling within the 1954 Act. Where the agreement is not a lease it matters not that it is a non-primary purpose agreement; Part 5 will apply to such a non-primary purpose agreement.

25.3.5

In determining whether the lease is one to which the 1954 Act applies, as is seen from the terms of paragraph 29(3), one ignores any exclusion agreement pursuant to s.38A of the 1954 Act. The fact that the security of tenure provisions contained within ss.24–28 of the 1954 Act may be excluded under s.38A does not, therefore, preclude it from being 'a lease to which Part 2 of the Landlord and Tenant Act 1954 ... applies'. The 1954 Act 'applies to any tenancy where the property comprised in the tenancy is or includes premises which are occupied by the tenant and are so occupied for the purposes of a business carried on by him or for those and other purposes': s.23(1) of the 1954 Act. The terms of s.38A of the 1954 Act simply provide for the exclusion of the provisions of ss.24–28 of the 1954 Act, which relate to the termination of a tenancy to which the 1954 Act applies.

25.3.6

Accordingly, the position in relation to a non-primary purpose *lease* which confers code rights is as follows:

(1) A non-primary purpose lease which is one to which the 1954 Act applies (being one falling within s.23 of the 1954 Act) is to be terminated in accordance with the 1954 Act. Removal of the electronic communications apparatus is required to be effected in accordance with Part 6 of the New Code (see paragraph 37(3)(d)).
(2) A non-primary purpose lease which is one to which the 1954 Act does not apply by reason of s.38A but which would otherwise be a lease to which the 1954 Act would apply, is terminated in accordance with the terms of the lease. There is no 1954 Act

protection and thus the lease is not required to be terminated in accordance with the 1954 Act and, equally, Part 5 of the New Code does not apply: paragraph 29(2) and (3) of the New Code. Removal of the electronic communications apparatus is, however, required to be effected in accordance with Part 6 of the New Code (see paragraph 37(3)(d)).

(3) Paradoxically, a non-primary purpose lease which is one to which the 1954 Act does not apply by reason of the fact that it does not fall within the terms of s.23 of the 1954 Act must, it would seem, be terminated in accordance with the terms of Part 5 of the New Code. Removal of the electronic communications apparatus is required to be effected in accordance with Part 6 of the New Code (see paragraph 37(3)(d)).

25.3.7

The New Code does not provide an example of what would constitute a 'non-primary purpose' agreement. In order to fall within the terms of the New Code the code rights must (1) be conferred upon an operator with respect to any land and (2) be conferred for 'the statutory purposes'. Thus, it is difficult to envisage circumstances where an operator takes a lease of premises which would not be a primary purpose agreement. For instance, if an operator agrees to be able to connect to an electricity supply on the landowner's land in connection with its network and that is the only right it seeks and has conferred, that is, nevertheless, a primary purpose agreement. The only purpose of the agreement is to grant code rights, namely the right falling within paragraph 3(g).[9]

25.3.8

In this regard, however, it may be that there are circumstances where the lease is more likely to be a 'non-primary purpose' agreement due to the definition of 'electronic communications apparatus'. Code rights, as defined in paragraph 3, all relate to 'electronic communications apparatus'. The terms of paragraph 5 define 'electronic communications apparatus' as meaning, amongst other things, 'other structures or things designed or adapted for use in connection with the provision of an electronic communications network'. In paragraph 5(3) it is provided that 'in this code… "structure" includes a building only if the sole purpose of that building is to enclose other electronic communications apparatus'. Thus, a building which houses other electronic communications apparatus is itself electronic communications apparatus. However, the code rights must be in relation to 'land': see the opening words of paragraph 3, the enumerated items in paragraph 3 and also the terms of paragraph 9; and 'land' excludes electronic communications apparatus: paragraph 108. In these circumstances it may be that the grant of a lease of a building to house electronic communications apparatus cannot be said to be a primary

9 In this example it probably matters little whether it is classified as a primary purposes agreement, for it is difficult to see how the operator could be said to be occupying premises for the purposes of a business carried on by it within the 1954 Act in any event. The right is probably an incorporeal hereditament and notwithstanding the decision of the Court of Appeal in *Pointon York Group plc v Poulton* [2007] 1 P& CR 115 (which held that 'premises' within the 1954 Act could include an incorporeal hereditament, and thus applied to car parking spaces occupied by the tenant during normal office hours in connection with its business), it is difficult to see what is the nature of the tenant's occupation of a connection to an electricity supply.

purpose agreement, because the primary purpose is not to provide code rights, e.g. to keep installed electronic communications apparatus on, under or over land (paragraph 3(b)), because the building itself is electronic communications apparatus and not land. Such an agreement would thus be either a non-primary purpose agreement (if, for instance, the grant provided for the some form of code right, e.g. installation of electronic communications apparatus on the land surrounding the building – an obvious example being the installation of a mast) or an agreement which has no form of code rights conferred under it at all. The lease of the building would, however, probably be one falling within the terms of s.23 of the 1954 Act and thus, at the very least, termination of such an agreement would be in accordance with the 1954 Act, not Part 5 of the New Code.

25.4 Transitional provisions

25.4.1

Schedule 2 to the New Code makes provision for dealing with the interrelationship of the New Code and the 1954 Act with respect to 'a subsisting agreement'. Paragraph 1 (4) of Schedule 2 to the 2017 Act defines 'subsisting agreement' as follows:

> A 'subsisting agreement' means –
> (a) an agreement for the purposes of paragraph 2 or 3 of the existing code, or
> (b) an order under paragraph 5 of the existing code,
> which is in force, as between an operator and any person, at the time the new code comes into force (and whose terms do not provide for it to cease to have effect at that time).

25.4.2

A 'subsisting agreement' has effect after the New Code comes into force as an agreement under Part 2 of the New Code subject to the modifications made by Schedule 2: paragraph 2 of Schedule 2 to the 2017 Act.

25.4.3

The words 'in force' are ambiguous. The Old Code contains no provision equivalent to that of paragraph 30 of the New Code which continues the rights of the operator under the code agreement notwithstanding the contractual termination of the term created by the lease. Under the Old Code a lease which has expired by effluxion of time is not 'continued', so that the relationship of landlord and tenant continues as between the parties, but it is provided simply that electronic communications apparatus remaining on land 'shall ... be deemed as against any person who was at any time entitled to require the removal of the apparatus, but by virtue of this paragraph [paragraph 21] is not entitled to enforce its removal, to have been lawfully so kept at that time': paragraph 21(9) of the Old Code.[10] Thus it is unclear whether a lease which has expired prior to the coming into force of the New Code can be said to be a 'subsisting agreement'.

10 The scheme under the Old Code envisages the operator claiming rights in these circumstances pending removal by serving a notice under paragraph 6 of the Old Code. However, it is rare, if at all, for any such notice to have been served by an operator once its contractual term had expired.

25.4.4

Paragraph 6 of Schedule 2 of the 2017 Act inserts, with respect to 'a subsisting agreement', in place of paragraph 29(2) to (4) of the New Code the following:

> (2) Part 5 of the new code (termination and modification of agreements) does not apply to a subsisting agreement that is a lease of land in England and Wales, if –
> (a) it is a lease to which Part 2 of the Landlord and Tenant Act 1954 applies, and
> (b) there is no agreement under section 38A of that Act (agreements to exclude provisions of Part 2) in relation the tenancy.
>
> (3) Part 5 of the new code does not apply to a subsisting agreement that is a lease of land in England and Wales, if –
> (a) the primary purpose of the lease is not to grant code rights (the rights referred to in paragraph 3 of this Schedule), and
> (b) there is an agreement under section 38A of the 1954 Act in relation the tenancy.

25.4.5

These provisions are similar but not identical to those contained in paragraph 29(2) and (3) of the New Code.[11] In particular it is to be noted that:

(1) albeit one could describe the primary purpose of a subsisting agreement as one to confer code rights as defined in the New Code, the subsisting agreement is, nevertheless, if the lease is one falling within the terms of the 1954 Act, one still governed by the 1954 Act. There is no exclusion of the potential application of the 1954 Act in relation to a subsisting agreement.

(2) If the subsisting agreement is one falling within the terms of the 1954 Act and there is no exclusion agreement of the security of tenure provisions pursuant section 38A of the 1954 Act, the 1954 Act governs termination of the subsisting agreement; Part 5 of the New Code does not apply. In these circumstances termination of the subsisting agreement will be in accordance with the 1954 Act, although there is nothing to disapply the application of Part 6 of the New Code dealing with the removal of the electronic communications apparatus from the land forming the subject matter of the subsisting agreement (see paragraph 37(3)(d)).

(3) If the subsisting agreement is excluded from the security of tenure provisions of the 1954 Act by reason of there being in existence an agreement under section 38A of the 1954 Act, Part 5 of the New Code does not apply only in circumstances where the primary purpose of the subsisting agreement *is not one* to grant code rights. Thus, it is only where there is an exclusion agreement under section 38A of the 1954 Act that one is concerned with the primary purpose/non-primary purpose distinction. However, that distinction is simply for the purposes of determining whether or not the termination provisions of Part 5 apply to the subsisting

11 Paragraph 29(4) makes special provision for Northern Ireland, where business tenancies are governed by different legislation.

agreement. Part 6 will continue to apply with respect to removal of the electronic communications apparatus from the land forming the subject matter of the subsisting agreement.

(4) It follows that where the subsisting agreement is one excluded from the 1954 Act by reason of there being in existence an agreement under section 38A of the 1954 Act, if the subsisting agreement is one the primary purpose of which is to grant code rights, Part 5 of the New Code will apply to termination of the subsisting agreement as well as Part 6 of the New Code with respect to removal of the apparatus.

25.5 Removal of apparatus

25.5.1

In the case of a New Code agreement having been entered into after the coming into force of the New Code, the 1954 Act will have little role to play other than in relation to non-primary purpose agreements. However, in circumstances where one has either a subsisting agreement which is subject to the 1954 Act (such that Part 5 of the New Code does not apply to termination of the subsisting agreement) or one has a New Code agreement which is a non-primary purpose agreement and subject to the 1954 Act, the 1954 Act may impact upon the timing of any application for removal of the electronic communications apparatus.

25.5.2

The effect of the 1954 Act on the ability of a landowner (as defined in paragraph 37 (1) of the New Code) to seek removal of electronic communications apparatus situated on the land forming the subject matter of the code agreement is an indirect one. The circumstances in which removal of apparatus may be sought to be effected has been dealt with in Chapter 24. Paragraph 37(1) provides that the landowner has the right to require removal of the apparatus 'on, under or over the land if (and only if) one or more of the following conditions are met'.

25.5.3

The conditions are discussed at length in Chapter 24. However, the second condition, contained within paragraph 37(3), provides the removal may be sought if 'a code right entitling an operator to keep the apparatus on, under or over the land has come to an end or has ceased to bind the landowner … (d) where the right was granted by a lease to which Part 5 of this code does not apply'. Part 5 will not apply to a subsisting agreement which is protected by the 1954 Act (discussed at paragraph 25.4.5) or to a New Code agreement where the agreement is a non-primary purpose agreement falling within the protection the 1954 Act (discussed at paragraph 25.3.2 onwards). It will be necessary in respect of such agreements for the agreement to be terminated in accordance with the 1954 Act. Accordingly, in the context of the second condition it is probably the case that the reference to the code right entitling the operator to keep the apparatus on, under or over the land, will only be said to have 'come to an end' when the tenancy has been effectively determined in accordance with the terms of the 1954 Act. Thus, until the 1954 Act court process been completed and, arguably, the tenancy has determined in accordance with s.64 of the 1954 Act,[12] the landowner will not be in a position to exercise its right to effect removal, at least in reliance upon the second condition set out in paragraph 37.

25.6 Renewal under the 1954 Act

25.6.1

As noted above, the New Code applies to a 'subsisting agreement'. A 'subsisting agreement' may be one to which the 1954 Act applies. What happens upon a renewal of the operator's lease in accordance with the 1954 Act?

25.6.2

If one assumes that the subsisting agreement is a primary purpose agreement, that is to say that its primary purpose is to grant code rights within the meaning of the New Code, any renewal pursuant to the 1954 Act will, by reason of the terms of section 43(4) of the 1954 Act, inserted by paragraph 4 of Schedule 3 to the 2017 Act, be one to which the 1954 not apply. It strikes one as slightly odd that the court should have jurisdiction under the 1954 Act, other than pursuant to an exclusion agreement in accordance with section 38A of the 1954 Act, to grant a new tenancy which will be one to which the 1954 Act will not apply.

25.6.3

Some interesting questions may arise particularly in relation to the assessment of rent in accordance with section 34 of the 1954 Act with respect to a renewal of what will, upon grant pursuant to the order of the court, be a primary purpose code agreement. The rent which is assessed in accordance with section 34 of the 1954 Act does not contain the detailed assumptions which one finds in paragraph 24 of the New Code as to the basis upon which the consideration for the grant of a code agreement is to be assessed and in particular there is nothing within section 34 to assess the rent by reference to a 'no scheme' world. Accordingly, it may be that the landowner on an assessment of rent under section 34 with respect to a renewal of a lease which will become a primary purpose code agreement upon grant, is better off than would be the case with respect to the consideration assessed in accordance with paragraph 24 of the New Code. Alternatively, it may be that the actual tenant argues that as in the real world the consideration for a primary purpose agreement is calculated in accordance with paragraph 24 of the New Code, a discount is required from the rent calculated in accordance with section 34 to reflect the much lower consideration which would otherwise be payable in the market if the rent were assessed in accordance with paragraph 24 of the

12 Which extends the tenant's tenancy for a period of three months in 21 days after 'final disposal' of any application for renewal/termination application.

New Code. These matters will no doubt be ironed out by agreement or, in default, by determination by the court.

25.6.4

Another problem area may be with respect to the terms of the new lease which are to be granted. The terms of the new lease are 'such as may be agreed between the landlord and tenant or as, in default of such agreement, may be determined by the court; and in determining those terms the court shall have regard to the terms of the current tenancy and to all relevant circumstances': s.35 of the 1954 Act.[13] No doubt the Court will, when considering the terms to be incorporated into the new tenancy, have regard to the terms of the New Code, e.g. the terms of Part 3 of the New Code dealing with assignment, upgrading apparatus and sharing.

25.6.5

A possibility which has been canvassed with the Authors is for the operator to avoid the 1954 Act altogether by allowing any s.25 notice under the 1954 Act to take effect (the effect of which will be to terminate the tenant`s current tenancy on the date specified in the notice) and then for the operator to seek an imposed agreement in accordance with Part 4 of the New Code so as to obtain the advantage, in particular, of the provisions as to consideration.[14]

25.7 Termination under the 1954 Act

25.7.1

As has been seen (sections 25.3 and 25.4 above) the 1954 Act may continue to apply to an agreement which confers code rights where:

(1) The code agreement conferred in accordance with the New Code is one the primary purpose of which is not to grant code rights. A non-primary purpose agreement is not excluded from protection of the 1954 Act and if it is one to which the 1954 Act applies, its termination is in accordance with the 1954 Act, not the New Code: paragraph 29(2).
(2) A 'subsisting agreement', being an agreement in force at the time the New Code came into force (paragraph 1 of Schedule 2 to the 2017 Act), is protected by the 1954 Act, remains one which continues to be so protected, albeit the primary purpose of any such subsisting agreement may be the granting of code rights as described in accordance with the New Code. Such a subsisting agreement accordingly remains

13 Authoritative guidance on s.35 of the 1954 Act was provided by the House of Lords in *O`May v City of London Real Property Co. Ltd* [1983] 2 AC 726.

14 What little experience has been had to date suggests that operators are avoiding the potential downside to a determination under s.34 of the 1954 Act by discontinuing any renewal proceedings if a satisfactory deal cannot be achieved with the landowner on term and rent for the renewal. The operator then immediately serves a paragraph 20 notice seeking an imposed agreement which, of course, brings with it the valuation assumptions provided for by paragraph 24.

one which is required to be terminated in accordance with the 1954 Act, for Part 5 does not apply to such a subsisting agreement: paragraph 6 of Schedule 2 to the 2017 Act.

25.7.2

It is no part of this work to consider in detail the termination of a lease in accordance with the terms of the 1954 Act.[15] However, as is well known, a landlord must, if he wishes to terminate the tenant's current lease, establish one of the seven grounds of opposition set out in s.30(1) of the 1954 Act. The most commonly invoked ground of opposition in the context of electronic communications apparatus is paragraph (f) of s.30(1) of the 1954 Act, namely, that the landlord intends to effect works of demolition or redevelopment (as elaborated in that sub-paragraph). The terms of s.30(1)(f) of the 1954 Act refers to the landlord possessing an intention 'on the termination of the current tenancy' to effect the works referred to within the sub-paragraph. The expression 'on the termination of the current tenancy' includes the continuation of such a tenancy pursuant to ss.24 and 64 of the 1954 Act, since the landlord cannot implement his intention until the tenancy has finally determined and the landlord is entitled to vacant possession. Thus, for instance, it has been said that an intention to effect work within three months after the termination of the tenancy is sufficient.[16]

25.7.3

Under the Old Code a concern arose in relation to the interrelation of the Old Code and the 1954 Act, in that it was often said that a landlord would be unable to establish the relevant intention for the purposes of section 30(1)(f) because until the landlord had obtained an order for removal of the electronic communications apparatus in accordance with paragraph 21 of the Old Code, he would not be able to establish that he would be capable of entering the demised property such as to be able to implement the stated intention. But the difficulty for the landlord is that it has been held that he cannot serve a paragraph 21 notice until the tenant's tenancy has been brought to an end, for until the tenancy has been brought to an end it cannot be said, within the terms of paragraph 21, that the landlord 'is for the time being entitled to require the removal' of the operator's apparatus.[17]

25.7.4

The terms of the New Code, as noted, avoids this complication. However, in relation to those code agreements to which the 1954 Act continues to apply (whether that be a New Code agreement or a subsisting agreement) the landlord will still be required to establish, in its opposition to any renewal under the 1954 Act under s.30(1)(f), that it intends to effect the work 'at the termination of the current tenancy'. But termination in

15 Reference can be made to Reynolds and Clark, *Renewal of Business Tenancies,* 5th ed. (2017).

16 *Livestock Underwriting Agency Ltd v Corbett and Newson Ltd* [1955] 165 EG 469.

17 *Crest Nicholson (Operations) Ltd v Arqiva Services Ltd* (2015) April 2015 (transcript available on Egi) and more recently, in Scotland, *Tait v Vodafone* (2017), 24 July 2017, Sheriff Kenneth J McGowan, Sheriffdom of Lothian and Borders at Edinburgh.

accordance with the 1954 Act simply terminates the tenant's tenancy in accordance with s.64 of the 1954 Act,[18] which provides for a continuation of three months and 21 days[19] after final disposal of the 1954 Act litigation.

25.7.5

An order for termination of the tenant's current tenancy without the grant of any new tenancy under the 1954 Act is not an order entitling the landlord to remove the operator's electronic communications apparatus from the demised premises. The landlord will still be required, once it has terminated the operator's lease in accordance with the 1954 Act, to implement the procedure under the New Code in accordance with Part 6 to remove the operator's apparatus. The procedure under Part 6 can be implemented only once the tenant's tenancy has come to an end: paragraph 37(3) of the New Code.[20] Removal requires service of a notice by the landowner requiring removal: paragraph 40. It would not appear possible, however, for the landlord to serve a removal notice to run concurrently with the termination proceedings under the 1954 Act. The landlord has to await the outcome of the 1954 Act litigation and the termination of the tenancy in accordance with s.64 of the 1954 Act until it can seek to implement the removal process under the New Code. If this is correct a landlord may be met with an argument in relation to any attempted termination of the operator's 1954 Act protected lease, that the process under Part 6 of the New Code is such that it cannot be said that the landlord intends 'on the termination of the current tenancy' to effect the relevant work. The removal process in accordance with Part 6 involves not simply service of a notice in accordance with paragraph 40 but, absent agreement between the landowner and the operator with respect to, inter alia, removal (which is unlikely to be forthcoming), necessitates an application to the court by the landowner for an order for removal: paragraphs 40(6) and 44(1).

25.7.6

It is considered that any attempt by the operator tenant to seek to undermine a landlord's ability to oppose renewal in reliance on s.30(1)(f) of the 1954 Act by serving at the last moment a notice pursuant to paragraph 20 of the New Code claiming code rights ought not to succeed given the terms of paragraph 21(5) of the New Code, which makes it clear that the 'court may not make an order under paragraph 20 if it thinks that the relevant person intends to redevelop all or part of the land to which the code Right would relate, or any neighbouring land, and could not really do so if the order were made'.

18 Which provides that the tenancy will expire 'at the expiration of the period of three months beginning with the date on which the application is finally disposed of' and 'the date on which an application is finally disposed of shall be construed as a reference to the earliest date by which proceedings on the application (including any proceedings on in consequence of an appeal) have been determined and any time for appealing of further appealing has expired': s.64(1) and (2) of the 1954 Act.

19 This being the period for appeal: CPR 52.12.

20 Which provides that 'The second condition is that a code right entitling the operator to keep the apparatus on, under or over the land has come to an end or has ceased to bind the landowner.' See the discussion in paragraph 25.5.3.

25.7.7

If there is doubt about whether or not the agreement conferring code rights complies with the mandatory requirements of paragraph 11, it may be advisable for a landowner to serve a precautionary notice pursuant to s.25 of the 1954 Act. If there is non-compliance with paragraph 11 there is no agreement which falls within the New Code but the agreement may, notwithstanding such non-compliance, be a lease to which the 1954 Act will apply and thus will need to be terminated in accordance with that Act.[21]

21 It will also require compliance with Part 6 in order to effect removal of the apparatus: see n. 1.

26 Transport land rights

26.1 Introduction

26.1.1

Part 7 of the New Code deals with so-called 'transport land rights'. These provisions correspond to paragraphs 12 to 14 of the Old Code ('linear obstacles', see Chapter 9). They deal with rights to traverse railways, canals and tramways.

26.1.2 General definitions

Paragraph 46 of the New Code defines 'transport land' as:

> land which is used wholly or mainly –
> (a) as a railway, canal or tramway, or
> (b) in connection with a railway, canal or tramway on the land.

26.1.3

Paragraph 46 defines a 'transport undertaker', in relation to transport land, as 'the person carrying on the railway, canal or tramway undertaking'.

26.1.4

Paragraph 47 states that an operator may exercise a transport land right for statutory purposes, in accordance with part 7 of the New Code. A 'transport land right' is defined in paragraph 48 as:

> (a) a right to cross any transport land with a line;
> (b) a right, for the purposes of crossing any transport land with a line –
>   (i) to install and keep the line and any other electronic communications apparatus on, under or over the transport land;
>   (ii) to inspect, maintain, adjust, alter, repair, upgrade or operate electronic communications apparatus on, under or over the transport land;
>   (iii) a right to carry out any works on the transport land for or in connection with the exercise of a right under sub-paragraph (i) or (ii);
>   (iv) a right to enter the transport land to inspect, maintain, adjust, alter, repair, upgrade or operate the line or other electronic communications apparatus.

26.1.5

As before, there are certain limitations on the route of a line installed pursuant to a transport land right. Paragraph 48(2)–(4) provide that:

> (2) A line installed in the exercise of a transport land right need not cross the transport land in question by a direct route or the shortest route from the point at which the line enters the transport land.
> (3) But the line must not cross the transport land by any route which, in the horizontal plane, exceeds that shortest route by more than 400 metres.
> (4) The transport land rights do not authorise an operator to install a line or other electronic communications apparatus in any position on transport land in which the line or other apparatus would interfere with traffic on the railway, canal or tramway.

26.1.6

As before, there is no statutory language limiting the vertical extent of the 'transport land' in question. Plainly, as discussed in relation to the Old Code, there might come a point where the vertical depth of the works is such that the right ceases to fall within this special regime, and falls instead within the general regime. The answer to that question may lie in the true construction of the long lease, Private Act of Parliament, or other document giving the transport undertaker its rights to run its undertaking, and the familiar presumptions as the vertical extent of any grant.[1]

26.2 Exercise of transport land rights in case of non-emergency

26.2.1

As under the Old Code, a distinction is drawn between emergency and non-emergency works. In a non-emergency cases, paragraph 49 requires the giving of a 'notice of proposed works' (paragraph 49(1)).[2] Non-emergency works are works which do not fall within the 'emergency works' provision under paragraph 51 (see paragraph 49(5)). Accordingly, it will be necessary to consider first whether any works to be undertaken could be said to fall within paragraph 51.

26.2.2

As before, pursuant to paragraph 49(2),

> Notice of proposed works must contain a plan and section of the works; but, if the transport undertaker agrees, the notice may instead contain a description of the works (whether or not in the form of a diagram).

1 *Bocardo SA v Star Energy UK Onshore Ltd* [2010] UKSC 35; *Lejonvarn v Cromwell Mansions Management Company Ltd* [2011] EWHC 3838 (Ch); *Stynes v Western Power (East Midlands) Plc* [2013] UKUT 214 (LC).

2 For notices and service, see Chapter 32.

26.2.3

Non-emergency works may not be commenced until the requisite notice period has ended (paragraph 49(3), the period being 28 days from the date on which the notice is given (paragraph 49(4)).

26.2.4

In the event that a notice of proposed works is given, then the transport undertaker has the right to give 'notice of objection' pursuant to paragraph 50(2), to which the right under paragraph 49 is expressly subject (see paragraph 49(4)). A notice if given, then either the operator or the transport undertaker are entitled to give an arbitration notice within the arbitration period (paragraph 50(3)). Under paragraph 50(7), 'arbitration notice period' means the period of 28 days beginning with the day on which objection notice is given. Paragraph 52 makes provision about arbitration.

26.2.5

The giving of any notice determines whether, when and how transport land rights can be exercised. Obviously in the absence of notice of objection, the transport land rights can be exercised. Otherwise, the position is governed by paragraphs 50(4)–(6):

> (4) In a case where notice of objection is given, the operator may exercise a transport land right in order to carry out the proposed works only if they are permitted under sub-paragraph (5) or (6).
> (5) Works are permitted in a case where –
> (a) the arbitration notice period has ended, and
> (b) no arbitration notice has been given.
> (6) In a case where arbitration notice has been given, works are permitted in accordance with an award made on the arbitration

26.3 Exercise of transport land rights in case of emergency

26.3.1

Where the works are emergency works, a different procedure applies, as there is obviously no time to comply with the notice and arbitration process under paragraph 50. Transport land rights can be exercised for emergency works (see paragraph 51(1)). 'Emergency works' are defined in paragraph 51(9) as meaning:

> works carried out in order to stop anything already occurring, or to prevent anything imminent from occurring, which is likely to cause –
> (a) danger to persons or property,
> (b) the interruption of any service provided by the operator's network, or
> (c) substantial loss to the operator,
> and any other works which it is reasonable (in all the circumstances) to carry out with those works.

26.3.2

In such a case, there is an abridged procedure put in place under paragraph 51(2) – (3):

> (2) If the operator exercises a transport land right to carry out emergency works, the operator must give the transport undertaker an emergency works notice as soon as reasonably practicable after starting the works.
> (3) An 'emergency works notice' is a notice which –
> (a) identifies the emergency works;
> (b) contains a statement of the reason why the works are emergency works; and
> (c) contains either –
> (i) the matters which would be included in a notice of proposed works (if one were given in relation to the works), or
> (ii) a reference to a notice of proposed works which relates to the works that are emergency works (if one has been given).

26.3.3

It therefore appears, by reason of paragraph 51(3)(c)(ii), that a notice under paragraph 49 for non-emergency works could become emergency works by reason of a change in circumstances. This is spelled out in paragraph 51(8):

> A reference in this paragraph to emergency works includes a reference to any works which are included in a notice of proposed works but become emergency works before the operator is authorised by paragraph 50 or 51 to carry them out.

26.3.4

In the case of emergency works, there is a right to claim compensation. This is catered for by paragraph 51(4), which gives the transport undertaker the right to give a 'compensation notice' within the 'compensation notice period'.[3] Under paragraph 51(9), 'compensation notice period' means the period of 28 days beginning with the day on which an emergency works notice is given.

26.3.5

The compensation recovered under paragraph 51(4) is limited to 'compensation for loss or damage sustained in consequence of the carrying out of emergency works', and does not extend to a price for the exercise of the right. That compensation will cover the disruption to the transport undertaker's undertaking, loss of profits and compensation for any damage caused by the works being undertaken. Lasting costs increases are expressly catered for in paragraph 52(7), which states that 'the arbitrator's power under sub-paragraph (3) or (4) to award compensation for loss includes power to award compensation for any increase in the expenses incurred by the transport undertaker in carrying on its railway, canal or tramway undertaking'. Although that is a special

3 For notices, see Chapter 32.

provision related to costs only, it is considered that this provision would not preclude a claim for future loss of income, which seems to be covered by paragraph 51(4) in any event. Presumably paragraph 52(7) is simply intended to make it explicit that increased costs are expressly covered in any compensation claim. Under paragraph 51(5) it is then provided that the operator must pay the transport undertaker any compensation which is required by a compensation notice (if given within the compensation notice period).

26.3.6

It is for the transport undertaken and the operator to agree the amount of compensation payable in the first instance (paragraph 51(6)). However, under paragraph 51(7) if –

(a) the compensation agreement period has ended, and
(b) the operator and the transport undertaker have not agreed the amount of compensation payable under sub-paragraph (6),

then the matter can be referred to arbitration under paragraph 52.

26.4 Arbitration

26.4.1

The statutory arbitration procedure is clearly set out in paragraph 52. It applies if notice is given under paragraph 50(3) or 51(7), which crystallises the 'matter in dispute' (paragraph 52(1)), namely either (a) an objection to proposed works or (b) a disagreement about an amount of compensation. The matter in dispute is referred to the arbitration of a single arbitrator appointed by agreement between the parties, or in the absence of such agreement, by the President of the Institution of Civil Engineers (paragraph 5(9)).

26.4.2

Although this could be read as allowing the transport undertaker to dispute the works under both emergency and non-emergency works, it seems that in fact, where the works are emergency works, the only issue that can be disputed is the level of compensation.

26.4.3

Where the arbitration is about an objection to proposed works, then paragraph 52(3) gives the arbitrator the following powers:

(a) power to require the operator to give the arbitrator a plan and section in such form as the arbitrator thinks appropriate;
(b) power to require the transport undertaker to give the arbitrator any observations on such a plan or section in such form as the arbitrator thinks appropriate;
(c) power to impose on either party[4] any other requirements which the arbitrator thinks appropriate (including a requirement to provide information in such form as the arbitrator thinks appropriate);

4 Paragraph 52(8) provides that '"party" means – (a) the operator, or (b) the transport undertaker'.

(d) power to make an award –
  (i) requiring modifications to the proposed works, and
  (ii) specifying the terms on which, and the conditions subject to which, the proposed works may be carried out;
(e) power to award one or both of the following, payable to the transport undertaker –
  (i) compensation for loss or damage sustained by that person in consequence of the carrying out of the works;
  (ii) consideration for the right to carry out the works.

26.4.4

In relation to 'consideration' under paragraph 52(3)(e)(ii), this is the exact phrase used in paragraph 13(2)(e)(ii) of the Old Code, so that, presumably, the limited extent of the consideration payable in light of the Court of Appeal's decision in *Bridgewater Canal Co. Limited v GEO Networks*[5] will apply under the New Code. In relation to the compensation element, the measure of damages will presumably be the same as for the emergency works, above. This seems to be enshrined in the statutory language of paragraph 52(8), which provides that 'An award of consideration under sub-paragraph (3)(e)(ii) must be determined on the basis of what would have been fair and reasonable if the transport undertaker had willingly given authority for the works to be carried out on the same terms, and subject to the same conditions (if any), as are contained in the award.'

26.4.5

In the event that the disagreement is about the amount of compensation following emergency works, the arbitrator has the following powers in paragraph 52(4) –

(a) power to impose on either party[6] any requirements which the arbitrator thinks appropriate (including a requirement to provide information in such form as the arbitrator thinks appropriate);
(b) power to award compensation, payable to the transport undertaker, for loss or damage sustained by that person in consequence of the carrying out of the emergency works.

This compensation is on the same basis as the compensation payable in relation to non-emergency works, above, including in relation to future costs under paragraph 52(7).

26.4.6

The arbitrator is given a power of compulsion in that he may make an award conditional upon a party complying with a requirement imposed under sub-paragraph (3)(a), (b) or (c) or (4)(a).[7] In making a determination pursuant to paragraph 52(6), the

5 [2010] EWCA Civ 1348.
6 Paragraph 52(8) provides that '"party" means – (a) the operator, or (b) the transport undertaker'.
7 Paragraph 52(5).

arbitrator is expressly directed that the matters to which the arbitrator must have regard include the public interest in there being access to a choice of high quality electronic communications services.

26.5 Transport undertaker's right to alter apparatus

26.5.1

The ability to have apparatus altered when it has been installed pursuant to a transport land right is governed by paragraph 53. Paragraph 53(1) provides that:

> A transport undertaker may give an operator notice which requires the operator to alter a line or other electronic communications apparatus specified in the notice ('notice requiring alterations') on the ground that keeping the apparatus on, under or over transport land interferes with, or is likely to interfere with –
> (a) the carrying on of the transport undertaker's railway, canal or tramway undertaking, or
> (b) anything done or to be done for the purposes of its railway, canal or tramway undertaking.

26.5.2

This is presumably intended to cover issues arising out of underground or overground cables limiting the ability of a canal operator to dredge, widen or deepen the canal, or preventing the passage of traffic above a certain height over the transport land.

26.5.3

The operator then has to power to give a counter-notice within the 'notice period' which is set at 28 days by paragraph 53(9). The counter-notice must specify the respects in which the operator is not prepared to comply with the notice requiring alterations (paragraph 53(2). It is not clear whether the operator is required to give reasons for non-compliance, rather than just identify the required works it is not prepared to comply with. The operator must, however, comply within a 'reasonable time' if the notice period has elapsed and no counter-notice has been given (paragraph 53(3)).

26.5.4

If a counter-notice has been given within the notice period, then the transport undertaker is entitled to apply to court[8] for an order in relation to the works in relation to the alteration of 'specified apparatus'.[9] There is no right for the operator to apply, and there is no specified time limit in which the application to court must be made – presumably because there is no prejudice to the operator in delay, as it will simply be able to maintain its works over the transport land, as before. That said, it may have been

8 See Chapter 33 for the meaning of 'court'.
9 Paragraph 53(8) provides that 'specified apparatus' means the line or other electronic communications apparatus specified in a notice requiring alterations.

beneficial to give the operator a right to permit it to resolve any uncertainty hanging over its network.

26.5.5

The court has the following powers upon the making of an application:

> (5) The court must not make an order unless it is satisfied that the order is necessary on one of the grounds mentioned in sub-paragraph (1).
> (6) In determining whether to make an order, the matters to which the court must also have regard include the public interest in there being access to a choice of high quality electronic communications services.
> (7) An order under this paragraph may take such form and be on such terms as the court thinks fit.
> (8) In particular, the order –
> (a) may impose such conditions, and
> (b) may contain such directions to the operator or the transport undertaker, as the court thinks necessary for resolving any difference between the operator and the transport undertaker and for protecting their respective interests.

26.6 Land ceasing to be transport land

26.6.1

If land ceases to be transport land, then, with the exception of paragraph 53,[10] the land continued to fall under Part 7 of the New Code: see paragraph 54(2). The operator may continue to exercise any transport land right in relation to the land as if it were still transport land). This is, however, expressly subject to sub-paragraphs (4) and (9) by reason of paragraph 54(3). The effect of those sub-paragraphs is that:

(1) Under paragraph 54(4), it is provided that 'references to the transport undertaker have effect as references to the occupier of the land'.
(2) By reason of paragraph 54(5),

> the application of this Part of this code to land in accordance with sub-paragraph (2) does not authorise the operator –
> (a) to cross the land with any line that is not in place at the time when the land ceases to be transport land, or
> (b) to install and keep any line or other electronic communications apparatus that is not in place at the time when the land ceases to be transport land.

(3) In relation to replacements on a life-for-like basis, paragraph 54(6) states that sub-paragraph (5):

> does not affect the power of the operator to replace an existing line or other apparatus (whether in place at the time when the land ceased to be transport land or a

10 That is, the power to alter apparatus under paragraph 53.

replacement itself authorised by this sub-paragraph) with a new line or apparatus which –

(a) is not substantially different from the existing line or apparatus, and
(b) is not in a significantly different position.

26.6.2

The effect of this is that new lines to be installed must thereafter be installed in accordance with the New Code's general regime (see the earlier chapters in this Part III of this book), as the land ceases, in that regard, to have any special status. In relation to existing transport land rights that have been implemented, these are preserved and replacement is permissible under paragraph 54(6).

26.7 Notice disapplying Part 7

26.7.1

However, the occupier is permitted to bring the application of Part 7 to an end by notice under paragraphs 54(7)–(9):

> (7) The occupier of the land may, at any time after the land ceases to be transport land, give the operator notice specifying a date on which this Part of this code is to cease to apply to the land in accordance with this paragraph ('notice of termination').
> (8) That date specified in the notice of termination must fall after the end of the period of 12 months beginning with the day on which the notice of termination is given.
> (9) On the date specified in notice of termination in accordance with sub-paragraph (8), the transport land rights cease to be exercisable in relation to the land in accordance with this paragraph.

26.7.2

The reference in those provisions to an occupier is a reference to the fact that the land must have ceased to be part of a transport undertaker's undertaking before that notice is given. The Law Commission considers that the effect of this notice is to deprive the operator of any right to remain on the land, unless it has come to another arrangement with the occupier:[11]

> We consider that in such cases, the Code Operator's rights should continue until the landowner, or the person with control of the land, gives the Code Operator 12 months' notice to terminate them. At the end of that 12 month period, the Code Operator will have no rights to retain the apparatus on the land, unless a new arrangement has been made with an appropriate Site Provider.

11 Law Com 336, paragraph 7.59.

26.8 Criminal offence

26.8.1

Part 7 of the New Code is given teeth by the imposition of a range of criminal sanctions under paragraph 55:[12]

> (1) An operator is guilty of an offence if the operator starts any works in contravention of any provision of paragraph 49, paragraph 50 or paragraph 51.
> (2) An operator guilty of an offence under this paragraph is liable on summary conviction to a fine not exceeding level 3 on the standard scale.
> (3) In a case where this Part of this code applies in accordance with paragraph 54, the reference in this paragraph to paragraph 49, paragraph 50 or paragraph 51 is a reference to that paragraph as it applies in accordance with paragraph 54.

26.8.2

Particular care will therefore have to be taken when deciding whether or not works can be classified as 'emergency works'.

12 Paragraph 104(9) provides that 'Paragraphs 55 (offence in relation to transport land rights) and 75(5) (offence in relation to notices on overhead apparatus) do not apply where this code applies in the case of the Secretary of State or a Northern Ireland department by virtue of section 106(3)(b).'

27 Street works, tidal waters and undertaker's works

27.1 Street works

27.1.1

As in the case of paragraph 9 of the Old Code, provision is made by the New Code for operator works to a street. As with the Old Code, the operator has the automatic right to carry out those works. In contrast to the Old Code, however, there is no provision for agreements to obstruct. Certain proposals made by the Law Commission in Law Com 336 were also not implemented.

27.1.2 General definitions

The following general definitions apply under paragraph 57:

> 'road' means –
> (a) a road in Scotland which is a public road;
> (b) a road in Northern Ireland;
>
> 'street' means a street in England and Wales which is a maintainable highway (within the meaning of Part 3 of New Roads and Street Works Act 1991), other than one which is a footpath, bridleway or restricted byway that crosses, and forms part of, any agricultural land or any land which is being brought into use for agriculture.

27.1.3

The definitions are therefore the same as under the Old Code, and mean that the rights do not apply to streets that are not maintainable highways, and do not apply to certain ways over agricultural land.

27.1.4 The street work rights

The operator can exercise a street work right for the 'statutory purposes' under paragraph 58, though subject to the provisions of the New Code. Under paragraph 59(1), a street work right is:

> (a) a right to install and keep electronic communications apparatus in, on, under, over, along or across a street or a road;

(b) a right to inspect, maintain, adjust, alter, repair, upgrade or operate electronic communications apparatus which is installed or kept by the exercise of the right under paragraph (a);
(c) a right to carry out any works in, on, under, over, along or across a street or road for or in connection with the exercise of a right under paragraph (a) or (b);
(d) a right to enter any street or road to inspect, maintain, adjust, alter, repair, upgrade or operate electronic communications apparatus which is installed or kept by the exercise of the right under paragraph (a).

27.1.5

In relation to the 'works' permitted under paragraph 59(1)(c), paragraph 59(2) provides that the works 'include':

(a) breaking up or opening a street or a road;
(b) tunnelling or boring under a street or a road;
(c) breaking up or opening a sewer, drain or tunnel.

27.2 Tidal water rights

27.2.1

As under paragraph 11 of the Old Code, provision is made for rights exercisable over tidal waters.

27.2.2 General definitions

Under paragraph 61, 'tidal water or lands' includes –

(a) any estuary or branch of the sea,
(b) the shore below mean high water springs, and
(c) the bed of any tidal water.

27.2.3 Conferral of tidal water rights

As with street works rights, the operator automatically receives a compendious set of rights under paragraph 62(1), again for the statutory purposes and again (by reason of paragraph 62(2) subject to the following provisions of this Part of this code.

27.2.4

A 'tidal water right', in relation to an operator, is expanded beyond the rights given by the Old Code, paragraph 63(1) providing

(a) a right to install and keep electronic communications apparatus on, under or over tidal water or lands;
(b) a right to inspect, maintain, adjust, alter, repair, upgrade or operate electronic communications apparatus on, under or over the tidal water or lands;

(c) a right to carry out any works on, under or over any tidal water or lands for or in connection with the exercise of a right under paragraph (a) or (b);[1]
(d) a right to enter any tidal water or lands to inspect, maintain, adjust, alter, repair, upgrade or operate electronic communications apparatus which is installed or kept by the exercise of the right under paragraph (a).

27.2.5 Exercise of tidal water rights: Crown land

Paragraph 64(1) provides that an operator may not exercise a tidal water right in relation to land in which a Crown interest subsists unless agreement to the exercise of the right in relation to the land has been given in respect of that interest by the appropriate authority in accordance with paragraph 104.

27.2.6

Paragraph 104 is the equivalent to the Old Code's paragraph 27, and provides

(1) This code applies in relation to land in which there subsists, or at any material time subsisted, a Crown interest as it applies in relation to land in which no such interest subsists.
(2) In this code 'Crown interest' means –
  (a) an interest which belongs to Her Majesty in right of the Crown,
  (b) an interest which belongs to Her Majesty in right of the Duchy of Lancaster,
  (c) an interest which belongs to the Duchy of Cornwall,
  (d) an interest which belongs to a government department or which is held in trust for Her Majesty for the purposes of a government department, or
  (e) an interest which belongs to an office-holder in the Scottish Administration or which is held in trust for Her Majesty for the purposes of the Scottish Administration by such an office-holder.
(3) This includes, in particular –
  (a) an interest which belongs to Her Majesty in right of Her Majesty's Government in Northern Ireland, and
  (b) an interest which belongs to a Northern Ireland department or which is held in trust for Her Majesty for the purposes of a Northern Ireland department.
(4) Where an agreement is required by this code to be given in respect of any Crown interest subsisting in any land, the agreement must be given by the appropriate authority.
(5) Where a notice under this code is required to be given in relation to land in which a Crown interest subsists, the notice must be given by or to the appropriate authority (as the case may require).
(6) In this paragraph 'the appropriate authority' means –

1 Paragraph 63(2) provided that the works that may be carried out under sub-paragraph (1)(c) include placing a buoy or seamark.

(a) in the case of land belonging to Her Majesty in right of the Crown, the Crown Estate Commissioners or the relevant person or, as the case may be, the government department or office-holder in the Scottish Administration having the management of the land in question;
(b) in the case of land belonging to Her Majesty in right of the Duchy of Lancaster, the Chancellor of the Duchy of Lancaster;
(c) in the case of land belonging to the Duchy of Cornwall, such person as the Duke of Cornwall, or the possessor for the time being of the Duchy of Cornwall, appoints;
(d) in the case of land belonging to an office-holder in the Scottish Administration or held in trust for Her Majesty by such an office-holder for the purposes of the Scottish Administration, the office-holder;
(e) in the case of land belonging to Her Majesty in right of Her Majesty's Government in Northern Ireland, the Northern Ireland department having the management of the land in question;
(f) in the case of land belonging to a government department or a Northern Ireland department or held in trust for Her Majesty for the purposes of a government department or a Northern Ireland department, that department.

(7) In sub-paragraph (6)(a) 'relevant person', in relation to land to which section 90B(5) of the Scotland Act 1998 applies, means the person having the management of that land.
(8) Any question as to the authority that is the appropriate authority in relation to any land is to be referred to the Treasury, whose decision is final.
(9) Paragraphs 55 (offence in relation to transport land rights) and 75(5) (offence in relation to notices on overhead apparatus) do not apply where this code applies in the case of the Secretary of State or a Northern Ireland department by virtue of section 106(3)(b).
(10) References in this paragraph to an office-holder in the Scottish Administration are to be construed in accordance with section 127(7) of the Scotland Act 1998.

27.2.7

A dispute resolution mechanism is put in place by paragraph 64(2) and (3):

(2) Where, in connection with an agreement between the operator and the appropriate authority for the exercise of such a right, the operator and the appropriate authority cannot agree the consideration to be paid by the operator, the operator or the appropriate authority may apply to the appointed valuer for a determination of the market value of the right.
(3) An application under sub-paragraph (2) must be made in writing and must include –
(a) the proposed terms of the agreement, and
(b) the reasoned evidence of the operator and of the appropriate authority as to the market value of the right.
(4) As soon as reasonably practicable after receiving such an application, the appointed valuer must –

(a) determine the market value of the tidal water right; and
(b) notify the operator and the appropriate authority in writing of its determination and the reasons for it.

(5) If the agreement mentioned in sub-paragraph (2) or an agreement in substantially the same terms is concluded following a determination under sub-paragraph (4), the consideration payable by the operator must not be more than the market value notified under sub-paragraph (4)(b).

(6) For this purpose the market value of a tidal water right is, subject to sub-paragraph (7), the amount that, at the date the market value is assessed, a willing buyer would pay a willing seller for the right –
(a) in a transaction at arm's length,
(b) on the basis that the buyer and seller were acting prudently and with full knowledge of the transaction, and
(c) on the basis that the transaction was subject to the proposed terms set out in the application.

(7) The market value must be assessed on these assumptions –
(a) that the right that the transaction relates to does not relate to the provision or use of an electronic communications network;
(b) that the right in all other respects corresponds to the tidal water right;
(c) that there is more than one site which the buyer could use for the purpose for which the buyer seeks the right.

(8) The appointed valuer may charge a fee in respect of the consideration of an application under sub-paragraph (4) and may apportion the fee between the operator and the appropriate authority as the appointed valuer considers appropriate.

(9) In this paragraph 'the appointed valuer' means –
(a) such person as the operator and the appropriate authority may agree;
(b) if no person is agreed, such person as may be nominated, on the application of the operator or the appropriate authority, by the President of the Royal Institution of Chartered Surveyors.

27.2.8

It therefore follows that a tidal water right over land with a Crown interest must be paid for – but not other tidal land. The valuation assumptions parallel those of the general regime under the New Code, and regard should be had to Chapter 30 in this regard.

27.3 Undertaker's works

27.3.1

Under Part 10 of the New Code, provision is made about the carrying out of undertaker's works by undertakers or operators.

27.3.2 Key definitions

Under paragraph 66(1), it is provided that

'undertaker' means a person (including a local authority) of a description set out in any of the entries in the first column of the following table;

'undertaker's works', in relation to an undertaker of a description set out in a particular entry in the first column of the table, means works of the description set out in the corresponding entry in the second column of the table.

'undertaker'	*'undertaker's works'*
A person authorised by any enactment (whether public general or local) or by any order or scheme made under or confirmed by any enactment to carry on any railway, tramway, road transport, water transport, canal, inland navigation, dock, harbour, pier or lighthouse undertaking	Works that the undertaker is authorised to carry out for the purposes of, or in connection with, the undertaking which it carries on
A person (apart from the operator) to whom this code is applied by a direction under section 106 of the Communications Act 2003	Works that the undertaker is authorised to carry out by or in accordance with any provision of this code
Any person to whom this Part of this code is applied by any enactment (whenever passed or made)	Works for the purposes of which this paragraph is applied to the undertaker

27.3.3

Paragraph 66(2) further states that

(a) a reference to undertaker's works which interfere with a network is a reference to any undertaker's works which involve, or are likely to involve, an alteration of any electronic communications apparatus kept on, under or over any land for the purposes of an operator's network;

(b) a reference to an alteration of any electronic communications apparatus is a reference to a temporary or permanent alteration of the apparatus.

27.3.4 Non-emergency undertaker's works

Under paragraph 67(1), it is provided that:

> Before carrying out non-emergency undertaker's works which interfere with a network, an undertaker must give the operator notice of the intention to carry out the works ('notice of proposed works').

27.3.5

Under paragraph 67(5), 'non-emergency undertaker's works' means any undertaker's works which are not emergency works under paragraph 71.

27.3.6

As to the content of such a notice, paragraph 67(2) further states that:

Notice of proposed works must specify –

(a) the nature of the proposed undertaker's works,
(b) the alteration of the electronic communications apparatus which the works involve or are likely to involve, and
(c) the time and place at which the works will begin.

27.3.7

The effect of this notice procedure is as follows, under paragraph 67(3):

(3) The undertaker must not begin the proposed undertaker's works (including the proposed alteration of electronic communications apparatus) until the notice period has ended.
(4) But the undertaker's power to alter electronic communications apparatus (in carrying out the proposed undertaker's works) is subject to paragraph 68.

27.3.8

Under paragraph 67(5), 'notice period' means the period of ten days beginning with the day on which notice of proposed works is given.

27.3.9

The operator is entitled to give a counter-notice under paragraph 68 within the notice period:

(1) This paragraph applies if an undertaker gives an operator notice of proposed works under paragraph 67.
(2) The operator may, within the notice period, give the undertaker notice ('counter-notice') stating either –
    (a) that the operator requires the undertaker to make any alteration of the electronic communications apparatus that is necessary or expedient because of the proposed undertaker's works –
        (i) under the supervision of the operator, and
        (ii) to the satisfaction of the operator; or
    (b) that the operator intends to make any alteration of the electronic communications apparatus that is necessary or expedient because of the proposed undertaker's works.
(3) In a case where counter-notice contains a statement under sub-paragraph (2)(a), the undertaker must act in accordance with the counter-notice when altering electronic communications apparatus (in carrying out the proposed undertaker's works).
(4) But, if the operator unreasonably fails to provide the required supervision, the undertaker must act in accordance with the counter-notice only insofar as it requires alterations to be made to the satisfaction of the operator.
(5) In a case where counter-notice contains a statement under sub-paragraph (2)(b) (operator intends to make alteration), the undertaker must not

alter electronic communications apparatus (in carrying out the proposed undertaker's works).

(6) But that does not prevent the undertaker from making any alteration of electronic communications apparatus which the operator fails to make within a reasonable time.

27.3.10

The undertaker is made financially liable to the operator under paragraph 69:

(1) This paragraph applies if an undertaker carries out any non-emergency undertaker's works in accordance with paragraph 67 (including in a case where counter-notice is given under paragraph 68).
(2) The undertaker must pay the operator the amount of any loss or damage sustained by the operator in consequence of any alteration being made to electronic communications apparatus (in carrying out the works).
(3) The undertaker must pay the operator any expenses incurred by the operator in, or in connection with, supervising the undertaker when altering electronic communications apparatus (in carrying out the works).
(4) Any amount which is not paid in accordance with this paragraph is to be recoverable by the operator from the undertaker in any court of competent jurisdiction.

27.3.11

An operator is able to make the alteration pursuant to paragraph 70:

(1) An operator may make an alteration of electronic communications apparatus if –
    (a) notice of proposed works has been given,
    (b) the notice period has ended, and
    (c) counter-notice has been given which states (in accordance with paragraph 68(2)(b)) that the operator intends to make the alteration.
(2) If the operator makes any alteration in accordance with this paragraph, the undertaker must pay the operator –
    (a) any expenses incurred by the operator in, or in connection with, making the alteration; and
    (b) the amount of any loss or damage sustained by the operator in consequence of the alteration being made.
(3) Any amount which is not paid in accordance with sub-paragraph (2) is to be recoverable by the operator from the undertaker in any court of competent jurisdiction.

27.3.12 Emergency undertaker's works

Under paragraph 71, provision is made for emergency undertaker's works as follows:

(1) An undertaker may, in carrying out emergency undertaker's works, make an alteration of any electronic communications apparatus kept on, under or over any land for the purposes of an operator's network.
(2) The undertaker must give the operator notice of the emergency undertaker's works as soon as practicable after beginning them.
(3) This paragraph does not authorise the undertaker to make an alteration of apparatus after any failure by the undertaker to give notice in accordance with sub-paragraph (2).
(4) The undertaker must make the alteration to the satisfaction of the operator.
(5) If the undertaker makes any alteration in accordance with this paragraph, the undertaker must pay the operator –
  (a) any expenses incurred by the operator in, or in connection with, supervising the undertaker when making the alteration; and
  (b) the amount of any loss or damage sustained by the operator in consequence of the alteration being made.
(6) Any amount which is not paid in accordance with sub-paragraph (5) is to be recoverable by the operator from the undertaker in any court of competent jurisdiction.
(7) In this paragraph 'emergency undertaker's works' means undertaker's works carried out in order to stop anything already occurring, or to prevent anything imminent from occurring, which is likely to cause –
  (a) danger to persons or property,
  (b) interference with the exercise of any functions conferred or imposed on the undertaker by or under any enactment, or
  (c) substantial loss to the undertaker,

  and any other works which it is reasonable (in all the circumstances) to carry out with those works.

27.3.13 Criminal offences

Paragraph 72 imposes criminal offences for non-compliance with Part 10:

(1) An undertaker, or an agent of an undertaker, is guilty of an offence if that person –
  (a) makes an alteration of electronic communications apparatus in carrying out non-emergency undertaker's works, and
  (b) does so –
    (i) without notice of proposed works having been given in accordance with paragraph 67, or
    (ii) (in a case where such notice is given) before the end of the notice period under paragraph 67.
(2) An undertaker, or an agent of an undertaker, is guilty of an offence if that person –
  (a) makes an alteration of electronic communications apparatus in carrying out non-emergency undertaker's works, and
  (b) unreasonably fails to comply with any reasonable requirement of the operator under this Part of this code when doing so.

(3) An undertaker, or an agent of an undertaker, is guilty of an offence if that person –
   (a) makes an alteration of electronic communications apparatus in carrying out emergency undertaker's works, and
   (b) does so without notice of emergency undertaker's works having been given in accordance with paragraph 71.

(4) A person guilty of an offence under this paragraph is liable on summary conviction to –
   (a) a fine not exceeding level 4 on the standard scale, if the service provided by the operator's network is interrupted by the works or failure, or
   (b) a fine not exceeding level 3 on the standard scale, if that service is not interrupted.

(5) This paragraph does not apply to a Northern Ireland department.

28 Overhead apparatus

28.1 Introduction

28.1.1

The New Code follows the Old Code[1] in making provision in Part 11, paragraphs 73 to 75 for conferring[2] upon an operator[3] the power to install and keep overhead apparatus on or over land for the purposes of the operator`s network.[4]

28.1.2

Part 11 also makes provision for the operator to affix a notice to certain overhead apparatus.[5]

28.1.3

The persons eligible to object to certain types of apparatus, the time periods within which objections may be made, the conditions required to be satisfied for any objection to be upheld and the orders which the court may make upon hearing any application to uphold the objection to the apparatus are provided for by Part 12 of the New Code.[6]

28.2 Power to fly lines

28.2.1

The power applies where any electronic communications apparatus[7] is kept on or over any land for the purposes of an operator`s network:[8] paragraph 74(1).

1 As to which see Chapter 9.
2 By paragraph 74.
3 An 'operator' is defined in paragraph 2 of the New Code. See Chapter 17.
4 A reading of only Part 11 would suggest that there is no right of objection. However, such a right is to be found in Part 12 of the New Code. The right of objection is considered further in section 28.4.
5 The terms of paragraph 75 replicate paragraph 18 of the Old Code.
6 See paragraph 28.4.
7 See paragraph 5 of the New Code. ECA is considered in sections 6.4 and 15.4.
8 Defined in paragraph 6 of the New Code. See sections 6.4 and 15.4.

28.2.2

Where such apparatus exists it is provided by paragraph 74(2) that the operator has the right[9] for statutory purposes,[10] to install and keep lines which:

(a) pass over other land adjacent to, or in the vicinity of, the land on or over which the apparatus is kept,
(b) are connected to that apparatus, and
(c) are not, at any point where they pass over the other land, less than three metres above the ground or within two metres of any building over which they pass.

28.2.3

Note that although Part 11 is headed 'Overhead Apparatus',[11] there is no definition of this phrase and, the power is limited in a number of respects, namely:

(1) the power is to pass such apparatus over 'land adjacent' to that land on which electronic communications apparatus is kept on or over;
(2) the apparatus which may be passed over such adjacent land is confined to 'lines'.
(3) the lines must be connected to the electronic communications apparatus on the land;
(4) when passing over the adjacent land the lines must not at any point pass over it at a height less than 3 metres above the ground or within 2 metres of any building over which they pass.

28.2.4 'Land adjacent'

It is clear that the land over which the lines may fly need not be immediately abutting the land upon which the apparatus, to which the lines are to be connected, is situated. This is not only the plain ordinary English meaning of the word 'adjacent' but paragraph 74(2)(a) makes it clear that the lines may be sought to be passed over land 'in the vicinity of the land' which contains the apparatus to which the lines are to be connected.

28.2.5 'Lines'

'Lines' are defined in paragraph 5(3) as meaning:

> any wire, cable, tube, pipe or similar thing (including its casing or coating) which is designed or adapted for use in connection with the provision of any electronic communications network or electronic communications service;

9 The right exists without the need for any Part 2 code agreement. But the absence of any code agreement binding the occupier or person who has an interest in the land so as to enable the operator to keep the lines over the land may lead to service of a notice of objection: see section 28.4.
10 Defined in paragraph 4 of the New Code. See section 15.4
11 The heading to paragraph 10 of the Old Code referred simply to 'Power to fly lines'.

28.2.6

This is identical to the definition provided for under the Old Code. There would seem to be no reason why a power cable, providing a power supply to the apparatus on the land to which the line is to be connected, should not fall within this definition.

28.2.7 No right to provide support

The power does not extend to authorising there to be placed 'any electronic communications apparatus used to support, carry or suspend a line installed under sub-paragraph (2)': paragraph 74(3)(a). This prohibition extends to 'any land'. Accordingly, having regard to how this paragraph is expressed, it would appear to exclude not only the provision of means of support, e.g. a pylon, to be placed upon the adjacent land over which the lines are intended to fly, but also exclude the possibility of there being any right, pursuant to the statutory power[12] to install a means of support on any land, including the land on which the apparatus is installed to which the lines are intended to be connected.

28.2.8 Height limitation

The 3 and 2 metre height restrictions[13] are no different to those provided for under the Old Code.[14] In order for the limitation to apply there must be a passing of the line *over* the ground or the building. Thus, if the line were to pass within 2 metres of a building but not actually pass over it, this would not appear to be caught by the prohibition.

28.2.9 No interference with business

No line is authorised to be installed or kept on or over any land if the line which, as a result of its position, interferes with the carrying on of any business carried on on that land: paragraph 74(3)(b). 'Business' includes a trade, profession or employment and includes any activity carried on by a body of persons (whether corporate or unincorporated): paragraph 74(4).[15] An obvious example would be the installation of lines which affect the carrying on of golf at a golf club.

28.2.10

It is to be noted that this limitation on the power applies irrespective of the fact that the line may in fact be at a height in excess of 3 metres.

28.3 Access to exercise power

28.3.1

No express provision is made within the terms of Part 11 or in the New Code as to how the power under paragraph 74 is to be exercised. Thus, no provision is made for notice to

12 There may of course be an express power under any agreement that the operator has with respect to the apparatus installed on or over the land to which the apparatus is to be connected.

13 See paragraph 28.2.3.

14 See Chapter 9.

15 This is the same definition as used under the Old Code. For a discussion on it see Chapter 9, section 9.5, and in particular paragraphs 9.5.4–9.5.6.

be served by the operator before entering upon the land to install the lines although one would expect this would be undertaken as a matter of courtesy. Both codes rather appear to provide that operators will proceed, with or without consent; attach the required notice (see paragraph 75 of the New Code); and wait for any objections to be made.[16]

28.3.2

The statutory power provides for the right, for statutory purposes, to 'install and keep lines'. Although there is no express language to this effect, it must surely be the case that this right will carry with it all rights to give effect to the right to install. This would therefore arguably include access to the land over which the lines are to be flown.

28.3.3

It is expressly provided by paragraph 13 of the New Code that a right under paragraph 74 may not be exercised so as to interfere with or obstruct any means of access to or from any other land (i.e. land other than that over which the lines are to be flown) unless, in accordance with the New Code, the occupier of the other land has conferred or is otherwise bound by a code right within the terms of paragraph 3(h) of the New Code.[17] A right of access includes that which is provided for use in emergencies.[18]

28.4 Objection to lines and other apparatus

28.4.1

Two rights of objection are given:[19]

(1) First, where lines have been installed over land pursuant to paragraph 74.[20] It would seem from the terms of that paragraph that the right of objection applies only once the line is present. There appears to be no right to object prior to the line being erected.[21]
(2) Second where an operator keeps on or over land electronic communications apparatus for the purposes of the operator's network, and the whole or part of it is at a height of 3 metres or more above the ground.[22] The right of objection under this category is (i) not confined to lines and (ii) need not relate to apparatus which has been erected pursuant to paragraph 74.

16 The form of notice required has been published by OFCOM and is included in Appendix E to this book. As to notices, see Chapter 32.
17 Paragraph 13(2). Paragraph 3(h) provides for a right 'to interfere with or obstruct a means of access to or from the land (whether or not any electronic communications apparatus is on, under or over the land)'.
18 Paragraph 13(3).
19 Paragraph 77(3) refers to 'A right to object under this Part of this code is available where …' Thus, the right of objection does not, it would seem, preclude a landowner falling within the description of paragraph 37 from seeking removal under Part 6.
20 Paragraph 77(3).
21 This seems to be consistent with the fact that there is no provision for the operator to provide notice that it is proposing to erect the line pursuant to the power contained in paragraph 74.
22 Paragraph 77(5).

28.4.2

It is to be noted that the rights of objection conferred by paragraph 77 say nothing about objection by reason of the line being below the relevant height restriction or the apparatus interfering with the carrying of any business on the land. The reason for this is that those limitations are limitation on the exercise of the power. Thus, if a person wishes to object that the line has been installed outside the terms of the statutory power, the means of objection is not a notice pursuant to paragraph 77 but, presumably, some form of injunctive relief for the lines to be removed on the basis that they have not been installed, and are not entitled to be kept, pursuant to the statutory power.

28.4.3 First category

Where the lines has been installed over land pursuant to paragraph 74 a person has a right to object if he is:

(i) an occupier of, or has an interest in the land;[23] and
(ii) is not bound by a code right enabling the operator to keep the apparatus installed over the land.[24]

28.4.4

The right of objection is thus given to an occupier of, or party having an interest in, *the land over which the lines pass.*

28.4.5

The power under paragraph 74 is not simply a power conferred on the operator to install the lines. Paragraph 74(2) confers power to install and keep. Thus, reference to 'is not bound' presumably is intended to cover something other than being bound by the statutory power. There is no doubt a difference between the right, to which any occupier or landowner is subject, conferred by paragraph 74 and the right, by way of a code agreement, binding the occupier or landowner to enable the operator to keep the lines installed.

28.4.6 Second category

Where electronic communication apparatus is kept on or over land for the purposes of the operator`s network and the whole or part of it is above a height of 3 metres or more from the ground, the right of objection is given to an occupier or person with an interest in any neighbouring land, i.e. land neighbouring that on or over which the apparatus is kept.[25]

23 Thus, a licencee in occupation would be able to object.
24 Paragraph 77(4).
25 Paragraph 77(5) and (6).

28.4.7

The person who has the right to object in these circumstances is the occupier of or has an interest in land in the neighbouring land.[26] Furthermore, it must be shown that because of the nearness of the neighbouring land to the land over which the apparatus is kept:

(1) the enjoyment of the neighbouring land is capable of being prejudiced by the apparatus,[27] or
(2) any interest in that land is capable of being prejudiced by the apparatus.[28]

28.4.8

No right of objection is possible under this Part of the New Code if the electronic communications apparatus to which objection is made:

(i) replaces any electronic communications apparatus which is not substantially different from the new apparatus, and
(ii) is not significantly in a different positon.[29]

28.5 Procedure with respect to objection

28.5.1

Where a person ('the objector') has a right to object in accordance with paragraph 77, he may exercise that right by giving notice[30] to the operator.[31]

28.5.2

The grounds upon which the Court may uphold an objection differ depending on whether the notice of objection is given before the end of the period of 12 months beginning with the date on which the installation of the apparatus was completed, or the notice is given after the end of that period.[32]

28.5.3

In *Jones v T Mobile (UK) Ltd*[33] (dealing with the terms of paragraph 17 of the Old Code, which provided a time-limited right for the occupier or owner of the affected land to object in certain circumstances, and the right was available only 'before the expiration of the period of three months beginning with the completion of the

26 Paragraph 77(6)(a).
27 Paragraph 77(6)(b)(i).
28 Paragraph 77(6)(b)(ii).
29 Paragraph 77(7).
30 As to the form of notice and method of service, see Chapter 32.
31 Paragraph 78(1).
32 Paragraph 78(2).
33 [2003] 3 EGLR 55.

installation of the apparatus'), the Court of Appeal considered what was meant by completion of the installation. Holman J indicated (at [40]) that he agreed with the trial judge that the 'installation was complete when physical installation was complete, and not when the apparatus was operational'. It was also held (at [13]) that the three months began whether or not the code operator had attached a notice pursuant to the code under paragraph 18 of the Old Code.

28.5.4

The basis upon which objection may be made 12 months after the installation was completed is very limited and, furthermore, the objector runs the risk of having to pay the expenses incurred by the operator in having to comply with any works required pursuant to any court order.

28.5.5 Notice of objection before expiry of 12-month period: paragraph 79

Once the objector has served notice of objection he may apply to the court to have the objection upheld. The application to the court must be made not earlier than two months and not later than four months after the date on which the notice of objection was given.[34] If the four-month time period has expired before any application has been made to the court, there would not, however, appear to be a restriction upon an objector serving a further notice of objection and proceeding to apply to the court pursuant to that notice.[35]

28.5.6

The court must uphold the objection if two conditions are met.[36] The first is that 'the apparatus appears materially to prejudice the objector's enjoyment of, or interest in, the land by reference to which the objection is made'.[37] The burden of proving this is clearly on the objector. The words would appear to cover the impact on the use of the land as well as any diminution in its value arising by reason of the apparatus being present. The prejudice must, however, be material.

34 Paragraph 79(2). This formula is similar to that in respect of 29(3) of Part II of the Landlord and Tenant Act 1954, before the amendments to that Act in 2004. Under the law as it stood prior to June 1, 2004, it was decided by the House of Lords in *Kammins Ballrooms Co Ltd v Zenith Investments (Torquay) Ltd* [1971] AC 850, HL, that a failure to observe the time limit contained within s.29(3) (in its then form) was an irregularity which could be waived by the landlord. Section 29(3) of the 1954 Act provided that 'No application under subsection (1) of section 24 of this Act shall be entertained unless it is made not less than two nor more than four months after the giving of the landlord's notice under section 25 of this Act or, as the case may be, after the making of the tenant's request for a new tenancy'.

35 That said, no doubt it may be argued that the New Code envisages only one notice and court application pursuant to it in respect of each installation of apparatus. Otherwise if multiple notice in respect of each installation were to be permitted, there would be no end to the attempts to object to the installation.

36 Paragraph 79(3) to (5).

37 Paragraph 79(4).

28.5.7

The second condition[38] is that the Court is not satisfied that the only possible alterations of the apparatus will:

(a) substantially increase the cost or diminish the quality of the service provided by the operator`s network to persons who have, or may in future, have, access to it:
(b) involve the operator in substantial additional expenditure (disregarding any expenditure caused solely by the fact that any proposed alteration was not adopted originally or, as the case may be, that the apparatus has been unnecessarily installed),[39] or
(c) give to any person a case at least as good as the objector has to have an objection under this paragraph upheld.[40]

28.5.8

Those sub-conditions are framed in the negative – 'the court is not satisfied' – and thus it would seem that (1) the burden of proof of satisfaction of any one of them is on the operator rather than the objector and (2) it is sufficient, in order to resist the Court upholding the objection, for the operator to establish only one of them.

28.5.9

If the Court upholds the objection it may by order:[41]

(a) direct the alterations of the apparatus to which the objection relates;
(b) authorise the installation (instead of the apparatus to which the objection relates) in a manner and position specified in the order, of any apparatus specified in the order;
(c) direct that no objection may be made under this paragraph in respect of any apparatus the installation of which is authorised by the court.

28.5.10

Where an objector has given both a notice of objection under paragraph 78 objecting to the apparatus and applied for compensation under any other provision in the New Code:

38 Paragraph 79(5).

39 In the Explanatory Notes to the Digital Economy Act, issued by the Department for Culture, Media and Sport, it was said that, para 472, 'It is not the cost of altering the apparatus that is taken into account in assessing additional expenditure, but rather the difference in cost between the original installation and the additional cost of the proposed altered installation if the latter had been installed from the outset. Similarly if the court considers that the apparatus has been unnecessarily installed, expenditure incurred in removing it is not to be taken into account.'

40 For this purpose the court has the power to give the objector directions for bringing the application to the notice of such interested person as it thinks fit: paragraph 78(8).

41 Paragraph 79(6). This is subject to the limitation contained in paragraph 81: see paragraph 28.5.12.

(a) the Court may give such directions as it thinks fit for ensuring that no compensation is paid until any proceedings under paragraph 79 have been disposed of, and
(b) if an order upholding the objection is made to order for some or all of the compensation otherwise payable to the objector is not to be payable or for some or all of the compensation paid to the objector be repaid to the operator.[42]

28.5.11 Notice of objection served after expiry of 12-month period: paragraph 80

If notice of objection is given after the expiry of 12 months beginning with the date on which the installation of the apparatus was completed, the objector may make an application to the court to have the objection upheld.[43] The application must be made not earlier than two months and not later than four months after the date on which the notice of objection was given.[44]

28.5.12

The court may uphold the objection only if it is satisfied that:[45]

(a) the alteration is necessary to enable the objector to carry out a proposed improvement[46] of the land by reference to which the objection is made, and
(b) the alteration will not substantially interference with any service which is or is likely to be provided using the operator`s network.

28.5.13

If the court upholds the objection it may by order direct the alterations of the apparatus to which the objection relates.[47] The order may provide for the alteration to be carried out with such modifications, on such terms and subject to such conditions as the court thinks fit.[48] The court must not include any such modifications, terms or conditions in its order without the consent of the objector and, if such consent is not given, may refuse to make an order.[49] An order must, unless the court otherwise thinks fit, require the objector to reimburse the operator in respect of any expenses which the operator incurs in or in connection with the execution of any works in compliance with the order.

28.5.14 Limitation on court's powers: paragraph 81

Whether notice of objection is given before or after the 12-month period, the Court, in considering any order to make under paragraph 79 (directing the alteration of apparatus

42 Paragraph 79(7).
43 Paragraph 80(2).
44 *Ibid.* See the discussion at n.34.
45 Paragraph 80(3).
46 Which is defined to include development and change of use: paragraph 80(9).
47 Paragraph 80(4). This is subject to the limitation contained in paragraph 81: see paragraph 28.5.12.
48 Paragraph 80(5).
49 Paragraph 80(6).

or authorising the installation of apparatus) or paragraph 80 (directing the alteration of any apparatus), must not make an order unless it is satisfied:[50]

(a) that the operator has all such rights as it appears to the Court appropriate that the operator should have for the purpose of making the alteration, or, as the case may be, installing the apparatus, or
(b) that:
    (i) the operator would have all those rights if the Court, on an application under paragraph 20 imposed an agreement on the operator and another person,[51] and
    (ii) it would be appropriate for the Court, on such an application, to impose an agreement.

28.5.15

The Court has power for the purposes of considering the requirements of paragraph 81 to give the objector directions for bringing the application to the notice of such other interested person as it thinks fit.[52]

28.6 Duty to attach notices

28.6.1

Where any electronic communications apparatus has been installed by an operator for the purposes of the operator`s network, and the apparatus is at a height of 3 metres or more above the ground, the operator must before the end of the period of three days beginning with the day after that on which the installation is completed, attach, in a secure and durable manner, a notice:

(1) to every major item[53] of apparatus installed; or
(2) if no major item of apparatus is installed, to the nearest major item of electronic communications apparatus to which the apparatus that is installed is directly or indirectly connected.[54]

28.6.2

The notice must be attached in a position where it is reasonably legible and must give the name of the operator and an address in the UK at which any notice of objection[55] may be given in respect of the apparatus in question.[56]

50 Paragraph 81(2).
51 The court has, for the purpose of avoiding the need for the agreement of any person to the alteration or the installation of any apparatus, the same powers as it would have if an application had been duly made under paragraph 20 of the New Code for an order imposing such an agreement: paragraph 81(3).
52 Paragraph 81(4).
53 There is no definition of this expression.
54 Paragraphs 75(1) and (2).The terms of paragraph 75 replicate paragraph 18 of the Old Code.
55 In accordance with paragraph 77; see section 28.4.
56 Paragraph 75(2). A person giving an address in respect of the apparatus is treated as having given that address for the purposes of paragraph 91(2): see paragraph 75(4). As to notices, see Chapter 32.

28.6.3

It is an offence for the operator to fail to provide the notice, save that it is a defence for the person charged to prove that they took all reasonable steps and exercised all due diligence to avoid committing the offence.[57]

28.7 Transitional provisions

28.7.1

It is provided by paragraph 8 of Schedule 2 to the 2017 Act that paragraphs 9 to 14 of the Old Code continue to apply in relation to anything in the process of being done when the New Code came into force.

28.7.2

It is further provided that anything installed pursuant to paragraph 10 of the Old Code is to be treated as installed under paragraph 74 of the New Code if it could have been installed under that provision if the provision had been in force or applied to its installation.[58]

57 Paragraphs 75(5) and (6).
58 Paragraph 8 of Schedule 2 to the 2017 Act.

29 Trees and vegetation

29.1 Introduction

29.1.1

This chapter considers the ability of an operator to remove overhanging trees or vegetation which obstruct or interfere with the working of relevant[1] electric communications apparatus.[2]

29.1.2

These rights are contained in paragraph 82, Part 13 of the New Code and are very similar to those contained in paragraph 19 of the Old Code, save that the right to cut back obstructing or interfering material has been extended to cover 'vegetation'. Under the Old Code, the right was limited to the lopping of 'any tree'.[3]

29.2 Paragraph 82

29.2.1

Paragraph 82 is said to apply where:

> (1) …
> (a) a tree or other vegetation overhangs a street in England and Wales or Northern Ireland or a road in Scotland and
> (b) the tree or vegetation –
> (i) obstructs or will or may obstruct, relevant electronic communications apparatus, or
> (ii) interferes with, or will or may interfere with, such apparatus.

1 Electronic communications apparatus which is 'relevant' is defined in paragraph 82(2) of the New Code. The limiting feature appears to be that the apparatus must be that which is used or is to be used 'for the purposes of the operator's network'.
2 As to the meaning of electronic communications apparatus see Chapters 6 and 15 of this book.
3 See Chapter 9, and in particular Section 9.8.

29.2.2

'Relevant electronic communications apparatus' means electronic communications apparatus which (a) is installed, or about to be installed, on land, and (b) is used, or to be used, for the purposes of an operator's network: para 82(2).

29.2.3

There is no definition of 'tree' or of 'vegetation'. It probably matters little as to what can be said to be a tree, or a hedge or a bush, given the extension of the New Code to 'vegetation'. A sapling of whatever size has been held to be a tree for the purposes of tree preservation regulations.[4] It might even be said to include small (dead) vegetation. The tree may be one protected by a tree preservation order. There is an exception in the tree preservation regime for 'statutory undertakers' which specifically includes code operators.[5] The exception also applies to trees in conservation areas.[6]

29.2.4

A 'street' has the same meaning as that in Part 3 of the New Roads and Street Works Act 1991.[7] By s.48(1) of the 1991 Act street means the whole or any part of the following, irrespective of whether it is a thoroughfare, namely:

(1) any highway, road, lane, footway, alley or passage;
(2) any square or court; and
(3) any land laid out as a way whether it is for the time being formed as a way or not.

29.2.5

It is to be noted from the definition that the highway does not have to be one maintainable at public expense. Where the street passes over a bridge or through a tunnel, references to a 'street' include that bridge or tunnel.[8]

29.2.6

It will be apparent from a consideration of the terms of paragraph 82 that the operator may seek to serve notice in anticipation of serving a notice pursuant to Part 4 of the New Code (notice requiring dispensation of the relevant person's agreement to the

4 *Palm Developments Ltd v Secretary of State for Communities and Local Government* [2009] 2 P & CR 16 (Admin).

5 See the Town and Country Planning (Tree Preservation) (England) Regulations 2012 SI 2012, No. 605, regs 14(1)(a))(iii) and 14 (3)(e). Under s.210 of the Town and Country Planning Act 1990 it is a criminal offence to breach a tree preservation order. However, this applies only where there is a breach 'in contravention of tree preservation regulations'.

6 See the Town and Country Planning (Tree Preservation) (England) Regulations 2012, SI 2012, No. 605, reg 15(1)(a)(i).

7 In relation to Northern Ireland, 'street' has the same meaning as in the Street Works (Northern Ireland) Order 1995 (SI 1995/3210 (NI 19)). In Scotland 'road' has the same meaning as in Part 4 of the New Roads and Street Works Act 1991.

8 Section 48(1) of the 1991 Act.

code agreement). Thus the lopping or cutting back may be effected before any electronic communications apparatus is installed or before the operator has a right to install it.

29.2.7

Although there was no judicial determination with respect to para 19 of the Old Code which may be of assistance, it would seem from a consideration of the terms of paragraph 82 that in order for notice to be served, the tree or vegetation must:

(1) overhang;
(2) overhang a 'street'; and
(3) obstruct or interfere with the operator's relevant electronic communications apparatus; or
(4) will or may obstruct or interfere with such apparatus.

29.2.8

Thus the operator does not have to establish on a balance of probabilities that there will be an obstruction or interference with the apparatus, simply that there may be. It may be, given the wording of paragraph 82, that there must be shown to be a physical disruption rather than e.g. interference with a wireless signal. The degree of obstruction or interference is not identified by the statutory wording. However, given that the 'relevant electronic communications apparatus' is apparatus used or to be used for the purposes 'of an operator's network', and given that even a minor obstruction or interference with the apparatus may, in connection with the operation of any such network, cause disruption, one would have thought that it would be sufficient for the operator to show a degree of obstruction on interference (or potential obstruction or interference) which is simply more than de minimis. It is clear from a consideration of the Law Commission Report[9] that the Commission was of the view that the terms of paragraph 82 enabled preventative measures to be effected, rather than wait for the tree, or other vegetation, to cause actual disruption to the apparatus

29.2.9

The interference or obstruction (or potential for such) must derive from a tree or vegetation which overhangs the street. Thus, it would not appear that any interference due to the impact of roots could be the subject of notice under paragraph 82.[10]

29.2.10

It would seem given the terms of paragraph 82(3) that although paragraph 82(2) refers to an obstruction or interference with apparatus 'of an operator's network', the person that can serve the notice requiring the lopping or cutting back would have to be the operator whose apparatus is being so obstructed or interfered with or will or may be obstructed or interfered with.

9 Law Com 336, paragraphs 8.61 and 8.62.

10 Contrast the more extensive wording in the Electricity Act 1989, Sched 4, paragraph 9 where the notice may require the occupier of the land 'to fell or lop the tree or cut back its roots'.

29.3 Procedure

29.3.1

If an operator is of the view that the overhanging tree or other vegetation is obstructing or interfering with or will obstruct or interfere with relevant electronic communications apparatus, the operator may serve a notice[11] to the occupier of the land[12] on which the tree vegetation is growing, requiring the tree to be lopped or the vegetation to be cut back to prevent the obstruction or interference: paragraph 82(3).

29.3.2

The occupier has 28 days beginning with the day on which the notice is given, to serve a counter-notice objecting to the lopping or cutting back: para 82(4). If such a counter-notice is served, the notice served by the operator has effect only if confirmed by an order of the court: paragraph 82(4).

29.3.3

If no counter-notice has been served or a counter-notice has been served and the court confirms the notice served by the operator, the operator may cause the tree to be lopped or the vegetation to be cut back: paragraph 82(5) and (6).

29.3.4

There is nothing in fact stopping the occupier of the land from complying with a notice served by the operator requiring a tree to be lopped or vegetation to be cut back. It is implicit from the terms of paragraphs 82(8) and (9) that the occupier may comply with the notice either upon its service or upon its confirmation by the court. In these circumstances the occupier may seek compensation.[13]

29.4 The work

29.4.1

There is no definition of 'lopping' or 'cutting back'. In the context of a provision under the highways legislation providing for the owner or occupier to, inter alia, 'prune or lop' a true excluding the sun or wind from a highway, it was held that the term 'lop' meant to cut off the branches of the tree laterally and did not comprise the cutting off of the top of the tree.[14] The tree cannot, it is considered, be felled.[15]

11 As to notices, see Chapter 32. The prescribed form of notice contains provision for details of the work to be provided, 'such as a map showing the location of the tree/vegetation and the precise works' which are considered are needed to be carried out.
12 As to the meaning of 'occupier' see Chapter 16.
13 As to which see section 29.5.
14 *Unwin v Hanson* [1891] 2 QB 115, CA. Lord Esher MR said at p.120 that the meaning of 'lopping', in s.65 of the Highways Act 1835, was well known with respect to trees. See now s.136(1) of the Highways Act 1981. It was said that if Parliament had intended that the tops of the trees could be cut, the words used would have been 'lopping and topping'.
15 Contrast the more extensive wording in the Electricity Act 1989, Sched 4, paragraph 9 where the notice may require the occupier of the land 'to fell or lop the tree or cut back its roots'.

29.4.2

There is nothing in the terms of paragraph 82 to indicate how, from a practical point of view, the operator is to effect the lopping or cutting back. It is unclear from the terms of paragraph 82 whether the operator has a right to enter upon the land upon which the tree or vegetation is growing to effect the necessary or any work. Paragraph 82 does not by its terms refer to the operator having access to or to encroach upon the occupier's land. If this is correct (1) the lopping or cutting back would, arguably, have to be effected from land other than that upon which the tree or vegetation is growing and (2) the extent of the work of lopping or cutting back cannot extend beyond the boundary of the occupier's land. On this basis the power conferred by paragraph 82 is, in essence, one only to remove that part of the tree or vegetation which 'overhangs'.

29.4.3

However, it may be said that it is implicit that the operator can access the occupier's land. The notice which is to be served upon the occupier is one which requires 'the tree to lopped or the vegetation to be cut back' to prevent the obstruction or interference'. It may be that the most effective way to undertake the work would be from the occupier's land and it may be, to comply with the duty to effect the work in a husband-like manner[16] that it would be inappropriate to lop or cut back just to the boundary of the occupier's land.

29.4.4

It would, it is considered, be sensible for the operator to notify the occupier as to when the operator intends to undertake the work, the equipment which the operator proposes to utilise, the length of time he expects the work to take and to provide a general description of work intended to be effected.

29.4.5

In undertaking the work the operator must do so in a husband-like manner[17] and in such a way as to cause the minimum damage to the tree or vegetation: paragraph 82(7). The operator will, of course, be subject to the normal duties at common law to exercise reasonable care and skill in effecting the work.

29.5 Compensation

29.5.1

An occupier who does comply with the notice may, whether compliance is effected upon service or upon confirmation by the Court, apply to the Court requiring the operator to pay compensation for loss or damage sustained in consequence of the lopping of the tree or the cutting back of the vegetation, or to pay expenses incurred in compliance with the notice.

16 See paragraph 29.4.5.

17 The draughtsman did not heed the suggestion of the Law Commission that 'the drafters of the revised code will wish to avoid this expression': para 8.54, n. 34. It is fully discussed in the standard texts on the subject – see e.g. Muir Watt and Moss, *Agricultural Holdings*, 15th ed. (2018).

29.5.2

There appears to be nothing in the terms of paragraph 82 which enables the operator to recover the expenses incurred in effecting the lopping of the tree or the cutting back of the vegetation.

29.6 Transitional provisions

29.6.1

Paragraphs 18 and 19 of Schedule 3 of the New Code make provision for what is to happen where a paragraph 19 notice under the Old Code had been served at a time of the coming into force of the New Code.[18] There are two scenarios to consider.

29.6.2

The first is where at the time when the New Code came into force a notice had been served under paragraph 19 of the Old Code but no application has been made to the court in respect of it, the notice and any counter-notice served shall take effect as if given under paragraph 82 of the New Code: paragraphs 18(1) and (2) of Schedule 3. Thus paragraph 82 will apply where:

(a) the 28-day period has elapsed without any counter-notice being served;
(b) a notice has been served but the 28-day period had not expired at the date of the coming into force of the New Code;
(c) a counter-notice was served prior to the New Code coming into force;
(d) the recipient of the notice has complied with it and has yet to make an application for compensation. Such an application can then be made in accordance with paragraphs 82(8) and (9) of the New Code.

29.6.3

The second is where at the time when the New Code came into force a notice had been served under paragraph 19 of the Old Code and an application had been made to the court in respect of it. In these circumstances, the Old Code continues to apply in respect of the application: paragraph 19 of Schedule 3.

18 For more detail, see Chapter 31.

30 Consideration and compensation under the New Code

30.1 Introduction

30.1.1

The terms of an imposed agreement are considered in Chapter 21 of this book. Paragraph 23(3) of the New Code provides that an imposed agreement must include terms as to the payment of consideration by the operator to the relevant person, and this is expressed to be for the relevant person's agreement to confer or be bound by the code right, as the case may be. Chapter 21 also summarises the effect of paragraph 24 of the New Code, which is headed 'How is consideration to be determined under paragraph 23?' This chapter examines in more detail the meaning and effect of the statutory definition of *consideration*.

30.1.2

Chapter 21 also summarises the position under paragraph 25 of the New Code – 'What rights to the payment of compensation are there?' This chapter examines the meaning and scope of the provisions of paragraph 25, by which the Court, in making an order for an imposed agreement, or for the removal of apparatus, may also order the operator to pay compensation to the relevant person for any loss or damage that has been sustained or will be sustained by that person as the result of the exercise of the code right to which the order relates. There are further compensation provisions in paragraph 84 of the New Code, where a court orders the payment of compensation for loss or damage, and in paragraph 85 of the New Code for the payment of compensation for injurious affection to any neighbouring land where a right conferred on an operator is exercised.

30.2 Consideration

30.2.1

Paragraph 24(1) of the New Code provides that the amount of consideration, payable by an operator to a relevant person under an agreement imposed by an order under paragraph 20 (an 'Imposed Agreement', as we refer to it), must be an amount or amounts representing the market value of the relevant person's agreement to confer or be bound by the code right (as the case may be). The relevant person is the person who is required to agree to confer a code right on a code operator, or is otherwise bound by a code right which is exercisable by the operator.[1] A code right in respect of land may only be conferred on an operator by an agreement between the occupier of the land and the operator.[2] It follows that the relevant person, for the purpose of New Code paragraph 24, will be the site provider, such as the freeholder in occupation of a corner of a field, where a mast is to be placed, or the freehold owner-occupier of a building, or part of a building, onto which apparatus can be fixed. The relevant person could also be a tenant entitled to occupation of the land or building in question.

30.2.2

Paragraph 24(1) of the New Code is expressed in terms as to the amount or amounts representing the *market value* of the relevant person's agreement to confer or be bound by the code right, as the case may be. This appears to embrace conflicting ideas.

30.2.3

First, it embraces the concept or idea of *market value*. Ordinarily, *market value* represents the price that a seller will obtain in the open market from a buyer in exchange for the sale by the seller, and the acquisition by the buyer, of land or rights in land. In the context of any other statutory hypothesis of market value, the question for the valuer is normally as to what price a seller can obtain from a buyer in the open market for the land or the rights in land that the seller is deemed to be selling.[3]

30.2.4

But New Code paragraph 24(1) is not so premised. It seeks to apply the concept of market value solely to the behaviour of the seller, namely the relevant person upon whom the agreement is imposed, and who will bear the burdens. The market value is that of the relevant person's agreement to confer or be bound by the code right. The relevant person is conferring on an operator certain rights, such as to erect a mast or erect apparatus, and is agreeing, or is deemed to be agreeing, to the imposition of the burden of those conferred rights. The concept underpinning New Code paragraph 24(1) is that there is a market in the grant of burdens, and that an analysis of that market will reveal a market value of the relevant person's agreement to confer or be bound by the rights granted. But a market in the conferment of burdens or obligations is hard to conceive, whilst a market in the conferment of rights, for which there may be competition between rival buyers to acquire those rights, is recognised.

30.2.5

The only sensible meaning of New Code paragraph 24(1) is that there is some notional market in which a relevant person is deemed to be prepared to accept the imposition of the imposed agreement for a certain price or consideration. But if that is right, it prompts more questions:

1 New Code paragraph 20(1).
2 New Code paragraph 9.
3 *Gray (Lady Fox's Executors) v Inland Revenue Commissioners* [1994] 2 EGLR 185.

(1) Who else is in that notional market?
(2) How should a valuer determine whether such other persons would demand more or less by way of consideration?

The answers to these questions are unclear from New Code paragraph 24(1) alone. On its own, the sub-paragraph is virtually meaningless, if its intention is in some way to establish some overriding principle behind the meaning of *consideration* to underpin the rest of paragraph 24. It might be better to understand this sub-paragraph as recognising that the conferment of an agreement to be bound by code rights is a burden, and that burden causes a negative value. That negative value is the value of the conferment of the code rights. The amount of the consideration is therefore the market value of the agreement to confer or be bound by the code rights, which market value removes that negative value. The consideration is then the value or price which the relevant person would be able to obtain in the open market in exchange for the conferment of the code rights; the price that should be payable in the market to be bound by the conferred burdens.

30.2.6

New Code paragraph 24(2) takes the matter a little further. For the purpose set out in paragraph 24(1), the market value of a person's agreement to confer or be bound by code right is, subject to sub-paragraph (3), the amount that, at the date the market value is assessed, the willing buyer would pay a willing seller for the agreement:

(a) in a transaction at arm's length,
(b) on the basis that the buyer and seller would be acting prudently and with full knowledge of the transaction, and
(c) on the basis that the transaction was subject to the other provisions of the agreement imposed by the order under paragraph 20.

30.2.7

Paragraph 24(2) is subject to two limitations. First, the provisions of sub-paragraph (3), considered below. Secondly, that the market value is assessed at a particular date, also considered below.

30.2.8

The introduction of 'a willing buyer', and what such a person would pay to 'a willing seller', accords with the usual notion of market value, on which there is a considerable jurisprudence derived both from compulsory acquisition compensation and from rent review cases.

30.2.9

Paragraph 24(2) embraces the three assumptions that are set out at (a), (b) and (c) above. Further, what the 'willing buyer' would pay 'the willing seller' for the agreement seems to turn New Code paragraph 24(1) on its head. We are now told that, after all, the consideration is what 'the willing buyer' would pay for the imposed agreement, and that is not the same as *the market value* of the relevant person's agreement to confer or be bound by the code right, under New Code paragraph 24(1).

30.2.10

New Code paragraph 24(3) then sets out four assumptions upon which market value is to be assessed.

30.2.11

The first is that the right that the transaction relates to does not relate to the provision or use of an electronic communications network. Where the code right in issue is, say, the right to install electronic communications apparatus, the phrase 'electronic communications apparatus' is defined as apparatus designed or adapted for use in connection with the provision of an electronic communications *network*.[4] That assumption immediately removes from the hypothetical market potential buyers of rights to place electronic communications apparatus on land and/or on a building, being such rights as would otherwise be code rights. It would not necessarily remove from the market buyers wanting to install the same type of apparatus, so long as it was not electronic communications apparatus. The valuer is therefore faced with the difficult concept of determining the market value of an agreement to confer, say, a right to install a mast on the roof of a building, where any potential buyers for such rights would, in most cases, be buyers wanting to provide or use the roof of a building to install apparatus as part of an electronic communications network. In such circumstances, where the acquisition of rights for the provision or use of an electronic communications network must be disregarded, there may be few if any buyers in the real market. However, the statutory hypothesis does assume that there would be a transaction conferring the code rights under an imposed agreement, or at least conferring rights that correspond to the code rights. It should therefore be assumed that the seller (the relevant person) would achieve in the hypothetical market a price (the consideration) which reflects the burdens conferred by the imposed agreement.[5] The valuer is therefore to assume that there is, or are, hypothetical willing buyers for the rights to be conferred by the imposed agreement. If demand from code operators, who may wish to place electronic communications apparatus, defined as apparatus designed or adapted for use in connection with the provision of an electronic communications network, is to be disregarded, it does not follow that the 'price' that would be payable in the notional market value would be nominal.[6] The hypothetical seller is likely to have the characteristics of the actual site provider, and would be concerned to obtain the best price reasonably obtainable in any negotiations that would lead to the conferment of the rights in the imposed agreement.[7]

4 New Code paragraph 5(1).

5 For decisions where there might be only one 'willing buyer' party, see *FR Evans (Leeds) Ltd v English Electric Co. Ltd* [1977] 36 P & CR 184, and *First Leisure Trading Ltd v Dorita Properties Ltd* [1991] 1 EGLR 133.

6 *Ibid.*

7 *Railtrack plc v Guinness Ltd* [2003] 1 EGLR 124, concerning rights acquired over a railway line.

30.2.12

The second assumption in paragraph 24(3) is that paragraphs 16 and 17 of the New Code do not apply to the right or any apparatus to which it could apply. These two paragraphs concern assignment, and upgrading and sharing rights. A number of points arise in connection with the second assumption.

30.2.13

First, it is unclear why paragraph 16 and 17 are specifically disregarded. Given that paragraph 24(3)(a) provides that the valuation exercise is to be on the assumption 'that the [code] right the transaction relates to does not relate to the provision or use of an electronic communications network', i.e. that one is valuing in a 'no scheme world', it would seem that there would be no need to expressly disregard paragraph 16 and 17. As one is valuing 'the right in all other respects' in a manner which 'corresponds to the code Right' (paragraph 24 (3)(c)), one is not valuing a code right at all. If one is not valuing a code right, arguably paragraphs 16 and 17 will not apply in any event.

30.2.14

The effect of this disregard would therefore seem to be as follows:

(a) As to the 'no scheme world' it may be thought that the disregard under paragraph 24(3)(a) only excludes one of the two 'statutory purposes' under paragraph 6 (as provision of a network is only one of two purposes, the other being the provision of an infrastructure system (see paragraph 7(2)); but in fact it covers both statutory purposes given the definition of network in paragraph 6. It follows that the valuation exercise relating to the 'corresponding right' requires the valuer to assume a 'no scheme world' in its true sense, that is, that none of the New Code rights will apply to it.
(b) In relation to paragraph 16 (assignment), the 'corresponding right' that is being valued is in principle freely assignable as a matter of law, but that is subject to any contractual limitation in the terms of the code agreement, which limitations are carried over to apply to the corresponding right. However, it is to be noted that the limitation carried over is not rendered void by paragraph 16, which is disapplied on the valuation exercise. The practical effect of this is that, if the parties have negotiated an exclusion of the right to assign, that will need to be reflected in the valuation, notwithstanding that the real code right will remain freely assignable by reason of paragraph 16.
(c) The effect of the disapplication of paragraph 17 seems different. As discussed in Chapter 20, paragraphs 20.1.4–20.1.7, it is considered that the better argument on the effect of paragraph 17 is that it confers a power which would not otherwise apply to the code agreement, unless separately negotiated or imposed by the Tribunal. In those circumstances the valuation exercise, given the express disregard of the paragraph 17 'power', would require one to treat the corresponding right to be valued as not including an automatic right to upgrade or share at all, even in the limited sense provided for by paragraph 17. This is of course subject to the terms

of the code agreement making express provision for upgrading or sharing, which express right would have to be taken into account under paragraph 24(2)(c).

30.2.15

If the rights of assignment under New Code paragraph 16, and the rights for an operator to upgrade or share apparatus under New Code paragraph 17, could be taken into account in determining the market value of the right to be bound by an agreement conferring code rights, the relevant person might be able to argue that the additional rights under New Code paragraphs 16 and 17 would confer additional burdens which the consideration should reflect. It seems that any such additional burdens are not to be taken into account in the assessment of the market value. For the reasons set out above, we do not consider that this analysis is likely to be the right one.

30.2.16

The third assumption in paragraph 24(3) is that the right in all other respects corresponds to the code right. This assumption does not seem to take any further the market value hypothesis in New Code paragraph 24(2), where at sub-paragraph (c) it is to be assumed that the hypothetical agreement is subject to the provisions of the agreement that is imposed by an imposed agreement under New Code paragraph 20. But, this assumption does introduce a degree of artificiality for the following reasons. The population of potential buyers requiring the rights to be included in the imposed agreement excludes those on the basis that the transaction does not relate to the provision or use of an electronic communications network. But the right which is actually conferred, for example, placing a mast on the roof of a building, together with the ancillary equipment, power supplies, etc., is the right that forms the subject matter of the transaction of which the market value is to be determined. It can therefore be said that the burden of the agreement that is to be valued is an agreement to confer rights that correspond to the code rights the subject of the actual agreement.

30.2.17

The fourth assumption in paragraph 24(3) is that there is more than one site which the buyer could use for the purpose for which the buyer seeks the right. The statutory hypothesis, having excluded buyers requiring the rights for the purposes relating to the provision or use of an electronic communications network, requires under this assumption that there is competition between potential site providers. Accordingly, the valuer has to consider a statutory hypothesis embracing the market value of a person's agreement to confer or be bound by a code right, where the right will not be used in relation to the provision or use of an electronic communications network, and further that there is more than one site available for the hypothetical grant of the right. Of course, in the real world, no such rights would be granted, for it is unlikely there would be any buyers for such rights under the statutory hypothesis. Nonetheless, it is to be assumed that such a right would be granted. The valuer will therefore have to consider the valuation from the point of view of the site provider, and what price the site provider could reasonably be expected to receive for the grant of the rights. Thus, in taking the example of a mast site on the top of a building, what price could reasonably be expected to be obtainable in the 'market' where the site provider is regarded as a willing seller, there are other potential sites, but in the hypothetical market, there would be no willing buyers? In an urban location, there may be other buildings suitable for the intended apparatus, and owned by parties other than the actual site provider. In such a case there could be competition between potential sellers of the rights under the imposed agreement, in the hypothetical market. In a rural location, where the site provider owns all the land, the site provider could be in a monopoly position for the 'sale' of the rights. As assumption (d) speaks of more than one site, not more than one seller, that monopoly position should be relevant to the assessment of the market value of the site provider's agreement to confer the rights under the imposed agreement.

30.2.18

New Code paragraph 24(4) provides that the imposed agreement may contain terms for the consideration to be payable:

(a) as a lump sum or periodically
(b) on the occurrence of a specified event or events, or
(c) in such other form or at such other time or times as the court may direct.

30.2.19

Subject to any special considerations in relation to a particular imposed agreement, periodic payments of 'rent' have been more attractive to site providers in the past, and may continue to be so.

30.2.20

The valuation date will, presumably, be the date of the court order in respect of the imposed agreement as it is the agreement itself which will contain the terms of the consideration.

30.2.21

There is one important matter, omitted from the New Code, in relation to the determination of the consideration. The underlying principles invoke the concept of market value, subject to the relevant assumptions. But, in most circumstances, the addition of a mast on a building, or the erection of a mast on land, will, in the ordinary way, amount to development requiring the grant of planning permission. In the case of the erection of masts in connection with the provision of an electronic communications network, the installation of such masts may well have permitted development rights under the Town and Country (General Permitted Development) (England) Order 2015, Part 16. On the market value assumptions set out above, it cannot be assumed that the right is required in connection with the use or provision of electronic communications network. Accordingly, the benefit of any permitted development rights under the 2015 Order is not available.

30.2.22

In the absence of any provisions in the New Code as to the assumption or otherwise of the planning status of the site the subject of an imposed agreement, the valuer will have to assume that the permitted development rights under the 2015 Order are not available, as these are generally limited to installation and similar works by code operators.[8] Whether there is any reasonable assumption or expectation of the grant of planning permission will depend entirely upon any relevant planning policies and other material considerations. In most cases it could well mean that the right to erect a mast will not have planning permission, will be unlawful and the site will have no value.

30.2.23

The potential argument, against the above conclusion that in the absence of any planning permission, the hypothetical 'sale' of the rights would have no value, is as follows. The underlying principle is the market value of the relevant person's agreement to confer the rights (paragraph 24(1)). There is also an underlying assumption that a transaction will take place conferring at the least corresponding rights on the imposed assumptions (paragraph 24(2) and (3)). It would be consistent with the statutory hypothesis established by paragraph 24, that the hypothetical transaction can occur lawfully, and therefore would not contravene planning control.[9] That point addresses the argument that the exercise of a right corresponding to a code right might otherwise be in breach of planning control, but it is not quite the same point as contending that such an exercise should be deemed as having the benefit of planning permission. Again, there is some assistance from the jurisprudence on rent reviews. In *Bovis Group Pension Trust Ltd v GC Flooring & Furnishing Ltd,*[10] the rent review provisions of a lease provided that the rent was to be the rent at which the demised premises 'might reasonably be expected to be let for office purposes'. It was held that the existence of planning permission for office use had to be assumed by the valuer on review. As the transaction conferring code rights, or at least rights equivalent to such rights, must be assumed under code paragraph 24, it is well arguable that planning permission for such rights must be assumed.

30.2.24

The separate claim for compensation, for loss or damage sustained by the exercise of code rights, includes a claim for the diminution in value of the land of the relevant person.[11] That raises the question as to whether the measure of the consideration should or should not reflect any diminution in value of the relevant person's land. On the face

8 In connection with the compulsory purchase of land or an interest in land, assumptions as to the prospect of planning permission may be available: Land Compensation Act 1961, ss.14–18.

9 See *Compton Group Ltd v Estates Gazette Ltd* [1978] 36 P & CR 148, where for the purposes of a rent review clause, the possibility of any future use of the subject property to be valued being in breach of planning control must be excluded from consideration.

10 [1984] 2 EGLR 123, followed in *Trust House Forte Albany Hotels Ltd v Daejan Investments Ltd* [1980] 2 EGLR 123, CA.

11 Paragraph 84(2)(b), as applied to the power to order compensation under paragraph 25(1) where an agreement is imposed under paragraph 20.

of the words of paragraph 24(1), the market value of the relevant person's agreement to confer the code rights could well reflect any diminution in value of his land. Such diminution in value might genuinely be caused by the exercise of the code rights of installing, keeping and maintaining apparatus. There could be damage in value by reason of the visual presence of the apparatus, and/or by the other code rights of works and entry. But, as the New Code makes express provision for compensation for the diminution in value of the relevant person's land, that suggests that there are two heads of claim: the claim for the consideration, under paragraph 24 and the separate and additional claim for diminution in value, under paragraph 84(2)(b). There is an analogy here in the legislation relating to the compulsory acquisition of land. There are separate claims for the value of the land taken, and for any injurious affection (diminution in value) of any retained land.[12] But that analogy cannot be pressed any further to the compulsory acquisition of rights in or over land, where the usual measure of compensation is limited to the diminution in value of the retained land, and it does not include any payment for the right, which is compulsorily acquired.[13]

30.2.25

The two potential claims can be reconciled. The claim for diminution in value, of the relevant person's land, only arises, to use the words of paragraph 84(2), 'depending on the circumstances'. Where the circumstances are that the assessment of the consideration plainly was on the basis that the measure included diminution in value of the land, it would not seem that there could be an additional claim for diminution in value under paragraph 84(2). Whereas, if the circumstances were clearly that the measure of the consideration did not include anything for the diminution in value of the land, then the additional claim under paragraph 84(2) could be maintained. It is certainly conceivable that the market value of a person's agreement to confer or be bound by a code right may reflect the inconvenience of the exercise of the code rights conferred. For example, where code rights confer rights to install and maintain apparatus on the roof of a multi-occupied building, the relevant person may be put to inconvenience in accommodating visits, inspections and maintenance work: the hassle factor. That inconvenience may not necessarily result in actual diminution in value of the land, for the purposes of code paragraph 84(2)(b), but the hassle factor might well go to the measure of the consideration and the market value of a person's agreement to confer or be bound by a code right.

30.2.26

A similar point can be made in relation to the measure of the consideration and a claim for compensation under code paragraph 25(1) for loss and damage expected to be sustained in the future. If the conferred code rights, in relation to, say, apparatus on a roof of a multi-occupied building, include a right to enter and maintain from time to time, the relevant person may incur actual costs in having to pay someone to unlock and provide security during a maintenance visit by the code operator. The hassle factor

12 Compulsory Purchase Act 1965, s.7.

13 Local Government (Miscellaneous Provisions) Act 1976, s.13, Sched 1, paragraph 6 modifying Compulsory Purchase Act 1965, s.7.

may go to the assessment of the measure of the consideration, as explained above. But such a claim would not necessarily negate a claim for compensation under code paragraph 25(1), assessed in some way, to reflect the future costs of such maintenance visits. Both claims could be made provided that the measure of the consideration, that is the market value of a person's agreement to confer or be bound by a code right, was limited to the hassle or similar factors, and did not reflect any measure of future actual costs or expenditure. Code paragraph 84(7) provides that where a person has a claim for compensation to which that paragraph applies, and a claim for compensation under any other provisions of the New Code in respect of the same loss, the compensation payable to that person must not exceed the amount of that person's loss.[14] If the word 'compensation' is used there in the broad sense of including the consideration, then the above discussion of the interrelationship of either a claim for compensation for diminution in value of land or for loss or damage, with the measure of consideration, is consistent with this direction against double-compensation for the same loss.

30.3 Compensation

30.3.1

New Code paragraphs 25 and 84 invoke principles that would otherwise have application to a claim for disturbance and other losses, in connection with the exercise of powers of compulsory acquisition of land, and rights in land, under rule (6) of section 5 of the Land Compensation Act 1961. In *Director of Lands v Shun Fung Ironworks Ltd*,[15] Lord Nicholls set out a number of principles. First, there must be a causal connection between the exercise of the power and the loss in question. Second, the losses must not be too remote. Third, the law expects those who claim recompense to behave reasonably. Such behaviour would include a duty to mitigate.

30.3.2

If the Court makes an order under paragraph 24, and there is an imposed agreement conferring a code right on an operator, the Court may also order the operator to pay compensation to the relevant person for any loss or damage that has been sustained or will be sustained by that person as the result of the exercise of the code right to which the order relates.[16]

30.3.3

Such an order may be made at any time the Court makes an order under New Code paragraph 20, or at any time afterwards, on the application of the relevant person.[17] An order for the payment of compensation may specify the amount of compensation to be paid by the operator, or give directions for the determination of any such amount.[18]

14 Paragraph 84(7).
15 [1995] 2 AC 111 at 126.
16 Paragraph 25(1).
17 Paragraph 25(2).
18 Paragraph 25(3).

Such directions may provide for the amount of compensation to be agreed between the operator and the relevant person, and for the dispute about that amount to be determined by arbitration.[19]

30.3.4

An order for the payment of compensation may provide for the operator to make a lump sum payment, to make periodical payments, to make a payment or payments on the occurrence of an event or events, or to make a payment or payments in such other form or such other times or times as the court may direct.[20]

30.3.5

Where an order has been made under New Code paragraphs 40 to 43, by which the operator must remove the electronic communications apparatus, and restore the land to its condition before the apparatus was placed on, under or over the land, then the order may include an order that the landowner may recover from the operator the costs of restoring the land to the condition before the code right was exercised and may require the operator to pay compensation to the landowner for any loss or damage suffered by the landowner as a result of the presence of the apparatus on the land during the period when the landowner had the right to require the removal of the apparatus from the land but was not able to exercise that right.[21]

30.3.6

New Code paragraph 84 applies to the powers of the court to order an operator to pay compensation to a person in two circumstances. First, where compensation is ordered under paragraph 25(1), and therefore associated with an order made under paragraph 20, imposing an agreement on a person conferring code rights. Second, where an order under paragraph 44(5) requires the payment of compensation in relation to the removal of the apparatus from the land.

30.3.7

Depending on the circumstances, the power of the Court to order the payment of compensation for loss or damage includes the power to order payment for:

(a) expenses (including reasonable legal and valuation expenses, subject to the provisions of any enactment about the powers of the Court by whom the order for compensation is made to award costs, or in Scotland, expenses);
(b) diminution of the value of land and,
(c) costs of reinstatement.[22]

19 Paragraph 25(4).
20 Paragraph 25(5).
21 Paragraph 44(1), (4).
22 Paragraph 84(2).

30.3.8

For the purposes of assessing such compensation for the diminution of the value of land, the following provisions apply with any necessary modifications as they apply for the purposes of assessing compensation for the compulsory purchase of any interest in land:

(a) in relation to England and Wales, rules (2) to (4) in section 5 of the Land Compensation Act 1961;
(b) in relation to Scotland, rules (2) to (4) set out in section 12 of the Land Compensation (Scotland) Act 1963; and
(c) in relation to Northern Ireland, rules (2) to (4) set out in Article 6(1) of the Land Compensation (Northern Ireland) Order 1982.[23]

30.3.9

New Code paragraph 84(4), (5) and (6) incorporate the provisions in the respective enactments relating to land compensation as they apply to the assessment of compensation in respect of mortgages, trusts of land and settled land (in Scotland restrictive interests). In England and Wales, the provisions are in section 10(1) to (3) of the Land Compensation Act 1973. In Scotland, they are in section 10(1) and (2) of the Land Compensation (Scotland) Act 1973. In Northern Ireland, they are in Article 13(1) to (3) of the Land Acquisition and Compensation (Northern Ireland) Order 1973.[24]

30.3.10

Where a person has a claim for compensation to which paragraph 84 applies, and a claim for compensation under any other provisions of the New Code in respect of the same loss, the compensation payable to that person must not exceed the amount of that person's loss.[25] This possible overlapping of claims for the consideration, and compensation for loss or damage and diminution in value of the land is discussed above under the heading 'Consideration'.

30.3.11

New Code paragraph 86 makes clear that, except as provided for by any of the provisions of Parts 2 to 13, or 14 of the New Code, an operator is not liable to compensate any person for, and is not subject to any other liability in respect of, any loss or damage caused by the lawful exercise of any right conferred by or in accordance with any provisions of those Parts of the New Code.

30.3.12

Paragraph 25(1) and paragraph 84 identify three principal heads of claim which deserve some discussion: a claim for loss or damage, a claim for the diminution of the value of the land and a claim for injurious affection.

23 Paragraph 84(3).
24 SI1973/1896 NI 21.
25 Paragraph 84(7).

30.4 Compensation for loss or damage

30.4.1

A claim for any loss or damage invokes the principles that apply to a claim for disturbance and other losses, in connection with the compulsory acquisition of land or rights in land, under rule (6) of section 5 of the Land Compensation Act 1961. As explained above, in *Director of Lands v Shun Fung Ironworks Ltd* (1995), Lord Nicholls set out a number of principles. First, there must be a causal connection between the exercise of the power and the loss in question. Second the losses must not be too remote. Third, the law expects those who claim recompense to behave reasonably. Such behaviour would include a duty to mitigate.

30.4.2

The scope of a claim for loss or damage will depend entirely upon the circumstances. In the case of the erection of a mast in the corner of a field, there may be minimal loss, although construction traffic accessing the work site may give rise to physical obstruction and perhaps loss of crops, and therefore include consequential loss of profits. The erection of apparatus on a building might have more onerous consequences and, again depending on the circumstances, there may be interferences with the conduct of business, which could give rise to loss of profits as a head of claim. It is conceivable that the grant of rights to erect apparatus might cause the total or partial extinguishment of a business, in certain cases. Compensation for such losses would be payable.[26] An order for compensation may be made when an order is made conferring code rights or at any time thereafter.[27] Further, in either case, an order for compensation may be made for any loss or damage that has been sustained or will be sustained.[28] Several difficulties follow from these provisions.

30.4.3

It is unlikely in most cases that a relevant person will have sustained any loss or damage as a result of the *exercise* of a code right prior to the date of the order under code paragraph 20, conferring code rights. At the date of any such order, the claim for compensation for loss or damage as a result of the exercise of a code right can only be for such loss or damage as 'will be sustained' by the relevant person. If any such future loss or damage is not reflected in the assessment of the consideration as part of the market value of the relevant person's agreement to confer or be bound by the code rights, then an assessment must be made of such future loss or damage. If, for example, the conferred code rights include a right to enter and maintain from time to time apparatus on a roof of a multi-occupied building, the relevant person may incur actual costs in having to pay someone to unlock any access and provide security during a maintenance visit by the code operator. Such costs could be assessed as a lump sum, or could

26 *Leonidis v Thames Water Authority* [1979] 2 EGLR 8.
27 New Code paragraph 25(2).
28 New Code paragraph 25(1).

be ordered to be paid on the occurrence of specified events.[29] The code also provides for an application for compensation for loss or damage at any time after the order conferring code rights is made.[30] There are two possibilities. First, and on the illustration above, it is conceivable that the code operator, acting within the code rights conferred, carries out a substantial work of repair or replacement to the apparatus on the roof of the building, beyond that anticipated at the time of any earlier order for compensation for loss or damage. There seems no reason why any such uncompensated loss should not be compensated on an application at any relevant time after the date of the order conferring the code rights. Second, an application can be made at any time after the order conferring code rights for any loss or damage which 'will be sustained' as a result of the exercise of a code right. Using the above illustration again, and where substantial work of repair or replacement is carried out in the exercise of the code right, it may be that in addition to the first claim above, there could be an expected future greater loss or damage, for the provision of security and access, if the apparatus will require more frequent inspections.

30.4.4

The claims for compensation under code paragraphs 25 and 84 are statutory claims for the purposes of the application of the six-year limitation period under the Limitation Act 1980.[31] Although code paragraph 25(2)(b) provides for an application being made for compensation at any time after an order under code paragraph 20 conferring code rights is made, it is suggested that the limitation period of six years must run from the date of the loss or damage, and not from the date of the paragraph 20 order.[32] That is because the claim is for loss or damage sustained by the *exercise* of a code right, and the conferment of a code right will not necessarily, or immediately, cause loss or damage. A code right conferred in 2018 to, among other things, upgrade apparatus, may not be exercised until 2030.

30.5 Diminution in value of the land

30.5.1

The claim for diminution in value, of the relevant person's land, only arises, to use the words of paragraph 84(2), 'depending on the circumstances'. Where the circumstances are that the assessment of the consideration, as explained above, plainly was on the basis that the measure included diminution in value of the land, it would not seem that there could be an additional claim for diminution in value under paragraph 84(2). Whereas, if the circumstances were clearly that the measure of the consideration did not include anything for the diminution in value of the land, then the additional claim under paragraph 84(2) could be maintained.

29 New Code paragraph 25(5).
30 New Code paragraph 25(2)(b).
31 S.9 – sums recoverable by statute.
32 See *Hillingdon London Borough Council v ARC Ltd* [1983] EGLR 18.

30.5.2

The entitlement to compensation for the diminution in value of the land on, under or over, which apparatus are installed pursuant to an agreement imposed under New code paragraph 20, may be significant in certain cases. The assessment of such compensation will normally proceed on the basis of a comparison of two valuations. The valuations will both be carried out, presumably on the date when the apparatus is first installed, with a valuation made of the land in question on the assumption that there is no apparatus installed and the second valuation on the basis that it is installed. The difference will represent the diminution in value for which compensation will be payable. This might be a significant claim where a mast or apparatus is to be placed on a building. The diminished value will reflect the presence and unsightliness of the apparatus, as well as the consequences of another party having rights of access for maintenance and servicing, perhaps sharing access to or through the building, with others. The principles underlying a claim for injurious affection in connection with the compulsory purchase of part of a claimant's land would appear to have application to this head of claim.[33] There is an analogy in the entitlement to compensation in relation to the installation of electric lines and a pylon. In *Turris Investments v CEGB*,[34] compensation was payable separately for the acquisition of rights over land and for injurious affection (diminution in value); in respect of the latter head of claim, the depreciation factors included the visual intrusion of the pylon and cables, the potential purchaser's fears, noise of corona discharge and the exercise of rights of entry without notice.

30.5.3

The second problem is the meaning of 'the land'. The New Code defines 'land' as not including electronic communications apparatus – see paragraph 108(1) – but does not otherwise define the term.[35] Section 5 of, and Schedule 1 to, the Interpretation Act 1978 defines 'land' as including any estate or interest in or over land.[36] 'The land' in paragraph 84(2)(b) would therefore include an estate in land, such as a freehold and/or a lease. That still leaves open the question as to whether 'the land' is physically limited to the footprint of the apparatus to be installed or kept, to the extent of the land of which the relevant person is in occupation, for the purposes of being the person deemed to be conferring code rights, or to the full extent of any land that the relevant person may actually own, whether wholly or only partly occupied, such as reversionary interests in other parts of a building. Having regard to the code rights in New Code paragraph 3(a) and (b), these speak of installing and keeping apparatus 'on, under or over the land'. But paragraphs 3(f) and (h) speak of entering the land and interfering or obstructing a means of access to 'the land'. Whilst sub-paragraphs (a) and (b) suggest a narrow physical definition of 'the land', sub-paragraphs (f) and (h) suggest that 'the land' is not limited to the footprint of the apparatus. Further, it is the occupier of 'the land' who confers an agreement: see paragraph 9. A court may construe the phrase 'the land' in the

33 *Cowper Essex v Acton Local Board* [1889] 14 App Cas 153; *Duke of Buccleuch v Metropolitan Board of Works* [1872] LR 5 HL 418.
34 [1981] 258 EG 1303.
35 See the detailed discussion of this topic in Chapter 18.
36 This approach was also adopted in *British Waterways Board v London Power Networks plc* [2003] 1 All ER 187, in relation to the Electricity Act 1989.

circumstances of an actual case. But, the most plausible interpretation is that 'the land' is the land of which the relevant person is the occupier.

30.5.4

A claim for the diminution in value of land may arise in relation to an imposed agreement conferring code rights for a defined term, rather than in perpetuity. It follows that the diminution in value is not necessarily a permanent feature. In such circumstances, where the land has a lesser value for the term of the imposed agreement, it is strongly arguable that the claim under paragraph 84(2)(b) is a claim for the capitalised diminished rental value for that period, as would be applied under the Compulsory Purchase Act 1965.[37]

30.5.5

The final problem is the valuation date. A claim for loss or damage under code paragraph 84 includes a claim for the diminution in value of the land. A claim for loss or damage is a claim where a relevant person has sustained or will sustain such loss or damage as a result of the exercise of a code right.[38] Unless or until any code rights are exercised, it is difficult to see that a relevant person has sustained or will sustain diminution in value of the land, even though it may be anticipated that loss will be sustained when the code rights are exercised. A New Code paragraph 20 order might be made conferring code rights some or all of which might not be exercised for some time. At the date of the order a diminution in value of the land might arise because of the fear that the code rights will be exercised, and that diminution in value might increase if and when the code rights are eventually exercised. It would be sensible for code operators to have the diminution in value claim settled earlier, and for landowners to have the claim settled when the rights are actually exercised and the diminution in value is real rather than merely reflective of fear.

30.6 Compensation for injurious affection

30.6.1

New Code paragraph 85 makes provision for the payment of compensation for injurious affection to neighbouring land. This applies where a right is conferred by or in accordance with Parts 2 to 4 of the New Code and is exercised by an operator. In the application of New Code paragraph 85 to England and Wales, compensation is payable by the operator under section 10 of the Compulsory Purchase Act 1965 as if that section applied in relation to injury caused by the exercise of such a right as it applies in relation to injury caused by the execution of works on land that has been compulsorily acquired.[39] There is a similar application of the equivalent provisions in Scotland under section 6 of the Railway Clauses Consolidation (Scotland) Act 1845. Further, there are equivalent provisions in relation to Article 18 of the Land Compensation (Northern Ireland) Order 1982 in respect of Northern Ireland.

37 Section 10, as applied in *Wildtree Hotels Ltd v Harrow LBC* [2001] 1.
38 New Code paragraph 25(1).
39 Paragraph 85(2).

30.6.2

The entitlement to compensation, under section 10 of the 1965 Act, has been the subject of interpretation by the House of Lords in *Wildtree Hotels Ltd v Harrow London Borough Council* (2000) where Lord Hoffmann restated a number of principles, originally known as the *McCarthy* rules.[40] The first rule is that the injury must have been done by reason of what is statutorily authorised. For example, if in the course of erecting a mast in the corner of a field, the contractor is careless or negligent and damage is caused in consequence of that, such negligently caused damage would not normally be statutorily authorised and the landowner would have to pursue a claim under the civil law against the contractor rather than a claim under the New Code against the operator. Second, there must be an injury which, but for the statutory authorisation, would have been actionable at law. Third, the damage must arise from a physical interference with some public or legal right, which the claimant has as owner of an interest in property is, by law, entitled to make use of. This usually means interference or an obstruction to a right of way, a right to light or a right of access onto a public highway. Fourth, the damage must arise from the execution of the authorised works, and not from their use. That means that it is only the consequence of the erection of apparatus which would give rise to a claim for injurious affection compensation. The use of that apparatus, in terms of any adverse effects from electrical fields or otherwise, would not be compensatable under s.10.

30.6.3

The measure of compensation for injurious affection is the diminution in value of the affected interest in land caused by the interference with or obstruction of the legal right affected. If there is a permanent diminution in value, such values are used. If the interference is only temporary, then the measure of compensation is diminution in the rental value of the affected property for the relevant period.[41]

30.6.4

New Code paragraph 85(8) provides that compensation is payable to a person under this paragraph irrespective of whether the person claiming the compensation has any interest in the land where a code right, which is conferred by or in accordance with any provision of Parts 2 to 9 of the New Code referred to in sub-paragraph, is exercised. Thus, where a code right has been granted by the person having an interest in the roof and airspace above a building, the owner of a lease of a flat or a floor in such a building, whose interest is not bound by the code right, may have a claim under s.10, if that person can satisfy the rules set out above.

40 *Wildtree Hotels Ltd v Harrow LBC* [2001] 1; *Metropolitan Board of Works v McCarthy* [1874] LR 7 HL 243.

41 *Wildtree Hotels Ltd v Harrow LBC* [2001] 1; *Ocean Leisure Ltd v Westminster City Council* [2004] 3 EGLR 9.

Part IV

Matters common to both codes

31 Transitional provisions

31.1 Introduction

31.1.1

Section 97(7) of the 2017 Act provides that Schedule 1, containing the New Code to be implanted into the 2003 Act, will be brought into force on a date to be appointed by the Secretary of State by statutory instrument.

31.1.2

The statutory instrument[1] in question was laid before Parliament on 14 December 2017, and provided for the New Code to come into force on 28 December 2017.

31.1.3

The New Code is not, however, retrospective, with the result that the many thousands of Old Code arrangements that were in place prior to 28 December 2017 will continue to subsist, in accordance with the detailed transitional provisions provided for by the New Code.

31.1.4

This chapter considers those transitional provisions, which are contained in Schedule 2 to the 2017 Act.

31.1.5

Of importance are:

(1) the application of the New Code to 'subsisting agreements';
(2) the retention of the termination of any subsisting code agreement in accordance with the provisions of Part II of the Landlord and Tenant Act 1954 ('the 1954 Act') where the subsisting code agreement is one to which the 1954 Act applies;

1 The Digital Economy Act 2017 (Commencement No. 3) Regulations 2017 (SI 2017 No. 1286) – www.legislation.gov.uk/uksi/2017/1286/made.

(3) the non-application to 'subsisting agreements' of certain beneficial provisions of the New Code which otherwise apply to New Code agreements;
(4) the preservation of rights of removal of electronic communications apparatus in accordance with paragraph 20 and 21 of the Old Code with respect to 'subsisting agreements';
(5) the continuing application of provisions affecting the special regimes contained within paragraphs 9–14 of the Old Code.

31.2 'Subsisting agreements'

31.2.1

The New Code, from the date of its coming into force, applies to 'subsisting agreements' and such a 'subsisting agreement' takes effect 'as an agreement under Part 2 of the New Code between the same parties, subject to the modifications made by this Schedule': paragraph 2(1) of Schedule 2 to the 2017 Act.

31.2.2

Paragraph 1(4) of Schedule 2 to the 2017 Act defines a 'subsisting agreement' as follows:

> A 'subsisting agreement' means –
> (a) an agreement for the purposes of paragraph 2 or 3 of the existing code, or
> (b) an order under paragraph 5 of the existing code,
>
> which is in force, as between an operator and any person, at the time the new code comes into force (and whose terms do not provide for it to cease to have effect at that time).

31.2.3 'In force'

The words 'in force' are ambiguous. The Old Code contains no provision equivalent to that of paragraph 30 of the New Code, which continues the rights of the operator under the code agreement[2] notwithstanding the contractual termination of the term created by the agreement. Under the Old Code a fixed term agreement which has expired by effluxion of time is not 'continued', so as to continue the relationship between the parties, but it is provided simply that the electronic communications apparatus remaining on land 'shall … be deemed as against any person who was at any time entitled to require the removal of the apparatus, but by virtue of this paragraph [paragraph 21] is not entitled to enforce its removal, to have been lawfully so kept at that time': paragraph 21(9) of the Old Code.

31.2.4

Accordingly, it may be said that a code agreement which has expired by the date of the coming into force of the New Code was not a 'subsisting agreement' and that a

2 Under the New Code a 'code agreement' is expressly defined as one to which Part 5 of the New Code (dealing with termination) applies: paragraph 29(5). However, in this chapter it refers to an agreement conferring code rights under the Old Code whether one which was consensual or imposed.

'subsisting agreement' applies only to one the contractual term of which is continuing at the date of the coming into force of the New Code.

31.2.5

There would appear to be no reason why an agreement to grant a code right subsisting at the date the New Code came into force could not be a 'subsisting agreement'. Paragraph 108(1) defines 'lease' as including 'any agreement to grant such a tenancy'. There is no reference to any such agreement being required to be unconditional at that date. However, given that there will, in equity, be no lease or agreement to confer the code right unless the agreement for lease was one susceptible to specific performance, it would seem to follow that the agreement must, in order to be a subsisting agreement 'in force' be one which, as at the date of the coming into force of the New Code, could be the subject of an order for specific performance.

31.2.6 Effect of subsisting agreement

The New Code applies to the subsisting agreement subject to the modifications made by schedule 2 to the 2017 Act: paragraph 2(1) of Schedule 2 to the 2017 Act. The New Code provides that the subsisting agreement 'has effect after the new code comes into force as an agreement under Part 2 of the new code between the same parties'.

31.2.7

Thus, the 'subsisting agreement' takes effect from the date of the coming into force of the New Code as a (new) Part 2 New Code Agreement. Accordingly:

(1) The subsisting agreement is, subject to the application of the terms of the Transitional Provisions in Schedule 2 of the 2017 Act, treated as a Part 2 Agreement under the New Code. The Transitional Provisions modify the application of the New Code in a number of respects as set out in this chapter e.g. by disapplying the favourable provisions as to assignment and sharing contained in Part 3 of the New Code: see section 31.5 below.
(2) The subsisting agreement is thus deemed to satisfy Part 2 of the New Code albeit it may not in fact, e.g. it may not satisfy all of the requirements of paragraph 11 of the New Code.
(3) The subsisting agreement is treated as a Part 2 code agreement (subject to the modifications provided for by Schedule 2 to the 2017 Act) 'between the same parties'. 'The same parties' are clearly the parties to the subsisting agreement immediately preceding the coming into force of the New Code, not the original parties to the subsisting agreement. Accordingly, where there has been an assignment of the interest of the conferring party or that of the operator, the Part 2 agreement will be one with whoever is the successor in title.
(4) As the 'subsisting agreement' is treated as a Part 2 code agreement subject only to the modifications made by Schedule 2 to the 2017 Act the following seem to follow.
    (a) The entirety of Part 6 of the New Code, making extensive provision for the removal of electronic communications apparatus, applies to the apparatus

installed under the subsisting agreement.[3] The transitional provisions only make modification to the application of Part 5, dealing with termination, and say nothing about Part 6.

(b) The non-application of any requirement to register any code agreement as a land charge in accordance with the Land Charges Act 1972 or to protect the same on the register at HM Land Registry in accordance with the Land Registration Act 2002, contained in paragraph 14 of the New Code, and which form part of Part 2 to the New Code, extend to the subsisting agreement. This would in any event reflect the terms of paragraph 2(7) of the Old Code.

(c) The anti-avoidance provisions of paragraph 100 of the New Code will apply to the subsisting agreement.

31.2.8 'Whose terms do not provide for it to cease to have effect at that time'

It is possible for code agreements entered into before the New Code came into force to anticipate the coming into force of the New Code, and provide expressly for the Old Code agreement to terminate on the date of the coming into force of the New Code. The Authors are unaware of code agreements having been entered into which make provision for termination on the New Code coming into force.

31.2.9 Paragraph 2(4) of the Old Code

A person who, in respect to subsisting agreement, is bound by it by virtue of paragraph 2(4) of the Old Code is, upon the New Code coming into force, treated as bound by it pursuant to Part 2 of the New Code: paragraph 2(2) of Schedule 2 to the 2017 Act.[4]

31.2.10

Paragraph 2(4) of the Old Code[5] provides in essence that where a landowner (whether or not at the time the occupier) agrees in writing to be bound by a code agreement, any successor in title to that person or anyone deriving title or a right of occupation from that person after the date of the relevant code agreement, will be bound by the code agreement.

31.2.11

There is a not dissimilar provision to paragraph 2(4) of the Old Code contained within paragraph 10 of the New Code. However, given that paragraph 2(1) of Schedule 2 to the 2017 Act treats, subject to the modifications provided for by it, a subsisting agreement as a Part 2 code agreement, it was obviously thought necessary and appropriate to

3 This may cause some difficulty where termination of the subsisting code agreement is required to be in accordance with the 1954 Act. See Chapter 25.

4 A different treatment is given to a person bound by a subsisting agreement only by reason of paragraph 2(3) of the Old Code. In that particular case it is provided that the subsisting agreement only continues to bind the person bound by paragraph 2(3) of the Old Code for so long as he would be bound by that paragraph assuming it continued to have effect post the coming into force of the New Code: see paragraph 31.3.10 of this chapter.

5 See Chapter 5.

extend the application of Part 2 to such other persons as were treated as bound by the subsisting agreement under the Old Code.

31.2.12

If the person bound by a subsisting agreement is bound by reason of paragraph 2(4) of the Old Code, that person may not, and in most cases would not, be a party to the code agreement. Given that paragraph 2(2) of Schedule 2 to the 2017 Act treats such a person as 'bound pursuant to Part 2 of the new code', presumably there is no reason why the terms of paragraph 10 of the New Code, and in particular paragraph 10(6), cannot apply, so as to treat a successor in title to someone who has agreed to be bound by the (subsisting) code agreement, albeit not a party to it, as a deemed party to it for the purposes of the New Code. This is important given that the termination provisions contained in Part 5 of New Code and in particular paragraph 31, apply only to 'a site provider[6] who is a party to a code agreement'.[7]

31.3 'Subsisting agreements': limitations with respect to code rights/ persons bound

31.3.1

Although the New Code applies to a 'subsisting agreement', there are a number of limitations with respect to the application of the New Code.

31.3.2 Limitation of code rights: limitation to existing description of 'code rights'

By the terms of paragraph 2(1) of the Old Code,[8] it is provided that a code agreement is one in writing whereby the occupier confers on the operator 'a right for the statutory purposes':

(a) to execute any works on that land for or in connection with the installation, maintenance, adjustment, repair or alteration of electronic communications apparatus; or
(b) to keep electronic communications apparatus installed on, under or over that land; or
(c) to enter that land to inspect any apparatus kept installed (whether on, under or over that land or elsewhere) for the purposes of the operator's network.

31.3.3

The definition of 'code right' in paragraph 3 to the New Code is much more extensive, consisting of nine code rights. Those within paragraphs 2(1)(a) to (c) of the Old Code

6 A 'site provider' is defined as someone who confers the code right or is otherwise bound by it: paragraph 30(1) of the New Code.
7 The draftsman was aware of this and has made provision to prevent a party bound by a code agreement only by reason of paragraph 2(3) of the Old Code from being treated as a party to the agreement and thus otherwise able to serve a termination notice in accordance with Part 5. See section 31.7 below.
8 For a discussion as to these rights, see Chapter 5.

correspond roughly to paragraphs 3(a) to (f) of the New Code, although the latter are phrased in slightly different and wider terms.

31.3.4

It is provided that in relation to a subsisting agreement, references in the New Code to a code right, are, in relation to an agreement for the purposes of paragraph 2 of the Old Code, references to a right for the statutory purposes to do the things listed in paragraphs 2(1)(a) to (c) of the Old Code, rather than the (corresponding) rights conferred by paragraph 3 of the New Code.

31.3.5

This may have some impact on the application of the New Code to the rights conferred on the operator by the subsisting agreement. By way of example, paragraph 12(2)[9] of the New Code provides that 'anything done by an operator in exercise of a code right conferred under this Part in relation to any land is to be treated as done in the exercise of a statutory power'. Thus, in relation to a subsisting agreement that immunity applies only to the exercise of rights falling within the terms of paragraph 2(1) of the Old Code.

31.3.6 Limitation of code rights: limitation with respect to obstructing access

By paragraph 3 of the Old Code[10] any right conferred by a code agreement 'shall not be exercisable so as to interfere with or obstruct any means of entering or leaving any other land unless the occupier for the time being of the other land conferred, or is otherwise bound by, a right to interfere with or obstructs a means of entering or leaving the other land'.[11]

31.3.7

Thus the consent of the occupier for the time being of the other land is required in order to enable the operator to obstruct the access to or from that other land. Provision is made by the terms of the Old Code to ensure that the agreement of the occupier should bind any person deriving title from or under that person.[12]

31.3.8

The New Code confers a similar code right, by the terms of paragraph 3(h), providing the right 'to interfere with or obstruct a means of access to or from the land (whether or not any electronic communications apparatus is on, under or over the land)'.

9 As to which, see Chapter 17.
10 See section 7.4 of Chapter 7.
11 This limitation also applies to any rights to be exercised under paragraph 9 (Street Works), paragraph 10 (power to fly lines) and paragraph 11 (tidal waters) of the Old Code: paragraph 3(1) of the Old Code.
12 See paragraph 3(4) of the Old Code, applying to such an agreement sub-paragraphs (2) to (7) of paragraph 2 of the Old Code.

31.3.9

The transitional provisions operate so that where there subsists an agreement entered into for the purposes of paragraph 3 of the Old Code, references in the New Code to a 'code right' is to the right mentioned in paragraph 3 of the Old Code, rather than, it would seem, the right referred to in the New Code under paragraph 3(h). However, there does not appear to be any material difference in the wording of the Old Code (paragraph 3(1)) and the New Code in this regard.

31.3.10 Limitation of code rights: limitation of persons bound

The terms of paragraph 4(1) of Schedule 2 to the 2017 Act make provision as to the continuing application of a party bound by a subsisting agreement by reason of paragraph 2(3)[13] of the Old Code. Paragraph 2(3) of the Old Code was a slightly unusual provision in that it dealt with the provision, by an agreement, of electronic communications services to an occupier of land. By paragraph 2(3) of the Old Code, provision was made for ensuring that any code agreement entered into by the occupier of the land, for purposes connected with the provision to the occupier from time to time of that land of any electronic communications services, bound persons other than the person who conferred the code agreement.

31.3.11

In particular (1) where the right is conferred by a person who was a freehold or leasehold owner or (2) where the right was conferred by someone who was not a freeholder or leasehold owner, but the freeholder or leasehold owner had in fact agreed in writing to be bound by the right, the code agreement had effect, where persons of the description falling within (1) and (2) were 'at any time' in fact in occupation,[14] as if every person for the time being owning an interest in the land had agreed in writing to the right being conferred 'for the said purposes' and 'to be bound by it'.

31.3.12

Paragraph 3 of Schedule 2 to the 2017 Act provides that any person bound by a code right by virtue only of paragraph 2(3) of the Old Code shall continue to be bound by it only for so long as they would be bound by it by reason of paragraph 2(3) of the Old Code.

31.3.13

It will be seen from the terms of paragraph 2(3) of the Old Code that the ambit of its operation was dependent upon occupation by the party who conferred the right or agreed to be bound by it. Accordingly, the extent of any party being bound by the New

13 See section 7.4 of Chapter 7.

14 And this would include any person deriving title from or under such a person: paragraph 2(4) of the Old Code.

Code by a code agreement in accordance with paragraph 2(3) of the Old Code may be lost if a party who conferred it or is otherwise bound by it ceases to be in occupation.

31.3.14

The transitional provisions also make an adjustment with respect to the compensation provisions which one finds contained in paragraph 4 of the Old Code with respect to any person bound by a code right by virtue only of paragraph 2(3) of the Old Code: see paragraph 4(2) of Schedule 2 to the 2017 Act.

31.4 'Subsisting agreements': 1954 Act protection

31.4.1

The New Code seeks, with respect to New Code agreements, to provide simply for there to be New Code protection without any security of tenure under the 1954 Act.[15] The New Code does this by providing that where the primary purpose[16] of the lease is to 'grant code rights within the meaning of Schedule 3 A of the Communications Act 2003 (the electronic communications code)' the 1954 Act will not apply to the lease: s.43(4) of the 1954 Act, inserted by paragraph 4 of Part 2 to Schedule 3 of the 2017 Act.

31.4.2

However, in relation to subsisting agreements the 1954 Act may still have a role to play with respect to the termination of the lease. The terms of paragraph 6 of Schedule 2 of the 2017 Act modifies the terms of paragraph 29 of the New Code with respect to 'a subsisting agreement', so as to provide that if the lease is one to which the 1954 Act applies, there being no agreement under s.38A of the 1954 Act excluding the security of tenure provisions of the 1954 Act to the lease, then termination of the lease is in accordance with the 1954 Act, not Part 5 of the New Code.

31.4.3

Paragraph 6 of Schedule 2 of the 2017 Act inserts, with respect to 'a subsisting agreement', in place of paragraph 29(2) to (4) of the New Code the following:

> (2) Part 5 of the new code (termination and modification of agreements) does not apply to a subsisting agreement that is a lease of land in England and Wales, if –
>
> (a) it is a lease to which Part 2 of the Landlord and Tenant Act 1954 applies, and
>
> (b) there is no agreement under section 38A of that Act (agreements to exclude provisions of Part 2) in relation the tenancy.
>
> (3) Part 5 of the new code does not apply to a subsisting agreement that is a lease of land in England and Wales, if –

15 See also Chapter 25 generally.
16 As to this expression, see Chapter 25, section 25.2.

(a) the primary purpose of the lease is not to grant code rights (the rights referred to in paragraph 3 of this Schedule), and
(b) there is an agreement under section 38A of the 1954 Act in relation the tenancy.

31.4.4

These provisions are similar but not identical to those contained in paragraphs 29(2) and (3) of the New Code.[17] In particular the following are to be noted.

(1) Albeit one could describe the primary purpose of a subsisting agreement as one to confer code rights as defined in the New Code, the subsisting agreement is, nevertheless, if the lease is one falling within the terms of the 1954 Act, one still governed by the 1954 Act: paragraph 6(2) to Schedule 2 of the 2017 Act.
(2) If the subsisting agreement is one falling within the terms of the 1954 Act and there is no exclusion agreement of the security of tenure provisions pursuant s.38A of the 1954 Act, the 1954 Act governs termination of the subsisting agreement; Part 5[18] of the New Code does not apply. In these circumstances termination of the subsisting agreement will be in accordance with the 1954 Act, not Part 5 of the New Code. It matters not whether the code agreement is or is not a primary purposes agreement.
(3) Albeit termination of a lease protected by the 1954 Act would have to be in accordance with the 1954 Act there is nothing in the transitional provisions of Schedule 2 to the 2017 Act to disapply the application of Part 6 of the New Code, dealing with the removal of the electronic communications apparatus from the land forming the subject matter of the subsisting agreement. Thus the procedure in Part 6 will still have to be adhered to in order to effect removal of the apparatus once the code agreement has been terminated in accordance with either Part 5 or the 1954 Act.
(4) If the subsisting agreement is excluded from the security of tenure provisions of the 1954 Act by reason of there being in existence an agreement under section 38A of the 1954 Act, Part 5 of the New Code does not apply only in circumstances where the primary purpose of the subsisting agreement *is not to grant* code rights: paragraph 6(3) of Schedule 2 to the 2017 Act. Thus, it is only where there is an exclusion agreement under s.38A of the 1954 Act that one is concerned with the primary purpose/non-primary purpose distinction. However, that distinction is simply for the purposes of determining whether or not the termination provisions of Part 5 apply to the subsisting agreement. Part 6 will continue to apply with respect to removal of the electronic communications apparatus from the land forming the subject matter of the subsisting agreement.
(5) It follows that where the subsisting agreement is one excluded from the 1954 Act by reason of there being in existence an agreement under s.38A of the 1954 Act, if the subsisting agreement is one the primary purpose of which is to grant code rights, Part 5 of the New Code will apply to termination of the subsisting agreement as well as Part 6 of the New Code with respect to removal of the apparatus.

17 Paragraph 29(4) applies to Northern Ireland, where the business tenancy statutory regime is governed by different legislation.

18 As to Part 5, see Chapter 22.

31.5 'Subsisting agreements': non-application of assignment and sharing provisions of New Code

31.5.1

Part 3[19] of the New Code confers upon an operator under New Code agreements very favourable anti-avoidance provisions[20] with respect to any term of a New Code agreement which seeks to restrict its ability to assign the code agreement, its ability to upgrade the electronic communications apparatus to which the agreement relates and to share the use of any such apparatus.[21]

31.5.2

By the terms of paragraph 5(1) of Schedule 2 to the 2017 Act, Part 3 does not apply to a subsisting agreement.

31.5.3

By paragraph 5(2) of Schedule 2, Part 3 of the New Code also does not apply in relation to a code right conferred under the New Code if, at the time when it is conferred, the exercise of the right depends on a right that has effect under a subsisting agreement. It is unclear as to the circumstances in which this paragraph will apply.

31.6 'Subsisting agreements': termination and adjustment of 18-month notice period

31.6.1

The New Code makes extensive provision as to the termination of the code agreement: Part 5[22] to the New Code.

31.6.2

The New Code introduces a minimum period of 18 months for termination of the code agreement: paragraph 31(3)(a) of the New Code. However, it is provided by paragraph 7(3) that:

> where the unexpired term of the subsisting agreement at the coming into force of the new code is less than 18 months, paragraph 31 applies (with necessary modification) as if the period of 18 months referred to in subparagraph (3)(a) there were

19 Contained in Part 3 (paragraphs 15–18 inclusive) to the New Code. Anything which was a 'relevant agreement' for the purposes of paragraph 29 of the Old Code is to be treated, after the coming into force of the New Code, as a relevant agreement for the purposes of paragraph 18 of the New Code: paragraph 22 of Schedule 2 to the 2017 Act.

20 Rendering the same void.

21 These provisions are discussed in Chapter 34.

22 See Chapter 22 generally.

substituted a period equal to the unexpired term of three months, whichever is greater.

31.6.3

To illustrate the operation of this provision:

(1) *Example 1*

O (the operator) has a fixed term subsisting agreement which, at the date of the coming into force of the New Code, had six months remaining unexpired. For the purposes of terminating that subsisting agreement in accordance with Part 5 (assuming that the 1954 Act does not apply to it – see section 31.4 above), in substitution for the period of 18 months, the relevant notice period is one of six months.

(2) *Example 2*

O has a fixed term subsisting agreement which, at the date of the coming into force of the New Code, had 24 months remaining unexpired. For the purposes of terminating that subsisting agreement in accordance with Part 5 (assuming that the 1954 Act does not apply to it – see section 31.4 above) the relevant notice period is one of 18 months, because at the date of the coming into force of the New Code the unexpired residue was not 'less than 18 months'.

(3) *Example 3*

O had a fixed term subsisting agreement which, at the date of the coming into force of the New Code, had already expired. For the purposes of terminating that subsisting agreement[23] in accordance with Part 5 (assuming that the 1954 Act does not apply to it – see paragraph 31.4 above), the relevant notice period is one of three months.

31.7 'Subsisting agreements': termination by 'site provider'

31.7.1

The provision for termination is conferred upon the person described as the 'site provider': paragraph 30(1)[24] and 31[25] of the New Code. The right to bring a code agreement to an end by giving notice is conferred upon the 'site provider': paragraph 31(1) of the New Code. Notice is given in accordance with paragraph 31 of the New Code 'to the operator who is a party to the agreement'.[26] The 'site provider' includes the

23 Assuming such an agreement can be described as a 'subsisting agreement': see paragraph 31.2.3.
24 Which contains the definition of 'site provider'.
25 Which deals with termination by the 'site provider'.
26 Note the extended reference as to who is a party to the agreement, by the terms of paragraph 10 of the New Code. See section 16.6 of Chapter 16.

person who confers the code right or a person upon whom the code right is otherwise binding: paragraph 30(1).

31.7.2

However, although the 'site provider' includes a person upon whom the code right is binding, albeit that person may not have the conferred code right, the transitional provisions make it clear that the 'site provider' does not include a person who is bound only by reason of another person who conferred it, or who has agreed in writing to be bound by it, being in occupation and thus bound pursuant to paragraph 2(3) of the Old Code: paragraph 7(2) of Schedule 2 to the 2017 Act.

31.8 'Subsisting agreements': factors to consider in determining order to be made on application in response to termination notice

31.8.1

In respect of any Code Agreement under the New Code, any site provider may implement the procedure for termination. Once the site provider has done so, it is then incumbent upon the operator to make application to the court to protect its code rights: paragraph 32 of the New Code. Where the Court considers such an application,[27] the Court, in determining the order to be made, must have regard to all the circumstances of the case, and in particular to the matters specified in paragraph 34(13) including '(d) the amount of consideration payable by the operator to the site provider under the existing code agreement'.

31.8.2

Where Part 5 applies to a subsisting agreement, the transitional provisions provide that sub-paragraph 13(d) shall not apply in determining the order to be made on the application. The reason, presumably, for this is that the consideration under the code agreement, being a subsisting agreement, will have been assessed by reference to criteria very different to that which one now finds in paragraph 24 of the New Code.[28]

31.9 Paragraph 5 and 6 notices

31.9.1

Paragraph 5 of the Old Code[29] conferred upon an operator the right to seek by way of notice on the relevant person,[30] an agreement conferring code rights and, in the absence of the required agreement, the right to apply to the court to confer the proposed right dispensing with the need for the agreement of the person to whom notice was given.

27 Or an application for modification of a code agreement requiring a change to the terms of the agreement which has expired: paragraph 33 of the New Code.

28 As to which see Chapter 30.

29 See Chapter 7 and, in particular, section 7.6. See also Chapter 32, which deals with the form of paragraph 5 notice.

30 The terms of paragraph 5 enabled the operator to require 'any person to agree for the purposes of paragraph 2 or 3 above that any right should be conferred on the operator, or that any right should bind that person or any interest in land'.

31.9.2

Where a paragraph 5 notice has been served at a time when the New Code came into force, but no application had been made to the court in relation to the notice, the notice takes effect as if given in accordance with paragraph 20(2) of the New Code (which is the corresponding provision of the New Code with respect to service of a notice in accordance with paragraph 5 of the Old Code): paragraph 11 of Schedule 2 to the 2017 Act.

31.9.3

Where a paragraph 5 notice has been served and an application has been made to the court with respect to the notice prior to the New Code coming into force, the Old Code continues to apply in relation to the court application save that any order made by the court has effect as an order under paragraph 20 of the New Code: paragraph 12 (3) of Schedule 2 to the 2017 Act.

31.9.4

Paragraph 5 enabled notice to be served not only with respect to the acquisition of a new code right but, furthermore, enables notice to be served with respect to electronic communications apparatus already installed: paragraph 6 of the Old Code. It is provided that the coming into force of the New Code does not affect any application or order made under paragraph 6 of the Old Code: paragraph 13 of Schedule 2 of the 2017 Act.

31.10 Special regimes

31.10.1

The Old Code contained provisions in paragraphs 9 to 13 for what, Lewison J referred to in *The Bridgewater Canal Co. Ltd v Geo Networks Ltd*[31] as 'Special Regimes' in respect of street works (paragraph 9 of the Old Code), a power to fly lines (paragraph 10 of the Old Code),[32] tidal waters (paragraph 11 of the Old Code) and linear obstacles (paragraph 12–14 of the Old Code).[33] The New Code has corresponding provisions.[34]

31 [2010] 1 WLR 2576, reversed [2011] 1 WLR 1487, CA.

32 The additional provisions relating to the right to object to overhead apparatus on the obligation to fix notices with respect to overhead apparatus, contained in paragraphs 17 and 18 respectively of the Old Code, are preserved by paragraph 15 of Schedule 2 to the 2017 Act 'in relation to anything whose installation was completed before the' repeal of the Old Code. The rights under the New Code to object to the installation of apparatus, contained in Part 12, do not apply in relation apparatus whose installation was completed before the New Code came into force: paragraph 17 of Schedule 2 to the 2017 Act. The special regimes are dealt with in Chapters 26 to 29 of this book.

33 These special regimes being expressly excluded from the general regime: paragraph 2(9) of the Old Code.

34 Part 8 deals with street works, see Chapter 27; Part 9 deals with tidal waters, see also Chapter 27; Part 11 deals with overhead apparatus, see Chapter 28.

31.10.2

The transitional provisions provide that 'paragraphs 9 to 14 *of the existing code … continue to apply in relation to anything in the process of being done when the new code comes into force*': paragraph 8(1) of Schedule 2 to the 2017 Act.[35]

31.10.3

Apparatus lawfully installed under any of those provisions, namely paragraphs 9 to 14 of the Old Code (before or after the time when the New Code came into force) is to be treated as installed under the corresponding provision of the New Code if it could have been installed under that provision if the provision had been in force or applied to its installation: paragraph 8(2) of Schedule 2 to the 2017 Act. The corresponding provisions are set out in paragraph 8(3) of Schedule 2 to the 2017 Act.[36]

31.11 Paragraph 20 of the Old Code: alteration of apparatus

31.11.1

The transitional provisions make provision for continuing the rights of a landowner to rely upon paragraph 20 of the Old Code to seek to alter apparatus installed on under or over land: paragraph 16 of Schedule 2 to the 2017 Act. This is discussed in detail in Chapter 10.

31.11.2

One matter which is unclear is whether in relying upon paragraph 20 of the Old Code, proceedings are to be instituted in the County Court (as was the position before the New Code came into force) or whether proceedings are to be in the Tribunals as directed by the Electronic Communications Code (Jurisdiction) Regulations 2017 (SI/1284). One possibility is that as a 'subsisting agreement' is treated as a paragraph 2 New Code agreement (see paragraph 2 of Schedule 2 to the 2017 Act) the new Regulations apply so as to ensure that Old Code procedures come within the remit of the Tribunals. Regulation 3 provides that 'the functions conferred by the code on the court are also exercisable by the following tribunals …'However, the terms of Regulations 2[37] and 4[38] of the Electronic Comunications Code (Jurisdiction) Regs 2017 (No. 1284), appear to enable Old Code paragraph 20 claims to be issued in the county Court. A paragraph 20 Old Code claim is not part of the 'electronic communications code set out in Schedule

35 Transitional provisions are also provided for in relation to undertaking work within a relevant conduit, e.g. a sewers, dealt with under paragraph 15 of the Old Code, and now to be found in paragraph 102 of the New Code: paragraph 10 of Schedule 2 to the 2017 Act.

36 Express reference is made to an agreement required by the Old Code to be given in respect of any Crown interest: see paragraph 9 of Schedule 2 to the 2017 Act.

37 '2. – (1) In these Regulations – "the code" means the electronic communications code set out in Schedule 3A to the Communications Act 2003; "relevant proceedings" means proceedings under any of the following provisions of the code – (a) Parts 4, 5, 6, 12 or 13, or (b) paragraph 53.'

38 '4. Relevant proceedings must be commenced – (a) in relation to England and Wales, in the Upper Tribunal, or … '

3A to the Communications Act 2003' within Regulation 2 nor is such a claim a claim for termination within Part 5 or removal within Part 6 of the New Code and thus not 'relevant proceedings' which 'must' be issued in the Upper Tribunal.

31.12 Paragraph 21 of the Old Code: removal of apparatus

31.12.1

The transitional provisions make provision for continuing the rights of a landowner to rely upon paragraph 21 of the Old Code to enforce the removal of apparatus on, under or over land: Paragraph 20 of Schedule 2 to the 2017 Act. This is discussed in detail in Chapter 11.

31.12.2

As to the appropriate court in which to bring paragraph 21 Old Code proceedings, see paragraph 31.11.2 above.

31.13 Tree lopping

31.13.1

The Old Code contains provisions enabling an operator with respect to any overhanging tree which obstructs or interferes with the working of any electronic communications apparatus or will obstruct or interfere with it, where it is about to be installed, to serve notice requiring the overhanging tree to be lopped so as to prevent the obstruction or interference: paragraph 19 of the Old Code.

31.13.2

The Old Code makes provision for service of a counter-notice by the person upon whom notice has been given by the operator and if such counter-notice has been served, the notice shall only take effect if confirmed by order of the Court: paragraph 19(2) of the Old Code.

31.13.3

The New Code deals with tree lopping under Part 13.[39] Where a notice pursuant to paragraph 19 of the Old Code has been served prior to the New Code coming into force, but no application has been made to the court in relation to that notice, the notice served by the operator and any counter-notice shall have effect as if given under paragraph 82 of Part 13 of the New Code: paragraph 18 of Schedule 2 to the 2017 Act.

31.13.4

If a notice has been served in accordance with paragraph 19 and an application has been made to the court in relation to it, the terms of the existing code continue to apply in relation to the application: paragraph 19 of Schedule 2 to the 2017 Act.

39 See Chapter 29.

31.14 Undertaker's works

31.14.1

The terms of paragraph 23 of the Old Code[40] make detailed provision with respect to where a relevant undertaker is proposing to execute any undertaker's works which involve or are likely to involve a temporary or permanent alteration of any electronic communications apparatus kept installed on, under or over any land for the purposes of the operator's network.

31.14.2

Paragraph 21 of Schedule 2 to the 2017 Act provides that the repeal of the Old Code does not affect the operation of paragraph 23 in relation to a notice given before the Old Code was repealed or in relation works which have otherwise been commenced before that time.

31.15 Compensation

31.15.1

The repeal of the Old Code does not affect paragraph 16 (compensation for injurious affection to neighbouring land) or any other right compensation with respect to the exercise of a right prior to the coming into force of the New Code: paragraph 14 of Schedule 2 to the 2017 Act.

31.15.2

A person entitled to compensation by virtue of the terms of Schedule 2 to the 2017 Act is not entitled compensation respect of the same matter under any provision of the New Code: paragraph 25 of Schedule 2 to the 2017 Act.

31.16 Notices

31.16.1

Part 15[41] of the New Code make detail provision as to the form of notice to be given by operators, by persons other than operators and the procedures for the giving of notice.

31.16.2

It is provided by paragraph 23 of Schedule 2 to the 2017 Act that part 15 of the New Code 'applies in relation to notices under this Schedule [i.e. Schedule 2] as it applies in relation to notices under that code'.

40 See Chapter 27.

41 As to notices, see Chapter 32 generally.

32 Notices under the codes

32.1 Introduction

32.1.1

Both the Old Code and the New Code contain provision for notice to be served by one party on another, sometimes in response to an initiating notice. The provisions of each code closely control the content of each such notice, the timing of their delivery and the mode of their delivery. It is obviously critical to landowners, operators and their advisers to ensure that the notice in the correct form is timeously given and appropriately served. This chapter examines the applicable criteria. The notice provisions under each code are not dissimilar, and raise similar issues.

32.1.2

Sections 32.2 and 32.3 below set out some general provisions of application to the consideration of form, content and service of notice under both codes.

32.1.3

Specific provisions for the form and service of notices under the Old Code are set out in paragraph 24 of the Old Code. These are examined in section 32.4 below.

32.1.4

The provisions concerning notice in the New Code are contained in Part 15 of that code, and are considered in section 32.5 below.

32.2 General principles concerning the form and content of notices

32.2.1

In very general terms, notices served pursuant to contractual provisions may be valid, even where they contain mistakes. To take an egregious example, in *TBAC Investments Ltd v Valmar Works Ltd*,[1] a notice to complete a contract for the sale of land was upheld

1 [2015] EWHC 1213 (Ch).

despite the fact that (a) it was given not by the seller or its original receivers, but by one original receiver and one replacement receiver; and (b) it contained several errors, including being wrongly dated and miscalculating the date by which completion was required. The test that the courts apply is whether a reasonable recipient of the notice would have been left in any doubt as to the purpose or effect of the notice.[2]

32.2.2

This relatively relaxed approach is in part due to the fact that it is relatively uncommon for contracts to specify what should be the *contents* of the notice; a typical notice provision will stipulate that in order to achieve a particular task, so many months' notice should be served upon the counterparty pursuant to that provision. It is usually superfluous for the notice provision to dictate that the notice should say anything in particular beyond those bare requirements.

32.2.3

This is exemplified by the approach of the House of Lords in *Mannai Ltd v Eagle Star Insurance Co. Ltd.*[3] In that case, a contractual break clause allowed a tenant to determine two leases of office premises by giving not less than six months' notice expiring on the third anniversary of the term commencement date (which had been 13 January 1992). The tenant gave notices to determine the leases on 12 January 1995. The Court of Appeal held that the notices were ineffective because they purported to expire one day early, but the House of Lords (by a majority of three to two) allowed the tenant's appeal and decided that the leases had been validly terminated.

32.2.4

The issue therefore concerned the validity of the notices, measured against the contractual requirements in the leases. Lord Steyn, Lord Hoffmann and Lord Clyde all approved the test for the validity of a notice which had been stated by Goulding J in *Carradine Properties Ltd v Aslam*: 'Is the notice quite clear to a reasonable tenant reading it? Is it plain that he cannot be misled by it?' Lord Steyn said 767D:

> The question is not how the [recipient] landlord understood the notices. The construction of the notices must be approached objectively. The issue is how a reasonable recipient would have understood the notices. And in considering this question the notices must be construed taking into account the relevant objective contextual scene.

Lord Clyde said of that test:

> The standard of reference is that of the reasonable man exercising his common sense in the context and in the circumstances of the particular case. It is not an absolute clarity or an absolute absence of any possible ambiguity which is desiderated. To demand a perfect precision in matters which are not within the formal requirements

2 See *Carradine Properties Ltd v Aslam* [1976] 1 WLR 442.
3 [1997] AC 749.

> of the relevant power would in my view impose an unduly high standard in the framing of notices such as those in issue here. While careless drafting is certainly to be discouraged the evident intention of a notice should not in matters of this kind be rejected in preference for a technical precision.

32.2.5

Lord Clyde's conclusion that 'the formal requirements of the relevant power' did not require technical precision reflected the views of Lord Steyn and Lord Hoffmann, both of whom pointed out that the particular contractual right to determine under consideration did not prescribe any 'indispensable condition for its effective exercise that the notice must contain specific information' or 'any particular form of words'.

32.2.6

As the test is an objective one, it matters not how the actual recipient actually treated the notice. The actual state of mind and understanding of the actual recipient is neither here nor there. *Lancrest Ltd v Asiwaju*;[4] *Lay v Ackerman*[5] at [80] per Neuberger LJ.

32.2.7

Where the notice does contain a mistake the notice may be construed in conjunction with the letter under cover of which it was served to validate the notice: *Stidolph v American School in London Education Trust Ltd*.[6]

32.2.8

If the form or content of the notice is in fact prescribed by the contractual provision, then a failure to comply with the requirement may well invalidate the notice. As Lord Hoffmann colourfully said in *Mannai*:

> If the clause had said that the notice had to be on blue paper, it would have been no good serving a notice on pink paper, however clear it might have been that the tenant wanted to terminate the lease.

4 [2005] 1 EGLR 40, CA.
5 [2005] 1 EGLR 139.
6 Although that case was concerned with the provisions of s.25 of Part II of the Landlord and Tenant Act 1954 Act and not a contractual provision, the general principle of construing the notice in conjunction with a covering letter holds good in the context of contractual documents. Lord Denning MR said:

> Any defect in the prescribed form can be made good by the covering letter or the stamped, addressed envelope. They can and should be read together. So long as the envelope contains the information which the Act requires, and is sufficiently authenticated, the notice is a good notice. The requirement of a notice should not be turned into a trap for the landlord. The documents must, of course, be served on the tenant. (p.805)

It is considered that the reference to 'envelope' in this passage was intended to be reference to 'the covering letter'. [1969] 20 P & CR 802, CA.

32.2.9

The approach to the validity of statutory notices (i.e. notices required by a statute to be given in a prescribed form) may however involve different considerations.[7]

32.2.10

First, in contrast to contractual provisions, it is common for statutory notice provisions to detail the content of the notices – and sometimes the very form of the notice. In those circumstances, the question that often arises is whether the notice contains the prescribed information and/or is in the correct form.

32.2.11

The proper approach in such circumstances is first to identify what information the statute requires the notice to contain in order to have the intended consequences; and secondly to ask whether the notice complies with those requirements. The first question is a matter of construction of the statute, and involves an assessment whether Parliament intended that the omission of a particular item of information from the notice would render it invalid.[8]

32.2.12

Secondly, the prescribing statute or statutory instrument will often contain a saving provision, using words to the effect that a notice must be in a particular form 'or substantially to the like effect'.[9] Such drafting may save an arguably deficient notice.

32.2.13

Thirdly, it may make a difference whether the prescription as to the form and/or content of a statutory notice is dictated by primary or secondary legislation. In considering the

7 The *Mannai* approach has, however, been applied in the following cases relating to statutory notices: *York v Casey* [1998] 2 EGLR 25 CA; *John Lyon's Free Grammar School v Secchi* [2000] 32 HLR 820. CA; *Burman v Mount Cook Ltd* [2002] Ch. 256 CA; *Ravenseft Properties Ltd v Hall* [2002] HLR 33 CA; *McDonald v Fernandez* [2003] 4 All ER 1033 CA; *Lay v Ackerman* [2004] L & TR 29; [2005] 1 EGLR 139 CA

8 See paragraph 44 of the judgment of the Upper Tribunal (Lands Chamber) in *St Stephens Mansions RTM Company Ltd v Fairhold NW Ltd* [2014] UKUT 0541 (LC). Neuberger MR applied a not dissimilar approach to all notices, whether contractual, statutory or at common law, in *Lay v Ackerman* [2004] L & TR 29, CA (concerning a statutory notice under the Leasehold Reform, Housing and Urban Development Act 1993) when he said:

> The correct approach on the basis of the decision and reasoning in *Mannai* is as follows. One must first consider whether there was a mistake in the information contained in the notice (as there was as to the date in *Mannai*, and there was as to the landlord, in the present case). If there was such a mistake, one must then consider how, in the light of the mistake, a reasonable person in the position of the recipient would have understood the notice in the circumstances of the particular case. Finally one must consider whether, as a result, the notice would have been understood as conveying the information required by the contractual, statutory or common law provision pursuant to which it was served. (At [40].)

9 See e.g. regs 3 and 4 of the Landlord and Tenant Act 1954, Part II (Assured Tenancies) (Notices) Regulations 1986/2181.

proper construction of a statutory provision governing the content of a notice, the fact that particulars are prescribed by regulation rather than by the statute itself has sometimes been taken as an indication that they are not intended to be crucial to the validity of the notice.[10]

32.2.14

Fourthly, the courts have exhibited a tendency to categorise statutory notice requirements as either matters of substance (which should therefore be followed faithfully) or mere machinery (which need not necessarily be reflected in the notice itself). This distinction was made in this way by Hobhouse LJ in *Belvedere Court Management Ltd v Frogmore Developments Ltd*:[11]

> I am strongly attracted to the view that legislation of the present kind should be evaluated and construed on an analytical basis. It should be considered which of the provisions are substantive and which are secondary, that is, simply part of the machinery of the legislation. Further the provisions which fall into the latter category should be examined to assess whether they are essential parts of the mechanics or are merely supportive of the other provisions so that they need not be insisted on regardless of the circumstances. In other words, as in the construction of contractual and similar documents, the status and effect of a provision has to be assessed having regard to the scheme of the legislation as a whole and the role of that provision in that scheme – for example, whether some provision confers an option properly so called, whether some provision is equivalent to a condition precedent, whether some requirement can be fulfilled in some other way or waived. Such an approach when applied to legislation such as the present would assist to enable the substantive rights to be given effect to and would help to avoid absurdities or unjustified lacunae.

32.2.15

Carnwath LJ (with whom Potter LJ agreed) adopted this approach in *Tudor v M25 Group Ltd*:[12]

> In my view, the judge was entirely correct in upholding the validity of the notice, and I would have been content to adopt his reasons. The same result may be arrived at by applying the analytical approach advocated by Hobhouse LJ in *Belvedere Court Management Ltd v Frogmore Developments Ltd.*[13] That involves considering which of the provisions are substantive and which are secondary or 'machinery'; and in relation to the latter, considering whether they are 'essential parts of the mechanics or merely supportive of the other provisions'. Here the substantive provisions are those conferring the right to acquire the freehold. The secondary (machinery) provisions include the notice requirements of section 11A itself, and the formal requirements of section 54, including the requirement for the addresses. The requirement for a notice

10 See e.g. the decision of the *Court of Appeal in Seven Strathray Gardens v Pointstar Shipping & Finance Ltd* [2004] EWCA Civ 1669.

11 [1997] QB 858.

12 [2004] 1 WLR 2319.

13 [1997] QB 858.

> is essential machinery, as no doubt is the requirement to indicate who is giving the notice. However, in agreement with the judge, I would hold that the requirement to state addresses in the notice is 'merely supportive'; and that accordingly a failure in this respect does not invalidate the notice.

32.2.16

To the same effect, in *Newbold v The Coal Authority*,[14] Sir Stanley Burnton said:

> In all cases, one must first construe the statutory or contractual requirement in question. It may require strict compliance with a requirement as a condition of its validity. … Against that, on its true construction a statutory requirement may be satisfied by what is referred to as adequate compliance. Finally, it may be that even non-compliance with a requirement is not fatal. In all such cases, it is necessary to consider the words of the statute or contract, in the light of its subject matter, the background, the purpose of the requirement, if that is known or determined, and the actual or possible effect of non-compliance on the parties. We assume that Parliament in the case of legislation, and the parties in the case of a contractual requirement, would have intended a sensible, and in the case of a contract, commercial result.[15]

32.3 General principles concerning the service of notices

32.3.1

Just as requirements as to the form and content of notices have provoked much litigation, so too have there been many disputes concerning contractual and statutory provisions prescribing service requirements for notices.

14 [2014] 1 WLR 1288.

15 That the courts have moved away from a simple mandatory/directory dichotomy can be seen from *Sumukan Ltd v Commonwealth Secretariat* [2007] EWCA Civ 1148, CA where Sedly LJ said:

> [43] In *Project Blue Sky Inc v Australian Broadcasting Authority* [1998] 194 CLR 355 the High Court of Australia abandoned this dichotomy. A majority held (at 390–391 (para 93)):
>
> '… The classification is the end of the inquiry, not the beginning. That being so, a court, determining the validity of an act done in breach of a statutory provision, may easily focus on the wrong factors if it asks itself whether compliance with the provision is mandatory or directory and, if directory, whether there has been substantial compliance with the provision. A better test for determining the issue of validity is to ask whether it was a purpose of the legislation that an act done in breach of the provision should be invalid. This has been the preferred approach of courts in this country in recent years, particularly in New South Wales. In determining the question of purpose, regard must be had to "the language of the relevant provision and the scope and object of the whole statute".'
>
> [44] It is true that the Blue Sky formula is also to an extent question-begging. Legislative purpose may be a mare's nest where it is probable that the legislature never considered the issue at all. Any such purpose has to be imputed to the legislature by reference to what the court judges to be the significance of the requirement. It may be, therefore, that whatever test is adopted – even a simple vernacular test such as whether the departure matters – it will return to two of the key questions canvassed under the old law: does the decision concern a procedural safeguard for persons affected by the scheme, and does it affect private rights (see Wade and Forsyth)?

See also *R v Soneji* [2006] 1 AC 340 at [14] and [15] per Lord Steyn, where it was said that the test was a more flexible approach of focusing on the consequences of non-compliance, and posing the question, taking into account those consequences, whether Parliament intended the outcome to be total invalidity.

32.3.2

The law as to service of notices generally was summarised thus by Lord Salmon in *Sun Alliance and London Assurance Co. Ltd v Hayman*:[16]

> According to the ordinary and natural use of English words giving a notice means causing a notice to be received. Therefore any requirement in a statute or contract for the giving of a notice can be complied with only by causing the notice to be actually received – unless the context or some statutory or contractual provision otherwise provides.

32.3.3

In all cases involving the service of notices, it is necessary to examine the question whether there are any contractual or statutory stipulations which must be observed in relation to mode of service. It will also be necessary to consider what happens if the party upon whom the notice is to be served cannot be located, because of death, insolvency and so forth. There are therefore a number of aspects of service which may need to be considered in any given case:

(a) What does the contract or statutory provision or regulation specify as to mode of service?
(b) Where is the notice to be served?
(c) Upon whom is the notice to be served?
(d) By whom is the notice to be given?
(e) What is the position if either the giver or recipient of the notice is unavailable?
(f) Are there any applicable statutory provisions imposing further constraints or offering assistance?

This section considers each of those questions in turn.

32.3.4 Mode of service

Whether proper service has been effected in accordance with contractual provisions depends upon the proper interpretation of those provisions and their application to the particular facts of the case. The contractual or statutory provision may specify many different modes of service: first class post, recorded delivery, hand delivery and so forth. Special attention needs to be paid to the time at which delivery is effected in accordance with these different modes. In *Wilderbrook Ltd v Olowu*,[17] the Court of Appeal cited the quotation above from (in paragraph 32.3.2) *Sun Alliance* with approval, and rejected an attempt by a tenant to contend that a notice that had arrived at the demised premises by recorded delivery, and been signed for, but that had not come into the tenant's hands until a later date, had not been 'received' by the tenant until that later date, within the meaning of the provisions in the lease. Further, in *Warborough Investments Ltd v Central Midland Estates Ltd*,[18] the judge felt bound by authority to reject the submission that there was a meaningful distinction between 'receipt' and 'service' of a notice.

16 [1975] 1 WLR 177.
17 [2006] 2 P & CR 4.
18 [2006] L & TR 10.

32.3.5

As with the form and content of notices, considered in section 32.2 above, where a clause prescribes a particular mode of service, the question often arises whether that prescription is mandatory (in which case service in some other fashion will not suffice); or directory (in which case service by some other fashion will be sufficient).

32.3.6

Examples of contractual service stipulations being construed as directory rather than mandatory are provided by the following decisions. First, in *Yates Building Co. Ltd v RJ Pulleyn & Son (York) Ltd*,[19] the Court of Appeal rejected the submission that an option notice sent by ordinary post, and arriving within the specified time limits, was nevertheless invalid, because of the wording of the option provision: 'such notice to be sent by registered or recorded delivery post'. Secondly, in *Midland Oak Construction Ltd v BBA Group Ltd* (an unreported decision given on 15 February 1983), the Court of Appeal held that a clause stipulating that a notice was to be posted by recorded delivery post was directory only, so that a notice could be validly served by hand.[20]

32.3.7 Place of delivery

Again, there is much variation to be found in contractual and statutory provisions: these may direct that the notice should be served at certain premises; at the last known place of abode of the recipient; or at its registered office. Again, care should be taken to ascertain which location is specified; and whether the stipulation is mandatory or directory only. For example, in the *Midland Oak* case, the Court of Appeal held that the further requirement in the clause that the notice should be addressed and sent to the tenant at the tenant's registered office was a mandatory requirement.[21]

32.3.8 Identity of recipient

The contract may expressly provide that a notice *must* be served on a specified agent, or (much more usually) that it *may* be served on a specified agent. In the absence of such an express provision, analysis of the contractual position will sometimes involve a consideration of the principles of agency. Unless the lease otherwise provides, notices should always be addressed to and served on the other party rather than on someone acting for him: solicitors, surveyors and others acting for a party as general agents will not normally have implied or ostensible authority to accept service of notices on his behalf.[22]

19 [1976] 1 EGLR 157.

20 The lease contained the following clause providing for notice to review the rent:

... it shall be such annual sum as shall be:

(a) specified by notice in writing signed by or on behalf of the lessor and posted by recorded delivery post in a pre-paid envelope addressed to the Tenant at its Registered Office at any time before [September 29, 1980] and such notice shall be deemed (unless the contrary shall be proved) to have been received by the Tenant in due course of post.

21 The provision that it be sent to the tenant's registered office, although mandatory, was overridden by the express incorporation of s.196 of the Law of Property Act 1925.

22 See most recently the decision of Popplewell J in *Glencore Agriculture BV v Conqueror Holdings Ltd* [2017] EWHC 2893 (Comm). Although a decision under the Arbitration Act 1996, the judgment analyses in depth the principles applicable to service upon an agent.

In the *Midland Oak* case (above) a notice addressed to a subsidiary company of the tenant was held to have been invalidly served, there being no sufficient evidence that the subsidiary had been authorised or held out as being authorised by the tenant to accept service of the notice on its behalf.[23] Of course, there may be circumstances from which it may be inferred that a general agent does have authority to accept notices on behalf of his principal. See, for example, *Townsends Carriers Ltd v Pfizer*[24] and *Peel Developments (South) Ltd v Siemens Plc*,[25] in which option notices were held to have been validly served upon the landlord where addressed to and served upon associated companies of the landlord. In those cases, the landlord had conferred on the associated company in question a general authority to manage the property and, upon the facts, it was held that such management included a capacity to receive such notices.[26]

32.3.9

If an agent has been authorised to accept service, then service on the agent will be good service, even if the intended recipient, e.g. the tenant, is able to show that the notice never came to its attention – see *Galinski v McHugh*[27] (notice addressed to tenant but served upon duly authorised solicitors). It is no concern of the sender, in the example landlord, and no matter for inquiry by the court, whether the agent did or did not fail to pass the notice to the intended recipient, the tenant.

32.3.10 Identity of server

In the case of service of a notice *by* an agent, the notice should be expressed to be served by or on behalf of the party to the contract, rather than just being given in the name of the agent; and the notice should be signed either by the party itself, or by the agent with the party's authority – see *Fox & Widley v Guram.*[28] If the name of the agent is stated as being the person giving the notice in the body of the notice itself, it may well be the case that the notice will be invalid, even if the addressee is aware that the server is the general agent of the addressor: see the decision of the Court of Appeal in *Dun & Bradstreet Software Services (England) Ltd v Provident Mutual Life Assurance Association,*[29] where this point was left open.

32.3.11

It will also be necessary to show, in any given case, that the purported agent serving the notice had authority to act as he did. A landlord failed to establish such authority on the facts in *Cordon Bleu Freezer Food Centre Ltd v Marbleace Ltd.*[30]

23 The provision that the notice be addressed to the tenant was mandatory and s.196 of the Law of Property Act 1925 did not assist since the notice under consideration in that case had not simply been addressed to 'the lessee' but to a company, which was not the tenant, by name.

24 [1977] 33 P & CR 36.

25 [1992] 2 EGLR 85.

26 Such general agency will arise where there has been a long course of dealing during which the agent has been treated or regarded as the principal: see in addition to the cases in the text, *Jones v Phipps* [1868] LR 2 QB 567; *Re Knight & Hubbard's Und*erlease [1923] 1 Ch 130; *Harmond Properties v Gajdzis* [1968] 1 WLR 1858; [1968] 3 All ER 263; *Dun & Bradstreet Software Services (England) Ltd v Provident Mutual Life Assurance* [1998] 2 EGLR 175 CA; *Lemmerbell Ltd v Britannia LAS Direct Ltd* [1998] 3 EGLR 67 CA.

27 [1988] 57 P & CR 359.

28 [1998] 1 EGLR 91.

29 [1997] 1 EGLR 57.

30 [1987] 2 EGLR 143.

32.3.12 Where recipient has died

Where a party dies testate, (that is to say, leaving a will providing for the division of his estate), notices should be served on its executors. Where the party dies intestate, then notices should be served upon the Public Trustee – see s.14 of the Law of Property (Miscellaneous Provisions) Act 1994.

32.3.13 Where recipient is insolvent

Where a party has been made bankrupt, its estate, including its leasehold interest, vests in its trustee in bankruptcy. Notices should therefore be served on the trustee in bankruptcy. On a company's insolvency, its assets remain vested in the company unless there is a disclaimer. Until that moment, therefore, notices should be served upon the company via the liquidator as its agent. Upon disclaimer, there will be no entity which can be served, and accordingly the counter-party should consider seeking to procure that the lease or other contract be vested in another party (if there is one).

32.3.14 Statutory provisions as to service of notices

Section 196 of the Law of Property Act 1925 sets out a number of detailed provisions with regard to the service of notices. Subsections (1) to (4) are as follows (subs.(5) being considered in paragraph 32.3.15 below):

(1) Any notice required or authorised to be served or given by this Act shall be in writing.
(2) Any notice required or authorised by this Act to be served on a lessee or mortgagor shall be sufficient, although only addressed to the lessee or mortgagor by that designation, without his name, or generally to the persons interested, without any name, and notwithstanding that any person to be affected by the notice is absent, under disability, unborn, or unascertained.
(3) Any notice required or authorised by this Act to be served shall be sufficiently served if it is left at the last known place of abode or business in the United Kingdom of the lessee, lessor, mortgagee, mortgagor, or other person to be served, or in the case of a notice required or authorised to be served on a lessee or mortgagor, is affixed or left for him on the land or any house or building comprised in the lease or mortgage, or, in the case of a mining lease, is left for the lessee at the office or counting house of the mine.
(4) Any notice required or authorised by this Act to be served shall also be sufficiently served, if it is sent by post in a registered letter addressed to the lessee, lessor, mortgagee, mortgagor, on the other person to be served, by name, at the aforesaid place of abode or business, office, or counting house, and if that letter is not returned through the post office undelivered; and that service shall be deemed to be made at the time at which the registered letter would in the ordinary course be delivered.
(5) The provisions of this section shall extend to notices required to be served by any instrument affecting property executed or coming into operation after the commencement of this Act unless a contrary intention appears.

32.3.15

It will be seen from s.196(5) that the section applies to 'notices required to be served by any instrument affecting property'. Section 196 may be and often is incorporated expressly into leases or other agreements relating to land with respect to the service of notices. Absent any express reference to the application of those provisions in the body of the instrument, the section will apply only where it can be said that the notice is 'required' by the instrument to be served. It has been held that it applies to an express contractual provision for service of notice: *Wandsworth LBC v Attwell*;[31,32] *Enfield LBC v Devonish*[33] which suggests, without deciding, that a contractual break clause is within s.196(5).

32.3.16

By virtue of the Recorded Delivery Service Act 1962, recorded delivery may be used instead of registered post, and s.196 applies equally to notices served by that means.

32.3.17

Section 23 of the Landlord and Tenant Act 1927 also contains provisions dealing with notices. However, this provision does not automatically apply to the code, unlike the 1954 Act, which expressly provides for its application to business tenancy renewals: s.66(4). The Authors have not encountered a lease or other agreement referring expressly to the application of s.23 to the service of notices. Although there is no reason why it should not be incorporated, it is probably safe to be assumed that, in practice, it will have no role to play in code cases.

32.4 Notices under the Old Code

32.4.1

There were three provisions in the Old Code that required action to be initiated by notice to be given by the operator. In summary, these included the following:

(a) a notice requiring agreement to a right proposed to be conferred upon it under paragraph 5;
(b) a notice of intention to execute works on land under paragraph 12 (linear obstacles);
(c) a notice requiring tree lopping under paragraph 19.

31 [1995] 27 HLR 536, CA.

32 The tenant was a periodic tenant. He left the premises to live in the United States. His half brother remained at the premises. The landlord served a notice to quit addressed to the tenant and left it at the premises. It was argued in possession proceedings against the half brother that the notice to quit did not take effect. The landlord sought to rely on s.196. The Court of Appeal held (per Glidewell LJ) that 'a tenancy agreement which makes no express provision for the service of a notice to quit to determine the tenancy does not "require" such a notice to be served. Thus in that situation section 196(5) does not apply.'

33 [1996] 74 P & CR 288 at 294.

32.4.2

In addition to these initiating notices, there were other notices that the operator could give in response to initiating notices given by others. These were called 'counter-notices'. Examples of these included:

(a) counter-notice under paragraph 14, where the person with control of any relevant land had given notice to the operator requiring it to alter apparatus crossing (or affecting) a linear obstacle;
(b) counter-notice under paragraph 20, when a person with an interest in land on which electronic communications apparatus had been installed has given notice requiring its alteration;
(c) counter-notice under paragraph 21, when a person entitled to require the removal of any of the operator's electronic communications apparatus had given notice requiring its removal;
(d) counter-notice under paragraph 23, where a relevant undertaker had given notice requiring the alteration of apparatus.

32.4.3

The Old Code had other provisions dealing with notices that could be given to operators by parties other than an operator. These included:

(a) a notice by the owner of a relevant interest in land claiming compensation under paragraph 4;
(b) a notice by a potential subscriber seeking access to an operator's network under paragraph 8;
(c) a notice by the Secretary of State requiring remedial works under paragraph 11;
(d) a notice by a person with control of relevant land seeking the alteration of apparatus affecting a linear obstacle under paragraph 14;
(e) a notice by a person occupying or claiming an interest in land seeking the alteration of apparatus which had been installed over that land under paragraph 17;
(f) a notice by a person with an interest in land seeking the alteration of apparatus which had been installed over that land under paragraph 20;
(g) a notice by a person entitled require the removal of apparatus seeking such a removal under paragraph 21;
(h) a notice by a relevant undertaker seeking the alteration of apparatus under paragraph 23.

32.4.4

Paragraph 24 of the Old Code also set out important provisions concerning notices under the Old Code. Some of these applied only to notices required to be given by operators, while others applied generally to all notices under the Old Code.

32.4.5

In the former category (notices required to be given by operators), paragraph 24(1) provided:

> Any notice required to be given by the operator to any person for the purposes of any provision of this code must be in a form approved by OFCOM as adequate for indicating to that person the effect of the notice and of so much of this code as is relevant to the notice and to the steps that may be taken by that person under this code in respect of that notice.

32.4.6

Paragraph 24(6) added:

> In any proceedings under this code a certificate issued by OFCOM and stating that a particular form of notice has been approved by them as mentioned in sub-paragraph (1) above shall be conclusive evidence of the matter certified.

32.4.7

Accordingly, all notices given by operators under the Old Code (including not only the initiating notices referred to in paragraph 32.4.1 above, but also the counter-notices under paragraph 32.4.2) were required to be in the OFCOM-approved form. OFCOM had approved model notices and counter-notices for use under the Old Code. Its website[34] directed that operators should fill in their details and those of the occupier of the land in question, together with other details as required in the areas in square brackets and highlighted in bold. Providing that the operator did not make any alteration to the text of a model notice, it stated that it would not be necessary to seek OFCOM's further approval for the form of words of the notice, as this would be taken as already having been given. If, however, an operator wished to alter the text of such a model notice, OFCOM's approval was to be sought in order for the notice to be considered valid.

32.4.8

In addition to this specific and mandatory requirement that applied only to notices to be given by operators, there were two other sets of provisions that applied to the service of all notices under the Old Code.

32.4.9

First, s.394 of the Communications Act 2003 sets out a number of stipulations governing how and upon whom 'notifications' (an expression which includes notices – see s.394(9)) and other documents can be served, and applied to all provisions in the Old Code or the original code in the Telecommunications Act 1984 which authorised or required notifications to be given or documents of any other description to be sent to any person. Section 394(9) also explains that references in the section to 'giving or sending' a notification or other document to a person include references to transmitting it to him and to serving it on him.

34 See www.ofcom.org.uk/phones-telecoms-and-internet/information-for-industry/policy/electronic-comm-code/notices.

32.4.10 Mode of service

S.394(3) provides:[35]

> The notification or document may be given or sent to the person in question –
> (a) by delivering it to him;
> (b) by leaving it at his proper address; or
> (c) by sending it by post to him at that address.

This drafting does not *mandate* service in the way specified: it merely prescribes three methods of service which, if successfully adopted, will be sufficient service. Other methods of service which result in the notice reaching its intended target will therefore also suffice.

32.4.11 Recipient of notice

S.394(4)–(6) deals with the proper person to serve where the intended recipient is other than a natural person, thus:

> (4) The notification or document may be given or sent to a body corporate by being given or sent to the secretary or clerk of that body.
> (5) The notification or document may be given or sent to a firm by being given or sent to –
> (a) a partner in the firm; or
> (b) a person having the control or management of the partnership business.
> (6) The notification or document may be given or sent to an unincorporated body or association by being given or sent to a member of the governing body of the body or association.

32.4.12 Place of service

S.394(7) explains the reference to the 'proper address' of a person in s.394(3), dealing again with the various different types of legal category of potential recipient as follows:

> For the purposes of this section and section 7 of the Interpretation Act 1978 (c. 30) (service of documents by post) in its application to this section, the proper address of a person is –
> (a) in the case of body corporate, the address of the registered or principal office of the body;
> (b) in the case of a firm, unincorporated body or association, the address of the principal office of the partnership, body or association;
> (c) in the case of a person to whom the notification or other document is given or sent in reliance on any of subsections (4) to (6), the proper address of the body corporate, firm or (as the case may be) other body or association in question; and
> (d) in any other case, the last known address of the person in question.

35 See the further discussion on these provisions in paragraph 32.5.6 below.

32.4.13

Section 394(8) adds provision for where the recipient is not based in the United Kingdom, as follows:

> In the case of –
> (a) a company registered outside the United Kingdom,
> (b) a firm carrying on business outside the United Kingdom, or
> (c) an unincorporated body or association with offices outside the United Kingdom,
>
> the references in subsection (7) to its principal office include references to its principal office within the United Kingdom (if any).

32.4.14

Section 394(10) stipulates that s.394 has effect subject to s.395 of the 2003 Act. That section adds the following digital overlay regarding mode of service:

> (1) This section applies where –
> (a) section 394 authorises the giving or sending of a notification or other document by its delivery to a particular person ('the recipient'); and
> (b) the notification or other document is transmitted to the recipient –
> (i) by means of an electronic communications network; or
> (ii) by other means but in a form that nevertheless requires the use of apparatus by the recipient to render it intelligible.
> (2) The transmission has effect for the purposes of the enactments specified in section 394(2) as a delivery of the notification or other document to the recipient, but only if the requirements imposed by or under this section are complied with.
> (3) Where the recipient is OFCOM –
> (a) they must have indicated their willingness to receive the notification or other document in a manner mentioned in subsection (1)(b);
> (b) the transmission must be made in such manner and satisfy such other conditions as they may require; and
> (c) the notification or other document must take such form as they may require.
> (4) Where the person making the transmission is OFCOM, they may (subject to subsection (5)) determine –
> (a) the manner in which the transmission is made; and
> (b) the form in which the notification or other document is transmitted.
> (5) Where the recipient is a person other than OFCOM –
> (a) the recipient, or
> (b) the person on whose behalf the recipient receives the notification or other document,
>
> must have indicated to the person making the transmission the recipient's willingness to receive notifications or documents transmitted in the form and manner used.

(6) An indication to any person for the purposes of subsection (5) –
   (a) must be given to that person in such manner as he may require;
   (b) may be a general indication or one that is limited to notifications or documents of a particular description;
   (c) must state the address to be used and must be accompanied by such other information as that person requires for the making of the transmission; and
   (d) may be modified or withdrawn at any time by a notice given to that person in such manner as he may require.

(7) An indication, requirement or determination given, imposed or made by OFCOM for the purposes of this section is to be given, imposed or made by being published in such manner as they consider appropriate for bringing it to the attention of the persons who, in their opinion, are likely to be affected by it.

(8) Subsection (9) of section 394 applies for the purposes of this section as it applies for the purposes of that section.

32.4.15

Put briefly, at its most obvious, s.395 enables the service of notices by email if, but only if the server of the notice has first cleared that means of communication with the intended recipient.

32.4.16

Section 396 of the 2003 Act gives the Secretary of State power to make regulations under s.395. No such regulations have yet been made.

32.4.17

The second set of provisions that governed the service of notices under the Old Code was found in the following provisions of paragraph 24 of the Old Code, and cut down (and in one case expanded) upon the relatively liberal regime of s.394 of the 2003 Act.

32.4.18 Mode of service

Paragraph 24(2) provides:

> A notice required to be given to any person for the purposes of any provision of this code is not to be sent to him by post unless it is sent by a registered post service or by recorded delivery.

32.4.19 Place of service

Paragraph 24(2A) provides:

> For the purposes, in the case of such a notice, of section 394 of the Communications Act 2003 and the application of section 7 of the Interpretation Act 1978 in relation to that section, the proper address of a person is –

(a) if the person to whom the notice is to be given has furnished the person giving the notice with an address for service under this code, that address; and
(b) only if he has not, the address given by that section of the Act of 2003.

32.4.20 Recipient of notice

Finally, paragraph 24(5) provides:

> If it is not practicable, for the purposes of giving any notice under this code, after reasonable inquiries to ascertain the name and address –
> (a) of the person who is for the purposes of any provision of this code the occupier of any land, or
> (b) of the owner of any interest in any land,
>
> a notice may be given under this code by addressing it to a person by the description of 'occupier' of the land (describing it) or, as the case may be, 'owner' of the interest (describing both the interest and the land) and by delivering it to some person on the land or, if there is no person on the land to whom it can be delivered, by affixing it, or a copy of it, to some conspicuous object on the land.

32.4.21

There are the following additional requirements to note about the notice provisions of the Old Code:

(a) First, many of the provisions spelt out the required contents of the notice (for example, in the case of a notice under paragraph 4, that it should give particulars of the land in respect of which the claim is made). Where the notice was to be given by an operator, that caused no difficulty, because the OFCOM prescribed form supplied the contents. However, in the case of a notice given by another, there was no prescribed form, with the result that there could be arguments as to whether what had been supplied by way of content was sufficient (and in turn possibly raising the issue discussed in section 32.2 above as to whether the provision is directory or mandatory).
(b) Secondly, in the case of notices requiring works (e.g. under paragraphs 11 or 12), the notice had to detail the works with sufficient particularity to enable the recipient to know what it had to do. Again, where the server was OFCOM, there would be no problem, because that would be spelt out in the standard form. However, where the server was another (e.g. the Secretary of State under paragraph 11), there was no standard form, and definitional problems could potentially arise.
(c) Thirdly, many of the provisions contained time limits – the most common example being that the recipient has 28 days to respond by counter-notice if it wishes to object (as in the cases of paragraphs 14, 19, 20 and 21, with just 10 days in the case of a counter-notice under paragraph 23). In other cases, the recipient must apply to court within a stated time to protect its rights if it wishes to object (as in the case of paragraph 17). In yet other cases, the operator may apply to

court after a stated time if it wishes to enforce the right its notice seeks (as in the case of paragraph 5). These are statutory time limits, and there is no leeway.

(d) Fourthly, paragraph 18 obliges operators who have erected electronic communications apparatus, the whole or part of which is at a height of 3 metres or more above the ground, to affix a notice to the apparatus:

> in a position where it is reasonably legible and shall give the name of the operator and an address in the United Kingdom at which any notice of objection may be given under paragraph 17 above in respect of the apparatus in question.

Any operator giving such a notice at that address in respect of that apparatus shall be deemed to have been furnished with that address for the purposes of paragraph 24(2A)(a) (see paragraph 32.4.19 above).

32.4.22

In summary, therefore, the notice provisions under the Old Code are explicit and detailed, particularly where operators are concerned (although even the general provisions that affect all servers are stringent).

32.4.23

The question that may arise in those circumstances is that canvassed in section 32.2 above: what if a notice is served other than in accordance with the requirements of paragraph 24 or s.394, but clearly reaches its target? For example:

(a) a notice is served by first class post; or
(e) a notice is served by email, but without the server validating that form of service with the recipient beforehand.

32.4.24

There is no authority on the point, and the question of validity, which is clearly a testing one, remains unresolved.

32.5 Notices under the New Code: notice by operators

32.5.1

As with the Old Code, there are a number of provisions of the New Code providing for service by operators of notices upon various persons, whether they be described as 'landowner' (as in the case of paragraph 37(1)), 'site provider' (paragraph 30(1)) 'relevant person' (paragraph 20(1)) or 'third party' (paragraph 41(1)). Notice may need to be given either to claim a code right[36] or in order to protect a code right by serving

36 For instance, where the operator requires a person to agree to confer a code right he is required to give the relevant person a notice in writing setting out the right that he seeks and requesting that person's agreement: paragraph 20(1) and (2).

appropriate counter-notice in response to a notice served upon the operator.[37] Notice may also need to be given by an operator under the special regimes.[38]

32.5.2

The detail concerning the form, service and timing of notices under the New Code is not significantly different in principle to the detail provided under the Old Code, considered in section 32.4 above. This section considers the comparable provisions of the New Code under the following headings:

(a) Form of notice;
(b) Service of notice;
(c) Proper address for service;
(d) The date of service.

32.5.3 Form of notice

Paragraph 88 of the New Code makes provision for a form of notice to be prescribed by OFCOM: paragraph 88(2). If OFCOM prescribes a form which may or must be given by an operator under a provision of the New Code, a notice given by an operator under that provision 'must be in that form'. A notice which does not comply with the prescribed form is not a valid notice for the purposes of the New Code: paragraph 88(3). However, if a notice is served by the operator other than in the appropriate form, the recipient may choose to rely on it: paragraph 88(4).

32.5.4

In the absence of any form being prescribed by OFCOM[39] the notice under the New Code by an operator must (paragraph 88(1)):

(a) explain the effect of the notice,
(b) explain which provisions of the code are relevant to the notice, and
(c) explain the steps that may be taken by the recipient in respect of the notice.

32.5.5

In relation to proceedings under the New Code, a certificate issued by OFCOM stating that a particular form of notice has been prescribed by them is conclusive evidence of that fact: paragraph 88(5).

37 A counter-notice is required in response to a termination notice served by a site provider: see paragraph 32 of the New Code. Equally a counter-notice is required to be served with respect to a notice requesting removal of apparatus by a third party: see paragraph 41 of the New Code.

38 For instance, an arbitration notice may be given in the context of the exercise of transport land rights, paragraph 50(3); paragraph 68(2) provides for service of a counter-notice by the operator upon 'the undertaker' (as defined in paragraph 66) where the undertaker proposes to carry out non-emergency undertaker's works which will interfere with the network.

39 Although OFCOM's practice has been to prescribe notices for almost all conceivable occasions: see section 41.7 of Chapter 41 of this book.

32.5.6 *Service of notice*

As with the Old Code, service of notices is governed by s.394 of the Communications Act 2003, but subject this time to paragraph 91 of the New Code.[40] Section 394(3) of the 2003 Act provides that, where notification is required to be given to a person or a document of any other description (including a copy) is to be sent to any person, under, inter alia, the New Code,[41] the 'notification or document' may be given or sent to the person in question by:

(i) delivering it to him;
(ii) leaving[42] it at his proper address; or
(iii) sending it by post to him at that address.[43]

32.5.7

For the purposes of s.394 'notification' includes 'notice' and 'document' includes 'anything in writing': s.394(9).

32.5.8

A notice to be given *by an operator* must not be sent by post unless it is sent by a registered post service or by recorded delivery: paragraph 91(1). This contrasts with the comparable provision in the Old Code, which applied generally to all notices, and not merely those served by operators.

32.5.9

The notification or document may be given or sent to a body corporate by being given or sent to the secretary or clerk of that body: s.394(4) of the 2003 Act.

32.5.10

Section 394(5) provides that the notification or document may be given or sent to a firm by being given or sent to –

(a) a partner in the firm; or
(b) a person having the control or management of the partnership business.

40 See s.394(11) of the 2003 Act, inserted by paragraph 46 of Schedule 3 to the 2017 Act.
41 See s.379(2) of the 2003 Act.
42 A requirement that a notice be 'left at' premises does not prescribe any particular method of leaving: *Warborough Investments Ltd v Central Midlands Ltd* [2007] 1 L & TR 10 Ch at [78]. Thus it may be hand delivered, or delivered by post, recorded delivery or some other method: *Kinch v Bullard* [1999] 1 WLR 423 per Neuberger J at 427. The method adopted must, however, be a proper one in the sense that it is one which a reasonable person, minded to bring the document to the attention of the person to whom the notice is addressed, would adopt: *Lord Newborough v Jones* [1975] Ch 90, Court of Appeal.
43 But note 'post' does not include ordinary post, where the server is an operator – paragraph 91(1). See paragraph 32.5.8.

32.5.11

The notification or document may be given or sent to an unincorporated body or association by being given or sent to a member of the governing body of the body or association: s.394(6) of the 2003 Act.

32.5.12 Proper address for service

Section 7 of the Interpretation Act 1978 provides that:

> Where an Act authorises or requires any document to be served by post (whether the expression 'served' or the expression 'give' or 'send' or any other expression is used) then, unless the contrary intention appears, the service is deemed to be effected by properly addressing, pre-paying and posting a letter containing the document and, unless the contrary is proved,[44] to have been effected at the time at which the letter would be delivered in the ordinary course of post.

32.5.13

The section makes a distinction between and provides for presumptions as to the fact of service and as to the time at which service has been effected. In the absence of proof to the contrary, service will be deemed to be effected on the day when in the ordinary course of post the document would be delivered. If the contrary can be proved, although the document will be deemed to have been served by reason of the statutory provision, it would have been proved not to have been served in time.[45]

32.5.14

It will have been seen in paragraph 32.4.12 above that s.394(3) refers to serving the notice at 'the proper address'. Section 7 of the 1978 Act refers to 'properly addressing' the letter containing the document. Section 394(7) makes provision as to what is the proper address in both cases. It provides that:

> (7) For the purposes of this section and section 7 of the Interpretation Act 1978 (service of documents by post) in its application to this section, the proper address of a person is –

44 As to the reference to 'contrary' in 'contrary being proved' in s.7 of the Interpretation Act 1978 see *Calladine-Smith v Saveorder Ltd* [2011] EWHC 2501 (Ch); [2011] 3 EGLR 55 considered in paragraph 32.5.20.

45 In *R. v County of London Quarter Session Appeals Committee Ex p. Rossi* [1956] 2 WLR 800; Parker LJ said:

> The section [i.e. s.26 of the Interpretation Act 1889, which in all material respects is identical to s.7 of the Interpretation Act 1978], it will be seen, is in two parts. The first part provides that the dispatch of a notice or other document in the manner laid down, shall be deemed to be service thereof. The second part provides that unless the contrary is proved that service is effected on the day when in the ordinary course of post the document would be delivered. The second part, therefore, concerning delivery as it does, comes into play and only comes into play in a case where under the legislation to which the section is applied the document has to be received by a certain time. If in such a case the 'contrary is proved', that is, that the document was not received by that time or at all, then the position appears to be that through under the first part of the section the document is deemed to have been served, it has been proved that it was not served in time.

(a) in the case of body corporate, the address of the registered or principal office of the body;
(b) in the case of a firm, unincorporated body or association, the address of the principal office of the partnership, body or association;
(c) in the case of a person to whom the notification or other document is given or sent in reliance on any of subsections (4)[46] to (6),[47] the proper address of the body corporate, firm or (as the case may be) other body or association in question; and
(d) in any other case, the last known address of the person in question.

32.5.15

Further provision is made in respect of the 'proper address' of the person to be served in paragraph 91(2), which stipulates that the 'proper address of a person' is, where the person to be served has given '*an address for service under this code, that address, and otherwise, the address given by* section 394' of the 2003 Act.[48]

32.5.16 Date of service

A number of provisions of the New Code provide for notice to be given within a certain time of receipt of an earlier notice.[49] Accordingly, the date when service is effected may be important. Section 7 of the Interpretation Act 1978 refers to the service of the document being deemed to have been effected at the time at which the letter would be delivered in the ordinary course of post, unless the contrary is proved. In *Stephenson v Orca Properties*,[50] it was held that where service is effected by recorded delivery, delivery cannot, in the ordinary course of post, be effected unless someone signs a receipt. If no one is available to sign or is willing to sign a receipt, delivery will not be effected. That decision was concerned with the interpretation of s.196(4) of the Law of Property Act 1925 which provides:

> (4) Any notice required or authorised by this Act to be served shall also be sufficiently served, if it is sent by post in a registered letter addressed to the lessee, lessor, mortgagee, mortgagor, or other person to be served, by name, at the aforesaid place of abode or business, office, or counting-house, and if that letter is not returned by the postal operator (within the meaning of the Postal Services Act 2000) concerned undelivered; and that service shall be deemed to be made at the time at which the registered letter would in the ordinary course be delivered.

32.5.17

The issue in *Stephenson v Orca Properties* was whether a notice initiating a rent review, which had to be served on the tenant before midnight on Sunday 30 June 1985, was

46 See paragraph 32.4.11.
47 See paragraph 32.4.11.
48 Notification to a landlord's solicitor of a change of the tenant's registered office was imputed to the landlord, even though the solicitor had forgotten that information and failed to pass it on to the landlord. Thus, service at the tenant's previous registered office was held to be ineffective: *Arundel Corpn v Khokher* [2003] EWCA Civ 1784.
49 See paragraphs 31(2) and (3), 32(1), 41(5), 50, 78(2) and 82(4), by way of example.
50 [1989] 2 EGLR 129.

deemed to have been validly served by that date, having been sent by recorded delivery on Friday 28 June. The evidence was that the post office had attempted to deliver the notice on 29 June but had found that the offices to which the letter was sent were not open. The letter was therefore returned to the delivery office and a successful delivery was effected on 1 July. Scott J held that the trigger notice had not been served in time. He said that in the case of delivery by recorded delivery:

> Delivery cannot, in the ordinary course of post, be effected unless someone signs a receipt. If no one is available to sign or is unwilling to sign a receipt, delivery will not be effected ... Delivery in the ordinary course of post requires, where recorded delivery letters are concerned, an available recipient; it cannot take place at a time when there is no available recipient.

32.5.18

Stephenson v Orca Properties was not followed by Patten J in *WX Investments Ltd v Begg*.[51] In that case Patten J held that, in determining for the purposes of s.196(4) of the Law of Property Act 1925 when delivery is treated as having taken place, it is necessary to decide when delivery in the ordinary course of post would have taken place. This does not require the presence of an available recipient. The fiction created by the deeming provision extends to what would in the ordinary course of post be the likely date of delivery. To make the presumption of delivery an effective one, s.196(4) requires an assumption to be made that an available recipient was present on that occasion. Once the statutory presumption comes into play with the operation of the deeming provision, the date on which the recorded receipt was in fact signed is irrelevant.

32.5.19

It is important to note that s.196(4) does not, unlike s.7 of the Interpretation Act 1978, contain any qualification as to the date of service by reference to the phrase 'unless the contrary is proved'. It thus provides for a statutory deeming as to service having been effected on the date when delivery in the ordinary course of post would have taken place. As s.7 of the Interpretation Act provides for a presumption as to service and, unless the contrary is proved, the date of service (see paragraph 32.5.11 et seq above), the need for a recipient may, from a practical evidential point of view, still be necessary for the purposes of the effective operation of s.7 in order to prevent the presumption as to the date of service being rebutted.

32.5.20

The reference to 'contrary' in 'contrary being proved' in s.7 of the Interpretation Act 1978 has been said to be a reference to the contrary being proved of the deeming provision that the document in question was delivered in the ordinary course of post. The reference to the contrary being 'proved' requires no more than evidence which supports a finding on the balance of probabilities that the letter was not delivered. There is no burden upon the addressee of the letter in question to lead positive evidence as to what

51 [2002] 3 EGLR 47.

happened to the letter and/or a burden upon the addressee to show that the sender of the letter was aware that it had not been delivered: *Calladine-Smith v Saveorder Ltd.*[52]

32.5.21 Substituted method of service

Where, for the purposes of giving a notice under the New Code, it is not practicable to find out after reasonable enquires[53] the name and address of the person who is the occupier of the land for the purposes of the New Code, a substituted form of service may be adopted: paragraph 91(3). Where reasonable enquiries have been undertaken the notice may be (1) addressed to a person by the description of 'occupier' of the land (and describing the land), and (2) by delivering it to a person who is on the land or, if there is no person on the land to whom it can be delivered, by fixing it, or a copy of it, to a conspicuous object on the land: paragraph 91(4).

32.5.22

Similar provision for a substituted form of service is contained in paragraphs 91(5) and (6) with respect to service of a notice on the 'owner of an interest in land' under the New Code.

32.6 Notices by others

32.6.1

The terms of paragraph 89 of the New Code make provision for service of a notice to be given by a person, other than an operator, under paragraphs 31(1), 33(1), 39(1) or 40(2) ('the specified paragraphs').

32.6.2

If OFCOM prescribes a form of notice to be given,[54] the notice must be in that form: paragraph 89(2). If it is not, it is not valid: paragraph 89(3). However, the operator may, notwithstanding the fact that the notice is not in a valid form, rely on it if it chooses to do so: paragraph 89(4).

32.6.3

In relation to any other notice given under any provision of the New Code, other than a notice given in accordance with the specified paragraphs, the notice will be valid even if not given in accordance with the prescribed form. It is provided that where a notice is served in response to a notice given by an operator, then albeit OFCOM may have prescribed the form of notice in response to that served by the operator, the notice is valid albeit not in the prescribed form: paragraph 89(5). The person who gives the notice other

52 [2011] 3 EGLR 55.

53 As to the obligation to make 'reasonable enquiries' see *Collier v Williams* [2006] 1 WLR 1945, CA. concerned with the service of court proceedings under the then provisions of CPR 6.5(6) (now to be found in CPR 6.9). See in particular the facts of *Marshall v Maggs*, one of the conjoined appeals in that case. See also *Cranfield v Bridgegrove Ltd* [2003] 1 WLR 2441, paragraphs 101–3.

54 As to which, see section 41.7 of Chapter 41 of this book.

than in the prescribed form must bear any costs incurred by the operator as a result of the note is not being in the prescribed form. It will, however, be necessary for the operator, in giving the notice which it has served, to draw the recipient's attention to the form prescribed by OFCOM for the notice to be given in response: paragraph 89(5) and (6).[55]

32.7 Service of notices by others

32.7.1

The provisions as to service of any notice upon an operator are those contained in s.394 of the 2003 Act and paragraph 91 of the New Code: paragraph 91. These provisions are discussed in paragraphs 32.2.5 et seq above. The provisions for substituted service contained within paragraph 91(3)–(6) do not apply given that they relate to service 'to the occupier' or 'to the owner' and thus do not apply to notices to be served upon the operator.

32.8 Service by email

32.8.1

The provisions of s.395 of the Communications Act 2003 will continue to apply to enable service of notices under the New Code to be effected by email. These provisions are considered in section 32.4 above.

32.9 OFCOM: prescription of notices

32.9.1

OFCOM must prescribe the form of notice to be given under each provision of the New Code that requires a notice to be given: paragraph 90(1). OFCOM may amend or replace a prescribed form from time to time: paragraph 90(2). Before prescribing a form for the purposes of the New Code, OFCOM must consult operators such other persons as they think appropriate: paragraph 90(3). The consultation requirements do not apply in relation to any amendment or replacement of an existing prescribed form: paragraph 90(4).

32.9.2

In practice, OFCOM has prescribed many forms of notice, including some where it had no obligation to do so: see section 41.7 of Chapter 41 of this book, and the template notices in Appendix E to this book.

55 Paragraph 89(6), which provides that the notice is valid albeit not in the prescribed form, applies where the provisions of sub-paragraph (5) are satisfied. One of the conditions contained in sub-paragraph (5) is that the operator has, in giving the notice it has served, drawn the recipient's attention to the OFCOM prescribed form of notice in response:: paragraph 89(5)(c). It cannot, it is considered, be the case that this condition contained in paragraph 89(5)(c) can be a condition to validity, but is a condition to the entitlement of the operator to the costs referred to in paragraph 89(6). It would be odd for the validity of a non-prescribed form of notice to depend upon the operator having notified the person serving the notice in response to the fact that there is a prescribed form.

33 Dispute resolution procedure under the codes

33.1 Introduction

33.1.1

Both the Old Code and the New Code (but especially the latter) provide for disputes arising under the codes to be determined by 'the court'.

33.1.2

In the case of the Old Code, disputes are primarily determined by the courts, rather than by tribunals. In practice, the number of disputes arising under the Old Code that have made their way into any forum is miniscule. This topic is discussed in more detail in section 33.2 below.

33.1.3

Under the New Code, although extensive reference is made to 'the court', the term is defined in a way that will allow the tribunals, with their relevant expertise and experience, to determine disputes. This is examined in sections 33.3 (which deals with Courts) and 33.4 (Tribunals) below. Section 33.5 compares and contrasts the two jurisdictions, while section 33.6 considers the practice in the Lands Chamber of the Upper Tribunal.

33.1.4

Costs of resolving disputes are of course at the forefront of the consideration of any dispute resolution mechanism. This topic is examined in section 33.7.

33.1.5

Section 33.8 considers the procedure for the resolution of code disputes in the Tribunals.

33.1.6

Quite apart from these traditional dispute resolution procedures, both Old and New Codes provide for arbitration in some limited circumstances, which section 33.9 reviews.

33.1.7

Section 33.10 looks at the procedures in the 2003 Act governing disputes that may be referred to OFCOM for resolution.

33.1.8

Finally, section 33.11 considers those disputes which might arise in connection with the use of telecommunications apparatus which do not trigger dispute resolution procedures under either code, but fall to be dealt with as part of the general law.

33.2 Dispute resolution procedure under the Old Code

33.2.1

Provision is made by the Old Code for disputes to be determined either by the Court, or by the Lands Chamber of the Upper Tribunal, or by arbitration.

33.2.2

The primary means of dispute resolution is the court. Paragraph 1(1) of the Old Code provides that 'the court' means:

(a) in relation to England and Wales and Northern Ireland, the county court: and
(b) in relation to Scotland, the sheriff.

In neither of these cases, therefore, is 'the court' the High Court. *Arqiva Ltd v Everything Everywhere Ltd*[1] considers what to do if code rights issues nevertheless arise in High Court proceedings. In *South v Chamberlayne*[2], Lightman J dealt with a case that should have been issued in the County Court, but which was instead listed in the High Court, by transferring it to the County Court and immediately thereafter transferring it back to the High Court, thus bestowing jurisdiction upon it (see ss.40 and 41 of the County Courts Act 1984). For appeals in Northern Ireland, see paragraph 25 of the Old Code.

33.2.3

Disputes to be resolved by the court include the following:

(1) Where a code operator wishes to install apparatus on a person's land, but that person does not agree to the installation: in such circumstances, the County Court has the power under paragraph 5 to dispense with the need for the agreement, and make a financial award, following an application by the code operator.
(2) Where a code operator already has apparatus installed on land in respect of which proceedings under paragraph 5 are pending, the County Court has the power under paragraph 6 to confer on the code operator temporary rights so as to ensure that its network is maintained pending determination of the proceedings.

1 [2011] EWHC 2016 (TCC).
2 [2002] L & TR 26.

(3) Where a code operator's apparatus is already installed on a person's land, that person may apply to the County Court for an order under paragraph 20 of the 2003 Code to require the operator to alter or remove the apparatus.

33.2.4

Provision is made for certain disputes as to compensation to be determined instead by the 'appropriate tribunal': see for example paragraphs 4 and 16. Paragraph 4(10A) defines 'the appropriate tribunal' as the Lands Chamber of the Upper Tribunal in the case of England and Wales, and the Lands Tribunal in the case of Scotland and Northern Ireland.

33.2.5

The Lands Chamber is the forum for the following issues:

(1) Where, on a right being conferred or varied in accordance with paragraph 2 of the Old Code, there is a diminution in value of a relevant interest in the land due to the security provisions of the code, the code operator is obliged to pay compensation under paragraph 4(4). The amount of compensation falls to be assessed by the Lands Chamber.
(2) Where a right conferred under the Old Code causes injurious affection to neighbouring land within the meaning of section 10 of the Compulsory Purchase Act 1965, the code operator must pay compensation under paragraph 16 of the code. The amount of compensation is determined by the Lands Chamber.

33.2.6

Finally, the Old Code also provides for some disputes to be resolved by arbitration, as follows:

(1) Disputes relating to the installation of apparatus, or emergency works to apparatus, crossing a linear obstacle are to be referred to arbitration under paragraphs 12 and 13 of the Old Code.
(2) In addition, where a landowner's agreement is dispensed with by the County Court making an order under paragraph 5 of the 2003 Code, the court is also obliged to make a financial award under paragraph 7. However, paragraph 7(4) allows the court to refer any questions arising as a consequence of making that award to an arbitrator.

33.2.7

In practice, neither the courts, the tribunal systems nor arbitrators have been overburdened by disputes arising under the Old Code. For the most part, the reluctance by landowners and operators to engage in formal dispute resolution has been attributable not to the lack of any contention in their dealings (for there has been much of this), but rather to the notorious degree of opaqueness in the drafting of the Old Code, leading to much uncertainty as to how the courts, in particular, would be likely to interpret the code provisions.

33.2.8

In addition to this, both code operators and site providers responded to the Law Commission consultation by expressing dissatisfaction with the County Court; its shortcomings were seen as expense, delay and lack of expertise.

33.2.9

The Law Commission noted this unsatisfactory feature of dispute resolution under the Old Code in paragraph 9.1 of their 2013 Report (Law Com 336):

> It was clear from many of our discussions with stakeholders before publication of the Consultation Paper that one of the principal sources of discomfort with the 2003 Code is its failure to provide for swift and effective dispute resolution. For many consultees a very important expectation of the revised Code is a solution to this problem …

33.2.10

The need for swift and effective dispute resolution is underpinned by the United Kingdom's obligations under the 2002 EU Framework Directive.[3] Article 11 of the Directive provides:

> Member states shall ensure that where a competent authority considers – an application for the granting of rights to install facilities on, over or under public or private property to an undertaking authorised to provide public communications networks … the competent authority: – acts on the basis of simple, efficient, transparent and publicly available procedures, applied … without delay, and in any event makes its decision within six months of the application, except in cases of expropriation, and – follows the principles of transparency and non-discrimination in attaching conditions to any such rights.

33.2.11

The Department for Business, Innovation and Skills described six months as a 'challenging timescale' where the competent authority is the County Court, but undertook to 'work with the Ministry of Justice and the courts' to meet it.[4]

3 Directive 2002/21/EC of 7 March 2002 on a common regulatory framework for electronic communications networks and services. See further Chapter 4.

4 Department for Business, Innovation and Skills, Implementing the revised EU Electronic Communications Framework: Overall approach and consultation on specific issues (September 2011), p.22, available at www.bis.gov.uk/Consultations/revised-euelectronic-communications-framework. Policy responsibility for these issues has now passed to the Department for Culture, Media and Sport.

33.2.12

Article 11 of the Directive was subsequently implemented in the United Kingdom by regulation 3(2) of the Electronic Communications and Wireless Telegraphy Regulations 2011.[5] This regulation applies where:

(a) a person authorised to provide public electronic communications networks applies to a competent authority for the granting of rights to install facilities on, over or under public or private property for the purposes of such a network, [or]
(b) a person authorised to provide electronic communications networks other than to the public applies to a competent authority for the granting of rights to install facilities on, over or under public property for the purposes of such a network …

33.2.13

The regulation requires that:

> except in cases of expropriation, the competent authority must make its decision within 6 months of receiving the completed application.

33.2.14

The next section examines the extent to which the New Code has satisfied this requirement, and attended to the perceived disadvantages of dispute resolution under the Old Code.

33.3 Dispute resolution procedure under the New Code: (a) the courts

33.3.1

Many provisions of the New Code also provide for disputes to be referred to 'the court': see for example Parts 4 (Power of the court to impose agreement), 5 (Termination and modification of agreements), 6 (Rights to require removal of electronic communications apparatus), 12 (Rights to object to certain apparatus), 13 (Rights to lop trees) and paragraph 53 (Alteration of apparatus at request of transport undertakers).

33.3.2

The impression that the draftsman of the New Code has opted in favour of the *courts* dealing with disputes under the code instead of *tribunals* appears at first sight to be confirmed by the drafting of paragraph 94 of Part 16 of the New Code, which deals with Enforcement and Dispute Resolution.

5 SI 2011 No. 1210.

33.3.3

Paragraph 94 defines the expression 'the court' to mean the County Court as far as England, Wales and Northern Ireland are concerned, and the Sheriff Court in Scotland. This drafting is testament to the fact that the United Kingdom does not have a single unified judicial system (save for the Supreme Court, and certain specialised areas such as some immigration appeals). England and Wales have one system, Northern Ireland a system that differs in some respects, but is broadly similar, and Scotland a substantively and procedurally different system. This is in contrast to the tribunal system considered in section 33.4 below, which covers England, Wales and in some cases Northern Ireland and Scotland.

33.3.4

In relation to England and Wales, although there are many county courts sitting in a great number of locations (with ten operating in Greater London alone), these courts are one court – see section A1 of the County Courts Act 1984 ('Establishment of a single county court'). Despite this, for all practical purposes, the courts are treated as separate litigation centres, with proceedings having to be intituled in the relevant county court, and with transfers from one court to another having to be applied for, where thought to be necessary, and adjudicated upon.

33.3.5

In Northern Ireland, the institution of the county court is established by the County Courts (Northern Ireland) Order 1980 and the Justice Act (Northern Ireland) 2015. The Lord Chief Justice of Northern Ireland gives directions under article 4 of the Order concerning the functioning of the county court and such other incidental, consequential, transitional or supplementary matters as appear to him to be necessary or proper. The jurisdiction of the county court in Northern Ireland is limited in ways that bear similarity with the restrictions imposed by the County Courts Act 1984 in England and Wales.

33.3.6

In Scotland, the sheriff court is administered by the Scottish Courts and Tribunals Service, established under the Judiciary and Courts (Scotland) Act 2008. Its jurisdiction is also limited by reference to the financial value of the claim, by virtue of the Court Reform (Scotland) Act 2014.

33.3.7

All such courts, in all parts of the United Kingdom, are presided over by judges who are lawyers rather than experienced surveyors (and sometimes also lawyers) of the sort that are to be found in the tribunal system. Those judges try the great majority of civil claims, with the result that their experience is broad, but necessarily leaving them with little opportunity to familiarise themselves with complex technical areas.

33.3.8

Some of the more complex value judgments required by the New Code (e.g. the balancing exercise test required under paragraph 21) may be said to be more suitable to an expert tribunal rather than a court. This is not because county court judges and sheriffs are not well able to assess the evidence and apply the law in those areas, but rather because the application of the test will require a familiarity with the concepts involved (in the case of paragraph 21, to take the same example, the weighing of the public benefit) that will be difficult to assemble with any degree of consistency when code cases heard by the same county court (let alone the same judge) will be relatively few. It will obviously be desirable for a course of dealing to emerge in the early years of the operation of the New Code so that practitioners and their clients may know what to expect.

33.3.9

In this respect, a parallel may usefully be drawn with the cases heard by the Lands Chamber of the Upper Tribunal, which specialises (apart from its appellate jurisdiction) in rating; compensation for the compulsory purchase of land, for land affected by public works, for damage to land caused by subsidence from mining and for blighted land; discharge or modification of restrictive covenants; tree preservation orders; the valuation of land or buildings for Capital Gains Tax or Inheritance Tax purposes; and right to light disputes.

33.3.10

Over many decades, that Tribunal and its predecessor the Lands Tribunal, have evolved a considerable expertise in those areas. The Upper Tribunal also has the following hallmarks (in comparison with the County Court).

(a) It is a superior court of record (s.3(5) of the Tribunals, Courts and Enforcement Act 2007), with the result that inferior courts and tribunals are bound to follow its decisions.
(b) Its Judges and Members include well qualified and experienced lawyers and chartered surveyors.
(c) It has its own procedural rules and practice directions (made by the Senior President of Tribunals, with the advice of the Tribunal Procedure Committee).
(d) It publishes all its decisions (more or less contemporaneously with the date upon which they are handed down to the parties): these are all available on BAILII[6] and on the Tribunal website,[7] which is administered by HM Courts & Tribunals Service.

33.3.11

It is to be hoped that a similar body of expertise concerning the New Code will be built up, so that consistency in the interpretation and application of that code emerges

6 www.bailii.org/form/search_cases.html
7 http://landschamber.decisions.tribunals.gov.uk//Aspx/Default.aspx

early on, guiding the expectations of owners and operators. As paragraph 33.3.9 above suggests, and as the next section explains, it is suggested that the tribunal system is more likely to produce those outcomes than the court service.

33.4 Dispute resolution procedure under the New Code: (b) the tribunals

33.4.1

Although paragraph 94 of the New Code defines 'the court' in such a way as to suggest at first sight that that is the primary – indeed the only – arena for the resolution of disputes arising under the code, that is not the case. Paragraph 94(2) adds that this definition is subject to 'provision made by regulations under paragraph 95'.

33.4.2

Paragraph 95 of the New Code is headed 'Power to confer jurisdiction on other tribunals', and provides:

> (1) The Secretary of State may by regulations provide for a function conferred by this code on the court to be exercisable by any of the following –
> (a) in relation to England, the First-tier Tribunal;
> (b) in relation to England and Wales, the Upper Tribunal;
> (c) in relation to Scotland, the Lands Tribunal for Scotland;
> (d) in relation to Northern Ireland, the Lands Tribunal for Northern Ireland.

33.4.3

The tribunal referred to in sub-paragraph (a) is dealt with in paragraph 33.4.6 below. The three tribunals referred to in sub-paragraphs (b), (c) and (d) carry out similar functions in their respective parts of the United Kingdom, adjudicating upon the property and compulsory purchase disputes referred to in paragraph 33.3.9 (although the Lands Tribunal for Northern Ireland also presides over business tenancy renewal and other disputes). Their respective jurisdictions are conferred and regulated by the following statutes:

(a) The Upper Tribunal (Lands Chamber) in England and Wales: The Tribunals, Courts and Enforcement Act 2007, and The Transfer of Tribunal Functions Order 2008 (2008) SI No. 2833;
(b) The Lands Tribunal for Scotland: The Lands Tribunal Act 1949;
(c) The Lands Tribunal for Northern Ireland: The Lands Tribunal and Compensation (Northern Ireland) Act 1964.

33.4.4

Although most of the Tribunals in England and Wales are administered by HM Courts and Tribunals Service[8] (and are headed by the Senior President of Tribunals),

8 The exception are the Valuation Tribunal for England and the Valuation Tribunal for Wales, which deal with business rates, council tax and land drainage rates.

each such Tribunal has its own procedural own rules (made by the Senior President of Tribunals, with the advice of the Tribunal Procedure Committee), governing such matters as how hearings are conducted, how the tribunal reaches its decisions, and how to appeal procedural rules and practice directions. The applicable rules are as follows:

(a) The Upper Tribunal in England and Wales: The Tribunal Procedure (Upper Tribunal) (Lands Chamber) Rules 2010 (as amended by The Tribunal Procedure (Upper Tribunal) (Lands Chamber) Rules 2013);
(b) The Lands Tribunal for Scotland: The Lands Tribunal for Scotland Rules 2003;
(c) The Lands Tribunal for Northern Ireland: The Lands Tribunal Rules (Northern Ireland) 1976.

33.4.5

In the case of Scotland, further, paragraph 106 of the New Code provides:

> The power to make rules under section 3(6) of the Lands Tribunal Act 1949 (Lands Tribunal for Scotland procedure rules) for the purposes of this code or regulations made under it is exercisable by the Scottish Ministers instead of by the Secretary of State (and any reference there to the approval of the Treasury does not apply).

33.4.6

Paragraph 95(a) of the New Code uniquely envisages in the case of England that the tribunal's functions may be exercised by the First-tier Tribunal ('the FTT', formerly known as Leasehold Valuation Tribunal) as well as the Upper Tribunal. In practice, within the Upper Tribunal in England, proceedings under the code are assigned to the Lands Chamber by the First-tier Tribunal and Upper Tribunal (Chambers) (Amendment No. 2) Order 2017. The Welsh Assembly has retained its Leasehold Valuation Tribunal, which will not deal with New Code matters.

33.4.7

All such tribunals are presided over by personnel who may be either legally qualified judges, called 'Tribunal Judges', or non-lawyers, called 'Members'. Most tribunal hearings are chaired by Tribunal Judges but they often sit with specialist, non-legal, members, depending on the subject matter of the hearing. Tribunal Members are not assessors who advise the tribunal (compare Rule 35.10 of the Civil Procedure Rules); they provide a practical, specialised view of the facts and evidence before the tribunal, and participate fully in the decision-making. In the case of the Upper Tribunal (Lands Chamber), moreover, such Members (who are required[9] to be chartered surveyors) frequently preside alone.

9 By The Qualifications for Appointment of Members to the First-tier Tribunal and Upper Tribunal Order 2008 (2008 SI No. 2692).

33.5 Court or tribunal?

33.5.1

Paragraph 95(2) of the New Code provides:

> (2) Regulations under sub-paragraph (1) may make provision for the function to be exercisable by a tribunal to which the regulations apply –
> (a) instead of by the court, or
> (b) as well as by the court.

Paragraph 95(3) adds:

> The Secretary of State may by regulations make provision –
> (a) requiring proceedings to which regulations under subparagraph (1) apply to be commenced in the court or in a tribunal to which the regulations apply;
> (b) enabling the court or such a tribunal to transfer such proceedings to a tribunal which has jurisdiction in relation to them by virtue of such regulations or to the court.

33.5.2

Given the advantages referred to in paragraph 33.3.9 above of using Tribunals rather than the County Court for dispute resolution in this specialist area, it was expected that the Secretary of State would not merely promulgate regulations providing for the functions of the court under the New Code to be exercised by the Tribunals, but that those functions would be exercised by the Tribunals *instead of* by the court, rather than *as well as* by the court.

33.5.3

Those expectations were realised in the Electronic Communications Code (Jurisdiction) Regulations 2017, which came into force on 28 December 2017.

33.5.4

The Regulations extend to England and Wales and Scotland.[10] They provide for the functions conferred by the New Code on the court to be conferred on and exercisable by all the Tribunals *as well as* by the court, rather than *instead of* by the court (see regulation 3).

10 At the time of writing, the decision has been taken to defer extending these regulations to Northern Ireland until a legislative executive is in place that can consider these provisions and ensure necessary supporting instruments are made.

33.5.5

However, this does not mean that there will be a confusing free-for–all where choice of forum is concerned. That is because regulation 4 of the Jurisdiction Regulations provides:

> Relevant proceedings must be commenced –
> (a) in relation to England and Wales, in the Upper Tribunal, or
> (b) in relation to Scotland, the Lands Tribunal for Scotland.

Regulation 2 defines "relevant proceedings" to mean proceedings under Parts 4, 5, 6, 12 or 13, or paragraph 53 of the New Code.

33.5.6

Taken together, the proceedings to which regulation 4 of the Jurisdiction Regulations relates comprise all the contentious areas which might arise under the New Code. Thus:

(a) parts 4, 5 and 6 deal with imposition of agreements, termination and modification of agreements), and rights to require removal of electronic communications apparatus;
(b) paragraph 53 deals with the circumstances in which a transport undertaker can require an operator to alter communications apparatus; and
(c) parts 12 and 13 deal with the rights to object to certain apparatus and to lop trees.

33.5.7

The regulation 4 list omits reference to part 14 of the New Code, which deals with compensation. However, paragraphs 84(3) and (4) and 85(2) provide for compensation to be determined respectively under the Land Compensation Act 1961, the Land Compensation Act 1973 and the Compulsory Purchase Act 1965 as far as England and Wales are concerned, for which disputes are already resolved by the Upper Tribunal (Lands Chamber), with cognate legislation having the same effect in relation to Scotland and Northern Ireland. In practice, therefore, disputes under the New Code will be dealt with, in the first instance at least, by the Tribunals, rather than by the Courts.

33.5.8

Regulation 5 of the Jurisdiction Regulations provides:

> (1) A relevant tribunal, and in England the First-tier Tribunal, may transfer relevant proceedings to the court, if the tribunal considers that the court is a more appropriate forum for the determination of those proceedings.
> (2) A transfer of proceedings in accordance with paragraph (1) may be made by the tribunal of its own motion or on the application of a party.

It is thought that the only circumstances in which proceedings might be transferred in this way would be if there were parallel related proceedings in court (for example an application to renew a tenancy of business premises under Part II of the Landlord and Tenant Act 1954), the resolution of which was intimately connected with the code proceedings, making it desirable therefore that both sets of proceedings should be heard together. The prospect of this happening in practice is, the Authors suggest, minimal. If that is right, then the instances of courts determining New Code cases will be rare indeed.

33.5.9

This shift in dispute resolution in favour of tribunals rather than courts meets the view expressed by the Law Commission in paragraph 9.46 of Law Com 336:

> We regard the Lands Chamber as the obviously better solution in view of its established jurisdiction and expertise.

33.6 A forecast of tribunal practice

33.6.1

Given the desirability addressed in paragraph 33.3.11 above for a body of expertise concerning the New Code to be built up, so that consistency in the interpretation and application of that code emerges early on, guiding the expectations of owners and operators, all cases under the New Code in England and Wales will in the first instance commence in the Upper Tribunal, and will be assigned by the Senior President of Tribunals to the Lands Chamber.

33.6.2

In paragraph 1.42 of their 2013 Report on the Electronic Communications Code (Law Com 336), the Law Commission commented, in relation to the employment of Lands Chamber:

> we recommend that the forum for almost all Code disputes should be the Lands Chamber of the Upper Tribunal. We think that this will be one of the most important changes brought in by the revised Code; many consultees regarded it as crucial for the revised Code to be backed by an adjudication system with more specialist expertise than the County Court can offer.

33.6.3

As time goes by, and practice becomes settled, there may possibly be a case for some cases to be transferred to the FTT. As the President of the Lands Chamber stated in his Practice Note dated 26 January 2018:

> Although the First-tier Tribunal has jurisdiction to determine Code disputes, such disputes may not be commenced in the First-tier Tribunal. By rule 5(k) of the Rules the Upper Tribunal may transfer proceedings to another court or tribunal if that other court or tribunal has jurisdiction in relation to the proceedings. For the time being it is not anticipated that Code disputes would normally be transferred by the

Tribunal to the First-tier Tribunal, but parties who agree that their dispute should be determined in the First-tier Tribunal may apply for transfer to be considered.

33.6.4

The practice in Scotland and Northern Ireland will no doubt be for New Code cases to be commenced in the Lands Tribunals.

33.7 Transitional Provisions

33.7.1

One matter which is unclear is whether in relying upon the Old Code grounds for alteration/removal which remain applicable to a 'subsisting agreement' (e.g. see paragraph 16 of the Schedule 2 to the 2017 Act, dealing with paragraph 20 of the Old Code and paragraph 20 of Schedule 2 of the 2017 Act dealing with paragraph 21 of the Old Code), proceedings are to be instituted in the County Court (as was the position before the New Code came into force) or whether proceedings are to be in the Tribunals as directed by the Electronic Communications Code (Jurisdiction) Regulations 2017 (SI/1284).

33.7.2

One possibility is that, as a 'subsisting agreement' is treated as a paragraph 2 New Code agreement (see paragraph 2 of Schedule 2 to the 2017 Act), the new Regulations apply so as to ensure that Old Code procedures come within the remit of the Tribunals. Regulation 3 provides that 'the functions conferred by the code on the court are also exercisable by the following tribunals …' However, the terms of Regulations 2[11] and 4[12] of the Electronic Communications Code (Jurisdiction) Regs 2017 (No. 1284), appear to enable Old Code paragraph 20 claims to be issued in the County Court. A paragraph 20 Old Code claim is not part of the 'electronic communications code set out in Schedule 3A to the Communications Act 2003' within Regulation 2 nor is such a claim a claim for termination within Part 5 or removal within Part 6 of the New Code and thus not 'relevant proceedings' which 'must' be issued in the Upper Tribunal.

33.8 Costs in the Tribunal

33.8.1

The costs of proceedings in the Tribunals will of course be a topic of considerable concern to those seeking to have disputes determined. Section 29 of the Tribunal, Courts and Enforcement Act 2007 provides that:

11 '2. – (1) In these Regulations – "the code" means the electronic communications code set out in Schedule 3A to the Communications Act 2003; "relevant proceedings" means proceedings under any of the following provisions of the code – (a) Parts 4, 5, 6, 12 or 13, or (b) paragraph 53.'

12 '4. Relevant proceedings must be commenced – (a) in relation to England and Wales, in the Upper Tribunal, or … '

> (1) The costs of and incidental to –
> (a) all proceedings in the First-tier Tribunal; and
> (b) all proceedings in the Upper Tribunal,
> shall be in the discretion of the Tribunal in which the proceedings take place.
> (2) The relevant Tribunal shall have full power to determine by whom and to what extent the costs are to be paid.

33.8.2

The exercise of this power is governed by rule 10 of the Tribunal Procedure (Upper Tribunal) (Lands Chamber) Rules 2010, which provides that the Tribunal may make a summary or detailed assessment of costs on its own initiative.

33.8.3

To this, paragraph 96 of the New Code adds, under the heading 'Award of costs by tribunal':

> (1) Where in any proceedings a tribunal exercises functions by virtue of regulations under paragraph 95(1), it may make such order as it thinks fit as to costs, or, in Scotland, expenses.
> (2) The matters a tribunal must have regard to in making such an order include in particular the extent to which any party is successful in the proceedings.

33.8.4

The Practice Directions of the Lands Chamber provide additional flexibility. Paragraph 2.2 of the Practice Directions provides that:

> In exercising its power to order that any or all of the costs of any proceedings incurred by one party be paid by another party or by their legal or other representative the Tribunal may consider whether a party has unreasonably refused to consider ADR when deciding what costs order to make, even when the refusing party is otherwise successful.

33.8.5

The general rule for costs is that the successful party ought to receive its costs. However, the Tribunal retains a discretion to depart from this rule in appropriate cases. In exercising this discretion, the Tribunal will have regard to:

> all the circumstances, including the conduct of the parties; whether a party has succeeded on part of their case, even if they have not been wholly successful; and admissible offers to settle. The conduct of a party will include conduct during and before the proceedings; whether a party has acted reasonably in pursuing or contesting an issue; the manner in which a party has conducted their case; whether or not they have exaggerated their claim; and the matters stated in paragraphs 2.2,

8.3(2) [written questions to experts], 8.4 [discussions between experts] and 10 [site inspections] above.[13]

33.8.6

A further important element in the Tribunal's costs regime is that in any proceedings before the Tribunal any party may make an offer to any other party to settle all or part of the proceedings or a particular issue on terms specified in the offer. Neither the offer nor the fact that it has been made may be referred to at the hearing if it is marked with 'without prejudice save as to costs' or similar wording.[14]

33.9 Procedure in the Tribunal

33.9.1

Most consultees to the Law Commission consultation agreed that the Old Code procedures for resolving code disputes could be improved. Many thought that a change of forum and clearer notice procedures would go most of the way to achieving this, but a few other suggestions were made.

33.9.2

These included a recommendation that rules akin to the pre-action protocols that regulate certain types of proceedings in the civil courts should be introduced, with appropriate sanctions for non-compliance. This would encourage the parties to 'negotiate and resolve their disputes reasonably, to adopt a "cards on the table" approach, and to encourage the early exchange of information'.

33.9.3

Ultimately, the Law Commission considered that the existing case management powers of the Lands Chamber deal with this concern. Rule 5 of the Tribunal Procedure (Upper Tribunal) (Lands Chamber) Rules 2010 confers a wide power on the tribunal to regulate its own procedure. Specifically, the tribunal may permit or require one party to provide documents, information, evidence or submissions to another party; deal with an issue in the proceedings as a separate or preliminary issue; hold a hearing to consider any matter, including a case management issue; and stay proceedings.

33.9.4

Timing will clearly be a big issue for some applications – particularly those for interim and temporary rights (see Chapter 21). Although Regulation 3 of the Electronic

13 Practice Directions of the Lands Chamber of the Upper Tribunal, paragraph 12.2.

14 Practice Directions of the Lands Chamber of the Upper Tribunal, paragraph 12.7(1). Paragraph 12(7)(2) gives further requirements as to the content of the offer.

Communications and Wireless Telegraphy Regulations 2011 requires that, except in cases of expropriation, the competent authority must make its decision within six months of receiving the completed application, even that (comparatively ambitious) time limit will be of little use in cases of emergency.

33.9.5

It will be interesting to see how the Lands Chamber and the other Lands Tribunals cater for such cases. The Lands Chamber currently operates a similar regime to the Commercial Court and Technology and Construction Courts, in that Fridays are kept clear for case management hearings, with the result that directions can be given at short notice (including at hearings over the telephone), while urgent contested hearings can be scheduled within a few weeks with cooperation or sufficient reason. However, even this helpful attitude may not be sufficient to meet the needs of the parties. It will be interesting to see whether the Lands Chamber and other Tribunals promulgate new practice directions specifically for code cases.

33.9.6

An early indication of the Lands Chamber's approach was provided by the Lands Chamber's President's Practice Note[15] dated 26 January 2018, in which he stated:

> (7) On receipt of a notice of reference in a Code dispute the Tribunal will fix an appointment for a case management hearing at which directions will be given to enable a final hearing to take place within 5 months of receiving the reference. Parties commencing a reference in a Code dispute should seek to agree in advance what directions will be required.
>
> (8) The first case management hearing in a Code dispute is likely to take place within 2 or 3 weeks of receipt of the notice of reference and will usually be held on a Friday. If it is more convenient to one or both of the parties, the case management hearing may be conducted over the telephone.

33.9.7

Paragraph 5 of the same Practice Note states that code disputes should be referred to the Tribunal using the forms and procedures applicable to references under Part 5 of the Tribunal Procedure (Upper Tribunal) (Lands Chamber) Rules 2010.[16]

33.10 Appeals from the Tribunal

33.10.1

The test for the grant of permission to appeal from the Upper Tribunal (Lands Chamber) when exercising an original jurisdiction is the ordinary first appeals test. Although the

15 This is not a formal practice direction, but is issued as guidance to parties at the commencement of the new jurisdiction.

16 A copy of the rules and the forms for use in references is available on the Tribunal's website at www.gov.uk/courts-tribunals/upper-tribunal-lands-chamber.

Appeals from the Upper Tribunal to the Court of Appeal Order 2008, prescribing the second appeals test, is apparently unqualified in its application to all appeals from the Upper Tribunal to the Court of Appeal, the statutory power under which that Order was made (section 13(6) of the Tribunal Courts and Enforcement Act 2007) is limited to appeals from decisions of the Upper Tribunal which were themselves decisions on appeal from the First Tier Tribunal: see *Nwankwo v Secretary of State for the Home Department.*[17]

33.10.2

Accordingly, in the case of an appeal from a case originating in the First Tier Tribunal to the Upper Tribunal, any further appeal will be governed by the 2008 Order; whereas if the case originated in the Upper Tribunal, the test for permission to appeal will be the ordinary first appeals test: see paragraph 4 of the judgment of Lewison LJ in *Mundy v The Trustees of the Sloane Stanley Estate.*[18]

33.10.3

The position in Scotland and Northern Ireland is the same in outcome, but with the difference being that the originating tribunals can only be the respective Lands Tribunals.

33.11 Arbitration under the codes

33.11.1

As noted above, some provisions in the Old Code allow for certain issues arising to be referred to arbitration:

(a) where a landowner's agreement is dispensed with by the County Court making an order under paragraph 5 of the Old Code, the Court was also obliged to make a financial award under paragraph 7, but with any questions arising as a consequence of making that award to an arbitrator, if the Court thought fit – see paragraph 7(4);

(b) disputes relating to the installation of apparatus, or emergency works to apparatus, crossing a linear obstacle (such as a railway line) – see paragraphs 12 and 13.

33.11.2

This limited provision for arbitration is replicated in the New Code, which provides similarly either for disputes to be referred by the court/tribunal to arbitration, or for operators and others to elect in certain circumstances for matters to be referred to arbitration, rather than continued through or commenced in the court (or, more likely, tribunal).

33.11.3

The category of disputes to which this applies are as follows:

17 [2018] EWCA Civ 5.
18 [2018] EWCA Civ 35.

(a) disputes about the amount of compensation any loss or damage that has been sustained or will be sustained as a result of the exercise of a code right under an imposed agreement – see paragraph 25;
(b) where an operator and transport undertaker cannot agree over proposed works – see paragraph 50;
(c) where an operator and transport undertaker cannot agree over the compensation that may be payable for loss or damage sustained in consequence of the carrying out of emergency works – see paragraphs 51 and 52.

33.11.4

Paragraph 107 of the New Code provides, under the heading 'Arbitrations in Scotland':

> Until the Arbitration (Scotland) Act 2010 is in force in relation to any arbitrations carried out under or by virtue of this code, that Act applies as if it were in force in relation to those arbitrations.

33.11.5

The way is therefore set for a limited number of arbitrations to be carried out under the New Code, whether in Scotland or in the remainder of the United Kingdom. It remains to be seen how useful this facility proves to be.

33.11.6

Some respondents to the Law Commission consultation urged upon it the advantages of more dispute resolution by arbitration. The Commission reviewed the competing advantages and disadvantages in Chapter 9 of Law Com 336, concluding that arbitration had a number of significant disadvantages: awards are not publicly available, they cannot be a source of valuation comparables and nor can they amount to legal precedents.

33.12 Dispute resolution by OFCOM

33.12.1

Section 185 of the 2003 Act allows three classes of dispute to be referred to OFCOM, essentially relating to disputes between communications and infrastructure providers.

33.12.2

Section 41.2 of this book analyses the procedure and the guidelines that have been published by OFCOM relating to such disputes.

33.13 Dispute resolution other than under the codes

33.13.1

This section considers those disputes which might arise in connection with the use of telecommunications apparatus, but which do not trigger dispute resolution procedures under either code, falling instead to be dealt with as part of the general law.

33.13.2

Not all disputes with code operators arise or will arise from the codes. Examples include actions in nuisance relating to apparatus installed under code rights; actions to enforce a leasehold covenant where the parties are site provider and code operator (for example where a code operator has upgraded its apparatus in breach of a leasehold covenant); actions for breach of repairing covenant and other general landlord and tenant matters; and claims for injunctions restraining development.

33.13.3

In each such example, although the parties' relationship may be governed by a code agreement, the disputes between them concern matters of general law, and do not engage the dispute resolution procedure under the codes.

34 Code avoidance

34.1 Introduction

34.1.1

This chapter considers the ability, if at all, of the parties being able to contract out of the provisions of the New Code.

34.1.2

At first sight, there would appear to be no obvious provision in the New Code which prohibits contracting out, in contrast to many comparable pieces of property legislation (see e.g. s.25 of the Landlord and Tenant (Covenants) Act 1995). Further, the drafter of the New Code has clearly been able, where necessary, to provide that certain provisions (e.g. restrictions on assignment, upgrading or sharing – see Chapter 19) should be void, which strongly suggests that the drafter was fully alert to the task of prohibiting certain contractual avoidance, but chose not to exercise it in relation to complete code avoidance.

34.1.3

In addition to this first impression, there would appear to be nothing preventing the parties arranging their affairs so as to engineer an agreement for the installation of apparatus that is not in fact a code agreement[1] (or achieving the same result through sheer inadvertence). For example, the parties could provide (genuinely) that the primary purpose of their agreement is not the grant of code rights; or they could contract orally; or they could enter into an agreement which lacks some of the prescribed formality for a code agreement, such as a signature by both parties.

34.1.4

Notwithstanding this first impression, and the relative ease with which parties may ensure that their agreement is not a code agreement, it was very much the intention of

1 Under the New Code a 'Code Agreement' is expressly defined as one to which Part 5 of the New Code (dealing with termination) applies: paragraph 29(5). However, in this chapter it is used in a more general sense as referring to an agreement conferring code rights whether one which is consensual or imposed. On the distinction between a conferring agreement and code agreement see Chapter 19 generally.

the drafters of the New Code that it should not be possible to contract out of it, save in the limited respects which this chapter explores. This intention is borne out by the paper issued by the Department for Culture, Media and Sport in May 2016, 'A New Electronic Communications Code',[2] which stated:

> We have given careful consideration to all the views expressed by stakeholders on this issue. We recognize that there is a divergence of views as to whether there should be measures in the reformed Code on the ability to contract out. We also recognize that the existing Code is unclear on the flexibility to contract out. The Government is bringing forward a new Code that makes significant policy changes in important areas in order to support investment in network growth and sustainability, and equip the UK with the best possible digital communications infrastructure. Given this, on balance, the Government considers that any attempts by one or more parties to gain advantage by circumventing the new Code's provisions must be prohibited if the Code is to be truly effective. We will therefore make provision in the revised Code to prohibit the ability to contract out and stop parties making private agreements capable of excluding Code provisions.

34.1.5

This legislative intention did in fact find its way into paragraph 100 of the New Code, as this chapter explores. It is however worth noting that if an agreement between parties resembling a code agreement is not after all subject to the New Code (e.g. because the telecommunications operator has not had a direction applied to it by OFCOM; or because it is not an agreement 'for the statutory purposes'; or for any of the reasons just considered in paragraph 34.1.3), then the anti-contracting out provisions which this chapter considers will not after all apply.

34.1.6

That last observation triggers another. If electronic communications apparatus is installed by a party who is not a code operator, the relationship that enables the installation will not be governed by the New Code. Suppose that that party subsequently becomes a code operator – what then? The Law Commission ventured the view in paragraph 2.25 of their Report 336 that:

> Agreements negotiated on the basis that Code Rights were not being granted should not have that basis changed because of a change in status of one of the parties.

This view has been taken up by the New Code: paragraph 9 provides that code rights are those which are granted to a code operator (see Chapter 17) – not those which are

2 www.gov.uk/government/uploads/system/uploads/attachment_data/file/523788/Electronic_Communications_Code_160516_CLEAN_NO_WATERMARK.pdf

subsequently assumed. Hence, it would appear that contracting-out provisions in an agreement that is not a code agreement from its inception will be effective throughout that agreement.

34.2 The Old Code

34.2.1

Before considering code avoidance in relation to the New Code it is worth, albeit briefly, considering the terms of the Old Code.

34.2.2

By paragraph 27(2) of the Old Code it was provided that:

> The provisions of this code, except paragraphs 8(5) and 21 and sub-paragraph (1) above, shall be without prejudice to any rights or liabilities arising under any agreement to which the operator is a party.

34.2.3

Paragraph 21 dealt with the right of removal.[3] Given the fact that only paragraph 21 was expressly referred to it was considered that there was no impediment to providing for contractual provisions with respect to paragraph 20 of the Old Code.[4]

34.2.4

Both paragraphs 20 and 21 were subject to a notice and counter-notice procedure. It was commonplace to find in many code agreements disincentives to an operator serving a counter-notice either by:

(1) expressly providing that it would not, thus providing for a contractual promise which could sound in damages;
(2) providing for some form of payment by the operator to the landowner to reflect loss and damage caused to the landowner by reason of the service of any counter notice by the operator;
(3) providing for payments for periods of occupation after expiration of a code agreement.

3 As to which see Chapter 11.
4 As to which see Chapter 10. Paragraph 20 conferred a right of alteration (including removal) where necessary to effect an improvement. See the recent decision of *PG Lewins Ltd v Hutchison 3G UK Ltd* [2018] Bristol CC, where it was held that paragraph 27(2) did not strike down a contractual entitlement to require the operator to effect an alteration not requiring complete removal from the land and that a failure to comply with the contractual obligation could sound in damages.

34.2.5

In the context of paragraph 21, albeit the Law Commission in their consultation paper[5] expressed the view that 'It is not clear what effect this has where, for example, a Code Operator covenants in an agreement not to seek to rely on the security provisions of paragraph 21', it seems to the Authors that it is likely that if tested such clauses would be likely to be struck down by paragraph 27(2).

34.3 Paragraph 100

34.3.1

Paragraph 100 of the New Code provides:

> (1) This code does not affect any rights or liabilities arising under an agreement to which an operator is a party.
> (2) Sub-paragraph (1) does not apply in relation to paragraph 99 or Parts 3 to 6 of this code.

34.3.2

This seems, given the terms of sub-paragraph 100(2), to be a departure from the view of the Law Commission. In their consultation paper, the view was expressed that 'We are attracted to the idea that parties should be able to contract out of the security provisions of a revised code.'[6] In their Report,[7] they stated that 'Turning to paragraph 21, we proposed that it should be possible to contract out of the provisions restricting a landowner`s right to remove apparatus (in other words, the successor, in the revised Code, to paragraph 21). This would enable Site Providers to agree to the installation of apparatus without giving security of tenure to the Code Operator.'[8]

34.3.3

It will be seen that there are two aspects to paragraph 100, (i) sub-paragraph (1) seeks to reinforce the sanctity of the parties' bargain; but (ii) sub-paragraph (2) seeks to ensure that there can be no contracting out of paragraph 99 and Parts 3–6 of the New Code.

34.3.4

Part 3 is a reference to the Assignment, Upgrading and Sharing provisions of the New Code; Part 4 is the Compulsory Acquisition section; Part 5 deals with Termination; and Part 6 deals with Removal of apparatus. Part 2, which deals with agreements conferring code rights, is not caught by paragraph 100.

5 Consultation Paper 205, paragraph 5.41, n. 40.
6 *Ibid.*, paragraph 5.45.
7 Law Com 336.
8 Paragraph 6.48.

34.4 Binding quality of the agreement

34.4.1 Sanctity of contract

On the face of it, the terms of paragraph 100(1) uphold whatever are the terms of the contract between the parties. The contract is sacrosanct.[9] Thus provisions in relation to e.g. an indemnity by the operator to the landowner concerning the costs of removal and a limitation as to any right of compensation to the landowner by reason of code rights being exercised, may take effect. However, these sorts of provisions are subject to the terms of paragraph 100(2), which is essentially a 'no contracting out' provision.

34.4.2

Paragraph 497 of the Explanatory Notes relating to the 2017 Act[10] state, in the context of paragraph 100(1), that:

> Agreements for code rights are final so that once agreed (see Part 2), it is not possible for either party to seek to apply to the court (see Part 4) to re-open the agreement. For example if an operator agrees to pay £5,000 for a (consensual) code agreement, the operator (or landowner) cannot then apply to the Court to get a better price or improve the accompanying terms that relate to the code right that has been agreed. It is possible to vary an agreement by further agreement of course (see paragraph 11 of the code.).[11]

34.4.3

What is interesting about this passage is that, to the extent that the contract does not fall foul of any of the provisions contained in Parts 3–6, the contract prevails. Of particular interest from the extract is the reference to the consideration payable. The criteria for determining the consideration payable in accordance with paragraph 24 of the New Code,[12] where the court imposes an agreement, has yet to be ironed out by the courts but the assumptions provided for by that paragraph are likely to lead to a reduction in market 'rent' for electronic communications sites. However, if the parties reach an agreement as to the consideration payable then, subject to any further consensual variation in accordance with paragraph 11(2) of the New Code, the parties are stuck with that consideration subject to any provision for rent review.

9 As was said in the recent decision of *PG Lewins Ltd v Hutchison 3G UK Ltd* [2018] Bristol CC, at [29]: 'Para 27(2) [of the Old Code] states that the provisions of the Code, except for paragraphs 8(5) . . . and para 21, shall be without prejudice to any rights or liabilities arising under any agreement to which the operator is a party. This suggests, in my judgement, that an agreement will take precedence over the provisions of the Code save where the Code forbids it.'

10 Prepared by the Department for Culture, Media and Sport – www.legislation.gov.uk/ukpga/2017/30/notes. These are referred to in this chapter as 'the Explanatory Notes'.

11 It would seem, however, that a proposal which alters the terms upon which a code right may be exercised may form the subject matter of an application to the Court under Part 4: see the extract from the Explanatory Notes in paragraph 34.4.8 below.

12 See Chapter 30.

34.4.4

In relation to the consideration, there is, in fact, nothing within the New Code that prevents the parties from agreeing a provision for a periodic review of the consideration payable for the lease or licence.[13] In fact, in passing, it is to be noted that it is not a necessary precondition to being a code agreement within the terms of the New Code for there to be a consideration at all and certainly at law rent is not a necessary prerequisite to the creation of a lease.[14] Where there is provision for a periodic review of the rent there appears to be nothing within the terms of the New Code requiring the review to be in accordance with paragraph 24. The parties would appear to be free to provide for their own assumptions and disregards. No doubt any operator will resist any attempt in negotiations over the rent review provisions to move away from the paragraph 24 terms.

34.4.5 Forfeiture and break clauses

It would seem, given the terms of Parts 3–6 of the New Code, that a consensual code agreement under the New Code can include:

(1) in the case of a lease, a provision for forfeiture or, in the case of a licence, provision for earlier termination. Such a forfeiture clause or earlier termination provision is not rendered void by the New Code.[15] It is simply the case that one cannot, for instance, override the requirements of Part 5 by e.g. forfeiting the lease. The incorporation of a forfeiture clause will still have practical effect. If there is a lease for, say, 10 years, paragraph 31 of the New Code, making provision for service of notice of 18 months to terminate the agreement, cannot bring the lease to an end earlier than it would otherwise expire by effluxion of time. This much is made clear by paragraph 31(3)(b) of the New Code. The contractual term will have to be brought to an end in order to be able to serve the 18 month notice.[16]

(2) a contractual break clause, enabling either party to bring the contractual term to an end earlier than it would expire by effluxion of time. The contractual

13 The agreement in order to be a code agreement must, of course, be 'an agreement between the occupier of the land and the operator': paragraph 9 of the New Code. See Chapter 19.

14 *Ashburn Anstalt v WJ & Co Arnold* [1989] Ch 1, CA.

15 Contrast, for instance, the terms of s.25 of the Landlord and Tenant (Covenants) Act 1995 which provides:
Agreement void if it restricts operation of the Act:
(1) Any agreement relating to a tenancy is void to the extent that –
(a) it would apart from this section have effect to exclude, modify or otherwise frustrate the operation of any provision of this Act, or
(b) it provides for –
(i) the termination or surrender of the tenancy, or
(ii) the imposition on the tenant of any penalty, disability or liability, in the event of the operation of any provision of this Act, or
(c) it provides for any of the matters referred to in paragraph (b)(i) or (ii) and does so (whether expressly or otherwise) in connection with, or in consequence of, the operation of any provision of this Act.

16 The forfeiture does not destroy the entitlement to exercise the code rights in the code agreement. The code agreement continues beyond the termination of the contractual term with the operator continuing to be able to exercise the code rights and the site provider continuing to be bound by such rights: paragraph 30(1) and (2) of the New Code.

arrangement would still apply, notwithstanding the terms of paragraph 100 and, as with the forfeiture clause, the 18-month period of notice contained within paragraph 31 does not bring the contractual term to an end (other than in the case of a periodic tenancy where it would appear that the statutory notice is all that is required[17]). Thus, the contractual break enables one to set an earlier expiry date than the date of the term for the purposes of giving the 18-month termination notice in accordance with paragraph 31.

34.5 Contracting out

34.5.1 Introduction

Paragraph 498 of the Explanatory Notes[18] states that the effect of paragraph 100(2) of the New Code is that it

> ensures that an operator cannot be required by a landowner to agree to exclude the effect of paragraph 99 and Parts 3 to 6 of the code. This is commonly known as a 'no contracting out' rule ... Paragraph 16(1)[19] and 17(5)[20] of the code prevent the possibility that operators could be required to 'contract out' of the upgrading and sharing rights provided by those paragraphs, and the reference in paragraph 100(2) to part 3 is to ensure consistency between paragraph 100 and those provisions. Part 5 provides the code agreements to continue, notwithstanding the terms of the agreement, and to be terminated only in accordance with the provisions of the code. Part 6 provides for apparatus to be removed only in accordance with the provisions of the code. Subparagraph 100(2) ensures that operators cannot be required to 'contract out' of Part 5 and 6 which are essential to protect networks.

34.5.2 Not rendered 'void'

Paragraph 100(2) is a qualification to the operation of paragraph 100(1). Paragraph 100(1) provides that the code does not affect rights and liabilities under the contract. Paragraph 100(2) says that that does not apply to the Parts of the code referred to. Thus, the effect of the two provisions read together is that Parts 3 to 6 do affect the rights and liabilities under the code agreement. What paragraph 100(2) does not say is how.

34.5.3

It is considered that, as noted above, the 'no contracting out' rule does not strike down the offending provision of the code agreement. It is not void,[21] it is simply that it will

17 See paragraph 31(3)(a) and (b).

18 See n. 1.

19 Paragraph 16(1) of the New Code, renders void any restriction on the ability of the operator to assign the code agreement. This is discussed in Chapter 20.

20 Paragraph 17(5) of the New Code, renders void any restriction on the ability of the operator to upgrade the apparatus or share the apparatus. This is also discussed in Chapter 20.

21 When the draftsman wanted to refer to something as void, he has done so e.g. in paragraphs 16(1) and 17(5) dealing with an assignment of the agreement conferring code rights and the right to share and upgrade apparatus respectively.

be of no effect, the statutory provisions overriding whatever is provided for in the contract which is inconsistent with the operation of those Parts of the code referred to in paragraph 100(2). Accordingly, by way of example the parties cannot, it is considered:

(1) Agree that the operator shall not serve a counter notice in accordance with paragraph 32 of the New Code in response to a paragraph 31 termination notice.
(2) Agree an entitlement on the part of the landowner to seek removal by reference to one of the conditions contained within paragraph 37, but with some form of dispensation of or variation of any of those conditions.
(3) Agree to some form of penal payment to be paid by the operator to the landowner where the operator exercises rights conferred on the operator by Parts 5 or 6.

34.5.4 What does not fall foul of 100(2)?

Effect will, it is considered, be given to terms of a code agreement which are not inconsistent with Parts 3 to 6. No doubt examples which arise in practice will come to highlight the circumstances in which the code agreement may make contractual provision without falling foul of paragraph 100(2). However, it seems to the Authors that the following, possibly quite important matters, may not fall foul of paragraph 100(2).

34.5.5(1) Lift and shift

The New Code says nothing about preventing a 'lift and shift' provision which was formerly said to be encapsulated in paragraph 20 of the Old Code, enabling apparatus to be moved to another location on the land forming the subject matter of the consensual code agreement. One often finds in an agreement as between landowner and occupier (other than in the context of code rights) a unilateral entitlement on the part of the landowner to move the occupier to another part of the landowner's premises.[22] It is to be noted that the provisions contained in Part 6 refer to 'removal of electronic communications apparatus'. Although 'alteration' is defined in paragraph 108(2) of the New Code as including references 'to the moving, removal or replacement of the apparatus', there is nothing which defines 'removal' as including 'alteration' by way of moving the apparatus to an alternative location within the land forming the subject matter of the code agreement and which location may be more convenient for the landowner (e.g. because the landowner wishes to undertake relevant work by way of repair to, say, a roof). It is to be noted that the 'lift and shift' provision postulated here is not one requiring removal to land outside the subject matter of the existing code agreement, but to a more convenient location (without the operator's consent) within the land forming the subject matter of the code agreement. It is to be noted that where the court imposes a code agreement pursuant to application made in accordance with Part 4, paragraph 23(8) provides that in determining the terms of the agreement to impose, the court must determine whether it should include the term 'enabling the relevant person to require the

22 By way of example see *Dresden Estates Ltd v Collinson* [1987] 1 EGLR 45, CA.

operator to reposition or temporarily to remove electronic communications equipment to which the agreement relates (and, if so, in what circumstances)' (paragraph 23(8)(b)).

34.5.6

A lift and shift provision would not appear to be inconsistent with Part 5, as it is not seeking to terminate the code agreement. On the contrary it is seeking to exercise a right which is consistent with the code agreement continuing, albeit to the altered site.

34.5.7

If it is right that paragraph 100(1) enables a 'lift and shirt' provision to be incorporated into the code agreement, there is no need for the landowner to be concerned with Part 5 or Part 6. Neither Part applies. The landowner is simply enforcing a contractual right which is not inconsistent with the operator`s code rights. There seems much to be said for this view. This appears also to be the view of the Law Commission who in their report[23] said:

> 6.65 ... Consultees felt that those who grant Code Rights have the opportunity to negotiate appropriate terms and should be relied upon to negotiate 'lift and shift' clauses, arrangements for structural repair of buildings on which apparatus is sited, break clauses in case of redevelopment, and so on. We agree.
>
> 6.66 Our agreement is fortified by the fact that we are recommending that those who grant Code Rights, or have Code Rights imposed upon them, are to receive consideration based on the market value of those rights. Along with market value consideration, and a revised Code based primarily upon the regulation of consensual arrangements, goes the responsibility to negotiate appropriate terms within the market place.

34.5.8(2) Surrender

There would appear to be nothing within the terms of the New Code to prevent a landowner and an operator agreeing terms within the code agreement for surrender of the code agreement. Part 5 makes provision for termination by the 'site provider'. Part 5 contains restrictions on the ability of the site provider to 'bring the agreement to an end'. It can do so only by serving notice and establishing one of the relevant conditions referred to in paragraph 37 of the New Code. However, 'surrender' is said to be 'a consensual transaction between the landlord and the tenant, and therefore dependent for its effectiveness on the consent of both parties'.[24] If the parties are unable to effect a surrender it seems difficult, logically, to see why, therefore, any compromise of court proceedings, which are commonplace, whereby the operator agrees to vacate on a specified date should not equally fall foul of the anti-avoidance provisions. However, it would seem, given the terms of paragraph 30(1)(b), that all that the surrender will do

23 Law Com 336.

24 Megarry and Wade, *The Law of Real Property,* 8th ed., paragraph 18-083.

is to accelerate the determination of the contractual term. It would seem the site provider will still need to serve notice in accordance with paragraph 31. On the other hand, paragraph 30 only continues the agreement, if '*under the terms of the agreement*, the right ceases to be exercisable over the site provider ceases to be bound by it, or the site provider may bring the code agreement to an end so far as it relates to that right' (paragraph 30(1)(b)). Thus, can it be said that a surrender is not one '*under the terms of the agreement*' (assuming that the surrender is not pursuant to a contractual provision providing for such) such that upon surrender the agreement will not in fact be continued requiring its termination in accordance with Part 5, and thus enabling the site provider to proceed straight to Part 6 to effect removal?

34.5.9(3) Guarantors

The New Code says nothing about guarantors or covenants of indemnity save with respect to the conditions which may be imposed on an assignment: see paragraph 16(2), (6), (7) and (9). Is it possible therefore to circumvent paragraph 100(2) by, having a third party to a code agreement who e.g. indemnifies the landowner for all losses if the operator were to serve a counter-notice to a paragraph 31 termination notice? Paragraph 100(2) undoubtedly applies to the code agreement and possibly to the third parties' liabilities under the code agreement. As noted paragraph 100(2) disapplies subparagraph (1) in relation to Parts 3 to 6. Paragraph 100(1) refers to 'an agreement to which an operator is a party'. Thus, in order for paragraph 100(2) to apply the operator does not have to be the only party – simply *a* party. Where there is a third party guarantor or covenant of indemnity the operator is a party as is the third party. But, notwithstanding that, in the example given, dealing with service of a counter-notice under Part 5 to service of a termination notice, there is nothing in Part 5 making reference to a third party.

34.5.10(4) Security arrangements

There would appear to be nothing in the code agreement to prevent the parties from making provision for security controls with respect to access to the land forming the subject matter of the code agreement. Many landowners whose land is the subject of code agreements wish, quite properly, to ensure that the land (e.g. a commercial office building) has appropriate controls in place in order to ensure that they know who is coming into the building at any given moment. The definition of code rights and the enumeration in paragraph 3 of the New Code of the rights which may form the subject matter of a code right say nothing about the manner in which they are to be exercised. Paragraph 12(1) provides that 'A code right is exercisable only in accordance with the terms subject to which it is conferred.' There seems no reason why, therefore, the landowner should not seek to impose, as a condition of the code right to inspect and maintain electronic communications apparatus (for example),[25] such conditions as it considers reasonable with respect to maintaining the security of its building. Of course if the operator does not wish to agree to those conditions, its option would be to seek

25 Paragraph 3(c) of the New Code.

a court imposed code agreement pursuant to Part 4, but it is to be noted in that context that an order of the court 'must require the agreement to contain such terms as the court thinks appropriate, subject to sub-paragraphs (3) to (8)'. The operator may yet therefore end up with precisely the same condition if the landowner has good cause for seeking it.

34.5.11 Part 2 of the New Code

In commentary upon the New Code, much has been made of the fact that the terms of paragraph 100(2) refer only to Parts 3 to 6 and do not make reference to Part 2.[26] It is not considered that this enables parties to avoid the application of the New Code altogether by agreeing e.g. that their code agreement 'is not a Code Agreement for the purposes of Part 2'.

34.5.12

Part 2 provides for who may confer code rights, what constitutes a code agreement, who is bound by code rights, the exercise of code rights, obstructing access pursuant to a code agreement and registration requirements. It is only once one has an agreement which satisfies Part 2 that one needs to consider the application of paragraph 100. Thus, it is ambitious to suggest that the parties may contract out of paragraph 11, which provides the mandatory requirements for determining whether one has a code agreement in the first place. It is difficult to identify, on a consideration of the terms of Part 2, what it may be said the parties could 'contract out' of. By way of example, paragraph 12 provides that 'A code right is exercisable only in accordance with the terms subject to which it is conferred.' The contract defines the manner in which the right may be exercised. Once the wording of the code right is agreed that is the term governing the exercise of the code right. There is no contracting out; the contract simply governs, by the words used, the manner in which the right may be exercised.

34.5.13 Part 4

The 'contracting out' rule enshrined in paragraph 100(2) applies equally to Part 4.[27] It is difficult to think of circumstances in which Part 4 may be the subject to the application of paragraph 100(2), given that Part 4 is dealing with the imposition of a code agreement by order of the court, dispensing with the need of the consent of the landowner. Part 4 is not consensual. True it is that paragraph 22 of the New Code provides that an agreement imposed by the court 'takes effect for all purposes of this code as an agreement under Part 2' but until the order is made there is no imposed Part 2 code agreement upon which paragraph 100(2) can bite. One cannot imagine that the court would impose terms inconsistent with Parts 3 to 6 when making their order.

26 As to Part 2, see Chapter 19.
27 As to Part 4, see Chapter 21.

34.5.14

The Explanatory Notes[28] at paragraph 499 provide some insight into the thinking behind why Part 4 should be subject to paragraph 100(2):

> Similarly an operator cannot be required to agree that it will forego its right to make an application to the court under Part 4 for a different or additional code right (for which you would have to pay additional consideration). For example an operator may have an agreement with a site provider under which the operator is entitled to enter the land to maintain apparatus on terms as it gives at least 72 hours' notice (see paragraph 3(f) of the code). The operator might later wish to seek a different code Right, i.e. to enter the land to maintain apparatus giving only 48 hours' notice. That would be a new right, more onus for the landowner, which would have to be agreed on further terms, including as to payment. Failing agreement, the operator could apply to the court to be granted the new right. The court would be required to apply the test under paragraph 20.

34.5.15

Thus, the Explanatory Notes suggest that the ambit of paragraph 100(2) to Part 4 is rather limited, namely, applying in relation to a pre-existing consensual code agreement which contains a term whereby the operator agrees to forego, during the term of the code agreement, from seeking any alternative code rights by an appropriate application to the court.

34.5.16 Special regimes

Paragraph 100(2) makes no reference to the various provisions of the New Code[29] which corresponded to the provisions of paragraphs 9–14 of the Old Code, referred to by Lewison J in *The Bridgewater Canal Co. Ltd v Geo Networks Ltd*[30] as special regimes. The reason for this is that none of those provisions requires any consensual agreement with a landowner for the rights conferred by the New Code to be exercised.

34.5.17

It was not uncommon for rights, for example in relation to crossings of linear obstacles, to be granted on a time-limited basis between the parties. It is considered that this was permissible under the Old Code, but that, at the end of the right, the undertaker-landowner was required to operate the paragraph 21 (or, as the case may be, paragraph 20) mechanism. It is considered that it is open to the parties to customise their rights in a similar way under the New Code, but there is a potentially significant structural difference. This is that, under the Old Code, paragraphs 20 and 21 came after both the general and special regimes, and could therefore readily be seen to apply for both. The

28 See note 10.

29 Part 7 deals with transport land rights, see Chapter 26; Part 8 deals with street works, see Chapter 27; Part 9 deals with tidal waters, see too Chapter 27; Part 11 deals with overhead apparatus, see Chapter 28.

30 [2010] 1 WLR 2576; reversed [2011] 1 WLR 1487, CA.

New Code changes the order, and Parts 5 (termination) and 6 (removal) come after the general, but before the special, regimes. Whether or not this is a significant change is unclear. On the one hand, it leads one to assume that those Parts apply to the new general regime only. On the other hand, if a special regime right can be time-limited, then special regime cases can readily be brought within the wording of Parts 5 and 6. On balance, it is considered that the restructuring of the order of regimes in the New Code is not significant.

35 Electronic communications and planning

35.1 Introduction

35.1.1

Planning permission is required for development, and development includes the carrying out of building, engineering or other operations in, on, over or under land, or the making of any material change in the use of any buildings or other land.[1] Building operations include structural alterations of or additions to buildings.[2] The importance of deciding whether activities amount to development of land is because, subject to the exceptions described below, development requires planning permission.

35.1.2

Certain operations or uses of land are expressly excluded from the definition of development, as otherwise defined above. Planning permission would not therefore be required for such matters. The matter excluded from the definition of development include the carrying out for the maintenance, improvement or other alteration of any building or works which only affect the interior of the building or do not materially affect the external appearance of the building.[3] Apart from these excluded operations, planning permission is required, and is granted for development following an application to the local planning authority and, if refused, on appeal to the Secretary of State.[4]

35.1.3

Planning permission may also be granted by a development order made by the Secretary of State. A development order may itself grant planning permission for development specified in the order, or for development of any class there specified.[5] Planning permission granted by a development order may be granted either unconditionally or subject to such conditions or limitations as may be specified in the order.[6]

1 Town and Country Planning Act 1990, s.55(1).
2 Town and Country Planning Act 1990, s.55(1A)(c).
3 Town and Country Planning Act 1990, s.55(2)(a).
4 Town and Country Planning Act 1990, ss.62 and 78–9.
5 Town and Country Planning Act 1990, s.59(1)–(2).
6 Town and Country Planning Act 1990, s.60(1).

35.1.4

In England, the Town and Country (General Permitted Development) (England) Order 2015 contains classes of development for which planning permission is granted by virtue of article 3 of the Order: permitted development rights, or known as 'permitted development'.[7] In Wales, the Town and Country Planning (General Permitted Development) Order 1995 has similar application.

35.1.5

For a detailed treatment of the law relating to town and country planning, reference should be made to more specialised works.[8] This chapter deals with the works and operations, relating to the electronic communications operated by code operators, for which there are permitted development rights under the 2015 Order, as it applies to England. The principles are the same in the other jurisdictions.

35.1.6

There are five separate classes of permitted development rights in Part 16, Schedule 2 to the 2015 Order. This chapter only addresses Class A, which makes provision for permitted development rights for electronic code operators.

35.2 Class A: electronic communications code operators

35.2.1

The permitted development is development by or on behalf of an electronic communications code operator for the purpose of the operator's electronic communications network in, on, over or under land controlled by that operator or in accordance with the electronic communications code, consisting of –

(a) the installation, alteration or replacement of any electronic communications apparatus,
(b) the use of land in an emergency for a period not exceeding six months to station and operate moveable electronic communications apparatus required for the replacement of unserviceable electronic communications apparatus, including the provision of moveable structures on the land for the purposes of that use, or
(c) development ancillary to radio equipment housing.

35.2.2

In relation to ground-based apparatus, and in relation to *the installation, alteration or replacement of any electronic communications apparatus mentioned at (a) in 35.2.1*

7 Article 3 of, and Sched. 2 to, the Order.
8 *Encyclopedia of the Law of Town and Country Planning*, subscription service.

above, paragraph A.1(1) of Class A sets out development that is not permitted by reference to heights and other limitations:[9]

(a) in the case of the installation of apparatus (other than on a building or other structure) the apparatus, excluding any antenna, would exceed a height of 15 metres above ground level;
(b) in the case of the alteration or replacement of apparatus already installed (other than on a building or other structure), the apparatus, excluding any antenna, would when altered or replaced exceed the height of the existing apparatus or a height of 15 metres above ground level, whichever is the greater; or
(c) in the case of the alteration or replacement of an existing mast (other than on a building or other structure, on article 2(3) land (National Parks, areas of outstanding natural beauty and conservation areas) or on any land which is, or is within, a site of special scientific interest) –
 (i) the mast, excluding any antenna, would when altered or replaced –
  (aa) exceed a height of 20 metres above ground level;
  (bb) at any given height exceed the width of the existing mast at the same height by more than one third; or
 (ii) where antenna support structures are altered or replaced, the combined width of the mast and any antenna support structures would exceed the combined width of the existing mast and any antenna support structures by more than one third.

35.2.3

In relation to building-based apparatus, and in relation to the installation, alteration or replacement of any electronic communications apparatus mentioned at (a) in 35.2.1 above, paragraph A.1(2) of Class A sets out development that is not permitted by reference to heights and other limitations:[10]

(a) in the case of the installation, alteration or replacement of apparatus on a building or other structure, the height of the apparatus (taken by itself) would exceed –
 (i) 15 metres, where it is installed, or is to be installed, on a building or other structure which is 30 metres or more in height; or
 (ii) 10 metres in any other case;
(b) in the case of the installation, alteration or replacement of apparatus on a building or other structure, the highest part of the apparatus when installed, altered or replaced would exceed the height of the highest part of the building or structure by more than –
 (i) 10 metres, in the case of a building or structure which is 30 metres or more in height;
 (ii) 8 metres, in the case of a building or structure which is more than 15 metres but less than 30 metres in height; or
 (iii) 6 metres in any other case;

9 Paragraph A.1(1).
10 Paragraph A.1(2).

(c) in the case of the installation, alteration or replacement of a mast on a building which is less than 15 metres in height, the mast would be within 20 metres of the highway (unless the siting remains the same and the dimensions of the altered or replaced mast are no greater);[11] or
(d) in the case of the installation, alteration or replacement of an antenna on a building or structure (other than a mast) which is 15 metres or more in height, or on a mast located on such a building or structure, where the antenna is located at a height of 15 metres or above, measured from ground level –
  (i) in the case of dish antennas, the size of any dish would exceed 1.3 metres or the aggregate size of all of the dishes on the building, structure or mast would exceed 10 metres, when measured in any dimension;
  (ii) in the case of antennas other than dish antennas, the development (other than the installation, alteration or replacement of a maximum of two small antennas or two small cell antennas) would result in the presence on the building or structure of –
    (aa) more than five antenna systems; or
    (bb) any antenna system operated by more than three electronic communications code operators; or
  (iii) the building or structure is a listed building or a scheduled monument.

35.2.4

In relation to apparatus on masts, and in relation to the installation, alteration or replacement of any electronic communications apparatus mentioned at (a) in 35.2.1 above, paragraph A.1(3) of Class A sets out development that is not permitted if, in the case of the installation, alteration or replacement of apparatus (other than an antenna) on a mast, the height of the mast would, when the apparatus was installed, altered or replaced, exceed any relevant height limit specified in respect of apparatus in paragraphs A.1(1)(a), (b) and (c), and A.1(2)(a) and (b), as are described above.[12] Although, for the purposes of applying the limit specified in paragraph A.1(2)(a), which concerns a mast on a building, the words '(taken by itself)' in that paragraph are omitted.

35.2.5

In relation to the installation, alteration or replacement of any electronic communications apparatus mentioned at (a) in 35.2.1 above, other than the following list, the ground or base area of the structure should not exceed 1.5 square metres:[13]

(a) a mast;
(b) an antenna;
(c) a public call box;
(d) any apparatus which does not project above the level of the surface of the ground; or
(e) radio equipment housing,

11 For the meaning of 'mast', see *R(Mawbey) v Lewisham LBC* [2018] EWHC 263 (Admin).
12 Paragraph A.1(3).
13 Paragraph A.1(4).

35.2.6

In the case of antennas installed, replaced or altered on Article 2(3) land (National Parks, areas of outstanding natural beauty and conservation areas) or on any land which is, or is within, a site of special scientific interest (SSSI), development within Class A(a) at 35.2.1 above is not permitted where it consists of:[14]

(a) in the case of development on any Article 2(3) land or any land which is, or is within, a site of special scientific interest, it would consist of –
  (i) the installation or alteration of an antenna or of any apparatus which includes or is intended for the support of such an antenna; or
  (ii) the replacement of such an antenna or such apparatus by an antenna or apparatus which differs from that which is being replaced,

unless the development is carried out in an emergency or is allowed by paragraphs A.1(5)(b), (9)(a), (9)(b) or (10)(b); or

(b) in the case of the installation of an additional antenna on existing electronic communications apparatus on a building or structure (including a mast) on Article 2(3) land (but not the site of a SSSI) –
  (i) in the case of dish antennas, the size of any additional dishes would exceed 0.6 metres, and the number of additional dishes on the building or structure would exceed three; or
  (ii) in the case of antennas other than dish antennas, any additional antennas would exceed 3 metres in height, and the number of additional antennas on the building or structure would exceed three.

35.2.7

Development is not permitted by Class A(a) if it would consist of the installation of a mast, on a building or structure which is less than 15 metres in height, and such a mast would be within 20 metres of a highway.[15]

35.2.8

In the case of radio equipment housing, development is not permitted by Class A(a) if –[16]

(a) the development is not ancillary to the use of any other electronic communications apparatus;
(b) the cumulative volume of such development would exceed 90 cubic metres or, if located on the roof of a building, the cumulative volume of such development would exceed 30 cubic metres; or
(c) on any Article 2(3) land, or on any land which is, or is within, a site of special scientific interest, any single development would exceed 2.5 cubic metres, unless the development is carried out in an emergency.

14 Paragraph A.1(5).
15 Paragraph A.1(7).
16 Paragraph A.1(8).

35.2.9

In the case of antennas installed, replaced or altered on a dwellinghouse, development is not permitted under Class A(a) if –[17]

(a) in the case of the installation, alteration or replacement on a dwellinghouse or within the curtilage of a dwellinghouse of any electronic communications apparatus, that apparatus –
  (i) is not a small antenna (defined as one use in connection with a telephone operating system operating on a point to fixed multi-point basis, does not exceed 0.5 metres by any linear measurement, and does not have an area exceeding 1,591 square centimetres);
  (ii) being a small antenna, would result in the presence on that dwellinghouse or within the curtilage of that dwellinghouse of more than 1 such antenna; or
  (iii) being a small antenna, is to be located on a roof or on a chimney so that the highest part of the antenna would exceed in height the highest part of that roof or chimney respectively; or
(b) in the case of the installation, alteration or replacement on Article 2(3) land of a small antenna on a dwellinghouse or within the curtilage of a dwellinghouse, the antenna is to be located –
  (i) on a chimney;
  (ii) on a building which exceeds 15 metres in height;
  (iii) on a wall or roof slope which fronts a highway; or
  (iv) in the Broads, on a wall or roof slope which fronts a waterway.

35.2.10

In the case of antennas installed, replaced or altered on a building other than a dwellinghouse, development is not permitted under Class A(a) if –[18]

(a) in the case of the installation, alteration or replacement of a small antenna on a building which is not a dwellinghouse or within the curtilage of a dwellinghouse –
  (i) the building is on Article 2(3) land (National Parks, areas of outstanding natural beauty or conservation areas);
  (ii) the building is less than 15 metres in height, and the development would result in the presence on that building of more than one such antenna; or
  (iii) the building is 15 metres or more in height, and the development would result in the presence on that building of more than two such antennas; or
(b) in the case of the installation, alteration or replacement of a small cell antenna on a building or structure which is not a dwellinghouse or within the curtilage of a dwellinghouse –

17 Paragraph A.1(9).
18 Paragraph A.1(10).

(i) the building or structure is on any land which is, or is within, a site of special scientific interest; or
(ii) the development would result in the presence on the building or structure of more than two such antennas.

35.3 Conditions to the permitted development rights under Class A

35.3.1

Paragraph A.2 sets out a number of conditions to the permitted developments rights for the three categories of Class A rights identified at 35.2.1 above.

35.3.2

First, Class A(a) and Class A(c) development is permitted subject to the condition that any antenna or supporting apparatus, radio equipment housing or development ancillary to radio equipment housing constructed, installed, altered or replaced on a building in accordance with that permission is, so far as is practicable, sited so as to minimise its effect on the external appearance of the building.[19]

35.3.3

Second, Class A(a) and Class A(c) development is permitted subject to the condition that any apparatus or structure provided in accordance with that permission is removed from the land, building or structure on which it is situated –[20]

(a) if such development was carried out in an emergency on any Article 2(3) land or on any land which is, or is within, a site of special scientific interest, at the expiry of the relevant period, or
(b) in any other case, as soon as reasonably practicable after it is no longer required for electronic communications purposes,

and such land, building or structure is restored to its condition before the development took place, or to any other condition as may be agreed in writing between the local planning authority and the developer.

35.3.4

Third, Class A(b) development is permitted subject to the condition that any apparatus or structure provided in accordance with that permission must, at the expiry of the relevant period, be removed from the land and the land restored to its condition before the development took place.[21]

19 Paragraph A.2(1).
20 Paragraph A.2(2).
21 Paragraph A.2(3).

35.3.5

Fourth, Class A development of the following descriptions is subject to the prior notification approval procedure set out in paragraph A.3 of Part 16, namely development –[22]

(a) on Article 2(3) land or land which is, or is within, a site of special scientific interest, or
(b) on any other land and consisting of the construction, installation, alteration or replacement of –
  (i) a mast;
  (ii) an antenna on a building or structure (other than a mast) where the antenna (including any supporting structure) would exceed the height of the building or structure at the point where it is installed or to be installed by 6 metres or more;
  (iii) a public call box;
  (iv) radio equipment housing, where the volume of any single development is in excess of 2.5 cubic metres,

There is an exception to this in case of emergency (for which there are special provisions).

35.4 Prior approval notification

35.4.1

In respect of the development listed at 35.3.5 above, the following conditions apply.

35.4.2

First, the developer must give notice of the proposed development to any person (other than the developer) who is an owner of the land to which the development relates, or a tenant, before making the application required by sub-paragraph (3) –[23]

(a) by serving a developer's notice on every such person whose name and address is known to the developer; and
(b) where the developer has taken reasonable steps to ascertain the names and addresses of every such person, but has been unable to do so, by local advertisement.

35.4.3

Second, where the proposed development consists of the installation of a mast within 3 kilometres of the perimeter of an aerodrome, the developer must notify the Civil Aviation

22 Paragraph A.2(4).
23 General Permitted Development Order 2015, Schedule 2, Part 16, paragraph A.3(1).

Authority, the Secretary of State for Defence or the aerodrome operator, as appropriate, before making the prior approval notification application explained below.[24]

35.4.4

Third, before beginning the development listed at 35.3.5 above, the developer must apply to the local planning authority for a determination as to whether the prior approval of the authority will be required as to the siting and appearance of the development. The application must be accompanied –[25]

(a) by a written description of the proposed development and a plan indicating its proposed location together with any fee required to be paid;
(b) by the developer's contact address, and the developer's email address if the developer is content to receive communications electronically;
(c) where the owner is required to be notified, by evidence that that requirement has been satisfied; and
(d) where the second condition above applies, by evidence that the Civil Aviation Authority, the Secretary of State for Defence or the aerodrome operator, as the case may be, has been notified of the proposal.

35.4.5

The local planning authority is then required to consult a number of defined consultees, and provide site notices and local advertisement of the application, in cases where the development does not accord with the provisions of the development plan or where it involves development on a site exceeding 1 hectare.[26]

35.4.6

The development, the subject of the prior approval notification application, must not begin before the occurrence of one of the following –[27]

(a) the receipt by the applicant from the local planning authority of a written notice of their determination that such prior approval is not required;
(b) where the local planning authority gives the applicant written notice that such prior approval is required, the giving of that approval to the applicant, in writing, within a period of 56 days beginning with the date on which they received the applicant's application;
(c) where the local planning authority gives the applicant written notice that such prior approval is required, the expiry of a period of 56 days beginning with the date on which the local planning authority received the application under sub-paragraph (4) without the local planning authority notifying the applicant, in writing, that such approval is given or refused; or
(d) the expiry of a period of 56 days beginning with the date on which the local planning authority received the application without the local planning

24 Paragraph A.3(2).
25 Paragraph A.3(4).
26 Paragraph A.3(5)–(6).
27 Paragraph A.3(7).

authority notifying the applicant, in writing, of their determination as to whether such prior approval is required.

35.4.7

For the purposes of a notification by the local planning authority before the expiration of the period of 56 days, the presumption, that posting can constitute the notification, can be rebutted if the notification is not received within that period.[28] Subject to any agreement in writing, where prior approval has been given, the development must be carried out in accordance with the details approved, and in any other case, in accordance with the details submitted with the application.[29] In any event the development must begin within five years of approval or where details were provided.[30] Once prior approval has been given, the planning permission under the General Permitted Development Order accrues or crystallises and is not lost by a later change in the status of the land, such as the creation of a conservation area: see *R(Orange Personal Communications Services Ltd) v Islington LBC*.[31]

35.4.8

The prior notification approval procedure does not infringe a claimant's rights under Article 6 of the European Convention of Human Rights, although the operation of it by the local planning authority may do so.[32]

35.5 Article 4 directions

35.5.1

Under Article 4 of the 2015 Order, the Secretary of State or the local planning authority may, if satisfied that it is expedient that development described in any part, class or paragraph in Schedule 2 (with certain exceptions) should not be carried out unless planning permission is granted for it on an application, may make a direction under Article 4 that the permission granted by Article 3 does not apply to any development specified in the direction.[33]

35.5.2

In the case of development mentioned in Class A of Part 16 of Schedule 2, which relates to development by a code operator, an Article 4 direction will not affect such development unless the direction specifically so provides.[34] Class A is explained below, but it includes the installation of electronic communications apparatus in accordance with the electronic communications code.

28 *Walsall MBC v Secretary of State for Communities and Local Government* [2012] JPL 1502.
29 Paragraph A.3(8).
30 Paragraph A.3(10).
31 [2006] JPL 1309.
32 *R(Nunn) v Secretary of State* [2005] Env LR 32.
33 Article 4(1).
34 Article 4(2)(c).

35.5.3

Article 4 directions may be found in such locations as conservation areas, national parks and areas of outstanding natural beauty. It follows that before a code operator considers whether it has permitted development rights under the General Permitted Development Order 2015, it should first ascertain whether there is an Article 4 direction in force in relation to the locality and the type of development it wishes to proceed with.

36 Compulsory purchase and entry for exploratory purposes

36.1 Introduction

36.1.1

Section 118 of, and Schedule 4, to the Communications Act 2003 makes provision for the compulsory purchase of land, and entry for exploratory purposes. The powers may be available to a provider of an electronic communications network, who is a 'code operator', by reason of a direction under section 106 of the 2003 Act. This chapter gives an outline of these powers. More specialised works should be consulted as to how these powers are obtained, the appropriate procedures, and the exercise of these powers.[1] The Schedule deals separately with the position in England and Wales, and in Scotland.

36.2 Duties of the Secretary of State in relation to compulsory purchase and entry for exploratory purposes

36.2.1

In exercising his powers under Schedule 4, to authorise a code operator to purchase compulsorily any land, or to enter any land for exploratory purposes, in England or Wales, it shall be the duty of the Secretary of State to have regard, in particular, to each of the following matters:

- (a) the duties imposed on OFCOM by sections 3 and 4 of the 2003 Act;
- (b) the need to protect the environment and, in particular, to conserve the natural beauty and amenity of the countryside;
- (c) the need to ensure that highways are not damaged or obstructed, and traffic not interfered with, to any greater extent than is reasonably necessary;
- (d) the need to encourage the sharing of the use of electronic communications apparatus.[2]

1 B. Denyer-Green, *Compulsory Purchase and Compensation*, 11th ed.; Roots, *Compulsory Purchase and Compensation*, subscription service.

2 Communications Act 2003, Schedule 4, paragraph 2.

36.2.2

For the purposes of Schedule 4, a 'code operator' is defined as meaning a provider of an electronic communications network in whose case the Electronic Communications Code is applied by a direction under section 106; and 'the operator's network', in relation to a code operator, means so much of the electronic communications network provided by the operator as is not excluded from the application of the Electronic Communications Code under section 106(5).[3]

36.2.3

It follows, that any code operator seeking compulsory powers, to acquire land or rights in land, or to enter land for exploratory purposes, must have regard to the general duties that the Secretary of State will take into account in considering whether or not to confirm any order conferring such powers.

36.3 Compulsory purchase of land

36.3.1

The Secretary of State may authorise a code operator to purchase compulsorily any land in England and Wales which is required by the operator:

(a) for, or in connection with, the establishment or running of the operator's network; or

(b) as to which it can reasonably be foreseen that it will be so required.[4]

However, no order is to be made authorising a compulsory purchase by a code operator except with OFCOM's consent.[5]

36.3.2

The power to purchase land compulsorily includes a power to acquire an easement or other right over land by the creation of a new right.[6]

36.3.3

In England and Wales, the Acquisition of Land Act 1981 applies to any compulsory purchase, under Schedule 4, as if the code operator were a local authority within the meaning of that Act.[7] The effect of incorporating the 1981 Act is that the provisions relating to the seeking, and making, of a compulsory purchase order will have application as if a code operator were a local authority seeking powers of compulsory purchase. In summary, the code operator will first prepare a draft compulsory purchase order, and then obtain the consent of OFCOM before proceeding any further. If

3 Communications Act 2003, Schedule 4, paragraph 1.
4 Communications Act 2003, Schedule 4, paragraph 3(1).
5 Communications Act 2003, Schedule 4, paragraph 3(2).
6 Communications Act 2003, Schedule 4, paragraph 3(3).
7 Communications Act 2003, Schedule 4, paragraph 3(4).

consent is given, it can then submit the order to the Secretary of State for confirmation. Part 2 of the 1981 Act prescribes certain requirements as to notification and publicity for the order, including the publication of a notice for two successive weeks in one or more local newspapers circulating in the locality, a notice to be fixed to the site, and a notice served on every 'qualifying person' explaining the effect of the order. In each case the notice must specify a time (at least 21 days from the service of the notice) within which objections can be made.[8] A 'qualifying person' is widely defined, and includes the owner of any interest in the affected land, and the owner of any right over it, such as an easement.[9]

36.3.4

In support of the making of a compulsory purchase order, the code operator must prepare a Statement of Reasons, setting out the details required in the *Guidance on Compulsory Purchase Process and the Crichell Down Rules* ('the Guidance').[10] The Guidance makes clear that a compulsory purchase order should only be made where there is a compelling case in the public interest. A code operator should ensure that the purposes for which the compulsory purchase order is made justify interfering with the human rights of those with an interest in the land affected, and particular consideration should be given to the provisions of Article 1 of the First Protocol to the European Convention on Human Rights and, in the case of a dwelling, Article 8 of that Convention.[11]

36.3.5

The Guidance also advises that undertaking negotiations in parallel to the preparation and making of a compulsory purchase order can help to build a good working relationship with those whose interests are affected by showing that the authority is willing to be open and to treat their concerns with respect. Talking to landowners will also assist the code operator to understand more about the land, or rights over it, it seeks to acquire, and any physical or legal impediments to development that may exist. It may also help in identifying what measures can be taken to mitigate the effects of the scheme underlying the compulsory acquisition on landowners and neighbours, thereby reducing the cost of a scheme. Code operators will be expected to provide evidence that meaningful attempts at negotiation have been pursued or at least genuinely attempted, save for lands where landownership is unknown or in question.[12]

36.3.6

If there are any objections or representations to the order, the Secretary of State may order the holding of a public inquiry.[13] Following any such public inquiry, the Inspector

8 Acquisition of Land Act 1981, s.12.
9 Acquisition of Land Act 1981, s.12.
10 Department of Communities and Local Government, February 2018.
11 See Guidance, paragraph 12.
12 See Guidance, paragraph 16.
13 Any public inquiry will be held in accordance with the Compulsory Purchase (Inquiries Procedure) Rules 2007 (SI 2007/3617).

will make a report to the Secretary of State. The Secretary of State will then consider whether to confirm the order or otherwise.

36.3.7

In respect of special descriptions of land, namely land belonging to local authorities or to statutory undertakers for the purposes of their undertaking, National Trust land held by the Trust inalienably, where an objection to the order has been made by the Trust and has not been withdrawn, and land forming part of a common, open space or field or field garden allotment, there are special procedures. The compulsory purchase order is subject to special parliamentary procedure.[14] Subject to certain exceptions,[15] the order must be laid before Parliament and petitions against the order may be presented within 21 days. Any petitions will be considered by a joint committee of the two Houses, who will then report as to whether the order shall be approved, and with or without amendments.

36.3.8

Certain provisions of the Town and Country Planning Act 1990 have effect in relation to land acquired compulsorily by a code operator as they have effect in relation to land acquired compulsorily by statutory undertakers.[16] The provisions of the 1990 Act are those concerned with the use and development of consecrated land and burial ground (special requirements to be observed), the use and development of land for open spaces (authorisation of use), the extinguishment of rights of way, and rights as to apparatus, of statutory undertakers (with safeguards).[17]

36.4 Compulsory purchase in Scotland and Northern Ireland

36.4.1

Schedule 4 deals with the position in Scotland and Northern Ireland only insofar as the legislation making provision for the making of compulsory purchase orders, and the limitation on the use of powers in relation to statutory undertakers, are different in those jurisdictions. In Scotland, compulsory purchase orders are made under the Acquisition of Land (Authorisation Procedure) (Scotland) 1947, and the relevant provisions relating to land acquired by statutory undertakers, as applied to code operators, are found in the Town and Country Planning (Scotland) Act 1997.

36.4.2

As to Northern Ireland, the powers to acquire land by a vesting order are found in Schedule 6 to the Local Government Act (Northern Ireland) 1972. As to the provision

14 Acquisition of Land Act 1981, ss.17–18.
15 Acquisition of Land Act 1981, s.19.
16 Communications Act 2003, Schedule 4, paragraph 3(5).
17 Communications Act 2003, Schedule 4, paragraphs (5) and (6), referring to ss.238–40, 241, and 271 to 274 of the Town and Country Planning Act 1990.

relating to inquiries, these are found in Schedule 8 to the Health and Personal Social Services (Northern Ireland) Order 1972.

36.5 Entry on land for exploratory purposes in England and Wales

36.5.1

A person nominated by a code operator, and duly authorised in writing by the Secretary of State, may at any reasonable time, enter upon and survey any land in England and Wales for the purpose of ascertaining whether the land will be suitable for use by the code operator for, or in connection with, the establishment or running of the operator's network.[18]

36.5.2

This power does not apply in relation to land covered by buildings or used as a garden or pleasure ground.[19]

36.5.3

The supplementary provisions, in the Country Planning Act 1990, as to rights of entry, apply to the powers of entry for exploratory purposes under the 2003 Act.[20] Those supplementary provisions include the following matters. A person authorised to enter any land shall, if so required, produce evidence of his authority and state the purpose of his entry before so entering, and shall not demand admission as of right to any land which is occupied unless 28 days' notice of the intended entry has been given to the occupier.[21] Any person who wilfully obstructs a person acting in the exercise of his powers shall be guilty of an offence and liable on summary conviction to a fine not exceeding level 3 on the standard scale.[22] If any person who, in compliance with the powers of entry, is admitted into a factory, workshop or workplace, discloses to any person any information obtained by him in it as to any manufacturing process or trade secret, he shall be guilty of an offence.[23] The authority of the Secretary of State is required for the carrying out of any works authorised if the land in question is held by statutory undertakers, and they object to the proposed works on the ground that the execution of the work would be seriously detrimental to the carrying on of their undertaking.[24]

36.5.4

The supplementary provisions referred to above are subject to two important modifications. In relation to the power to search and bore for the purposes of ascertaining the nature of the subsoil, this cannot include searching and boring for the presence of

18 Communications Act 2003, Schedule 4, paragraph 6(1).
19 Communications Act 2003, Schedule 4, paragraph 6(2).
20 Communications Act 2003, Schedule 4, paragraph 6(3)–(4).
21 Town and Country Planning Act 1990, s.325(1).
22 Town and Country Planning Act 1990, s.325(2).
23 Town and Country Planning Act 1990, s.325(3).
24 Town and Country Planning Act 1990, s.325(9).

minerals.[25] Second, in place of the 24-hour notice to enter, required under section 325 of the 1990 Act, 28 days' notice is required.[26]

36.5.5

Where, in the exercise of the powers of entry, any damage is caused to land or to chattels, the code operator must make good the damage or pay compensation in respect of the damage to every person interested in the land or chattels.[27] Further, where in consequence of an exercise of the power of entry, a person is disturbed in his enjoyment of land or chattels, the code operator must pay that person compensation in respect of the disturbance.[28] A person will be disturbed in his enjoyment of land if, for example, business is interfered with or obstructed for a period of time and profits or other losses are incurred. That may apply to agricultural as well as business premises.

36.6 Entry on land for exploratory purposes in Scotland or Northern Ireland

36.6.1

The provisions relating to entry on land for exploratory purposes in Scotland and Northern Ireland are, broadly the same. However, separate provision is made by virtue of the relevant legislation in each jurisdiction. In Scotland, the supplementary provisions relating to the exercise of powers of entry are found in ss.269(6) and 270(1) to (5), (8) and (9) of the Town and Country Planning (Scotland) Act 1997. In Northern Ireland the supplementary provisions relating to the exercise of powers of entry are contained in sub-sections (2) to (5) and (8) of s.40 of the Land Development Values (Compensation) Act Northern Ireland 1965.

36.7 Acquisition of land by agreement

36.7.1

Where land is acquired by agreement by a code operator in England and Wales, the provisions of Part 1 of the Compulsory Purchase Act 1965, with three exclusions, apply as they would otherwise apply to a compulsory purchase.[29] The excluded provisions are sections 4 to 8 of the 1965 Act (time limits, notices to treat, etc.) and s.31 (ecclesiastical property). It is not certain that the application of the 1965 Act necessarily incorporates all the relevant provisions relating to the assessment of, and entitlement to, compensation for the acquisition of land or any rights in land, to an acquisition by agreement. It follows that if an agreement is made to acquire land by a code operator, or any rights in land, any contract should satisfy the requirements of s.2 of the Law of Property (Miscellaneous Provisions) Act 1989 and clearly make provision for the measure of the compensation, and of the additional sums that may be payable for disturbance and other losses.

25 Communications Act 2003, Schedule 4, paragraph 6(4)(a).
26 Communications Act 2003, Schedule 4, paragraph 6(4)(b).
27 Communications Act 2003, Schedule 4, paragraph 6(5).
28 Communications Act 2003, Schedule 4, paragraph 6(6).
29 Communications Act 2003, Schedule 4, paragraph 9(1).

36.8 Other matters

36.8.1

Where a compulsory purchase order is made under the Communications Act 2003, reference should be made to more specialised works in relation to the following matters: the procedures for the exercise of powers of compulsory purchase of land, or any rights in land, the manner of taking of possession of land, and as to the measure of compensation, the heads of claim for disturbance and other losses.[30]

30 B. Denyer-Green, *Compulsory Purchase and Compensation*, 11th ed.; Roots, *Compulsory Purchase and Compensation*, subscription service.

37 Telecommunications and non-domestic rates

37.1 Scope of this chapter

37.1.1

Non-domestic rates are a property tax. Many of the fundamental concepts found in the law of rating are also found in or overlap with those found in property law. The arrangement and configuration of apparatus can affect how and who is liable to pay rates. While taxation generally is outside the scope of this book, the nature of rating means that it can usefully be considered together with other aspects of property law relating to telecommunications. A comprehensive review of the law of rating, however, is outside the ambit of this chapter.[1] It will be limited to introducing the legal framework of rating in very general terms but highlighting the peculiarities which apply to telecommunications apparatus and sites.[2] As will be seen, the nature of telecommunications and how the industry operates have required certain assumptions to be made by statute so that the industry can be shoe-horned into a taxation system devised for different times and different sorts of interests in land.

37. 2 Historical matters

37.2.1

The origins of the law of rating were in the Poor Relief Act 1601.[3] The persons made chargeable under this Act were 'every inhabitant, parson, vicar and every occupier of land'. The charge theoretically extended to both real and personal property although in practice personal property appears to have been left out of account. The liability of every 'inhabitant' was abolished by the Poor Rate Exemption Act 1840. Thereafter the tax became a tax on the occupation of land.

1 For a full treatment of the law rating see *Ryde on Rating and Council Tax* (Butterworths). See too the Valuation Office Agency's Rating Manual.

2 In this chapter we have referred to 'telecommunications apparatus' rather than 'electronic communications equipment' or 'electronic communications apparatus' used elsewhere in the book. This follows the terminology of Non-Domestic Rating (Telecommunications Apparatus) (England) Regulations 2000 SI 2000/2421 and Non-Domestic Rating (Telecommunications Apparatus) (Wales) Regulations 2000 SI 2000/3383: see para 37.7.4 below.

3 43 Eliz I.

37.2.2

The law was subsequently contained in the General Rate Act 1967. The Local Government Finance Act 1988 repealed this earlier Act with effect from 1 April 1990. From that date, only relevant *non-domestic* property was subject to rates. Nevertheless the 1967 Act remains relevant and some important aspects of the previous law of rating have been continued in the current legislation.

37.3 The main legislation

37.3.1

The main source of the law of rating is now the Local Government Finance Act 1988 ('the LGFA 1988'). The primary provision governing the collection of rates generally is the Non-Domestic Rating (Collection and Enforcement) (Local Lists) Regulations 1989.[4] As will be explained below, it is also necessary to consider the Central Rating Lists (England) Regulations 2000[5] and the Central Rating Lists (Wales) Regulations.[6] Specific adaptations of the law for telecommunications are found in the Non-Domestic Rating (Telecommunications Apparatus) (England) Regulations 2000,[7] the Non-Domestic Rating (Telecommunications Apparatus) (Wales) Regulations 2000.[8] Between 1 April 2017 and 1 April 2022, specific relief where telecommunications apparatus is to be provided under amendments to the Local Government Finance Act 1988 introduced by the Telecommunications Infrastructure (Relief from Non-Domestic Rates) Act 2018.[9]

37.4 Key concepts

37.4.1

In essence, rates are levied on ratepayers in respect of non-domestic property (hereditaments) shown either on local lists or central lists. Rates were originally levied where ratepayers occupied property, but they are also now levied on unoccupied property. The level of rates charged depends upon a statutory formula which is underpinned by the 'rateable value' of each hereditament. In considering the peculiarities of the rating of property used for telecommunications equipment, it is necessary to be familiar with the key concepts found in the law of rating generally. These can be considered under the following headings.

(1) rateable property and the 'hereditament' – see section 37.5 below;
(2) the lists – see section 37.6 below;

4 SI 1989/1058.
5 SI 2000/535.
6 SI 2005/442 W40.
7 SI 2000/2421.
8 SI 2000/3383.
9 See paragraph 37.7.2 below.

(3) when liability for rates arises for rateable occupation, unoccupied property, central lists and for shared telecommunications sites – see section 37.7 below;
(4) the formulae for the calculation of rates and 'rateable value' – see section 37.8 below.

37.5 Rateable property and the 'hereditament'

37.5.1 Generally

Rateable property comprises every relevant non-domestic hereditament. Every such hereditament must be entered in a local rating list unless it is exempt or entered in a central rating list.[10]

37.5.2 'Hereditament'

In the LGFA 1988, a hereditament is 'anything which, by virtue of the definition in section 115(1) of the 1967 Act, would have been a hereditament for the purposes of that Act' if the 1988 Act had not been passed.[11] Section 115(1) of the 1967 Act provided that

> 'hereditament' means property which is or may become liable to a rate, being a unit of such property which is or would fall to be, shown as a separate item in the valuation list'

A hereditament is not a hereditament unless it is liable to a rate shown as a separate item in the valuation list but an item will only be shown as a separate item in the valuation list if it is a hereditament. The circularity of this definition has led the Court of Appeal to describe these provisions as 'legislative gobbledegook'.[12]

Accordingly, to make sense of the law of rating, it is necessary to consider case law to determine what constitutes a hereditament. This establishes that the hereditament is equated with the unit of occupation.[13] In most instances, therefore, the hereditament will be defined by reference to the extent of a person's rateable occupation since there cannot be more than one rateable occupier of a hereditament. Telecommunications hereditaments, however, are subject to special rules as will be explained below.[14]

The leading case explaining the general position, however, is that of *Woolway (VO) v Mazars LLP*.[15] In that case, the Supreme Court identified three tests for what constitutes a hereditament: (1) a geographical test, (2) a functional test and (3) a test of 'effectual enjoyment'. Lord Sumption JSC said the following:[16]

10 LGFA 1988 s.42. Agricultural land or agricultural buildings are amongst the exemptions: see LGFA Schedule 5.
11 LGFA 1988 s.64(1).
12 *Reeves (LO) v Northrop* [2013] EWCA Civ 362 at [9].
13 *Vtesse Networks Ltd v Bradford (VO)* [2006] EWCA Civ 1339.
14 See paragraphs 37.6.4 et seq.
15 [2015] AC 1862.
16 [2015] AC 1862 at [12].

> I derive ... three broad principles relevant to cases like this one where the question is whether distinct spaces under common occupation form a single hereditament. First, the primary test is, as I have said, geographical. It is based on visual or cartographic unity. Contiguous spaces will normally possess this characteristic, but unity is not simply a question of contiguity, as the second Bank of Scotland case illustrates. If adjoining houses in a terrace or vertically contiguous units in an office block do not intercommunicate and can be accessed only via other property (such as a public street or the common parts of the building) of which the common occupier is not in exclusive possession, this will be a strong indication that they are separate hereditaments. If direct communication were to be established, by piercing a door or a staircase, the occupier would usually be said to create a new and larger hereditament in place of the two which previously existed. Secondly, where in accordance with this principle two spaces are geographically distinct, a functional test may nevertheless enable them to be treated as a single hereditament, but only where the use of the one is necessary to the effectual enjoyment of the other. This last point may commonly be tested by asking whether the two sections could reasonably be let separately. Third, the question whether the use of one section is necessary to the effectual enjoyment of the other depends not on the business needs of the ratepayer but on the objectively ascertainable character of the subjects. The application of these principles cannot be a mere mechanical exercise. They will commonly call for a factual judgment on the part of the valuer and the exercise of a large measure of professional common sense. But in my opinion they correctly summarise the relevant law. They are also rationally founded on the nature of a tax on individual properties. If the functional test were to be applied in any other than the limited category of cases envisaged in the second and third principles, a subject (or in English terms a hereditament) would fall to be identified not by reference to the physical characteristics of the property, but by reference to the business needs of a particular occupier and the use which, for his own purposes, he chose to make of it.

Applying this approach, where an occupier uses two or more separated spaces within a single building which are accessed through a communal area, the separate areas of occupation are treated as separate hereditaments for business rates purposes.

By reason of the Non-Domestic Rating (Miscellaneous Provisions) Regulations 1989 there is no longer any rule that a single hereditament crossing the boundary between rating districts should be treated as two separate hereditaments.[17]

37.5.3 'Relevant hereditament'

A hereditament is a 'relevant hereditament' if it consists of lands, coal mines, mines of any other description, other than a mine of which the royalty or dues are for the time being wholly reserved in kind, or certain rights for the purposes of exhibiting advertising or to use land for operating meters to measure the supply of gas or electricity and, by statutory instrument, to other meters.[18]

17 SI 1989/1060 reg 6.
18 LGFA 1988 s.64(4).

37.5.4 'Non-domestic' hereditament

A hereditament is non-domestic if either (a) it consists entirely of property which is non-domestic or (b) it is a composite hereditament.

Property is non-domestic if (a) it is used wholly or mainly for the purposes of living accommodation, (b) it is a yard, outhouse or other appurtenance belonging to or enjoyed with such property, (c) it is a private garage which either has a floor area of 25 sq metres or less or is used wholly or mainly for the accommodation of a private vehicle or (d) it is private storage premises used wholly or mainly for the storage of articles of domestic use.[19]

37.5.5 Composite hereditaments

A hereditament is composite if part only of it consists of domestic property. If a hereditament is a composite hereditament, it is the whole hereditament which falls to be entered in the local rating list, notwithstanding the domestic part.[20]

37.5.6 Incorporeal rights

Incorporeal rights such as easements cannot in themselves constitute rateable property. They will only become rateable property if in practice the grantee of the right can be said to be in rateable occupation of the land.[21] The question is whether the grantee's occupation is 'paramount' to that of the grantor.[22] Rateable occupation is considered briefly below.[23]

37.5.7 Chattels and equipment

As explained above, personal property is outside the scope of rates.[24] Accordingly, chattels themselves are not generally rateable.[25] On the other hand, rateable occupation might occur by reason of the presence of chattels.

Whether or not an item forms part of the land comprising the hereditament itself or whether it remains a chattel is a matter of fact dependent on the intention of the person placing the chattel on the land, the elements of annexation, period, size quality, amenities and purposes of enjoyment are all relevant.[26] It seems clear, however, that cables and telegraph posts can themselves constitute rateable subjects.[27] In the case of *Vtesse Networks Limited v Alan Roy Bradford (Valuation Officer)*[28] the Court of Appeal

19 LGFA 1988 s.66; see this section too for the exceptions.

20 LGFA 1988 ss. 42(1), 64(4) and 64(8). The amount of rates payable, however, is calculated by reference to the value reasonably attributable to non-domestic use: LGFA 1988 Sched 6 paragraph 2(1A).

21 *Holywell Union and Halkyn Parish v Halkyn District Mines Drainage Co.* [1895] AC 117 at 126.

22 *Westminster City Council and Kent Valuation Committee v Southern Rly Co.* [1936] AC 511, 529.

23 See paragraph 37.7.

24 See paragraph 37.2 above.

25 But plant and machinery may be made rateable under the Valuation for Rating (Plant and Machinery) (England) Regulations 2000 SI 2000/540; Valuation for Rating (Plant and Machinery)(Wales) Regulations 2000 SI 2000/1097.

26 *LCC v Wilkins (Valuation Officer)* [1959] AC 362.

27 *Ibid.*, at 378 per Lord Radcliffe.

28 [2006] EWCA Civ 1339 at [18].

confirmed that fibre cables can comprise a rateable hereditament. Accordingly, quite apart from the special rules applicable to telecommunications equipment, land used for telecommunications purposes and the equipment affixed to it can constitute a rateable hereditament.

37.5.8 Combined hereditaments and artificial hereditaments

Specific statutory provisions require property to be treated as separate or combined hereditaments. As will be explained below, such provisions have been applied to telecommunications equipment attached to a hereditament.[29]

37.5.9 Time for deciding nature of hereditament

A right or other property is a hereditament on a particular day if (and only if) it is a hereditament immediately before the day ends.[30] Similarly, a hereditament is relevant, non-domestic, composite, unoccupied or wholly or partly occupied on a particular day if (and only if) it is a relevant, non-domestic, composite, unoccupied or wholly or partly occupied (as the case may be immediately before the day ends.[31] Furthermore, for the purpose of deciding the extent (if any) to which a hereditament consists of domestic property on a particular day, the state of affairs existing immediately before the day ends is treated as having existed throughout the day.[32]

37.6 The lists

37.6.1 Generally

It follows from the definition of hereditament[33] that to constitute a rateable hereditament it must be entered on a valuation list. There are two lists. There is a local list for each billing authority area and a central list.[34] All relevant non-domestic hereditaments must be shown in the appropriate local list unless they are required to be included in the central list.[35]

37.6.2 Local non-domestic rating lists

The valuation officer for a billing authority must compile and maintain lists for the authority called local non-domestic rating lists.[36] Local lists were initially to be compiled

29 See section 37.6 below. Non-Domestic Rating (Telecommunications Apparatus) (England) Regulations 2000 SI 2000/2421, Non-Domestic Rating (Telecommunications Apparatus) (Wales) Regulations 2000 SI 2000/3383 SI 2000/3383.

30 LGFA 1988 s.67(3).

31 LGFA 1988 s.67(4).

32 LGFA 1988 s.67(5).

33 See paragraph 37.5.2 above and LGFA 1988 s.64(1) and General Rate Act 1967 s.115(1).

34 In England, there is also a rural settlements list in respect of which certain hereditaments are entitled to relief: see LGFA 1988 s.42A(1).

35 LGFA 1988 s.42(1). For the procedure for compiling, altering and challenging rating lists see *Ryde on Rating*.

36 LGFA 1988 s.41(1).

on 1 April 1990 and thereafter on 1 April in every fifth year.[37] The list that was due to be compiled on 1 April 2015 was postponed until 1 April 2017.[38] The list must show, for each day in each chargeable financial year for which it is in force, each hereditament which fulfils on the day concerned the following four conditions, namely:[39]

(a) it is situated in the billing authority's area;
(b) it is a relevant non-domestic hereditament;
(c) at least some of it is neither domestic property nor exempt from non-domestic rating; and
(d) it is not a hereditament which must be shown for the day in a central non-domestic rating list.

37.6.3 Central non-domestic rating lists

It is the duty of the central valuation officer to compile and then maintain central non-domestic rating lists.[40] As with local non-domestic rating lists, central lists were initially to be compiled on 1 April 1990 and thereafter on 1 April in every fifth year.[41] The list that was due to be compiled on 1 April 2015 was postponed until 1 April 2017.[42]

The contents of any central non-domestic rating list compiled on or after 1 April 2005 is prescribed by the Central Rating Lists (England) Regulations 2005[43] and the Central Rating Lists (Wales) Regulations 2005.[44] The Schedules to these regulations designate a number of corporate bodies and prescribe in relation to each of them a group or series of 'relevant hereditaments'.[45] As described in the next paragraph hereditaments of British Telecommunications plc or which are licensed or let by British Telecommunications plc fall within the scope of these regulations. Other telecommunications hereditaments fall within local lists.

The central lists must show for each day in each year for which the list is in force, the name of each 'person' so designated and against that name each hereditament which on that day is occupied or (if unoccupied) owned by that person and which falls within the description contained in the schedule.[46] The list must also show against the name of the designated person the rateable value (as a whole) of the hereditaments and (if after 1 April 2005) the first day for which that rateable value took effect.[47]

37 LGFA 1988 s.41(2). In Wales amalgamated lists were prepared under s.421A and were deemed to come into force on 1 April 1995: s.41A(4).
38 LGFA 1988 s.41(2) and (2A) as amended and inserted by Growth and Infrastructure Act 2013 s.29. In relation to Wales see LGFA 1988 s.54 and the Rating Lists (Postponement of Compilation) (Wales) Order 2014 SI 2014/1370.
39 LGFA 1988 s.42(1).
40 LGFA 1988 s.52(1). Separate lists are compiled and maintained for England and for Wales: see LGFA 1988 s.140(2)(a).
41 LGFA 1988 s.41(2). In Wales amalgamated lists were prepared under s.421A and were deemed to come into force on 1 April 1995: s.41A(4).
42 LGFA 1988 s.52(2) and (2A) as amended and inserted by Growth and Infrastructure Act 2013 s.29. In relation to Wales see LGFA 1988 s.54 and the Rating Lists (Postponement of Compilation) (Wales) Order 2014 SI 2014/1370.
43 SI 2005/551 made under LGFA 1988 s.53.
44 SI 2005/422 made under LGFA 1988 s.53.
45 SI 2005/551 reg 3; SI 2005/422 reg 4.
46 SI 2005/551 reg 4; SI 2005/442 reg 5.
47 SI 2005/551 reg 5; SI 2005/442 reg 6; LGFA 1988 s.53(3).

A hereditament is treated as shown in a central non-domestic rating list if on the day in question it falls within a class of hereditament shown for that day in the list; and for that purpose a hereditament is taken to fall within a class on a particular day only if it does so immediately before the day ends.[48]

37.6.4 The central list and telecommunications: treatment of hereditaments as one hereditament

The effect of the Central List Regulations[49] is that where

(a) British Telecommunications plc occupies or, if it is unoccupied, owns any hereditament which comprises posts wires, fibres, cables, ducts, telephone kiosks, towers, masts, switching equipment, or other equipment, or easements or wayleaves, being property used for the monitoring, processing or transmission of telecommunications or other signals for the provision of electronic communications services; or
(b) any person occupies, or if it is unoccupied, owns any hereditament which is an 'unbundled local loop',

and which would apart from the Regulations be more than one hereditament, those hereditaments are to be treated as one hereditament.[50]

For these purposes, an 'unbundled local loop' means:[51]

(a) cables, fibres, wires and conductors (or any part of them) used or intended to be used for carrying communications or other signals between the network terminating equipment on the premises of the end-users and premises (or any part of them) used for the processing of the communications or other signals, and the land occupied therewith; and
(b) poles, posts, towers, masts, mast radiators, pipes, ducts, conduits and any associated supports and foundations (or any part of them) used or intended to be used in connection with any of the items listed in sub-paragraph (a) and any land occupied therewith,

which British Telecommunications plc lets or licenses to any person.

37.7 Liability for rates: rateable occupation, unoccupied property, shared telecommunications sites, telecommunications equipment used for domestic or business purposes

37.7.1 Occupied property

A person becomes liable to non-domestic rate in respect of any day in the year if (a) on that day he is in occupation of all or part of the hereditament and (b) the hereditament is shown for the day in a local non-domestic rating list in force.[52] Thus, in general, it is

48 LGFA 1988 s.67(9).
49 SI 2005/551 and SI 2005/422.
50 In England, SI 2005/551 reg 8; in Wales, SI 2005/422 reg 8.
51 *Ibid.*
52 LGFA 1988 s.43(1).

the occupier of a hereditament who is liable for rates. In essence, for occupation to be rateable it must be (1) actual, (2) exclusive, (3) beneficial and (4) actual.[53] The concept of 'occupation' is perhaps difficult to comprehend when applied to an object as narrow as a fibre optic cable running through land. In the case of *Vtesse Networks Limited v Alan Roy Bradford (Valuation Officer)*[54] the Court of Appeal confirmed not only that fibre cables can comprise such a hereditament but their use could be sufficient to amount to exclusive occupation where no third party had control over the use which the ratepayer made of the cable. So, the ratepayer was in rateable occupation of a hereditament comprising fibre optic cables even though the ratepayer had no separate access to the cables. It appears that as a general rule of thumb if a person owns and/or leases from a provider dark (unlit) fibre and lights it themselves, they will be in rateable occupation. Under the Telecommunications Infrastructure (Relief from Non-Domestic Rates) Act 2018 provision is made for relief from non-domestic rates for an occupied hereditament where the hereditament is wholly or mainly used for the purposes of facilitating the transmission of communications by any means involving the use of electrical or electromagnetic energy, the chargeable day concerned falls before 1 April 2022 and conditions provided for by regulation are satisfied. The amendments will have effect in relation to financial years beginning on or after 1 April 2017: accordingly, the power to make regulations includes a power to make provision in relation to times before the coming into force of the Act. This is considered further below.[55]

37.7.2 Unoccupied property

In some circumstances a person, the owner, may be liable in respect of unoccupied hereditaments.[56] A person becomes liable to non-domestic rate in respect of any day in the year if on that day (a) none of the hereditament is occupied, (b) he is the owner of the whole hereditament, (c) the hereditament is shown in a local non-domestic rating list and (d) falls within a class prescribed by the Secretary of State by Regulations.[57]

The class of non-domestic hereditaments now prescribed in respect of which rates can be charged on unoccupied property consists of all 'relevant non-domestic hereditaments' subject to 13 exceptions.[58] Under the Telecommunications Infrastructure (Relief from Non-Domestic Rates) Act 2018 provision is made for relief from non-domestic rates for an unoccupied hereditament where the hereditament is wholly or mainly used for the purposes of facilitating the transmission of communications by any means involving the use of electrical or electromagnetic energy, the chargeable day concerned falls before 1 April 2022 and conditions provided for by regulation are satisfied. The amendments will have effect in relation to financial years beginning on or after 1 April 2017: accordingly, the power to make regulations includes a power to make provision in relation to times before the coming into force of the Act. This is considered further below.[59]

53 *John Laing & Son Ltd v Kingswood Assessment Area Assessment Committee* [1949] 1 KB 374, CA; *LCC v Wilkins (Valuation Officer)* [1957] AC 362.
54 [2006] EWCA Civ 1339 at [18].
55 See paragraph 37.8.2.
56 LGFA 1988 s.45(1), (9), (10).
57 LGFA 1988 s.45(1).
58 See the Non-Domestic Rating (Unoccupied Property) (England) Regulations 2008 SI 2008/386.
59 See paragraph 37.8.2.

37.7.3 Hereditaments entered in the central rating list

A person becomes liable to non-domestic rates in respect of any day if his name is shown in a central rating list.[60]

37.7.4 Specific provision for telecommunications sites

The nature of telecommunications sites and apparatus, however, means that applying the usual rules makes it particularly difficult to discern distinct hereditaments and occupation where sites are shared. Thus, telecommunications sites are one of the classes of hereditaments that may fall outside the normal rules. Special rules are provided under the Non-Domestic Rating (Telecommunications Apparatus) (England) Regulations 2000[61] and the Non-Domestic Rating (Telecommunications Apparatus) (Wales) Regulations 2000.[62] For these purposes a 'telecommunications hereditament' means

(a) a site forming a hereditament occupied exclusively by telecommunications apparatus, or
(b) a site which would constitute a single hereditament but for the occupation of the whole of it exclusively by telecommunications apparatus operated or owned by more than one operator,

and for the purposes of determining whether a hereditament is occupied exclusively by telecommunications apparatus, the presence on the site of, and of accommodation for, personnel for the sole purpose of maintaining, repairing, operating or safeguarding that apparatus shall be disregarded.[63]

'Telecommunications apparatus' includes –

(a) telecommunications apparatus within the meaning given by Schedule 2 to the 1984 Act which is used, or designed for use, for –
  (i) wireless telegraphy within the meaning given by section 19 of the 1949 Act, or
  (ii) broadcasting; and
(b) structures in the nature of huts or other buildings (including structures forming part only of a building) used, or designed for use, solely to house apparatus within the description in paragraph (a);

together with any ancillary equipment occupied exclusively for the purposes of an operator.[64]

These definitions have not yet been amended so that they dove-tail with the New Code. The repeal of Schedule 2 of the 1984 Act cannot affect these definitions.

60 LGFA 1988 s.54.
61 SI 2000/2421.
62 SI 2000/3383.
63 SI 2000/2421 reg 2; SI 2000/3383 reg 2.
64 *Ibid.*

37.7.5 Telecommunications sites where there is one occupier or operator

Where there is only one occupier or operator of a hereditament used for telecommunications purposes, then no problem arises and liability for rates will be assessed to the operator of the site (even where the operator merely provides infrastructure and is not themselves broadcasting). Where a telecommunications hereditament is shown on the central list,[65] then the list will identify the occupier who will be liable for rates.

37.7.6 Treatment of telecommunications apparatus on shared sites

In practice, however, many telecommunications sites are shared. Special provisions allow the aggregation of site sharers into a single hereditament. The relevant provisions are found in the Non-Domestic Rating (Telecommunications Apparatus) (England) Regulations 2000[66] which came into force on 1 October 2000 and the Non-Domestic Rating (Telecommunications Apparatus) (Wales) Regulations 2000[67] which came into force on 1 April 2001.

The Telecommunications Apparatus Regulations apply to 'telecommunications hereditaments'. These comprise

a. a site forming a hereditament occupied exclusively by telecommunications apparatus or
b. a site which would constitute a single hereditament but for the occupation of it exclusively by telecommunications apparatus operated or owned by more than one operator.

For the purposes of determining whether a hereditament is occupied exclusively by telecommunications apparatus, the presence on the site of, and accommodation for, personnel for the sole purpose of maintaining, repairing, operating or safeguarding that apparatus is disregarded.[68]

Where telecommunications apparatus is attached to or situated in or on a 'telecommunications hereditament' and is occupied or owned so that it would otherwise constitute one or more separate hereditaments, then the telecommunications apparatus is treated as a single hereditament in the occupation or ownership of the 'host'.[69] A host is defined in the Telecommunications Apparatus Regulations as 'An individual who has the right to receive payment in respect of the use of any part of the site by any other who is an operator occupying telecommunications apparatus, or would have such a right if any such part were so used.'[70] So, a telecommunications site provider or operator who receives the site share payments (or would be entitled to receive such payments) is the ratepayer for the whole telecommunications hereditament.

65 See paragraphs 37.6.3 et seq.
66 SI 2000/2421.
67 SI 2000/3383.
68 SI 2000/2422 reg 2. For convenience, reference will be made to the provisions of the English regulations. There is no material difference in Wales.
69 2000/2421 reg 3(1), (2).
70 SI 2000/2421 reg 2.

It is important to distinguish the situation where a host within the definition shares a telecommunications site which is a 'telecommunications hereditament' from that where the site is not within the definition. So, a mast on a stand-alone site exclusively occupied by telecommunications apparatus which is shared will comprise a telecommunications hereditament and the host who receives payments from sharers will be liable for rates under the Telecommunications Apparatus Regulations.

On the other hand, if the apparatus forms part of a larger site which is not exclusively occupied by telecommunications apparatus, then it will not comprise a separate 'telecommunications hereditament' and each sharer will be assessed on the parts of the site of which they have exclusive use. If the sharers share non-exclusive accommodation with the central list host they will be included in the central list assessment as there is no separately identifiable hereditament and the paramount control of the shared accommodation will be with the designated central list occupier. The central list host's site will form part of the occupier's central rating list assessment (unless the sharer is itself designated under the central list, in which case its liability will be under its own central list assessment).

37.7.7 Excepted apparatus: central rating list 'designated persons'

Apparatus is 'excepted apparatus' and will not be included in this aggregation if it is occupied by a person whose name is for the time being shown in a central non-domestic rating list and who is not the host in relation to the site on which the apparatus is situated or to which it is attached.[71] So, telecommunications apparatus of central rating list occupiers[72] who are *not* the host are treated as 'excepted apparatus' and are excluded from the aggregation. The person shown on the central list will be liable for the rates in respect of such apparatus.

37.7.8 Exclusion from aggregation where building or structure used for other purposes

The provisions requiring aggregation also do not apply where a telecommunications hereditament is on or forms part of a building or structure all or any part of which is owned or occupied by the host for any purpose other than the provision of or operation of a site for telecommunications apparatus.[73] So, the Telecommunications Apparatus regulations only apply to sites used exclusively for telecommunications. Thus, if a mast operated by a designated central list occupier is part of a larger non-telecommunications hereditament, the regulations will not apply and the sharers will be separately assessed where they have their own exclusive room, cabin or compound. If they share non-exclusive accommodation with the central list host they will be included in the central list assessment as there is no separately identifiable hereditament since paramount control of the shared accommodation will be with designated central list occupier.[74] The site will form part of the central list occupiers' central list assessment.

71 2000/2421 reg 2, 3(1).
72 See paragraphs 37.6.3 et seq.
73 2000/2421 reg 3(3).
74 See paragraphs 37.5.2 and 37.7.6.

37.7.9 Position of landlords

It is also important to note that these regulations do not make a landowner or landlord who is not in the telecommunications business responsible for the rates liability where they are in occupation of land, building or structures for purposes other than telecommunications. So, where land (such as a hospital, school, water tower or a farmer's field) is let to a telecommunications operator, the distinct telecommunications use creates a separate hereditament. This is made clear because the regulations specifically apply to a 'telecommunications hereditament' (i.e. a site forming a hereditament *occupied exclusively* by telecommunications apparatus of either a single operator or more than one operator).

37.7.10 Domestic installations

Dwellings are subject to Council Tax and are not liable to non-domestic rates.[75] Thus, wireless routers and other equipment used for domestic purposes, whether internal or external, will be considered part of the domestic hereditament. Residents can connect to telecommunications services and use them without becoming liable to non-domestic rates.

37.7.11 Telecommunications apparatus used by business premises

On the other hand, where occupiers of business premises such as shops or offices have paramount control of telecommunications equipment such as a wifi installation and use it on their premises in connection with their business (including the enjoyment of their customers – such as in a coffee shop), then this forms part of the hereditament comprising the business premises and will be reflected in the assessment for that hereditament.

37.8 Chargeable amount

37.8.1 Generally

The chargeable amount[76] of rates levied on a hereditament is assessed by the application of a multiplier to the rateable value of the property in accordance with a statutory formula.

37.8.2 The statutory formulae and relief of telecommunications hereditaments

Where a ratepayer is liable for non-domestic rates by reason of his occupation of a hereditament shown in the local rating list, subject to regulations mentioned below, the chargeable amount is calculated by ascertaining the chargeable amount for each chargeable day (i.e. each day of occupation) and aggregating such amounts. For most occupied hereditaments the amount is calculated in accordance with the following formula:

$$\frac{A \times B}{C}$$

75 See paragraph 37.2.2.
76 [1998] RA 319 at 415.

A is the rateable value shown for the day in the local rating list; B is the non-domestic multiplier for the financial year; C is the number of days in the financial year.[77] Separate formulae are provided where the ratepayer is a charity[78] or where the hereditament is a qualifying post office, general store, public house or petrol filling station.[79]

Since 2008 the amount chargeable to a person liable to rates for an unoccupied hereditament is calculated in a similar way, except, most notably, that where the ratepayer is a charity and it appears that when next in use the hereditament will be wholly or exclusively used for charitable purposes, then the chargeable amount is zero.[80]

Subject to regulations, where hereditaments are entered in a central rating list the ratepayer whose name is shown in the list pays rates in accordance with a similar formula except the rateable value is that shown in the central list.[81]

Detailed consideration of how these multipliers are calculated is outside the scope of this work. Under the Telecommunications Infrastructure (Relief from Non-Domestic Rates) Act 2018, however, there is specific power to provide relief from non-domestic rates where the hereditament is wholly or mainly used for the purposes of facilitating the transmission of communications by any means involving the use of electrical or electromagnetic energy, the chargeable day falls before 1 April 2022 and conditions provided for by regulation are satisfied by amendment of the statutory formula. In essence, there is power to provide a new formula for calculating chargeable amounts for hereditaments shown on local non-domestic rating lists wholly or mainly used for the purpose of facilitating the transmission of communication by any means involving the use of electrical or electromagnetic energy. The amendments introduced by this legislation give the Secretary of State, or in relation to Wales, the Welsh Ministers, powers to prescribe further conditions that must be met for the relief to apply. The legislation also provides the Secretary of State, or in relation to Wales, the Welsh Ministers, with powers to set the level of the relief by prescribing the amount of relief in regulations. Regulations may also impose duties or confer powers on valuation officers. The amendments will have effect in relation to financial years beginning on or after 1 April 2017: accordingly, the power to make regulations includes a power to make provision in relation to times before the coming into force of the Act.[82]

The formula set out above in relation to occupied telecommunications hereditaments within the Act is modified to

$$\frac{A \times B \times F}{C}$$ [83]

77 LGFA 1988 s.44.

78 LGFA 1988 s.43: there is an 80% reduction.

79 LGFA s.43(6C)–(6L): there is a 50% reduction.

80 LGFA 1988 s.45.

81 LGFA 1988 s.54.

82 Telecommunications Infrastructure (Relief from Non-Domestic Rates) Act 2018 s.6(1).

83 See LGFA 1988 s.43(4E)-(4H) as inserted by Telecommunications Infrastructure (Relief from Non-Domestic Rates) Act 2018 s.1.

F is the amount prescribed by the regulations.[84]

The formula set out above in relation to unoccupied hereditaments within the Act is modified to

$$\frac{A \times B \times T_{85}}{C}$$

The formula set out above in relation to central non-domestic rates is similarly modified with a multiplier 'T' being prescribed or calculated in accordance with provision prescribed by the appropriate national authority by regulation.[86] The relief occurs in relation to central non-domestic rates (a) in a case where there is only one hereditament falling within the description of the hereditament shown against the person's name in the list, the hereditament is wholly or mainly used for the purposes of facilitating the transmission of communications by any means involving the use of electrical or electromagnetic energy, or (b) in a case where there is more than one hereditament falling within the description, those hereditaments are, taken together, wholly or mainly so used.[87]

37.8.3 The 'F' and 'T' multipliers and relief for 'new fibre'

Detailed provisions for how the multipliers are generally applied under these statutory formulae are found in statutory instruments. These are complex and reference is made to the relevant statutory instruments which are beyond the scope of this work.[88] As explained in paragraph 37.8.2 above, however, specific relief is provided in relation to telecommunications hereditaments and statutory instruments make detailed provision for the 'F' and 'T' multipliers described above.

The condition for relief is that 'new fibre' is part of the hereditament. 'New fibre' means fibre that was not laid, flown, blown, affixed or attached before 1 April 2017.[89] Where in relation to a description of hereditament this condition is satisfied, the appropriate valuation officer must certify the proportion of rateable value shown for the hereditament in the local list or against the ratepayers name in the central list (as the case may be) which appears to that officer to be attributable to (a) new fibre, (b) any plant and machinery used or intended to be used in connection with new fibre; and the proportion of the hereditament or hereditaments which is exclusively occupied by (a) or (b).[90] At the date of writing the methodology of certificates is the subject of discussion

84 See paragraph 37.8.3.

85 See LGFA 1988 s45 (4C)-(4G) as inserted by Telecommunications Infrastructure (Relief from Non-Domestic Rates) Act 2018 s.1.

86 See paragraph 37.8.3.

87 See LGFA 1988s.54ZA.

88 The Non-Domestic Rating (Chargeable Amounts) (England) Regulations 2016 SI 1265 as amended by The Non-Domestic Rating (Telecommunications Infrastructure Relief) (England) Regulations 2018 SI 2018/425; at the time of writing, provision has not yet been made for Wales.

89 See Non-Domestic Rating (Telecommunications Infrastructure Relief) (England) Regulations 2018 SI 2018/425 reg 3. At the time of writing, provision has not yet been made for relief in Wales.

90 Non-Domestic Rating (Telecommunications Infrastructure Relief) (England) Regulations 2018 SI 2018/425 reg 4(2), 5(2), 5(2).

between the Valuation Office Agency and the telecom sector.[91] Procedures are set out for certification[92] and for appeals against certificates.[93] Provision is made for relief to be awarded against this disaggregated rateable value for the fibre as distinct from the rest of the property.

The amount of F in the prescribed formula for occupied hereditaments in a local list is the amount calculated in accordance with the formula

$$1-\frac{CRV}{RV}$$

where CRV is the proportion of the rateable value shown for the hereditament in a local list that is certified by the appropriate valuation officer; and RV is the rateable value shown for the hereditament in a local list for the day.[94]

The amount prescribed of T for unoccupied hereditaments in a local list is the amount calculated in accordance with the formula

$$1-\frac{CRV}{RV}$$

where CRV is the proportion of the rateable value shown for the hereditament in a local list that is certified by the appropriate valuation officer; and RV is the rateable value shown for the hereditament in the local list for the day.[95]

The amount prescribed of T for descriptions of hereditaments in a central list is the amount calculated in accordance with the formula

$$1-\frac{CRV}{RV}$$

where CRV is the proportion of the rateable value shown against the ratepayer's name in the central list that is certified by the appropriate valuation officer; and RV is the rateable value shown against the ratepayer's name in the central list for the day.[96]

37.8.4 Rateable value

The rateable value by reference to which non-domestic rates are calculated is defined by statute:

91 See Business Rates: Relief for New Fibre on Telecommunications Hereditaments, Summary of Responses and Government Response paragraph 11 (March 2018).

92 Non-Domestic Rating (Telecommunications Infrastructure Relief) (England) Regulations 2018 SI 2018/425 reg 7. Separate Welsh provision is yet to be made at the time of writing.

93 Non-Domestic Rating (Telecommunications Infrastructure Relief) (England) Regulations 2018 SI 2018/425 reg 8. Separate Welsh provision is yet to be made at the time of writing.

94 Non-Domestic Rating (Telecommunications Infrastructure Relief) (England) Regulations 2018 SI 2018/425 reg 4. Separate Welsh provision is yet to be made at the time of writing.

95 Non-Domestic Rating (Telecommunications Infrastructure Relief) (England) Regulations 2018 SI 2018/425 reg 5. Separate Welsh provision is yet to be made at the time of writing.

96 Non-Domestic Rating (Telecommunications Infrastructure Relief) (England) Regulations 2018 SI 2018/425 reg 6.

> The rateable value of a non-domestic hereditament … shall be taken to be an amount equal to the rent at which it is estimated the hereditament might reasonably be expected to let from year to year on these three assumptions –
> (a) the first assumption is that the tenancy begins on the day by reference to which the determination is to be made;
> (b) the second assumption is that immediately before the tenancy begins the hereditament is in a state of reasonable repair, but excluding from this assumption any repairs which a reasonable landlord would consider uneconomic;
> (c) the third assumption is that the tenant undertakes to pay all usual tenant's rates and taxes and to bear the cost of the repairs and insurance[97]

Viscount Maughan in *Townley Mill (1919) Ltd v Oldham Assessment Committee*[98] explained that the effect of the formula was that the rent was to be assessed by reference to 'a hypothetical tenant and a hypothetical rent, but … a real and concrete hereditament'. Subject to the hypotheses which must be made to value such a letting the valuation will accord as far as possible to give effect to 'the principle of reality.' In *Hoare (Valuation Officer) v National Trust*[99] Peter Gibson LJ described the principle:[100]

> In particular I would emphasise the necessity to adhere to reality subject only to giving full effect to the statutory hypothesis, so that the hypothetical lessor and lessee act as a prudent lessor and lessee. I would call this the principle of reality

Schiemann LJ made clear the importance of this:[101]

> 'The statutory hypothesis is only a mechanism for enabling one to arrive at a value for a particular hereditament for rating purposes. It does not entitle the valuer to depart from the real world further than the hypothesis compels.'

37.8.5 Notional parties to hypothetical letting

The rent is assessed by reference to a notional letting from year to year between hypothetical landlord and hypothetical tenant.[102] Because the statute requires the hereditament to be let, the fact that the premises are occupied is immaterial[103] but all possible occupiers (including the actual occupier) must be taken into account as possible tenants in the market when assessing what the hypothetical parties would agree as the rent.[104] The hypothetical landlord for its part must be assumed to be in a position to let the entire hereditament and not to have let any portion of it already.[105] So, even if the

97 LGFA 1988 Sched 6 paragraph 2(1).
98 [1937] AC 419 at 436.
99 [1998] RA 319.
100 *Ibid.*, at 415.
101 *Ibid.*, at 408.
102 *R v West Middlesex Waterworks Co.* [1859] 28 LJMC 135, at 137; *R v Sheffield United Gaslight Co.* [1863] 32 LJMC 169 at 173.
103 See generally *R v London School Board* [1886] 17 Q.B.D. 738.
104 *LCC v Erith Parish (Churchwardens); West Ham Parish (Churchwardens) v LCC* [1893] AC 562, 596; see too *R v London School Board* [1886] 17 QBD 738, CA; *Davies v Seisdon Union* [1908] AC 315.
105 *Dawkins (VO) v Ash Bros and Heaton Ltd* [1969] 2 AC 366, 391.

hereditament is in fact let by more than one landlord, it must be assumed to be let as a whole. There is no requirement, however, to assume that the hypothetical landlord of separate hereditaments which complement one another must be the same person.[106]

37.8.6 The hypothetical term

The hypothetical tenancy must be assumed to be one from year to year and the rateable value will be assessed on that basis and not by reference to a fixed term of years.[107] On the other hand, it is not to be assumed that the tenancy will only last for one year but rather it must be assumed that there will be a reasonable expectation that it will continue: in many cases it would be unreasonable to suppose that a tenancy would be agreed if it were certain that the tenancy would expire after one year (for instance, if substantial plant and machinery were required).[108]

37.8.7 Hereditament vacant and available to let

It is also to be assumed that at the date of valuation the herediatement is vacant and available to be let.[109]

37.8.8 Effect of actual restrictions on terms?

It follows from these assumptions that restrictive covenants and private arrangements which restrict letting are to be ignored.[110] On the other hand, where the hypothetical tenant would be affected by statutory restrictions these will be taken into account.[111] Where, however, no rent can actually be charged because of statutory provisions, this does not mean that the rateable value will be nil: the statutory right is not determinative of value just as the actual rent is not determinative.[112] So, in the case of *Orange PCS Ltd v Bradford (VO)*[113] it was held that land occupied by a telecommunications operator's apparatus was to be valued at an amount which reflected the value of the occupation even though the land in question was part of a highway and no rent was payable by virtue of the telecommunications code. The occupation of the land on which a mast was erected was a right that would ordinarily be of value to the occupier; that followed from the operation of any market. The value of the right was illustrated by the fact that telecommunications operators had to pay for the right to occupy land in private ownership on which they erected masts. For the purpose of determining the value of the occupation under the statutory hypothesis, the rent payable by the tenant should be that value. The statutory right that Orange had to occupy the land without payment

106 *Coppin (VO) v East Midlands Airport Joint Committee* [1971] 17 RRC, 31, CA.

107 *Staley v Castleton Overseers* [1864] 33 LJMC 178.

108 *Great Eastern Rly Co. v Haughley* [1866] LR 1 QB 666, 679, 370; *Railway Assessment Authority v Southern Rly Co.* [1936] AC 266. See too *Dawkins (VO) v Ash Bros and Heaton Ltd* [1969] 2 AC 366.

109 *LCC v Erith Parish (Churchwardens); West Ham Parish (Churchwardens) v LCC* [1893] AC 562 at 588.

110 See e.g. *Robinson Brothers (Brewers) Ltd v Durham County Assessment Committee* [1938] AC 321, 336–7.

111 *Port of London Authority v Orsett Union* [1920] AC 273, 305; *Clement (VO) Addis Ltd* [1988] 1 WLR 301, HL. For example, this would mean that the provisions of New Code paragraphs 16 and 17 in principle would be taken into account.

112 *Poplar Assessment Committee v Roberts* [1922] 2 AC 93.

113 [2003] RA 141, LT upheld at [2004] EWCA Civ 155. See too *Re Coll (VO)'s appeal* [2012] UKUT 5 (LC).

was not determinative of that value in the same way as an actual rent is not determinative. The Old Code operated to determine that the price paid for the occupancy was nil; but it did not operate to determine the value of the occupation. Moreover, the right to occupy free of charge was a right that pertained to Orange as a personal right or qualification derived from the Telecommunications Code; it was not a matter that affected the hereditament as any other occupier (other than a telecommunications operator) would have had to pay for that right.[114]

37.8.9 Rebus sic stantibus

LGFA 1988 expressly requires reality to be taken into account in respect of the physical state and enjoyment of the hereditament at the valuation date. It requires the hereditament to be valued on the basis that the following matters are to be taken into account as they were on the material day:

(a) matters affecting the physical state or physical enjoyment of the hereditament;
(b) the mode or category of occupation of the hereditament.[115]

These provisions enact what had been established in case law, namely that premises are to be valued as things stood at the valuation date, i.e. *rebus sic stantibus.* In *Robinson Bros (Brewers) Ltd v Houghton and Chester-le-Street Assessment Committee*[116] Scott LJ said:

> The hereditament to be valued … is always the actual house or other property for the occupation of which the occupier is to be rated, and that hereditament is to be valued s it in fact is – *rebus sic stantibus*.

The LGFA 1988, however, makes assumptions about the state of repair as described above.[117]

37.8.10 Valuation methodology

In the case of *Vtesse Networks Ltd v Bradford (VO)*[118] the Court of Appeal summarised the four usual methods adopted in assessing rateable value of a given hereditament. These are as follows.

(1) The rental method uses direct and actual evidence of the rent of the actual hereditament. It is not available in the present case because the hereditament cannot be let as a whole.
(2) The tone of the list uses a comparison with the rateable value of other comparable hereditaments in order to derive a unit cost which is consistent

114 *Orange PCS Ltd v Bradford (VO)* [2004] EWCA Civ 155 at [22]–[24]
115 LGFA 1988 Sched 6 paragraph 2(7).
116 [1937] 2 KB 445 at 468, CA. See too *Dawkins (VO) v Ash and Heaton Ltd* [1969] 2 AC 366, HL at 385 per Lord Wilberforce.
117 See paragraph 37.8.4 above.
118 [2006] EWCA 1339 at [33].

with that of other hereditaments.[119] It is concerned with the weight to be given to assessments in the rating list. It is settled law that assessments of comparable hereditaments are admissible as evidence of value.[120]

(3) The receipts and expenditure method estimates the future receipts of a putative tenant and deducts its future costs and the return it would require, thereby leaving the maximum rent which it would be prepared to pay.

(4) The contractor's method is based on the replacement cost of a business property, on the basis that a putative tenant would not be willing to pay more in annual rent for the hereditament than it would cost in terms of annual interest on the capital sum needed to build a similar hereditament.

The Valuation Office Agency has published the methodology which it adopts when assessing various sorts of telecommunication sites in its Rating Manual.

37.9 Procedure and recovery of rates

The procedure for challenging inclusion in the local and central lists, for challenging assessment of rateable value and the procedure for the recovery of rates of (and defending claims) are outside the scope of this work. The reader is referred to the standard works on Non-Domestic Rating.

119 *K Shoe Shops Limited v Hardy (VO)* [1983] RA 145 at 154; *Burroughs Machines Limited v Mooney (VO)* [1977] RA 45 at 55; *Marks v Eastaugh (VO)* [1993] RA 11 at 20–23; *Jafton Properties Limited v Prisk (VO)* [1997] RA 137 at 166–7.

120 *Pointer v Norfolk Assessment Committee* [1922] 2 KB 471.

38 Land registration

38.1 Introduction

38.1.1

This chapter is concerned with the way in which code rights affect successors. It is divided into three broad sections:

(a) An introduction to land registration concepts. The reader who is familiar with the land registration regime can skip over this section.
(b) The Old Code and land registration.
(c) The New Code and land registration.

38.2 Introduction to land registration

38.2.1

Land registration has existed in England and Wales, in one guise or another, for many years. The rules about land registration are now set out in the Land Registration Act 2002. Most freehold estates in England and Wales are now registered at Land Registry. Many long leasehold estates are also registered with their own title at Land Registry.

38.2.2

A new lease will require registration as a new leasehold title if it is:

(a) for a term of more than 7 years;
(b) a reversionary lease taking effect more than 3 months from the date of grant; or
(c) a discontinuous lease, which is granted out of a registered title or granted for a term of more than seven years out of unregistered freehold or leasehold estates: Land Registration Act: 2002 s.27(2)(b).

38.2.3

The register[1] for each title (i.e. each registered estate) is divided into 3 parts:

1 Two additional types of leases are listed in the statute, but these are unlikely to be relevant to telecoms work.

(a) a description of the property so that the parcel in question can be identified, almost invariably by reference to a title plan, and the estate which the proprietor holds;
(b) the identification of the owner of that title; and
(c) the charges register, where legal charges, and notices or restrictions relating to the title appear.

38.2.4

Interests in land, such as easements (rights of way, rights of light and so on) and restrictive covenants (which prevent use of land in a particular way), which burden the title are generally protected by notice in the charges register. A notice can only be registered 'in respect of the burden of an interest affecting [the title]'.[2] A notice cannot be registered if the applicant does not have an interest in the title. A notice also cannot be registered if the applicant's interest is one of the excluded interests listed in s.33 of the Land Registration Act 2002. The list of excluded interests includes leasehold estates which are granted for a term of three years or less from the date of the grant.

38.2.5

There are two types of notices: agreed notices and unilateral notices. An agreed notice can only be sought where the registered proprietor consents. If he does, the application should be made on Form AN1. If the registered proprietor does not consent, an application for a unilateral notice should be made on Form UN1.

38.2.6

Restrictions typically prevent dispositions being registered without the consent of the person registering the restriction. A restriction can only be entered if it is necessary or desirable for:

(a) preventing invalidity or unlawfulness in relation to dispositions of the title;
(b) securing that interests which are capable of being overreached on a disposition of a registered estate or charge are overreached; or
(c) protecting a right or claim in relation to a registered estate or charge, which cannot be protected by notice.[3]

38.2.7

Section 29(1) of the Land Registration Act 2002 provides:

> If a disposition is made for valuable consideration, completion of the disposition by registration has the effect of postponing to the interest under the disposition any interest whose priority is not protected at the time of the disposition.

2 Land Registration Act 2002 s.32.
3 Land Registration Act 2002 s.42.

Sub-section (2) provides that an interest is protected if, in broad terms, it is within the category of overriding interests set out in Schedule 3, or it is protected by notice on the register. In summary, therefore, a purchaser can expect to take free of any interests affecting the land, apart from a limited list of overriding interests and those noted on the register.

38.2.8

An interest which belongs to a person in actual occupation is, *prima facie*, an overriding interest. However, the interest of a person in actual occupation will not be overriding if either:

(a) enquiry was made of the person and the interest was not disclosed when it could reasonably have been expected to be; or
(b) the occupation would not have been obvious on a reasonably careful inspection of the land (and the purchaser did not in fact know of the interest).[4]

38.2.9

The entry of a restriction does not render the underlying right binding on successors. All it does is give the person who registered the restriction the ability to prevent the disposition taking place until it has had the opportunity to protect his rights.

38.3 The Telecommunications Act 1984

38.3.1

The 1984 Act was the first legislation to create a code ('the Telecommunications Code') regulating the installation and operation of telecommunications apparatus upon land. At that time, telecommunications were in their infancy. As the Law Commission said in paragraph 4.49 of their Report No. 336:

> The Telecommunications Act 1984 was enacted in order to create and regulate the market in landline telephone services, and it is likely that it was drafted with only wayleaves in mind.

38.3.2

Indeed, the original draft of the code was based on several 19th- and early 20th-century statutes dealing with telephone wayleaves. It is clear that the drafters of that code (and, to an extent, the 2003 code considered next) did not contemplate that code operators might require a leasehold estate in land in order to site their apparatus on it. It seems likely that the original drafters would have considered that they were creating a system of *sui generis* rights governed by a closed, separate code.

4 Land Registration Act 2002 Schedule 3, paragraph 2.

38.3.3

Wayleaves are contractual arrangements, and are not property rights, and are not therefore registrable interests. Nevertheless, they have enduring popularity, not merely under the 1984 legislation, but also through the 2003 legislation and to the present day. That is because practitioners are familiar with the concept, and the drafting is usually simple, based upon tried and tested standard forms.[5]

38.3.4

Wayleaves are suitable primarily for agreements conferring the right to run cables in or across land. However, given that code rights might include the right to install a mast on land, para 4.49 of Law Com 336 notes:

> It is likely that in some cases the Code Rights, together with the terms and conditions imposed by the court, give the Code Operator exclusive possession of the land concerned, for a term, and therefore amount to a lease. Indeed, in many cases the grant of a lease will be what the Code Operator wants; and the landowner too may take the view that if Code Rights are to be imposed, he or she wishes to be in the position of landlord (with the associated rights, in particular to remedies for breach of leasehold covenants) rather than the Code Operator having a simple wayleave.

38.3.5

Although wayleaves are in common use for telecommunications agreements, there are risks associated with their use. In particular, it is not clear in some cases whether what is being created is a genuine licence, as opposed to a tenancy or easement, both of which are registrable interests in land. There was nothing in the original Telecommunications Code which assisted with that problem. As the next section shows, the revision in 2003 of the Telecommunications Code provided only limited assistance.

38.4 The Old Code

38.4.1

The Communications Act 2003 extensively revised the Telecommunications Code to provide the Electronic Communications Code (in this chapter, 'the Old Code', to distinguish it from its 2017 successor, 'the New Code'). However, from a land registration point of view, the Law Commission said in paragraph 1.9 of their Report No. 336 that it was 'difficult to discern the relationship of the 2003 Code with other elements of the law, such as the Land Registration Act 2002'. This section seeks to explain that relationship, mindful however of the Law Commission's remark.

5 An example is provided by the Standardised Wayleave Agreement devised and provided by the City of London in partnership with the City of London Law Society, and supported by DCMS and major industry groupings: www.cityoflondon.gov.uk/business/commercial-property/utilities-and-infrastructure-/Documents/standardised-wayleave-agreement.pdf.

38.4.2

Paragraph 2 of the Old Code contains detailed provisions as to who is to be bound by the rights set out in paragraph 2(1) ('Old Code rights'). The Old Code rights are rights conferred on an operator for the statutory purposes to:

(a) execute any works on that land for or in connection with the installation, maintenance, adjustment, repair or alteration of electronic communications apparatus; or
(b) to keep electronic communications apparatus installed on, under or over that land; or
(c) to enter that land to inspect any apparatus kept installed (whether or, under over that land or elsewhere) for the purposes of the operator's network.

It is noteworthy that nothing is said about the nature of the Old Code rights being conferred, and in particular whether they take the form of simple licences, tenancies or anything else.

38.4.3

The Old Code provisions about who is to be bound by the rights are considered above. The question for consideration in this section is the extent to which any registration requirements apply to alter what might otherwise appear to be the result from reading paragraph 2 of the Old Code.

38.4.4

Paragraph 2(7) of the Old Code provides:

> It is hereby declared that a right falling within sub-paragraph (1) above is not subject to the provisions of any enactment requiring the registration of interests in, charges on or other obligations affecting land.

It is therefore clear that the rights falling within sub-paragraph (1) do not need to be registered in order to bind successors as set out in paragraph 2.

38.4.5

But what if the agreement by which the rights are created also conferred additional (i.e. non-code) rights on the operator, such as an option to renew its lease? Do these additional rights bind successors without registration because they are conferred by an agreement which creates code rights? The position is unclear.

38.4.6

There are two potential answers:

(a) The first option is full exemption from registration. The parties are free, subject to paragraph 27 of the Old Code, to insert such terms as they wish into an agreement under paragraph 2 conferring code rights. Code rights will then bind successors and others in accordance with the special rule of priorities contained in paragraph 2. Under that priorities scheme, the right conferred will only be exercisable in accordance with those terms by virtue of paragraph 2(5). Therefore, all of the terms of an agreement conferring a code right are land registration exempt. This analysis is by no means a perfect 'fit' with the drafting of paragraph 2, however. First, paragraph 2(7) only exempts the 'right' from registration. The answer to this may be that the right, and the terms on which it can be exercised, are one and the same thing. Secondly, although the first part of paragraph 2(5) is consistent with this interpretation, the second part is harder to reconcile, as it expressly talks only of the benefit of those terms passing to any person bound by the code rights. It is unclear whether the second part of paragraph 2(5) is intended to cut down the generality of the first part, or whether it is simply an 'avoidance of doubt' provision. On the other hand, paragraph 2(6) talks of variations of the terms on which code rights are exercisable as binding, which might suggest that the burden of such terms would otherwise pass. If this is the correct view, then not just land registration, but also other aspects of property and contract law – such as the rule that contractual licences (if that is what an agreement with an operator would be, absent the code) do not bind successors in title[6] – would also seem to be irrelevant.

(b) The second option is a hybrid position. To the extent (and only to the extent) that what has been conferred is a code right is there exemption from registration. Again, reference is made to paragraph 2(7) which focuses on the 'right'. Insofar as what is conferred are not terms regulating the exercise of that right, but, in truth, additional rights, those are not 'terms'. It would follow that the code right alone is binding under paragraph 2 as provided, but, insofar as the operator would wish to ensure that the agreement also takes effect as 'full' property right, like a lease which requires registration to take effect at law, then the requirement of the Land Registration Act 2002 would need to be complied with (though one would expect that in most cases the lease would override by reason of actual occupation in any event).

38.4.7

It is considered that the better view is that those rights would not be saved from any consequences of non-registration by paragraph 2(7), because:

(a) that appears to be the more natural reading of clause 2(7);
(b) a provision which disapplies another statute should be read restrictively;
(c) the statutory purpose was to ensure that operators were able to obtain the specific listed rights necessary for them to run their networks, not other rights;
(d) it is not onerous to expect operators to register additional rights that they bargain for, beyond the statutory rights;

6 *National Provincial Bank v Ainsworth* [1965] AC 1175; the general law as stated in *Terunnanse v Terunnanse* [1968] AC 1086 that a licence also terminates on death of the licensor must, on this analysis, also have no application.

(e) purchasers can be told that operators might have unregistered rights to do the things set out in paragraph 2(1) for the statutory purposes, and can reasonably be expected to take a view about that. However, it is unsatisfactory for purchasers to have to accept a risk that an operator might have any other right over the land which it would not know about when it purchased. To take an extreme example, if an operator had an option to buy the land for £1, it cannot really have been intended that the purchaser would be bound by this right even if it was not registered and the purchaser had no way of finding out about it.

38.4.6

If it is right that additional rights are not protected by paragraph 2(7), an operator who wants such rights to bind third parties must adopt conventional conveyancing techniques to do so. In simple terms, this means that an estate or interest in land must be created, and any registration formalities must be complied with. The simplest way (and a common way) is for a lease to be granted. A lease is created if exclusive possession is granted, for a term, generally at a rent.[7] Whether a lease is created is not determined by the name which the parties have given to the agreement, or themselves in it. It is determined by an analysis of the rights granted to the occupier. Specialist landlord and tenant works should be consulted for more detailed commentary as to when an agreement will be held to create a tenancy.

38.4.7

If a tenancy is created, whether as a technique to ensure other rights bind successors, or because the operator wants exclusive possession of an area of land for a term:

(a) *If it is for a term of seven years or more* (or if it takes effect in possession more than three months from the date of the lease; or the right to possession is discontinuous), it must be registered: Land Registration Act 2002 s 27(2)(b). If it is not registered, successors in title will not be bound by any rights beyond the code right. The application should be made to Land Registry on Form AP1.
(b) If it is for a term of less than seven years (and it does not otherwise require registration), successors in title will be bound by the lease without the need for registration because such leasehold estates are overriding interests: Land Registration Act 2002 Schedule 3 paragraph 1.
(c) However, if the lease is for a term of at least three years and does not require registration,[8] the operator can apply to register a notice against the landlord's title. The application should be made on form AN1 if the landlord agrees, or form UN1 if the operator is making the application unilaterally.

7 The reservation of a rent is not in fact essential in order that a lease be granted: *Ashburn Anstalt v WJ & Co Arnold* [1989] 1 Ch 1, CA.

8 Notices cannot be registered to protect leases of less than three years term which are not required to be registered: Land Registration Act 2002 s 33(2)(b).

38.4.8

It is not the practice to register Old Code rights which are granted other than by lease. If the agreement created only Old Code rights, there is of course no need to do so, because of paragraph 2(7). Indeed, it is unclear whether such rights are registrable at all. Rights can only be registered if they amount to an interest in land. It remains unclear whether Old Code rights themselves were interests in land.

38.4.9

However, agreements creating Old Code rights could, in theory, also create other types of interest in land, such as an easement, if the operator had a dominant tenement which was benefited by the rights. If rights beyond Old Code rights were expressly granted, which were capable in nature of being an easement and which benefited the operator's dominant tenement, they would not bind successors without registration unless they were an overriding interest.

38.5 The New Code

38.5.1

The introduction of the New Code was preceded by two detailed consultations calling for suggestions for improvements to the Old Code, amongst which the question of registrability was raised, albeit perhaps not expressly enough. In paragraph 4.47 of their Report No. 336, the Law Commission say:

> One issue that we did not raise explicitly in the Consultation Paper is whether the imposition of Code Rights by the court in the 2003 Code can result in, or amount to, the grant of a lease or easement, or whether it can only ever amount to a wayleave (that is, a bare right to do something or keep something on land, without any interest in the land).

As this section establishes, it is moot whether the New Code has conclusively resolved the issues under the Old Code.

38.5.2

The starting point when considering whether a party other than the person who made the agreement conferring the code rights is bound by them is paragraph 10 of the New Code. This says:

... (2) If O has an interest in the land when the code right is conferred, the code right also binds –
- (a) the successors in title to that interest,
- (b) a person with an interest in the land that is created after the right is conferred and is derived (directly or indirectly) out of –
  - (i) O's interest, or
  - (ii) the interest of a successor in title to O's interest, and

(c) any other person at any time in occupation of the land whose right to occupation was granted by –
   (i) O, at a time when O was bound by the code right, or
   (ii) a person within paragraph (a) or (b).

(3) A successor in title who is bound by a code right by virtue of sub-paragraph (2)(a) is to be treated as a party to the agreement by which O conferred the right.

(4) The code right also binds any other person with an interest in the land who has agreed to be bound by it.

(5) If such a person ('P') agrees to be bound by the code right, the code right also binds –
   (a) the successors in title to P's interest,
   (b) a person with an interest in the land that is created after P agrees to be bound and is derived (directly or indirectly) out of –
      (i) P's interest, or
      (ii) the interest of a successor in title to P's interest, and
   (c) any other person at any time in occupation of the land whose right to occupation was granted by –
      (i) P, at a time when P was bound by the code right, or
      (ii) a person within paragraph (a) or (b).

(6) A successor in title who is bound by a code right by virtue of sub-paragraph (5)(a) is to be treated as a party to the agreement by which P agreed to be bound by the right.

38.5.3

In summary:

(1) If the occupier granting the code right had an interest in land, his successors will be treated as if they were a party to the agreement. So too the successors of anyone else with an interest in the land who has agreed to be a party to it.

(2) Those with derivative rights (including rights of occupation) whose rights were created *after* the code right was granted are also bound. However, they are not, apparently, to be treated as parties to the agreement.[9] Those with derivative interest are, presumably, bound by force of statute.

38.5.4

How does the land registration regime fits in with the provisions of paragraph 10? Must an operator register his code rights in order to take advantage of paragraph 10? Paragraph 14 of the New Code states:

9 This means that they are unable to terminate the agreement. Paragraph 31 only permits a 'site provider who is party to a code agreement' to serve notice. However, the party with a derivative interest will be subject to the restriction on the right to seek removal to electronic communications apparatus under Part 6. Any successor to such a person will be equally bound. See section 22.4 of Chapter 22, where this is dealt with in more detail.

> Where an enactment requires interests, charges or other obligations affecting land to be registered, the provisions of this code about who is bound by a code right have effect whether or not that right is registered.

This makes it clear that the provisions in the New Code which provide for others to be bound by code rights trump the provisions of the Land Registration Act 2002 which provide for rights to be lost if they are neither overriding interests nor noted on the register. Code rights will bind in accordance with paragraph 10 of the New Code even if they are not registered. Thus code rights can be considered an additional type of overriding interest which is not listed in Schedule 3 of the Land Registration Act 2002.[10]

38.5.5

However, there is an argument that paragraph 14 does not provide the wide protection which, at first glance, it appears to provide for operators. Paragraph 14 provides that 'the provisions of this code about who is bound by a code right' trump the land registration regime. What are those provisions? Paragraph 10 provides for code rights to bind successors and those with derivative interests. Code rights are the rights set out in paragraph 3 of the New Code. As before, operators might have additional rights, like, for example, the right to exclusive possession of an equipment room. That right is not in itself a code right.[11] Paragraph 10 appears to say that successors would be bound by rights beyond code rights (because they are to be treated as though they were a party to the agreement by which the rights are conferred), whereas the holders of other derivative interests would not be bound, by virtue of paragraph 10, by such additional rights. But paragraph 14 only exempts 'code rights' from registration, not other rights. This does not fit well with the provision in paragraph 10 deeming successors to be parties to the agreement. It follows that where the agreement creates rights beyond code rights, paragraph 10 does not provide for all relevant third parties to be bound by the non-code rights, and paragraph 14 will not save those non-code rights from being defeated by a failure to register. In the circumstances, wherever any rights beyond code rights are created, it is sensible to make provision for those rights to bind all necessary third parties by ordinary conveyancing techniques.

10 A code right can only truly be an overriding interest if it is an interest in land. There is some doubt about this, as mentioned in paragraph 38.4.8 above, and discussed in more detail below. If they are not property interests, they are *sui generis* rights like overriding interests. It was a proposal of the Law Commission that there should be an amendment to the Land Registration Act 2002 to provide for code rights, which amounted to an interest in land conferred otherwise than by a lease, to be an overriding interest so as to be enforceable against purchasers of registered land despite not being registered: Law Commission, Report No. 336, paragraph 2.132. However, this proposal has not found its way into the amendments to legislation to be found in Schedule 3 to the 2017 Act. Nonetheless this appears to be the intended effect of paragraph 14.

11 Note though, that if the equipment room is a cabin which is installed by the operator, the operator would typically have exclusive possession of the land on which it is placed, and the cabin itself could be apparatus, rather than land in such circumstances (as discussed in Chapter 18). If another operator is granted rights to install equipment over the cabin, that could not be a code right, because such rights can only be granted over land, and not apparatus. However, even if this is the case, one would expect there to be ancillary rights of access (e.g. a right to access the landowner's land to e.g. inspect and maintain the cabin). Such a right *is* a code right: paragraph 3(c).

38.5.6

As under the Old Code, non-code rights will only bind if an estate or interest in land is created, and all appropriate registration formalities are completed. It is still the case that the simplest way to do this is to grant a lease. (It is clear from paragraph 29 of the New Code that it was envisaged that code rights can still be granted within leases.) The registration requirements have not changed, so if a lease is created, the same provisions as applied under the Old Code will apply.

38.5.7

The view of the Law Commission was that no special provision was needed with respect to leases. As the Law Commission stated:

> leases that confer Code Rights should – to put it colloquially – look after themselves. The general law will apply; normal registration requirements should apply. Accordingly, there is nothing that the revised Code need say about the registration of leases.[12]

What paragraph 14 is aimed at are code rights conferred otherwise than by leases but which are in fact (whether or not the parties appreciate it) property rights, because they amount to an easement. It is here that the priority rules in the code prevail over those of the Land Registration system.

38.5.8

Occasionally, operators may wish to register code rights, even if no additional rights are created – although they do not need to in order to bind third parties. From an operator's point of view there is a practical advantage in doing so: disputes with successors and others coming later to the land are likely to be minimised. Paragraph 14 appears to suggest that, although there is no need to do so, code rights themselves can be registered.

38.5.9

It may be that this provision is intended to do no more than recognise that code rights might be capable of registration, for example, if an easement (which is an interest in land) is created, and was intended to avoid such rights being lost due to the parties not appreciating the legal effect of their agreement.[13] However, for an easement to exist, there must be a dominant tenement which is benefited by the easement. It will be comparatively rare that there will be a dominant tenement in the case of a standard telecommunications access agreement. Far more frequently, if an operator does not

12 Law Com 336, paragraph 2.123.

13 The Law Commission specifically recognised this possibility. It said that it 'will often be unclear how a particular agreement is to be categorised [i.e. as to whether it created an easement and thus a property interest or a wayleave which is not] and we do not think that it is appropriate for the parties to have to struggle to ascertain whether or not a Code Right (giving access for example) is in fact an easement and needs to be noted at [the] Land Registry'. Law Com 336, paragraph 2.124.

have a lease, it will have a mere licence, or wayleave as it is sometimes called in this context. Is such an agreement registrable?

38.5.10

As a matter of first principle, the starting point must be to ascertain whether a code right is an interest in land. Licences are not interests in land. However, rights created by statute could be interests in land even if the ordinary incidents required for the creation of an interest in land are not present.[14] Whether a particular statute does so is a question of construction of that statute. It is unclear whether Parliament intended code rights to be interests in land. However, it is considered that the better view is that they are not interests in land, because 'the general rule is that no greater rights or interests should be treated as conferred upon the undertakers than are necessary for the fulfilment of the object of the statute'.[15] Given the terms of paragraph 10, code rights do not need to be interests in land.

38.5.11

Given the doubt, however, an operator who wishes to register a particular code right should attempt to register both a notice and a restriction. If the Land Registry takes the view that code rights are not interests in land, it will decline to register the notice, but should register the restriction. A restriction can be registered even if the code rights are not interests in land. All that is required is a 'a right or claim in relation to' land. As set out above, entering a restriction will merely result in the operator having an opportunity to inform the purchaser of his code rights before any disposition can be registered; it does not amount to registration of the code rights. However, given that the purpose of registering the code rights is simply to ensure that those who come to the land subsequently are aware of the rights, a restriction may provide adequate protection. Note, however, that if the code rights are interests in land, it is not possible simply to apply for a restriction instead of a notice.

38.6 Existing agreements and the New Code

38.6.1

No action will need to be taken as a result of the New Code coming into force in respect of many existing agreements. Any agreement which is already registered will remain registered. Anything which was an overriding interest will still be an overriding interest. Further, any leases which should have been registered and are not will not be cured by the change to the New Code: they will still require registration.

38.6.2

However, an interesting question does arise about the extent to which code rights in existing agreements bind successors under paragraph 10, and are exempted from registration requirements under paragraph 14. Although existing agreements will be

14 *Newcastle under Lyme Corporation v Wolstanton* [1947] 1 Ch 427 at 455–7.
15 *Newcastle upon Lyme Corporation v Wolstanton* (supra) at 455.

governed by the New Code when it comes into force, the definition of 'code rights' is modified in relation to those agreements. By Schedule 2, paragraph 3 of the Digital Economy Act 2017:

> In relation to a subsisting agreement, references in the new code to a code right are –
> (a) in relation to the operator and the land to which an agreement for the purposes of paragraph 2 of the existing code relates, references to a right for the statutory purposes to do the things listed in paragraph 2(1)(a) to (c) of the existing code …

38.6.3

This means that only Old Code rights are exempted from the requirements of registration, even after the New Code comes into force. Rights beyond Old Code rights in existing agreements will still not bind unless they amount to an interest (or estate) in land, and that interest or estate is registered where appropriate or is an overriding interest.

38.6.4

A further interesting question is whether, if the Old Code rights were not interests in land, and if (contrary to the view expressed above) New Code rights are interests in land, existing rights become interests in land when the New Code comes into force. Schedule 2, paragraph 2(1) of the Digital Economy Act 2017 provides:

> A subsisting agreement has effect after the new code comes into force as an agreement under Part 2 of the new code …

Although the position is not clear, the better view is probably that Old Code rights do become interests in land. A notice could therefore be registered so that their existence is plain to third parties on the face of the register once the New Code comes into force.

38.7 Sharers and assignees of code rights

38.7.1 Registration of sharer's rights

There are two types of sharers: those who have no apparatus of their own on the land, who simply share the main operator's equipment; and those who bring their own apparatus onto the land. The first category have no code rights and no interest in land, and no question of registration arises.

38.7.2

The second category of sharers, however, might have code rights and/or an interest in land.[16] Their rights can be either limited to code rights only, or could be more extensive (if the main operator had more extensive rights).

16 Although it should be noted that the main operator's apparatus cannot be 'land' for these purposes, with the result that the main operator cannot grant code rights to a sharer of its apparatus. This is described in more detail in Chapter 20.

38.7.3

If the main operator had a lease, a sharer could have a sub-lease. In that event, the registration regime will depend on the length of the term (see above).

38.7.4

If the sharer does not have a sub-lease, but does have code rights then its rights will be protected even if nothing is registered, by paragraph 14. If a sharer wanted to voluntarily register its rights, the position is the same as for the main operator.

38.7.5 Registration of assignees

If the code rights are contained in a lease which is registered as a leasehold title, the assignee will not have legal title, or be able to exercise 'owner's powers', until he has been registered: Land Registration Act 2002 s.27(1); *Skelwith (Leisure) Ltd v Armstrong.*[17] The assignment should be on Land Registry Form TR1. Land Registry Form AP1 will also be needed to register the assignment.

38.7.6

If simple code rights are assigned,[18] there is no requirement to register the assignment. However, if a restriction had been registered, the assignee will want to ensure that the assignor joins in an application under Land Registration Act 2002 s.41(2)(b) for a modification of the restriction to substitute the assignee's name as the person whose consent should be sought before any disposition is made. Similarly, the assignee can apply to be registered with the benefit of any existing unilateral notice, by application on form UN3. If consent from the assignor is not provided at the time the application is made, notice of the application will be served on the assignor and it will have an opportunity to object, before the application is completed. There is no equivalent procedure if an agreed notice has been registered: in these circumstances, the assignee will have to make a fresh application for an agreed notice or a unilateral notice.

17 [2016] Ch 345.
18 See Chapter 20.

39 Electronic communications and competition law

39.1 Introduction

39.1.1

Terrestrial broadcasting in the UK is delivered over a network of masts and sites spread across the country. These masts and sites are used to broadcast (television and radio) signals to end-consumers, that is, viewers and listeners. Those involved in this delivery include primarily:

(a) landowners who own or lease the sites;
(b) entities who provide the electronic communications equipment (including masts, buildings and shared equipment) suitable for the transmission of terrestrial broadcast services including in the wireless communications sector, commonly referred to as wholesale infrastructure providers ('WIPs');
(c) operators of the equipment, including mobile network operators ('MNOs'), and mobile virtual network operators ('MVNOs') which typically do own the wireless network infrastructure over which they provide services to their customers.

This grouping should not be taken to suggest that the activities are distinct or discrete. It is quite possible for an MNO to acquire and build its own site; or lease a site from another MNO. Thus an MNO could in principle lease or own both site and infrastructure, and operate from them, and hence belong to all three groups.

39.1.2

The market for electronic communications in the United Kingdom has become dominated over the decades by agglomerations of members within these groups. Thus:

(a) Although many landowners are farmers with perhaps one or two sites, many of these are members of, or form associations with, or are represented by, bodies such as the National Farmers Union ('NFU'), the Central Association of Agricultural Valuers ('CAAV') and the Country Land and Business Association ('CLA'). In the cities, large estates and commercial groupings will have substantial power in their local markets.
(b) WIPs within the telecommunications industry are dominated by two groups: Arqiva Ltd and Wireless Infrastructure Group. According to its

response to the 2017 OFCOM consultation, Arqiva's portfolio includes over 8,000 active mobile, radio and television sites.

(c) The UK's MNOs account for almost 70 per cent of supply of UK shareable active sites. In 2007, there were five MNOs operating in the UK, namely: Three; O2; Orange; Vodafone and T-Mobile, and just nine MVNOs. Although the number of MNOs has remained small (currently numbering just four: Vodafone, O2 (Telefónica), 3UK and T-Mobile (now EE)), the number of MVNOs has increased to over 20. Vodafone and Telefónica have formed a joint team known as Cornerstone Telecommunications Infrastructure Ltd; while 3UK and EE have a joint venture known as Mobile Broadband Network Limited.

39.1.3

These industry and landowning groupings can create the potential for practices to arise which are contrary to domestic and European competition law. Although the potential for competition issues may seem obvious in the case of the dominance of the market by some operators, the problem may also arise where landowners in the cities and in rural areas form associations,[1] with a view to improving their negotiating stance with electronic communications operators, or otherwise engaging in economic activity.[2] This is the case even where the cooperation between them does not relate to their principal business activity and is not necessarily motivated by the pursuit of profits.[3]

39.1.4

This chapter summarises the relevant legislation, and sets out some guiding principles which those who are unfamiliar with it may wish to bear in mind in ordering their affairs – although where a competition issue does arise, a standard text on the subject should be consulted, or specialist advice sought.

39.2 Competition legislation

39.2.1

The competition regime governing the telecommunications market is set out in the Competition Act 1998 ('the 1998 Act', as amended by the Enterprise and Regulatory Reform Act 2013), the EU Common Regulatory Framework and the Communications Act 2003.

1 See generally OFT1289 *Land Agreements: The Application of Competition Law Following the Revocation of the Land Agreements Exclusion Order* (March 2011) and OFT740rev *How Competition Law Applies to Co-operation between Farming Businesses: Frequently Asked Questions* (November 2011).

2 *Case C-41/90 Hofner and Elser v Macrotron GmbH* [1991] ECR1-1979 at paragraph 21 and Case C-180/98 *Pavlov v Stichting Pensioenfonds Medische Specialisten* [2000] ECR I-6451 at paragraph 75: 'The notion of economic activity can encompass any activity consisting in offering goods or services'.

3 Case C-209/78 *Van Landewyck v Commission* [1980] ECR 3125 at paragraph 88.

39.2.2 The Chapter I prohibition

Section 2 of the 1998 Act prohibits agreements or concerted practices between undertakings (collectively 'Agreements') which have as their object or effect an appreciable prevention, restriction or distortion of competition within the UK or a part of it and which may affect trade within the UK or a part of it, unless they fall within an excluded category or are exempt. Section 2(4) of the 1998 Act provides that any agreement which falls within this prohibition (called 'the Chapter I prohibition'), and is not exempt, is void and unenforceable.[4]

39.2.3

The term 'undertaking' is not defined in either European legislation or the 1998 Act, but its meaning has been set out in Community law. It covers any natural or legal person engaged in economic activity, regardless of its legal status and the way in which it is financed. It includes companies, firms, businesses, partnerships, individuals operating as sole traders, agricultural co-operatives, associations of undertakings (e.g. trade associations), non-profit-making organisations and (in some circumstances) public entities that offer goods or services on a given market. The key consideration in assessing whether an entity is an undertaking for the application of Article 81 and/or the Chapter I prohibition (see below) is whether it is engaged in economic activity. An entity may engage in economic activity in relation to some of its functions but not others.

39.2.4

The Chapter I prohibition is considered further in section 39.3 below.

39.2.5 The Chapter II prohibition

Section 18 of the 1998 Act introduces a prohibition upon any conduct on the part of undertakings which amounts to the abuse of a dominant position if it may affect trade within the United Kingdom. This prohibition is called 'the Chapter II prohibition'.

39.2.6

The Chapter II prohibition is considered further in section 39.4 below.[5]

39.2.7

If an agreement is incompatible with either chapter prohibition, there are two possible consequences:

4 A clear guide to this subject is provided by the OFT publication 'Agreements and Concerted Practices – Understanding Competition Law' – www.gov.uk/government/uploads/system/uploads/attachment_data/file/284396/oft401.pdf.

5 A clear guide to this subject is provided by the OFT publication 'Abuse of a Dominant Position – Understanding Competition Law' – www.gov.uk/government/uploads/system/uploads/attachment_data/file/284422/oft402.pdf.

(a) The offending provision in the agreement is treated as being null and void (see s.2(4) of the 1998 Act). In *Gibbs Mews plc v Gemmell*,[6] the Court of Appeal said that an infringing agreement was not merely void and unenforceable, but also illegal, with the result that monies paid pursuant to it are not returnable. The ECJ has allowed severance in certain circumstances.[7]

(b) Any person who has been harmed by the prohibited provision can sue for damages for breach of statutory duty.

39.2.8

In the UK, competition law is applied and enforced by the CMA, acting concurrently with OFCOM. The 1998 Act gives the CMA powers to apply, investigate and enforce the Chapter I and Chapter II prohibitions in the CA98. The CMA's powers and duties are analysed in more detail in section 39.6 below.

39.2.9

The European equivalents of both these prohibitions are Articles 101 and 102 (formerly Articles 81 and 82) of the Treaty on the Functioning of the European Union (the Treaty of Rome – 'TFEU'). Under EU legislation, as a 'designated national competition authority', when the Competition and Markets Authority ('CMA'; previously the Office of Fair Trading – 'OFT') applies national competition law either to agreements which may affect trade between member states or to abuse prohibited by Article 102, the CMA is also required to apply Articles 101 and 102 of the TFEU.[8]

39.2.10

On 1 May 2004, EC Regulation 1/2003 (the Modernisation Regulation)[9] came into force. The Modernisation Regulation substantially changes the framework for enforcement of European competition law. It requires the designated national competition authorities of the member states and the courts of the member states to apply and enforce Articles 101 and 102 when national competition law is applied to agreements which may affect trade between member states or to abuse prohibited by Article 102. It also establishes a 'legal exception' regime. These changes are referred to in this guideline as 'modernisation'.

39.2.11

The Modernisation Regulation is just one part of the EU Common Regulatory Framework, comprising the Framework Directive (2002/21/EC), the Access Directive (2002/19/EC), the Authorisation Directive (2002/20/EC) and certain other directives, all of which govern the telecommunications market. These directives are considered further in section 39.5 below.

6 [1998] Eu LR 588.
7 See Chitty on Contracts, paragraph 16.188/197.
8 Article 3 of Regulation 1/2003, 2003 OJ L1/1, at p.8.
9 Council Regulation (EC) No 1/2003 of 16 December 2002 on the implementation of the rules on competition laid down in Articles 81 and 82 of the Treaty (OJ L1, 4.1.03, p1).

39.2.12

As a sectoral regulator, OFCOM also has regulatory powers and duties under the 2003 Act, which are examined in section 39.7 below.

39.3 The Chapter I prohibition

39.3.1

The European Courts have held that certain forms of agreement between undertakings ('Agreements') can be regarded, by their very nature as being injurious to the proper functioning of competition.[10] These restrictions, which include price fixing, market sharing and restrictions of output, are classed as restrictions of competition by object. The CMA will generally regard any Agreement which directly or indirectly fixes prices as being likely appreciably to restrict competition,[11] regardless of the parties' market shares.[12]

39.3.2

In its 'Guidance on the application of Article 81(3) [now Article 101(3)] of the Treaty',[13] the European Commission explains:

> Agreements between undertakings are caught by the prohibition rule of Article 81(1) when they are likely to have an appreciable adverse impact on the parameters of competition on the market, such as price, output, product quality, product variety and innovation. Agreements can have this effect by appreciably reducing rivalry between the parties to an agreement or between them and third parties.

39.3.3

Under s.60 of the 1998 Act, when determining a question arising under Part 1, the CMA must ensure that, having regard to any relevant differences between the provisions concerned, there is no inconsistency with the principles laid down by any relevant decision of the European Courts.[14] It must also have regard to any relevant decision or statement of the European Commission.

39.3.4

In a case thought to engage competition aspects, the CMA would ordinarily consider it appropriate to have regard to the guidelines set out in two Commission Notices:

10 Case C-8/08 *T-Mobile Netherlands v NMa*, CJ judgment of 4 June 2009 at paragraphs 28 to 30.
11 OFT408 *Trade Associations, Professions and Self-Regulating Bodies* (December 2004) at paragraph 3.3.
12 Commission Notice on agreements of minor importance which do not appreciably restrict competition under Article 101(1) TFEU (de minimis) 2001 OJ C 368/07 at paragraph 11. OFT401 *Agreements and Concerted Practices* (December 2004), at paragraph 2.17.
13 OJ 2004 C101/97.
14 The European Courts are together the Court of Justice and the General Court (formerly the Court of First Instance).

- Guidelines on the applicability of Article 101 TFEU to horizontal cooperation agreements (EU Horizontal Guidelines);[15] and
- Guidelines on the application of Article 101(3) of the Treaty (Article 101(3) Guidelines).[16]

39.3.5

When applying the Chapter I prohibition and/or Article 101, the CMA is only obliged to define the relevant market where it is not possible, without such a definition, to determine whether an Agreement is liable to affect trade in the UK and/or between member states, and whether it has as its object or effect the prevention, restriction or distortion of competition.[17]

39.3.6

The task for the CMA is to determine whether an agreement affects competition to an appreciable extent. This ingredient is not spelt out in the 1998 Act, but appears from the way in which the European Commission has approached the identically worded Article 81. The CMA has indicated[18] that it will have regard to the EU Commission's approach as set out in the Notice on Agreements of Minor Importance.[19] This is summarised as follows in paragraph 2.16 of the Guidelines:

> The European Commission considers that agreements between undertakings which affect trade between Member States do not appreciably restrict competition within the meaning of Article 81 if:
>
> - the aggregate market share of the parties to the agreement does not exceed 10 per cent on any of the relevant markets affected by the agreement where the agreement is made between competing undertakings (i.e. undertakings which are actual or potential competitors on any of the markets concerned), or
> - the market share of each of the parties to the agreement does not exceed 15 per cent on any of the relevant markets affected by the agreement where the agreement is made between non-competing undertakings, (i.e. undertakings which are neither actual nor potential competitors on any of the markets concerned).

39.3.7

The relevant market has two ingredients: the product market and the geographic market. In practice, therefore the task is to determine the relevant product market and the relevant geographic market, and then to enquire what is the competitive effect on that market of the agreement in question.

15 2011 OJ C 3/02.

16 2004 OJ C 101/97.

17 Case T-199/08 *Ziegler v Commission* [2011] 5 CMLR 261 paragraphs 41–5, Case T-62/98 *Volkswagen AG v Commission* [2000] ECR II-2707, paragraph 230 and Case T-29/92 *SPO v Commission* [1995] ECR II-289, paragraph 74.

18 'Agreements and Concerted Practices', OFT Guidelines 401, paragraph 2.18.

19 OJ [2001] C 368/13.

39.3.8

In carrying out this task, the CMA and the EU use the concept of the 'hypothetical monopolist'.[20] Paragraph 2.5 of the OFT Guideline 403, December 2004, explains:

> The process of defining a market typically begins by establishing the closest substitutes to the product (or group of products) that is the focus of the investigation. These substitute products are the most immediate competitive constraints on the behaviour of the undertaking supplying the product in question. In order to establish which products are 'close enough' substitutes to be in the relevant market, a conceptual framework known as the hypothetical monopolist test (the test) is usually employed.

39.3.9

The stress is therefore upon the 'relevant product market'. There are complicated tests which are applied to determine what is the relevant market. One such test is based upon the hypothetical monopoly: if an operator started raising prices by 5 per cent, would customers still come, or would they look elsewhere? If elsewhere, then the place they go to instead might indicate the bounds of the market.

39.3.10

Having determined the likely scope of the relevant product market, assessing the scope of the relevant geographic market requires a consideration of whether each party faces actual or potential competition from neighbouring landowners or operators as the case might be.

39.3.11

Undertakings are treated as actual competitors if they are active on the same relevant market. An undertaking may be treated as a potential competitor of another undertaking if, in the absence of the cooperation between them, it would enter into the same relevant market in response to a small but significant non-transitory increase in price.[21]

39.3.12

Having ascertained the relevant market, the next step is to ascertain what is the competitive effect on that market of the Agreement in question. This is often referred to as the 'de minimis' question. Agreements where the parties are below certain market share thresholds (10 to 15 per cent) will generally be regarded as de minimis and therefore outside the ambit of the Chapter I prohibition.

20 See the EU's *Notice on the Definition of Relevant Markets for the Purposes of Community Competition Law*, OJ [1997] C 372/5; [1998] 4 CMLR 177 – europa.eu.int/com/competition/mergers/legislation; and see the OFT Paper 'Market Definition', OFT Guideline 403, December 2004.

21 EU Horizontal Guidelines, paragraph 10.

39.3.13

The parties to an Agreement which is deemed to create an appreciable restriction of competition within the meaning of the Chapter I prohibition may yet be able to demonstrate that their Agreement is eligible for individual exemption under s.9 of the 1998 Act.[22]

39.3.14

If the effect of the Agreement is not de minimis (i.e. competition is affected to an appreciable extent), the Agreement may nevertheless merit exemption under s.9 of the 1998 Act, if it produces economic and consumer benefits that outweigh any impact on competition. Provided that the restrictions are indispensable to achieving the claimed efficiency, consumers share in the resulting benefit and competition is not thereby eliminated.

39.3.15

Section 9(2) of the 1998 Act states that the parties claiming the benefit of exemption under s.9(1) bear the burden of proving that its conditions are satisfied.

39.3.16

Examples of the sorts of competition issues engaging code and property law are examined in sections 39.6 and 39.8 below.

39.4 The Chapter II prohibition

39.4.1

As previously remarked, s.18 of the 1998 Act introduces a prohibition called 'the Chapter II Prohibition' upon any conduct on the part of undertakings which amounts to the abuse of a dominant position if it may affect trade within the United Kingdom. In essence, according to established case law, a dominant position exists when an undertaking has the power to behave to an appreciable extent independently of its competitors and its customers and thus is capable of preventing effective competition being maintained in the relevant market. Section 18(2) gives as examples of such conduct, 'directly or indirectly imposing unfair purchase or selling prices or other unfair trading conditions' and 'applying dissimilar conditions to equivalent transactions with other trading parties, thereby placing them at a competitive disadvantage'.

39.4.2

The Chapter II Prohibition is therefore concerned primarily with controlling single undertakings which are so powerful in the market that they are able to behave with

22 OFT401 *Agreements and Concerted Practices* (December 2004) at paragraphs 5.1–5.5 and Article 101(3) Guidelines at paragraph 41.

a degree of impunity, because there is no real competition that would enable their customers to obtain similar services elsewhere. Such undertakings are referred to as 'dominant'.

39.4.3

The holding of a dominant position is not prohibited. It is the abuse of a dominant position that is prohibited. The Chapter II Prohibition prevents such undertakings from abusing their dominance, either by exploiting their customers, or by taking steps to shut out possible competitors.

39.5 The (European) Common Regulatory Framework

39.5.1

The 2003 Act implements in the UK the regulation of the telecommunications sector under the Common Regulatory Framework ('the CRF') which applies across the European Union.

39.5.2

The CRF imposes on member states the obligation to designate independent national regulatory authorities (NRAs); sets out objectives and principles that the NRAs are to be guided by in carrying out their functions; obliges them to carry out market reviews; and empowers them to impose certain obligations on undertakings with significant market power ('SMP'[23]), including price controls. The designated NRA for the UK is OFCOM (see section 39.7 below).

39.5.3

The CRF consists of a number of directives,[24] the most relevant of which are:

(a) Directive 2002/21/EC on the common regulatory framework for electronic communications networks and services (as amended) ('the Framework Directive');
(b) Directive 2002/20/EC on the authorisation of electronic communications networks and services (as amended) ('the Authorisation Directive'); and
(b) Directive 2002/19/EC on access to, and interconnection of, electronic communications networks and associated facilities (as amended) ('the Access Directive').

23 Pursuant to Article 14(2) of the Framework Directive, an undertaking is deemed to have significant market power if, either individually or jointly with others, it enjoys a position equivalent to dominance, that is to say a position of economic strength affording it the power to behave to an appreciable extent independently of competitors, customers and ultimately consumers.

24 Explained in more detail in Chapter 4 of this book.

39.5.4

The Framework Directive establishes a harmonised framework for the regulation of electronic communications services and networks and associated facilities and services. It also establishes a set of procedures to ensure the harmonised application of the regulatory framework throughout the European Community.

39.5.5

In particular, Article 8 of the Framework Directive sets out the objectives which national regulatory authorities (such as OFCOM) must take all reasonable steps to achieve. These include:

- the promotion of competition in the provision of electronic communications networks and services by, amongst other things: ensuring that there is no distortion or restriction of competition in the electronic communications sector and encouraging efficient use of radio frequencies; and
- contributing to the development of the internal market by, amongst other things, removing obstacles to the provision of electronic communications networks and services at a European level and encouraging the interoperability of pan-European services.

In pursuit of these objectives, Article 8 requires national regulatory authorities to apply objective, transparent, non-discriminatory and proportionate regulatory principles by, among others: (i) ensuring that, in similar circumstances, there is no discrimination in the treatment of undertakings providing electronic communications networks and services; and (ii) promoting efficient investment and innovation in new and enhanced infrastructures. Article 8 also stipulates that member states must ensure that, in carrying out their regulatory tasks, national regulatory authorities take the utmost account of the desirability of making regulations technologically neutral.

39.5.6

Article 9 of the Framework Directive requires member states to ensure the effective management of radio frequencies for electronic communications services in accordance with Article 8. It also requires them to ensure that spectrum allocation used for electronic communication services and the issuing of general authorisations or individual rights of use of such radio frequencies are based on objective, transparent, non-discriminatory and proportionate criteria. In addition, it requires member states to promote the harmonisation of use of radio frequencies across the Community, consistent with the need to ensure effective and efficient use of frequencies. It further requires member states to ensure technology and service neutrality.

39.5.7

Article 5 of the Authorisation Directive states that where necessary to grant individual rights of use of radio frequencies, member states must grant such rights through open, objective, transparent, non-discriminatory and proportionate procedures (and in accordance with the provisions of Article 9 of the Framework Directive). When granting such rights, member states are required to specify whether they can be transferred by the holder, and if so, under which conditions.

39.5.8

Article 7 of the Authorisation Directive provides that where member states limit the number of rights of use to be granted for radio frequencies, they must, inter alia, give due weight to the need to maximise benefits for users and to facilitate the development of competition.

39.5.9

The Access Directive harmonises the way in which member states regulate access to, and interconnection of, electronic communications networks and associated facilities. It deals with the imposition of obligations by NRAs (national regulatory authorities) on operators designated as having significant market power.

39.6 The role of the CMA

39.6.1

The CMA is a non-ministerial department formed from a merger between the OFT and the Competition Commission, which was announced by the Department for Business, Innovation and Skills on 15 March 2012; created under Part 3 of the Enterprise and Regulatory Reform Act 2013; and brought into operation on 1 April 2014. Guidance on its procedures and the way in which it conducts investigations into suspected competition law infringements, is set out in its March 2014 publication *Guidance on the CMA's Investigation Procedures in Competition Act 1998 Cases.*[25]

39.6.2

The CMA's primary areas of responsibility in relation to telecommunications all concern situations where competition could be unfair or consumer choice may be affected, including:

(a) investigating mergers;
(b) conducting market studies;
(c) investigating possible breaches of prohibitions against anti-competitive agreements under the Competition Act 1998;
(d) encouraging regulators to use their competition powers; and
(e) considering regulatory references and appeals.

25 www.gov.uk/government/uploads/system/uploads/attachment_data/file/537006/CMA8_CA98_Guidance_on_the_CMA_investigation_procedures.pdf

39.6.3

So far as mergers are concerned (which form a substantial part of the work of the CMA), the CMA assesses the competitive effects of a merger by comparing the prospects for competition with the merger against the competitive situation absent the merger. The CMA will not take the matter further if, upon investigation, it does not believe that it is or may be the case that the merger may be expected to result in a substantial lessening of competition within a market or markets in the United Kingdom.

39.6.4

In 2012, the OFT (the predecessor to the CMA) was requested by the NFU and the CLA, to provide a Short-form Opinion in circumstances where those parties proposed to recommend a reference rate to their members for the grant of wayleaves for broadband in rural areas (the Rate Recommendation), the purpose of which was to provide certainty and consistency in the market place. The Short-form Opinion provides guidance as to the application of competition law to the Rate Recommendation, to facilitate the parties' self-assessment, and sets out general principles which may assist other associations of undertakings considering making similar recommendations to their members.[26]

39.6.5

There is a right of appeal from appealable decisions of the CMA to the Competition Appeal Tribunal ('CAT') by parties and third parties with a sufficient interest in appealable decisions, pursuant to Chapter IV of the 1998 Act. Where such an appeal is unavailable, an application for judicial review may be brought in certain circumstances.

39.6.6

The discussion in section 39.7 below concerning appeals from OFCOM decisions to the CAT or CMA under the 2003 Act applies equally to appeals from decisions of the CMA under the 1998 Act.

39.7 The role of OFCOM

39.7.1

The Office of Communications (OFCOM), the sectoral regulator for the communications sector, has concurrent powers and duties under the 2003 Act and TFEU to deal with anti-competitive behaviour in broadcasting, spectrum and telecommunications.[27] OFCOM

26 http://webarchive.nationalarchives.gov.uk/20140402165731/http://oft.gov.uk/shared_oft/SFOs/wayleave.pdf

27 Communications Act 2003 c.21, ss.369 to 372. OFT405 *Concurrent Application to Regulated Industries* (December 2004) paragraph 1.1, *Guidelines for the Handling of Competition Complaints, and Complaints and Disputes about Breaches of Conditions Imposed under the EU Directives*, OFCOM, July 2004, paragraph 1.

also has powers to investigate complaints about breaches of conditions imposed on providers and a duty to resolve disputes relating to conditions imposed under the EU Directives.[28]

39.7.2

These general powers and duties include:

(a) the duty to further the interests of citizens in relation to communications matters and to further the interests of consumers in relevant markets, where appropriate by promoting competition: s.3;

(b) duties relating to the fulfilment of OFCOM's EU obligations, which, insofar as are relevant, include a requirement to promote competition in relation to the provision of electronic communications networks and services; an obligation to encourage the provision of network service and interoperability for the purpose of securing efficient investment and innovation; and a requirement to take account of the desirability of it carrying out its functions in a manner which, so far as practicable, does not favour one form of electronic communications network, service or associated facility over another or one means of providing or making available such a network, service or facility over another: s.4;

(c) a duty to take due account of all applicable recommendations issued by the Commission under the Framework Directive: s.4A;

(d) the power to set binding conditions, including significant market power (SMP) conditions: s.45. An SMP condition can be applied to a communication provider that OFCOM has determined as having SMP in a specific market (s.46), but only if OFCOM is satisfied that the condition is:

   (i) objectively justifiable in relation to the networks, services, facilities, apparatus or directories to which it relates;

   (ii) not such as to discriminate unduly against particular persons or

   (iii) proportionate to what it is intended to achieve; and

   (iv) in relation to what it is intended to achieve, transparent (see s.47);

(e) Ofcom has a specific power to set SMP conditions that impose price controls: s.87(9). The imposition of such controls is subject to s.88, which introduces limitations upon the exercise of the power, such as a need to find both that there is an apparent risk of adverse effects arising from price distortion; and that the conditions must be appropriate to promote efficiency, sustainable competition and confer the 'greatest possible benefits on the end-users of public electronic communications services'.

39.7.3

Specific examples of the use by OFCOM of its powers under the 2003 Act are given in section 39.8 below.

28 www.ofcom.org.uk/static/archive/oftel/ind_info/eu_directives/index.htm

39.7.4

An appeal from a decision by OFCOM lies to the Competition Appeal Tribunal (CAT) under s.192 of the 2003 Act, on the ground that OFCOM's decision was based on an error of fact or was wrong in law, or both, or an erroneous exercise of discretion by OFCOM.

39.7.5

Under s.195(2) of that Act, the CAT must decide an appeal on the merits and by reference to the grounds of appeal set out in the notice of appeal.

39.7.6

Cases may also be referred by the CAT to the CMA (previously the Competition Commission), in cases where the appeal raises specified price control matters. The CMA conducts appeals in accordance with the guidance set out in the Competition Guidelines, which it has adopted.[29] Once the CMA has notified the CAT of its determination of the specified price control matters referred to it, the CAT must decide the appeal on the merits.

39.7.7

The approach of the CAT and CMA to appeals before them is identical, and has been elucidated by a series of cases.

39.7.8

First, in *T-Mobile (UK) Limited v Office of Communications*,[30] the CAT stated:

> It is also common ground that there may, in relation to any particular dispute, be a number of different approaches which OFCOM could reasonably adopt in arriving at its determination.
>
> There may well be no single 'right answer' to the dispute. To that extent, the Tribunal may, whilst still conducting a merits review of the decision, be slow to overturn a decision which is arrived at by an appropriate methodology even if the dissatisfied party can suggest other ways of approaching the case which would also have been reasonable and which might have resulted in a resolution more favourable to its cause.

39.7.9

Secondly, in *T-Mobile (UK) Ltd v Office of Communications*[31], the Court of Appeal confirmed that the s.192 appeal process is not intended to duplicate, still less, usurp, the functions of OFCOM:

29 *Price Control Appeals under Section 193 of the Communications Act 2003: Competition Commission Guidelines (CC13).*

30 [2008] CAT 12.

31 [2008] EWCA Civ 1373.

> After all it is inconceivable that Article 4 [of the Framework Directive], in requiring an appeal which can duly take into account the merits, requires Member States to have in effect a fully equipped duplicate regulatory body waiting in the wings just for appeals. What is called for is an appeal body and no more, a body which can look into whether the regulator has got something materially wrong. That may be very difficult if all that is impugned is an overall value judgment based upon competing commercial considerations in the context of a public policy decision.

39.7.10

Thirdly, in *British Telecommunications plc v Office of Communications*,[32] the CAT stated:

> By section 192(6) of the 2003 Act and rule 8(4)(b) of the 2003 Tribunal Rules, the notice of appeal must set out specifically where it is contended OFCOM went wrong, identifying errors of fact, errors of law and/or the wrong exercise of discretion. The evidence adduced will, obviously, go to support these contentions. What is intended is the very reverse of a *de novo* hearing. OFCOM's decision is reviewed through the prism of the specific errors that are alleged by the appellant. Where no errors are pleaded, the decision to that extent will not be the subject of specific review. What is intended is an appeal on specific points.

39.7.11

Fourthly, in *TalkTalk v Office of Communications*,[33] the CAT noted that the appropriate level of scrutiny in such appeals was 'profound and rigorous' and said that 'the question is whether Ofcom's determination was right, not whether it lies within the range of reasonable responses for a regulator to take'. It added:

> [w]here a decision can be challenged by way of a merits appeal, it is incumbent upon an appellant to show – if necessary by way of new evidence – that the original decision was wrong 'on the merits'. It is not enough to suggest that, were more known, the Tribunal's decision might be different.

39.7.12

Fifthly, in *British Telecommunications plc v Ofcom*,[34] the CAT held:

> 201. The proposition that an administrative decision-maker should ask himself the right question and take reasonable steps to acquaint himself with the relevant information to enable him to answer it correctly is well-established and uncontroversial: see, for example, *Secretary of State for Education and Science v Tameside MBC*.[35]
>
> 202. It is absolutely clear that when the Commission is exercising its original and investigative jurisdiction, it is under precisely such a duty: see, for example, *Tesco plc v Competition Commission*;[36] *BAA Limited v Competition Commission*.[37]

32 [2010] CAT 17.
33 [2012] CAT 1.
34 [2012] CAT 11.
35 [1977] 1 AC 1014 at 1065.
36 [2009] CAT 6 at paragraph [139].
37 [2012] CAT 3 at paragraph [20(3)].

203. In this case, however, the Commission is not exercising any kind of original or investigative jurisdiction. As we made clear in paragraph 118 above, that is the function of OFCOM. The Commission's role is confined to determining the questions referred to it by the Tribunal. The Commission is not investigating anything – it is determining whether OFCOM erred in its decision for the reasons set out in the notice of appeal. As we noted in paragraph 118 above, the Commission is acting as an administrative appeal body.

204. Accordingly, we hold that the duty on an administrative decision-maker to investigate and seek out the relevant information to enable him to answer the question before him correctly does not apply to the Commission when determining reference questions pursuant to section 193 of the 2003 Act. Rather, the Commission's duty is to discharge its functions under this section in a more judicial manner: it is not investigating with a view to making a decision; it is considering specific complaints about the decision of another. In short, the nature and quality of the scrutiny that the Commission gives to Ofcom's decisions is altogether different (to say 'lower' or 'higher' would be to compare qualitatively different functions) from exercises conducted by the Commission as administrative decision-maker.

39.7.13

Sixthly, in *Everything Everywhere Ltd v Competition Commission*,[38] the Court of Appeal said:

22. ... If the appellant can do no more than show that there is a 'real risk that the decision was wrong' then it has not shown that Ofcom's decision was wrong and the appeal should be dismissed. But there remains scope for dispute as to what is meant by showing that an original decision is wrong 'on the merits'.

23. It is for an appellant to establish that Ofcom's decision was wrong on one or more of the grounds specified in s.192(6) of the 2003 Act: that the decision was based on an error of fact, or law, or both, or an erroneous exercise of discretion. It is for the appellant to marshal and adduce all the evidence and material on which it relies to show that Ofcom's original decision was wrong. Where, as in this case, the appellant contends that Ofcom ought to have adopted an alternative price control measure, then it is for that appellant to deploy all the evidence and material it considers will support that alternative.

24. The appeal is against the decision, not the reasons for the decision. It is not enough to identify some error in reasoning; the appeal can only succeed if the decision cannot stand in the light of that error. If it is to succeed, the appellant must vault two hurdles: first, it must demonstrate that the facts, reasoning or value judgments on which the ultimate decision is based are wrong, and second, it must show that its proposed alternative price control measure should be adopted by the Commission. If the Commission (or Tribunal in a matter unrelated to price control) concludes that the original decision can be supported on a basis other than that on which Ofcom relied, then the appellant will not have shown that the original decision is wrong and will fail.

38 [2013] EWCA Civ 154.

25. Usually an appellant will succeed by demonstrating the flaws in the original decision and the merits of an alternative solution. But that is not necessarily so. I would not rule out the possibility that there could be a case where an appellant succeeds in so undermining the foundations of a decision that it cannot stand, without establishing what the alternative should be. In such a case, if there is no other basis for maintaining the decision, the Commission or Tribunal would be at liberty to conclude that the original decision was wrong but that it could not say what decision should be substituted. The Tribunal would then be required to allow the appeal under s.195(2) and direct Ofcom to make a fresh decision with such directions as the Tribunal thinks are necessary to reach a properly informed conclusion. The Tribunal may wish to specify the steps to be taken by Ofcom to make good any deficit in evidence and material so as to reach a fresh decision, or leave it to Ofcom to act as it sees fit in the light of the Commission's conclusion.

39.7.14

Lastly, in determining whether OFCOM erred, OFCOM is not held to be wrong simply because it considers there to be some error in its reasoning on a particular point – the error in reasoning must have been of sufficient importance to vitiate OFCOM's decision on the point in whole or in part. This check on the appellate function was expressed in this way in a determination by the CMA in 2016 in two references by the CAT concerning superfast broadband price control appeals by BT plc and TalkTalk Telecom Group plc:[39]

> Ofcom is a specialist regulator whose judgement should not be readily dismissed. Where a ground of appeal relates to a claim that Ofcom has made a factual error or an error of calculation, it may be relatively straightforward to determine whether it is well founded. Where, on the other hand, a ground of appeal relates to the broader principles adopted or to an alleged error in the exercise of a discretion, the matter may not be so clear. In a case where there were a number of alternative solutions to a regulatory problem with little to choose between them, we do not think it would be right for us to determine that Ofcom erred simply because it took a course other than the one that we would have taken. On the other hand, if, out of the alternative options, some clearly had more merit than others, it may more easily be said that Ofcom erred if it chose an inferior solution. Which category a particular choice falls within can necessarily only be decided on a case-by-case basis.

39.8 Examples of OFCOM involvement in competition matters

39.8.1

Specific examples of the use by OFCOM of its powers under the 2003 Act are as follows.

39 Reference under s.193 of the Communications Act 2003 – *British Telecommunications plc v OFCOM* Case 1238/3/3/15; *TalkTalk Telecom Group plc v OFCOM* Case 1237/3/3/15 – https://assets.publishing.service.gov.uk/media/5767bd34ed915d3cfd0000a2/bt-talktalk-final-determination.pdf.

39.8.2

First, OFCOM is entitled to request the separation of the provision of managed transmission services (the combination of network access together with the supply of transmission services, involving management and maintenance services including field maintenance and support, electricity, insurance, service management and monitoring) and network access (including site access, site accommodation, shared combiner systems, shared feeder systems, shared antenna systems, provision of bespoke antennas and feeders and main power or diesel generator infrastructure). It can impose an obligation to permit network access to third parties on the basis of published Reference Offers. These Offers must be non-discriminatory and cost-orientated, with the aim of encouraging competition at the managed transmission services level.

39.8.3

Secondly, in its 2004 final form Leased Lines Market Review, OFCOM identified BT as having significant market power in the interface symmetric broadband origination market, and imposed cost orientation controls. In 2012, OFCOM found that BT had failed to comply with these controls, and ordered BT to make substantial payments to other communications providers which had been overcharged. Appeals against different aspects of the determination were made by all parties to the CAT, and thence to the Court of Appeal, whose judgment[40] gives a good overview of the relevant legislation and of the issues arising.

39.8.4

Thirdly, in 2015, OFCOM imposed a pricing rule requiring BT to maintain a sufficient margin between its wholesale and retail superfast broadband charges, in order to allow other providers profitably to match its prices. In imposing this rule, in accordance with s.88 of the 2003 Act, it appears to OFCOM from its analysis in the Fixed Access Market Reviews that there was a relevant risk of adverse effects arising from price distortion.

39.8.5

Fourthly, in 2015 OFCOM set annual licence fees for the 900 MHz and 1800 MHz bands of radio spectrum used for mobile communications, in order to comply with a Direction made in 2010 by the Secretary of State for Culture, Media and Sport under the powers contained in s.5 of the Wireless Telegraphy Act 2006. In choosing the level of fees, OFCOM was required by Article 13 of the Authorisation Directive to take into account the objectives under Article 8 of the Framework Directive, such as promoting competition and investment in new technology. A challenge was mounted by EE Ltd on the basis that, in setting the amount of the fees, OFCOM had deliberately left out of account any considerations other than the market value of the spectrum in question, and had therefore failed to have regard to the Article 8 considerations. Its challenge failed upon judicial review, but succeeded in the Court of Appeal – see *EE Ltd v Ofcom*.[41]

40 *British Telecommunications Plc v OFCOM* [2017] EWCA Civ 330.
41 [2017] EWCA Civ 1873.

39.8.6

Fifthly, in 2016, OFCOM issued a statement following its review of competition in the provision of leased lines, in which it defined a single product market for contemporary interface symmetric broadband origination services of all bandwidths; and four separate relevant geographic markets (namely the Central London Area; the London Periphery; Hull; and the Rest of the UK ('RoUK')). It also made determinations concerning the extent of BT's core network. OFCOM proposed a package of remedies including a so-called passive remedy allowing communications providers to lease only the fibre element of the leased lines from BT, thus enabling them to attach equipment of their own choosing at either end to 'light' the fibre. BT appealed against the determinations on a number of grounds, including alleged errors concerning market definition and the remedies imposed. The CAT held, quashing the determinations,[42] that OFCOM had erred in a number of respects, by concluding that it was appropriate to define a single product market for CISBO services of all bandwidths; by concluding that the RoUK comprises a single geographic market; and in its determination of the boundary between the competitive core segments and the terminating segments of BT's network.

39.8.7

Sixthly, in 2017, OFCOM announced its intention to auction wireless telegraphy licences in the 2.3 GHz and 3.4 GHz bands, setting out its decision regarding the regulation of the auction in a decision paper dated 11 July 2017. The regulations imposed restrictions capping the amount of spectrum that any MNO could hold. These restrictions and others were opposed, leading to claims for judicial review of OFCOM's decision brought by Hutchison 3G (UK) Ltd, British Telecommunications Plc and EE Ltd, with Telefónica UK Ltd and Vodafone Ltd appearing as interveners to oppose both claims. The decision of Green J[43] rejecting all the claims is required reading for those with a thirst for knowledge of competition issues to a depth that this book is unable to cover.

39.9 Land agreements

39.9.1

The Competition Act 1998 came into force on 1 March 2000. It replaced the Restrictive Trade Practices Acts 1976 and 1977. Under those Acts, although a lease was capable of being 'an agreement' to which the Acts applied, it was not a registrable agreement, unless there was a trading nexus between landlord and tenant.[44]

39.9.2

Section 50 of the 1998 Act provides that the Secretary of State may order any provision of Part I of the Act to apply to a 'land agreement'. The Competition Act 1998 (Land

42 See BT plc v Ofcom [2017] CAT 17.

43 *The Queen on the application of Hutchison 3G UK Ltd v Office of Communications v Telfónica UK Ltd, EE Ltd, British Telecommunications Plc, Vodafone Ltd; The Queen on the application of British Telecommunications Plc v Office of Communications Telefónica UK Ltd, Vodafone Ltd, Hutchison 3G UK Ltd* [2017] EWHC 3376 (Admin).

44 See Ravenseft Properties Ltd v Director of Fair Trading [1978] QB 52.

Agreements Exclusion and Revocation) Order 2004 ('the Exclusion Order') defines 'a land agreement' as an agreement creating, altering, transferring or terminating any interests in land. This clearly therefore included leases. Article 4 of that Order provided that the Chapter 1 prohibition was not applicable to such land agreements.

39.9.3

In January 2010, the Department for Business Innovation and Skills decided to revoke the Exclusion Order with effect from 6 April 2011. This was confirmed by the new Government on 29 June 2010, when it made the Competition Act 1998 (Land Agreements Exclusion and Revocation) Order 2010 ('the Revocation Order'). The effect was therefore that, as at 6 April 2011, the Chapter 1 prohibition would apply to leases, with retrospective effect.

39.9.4

The Government's impact assessment concerning the Revocation Order provided that:

> It has always been the case that agreements concerning land must be compatible with Chapter I of the Competition Act 1998. The exclusion only meant that such agreements did not need to be notified to the OFT for approval. They would be assumed to be compatible with the Chapter I prohibition unless and until found not to be – at which point the benefit of the exclusion would be withdrawn from the relevant agreement. Removing the exclusion will make it clear that land agreements must be assessed for compatibility with the Chapter I prohibition in the same way as must all other types of agreement. The great majority of agreements are compatible with the prohibition. It is only agreements which have the effect of restricting competition in markets that are prohibited.

39.9.5

Leases and other forms of land agreements to operators are therefore to be treated in the same way as any other form of agreement, as far as competition law is concerned.

40 The position in Scotland

40.1 Introduction

40.1.1

The territorial scope of both the Old Code and the New Code is England, Wales, Scotland and Northern Ireland – see generally the Telecommunications Act 1984 (which operates by including parallel texts relating to England and Wales or Scotland) and s.119 of the Digital Economy Act 2017.

40.1.1

The substantive law in England and Wales is identical for all material purposes, although procedural differences (essentially matters of nomenclature) are gradually opening up. For example, Wales has chosen not to implement the Tribunal changes effected in England on 1 July 2013 under the Tribunal Procedure (First-tier Tribunal) (Property Chamber) Rules 2013, whereby the former Leasehold Valuation Tribunal became the First-tier Tribunal (Property Chamber).

40.1.2

The substantive law in Northern Ireland is similar, so far as the issues with which this book are concerned, although there is some difference in courts and tribunal procedure, which are examined in Chapter 33.

40.1.3

In Scotland, however, not merely is the language of the law different from that employed in the remainder of the United Kingdom in a number of respects, but there are also substantive differences in the law itself, and in practice and procedure. This chapter examines the differences in labelling and in substance, while Chapter 33 looks at the different courts and tribunals found in the United Kingdom that adjudicate upon telecommunications disputes.

40.2 Labelling differences

40.2.1

There are similar labelling differences in both the Old Code and the New Code. It suffices to treat with the New Code to bring out the examples.

40.2.2

Paragraph 12(5) of the New Code (which deals with the exercise of code rights), provides:

> In the application of sub-paragraph (3) to Scotland the reference to a person who is the owner of the freehold estate in the land is to be read as a reference to a person who is the owner of the land.

This drafting accommodates the fact that, in Scotland, there is a system of outright ownership of land.[1]

40.2.3

Paragraph 13(5) of the New Code makes the point that a street in England and Wales or Northern Ireland is referred to as a road in Scotland. Each expression has a well-known statutory meaning, which amounts to the same thing, and which is explained by paragraph 108(1). In Scotland, a 'road' is defined as a public road (see paragraph 57(a)), and the further explanation in paragraph 108(1) by reference to the meaning in Part 4 of the New Roads and Street Works Act 1991; while in England and Wales, 'street' receives its definition from paragraphs 57(b) and 108(1) by reference to the New Roads and Street Works Act 1991. The same difference is also noted in other parts of the New Code that deal with streets and roads (paragraphs 42, 82 and 105, and Part 8 generally). Paragraph 105(4) adds that references to an occupier of land in relation to a road in Scotland are to the road managers within the meaning of Part 4 of the New Roads and Street Works Act 1991.

40.2.4

Paragraph 16 makes the point that in Scotland, an 'assignment' of an agreement is referred to as an 'assignation'. Language aside, there is no substantive difference in concept. However, as is more fully explained in paragraph 40.4.2 below, the concept of a 'guarantee agreement', whereby an Assignor guarantees future performance by an Assignee (given statutory footing by paragraph 16) is new. It has not previously formed part of the law and practice in Scotland.

1 Following the coming into force, on 28 November 2004, of the Abolition of Feudal Tenure etc. (Scotland) Act 2000.

40.2.5

Paragraphs 37, 38 and 43 similarly note that in Scotland a 'covenant' in the context of a land obligation is referred to as a contractual term, enforceable in the same way. There is much substantive law regarding enforceability of covenants affecting land in the other parts of the United Kingdom, and in Scotland the substantive law is, broadly speaking, the same, even if the nomenclature is different. It is worth noting however that, generally speaking, a contractual term will more readily be enforced by the Scottish courts by way of an order for performance rather than damages.[2]

40.2.6

Paragraph 84 makes the point that costs awarded by a court are known as 'expenses' in Scotland. This is repeated for good measure in paragraphs 85 and 96.

40.2.7

Paragraph 94 refers to the difference in nomenclature between the various courts and tribunals in the United Kingdom. Reference should be made to the explanation provided in Chapter 33.

40.2.8

Paragraph 104 stipulates that, in its application to the Crown:

(a) references to a Crown interest include an interest which belongs to an office-holder in the Scottish Administration or which is held in trust for Her Majesty for the purposes of the Scottish Administration by such an office-holder;
(b) 'the appropriate authority', for the purposes of considering whether a valid agreement required by this code is given in respect of any Crown interest subsisting in any land, in the case of land belonging to Her Majesty in right of the Crown, includes the relevant person or, as the case may be, the government department or office-holder in the Scottish Administration having the management of the land in question; and in the case of land belonging to an office-holder in the Scottish Administration or held in trust for Her Majesty by such an office-holder for the purposes of the Scottish Administration, includes that office-holder;
(c) 'the relevant person' in relation to land to which section 90B(5) of the Scotland Act 1998 applies, means the person having the management of that land;
(d) references to an office-holder in the Scottish Administration are to be construed in accordance with s.126(7) of the Scotland Act 1998.

2 See e.g. the position in relation to 'keep open' clauses in commercial leases e.g. *Highland and Universal Ltd v Safeway Properties* (No. 2) [2000] SLT 414 and *Oak Mall Greenock Ltd v McDonald's Restaurants Ltd* GWD 17–540.

40.2.9

Paragraph 108(1) contains a series of terms which set out further linguistic differences between the jurisdictions, including, so far as relevant to this chapter:

'agriculture' and 'agricultural' –
- (a) in relation to England and Wales, have the same meanings as in the Highways Act 1980,
- (b) in relation to Scotland, have the same meanings as in the Town and Country Planning (Scotland) Act 1997, and
- (c) in relation to Northern Ireland, have the same meanings as in the Agriculture Act (Northern Ireland) 1949;

'bridleway' and 'footpath' –
- (a) in relation to England and Wales, have the same meanings as in the Highways Act 1980,
- (b) in relation to Scotland, have the same meanings as Part 3 of the Countryside (Scotland) Act 1967, and
- (c) in relation to Northern Ireland, mean a way over which the public have, by virtue of the Access to the Countryside (Northern Ireland) Order 1983 (SI 1983/ 1895 (NI 18)), a right of way (respectively) on horseback and on foot;

'enactment' includes –
- (a) an enactment comprised in subordinate legislation within the meaning of the Interpretation Act 1978,
- (b) an enactment comprised in, or in an instrument made under, a Measure or Act of the National Assembly for Wales,
- (c) an enactment comprised in, or in an instrument made under, an Act of the Scottish Parliament, and
- (d) an enactment comprised in, or in an instrument made under, Northern Ireland legislation;

'lease' includes –
- (a) in relation to England and Wales and Northern Ireland, any leasehold tenancy (whether in the nature of a head lease, sub-lease or underlease) and any agreement to grant such a tenancy but not a mortgage by demise or sub-demise, and
- (b) in relation to Scotland, any sub-lease and any agreement to grant a sub-lease.

40.3 Substantive differences: (a) form of contract

40.3.1

Neither the Old Code nor the New Code is prescriptive as to the legal form which an agreement conferring code rights must take. Paragraph 2(1) of the Old Code provided that an agreement must be 'in writing'. Paragraph 11(1) of the New Code provides that an agreement:

(a) must be in writing,
(b) must be signed by or on behalf of the parties to it,
(c) must state for how long the code right is exercisable, and
(d) must state the period of notice (if any) requited to terminate the agreement.

40.3.2

This flexibility gives parties the freedom to choose the legal form of an agreement by which code rights are conferred. In practice, under the Old Code, most agreements took the form of a lease or, alternatively, a wayleave or other contract (often labelled a 'site access agreement' or similar). This is likely to continue to be the case under the New Code (although the use of leases may well diminish). However, under the Old Code and the New Code, code rights can be conferred using other legal forms such as a servitude (easement), if the operator owns property which can act as the 'benefited property', as defined in section 1(2) of the Title Conditions (Scotland) Act 2003.

40.4 Substantive differences: (b) leases

40.4.1 Legislative intervention

Scots law as it applies to commercial leases has developed with very little in the way of legislative intervention. Notable exceptions include irritancy[3] (forfeiture) and limited rights of lease extension for tenants of shops.[4] As such, the Old Code introduced and, to a far greater extent, the New Code introduces significant changes in certain aspects of the law of landlord and tenant in Scotland (in respect of leases conferring code rights).

40.4.2 Assignation

The prohibition, introduced by paragraph 16(1) of the New Code, of any provision in a lease preventing, or limiting assignation of an agreement conferring code rights is one such change.[5] Whilst practitioners in England and Wales will be familiar with authorised guarantee agreements[6] and, indeed, the concept of an assignor remaining liable to a landlord until expiry,[7] that is not so in Scotland. The long established position under Scots law is that an assignor steps out of the picture entirely and that its obligations are assumed by the assignee alone. This is not to say that no leases in Scotland ever make provision for such guarantees (for they sometimes do, for example, in relation to permitted transfers between group companies) but, in practice, the use of clauses of this

3 Law Reform (Miscellaneous Provisions) (Scotland) Act 1985, ss.4–5.
4 The Tenancy of Shops (Scotland) Act 1949.
5 The prohibition does not apply to any subsisting agreement in force when the New Code came into force on 28 December 10167. See Digital Economy Act 2017, Schedule 2, Paragraph 5.
6 Introduced by the Landlord and Tenant (Covenants) Act 1995, s.16.
7 Since this was the law prior to 1 January 1996.

type is rare. It remains to be seen whether a provision requiring a 'guarantee agreement' will in future become a standard term in Scottish leases conferring code rights.

40.4.3 Termination

Since the Landlord and Tenant Act 1954 does not apply in Scotland, termination of leases (and other agreements) conferring code rights is always governed by Part 5 of the New Code (even if the primary purpose of the agreement is not to grant code rights).[8]

40.4.4

Paragraph 31(3) of the New Code provides that the notice period for bringing a code agreement to an end is 18 months and that the expiry of the notice period must fall on a date when the agreement would otherwise have terminated (by virtue of expiry or the exercise break). This has several consequences.

40.4.5

First, it provides helpful clarification of when a termination notice may be served (and, in particular, that it may be served prior to the expiry of a lease). Paragraph 21 of the Old Code was far from clear and many operators sought to argue that a termination notice could only be lawfully given, in the case of a Scottish lease, after a valid notice to quit had been given by landlord or tenant (or both) and lease expiry had actually occurred.

40.4.6

Secondly, since a termination notice will often serve the dual function of constituting a notice to quit in respect of a lease in Scotland, the de facto period (as a matter of practice rather than law) for a notice to quit on behalf of a landlord is 18 months rather than 40 days.[9] A termination notice which complies with the terms of paragraph 31 of the New Code will almost certainly be sufficient to satisfy the criteria for a notice to quit.[10]

40.4.7

Thirdly, an asymmetry is created since an operator can still (unless a lease provides otherwise) terminate a lease, which confers code rights, by serving a notice to quit not less than 40 days prior to expiry, whereas a site provider must give not less than 18 months' notice. It seems likely that the drafters of leases conferring code rights will seek to address this imbalance by requiring operators to give notice of their intentions far sooner than 40 days prior to lease expiry.

8 New Code, paragraph 29.

9 At common law most commercial leases in Scotland terminate at expiry only if landlord or tenant serves a notice to quit not less than 40 days prior to lease expiry. Otherwise tacit relocation operates; see paragraph 40.4.8.

10 Being an unequivocal declaration that the lease will come to an end at the expiry date and that the tenant must remove.

40.4.8 Tacit relocation and termination

It is common for leases conferring code rights to continue beyond their contractual expiry date by virtue of the operation of the Scots law doctrine of tacit relocation. In the event that neither landlord or tenant serves a notice to quit the requisite period prior to expiry[11] then the lease continues for a period of 12 months (or the period of the original lease, if less than 12 months) on the same terms and conditions as before.[12] The effect of paragraph 31(3) of the New Code is that, whilst it will still be possible (under the general law) to terminate a lease continuing by tacit relocation by giving a notice to quit the requisite period prior to any anniversary of the contractual expiry date, termination pursuant to the New Code cannot occur without 18 months' notice.

40.4.9

Subsisting leases conferring code rights and continuing on tacit relocation at the date on which the New Code came into force can be terminated on three months' notice or a period equal to the unexpired term, whichever is greater.[13]

40.4.10 Irritancy

As is the position in other parts of the United Kingdom, a risk for site providers is that a lease may be subject to irritancy[14] (forfeiture) and therefore terminated, within as little as 14 days or so (in the event of, for example, non-payment of rent), but can only be terminated for the purposes of the New Code on 18 months' notice.[15]

40.4.11 Continuation after contractual termination

Whilst practitioners in other parts of the UK will be well familiar with the concept of leases continuing automatically, as a matter of statute, notwithstanding that they may have come to an end in accordance with their contractual terms,[16] this is an entirely new concept in Scotland – introduced by paragraph 30 of the New Code.

40.4.12 Grounds for termination

A number of the grounds for termination, set out in paragraph 31(4) of the New Code, will be well known to practitioners in England and Wales. The grounds set out at paragraph 31(4)(a)–(c) (being: substantial breaches of the agreement by the operator; persistent delays in payment by the operator; and intention of the site provider to redevelop) are substantially similar to the grounds on which a landlord can oppose an application for a new tenancy, as set out at section 30(1)(b), (c) and (f) of the Landlord and Tenant

11 Usually 40 days.
12 Save for duration (term) and any other contract terms which are inconsistent with a 12-month lease.
13 Digital Economy Act 2017, Schedule 2, paragraph 7(3).
14 Pursuant to the Law Reform (Miscellaneous Provisions) (Scotland) Act 1985, s.4 or s.5.
15 New Code, paragraph 31(3).
16 Since the Landlord and Tenant Act 1954 contains provisions which apply to business tenancies generally, and there is similar legislation in Northern Ireland (the Business Tenancies (Northern Ireland) Order 1996).

Act 1954. If any of these grounds are used in Scotland then the Lands Tribunal for Scotland (or the Sheriff Court) will no doubt have regard to the body of case law which has built up in England and Wales in relation to the interpretation and application of these grounds, but will not be bound by it. It seems to the Authors to be unlikely that there will be divergence in the general approach taken by the Tribunals (and Courts) in the various parts of the United Kingdom to the interpretation and application of paragraph 31 of the New Code.

40.5 Substantive differences: (c) registration and the creation of real rights in land

40.5.1

The New Code contains detailed provisions, at paragraphs 9 and 10, governing who may confer code rights and who else is bound by code rights. These provisions are examined in Chapter 16.

40.5.2

There is no requirement to register, in the Land Register of Scotland (or elsewhere), code agreements under the New Code (Paragraph 14), as was the position under the Old Code.

40.5.3

For present purposes, it suffices to observe that an agreement to confer code rights will bind not only the granter but also its successors and any party taking derivative rights from the granter's successors. In this sense, the conferral of a code right creates a real right[17] rather than a personal right.[18] This is an important distinction under Scots law (as it is in many other legal systems). For instance, if a code right is conferred by the owner of land then all successive owners (as well as any future acquirer of a subordinate right from the owner, such as a lessee or the grantee of a standard security) will also be bound. If a code right is conferred by a lessee then the code rights will bind successive lessees but not the owner (unless the owner agrees to be so bound).

40.5.4

What emerges therefore is that the New Code allows for the creation of a new type of real right in land – but one that does not have to be registered. This is true even if the legal form chosen by the parties would ordinarily require registration in the Land Register in order to create a real right. So, for example, an unregistered lease of more than 20 years which ordinarily requires to be registered in order to confer a real right upon a tenant[19] would nevertheless create code rights which would bind successor

17 i.e. a right in the land.

18 i.e. a right enforceable against a person.

19 Registration of Leases (Scotland) Act 1857, s.20B.

owners. Similarly code rights conferred by way of deed of servitude,[20] or even by contractual wayleave, will have a real effect irrespective of registration. To mitigate against this issue, a landowner or occupier of neighbouring land (but no other party) is entitled, in terms of paragraph 39, to require an operator to disclose ownership of apparatus and the existence of code rights.

40.6 Substantive differences: (d) compensation

40.6.1

Paragraph 84(3) of the New Code provides:

> For the purposes of assessing such compensation for diminution in the value of land, the following provisions apply with any necessary modifications as they apply for the purposes of assessing compensation for the compulsory purchase of any interest in land –
> (a) in relation to England and Wales, rules (2) to (4) set out in section 5 of the Land Compensation Act 1961;
> (b) in relation to Scotland, rules (2) to (4) set out in section 12 of the Land Compensation (Scotland) Act 1963;
> (c) in relation to Northern Ireland, rules (2) to (4) set out in Article 6(1) of the Land Compensation (Northern Ireland) Order 1982 (SI 1982/712 (NI 9)).

40.6.2

The position in relation to the Land Compensation Act 1961 (which is broadly similar to the position in Northern Ireland) is discussed in Chapter 30.

40.6.3

The position in relation to s.12 of the Land Compensation (Scotland) Act 1963 is identical to that in England and Wales save that section 2A of the Land Compensation Act 1961[21] (value to be assessed in light of the 'no-scheme' principle) is not replicated in the Scottish Act. This reflects the fact that, prior to the coming into force of the Scotland Act 1998,[22] the compulsory purchase systems were more or less identical north and south of the border.[23] However, the changes introduced by Part 9 of the Localism Act 2011[24] and, most recently, by Part 2 of the Neighbourhood Planning Act 2017 have not been followed in Scotland. As such, the position in Scotland is, in simple terms, as it was in England and Wales prior to the coming into force of 2011 and 2017 Acts.

20 Which require to be created as a real burden and registered in terms of the Title Conditions (Scotland) Act 2003, s.75.
21 Added by Neighbourhood Planning Act 2017, s.32(2).
22 On 1 July 1999.
23 Scottish Law Commission Discussion Paper on Compulsory Practice (No. 159), December 2014, paragraphs 13–16.
24 Amending the Land Compensation Act 1961, ss.14–16 and establishing new rules for 'taking account of planning permission when assessing compensation'.

40.6.4

The Scottish Law Commission published a Discussion Paper on Compulsory Purchase in December 2014,[25] proposing a comprehensive rewriting of the law of compulsory purchase (and compensation, in particular). The Scottish Law Commission looked favourably upon the reforms which had (at that time) been implemented in England and Wales. The Commission subsequently submitted a Report to the Scottish Government on 22 September 2016. The Scottish Government's Programme for 2017–18 includes a statement to the effect that it intends to introduce a Planning Bill dealing with, amongst other things, 'modernising Compulsory Purchase Orders'. The Planning (Scotland) Bill introduced on 4 December 2017[26] does not contain any provisions concerning compulsory purchase. Reform does, however, appear inevitable in the short to medium term. This may well bring the Scottish legislation back into line with England and Wales.

40.6.5

Paragraph 84(3) to (5) also note the following further differences in compensation assessment as between the jurisdictions:

(a) England and Wales: s.10(1) to (3) of the Land Compensation Act 1973 (compensation in respect of mortgages, trusts of land and settled land) applies;
(b) Scotland: s.10(1) and (2) of the Land Compensation (Scotland) Act 1973 (compensation in respect of restricted interests in land) applies;
(c) Northern Ireland: Article 13(1) to (3) of the Land Acquisition and Compensation (Northern Ireland) Order 1973 (SI 1973/1896 (NI 21)) (compensation in respect of mortgages, trusts for sale and settlements) applies.

The position in England and Wales (and the similar approach in Northern Ireland) is also dealt with in Chapter 30.

40.6.6

The foregoing provisions are concerned, first, with the interest of a heritable creditor (mortgagee). The position of a heritable creditor in Scotland is substantially similar to a mortgagee in England and Wales, save for nomenclature.

40.6.7

Secondly, the provisions are concerned with the application of compensation funds in certain prescribed circumstances. In England and Wales, this is limited to circumstances where the interest is subject to a 'trust in land' or is 'settled land' (which has no direct counterpart in Scots law). Section 67 of the Land Clauses Consolidation (Scotland) Act 1948 sets out: (a) a far broader range of circumstances in which the provision is

25 Discussion Paper No. 159.
26 SP Bill 23.

engaged; and (b) detailed rules governing the application of compensation funds in circumstances where the provision is engaged.[27]

40.6.8

Paragraph 85(2) to (6) also note the following further differences in assessment as between the jurisdictions, in relation to compensation payable by the operator for injurious affection to neighbouring land:

(a) England and Wales: compensation is payable under s.10 of the Compulsory Purchase Act 1965;
(b) Scotland: compensation is payable under s.6 of the Railway Clauses Consolidation (Scotland) Act 1845, and is to be determined, in default of agreement, by the Lands Tribunal for Scotland;
(c) Northern Ireland: compensation is payable under Article 18 of the Land Compensation (Northern Ireland) Order 1982 (SI 1982/712 (NI 9)), and is to be determined, in default of agreement, by the Lands Tribunal for Northern Ireland.

The position in England and Wales (and the similar approach in Northern Ireland) is also dealt with in Chapter 30.

The position in relation to s.6 of the Railway Clauses Consolidation (Scotland) Act 1845 is substantially the same as in England and Wales. The 'rules' concerning claims for compensation arising from injurious affection as set out in *Metropolitan Board of Works v McCarthy*[28] apply equally in Scotland.[29]

40.7 Substantive differences: (e) rating

40.7.1

The law of non-domestic rates as it applies to telecommunications is examined in Chapter 37. A detailed examination of the differences between the law of rating in Scotland, as compared to England and Wales, is beyond the scope of this book.[30]

40.7.2 The main legislation

The main source of the law in Scotland is the Lands Valuation (Scotland) Act 1854 and the Valuation and Rating (Scotland) Act 1956. The primary provision governing the collection of rates generally is the Non-Domestic Rating Contributions (Scotland) Regulations

27 Broadly, that the funds must be paid into an incorporated or Chartered Bank in Scotland and can only be applied, with the authority of the Court of Session, for certain specified purposes.

28 [1874] LR 7 HL 2463.

29 The Scottish Courts and Tribunals will also have regard to, amongst other authorities on the subject: *Cliff v Welsh Office* [1999] 1 WLR 796; *Wildtree Hotels Ltd v Harrow London Borough Council* [2001] 2 AC 1; and *Moto Hospitality Ltd v Secretary of State for Transport* [2008] 1 WLR 2822.

30 For a full treatment of the law of rating in Scotland see Armour on *Valuation for Rating*, 5th ed. (W. Green & Son).

1996.[31] Specific adaptations of the law for telecommunications are found in the Non-Domestic Rating (Telecommunications and Canals) (Scotland) Order 1995[32] and the Non-Domestic Rates (Telecommunication Installations) (Scotland) Regulations 2016.[33]

40.7.3 Valuation authorities, assessors and the valuation rolls

For the purposes of valuation, the whole of Scotland is divided into valuation areas. The valuation authority is the local authority for each local authority area in Scotland,[34] subject to the Secretary of State's (now the Scottish Government's) right to establish a joint board of two or more valuation authorities.[35] In terms of the Valuation Joint Boards (Scotland) Order 1995[36] ten joint boards were established. Four local authorities are not involved in joint boards. These authorities are: (i) Borders Council; (ii) Dumfries and Galloway Council; (iii) Fife Council; and (iv) Glasgow City Council.

40.7.4

In terms of s27(2) of the Local Government etc. (Scotland) Act 1994 Act, every valuation authority is required to appoint an assessor. There are currently 14 assessors for the 32 local government areas in Scotland. The assessor for each valuation area is responsible for making up the valuation roll (the Scottish equivalent to the valuation list) for that area.[37]

40.7.5 Treatment of telecommunications subjects

The Non-Domestic Rating (Telecommunications and Canals) (Scotland) Order 1995 provides that all telecommunications subjects within a local government area occupied by (or, if unoccupied, owned by) the same person shall be treated as justifying only one entry in the relevant valuation roll.[38] From a practical perspective, this means that each and every telecommunications subject must be valued as if it existed in isolation but, for administrative convenience, the individual values are aggregated within a cumulative valuation for each host and operator in each valuation area.[39]

40.7.6 Treatment of shared sites

There is no equivalent in Scotland to the provisions found within the Non-Domestic Rating (Telecommunications Apparatus) (England) Regulations 2000[40] concerning

31 SI 1996/3070.

32 SSI 1995/239.

33 SSI 2016/122 as amended by the Non-Domestic Rates (Telecommunication Installations) (Scotland) Amendment Regulations 2018, SSI 2018/63.

34 Local Government etc. (Scotland) Act 1994 s.27(1).

35 Local Government etc. (Scotland) Act 1994 s.27(7).

36 SI 1995/2589.

37 Local Government etc. (Scotland) Act 1994 s.28.

38 SI 1995/239, Art. 2.

39 Scottish Assessors Association Utililities Committee Practice Note No. 1 of 2017, Valuation of wireless telecommunications subjects.

40 SI 2000/2421 and the Non-Domestic Rating (Telecommunications Apparatus) (Wales) Regulations 2000 SI 2000/3383.

the treatment of telecommunications apparatus on shared sites. The position in England and Wales is discussed at paragraphs 37.7.7–37.7.11 above. In Scotland, the rateable value of an occupier's interest in telecommunications subjects always depends upon the nature of that occupation (i.e. whether occupation is as a host or as a sharer).

40.7.7 Relief from non-domestic rates for telecommunications subjects

There is no equivalent in Scotland to the Telecommunications Infrastructure (Relief from Non-Domestic Rates) Act 2018. The provisions of the 2018 Act are discussed at paragraph 37.8.2 of this book.

40.7.8

There is, however, in Scotland, the Non-Domestic Rates (Telecommunication Installations) (Scotland) Regulations 2016,[41] which initially provided 100 per cent non-domestic rates relief (from 1 April 2016 until 31 March 2021) for new mobile telecommunications masts located in three eligible pilot areas.[42] The purpose of the 2016 Regulations was to incentivise infrastructure investment in non-commercial areas (so called 'notspots') in order to extend the breadth of 2G/3G/4G coverage across Scotland. The 2016 Regulations were amended by the Non-Domestic Rates (Telecommunication Installations) (Scotland) Amendment Regulations 2018[43] which had the effect, first, of extending the relief period from 31 March 2021 until 31 March 2029[44] and, secondly, of extending the eligible pilot areas so as to include additional areas identified in a document entitled 'Non-Domestic Rates Relief – Mobile Masts Pilot Extension – Eligible Grid References' and dated 12 February 2018.

40.7.9

For the time being, therefore, whereas the new 2018 Act in England and Wales seeks to facilitate the roll-out of 5G by incentivising providers who deploy new fibre, the Scottish Regulations are focused on achieving 2G/3G/4G coverage in 'notspots' by incentivising the construction of new mobile masts.

40.8 Procedural differences: The Lands Tribunal for Scotland

40.8.1

As explained at section 33.5 above, the Electronic Communications Code (Jurisdiction) Regulations 2017 confers jurisdiction of all of the functions conferred by the New Code on the Court, in Scotland, on the Lands Tribunal for Scotland.[45]

41 SSI 2016/122.
42 Regulation 3.
43 SSI 2018/63.
44 Regulation 2(3).
45 The Lands Tribunal will become the Lands Chamber in the First Tier Tribunal for Scotland, under the Tribunals (Scotland) Act 2014 on a date yet to be specified.

40.8.2

Proceedings under the New Code must be raised in the Lands Tribunal, subject to the possibility, provided for a Regulation 5 of the Jurisdiction Regulations,[46] that proceedings may be transferred to the relevant Sheriff Court. It remains to be seen how the Lands Tribunal will exercise this power.

40.8.3

The Act of Sederunt (Summary Applications, Statutory Applications and Appeals etc. Rules Amendment) (Transfer from Lands Tribunal for Scotland) 2017, which came into force on 19 January 2018,[47] sets out the procedure to be followed by a Sheriff upon transfer to the Sheriff Court pursuant to Regulation 5 of the Jurisdiction Regulations.

40.8.4

Pursuant to the power provided at paragraph 106 of the New Code, more fully set out at paragraph 33.4.5 above, the Scottish Ministers made the Lands Tribunal for Scotland Amendment Rules 2017[48] which amended rule 28(1) (expenses) of the Lands Tribunal for Scotland Rules 2003 so as to read:

> For the purposes of determining applications under schedule 3A (electronic communications code) of the Communications Act 2003, expenses shall be determined in accordance with paragraph 96 of that Schedule.

40.8.5

Paragraph 96 of the New Code provides:

> (1) Where in any proceedings a tribunal exercises functions by virtue of regulations under paragraph 95(1), it may make such order as it thinks fit as to costs, or, in Scotland, expenses.
> (2) The matters a tribunal must have regard to in making such an order include in particular the extent to which any party is successful in the proceedings.

In other words, the Lands Tribunal will apply the general rule that expenses follow success but retains a broad discretion.

40.8.6

Absent agreement on expenses between the parties, expenses awarded by the Lands Tribunal are taxed (assessed), at the discretion of the Tribunal, either by the Auditor of the Court of Session according to the fees payable in the Court of Session or by the Auditor of the Sheriff Court specified by the tribunal according to the Sheriff Court Table of Fees.[49]

46 See paragraph 33.5.8 above.
47 SSI 2017/459.
48 SSI 2017/427.
49 Lands Tribunal for Scotland Rules 2003, Rule 28(3).

40.8.7

The Lands Tribunal for Scotland Amendment (Fees) Rules 2017[50] sets out the fees payable in respect of the various applications that can be made under the New Code.

40.8.8

It remains to be seen where there will be further changes to the Lands Tribunal Rules in respect of code applications.

40.8.9 Arbitration

As noted at paragraph 33.9.4 above, paragraph 107 of the New Code provides that:

> until the Arbitration (Scotland) Act 2010 is in force in relation to any arbitrations carried out under or by virtue of this code, that Act applies as if it were in force in relation to those arbitrations.

This reflects the fact that the provisions of the 2010 Act are not yet in force in relation to 'statutory arbitrations' as defined by section 16(1) of the 2010 Act.[51] As such, the effect of paragraph 107 of the New Code is to, nevertheless, apply the provisions of the 2010 Act to the categories of dispute which may be referred to arbitration under the New Code.[52]

40.8.10

The 2010 Act draws upon the Arbitration Act 1996 and introduces substantially similar (but by no means identical) rules for arbitrations in Scotland. It is beyond the scope of this book to explore the material differences between these two sets of rules.

50 SSI 2017/426.
51 Being: 'arbitration pursuant to an enactment which provides for a dispute to be submitted to arbitration'.
52 See section 33.9 above.

41 The role of OFCOM

41.1 Introduction

41.1.1

This chapter considers the role of OFCOM under the codes, with specific regard to the New Code. It examines that material in the following order:

(a) OFCOM's functions;
(b) OFCOM's role under the Old Code;
(c) OFCOM's role under the New Code;
(d) OFCOM documents under the New Code;
(e) The 2017 OFCOM consultation;
(f) The OFCOM notices – background;
(g) The OFCOM notices – critique;
(h) The OFCOM Code of Practice – background;
(i) The OFCOM Code of Practice – critique;
(j) The OFCOM standard terms – background;
(k) The OFCOM standard terms – background;
(l) OFCOM as a forum for disputes.

41.2 OFCOM's general functions

41.2.1

The Office of Communications ('OFCOM') was established by the Office of Communications Act 2002.[1] It is the statutory regulator for the electronic communications sector, and also the competition authority, using powers flowing from provisions under the Communications Act 2003 and the Wireless Telegraphy Act 2006 which transpose the European Framework Directive and the Authorisation Directive into national law.[2] Its functions are set out in the Communications Act 2003, which has been extensively amended by Part 6 of the 2017 Act, so as to extend the range of reports on infrastructure and other matters which OFCOM may require.

1 For a full treatment of OFCOM, see *Halsbury's Laws of England,* vol. 97.
2 See Chapter 39 of this book.

41.2.2

As part of its responsibility for applying the Old Code, and now the New Code, to Code Operators, OFCOM must secure 'the optimal use for wireless telegraphy of the electro-magnetic spectrum' and 'the availability throughout the United Kingdom of a wide range of electronic communications services' (s.3(2) of the Communications Act 2003).

41.2.3

Quite apart from its code powers and duties, OFCOM has many other obligations and powers with regard to telecommunications under the 2003 and 2017 Acts, the nature and extent of which are outside the scope of this book.

41.2.4

OFCOM's functions are to be generously construed: s.405(3) of the 2003 Act provides:

> References in this Act to OFCOM's functions under an enactment include references to their power to do anything which appears to them to be incidental or conducive to the carrying out of their functions under that enactment.

41.3 OFCOM's role under the Old Code

41.3.1

The Old Code provides for network operators to obtain rights to install and maintain their apparatus on public and private land. Only those operators that apply for and receive approval from OFCOM under s.106(3)(a) of the Communications Act 2003 are able to benefit from, and be subject to, the Old Code. In practice, all network operators have such approval, without which they would not be able to operate.

41.3.2

The procedure and criteria for a direction by OFCOM applying the code in any person's case are set out in s.107 of the 2003 Act, and are described in detail in section 17.2 of Chapter 17 of this book.

41.4 OFCOM's role under the New Code

41.4.1

OFCOM's role under the New Code remains governed by the 2003 Act, and is little different to its role under the Old Code, although the 2017 Act has introduced some amendments to OFCOM's procedures, the nature of which are explained in detail in this chapter.

41.4.2

For present purposes, it suffices to explain that OFCOM has the following core functions in relation to the New Code:

(a) It makes directions under which entities can secure operator status under s.106 of the Communications Act 2003 (see Chapter 17).
(b) It maintains a list of entities with operator status (see too Chapter 17).
(c) It regulates the activities of operators under the 2003 Act.
(d) It collates information and reports on infrastructure under ss.134A–D of the 2003 Act (as amended by the 2017 Act).
(e) It prescribes standard forms and notices, and publishes a Code of Practice for performance under the New Code (see sections 41.9 and 41.10 below).
(f) It has important dispute resolution functions relating to disputes relating to electronic communications networks under ss.185 to 191 (see section 41.13 below).

41.4.3

OFCOM also has other powers with regard to digital infrastructure which may impact upon operators and owners/occupiers of land. For example, under regulation 16 of The Network and Information Systems Regulations 2018 (SI 2018/506) (in force from 10 May 2018), as designated competent authority, OFCOM has the right to inspect premises in order to assess whether an operator of essential services has fulfilled the duties imposed on it by regulations 10 and 11.[3]

41.5 OFCOM documents under the New Code

41.5.1

The New Code specifies three categories of documents with which OFCOM is to be concerned under the New Code.

41.5.2

First, paragraph 90 of the New Code provides that OFCOM must 'prescribe the form of a notice to be given under each provision of this code that requires a notice to be given'. In relation to operator notices to be used under the New Code provision is made in paragraph 88. In relation to notices to be given by any other person, provision is made in paragraph 89.

3 These duties include to take appropriate and proportionate technical and organisational measures to manage risks posed to the security of the network and information systems on which their essential service relies.

41.5.3

Secondly, paragraph 103(1) of the New Code requires OFCOM to prepare and publish a code of practice dealing with:

(a) the provision of information for the purposes of the New Code by operators to persons who occupy or have an interest in land;
(b) the conduct of negotiations for the purposes of the New Code between operators and such persons;
(c) the conduct of operators in relation to persons who occupy or have an interest in land adjoining land on, under or over which electronic communications apparatus is installed; and
(d) such other matters relating to the operation of the New Code as OFCOM think appropriate.

41.5.4

Thirdly, paragraph 103(2) of the New Code requires OFCOM to prepare and publish standard terms which may (but need not) be used in agreements under the New Code.

41.6 The 2017 OFCOM consultation

41.6.1

Working with what it referred to as 'a balanced cross-section of representatives nominated by different stakeholder groups, in order to take full account of a broad range of interests and concerns', OFCOM prepared drafts of each of the documents referred to above.

41.6.2

On 24 March 2017, OFCOM published a consultation[4] seeking views on those drafts, in accordance with paragraphs 90(3) and 103(4) of the New Code, and setting 2 June 2017 as the closing date for responses.

41.6.3

OFCOM received 34 responses to its consultation, from a broad range of stakeholders expressing a variety of views, primarily on the draft Code of Practice. Four of these responses were confidential; the remaining 30 responses are published on OFCOM's website.

4 Electronic Communications Code – Digital Economy Bill: Proposed Code of Practice, Standard Terms of Agreement and Standard Notices.

41.6.4

After the consultation closed, OFCOM reviewed all the submitted responses, and published finalised versions of the Code of Practice and accompanying standard terms and notices in a Final Statement on 15 December 2017. In that Statement,[5] OFCOM note that 'Having taken account of stakeholder responses', they have made 'a small number of changes to these documents where we considered appropriate in line with stakeholder comments'. The changes are indeed minor (consisting for the most part of a softening of the language to replace mandatory with advisory terms, and imposing limitations as to reasonableness), and are considered in each category of document examined below. OFCOM add in the Statement that:

> We consider it may be prudent, depending on parties' experience of using the documents we are publishing, for Ofcom to review the Code of Practice, the standard terms and the notices, if necessary after an appropriate period, to consider their effectiveness. We would envisage working with relevant parties in carrying out this exercise if we proceeded to do so.

41.7 The OFCOM notices: background

41.7.1

As explained in section 41.4 above, paragraph 90 of the New Code provides that, where a provision in the New Code 'requires a notice to be given', OFCOM *must* prescribe the form of that notice.

41.7.2

It is however difficult always to ascertain when a notice is 'required' to be given, because although the New Code uses terms such as 'may' or 'must' in relation to notices, it is not always clear whether the situation itself is one which engages the mandatory provisions. This topic is discussed in detail in Chapter 32.

41.7.3

The following are the paragraphs of the New Code which deal with notice provisions, and which may or must entail prescribed form notices: 20(2) ('the operator may'); 31(1) ('the site provider may'); 32(1) ('the operator may'); 33(1) ('the site provider may'); 39(1) ('a landowner may'), (2) ('a landowner or occupier may') and (4) ('the operator may'); 40(2) ('the landowner or occupier may'); 41(2) ('the third party may') and (5) ('the operator may'); 43(5) ('a person may'); 49(1) ('the operator must'); 50(2) ('the transport undertaker may'); 50(3) ('[either party] may'); 51(2) ('the operator must'); 51(4) ('the transport undertaker may'); 51(7) ('[either party] may'); 53(1) ('a transport undertaker

5 See www.ofcom.org.uk/__data/assets/pdf_file/0027/108792/ECC-Statement.pdf.

may'); 53(2) ('the operator may'); 54(7) ('the occupier may'); 67(1) ('an undertaker must'); 68(2) ('the operator may'); 71(1) ('the undertaker must'); 78(1) ('the objector may'); 82(3) ('the operator may'); 82(4) ('the occupier may').

41.7.4

Where the form of a notice is prescribed by OFCOM, paragraphs 88(2) and 89(2) of the New Code require that, to be valid, notices given by code operators and certain notices given by other parties *must* be in the prescribed form. However, paragraphs 89(5) and (6) envisage that certain other notices *may* be given in a form other than that prescribed by OFCOM (subject to the party giving the notice bearing the other party's resulting costs, if any, attributable to the notice not being in the correct form).

41.7.5

OFCOM interpreted its obligation under paragraph 90 of the New Code expansively, and has sought to provide drafts of standard notices even where they are not strictly required by the New Code, in the hope that code operators and landowners/occupiers alike would consider this useful, and that this would facilitate a smooth transition to the New Code regime. Although this prescription was intended to be beneficial, it has had the consequence that notices which were not *required* to be given, *must* now be in the prescribed form if they are given.

41.7.6

In a very limited number of cases, OFCOM took the view that there was likely to be limited (if any) value in it prescribing the form of a discretionary notice. In particular, it considered that there would be little value in prescribing the form of notices under paragraphs 32(1)[6] and 39(4)[7] of the New Code, as the contents of any such notices would be highly fact-specific and code operators were expected to be able to prepare these easily. OFCOM did not therefore prescribe any standard form notices under these specific paragraphs of the New Code.

41.7.7

In preparing the standard notices, OFCOM was mindful of the need to ensure that they are as clear and concise as possible, as well as the need for code operators to comply with paragraph 89(1) of the New Code when giving notice (i.e. that they explain the effect of the notice and the steps that may be taken by recipients). OFCOM was also mindful of the fact that, whilst code operators may be familiar with the New Code, this may not necessarily be the case for landowners/occupiers. It therefore sought to provide helpful 'Notes' or guidance, particularly for landowners/occupiers, at the end of a number of the notices which should assist them when sending or receiving notices under the New Code.

6 i.e. a counter-notice from a code operator regarding the termination of a code agreement.
7 i.e. a notice from a code operator disclosing whether apparatus is on land pursuant to a code right.

41.7.8

The standard form OFCOM notices are included in Appendix E to this book.

41.7.9

The delivery of such notices is discussed in Chapter 32.

41.8 The OFCOM notices: critique

41.8.1

The consultation exercise carried out by OFCOM in 2017 drew a large number of comments from stakeholders, ranging from those of general application to detailed comments on individual notice templates. For the most part, OFCOM refrained from incorporating the alternative or additional drafting proposed by stakeholders, unless it agreed that further clarity was required. Instead, OFCOM indicated that it considered that:

> it would be more appropriate to allow stakeholders to use the notices and then, depending on their experience of using them, for OFCOM to review them, if necessary after an appropriate period, to consider their effectiveness.

41.8.2

The remainder of this section sets out perhaps the two most substantive areas of criticism of the template notices, together with OFCOM's response. The many other more detailed criticisms that were made were typically met with the response that, having had regard to the requirements of the provisions of the relevant paragraph of the New Code, OFCOM considered the drafting of the template notice met those requirements and was sufficiently clear; and consequently had decided that it would not be necessary to make any changes.

41.8.3 Name and address blocks

A number of the respondents to the consultation exercise stated that it would be helpful if each of the prescribed notices had a clear section at the top which sets out who the notice is from, and to whom it is being sent. This would avoid the current inconsistency whereby some of the notices require this information to be provided in the main body of the notice, while some do not require it at all, and do not set out the sender's details, which is unhelpful where notices require a counter-notice or response to be served. To cure these deficiencies, the template might begin:

> To: Operator/Landowner (name and Co. no.) as appropriate
> Of: [address] Please quote formal Address for Service
> From: Operator/Landowner (name and Co. no.) as appropriate
> Of: [address] Please quote formal Address for Service

This is a common expedient in leasehold cases, but not one which the template notices use. OFCOM rejected this suggestion, considering that the template notices are sufficiently clear in identifying the purpose for which they are being given, as well as the identity of the sender and of the recipient.

41.8.4 Notice seeking agreement to the conferral of code rights

The drafting of this template notice generated a great deal of controversy among respondents, many of whom considered it lengthy and confusing, and argued for separate notices distinguishing between situations where (a) code rights are to be conferred on a new site; and (b) code rights are to be conferred in respect of an existing site where the led contractual term has ended. These comments persuaded OFCOM to make a substantial amount of revision, albeit retaining the dual notice.

41.8.5

Unhappily, OFCOM did not accede to the further request from respondents to make clear that the landowner cannot be compelled by a tribunal to accept additional rights that are not code rights. It also rejected the proposition that the 28-day deadline in clause 9 concerning the paragraph 20(2)/27(1) notice was unreasonably short.

41.9 The OFCOM Code of Practice: background

41.9.1

The 2013 Law Commission report (Law Com 336) proposed a possible code of practice, to be consulted on and agreed by OFCOM with code operators, covering issues such as the provision of information to landowners, conduct in negotiations with landowners, the content of agreements granting code rights and relationships with those whose property adjoins land where apparatus is sited (including highways) – see paragraphs 9.133 et seq.

41.9.2

In its document announcing the New Code,[8] the Government explained that the proposed OFCOM Code of Practice would enable smooth implementation of its long-term reforms, and that it would like to see all stakeholder groups in the industry work together with OFCOM in its development.

41.9.3

With that in mind, OFCOM developed an initial draft of the proposed Code of Practice, working with what it described in paragraph 2.18 of its 2017 consultation paper as 'a

8 'A New Electronic Communications Code'; May 2016 – www.gov.uk/government/uploads/system/uploads/attachment_data/file/523788/Electronic_Communications_Code_160516_CLEAN_NO_WATERMARK.pdf.

wide spectrum of stakeholders including representatives from the fixed and mobile operator community, communications infrastructure providers and representatives from the National Farmers Union (NFU), the Country Land & Business Association (CLA), the British Property Federation (BPF) and the Central Association of Agricultural Valuers (CAAV)'.

41.9.4

On 28 July 2016 OFCOM held an initial scoping meeting with stakeholders setting out its approach to the Code of Practice drafting process, and invited different stakeholder communities to nominate representatives to serve on a Code of Practice Drafting Group.

41.9.5

In September 2016, OFCOM confirmed the membership of the Code of Practice Drafting Group, composed of eight specialist practitioners representing landowners, communications network operators and infrastructure providers.

41.9.6

Between September and December 2016, the Drafting Group prepared successive versions of the draft Code of Practice document, which was reviewed at monthly meetings hosted by OFCOM, designed to capture additional input from a wider group of cross-sector stakeholders.

41.9.7

The Drafting Group submitted a finalised version of the draft Code of Practice to OFCOM on 16 December 2016. Having reviewed their output, OFCOM believed that, with some minor drafting amendments, it met the requirements specified in paragraph 103(1) of the New Code.

41.9.8

OFCOM then put out the amended draft for consultation as described in section 41.6 above.

41.9.9

OFCOM stated in paragraph 2.15 of its March 2017 consultation paper:

> It is important to note that the purpose of this proposed Code of Practice is to set out expectations for the conduct of parties to agreements made under the New Code. It does not represent a guide to the New Code nor does it replace the provisions of the New Code. Instead it is designed to complement the New Code by

> facilitating positive and productive engagement between all parties across a range of issues, roles and responsibilities.

It should be emphasised that the Code of Practice merely regulates behaviour; it does not explain or otherwise supplement the New Code.

41.9.10

The OFCOM Code of Practice is non-binding (in the sense that there is no statutory obligation on operators or landowners to comply with its provisions). This aspect of the Code of Practice caused disquiet among landowner representatives during the legislative passage of the Digital Economy Bill. The only response to these concerns was this suggestion by Lord Ashton during a debate on the Bill:

> I understand the desire to ensure that Ofcom's code of practice effects real change in behaviour within industry. It will have weight. Indeed, failure to abide by it could be taken into account by a court or tribunal in the event of a dispute.[9]

It is right to say that the courts may take account of compliance with relevant codes of practice when assessing conduct in awarding costs (although there is nothing in the Civil Procedure Rules that currently mandates this, in contrast with the position under the Pre-action Protocols). However, there will be many occasions where litigation is unlikely to ensue, but where one party or the other (but typically a landowner) will wish that the Code of Practice had greater force. Accordingly, although OFCOM expects parties to seek to act in accordance with the principles set out in the Code of Practice, and although the courts may also expect compliance, the comment by OFCOM in paragraph 2.16 of its 2017 consultation that these factors alone 'should ensure that the Code of Practice has real impact on ensuring best practice in the deployment of digital communications infrastructure' may be doubted.

41.9.11

Paragraph 103(3) of the New Code enables OFCOM to amend or replace elements of the Code of Practice and accompanying standard terms at the time of its choosing, and to publish such amendments. OFCOM must, under paragraph 103(4) and (5), consult operators and other appropriate stakeholders before publishing a code of practice or standard terms, though this does not apply to publication or amendments to either type of document, or a replacement. OFCOM has stated that it is 'committed to monitoring and reviewing the Code of Practice to ensure that its content remains fit for purpose, appropriate and proportionate in light of on-going developments'. The authors of this work will seek to ensure that any revisions are communicated as soon as they are disseminated.

41.9.12

The OFCOM Code of Practice is included in Appendix D to this book.

9 Lords, Hansard, 31 January 2017, col. 1183.

41.10 The OFCOM code of practice: critique

41.10.1

This section does not set out to explain every aspect of the OFCOM Code of Practice, which for the most part speaks for itself. It simply draws attention to aspects of the Code of Practice which may require discussion, before drawing attention to a number of respects in which the respondents to the consultation exercise felt that the code fell short, but whose comments have not been reflected in the final draft.

41.10.2 Terms

The OFCOM Code of Practice uses the expression 'Site' to refer to the New Code term 'land' – that is to say, the place where electronic communications apparatus is installed, be that on open land, the rooftop of a building, a tunnel or a lamp-post. Chapter 18 of this book explains the distinction between land and apparatus in more detail.

41.10.3 Purpose

The OFCOM Code of Practice is more of a behavioural code than anything else, enjoining parties to 'treat each other professionally and with respect, remembering always that the goal is to improve and maintain essential communications services for all'.[10]

41.10.4 Scope

The OFCOM Code of Practice 'covers a wide range of scenarios, from the construction of a full mobile mast to the installation of just one telegraph pole or a very small length of cable'.[11] It is therefore to be noted that not all its procedural elements will be required in each and every case.

41.10.5 Effect

As section 41.9 above has made clear, the OFCOM Code of Practice is not binding, and its role in practice may therefore be comparatively limited, unless it is regarded by the telecommunications industry as a gold standard to which all should aspire. The Code of Practice set out a useful sequence of steps to be followed in relation to the installation, access to and decommissioning of apparatus, which should be useful as an indication of good practice that should be followed unless there is good reason to depart from that practice.

41.10.6 Installation of apparatus

This part of the OFCOM Code of Practice sets out a sequence of 'stages', depending on the nature of the apparatus to be installed, which OFCOM proposes that operators

10 See paragraph A1.5 of the OFCOM Code of Practice.
11 *Ibid.*, paragraph A1.8.

seeking to install new apparatus '*will* follow'. This language is instructive: the remainder of the code generally uses the expressions 'should' and 'ought' rather than 'will' or 'must' (largely as a result of amendments to the drafting made during the final stage), and it is therefore to be expected that operators will abide by what follows.

41.10.7

The Code of Practice makes the point that, save where the apparatus is minor, a site visit may be required. It then goes into some detail concerning the nature of the request for access, and the information that will or may be required. It then turns to consider the nature and extent of the consultation and agreement that may be required in order to engage the landowner and operator in the most productive way. Following agreement, it goes on to suggest the most effective way in which the deployment of the apparatus may be carried out, involving dealings with the landowner and any affected neighbours. Reference is usefully made to existing codes of practice concerning such matters.[12] Annexes A and B to the Code of Practice summarise the key points to be borne in mind in requesting and arranging access, in a way that should provide not merely a good aide memoire for the operator, but also a clear guide for the landowner to the key points to bear in mind.

41.10.8 Access to existing apparatus

This part of the OFCOM Code of Practice sets out a further sequence of steps to be taken where apparatus has already been installed, and access is needed for various purposes, including maintenance and possible upgrading. The point is made that access may be needed in case of emergency, and the Code of Practice gives useful advice concerning such matters as identification of contractors and the provision of emergency contact details.

41.10.9

The OFCOM Code of Practice says little concerning upgrading existing apparatus, notwithstanding the representations to the consultation exercise by a number of landowner groups, proposing operators should notify landowners at least 28 days in advance of carrying out any proposed upgrade works that would change the external appearance of a site, and providing sufficient detail of the proposed upgrade (e.g. specifications, drawings and plans) to enable landowners to make their own assessment as to whether the upgrade would be likely to have these effects. The same respondents also proposed that where a landowner feels that a proposed upgrade will have more than a minimal impact on appearance or impose an additional burden, he may advise the operator, who should consider the landowner's concerns and address them before the upgrade is carried out. This proposed revision was also rejected by OFCOM.

12 Cabinet and pole siting Code of Practice, and The Code of Best Practice on Mobile Network Development in England, www.mobileuk.org/codes-of-practice.html.

41.10.10 Decommissioning sites

This part of the OFCOM Code of Practice makes a couple of brief practical points concerning redundant apparatus, which will serve as a benchmark for appropriate conduct in such cases.

41.10.11 Renewing agreements

The Code of Practice also briefly covers renewal, while making the point that time for agreeing new terms will usually be short.

41.10.12 Repairs by landowners

The Code of Practice notes that, where landowners need to carry out essential repairs that will need apparatus to be moved temporarily, a balance will have to be struck that accommodates the need for any interruption in service to be kept to a minimum. This paragraph of the Code of Practice (paragraph 1.47) was revised in the operators' favour, perhaps more than any as a result of the responses to the consultation exercise.

41.10.13 Redevelopment

Again, the Code of Practice emphasises the balance to be struck where a site is needed for redevelopment, stressing that landowners should give as much notice as possible; while equally operators should be proactive in seeking alternative sites, if not on the redeveloped land.

41.10.14 Disputes

Finally, the Code of Practice offers advice where disputes materialise, while stressing the interest each party should have in resolving disputes informally.

41.10.15 Shortcomings of the OFCOM Code of Practice?

Many of the respondents to the consultation exercise will feel disappointed at the extent to which their comments have not been reflected in the final form of the OFCOM Code of Practice. Particular matters which were voiced by respondents but which did not find their way into the code are as follows:

(a) The Code of Practice does not set out sanctions for failure to comply with its terms.
(b) The code does not provide any guidance on how the court will take it into account when assessing conduct and costs. Again, OFCOM's response was to stress party autonomy. It might be added that the Court is perfectly capable of taking conduct into account without such guidance.
(c) The point has already been made in paragraph 41.10.9 above that the code adopts a very light touch in relation to upgrading, suggesting little that the operator might usefully do in order to inform the landowner of its plans. This

provision was responsible for the greatest part of the controversy generated by the drafting of the OFCOM Code of Practice.

(d) The code does not require an operator to notify the landowner/occupier in the event that it assigns the agreement.

41.10.16

OFCOM's response to points like these was that the Code of Practice is based on the core principle of agreement directly between parties, and that it is not intended to impose any rights or obligations on either party beyond those detailed in the 2017 Act. It adds:

> It is also important to note that the Code of Practice does not represent a guide to the new Code nor does it replace or supplement its provisions by imposing any new rights or obligations on the respective parties. The Code of Practice is not binding and cannot change the balance that the law delivers under the Code. Instead it is designed to complement the new Code by suggesting best practice to facilitate positive and productive engagement between all parties across a range of issues, roles and responsibilities. Whilst the Code of Practice provides some examples of best practice these are not intended to be exhaustive.

41.11 The OFCOM standard terms: background

41.11.1

As noted above, paragraph 103(2) of the New Code requires OFCOM to prepare and publish standard terms which may (but need not) be used in agreements under the New Code. In parallel to the mainstream stakeholder engagement process concerning the Code of Practice referred to in section 41.9 above, OFCOM prepared a number of supporting templates for standard terms, based on material submitted by the wider stakeholder group. These included some template agreements used by members of the Code of Practice drafting group, based upon their standard practice under the Old Code.

41.11.2

In preparing its draft standard terms, OFCOM bore in mind the views and recommendations expressed by the Law Commission.[13] In particular, the Commission had explained that standard terms would be useful on the basis that they could give a starting point for negotiations, but could be amended as necessary to meet particular circumstances. It considered that, at a most basic level, standard terms could assist parties, particularly landowners, to ensure that important terms are not forgotten. In effect, the standard terms might best operate as a useful aide memoire.

41.11.3

OFCOM recognised that some parties might consider that it would be useful if it were to prepare more than one set of standard terms. However, OFCOM took the view that

13 The Law Commission, *The Electronic Communications Code*, 27 February 2013, pp.203–5.

this was not necessary, and that the value (if any) of it preparing more than one set of terms would be limited. Code agreements would, in practice, cover an extremely wide range of circumstances; the technology to be installed, the physical characteristics of the site and the preferred approach and sophistication of the parties to the agreement will often differ significantly. For OFCOM to prepare a variety of standard terms which would suit each type of technology, site, operator and landowner/occupier would be a significant task, and it was not clear to it that this would be of benefit to code operators and landowners/occupiers.

41.11.4

The purpose of the standard terms is therefore to provide parties with a starting point for their negotiations, rather than to provide a final set of terms for all parties. OFCOM anticipated that many experienced site providers and code operators may prefer to use their own terms and that, for more complex transactions, parties are likely to seek independent legal advice in order to ensure that their code agreement is properly tailored to their specific circumstances. OFCOM was also aware that there are other sources of standard terms and conditions that have been developed through consultation between interested parties. It cites as a good example of this the multi-occupant office building wayleave agreement developed for the City of London.[14]

41.11.5

Paragraph 102(3) of the New Code enables OFCOM to amend or replace elements of the Code of Practice and accompanying standard terms at the time of its choosing. OFCOM states that it is committed to monitoring and reviewing the Code of Practice to ensure that its content remains fit for purpose, appropriate and proportionate in light of ongoing developments, and the same presumably goes for its standard terms. The Authors of this work will seek to ensure that any revisions are communicated as soon as they are disseminated.

41.11.6

The OFCOM standard terms are included in Appendix F to this book. They take the form of a generic agreement containing a number of standard term templates, in which the landowner/occupier is referred to as 'the Grantor' and the operator as 'the Operator'.

41.12 The OFCOM standard terms: critique

41.12.1

As in the case of the notices and Code of Practice, the consultation exercise carried out by OFCOM in 2017 drew a large number of comments from stakeholders, ranging from those of general application to detailed comments on individual standard term clauses

14 City of London, Digital Infrastructure Toolkit, standardised wayleave agreement; this is reviewed and updated on a regular basis.

templates. The respondents tended to fall into two camps: (a) those who considered that detailed terms were inappropriate, given the wide diversity of situations encountered, and who proposed a general 'heads of terms' or principles approach, signposting other standard terms created by other stakeholders; and (b) those who were concerned that OFCOM had not chosen to draft different types of standard terms, and who considered that adopting a 'one-size-fits-all' approach for the standard terms would not be appropriate. For the most part, OFCOM refrained from incorporating the alternative or additional drafting proposed by stakeholders, unless it agreed that further clarity was required. Instead, OFCOM decided that:[15]

- creating standard 'heads of terms', as opposed to actual terms themselves, would not be sufficient for OFCOM to meet its requirement under paragraph 103(2) of the new Code;
- publishing more than one set of standard terms is not necessary in the sense that it is not required by the new Code;
- to prepare a variety of standard terms which suit each type of technology site, operator and landowner/occupier would be a significant task and stakeholders' responses were mixed in their respective views as to the benefit of carrying out such a task; and
- it would not be appropriate to seek some sort of compromise by preparing a limited number of additional standard terms since this would necessitate a judgment call on which standard terms to include, which would be a call we would not be in an informed position to take without further, and not insignificant, consultation with stakeholders.

41.12.2

The details of the standard terms, together with a brief commentary upon each drawn from a summary of the responses to the OFCOM consultation exercise now follows. It is to be emphasised that OFCOM's own summary of the responses in its Statement does not reflect the full range of cogent comments made by the 34 respondents to the consultation.[16]

41.12.3 Template recitals

There are four recitals to the template agreement, (a) referring to the New Code; (b) noting that the New Code has been applied to the operator intending to enter into the agreement; (c) the Grantor is 'the occupier of certain land'; and (d)

> This Agreement is an agreement pursuant to paragraph 9 of Part 2 of the Code. It sets out the contractual basis upon which the Grantor is willing to confer code rights in respect of that land on the Operator.

15 See paragraph 4.6 of its 15 December 2017 statement at www.ofcom.org.uk/__data/assets/pdf_file/0027/108792/ECC-Statement.pdf.

16 The full range of comments (save for those who wished to remain confidential) may be found at www.ofcom.org.uk/consultations-and-statements/category-1/electronic-communications-code.

41.12.4

One of the respondents to the consultation exercise considered that:

> It should be added that the primary purpose of the Agreement is to confer Code Rights (so that in the event of an Agreement having the potential to be caught by the provisions of the Landlord and Tenant Act 1954, it is clear that the Code applies and the 1954 Act does not).

41.12.5

OFCOM did not refer to this comment in its Statement. The context for the comment is discussed in section 25.2 of Chapter 25 of this book.

41.12.6 Template clause 1: definitions and interpretation

This first part of this clause briefly describes a handful of critical expressions in the following way:

(a) '*Act*': the 2003 Act;
(b) '*alter*', '*Code*' and '*Electronic Communications Apparatus*': by reference to their definitions in the New Code;
(c) '*Apparatus*': the equipment described in Schedule 1 to the draft agreement;
(d) '*Code Rights*': the rights set out in clause 2.1 of the draft agreement;
(e) '*Land*': the physical land so described in in Schedule 2 to the draft agreement;
(f) '*Term*': the period of time during which the draft agreement is in force.

41.12.7

The second part of the clause contains a very few boiler plate terms.

41.12.8

The respondents to the consultation exercise made two broad points about the definitions in the first part of the clause (amongst a range of more minor points):

(a) There was no clarity as to what was the landowner/occupier's retained land to which the code rights might apply (compare the 'reversion' in leasehold parlance), and in particular whether the land was simply the area on which the apparatus was to be sited, or a larger area.
(b) The definition of 'term' was such that the agreement would run indefinitely until terminated, whereas many landowners, it was thought, would usually want to grant the rights for a specific fixed term, with successive renewals thereafter.

41.12.9

OFCOM's response to these points was as follows:

(a) As to the definition of 'Land', it was not considered necessary to make changes, since the definition merely refers to what would be attached at Schedule 2, which would be for the parties to provide.
(b) OFCOM amended the definition of 'Term' by adding the following at the end (which may, but need not, be used by the parties): 'Term' means the period of time during which this Agreement is in force[, which shall be a period starting on [insert date] and ending on [insert date]]'.

41.12.10 Template clause 2: rights of the operator

Sub-clause 2.1 simply sets out the full list of code rights, more or less repeating the language of paragraph 3 of the New Code. Sub-clause 2.2 provides that the operator may also share the use of the apparatus with another operator, and carry out any works to the apparatus to enable that sharing to take place. This very wide provision in the operator's favour is however subject to the restrictions in sub-clause 2.3, which repeats the New Code paragraph 17 safeguards (see section 20.3 of Chapter 20 of this book). Finally, sub-clause 2.4 provides that the operator's rights of entry may be exercised by with or without workmen, vehicles (where appropriate), plant equipment or machinery.

41.12.11

The respondents to the consultation exercise made a number of points about this drafting, principal amongst which were the following:

(a) Some considered that it should be made clear that it would not be necessary in all cases for the whole panoply of rights to be conferred.
(b) Others commented that there were no restrictions on the time or manner within which the operator might access the land (noting that on shared or sensitive sites there may be security protocols in place).
(c) One respondent made the points that clause 2.1(h) does not (i) expressly place any restrictions on the extent to which the operator can interfere with or obstruct a means of access to or from the land (for example, upon first giving reasonable written notice to the landowner save in the case of an emergency); or (ii) reflect the fact that such access may not be in the landowner's control in the first place. Although clause 4.1 does deal with this point in more detail, the respondent considered that it would be helpful to cross refer to clause 4.1 in clause 2.1(h).
(d) Clause 2.3 was a particular cause of concern, with one respondent making the following criticism:

> it is questionable whether a requirement that the Operator notify the Landowner after the event of the changes to the Apparatus would fall foul of this restriction in the Code and some form of notification would be helpful so that a Landowner can keep track of the changes and the extent of the Apparatus on a site – for example when considering the maximum load restrictions on a lattice tower. An accumulation of Apparatus may over time create a 'burden' which previously did not exist so a Landowner should be entitled to be able to keep track of incremental changes.

This important topic is discussed in more detail in section 20.3 of Chapter 20 of this book.

(e) Others were concerned to stress that the code rights alone would be insufficient in some cases, and that additional or ancillary rights (such as the right to install a power supply) should be included expressly.
(f) With further reference to the power supply, some pointed out that the clause does not clarify whether the operator's right is limited to the operator installing and providing its own power supply at its own cost, or to an existing power supply to be provided by the grantor, with costs to be recharged to the operator.
(g) Others wished the obligation to remove decommissioned apparatus (included in clause 10.3, considered in paragraph 41.12.39 below) to be accompanied by a right to remove.

41.12.12

OFCOM was largely unmoved by these points, noting that any such rights or further detail would be a matter for negotiation between the respective parties, reflecting the circumstances pertaining to the relevant agreement.

41.12.13 Template clause 3: payment

The template payment clause takes the form of a simple requirement to pay a flat amount, expressed as a fixed yearly sum.

41.12.14

The respondents to the consultation exercise made a number of points about this drafting, including the following:

(a) It contains no provision for review of the amount payable.
(b) There is no mechanism indicating when, how often, and by what means payment should be made.
(c) There is no mechanism providing for interest to be paid in the event of late payment.
(d) VAT should be mentioned.

41.12.15

Despite these comments, OFCOM left its drafting unchanged, stating that:

> the precise drafting of the term, including provision for review and amount, if indeed any, of payment and compensation should be regarded as a matter for negotiation between parties.

41.12.16 Template clause 4: operator's obligations

This template clause deals expressly with a number of tenant-like obligations, including requirements: (a) to give seven days' notice of its intention to enter the land (save in case of emergency); (b) to provide information where emergency access is required or has been taken; (c) otherwise to exercise access in accordance with Schedule 3 to the draft agreement; (d) exercise its code rights in a proper and workmanlike manner; (e) to do as little damage as possible in the exercise of those rights, and to make good any such damage; (f) comply with legislation; (g) obtain any necessary consents; (h) maintain the apparatus; (i) pay rates; (j) maintain insurance. There is no express limitation upon sub-letting.

41.12.17

The respondents to the consultation exercise made a number of points about this drafting, principal amongst which were that:

(a) The access restrictions were onerous, and exceeded those laid down in the New Code itself.
(b) Many obligations commonly accepted by the industry (e.g. to pay costs if in default; keep the site in good repair or clean and tidy; prevent encroachment or acquisition of rights; deal with interference with pre-existing equipment; grant reservations or access to the site provider; make contributions to shared facilities) had been omitted.
(c) Agreements for rooftop sites commonly required denser drafting to deal with such matters as out-of-hours access and security arrangements – and all these were lacking.

41.12.18

Again, OFCOM was unmoved by these points, saying in paragraph 4.54 of the Statement:

> the purpose of the new Code is not to set out all the specific obligations that would be contained in agreements between code operators and landowners/occupiers. And the Code of Practice does not seek to cover all the standard terms, but instead sets out best practice principles in accordance with which Ofcom would expect parties to act. Consequently, having considered stakeholders' comments, we have decided to keep clause 4 unaltered …

41.12.19 Template clause 5: grantor's obligations

This clause imposes three specific obligations upon the landowner/occupier:

(a) Not to build or place anything on the Land, or permit any third party to do the same, that makes it more difficult for the Operator to access the Apparatus, or which might interfere with the Apparatus, without the Operator's express written consent (which should not be unreasonably withheld);
(b) Not to cause damage to or interfere with the Apparatus or its operation and not permit any third party to do the same; and
(c) To give reasonable prior written notice to the Operator of any action that it intends to take that would or might affect the continuous operation of the Apparatus, including (but not limited to) causing an interruption to any power supply to which the Apparatus is connected.

41.12.20

The substantive criticisms of this clause were:

(a) That it went beyond the New Code, and gave the grantor the ability to disrupt the operator's operation, which was not acceptable.
(b) That it lacked reference to confirmation that the grantor has either obtained the consent of a superior landlord, mortgagee or other required person to the agreement, or alternatively does not require such consent.

41.12.21

Yet again, OFCOM was unmoved by the first point (and neglected to refer to the second), saying in paragraph 4.57 of the Statement:

> We consider stakeholders may have misunderstood the purpose of clause 5.1(c) which would be to merely ensure that the code operator is always aware of any action the consequences of which, as the clause says, 'would or might affect the continuous operation of the Apparatus'. Importantly, the clause is not designed to give landowners/ occupiers the right to take action in order to affect the continuous operation of the apparatus.

41.12.22 Template clause 6: ownership of the apparatus

Clause 6 simply states: 'The Apparatus shall remain the absolute property of the Operator at all times'.

41.12.23

An objection was made to this drafting by a respondent to the consultation exercise on the ground that the apparatus might in fact belong to a sharer of the operator, particularly where the grantee was an infrastructure provider.

41.12.24

OFCOM left this drafting as it was, commenting that it was consistent with paragraph 101 of the New Code, and that:

> the parties to the particular agreement would be best placed to determine whether further explanation would be warranted to reflect the circumstances pertaining to that agreement.

41.12.25 Template clause 7: general

This clause collects together three unrelated concepts. The first sets out the parties' agreement 'that no relationship of landlord and tenant is created by this Agreement'. This provision is not foolproof, for if as a matter of substance the agreement amounts to a tenancy, then no amount of such labelling would distract a tribunal from so finding.[17] It is, however, a term to which the tribunal will be bound to give weight, as an indication of the characterisation of the relationship which the parties clearly intended.

41.12.26

The second is the declaration in clause 7.2 that the agreement will not apply to any part of the land which is or (from the date of such adoption) becomes adopted as highway maintainable at the public expense. There is no reason why this provision would not be given its full effect (in possible contrast to clause 7.1).

41.12.27

The third is the stipulation in clause 7.3 that 'In the event of any inconsistency between this Agreement and any provision of the Code, the Code will prevail'. This recognises the intended supremacy of the New Code – although it remains to be seen to what extent a provision purporting to oust a provision of the code would prevail.[18]

41.12.28

The respondents to the consultation exercise made a number of points about this drafting:

(a) First, some made the point concerning clause 7.1, referred to in paragraph 41.12.23 above.
(b) Secondly, some considered that in many cases it would be necessary to create a tenancy. A respondent put the point in these terms:

> This is only appropriate for certain agreements – for example Apparatus which is to be laid underground. In a lot of cases the Operator will be taking a roof top site or erecting a lattice tower or creating a compound within which to locate its Apparatus and in those scenarios the relationship of Landlord and Tenant will almost certainly be required as the Operator will have legal possession and control of an area of land or building which can and should be demised to it.

(c) Others also considered that clause 7.3 was unnecessary and potentially confusing, and that it might prevent negotiation of broader rights beyond those provided for in the New Code.

41.12.29

As with so many of the comments, OFCOM rejected the criticism, asserting that the drafting was intended to provide baseline clarity, while leaving the parties free to craft their own bespoke agreements.

17 See the discussion of this topic in section 14.3 of Chapter 14.
18 See the discussion of this topic in Chapter 34.

41.12.30 Template clause 8: indemnity for third party claims

This clause provides for the operator to indemnify the grantor up to a suggested series of ceilings (£1m–£5m) each year against third party claims caused by the operator, subject to three caveats (timeous notification by the grantor; no compromise of the claim without the operator's consent; subrogation rights).

41.12.31

The respondents to the consultation exercise made a number of points about the scope and suitability of this indemnity provision, noting in particular that there should be provision for the grantor to use reasonable endeavours to mitigate the liabilities for which it seeks indemnity.

41.12.32

OFCOM's response to these comments in paragraph 4.71 of the Statement was characteristically negative:

> We consider the scope of the indemnity, as well as the suitability of the indemnity for any site, should be matters for the parties to the particular agreement to negotiate and agree on. Consequently we have not made any changes to clause 8.

41.12.33 Template clause 9: limitation of liability

In contrast to clause 8, clause 9 deals more extensively with the general liabilities of each party to the other, imposing limits upon such liabilities, but subject to the further limitation prohibiting the application of such limits (a) in the event of death or personal injury caused by negligence; (b) in the event of damage caused by fraud; and (c) where that would be unlawful.

41.12.34

Objections were made by respondents to the consultation on the grounds that (i) operators should not be liable for consequential or economic loss; and (ii) it is unusual for there to be a provision for site provider liability.

41.12.35

These objections were dismissed by OFCOM, on the ground that the extent of and limitations upon party liability 'would be a matter for the parties to the particular agreement to negotiate and agree on'.

41.12.36 Template clause 10: termination

Clause 10 sets out three separate provisions concerning termination of the agreement, all of which have as their source or parallel provisions in the New Code.

41.12.37

Clause 10.1 deals with termination by the grantor in four specified sets of circumstances, each of which must be preceded by 30 days' written notice. The circumstances, three of which are modelled upon paragraph 31(4) of the New Code,[19] are (a) material breach of obligation which is either irremediable, or has not been remedied despite notice; (b) persistent non-payment of rent (a non-code ground, unless it is simply regarded as a material breach falling within (a)); (c) redevelopment; and (d) the prejudice/public benefit balance favours the grantor.

41.12.38

Clause 10.2 allows for termination by the operator, by written notice of an agreed duration.

41.12.39

Clause 10.3 provides that on termination of the agreement (except where it continues in accordance with paragraph 30(2) of the New Code), the operator will as soon as reasonably practicable remove the apparatus from the land and make good any damage to the land caused by its removal to the reasonable satisfaction of the grantor.

41.12.40

As might have been expected, this drafting generated much controversy, reflected in the comments made in response to the consultation exercise. These fell into four main groups:

(a) The most substantial criticism was directed at the length of notice to be given by the grantor in redevelopment cases, which was said to be unrealistically short, especially given (i) the length of time an operator needed to relocate ('it usually takes a minimum of 24 months for a Code operator to be able to find a replacement site'); and (ii) the 18-month notice period stipulated in paragraph 31(3) of the New Code.

(b) There is no provision for termination in the event of loss by the operator of its operating licence; damage to the site to such an extent that operation has to cease; or abandonment by the operator.

(c) The prejudice/public benefit balance ground is only applied under the New Code where the court is deciding to order the grant of code rights (paragraphs 19/20 of the code) or at the end of a contractual term (paragraph 31(4)(d) of the code). Its incorporation continuously throughout the term of a negotiated agreement was not the intention of the New Code.

(d) The apparatus removal provision should apply during the agreement rather than at its end, when the parties will no longer be in a legal relationship.

(e) One respondent questioned whether there should not be scope to include a lift and shift provision in the agreement, regarded as very important by landowners to cater for example for development situations.

19 See Chapter 22 for a full discussion of termination under the New Code.

41.12.41

OFCOM was not persuaded that any of these comments required changes to be made to clause 10. They commented (without dealing with each in turn):

(a) The criticism as to notice length betrayed a possible misunderstanding of paragraph 31(3) of the New Code, while the footnote to the clause referred to the 18 month Code termination period (see paragraphs 4.88 to 90 of the Statement).
(b) Any additional grounds for termination could be negotiated by the parties if they so chose.
(c) Removal provisions were a matter for negotiation.

41.12.42 Template clause 11: assignment

Clause 11 contains two provisions affecting assignment by the operator. The first – providing that the operator may assign the agreement to another operator who will be bound by its terms with effect from the date of the assignment – triggers a number of observations, which are considered in detail in section 20.2 of Chapter 20 of this book. In essence, this provision reflects the stipulation in paragraph 16(1) of the New Code that assignment by and to an operator may not be prevented.

41.12.43

Clause 11.2 provides that the operator will not be liable for any breach of the agreement occurring on or after the date of the assignment – provided that (a) the grantor is given written notice of the name of the operator assignee and its address for the purposes of clause 15.2 (see clause 41.12.47 below); and (b) notice was given prior to the breach occurring. Again, this drafting merely repeats the provisions set out in paragraph 16(4) and (5) of the New Code.[20]

41.12.44

Respondents argued for extra protection for the grantor, including requirements for the operator (a) to give pre-notification of an intention to assign; and (b) to give notice to the grantor of the date of assignment and the details of the assignee. They also noted that the template does not include provision for the site provider to seek an authorised guarantee agreement (AGA), which is permitted by paragraph 16(6)–(9) of the New Code.

41.12.45

Such provisions are of course customary and regarded as reasonable in commercial leases. However, OFCOM considered that such changes were not necessary to ensure that this clause 16 remains a term that may (but need not) be used in agreements under the New Code.

20 See again section 20.2 of Chapter 20.

41.12.46 Template clause 12: Contracts (Rights of Third Parties) Act

This clause repeats the customary default provision in leases ousting the application of the Contracts (Rights of Third Parties) Act 1999.

41.12.47

There were no substantive comments raised in relation to this clause, save for the observation that the 1999 Act does not have effect in Scotland.

41.12.48 Template clause 13: entire agreement

This boilerplate provision did not generate any comments.

41.12.49 Template clause 14: severance

The same observation may be made about this provision.

41.12.50 Template clause 15: notices

Clause 15 contains four separate provisions. Sub-clauses 15.1 and 15.2 simply repeat the general provisions applicable to the format and delivery of notices set out in paragraphs 88, 89 and 91 of the New Code and s.394 of the 2003 Act.[21] Sub-clause 15.3 helpfully provides that 'Following the execution of this Agreement, either party may amend its address for the purposes of clause 15.2 by notice to the other party'. Sub-clause 15.4 adds the parties' agreement that the address set out in clause 15.2 (as it may be subsequently amended under clause 15.3) will also constitute their address for service for the purposes of paragraph 91(2)(a) of the New Code.

41.12.51

Respondents to the consultation suggest that this clause should provide that notices could be in any form required by the New Code. This prompted OFCOM to add a sentence to that effect at the end of sub-clause 15.1.

41.12.52 Template clause 16: mediation

This clause allows, alternatively requires, the parties to seek to mediate any dispute arising before turning to other forms of resolution. Respondents to the consultation exercise objected that this drafting imposed a constraint on other forms of dispute resolution, and had the potential to delay the institution of proceedings.

41.12.53

Again, OFCOM was unmoved by these comments, noting that:

21 Considered in detail in Chapter 32.

> whether parties should be obliged to first try some means of alternative dispute resolution before recourse to the courts, should be a matter for the parties to particular agreement to negotiate and agree on.

41.12.54 Template clause 17: governing law and jurisdiction

Clause 17 provides for the domestic courts of the United Kingdom to have jurisdiction in their own domains, and for the parties to submit to the exclusive jurisdiction of those courts (by which, as its footnote to this clause states, OFCOM meant the tribunals, that being the principal forum for dispute resolution under the New Code[22]). Aside from observations concerning the differing practices in the three jurisdictions, there were no substantive comments on this provision.

41.12.55 Template clauses 17: conclusion

These template terms may be said to represent an unsatisfactory compromise between safe and adventurous drafting: summarising some of the terms of the New Code, while failing to take the opportunity to provide a helpful drafting guide for parties. As one respondent commented:

> this is a very basic form of Code Agreement which will not be appropriate for green field mast sites, roof tops or multi-site providers.

Perhaps the fairest comment to make is that it is always easier to criticise than to draft. While OFCOM's responses to the consultation exercise comments perhaps betray an element of bias in favour of the text it first proposed, its task of reconciling the views emanating from the diametrically opposed camps of landowner and operator in a diverse operating environment was never going to be trouble-free.

41.12.56

None of the criticism noted in this section may matter very much, in the sense that, as OFCOM said repeatedly in its Statement, the parties are (relatively) free to enter into whatever agreement they wish. Nevertheless, the template clauses may come to have a rather more enduring role than OFCOM perhaps intended, for two reasons.

41.12.57

First, less well-resourced parties may treat the template terms as officially sanctioned and exhaustive terms which represent all that they could reasonably need, and may (a) look with suspicion upon any attempt to renegotiate the terms; and (b) consider that no further protective terms are required. If this is what transpires, it will slow negotiations, and risk landowners agreeing to sub-optimal terms.

22 See in this respect Chapter 33.

41.12.58

Secondly, the tribunal that is asked to impose terms (as to which, see Chapter 21) may be tempted too to consider that the template terms are the gold standard, when that is of course far from the case.

41.12.59

On the other hand, it is possible that the template terms will simply be ignored in practice, given the shortcomings identified by the respondents to the consultation exercise, virtually none of which were attended to by OFCOM.

41.13 OFCOM as a forum for disputes

41.13.1 Nature of disputes covered

Section 185 of the 2003 Act allows three classes of dispute to be referred to OFCOM:

> (1) This section applies in the case of a dispute relating to the provision of network access if it is –
> (a) a dispute between different communications providers;
> (b) a dispute between a communications provider and a person who makes associated facilities available;
> (c) a dispute between different persons making such facilities available;
> (1A) This section also applies in the case of a dispute relating to the provision of network access if –
> (a) it is a dispute between a communications provider and a person who is identified, or is a member of a class identified, in a condition imposed on the communications provider under section 45; and
> (b) the dispute relates to entitlements to network access that the communications provider is required to provide to that person by or under that condition.
> (2) This section also applies in the case of any other dispute if –
> (a) it relates to rights or obligations conferred or imposed by or under [a condition set under section 45, or any of the enactments relating to the management of the radio spectrum];
> (b) it is a dispute between different communications providers; and
> (c) it is not an excluded dispute.

41.13.2

Section 45 conditions can be applied to any person identified in s.46 of that Act, which identifies different kinds of conditions which can be applied in a particular instance. For present purposes, it is enough to note that general conditions can be applied to any person providing an electronic communications network or service. Section 45(2) provides that

(2) A condition set by OFCOM under this section must be either –
  (a) a general condition; or
  (b) a condition of one of the following descriptions –
    (i) a universal service condition;
    (ii) an access-related condition;
    (iii) a privileged supplier condition;
    (iv) a significant market power condition (an 'SMP condition').

41.13.3

Section 45(3) provides that a 'general condition' is a condition which contains only provisions authorised or required by one or more of ss.51, 52, 57, 58 or 64.

41.13.4

Section 45(5) provides that an 'An access-related condition' is a condition which contains only provisions authorised by s.73. Section 73(3) provides that:

(3) Access-related conditions may include conditions appearing to OFCOM to be appropriate for securing that persons to whom the electronic communications code applies participate […][4] in arrangements for –
  (a) sharing the use of electronic communications apparatus; and
  (b) apportioning and making contributions towards costs incurred in relation to shared electronic communications apparatus.

41.13.5

Section 46(6) makes provision for the persons to whom an access-related condition can be applied.

41.13.6 Means of referring a dispute

Section 185 makes provision for the mode of referring a dispute:

(3) Any one or more of the parties to the dispute may refer it to OFCOM.
(4) A reference made under this section is to be made in such manner as OFCOM may require.
(5) The way in which a requirement under subsection (4) –
  (a) is to be imposed, or
  (b) may be withdrawn or modified,

is by a notice published in such manner as OFCOM consider appropriate for bringing the requirement, withdrawal or modification to the attention of the persons who, in their opinion, are likely to be affected by it.

(6) Requirements imposed under subsection (4) may make different provision for different cases.
(7) A dispute is an excluded dispute for the purposes of subsection (2) if it is about –

(a) obligations imposed on a communications provider by SMP apparatus conditions

(8) For the purposes of this section –

(a) the disputes that relate to the provision of network access include disputes as to the terms or conditions on which it is or may be provided in a particular case; and

(b) the disputes that relate to an obligation include disputes as to the terms or conditions on which any transaction is to be entered into for the purpose of complying with that obligation.

41.13.7 How disputes are handled

On 7 June 2011, OFCOM published guidelines in relation to its handling of regulatory disputes.[23] This guidance explains that:[24]

2.7 In accordance with s.186(2) of the 2003 Act, OFCOM must decide whether or not it is appropriate for it to handle the dispute.

2.8 In the case of a dispute falling within s.185(1A) or (2) of the 2003 Act, OFCOM will decide that it is appropriate for it to handle the dispute unless there are alternative means to resolve the dispute promptly and satisfactorily, in line with the Community requirements set out in s.4 of the 2003 Act.

2.9 In some cases, for example, OFCOM may consider that it would be appropriate to send the dispute for alternative dispute resolution ('ADR'). It would make this assessment on the particular facts of each individual case.

2.10 If the dispute is not resolved by alternative means before the end of the four months after the day of OFCOM's decision not to accept the dispute, one or more of the Parties may refer the dispute back to OFCOM.

2.11 In the case of a dispute falling with s.185(1) of the 2003 Act, OFCOM has discretion whether to decide that it is appropriate for it to handle the dispute. In exercising that discretion, OFCOM may in particular take into account its priorities and available resources at the time (s.186(2A) of the 2003 Act).

2.12 OFCOM's powers when resolving disputes are set out in s.190 of the 2003 Act.

2.13 OFCOM must resolve disputes within the four month statutory deadline, except in exceptional circumstances (s.188(5) of the 2003 Act).

2.14 In all cases Parties should have realistic expectations of the depth of analysis OFCOM is able to carry out within the four month statutory deadline. Whilst dispute resolution is a separate regulatory function to be used in parallel to OFCOM's other regulatory powers, in the time available OFCOM is clearly not able to undertake the type of analysis it would normally carry out in exercising its ex ante regulatory powers or its powers under the Competition Act 1998 (as to which, see Chapter 39 of this book)[25].

23 Ofcom's guidelines for the handling of regulatory disputes – www.ofcom.org.uk/__data/assets/pdf_file/0020/71624/guidelines.pdf.

24 See also *Halsbury's Laws,* vol. 97, paragraph 217 and following.

25 And see *Guidelines for the Handling of Competition Complaints, and Complaints and Disputes about Breaches of Conditions Imposed under the EU Directives,* OFCOM, July 2004 – www.ofcom.org.uk/__data/assets/pdf_file/0029/37946/guidelines.pdf.

2.15 Where necessary, OFCOM will consider exercising any of its regulatory powers listed in s.190(4) of the 2003 Act, or any other of its regulatory powers as the sectoral regulator, instead of or at the same time as resolving the dispute.

41.13. 8 Relevance of disputes

As discussed in Chapter 18, one of the key changes which the New Code seeks to implement is that operator–operator disputes will no longer be capable of being brought under the New Code. This is because no code rights can be acquired over electronic communications apparatus. Instead, such disputes will need to be dealt with through other mechanisms.

41.13.9

During the development of the package of reforms to the code that led to the amendments proposed in the 2017 Act, stakeholders provided a range of inputs to DCMS and OFCOM with regard to whether the relationships between wholesale infrastructure providers (WIPs) and communications providers (particularly mobile network operators – MNOs) were effectively governed by the code and, if not, what reforms to the code might be necessary.

41.13.10

Ultimately, the Government concluded that specific changes to the code were unnecessary, as explained by the DCMS in May 2016:[26]

> The Code provides for a series of rights which will be binding on site providers – there has been considerable debate on the definition of land within the Code, and in particular whether 'apparatus' should be regulated under the new Code.
>
> The Government received a number of responses on this issue, and there were strongly opposing views on all sides, suggesting the legal position under the current Code to be ambiguous. However, the original purpose of the Code was to allow access to land so that communications infrastructure could be installed rather than to allow access to the infrastructure itself. That rationale has not changed, and Government does not want to increase regulation and risk disruption of market incentives for investment in passive infrastructure. There is an existing and well understood legal framework in place to provide for access to apparatus in cases where there is significant market power and / or anticompetitive behaviour. As the UK's independent regulator for telecommunications, Ofcom is responsible for ensuring effective competition in telecommunications markets. Given this, the Government will exclude apparatus from the scope of land within the Code and avoid 'gold-plated' regulation.

41.13.11

Since DCMS published this statement, some progress towards the development of a voluntary Code of Practice between WIPs and MNOs has been made, which OFCOM

26 DCMS, *A New Electronic Communications Code*, May 2016.

is supporting, although it recognises that industry stakeholders wish for confirmation that, in the event that commercial negotiation and any subsequent arbitration fails, the terms on which WIPs grant access to their infrastructure can be regulated.

41.13.12

In paragraph 2.28 of their 24 March 2017 consultation paper,[27] OFCOM dealt with this matter as follows:

> Whilst it is difficult to provide a view on this issue in the abstract, we can confirm that Ofcom has a number of statutory powers which could enable us, in principle, to regulate the terms on which WIPs grant access to their infrastructure. For example, Ofcom has certain powers to regulate access to infrastructure under the following legislation:[28]

- regulation 3(4) of the Electronic Communications Code (Conditions and Restrictions) Regulations 2003/2553;[29]
- section 73 of the Communications Act 2003 (the '2003 Act') (i.e. by imposing an access-related condition);
- section 87 of the 2003 Act (i.e. by imposing a significant market power (SMP) condition); and
- section 185 of the 2003 Act (i.e. by resolving a dispute relating to the provision of network access).

41.13.13

OFCOM added in paragraph 2.29 of the same paper:

> We note that, in the first instance, we would generally seek to resolve problems using the most appropriate and least intrusive approach, with recourse to more formalised regulatory interventions only where necessary.

41.13.14

The extent to which WIP/MNO disputes arise under the New Code and are referred to OFCOM for resolution remains to be seen.

27 Electronic Communications Code – Digital Economy Bill: Proposed Code of Practice, Standard Terms of Agreement and Standard Notices.

28 OFCOM also has certain powers under general competition law to address abuse of dominance or anti-competitive agreements.

29 Regulation 3(4) specifies that a code operator 'where practicable, shall share the use of electronic communications apparatus'. Ofcom have powers under s.110 of the 2003 Act to enforce this requirement.

Part V

Drafting

42 Drafting considerations for code agreements

Part 1 The nature of agreements

42.1 Introduction

42.1.1

Telecommunications agreements are the subject of statutory intervention by the New Code and, formerly, the Old Code. As a result (and to the surprise of many landowners), the deal which they have made with the operator and which has been duly recorded in the written contract between them, invariably does not reflect the true scope and extent of the operator's rights, nor the impact those rights will have upon the landowner's property.

42.1.2

To take one example, operators do not commonly simply abandon their apparatus on-site when they cease operating. Often (in the case of removal for redevelopment, for example), the costs of enforcing the removal of telecommunications equipment will exceed the income which the landowner will have received over the entire lifetime of the agreement, in some cases by many multiples.

42.1.3

The point just made underlines the need for drafters of code agreements to be alert not merely to the consequences of the text they draft, but also to the statutory overlay imposed by the New Code, which may – and commonly will – have the effect that the text does not dictate the eventual outcome. This chapter discusses the various drafting considerations for landowners and operators that may help to ensure that the various outcomes are foreseen, if not always mitigated to every party's satisfaction.

42.1.4

This chapter is divided into a number of parts, as follows:

- This first part considers the nature of agreements and in particular whether they constitute licences or tenancies, as well as the ramifications of that distinction under each code.
- Part 2 analyses the ingredients a telecommunications agreement has to satisfy in order to engage the New Code.

- Part 3 considers agreements to use apparatus, rather than land – where the New Code is not after all engaged.
- Part 4 looks at other drafting issues to consider when granting a telecommunications agreement.
- Part 5 considers the New Code documents (Code of Practice, standard terms and notices) for which OFCOM has been responsible.
- Part 6 deals with the drafting of the notices required under the New Code.
- Part 7 considers some dispute resolution issues that may arise under the New Code.
- Finally, Part 8 looks at other practical drafting issues that may arise.

42.2 *Old Code agreements: overview*

42.2.1

In general terms, the Old Code applied to any written agreement to grant operators rights to install telecoms apparatus (that is, before 28 December 2017 when the New Code came into force). As discussed in more detail in Chapter 14, Old Code agreements faced a potential double overlay of statutory regimes.

42.2.2

The rights granted by the Old Code, over and above those set out in the written contract, were in reality however, fairly limited (which is very much not the case under the New Code, considered later), although they did, and, in the case of subsisting agreements (as to which see Chapter 31, Section 31.2), do go to the fundamental issue of when an operator is required to give up possession of the property occupied by its apparatus. That is, not when the contract says, but only if and when the tests in paragraphs 20 and/or 21 of the Old Code have been proven and overcome.

42.2.3

Over and above this, there is also the security of tenure regime imposed by the Landlord and Tenant Act 1954 ('the 1954 Act'), where applicable (see further Chapter 25). Again, this goes to the question of exactly when an operator can be required to give up possession of its site, which is (again) not when the contract says, but only once the various hurdles set out in the Act have been dealt with.

42.2.4

In summary, therefore, when considering the effect of an agreement under the Old Code which is a subsisting agreement under the terms of the New Code (see Chapter 31, Section 31.2), regard must be had to:

(a) the terms of the agreement itself setting out the bargain agreed between the parties; and
(b) the provisions of the Old Code; and
(c) where applicable, the 1954 Act.

42.2.5

Whilst the provisions and effect of the 1954 Act are reasonably well known and understood by most commercial landlords, that is significantly less the case in terms of the Old Code. Landlords of a code agreement which has been duly contracted out of the 1954 Act, reasonably expecting to take back possession of their property, for example, to carry out a redevelopment, may ultimately discover that, due to the provisions of the Old Code, the vacant possession date is likely to be pushed back, in some cases (given the terms of the New Code and in particular Part 5 – see Chapter 22) for another 18 months or more. It is therefore critical in any case where a landlord is looking to terminate an Old Code agreement that the time line for finally securing possession from an operator is factored in. This problem will in future only arise in the context of the Old Code, in respect of notices served pursuant to the transitional provisions of the Digital Economy Act 2017 (as to which see Chapter 31 of this book).

42.3 New Code agreements: overview

42.3.1

The New Code alters the position substantially, and drafting practice for New Code agreements will have to change significantly from the approach adopted hitherto in relation to the Old Code. Whilst the New Code expressly dis-applies the application of the 1954 Act (in relation to what are termed 'primary purpose' agreements, as to which see Chapter 19) and therefore removes that layer of complexity and sometimes downright incompatibility with the telecommunications legislation, the mandatory intervention of the New Code upon the bargain which the parties make and the agreement they enter into is expanded very significantly indeed.

42.3.2

Whereas the Old Code was concerned primarily with security of tenure for telecoms operators, the New Code prescribes (in addition to security of tenure, as before), for example, rights to share apparatus, rights to assign the agreement, rights to upgrade the apparatus and detailed provisions regarding how (on compulsory acquisition of a code agreement) the consideration for the agreement is to be calculated.

42.3.3

As with the Old Code, it is not possible for the parties to contract out of these provisions of the New Code – see paragraph 100. The anti-avoidance provisions are considered in Chapter 34 of this book.

42.3.4

When considering the nature and effect of any New Code agreement, therefore, one needs to have regard to:

(a) the terms of the agreement itself; and
(b) the extensive provisions of the New Code.

42.4 The nature of agreements: lease v licence

42.4.1

As above, for the Old Code to apply to an agreement (pre-28 December 2017), all that was required was an agreement in writing; this limited level of formality brought within the Old Code's remit both formal leases and also agreements which were, on their terms, no more than licences.

42.4.2

The lease/licence distinction is dealt with in detail in Chapter 14 of this book. The distinction is important in relation not merely to the Old Code, but also the New Code.

42.4.3

A telecoms agreement under the Old Code, whether one which expired contractually before the New Code came into force or which was continuing at that date, may, in addition to the security of tenure provided by the Old Code, also benefit from the additional rights conferred by Part II of the Landlord and Tenant Act 1954 if it is a tenancy (unless the contracting out procedures have been followed), but not if it is a licence.

42.4.4

Under the New Code, by contrast, while a licence will not, as before, benefit from 1954 Act protection, even a tenancy will fall outside that protection, unless it is a non-primary purpose agreement to which Part 5 of the New Code is not applied.

42.4.5

In general terms, therefore, the 1954 Act is expressly dis-applied from the New Code regime, although many aspects of the security of tenure model used by the 1954 Act have been imported into the New Code. This is discussed in Chapter 22.

42.5 Formalities of an agreement

42.5.1

Any new agreements (relating either to fresh installations or to the renewal of existing sites) will as from 28 December 2017 fall under the jurisdiction of the New Code, and the remainder of this chapter is predicated on that basis.

42.5.2

In drafting future agreements, the drafter will be required to keep in mind a series of points. The first point (as before) to check is whether one is actually dealing with a code-protected operator; that is one which has been granted code powers by OFCOM under the Communications Act 2003 and is as a result listed on the OFCOM register of operators (which can be easily accessed on the OFCOM website). If not, then the New Code will simply not apply.

42.5.3

For the avoidance of any doubt, transitional provisions[1] made as part of the implementation of the New Code regime mean that those operators who currently have protection under the Old Code will also benefit from that protection going forward.

42.5.4

Once it is established that the relevant counterparty to an agreement is a code-protected operator, then paragraph 11 of the New Code sets out the formalities which must be complied with in order to fall within the new regime. These are considerably more developed than the provisions under the Old Code, and introduce a level of formality closer to section 2 of the Law of Property (Miscellaneous Provisions) Act 1989, but also imposing content requirements on the agreement. In essence paragraph 11 of the New Code requires that the agreement:

(a) be in writing;
(b) be signed by or on behalf of parties to it;
(c) state for how long the code right is exercisable; and
(d) state the period of notice (if any) required to terminate the agreement.

42.5.5

Compliance with these requirements is strict. If, for example, only one party has signed the agreement, then the agreement will not benefit from New Code protection. Some have pointed out that this is somewhat incompatible with the mandatory nature of the New Code and therefore a possible way in which to circumvent the prohibition on contracting out: by not following one or other of the requirements of paragraph 11, the parties simply do not get into the New Code in the first place.

Part 2 Ingredients of the New Code agreement

42.6 'Bare' and fuller agreements

42.6.1

This part examines the terms which a New Code agreement should contain.

1 Contained in Schedule 2 to the 2017 Act. See Chapter 31.

42.6.2

It is important at the outset to make the distinction between what may be called a 'bare' New Code agreement and one which is supplemented by other provisions, through the agreement of the parties.

42.6.3

The overriding principle at the heart of the New Code is the agreement of the parties. They are expected to first see whether an agreement can be reached by consent. Only in the absence of such agreement is it envisaged that the operators will then fall back on their right to seek the imposition of an agreement under the New Code.

42.6.4

Whilst the parties are (subject to some express caveats in the New Code, which are considered in more detail below) free to agree whatever they want in terms of what goes into a New Code agreement (as to which, see section 42.8 below), the rights which are available to an operator seeking to *impose* rights using the New Code are much more restricted. The expression for such an agreement (although not one used in the code itself) may be said to be a 'bare agreement'.

42.6.5

The expression 'bare agreement' may, however, not be as 'minimalist' as the expression implies and may be considered to be a little misleading. Notwithstanding the fact that the terms of Part 4 of the New Code are dealing with the imposition of an agreement with respect only to code rights (as set out in paragraph 3 – as to which see section 42.7 below), there are a number of provisions in Part 4 which confer upon the court, it would seem, a degree of flexibility with respect to the terms to be imposed in the agreement. In particular:

(1) First, the starting position is paragraph 20(2)(a), which provides that the operator is to give the relevant person a notice in writing 'setting out the code right, and all of the other terms of the agreement that the operator seeks …' It would appear, therefore, that the operator is not limited simply to terms reflecting the menu of code rights contained within paragraph 3. Of course the relevant person need not agree to all the other terms which the operator seeks.
(2) Second, paragraph 23(1) provides that 'an order under paragraph 20 may impose an agreement which gives effect to the code rights sought by the operator with such modifications as the court thinks appropriate'. Furthermore, paragraph 23(2) provides that 'an order under paragraph 20 must require the agreement contains such terms as the court thinks appropriate, subject to subparagraph (3) and (8)'. Those two provisions would appear to confer upon the court a discretion to impose appropriate terms not only to give effect to the code right under paragraph 3 but such other terms as are considered appropriate.

Presumably, in considering what terms are 'appropriate', the court will have regard to those which the operator seeks in its notice. This would suggest that operators should consider very carefully what other terms going beyond simply the paragraph 3 code rights they would like to find incorporated into the agreement to be imposed upon the relevant person.

(3) Third, paragraph 23(5) provides that there may be further terms incorporated into the agreement which the 'court thinks appropriate for ensuring that the least possible loss and damage is caused by the exercise of the code right to persons who (a) occupy the land in question, (b) own interest in that land, or (c) are from time to time on that land'.

(4) Finally, paragraph 23(8) makes provision for the incorporation of terms as to termination (which could, of course, include not only a break clause but also a forfeiture clause or any other formulation for termination), together with a right to 'lift and shift'.

42.7 The 'menu' of code rights

42.7.1

The New Code provides for nine express rights (and nothing more) which can be 'Code Rights'. These are set out at paragraph 3 (and see the discussion at Chapter 15 of this book). Those rights are:

(a) To install electronic communications apparatus on, under or over the land
(b) To keep installed electronic communications apparatus which is on, under or over the land
(c) To inspect, maintain, adjust, alter, repair, upgrade or operate electronic communications apparatus which is on, under or over the land
(d) To carry out any works on the land for or in connection with the installation of electronic communications apparatus on, under or over the land or elsewhere
(e) To carry out any work on the land for or in connection with the maintenance, adjustment, alteration, repair, upgrading or operation of electronic communications apparatus which is on, under or over the land
(f) To enter the land to inspect, maintain, adjust, alter, repair, upgrade or operate any electronic communications apparatus which is on, under or over the land
(g) To connect a power supply
(h) To interfere with or obstruct a means of access to or from the land (whether or not any electronic communications apparatus is on, under or over the land)
(i) To lop or cut back, or require another person to lop or cut back, any tree or other vegetation that interferes or will or may interfere with electronic communications apparatus

42.7.2

These rights are self-explanatory, although some further consideration of their true scope and extent, together with some drafting recommendations, are set out below.

42.7.3

It is implicit in the New Code that the parties can pick and choose which combination of these nine paragraph 3 code rights they wish to include in any New Code agreement. It is therefore not the case that any New Code agreement will automatically benefit from all nine rights, as a matter of course, though there is no reason why an agreement should not refer to them as being conferred compendiously. However, absent a compendious conferral of code rights (and a clause seeking to grant all rights in one compendious grant might lead to construction arguments at a later stage), operators will need to take care in their negotiations with the landowner to specify which of the nine rights they wish to be incorporated in the agreement. It is considered preferable that the individual code rights are expressly referred to, in order to avoid any argument as to which of them has been conferred.

42.7.4

It is also considered prudent, in parallel to such negotiations, that the operator should satisfy itself that it could, if pressed, meet the test at paragraph 21(2) and (3) of the New Code; i.e. that there is a public interest likely to derive from the imposition of the specific rights, which both outweighs the interests of the landowner and also is capable of being adequately compensated in money.

42.7.5

Absent further agreement between the parties, a New Code agreement will not of itself constitute a lease, that is, the demise of a right to exclusive possession of land to an operator to the exclusion of all others, including the landowner. At heart, it need be no more than the grant of a right to install electronic communications apparatus on, under or over land; that is, a mere permission to place something on land which is much more akin to a licence, i.e. making something legal which would otherwise be a trespass. It is considered that, going forward, the lease/licence distinction which has preoccupied litigators under the Old Code will lose its relevance given the provisions under the New Code for agreements to be outside the scope of the Landlord and Tenant Act 1954 (see Chapter 14).

42.8 Supplementing code rights

42.8.1

Although the bare code rights offer a greater range of rights than was available under the Old Code, it is to be expected (as before) that the parties will seek to supplement those rights with qualifications or additional rights, or that they will wish to ensure that the agreement conferring code rights is in a form, and uses language, already familiar to the parties, their property or estate managers, and those advising them. Further, it is to be anticipated that parties to the agreement will seek to have the comfort of familiar clauses, such as contractual break clauses or 'lift and shift' provisions (that is to say, provisions giving the landowner and/or the operator the right to insist upon the apparatus being relocated to a different location upon the land). Additionally, the operator may require separate rights in the nature of easements to access apparatus, rights to park when inspecting the apparatus or the right to keep equipment near the electronic communications apparatus which is not, itself, electronic communications apparatus.

42.8.2

On a practical note, the extent to which the operator will be able to obtain a full agreement with all necessary additional rights by negotiation is likely to be subject to the usual constraint of time and commercial pressures on operators to secure rights to sites. The pressure on the operator to get onto the site may be such that it is simply not possible to negotiate all the desired rights at leisure.

42.8.3

Although the operator has, as a well-advised landowner will know, compulsory acquisition powers in the background under paragraph 20 of the New Code (as to which see Chapter 21), there must be at least some doubt as to the nature of the Upper Tribunal's jurisdiction as to terms in two respects:

(a) as to whether the powers of the Upper Tribunal to impose terms will extend to a power to provide all such rights and advantages as the parties might be able to negotiate freely; and
(b) as to the effect of the consideration/compensation provisions, and how they operate.

42.8.4

Further, the reality is that operators require access to sites relatively quickly. Although the underlying European materials (see Chapter 4) make it clear that member states should provide mechanisms for the quick roll-out of networks and easy access to land, in practice the acquisition of rights through the New Code regime is unlikely to be quick or straightforward. Particularly in the early days of the New Code, the potency in negotiations of a background threat of a paragraph 20 acquisition is likely to be diluted.

42.8.5

In addition to the doubt over whether the Upper Tribunal using its powers to impose agreement under paragraph 20 can, in all respects, replicate a freely negotiated agreement between the parties, there are likely to be two practical difficulties that arise in operating the paragraph 20 jurisdiction:

(a) First, the test at paragraph 20 is not a tick-box exercise. Rather, it will require the submission of expert and other evidence to demonstrate exactly why the imposition of rights is necessary – that there is a public interest which both outweighs the interests of the landowner and also is capable of being adequately compensated in money. The compilation, submission and consideration of such evidence will obviously take time and money.

(b) Secondly, the Upper Tribunal may be unable (in the early days at least) to deal with applications under the New Code within a time frame that is likely to suit the operators, at least in the case of an urgently required site. It seems likely that temporary rights would need to be sought on an interim basis. However, absent any clear guidance as to procedure that the Upper Tribunal is likely to adopt, there will doubtless again be delays as this new jurisdiction 'beds in'.

42.8.6

For all of these reasons, it may well be the case that a negotiated agreement by the consent of the parties (i.e. outside of the strict confines of the New Code, in so far as that is possible), will in fact be the only practical way forward, at least in the early days of New Code operation.

42.9 The individual code rights

42.9.1

Chapter 15 considers the meaning and effect of individual code rights. This part is concerned not so much with their content but with how, as a matter of drafting practice, they ought to be reflected in the agreement and what the agreement ought to provide for, in addition to the bare right itself.

42.9.2

Each of the following sections in the remainder of this part takes the nine New Code rights at paragraph 3 in turn.

42.10 Code rights (a) To install electronic communications apparatus on, under or over the land; and (b) To keep installed electronic communications apparatus which is on, under or over the land

42.10.1

Both sub-paragraphs (a) and (b) are self-explanatory. The key to the clear and effective operation of the agreement, however, will be defining in as much detail as possible exactly what apparatus the operator will be entitled to install and subsequently keep installed on the land in question.

42.10.2

With this in mind, a detailed apparatus schedule and plans will be required for every agreement – setting out exactly what apparatus will be installed and where, along with the exact dimensions of that apparatus, the manufacturer, serial numbers and so on.

42.10.3

A clear distinction should also be drawn between the electronic components – the antennae, dishes (if any), cables and supporting electronic components and, separately, the physical infrastructure – principally the mast itself (for which exact details of its proposed location, dimensions, construction, and height should be expressly specified) and also the cabinet(s) (again for which manufacturer, dimensions, location etc. should all be listed).

42.10.4

It should be reiterated at this point that a 'bare' New Code agreement will probably not take the form of a lease. Under no circumstances therefore should such an agreement contain plans with red lines around a demise of land, to the exclusion of the landowner, unless it is expressly intended that the agreement should take effect as a lease. If a fenced compound is required, or a locked equipment room, then this looks closer to a demise of land. If that is what the operator requires, then it will need to be dealt with by an agreement of the parties over and above what is provided for in the New Code.

42.10.5

Consideration should also be given to a clause in the agreement making plain what the parties consider is the legal effect of the agreement (see Chapter 14 as to the effect of such clauses), and additionally, consideration should be given (particularly in cases where the agreement confers a mixture of code rights and other rights) to a statement as to what is the 'primary purpose' of the agreement (as to the effect of which see Chapter 19).

42.10.6

If there is genuine doubt as to primary purpose, then consideration should be given to contracting out of the Landlord and Tenant Act 1954, albeit that now, termination under that regime (with a six to twelve-month notice period) is in fact potentially more favourable to the landowner than the position under Part 5 of the New Code (though again this depends on the swiftness with which the Part 5 process is operated, and the relative merits and efficiency of the County Court for 1954 Act claims versus the Upper Tribunal for code termination claims).

42.10.7

The site plans to the agreement should mark in simple terms the exact location of the equipment cabinet, along with the exact location of the points at which the mast legs will be touching the ground. Further plans can then record exactly where the various electronic apparatus will sit on that mast and within that cabinet, along with any connecting cables etc.

42.10.8

It is relevant at this stage to note the definition of 'electronic communications apparatus', at paragraph 5(1) of the New Code:

> In this Code 'electronic communications apparatus' means-
> (a) apparatus designed or adapted for use in connection with the provision of an electronic communications network,
> (b) apparatus designed or adapted for a use which consists of or includes the sending or receiving of communications or other signals that are transmitted by means of an electronic communications network,
> (c) lines, and
> (d) other structures or things designed or adapted for use in connection with the provision of an electronic communications network.'

42.10.9

As paragraph 403 of the Explanatory Notes to the 2017 Act state:

> Whether a particular structure or thing has been adapted to a point at which it can properly be considered as electronic communications apparatus is a question of fact, which will depend on the specific circumstances, including what the parties have agreed, the nature of the installation and the extent of the adaptation.

42.10.10

In the light of the provisions of paragraph 5(1), as so explained, the drafter must consider whether, for example, a compound fence, a concrete base, or alternatively a generator, constitute electronic communications apparatus, being 'designed or adapted for use' in connection with the provision of an electronic communications network, as defined. If the operator wishes to bring onto site apparatus which falls outside this definition, then that is not within the remit of a 'bare' New Code agreement and the agreement of the landowner will be required.

42.10.11

A distinction should be drawn between cases where (i) a fresh site is being put into operation, where the parties start with in effect a blank sheet of paper, and (ii) instances where an existing site is being renewed, which may already have apparatus on site.

42.10.12

In the case of a *new site*, the parties ought to apply the drafting principles set out above.

42.10.13

Matters are somewhat different in relation to *existing sites*, where a renewal is being negotiated. Many existing sites (particularly for example historic, large television mast-type installations) will have a quantity of equipment on site which may not fall into the definition of 'electronic communications apparatus' and in relation to which code rights (whether freely negotiated or imposed by the Upper Tribunal) may not apply.

42.10.14

There will, therefore, need to be an inspection to determine what, if anything, needs to be removed from site if a 'bare' New Code agreement is all that will be in place going forward. In practical terms, therefore, a full site audit of any and all equipment at site is likely to be required for these purposes from the outset.

42.11 Code right (c): To inspect, maintain, adjust, alter, repair, upgrade or operate electronic communications apparatus which is on, under or over the land

42.11.1

The power at paragraph 3(c) to both upgrade and/or to subsequently share that (specified) apparatus is elaborated upon at paragraph 17 of the New Code.

42.11.2

The first point to make here, however, is that it is not possible for the parties to contract out of paragraph 17, by agreement or otherwise – see 17(5) and Chapter 34:

> Any agreement under Part 2 of this Code is void to the extent that-
> (a) it prevents or limits the upgrading or sharing, in a case where the conditions in sub-paragraphs (2) and (3) are met, of the electronic communications apparatus to which the agreement relates; or
> (b) it makes upgrading or sharing of such apparatus subject to conditions to be met by the operator (including a condition requiring the payment of money).

42.11.3

In any new telecoms agreement going forward (be that a 'bare' agreement, or one that is supplemented by the agreement of the parties), express rights in the terms set out at paragraph 17 should therefore be included.

42.11.4

As a minimum, that would include express reference to sub-paragraphs (1)–(3), inclusive of not only the right to share and upgrade the apparatus as permitted, but also the limitations qualifying these rights at sub-paragraphs (2) and (3).

42.11.5

The key point to note, however, is that these upgrading and sharing rights (see paragraph 17(1)) relate to the specified electronic communications apparatus only and, crucially, do not include the addition of any new equipment – hence it is advisable for landowners to define in as much detail as possible (see paragraphs 42.10.1–4 above) exactly what apparatus is permitted to be installed at the site from the outset.

42.11.6

If an operator wishes to share a site with one of its counterparts, but that would require the installation of fresh apparatus over and above that listed in the agreement specifications/plans, then paragraph 17 is therefore unlikely to be of use to that operator. Rather, a fresh application to the Upper Tribunal (in the absence of a negotiated agreement) for additional apparatus rights will be required.

42.11.7

In an age of increased roaming-based sharing, this may, however, be less of an issue for the operators going forward. As technology advances, the need for additional physical apparatus is expected to decrease over time.

42.11.8

All this will no doubt inevitably lead to further debate on exactly what constitutes the upgrading of the apparatus. Whereas adding fresh additional antennae onto the site, for example, might be said to fall outside the scope of the upgrade rights permitted by paragraph 17, adding new electronics into the existing antennae is, perhaps, less clear cut. Again, each instance of upgrading will and should need to be considered on its own facts. Drafters of agreements will need to be astute to technological advances, and will need to keep precedents updated. The historic practice, sometimes encountered, of simply applying standard lease precedents without taking into account the idiosyncrasies of electronic communications, and without drafting with a view to, for instance, technological innovations in the sector, caused considerable difficulties under the Old Code, and is likely to cause even more substantial problems if that practice is followed in relation to the New Code.

42.11.9

It seems advisable to include a contractual provision in the agreement (despite the lack of OFCOM guidance to this effect[2]) to require the operator to give prior written notice of any planned upgrade works, so as to give the landowner an opportunity to consider and assess:

(a) whether that upgrade is indeed permitted under the confines of the existing apparatus specification, or whether further additional equipment rights are going to be required; and also

2 See paragraph 41.10.9 of Chapter 41 of this book.

(b) what the likely impact upon the landowner will be, in terms of the limitations imposed upon the operators at paragraphs 17(2) and (3).

42.11.10

Finally, for the reasons set out above, the Authors suggest that prudent landowners will wish to be proactive in managing or policing the operation of telecoms agreements under the New Code. The days of completing the agreement, putting that away in a drawer and collecting the rent each year may well be at an end. What apparatus and which operators are on site, upgrade and access requests etc. are all going to need close and careful monitoring and management going forward under the New Code.

42.11.11

In this regard it will also be sensible, therefore, to include a power similar to s.40 of the Landlord and Tenant Act 1954 to require any operator to disclose who else is using the site. It is probably wise to avoid (or at any rate to not limit the clause to) old and uncertain concepts like 'occupation' in such a clause. It is probably a better idea to require the operator to disclose (a) anyone with a proprietary or contractual interest in the site and (b) anyone who is using any electronic communications apparatus on the site in connection with their electronic communications network. It will also be sensible to confer rights to inspect the site (though care should be taken in drafting this clause that it does not amount to a reservation of a right to enter consistent with the operator's rights under the agreement being a lease).

42.12 Code right (d) To carry out any works on the land for or in connection with the installation of electronic communications apparatus on, under or over the land or elsewhere

The concepts associated with this right are discussed in relations to code rights (a), (b), (c) at sections 15.3 and 42.9 above.

42.13 Code right (e) To carry out any work on the land for or in connection with the maintenance, adjustment, alteration, repair, upgrading or operation of electronic communications apparatus which is on, under or over the land

The concepts associated with this right are discussed in relations to code rights (a), (b), (c) at sections 15.3 and 42.9 above.

42.14 Code right (f) To enter the land to inspect, maintain, adjust, alter, repair, upgrade or operate any electronic communications apparatus which is on, under or over the land

42.14.1

Site access has been the source of a good deal of friction between landowners and operators over the years. Many older telecoms agreements will provide for 24-hour, seven day a week unlimited and unregulated access, regardless of whether that is appropriate in the specific circumstances and also without due consideration being given to, for example, specific security concerns, access to locked land and/or buildings, business interruption and crop management/livestock/shooting seasons. In that sense, these agreements are simply not fit for purpose and have therefore inevitably led to access disputes.

42.14.2

Care should therefore be taken when drafting to take account of the nature of the land over which access is to be had. It has not been uncommon to find identically drafted rights of way conferring rights in both an agreement conferring rights over the common parts of a residential block of flats with a locked front door, and in an agreement conferring rights over an open field which has not been made up for vehicular use. Particularly in relation to agreements relating to rural land, it is a not uncommon complaint that gates are left open or vans are used to drive over fields which are suitable only for tractors. Careful consideration therefore needs to be given to the timing of access, the means by which access is exercised having regard to the land, and the duties of the operators and their employees, agents and contractors when exercising such rights.

42.14.3

To compound these issues, site visits are often conducted on the behalf of the operators by numerous different contractors. Not only does this mean that the landowner does not know who it has on site at any given time, but it also means that a landowner may find itself confronted with a large volume of access requests from companies and individuals it has never contracted with or even heard of. This can make it difficult to identify who is responsible if something occurs on such a visit. Consideration should be given to a clause requiring an operator to disclose a list of pre-approved contractors, along with a description of their function and an emergency contact number. Further, routine inspection clauses should make provision for a reasonable prior notice period, identifying who will be attending the site, in what capacity (and as whose contractor) and for what purpose.

42.14.4

It is quite common, in addition to access rights on notice, for there to be an 'emergency access' provision. In some cases experience shows that a rather too flexible approach has been taken to what amounts to 'emergency access', capable of being exercised without notice. It is difficult to draft an agreement in such a way as to stop this altogether. An operator could be required, after the event, to identify what the emergency was (as occurs under some of the New Code special regimes – see Chapter 28).

42.14.5

When disputes of these types arise, it is not uncommon for the landowner to simply block any access to the site altogether, often in breach of the terms of the agreement.

This outcome clearly causes operational issues for the providers and, at the end of the day, is in none of the parties' ultimate best interests. It is therefore important for those advising landowners about access rights to explain the position in relation to those rights.

42.14.6

For these reasons, access provisions in any new agreement should be comprehensively dealt with in as much detail as possible. For example:

(a) It goes without saying that the access route should be clearly shown on the plans to the agreement.
(b) Access should be by prior notification only, save in cases of genuine emergency (see further below).
(c) Access requests should include the details of who (which individuals) will attend.
(d) When will those individuals attend (within a specific window)?
(e) Can, for example, lifts be used within the building? If so, can they be used to transport equipment to the upper floors? Is there a weight restriction?
(f) Can residential common parts be used?
(g) Can any part of the land or the building be used for temporary storage? What conditions should be imposed on storage?
(h) If heavy equipment needs to be installed using external pulleys, cherry pickers or similar apparatus, or if scaffolding is required, should express provision be made for that? If such rights (as might be the case) are included within paragraph 3(d), ought restrictions on their use to be imposed? If external parts of the building (flat roofs or in the case of some older residential leases, the individual external walls of flats) have been demised away, has the operator acquired all relevant rights from the right persons in order to exercise access rights?
(i) By what means, and whether that is to include vehicles, and if so, what type of vehicle (four wheel drive, no vehicle over a particular weight, and so on), with registration numbers? Should there be parking on site, and if so subject to what conditions?
(j) Are there site-specific security measures (locked gates, keypad entry, registration with a porter, livestock) that need to be addressed?
(k) Will accompanied access be required, in which case what are the specific arrangements for that? If this sort of access is likely to cause the landowner to incur fees, then there should be specific provisions for the operator to meet those costs, including a mechanism for payment etc.
(l) Although addressed above, it is particularly important to note if there are any building specific restrictions (which there almost certainly always will be in the case of rooftop sites – out of hours access, for example), then these should also be dealt with.
(m) There ought to be a requirement that the notice of access furnishes the details of what work is to be carried out.

42.14.7

As already indicated above, in relation to emergency access, some consideration should be given to restrictions that control the operator's ability to invoke such a clause. Defining exactly what a qualifying emergency may be can difficult, if not impossible. As a minimum, however, there should be an obligation upon the operator to advise each time access is undertaken on this basis exactly what the emergency is/was and what works will be/were required to resolve the position. There should be a notification obligation unless that is not reasonably practicable. If access to the site on an emergency basis causes the landowner to incur additional costs/liabilities, then the operator should pay for those.

42.15 Code right (g): to connect a power supply

42.15.1

The New Code and the associated OFCOM guidance (see section 41.12 of Chapter 41) unfortunately do not elaborate exactly what is envisaged by the right 'to connect to a power supply'.

42.15.2

The New Code does not, for example, state whose power supply can/will be connected into – does it envisage a stand-alone supply, or the landowner's own supply? Alternatively is the landowner expected to enter into wayleaves with third party power supply companies for a stand-alone supply to be installed to the site? Such wayleaves are common, as operators tend not to want to be dependent upon the landowner's supply, particularly if that is a supply serving a residential block or offices. Such a supply can be vulnerable or of insufficient power or quality to satisfy the operator that it is resilient. Nor is anything said about the ability of the operator, once connected, to share the supply. There would appear to be no reason why the code agreement should not seek to restrict the use to which the supply of power can be put. Any such restriction, e.g. to prevent sharing of the power, would probably not fall foul of either paragraph 17 (conferring upon the operator the right to share apparatus) or paragraph 100 (anti-avoidance).

42.15.3

In the case of paragraph 17, any such arrangements will need to be carefully documented in terms of the installation of a metered supply (at the operator's cost) and detailed provisions for the reading of that meter and, in turn, the payment of periodic payments to the landowner to cover the electricity used.

42.15.4

In terms of a separate power supply to be installed under a wayleave with a third party, the landowner's obligation to enter into any such agreement has to be limited to using reasonable endeavours (or similar) to agree the terms of the wayleave.

Provision for the payment of the landowner's professional costs will also need to be dealt with. Consideration should also be given, where appropriate, to a limit on the number of such wayleave agreements which the landowner is obliged to enter into. Provision should also be made for the removal of the electrical cabling in the event that it ceases to power any electronic communications apparatus. A lift and shift provision and clauses mirroring the termination and removal provisions of the New Code should also be considered so that the wayleave and the agreement under the New Code are, as far as possible, parallel.

42.15.5

Finally, it goes without saying that the route of any proposed electricity supply should be expressly marked on the plans to the agreement. It will be necessary for the parties to ensure that any electricity supply is over land retained by the landowner (or other land over which the necessary consents can be obtained), and that other consents, for example from the relevant highways authority, are obtained.

42.16 Code right (h): To interfere with or obstruct a means of access to or from the land (whether or not any electronic communications apparatus is on, under or over the land)

See paragraph 15.3.7 of Chapter 15.

42.17 Code right (i): To lop or cut back, or require another person to lop or cut back, any tree or other vegetation that interferes or will or may interfere with electronic communications apparatus

42.17.1

The point to make in relation to this code right is that, in some instances, the rights granted by paragraph 3(i) could result in a potentially very significant and detrimental impact upon the landowner's property. The right to lop or cut back trees and/or other vegetation may apply to the entirety of the landowner's property holding: it is important therefore to consider the impact upon the owner of a large forestry operation or the like. The extent to which an operator may lop and cut back is therefore potentially significant. Chapter 29 provides further discussion on this subject.

42.17.2

In order to mitigate this risk, the specification of lines of transmission from the site should be identified and documented from the very outset, such that the potential impact on the landowner (including with specific reference to the test at paragraph 21 as required, but also in terms of what compensation/consideration should be payable under paragraph 24) can be properly assessed.

Consideration ought also to be given to (1) the means if any by which access is to be obtained to undertake any lopping and (2) the restrictions, if any, as to the type of equipment which may be utilised to effect the same.

Part 3 Agreements to use apparatus

42.18 Apparatus-only agreements

42.18.1

The New Code applies only to code rights (elaborated upon as discussed above in relation to paragraph 3 of the New Code). Those rights must be 'in relation to an operator and any land': paragraph 3. If the rights do not relate to land but only to electronic communications apparatus ('ECA'), the agreement bestowing those rights cannot take effect as a code agreement, for the reason that 'land' is defined so as to exclude ECA: paragraph 108(1). ECA is elaborated upon in paragraph 5 of the New Code, which is discussed in detail in Chapter 18.

42.18.2

Paragraph 403 of the Explanatory Notes to the 2017 Act supports this explanation, by providing:

> Paragraph 5 defines electronic communications apparatus i.e. the apparatus which can be installed on, under or over land. This definition is important for two reasons. Firstly, it defines the scope of what can be installed and kept on land under the provisions of the code. Secondly, because paragraph 108(1) specifically provides that land does not include electronic communications apparatus, anything that falls within the definition of 'electronic communications apparatus' (as it is set out in paragraph 5) cannot have code rights imposed against it. This has the practical effect of ensuring that one code operator cannot seek to exercise code rights against the apparatus of another (or indeed against the apparatus of a person who is not a code operator).

42.18.3

Accordingly, *there is no code protection for a mere right to use something which qualifies as ECA*. It matters not whether or not it was installed or utilised by an operator; in either case the agreement is not a code agreement. Where there are no code rights conferred by the grant of a right simply to use ECA, any ECA-only agreement will either be governed by the general law of property, or may be a protected 1954 Act tenancy of the ECA.

42.18.4

Thus, it may be that some of the rights conferred under the relevant agreement (e.g. a right to access the apparatus, a right to house an equipment cabin on bare land), if exercisable over land, are code rights. Other rights, such as the right to affix an antenna on a mast which has been installed by another operator (or indeed anyone else), is not a code right, as it is a right over ECA and not land, so long as it can be said that the mast is e.g. 'apparatus … designed or adapted' or 'a structure or thing designed or adapted' for use in the provision of an electronic communications network within paragraph 5 of the New Code.

42.18.5

This dichotomy between 'land' and 'ECA' may find itself reflected in the drafting of appropriate agreements providing for the use of ECA only, making no reference to the right to use it on, in or under land. It may be possible to draft an agreement conferring the right to use ECA per se with a separate agreement providing 'ancillary land rights' which fall within paragraph 3, and which would thus be code rights. If it is possible as a matter of drafting to adopt the dichotomy of ECA and 'ancillary' code rights in connection with such ECA,[3] it has an obvious practical advantage: the ECA agreement can be terminated free of code protection. The 'ancillary land rights' agreement is code-protected but it matters not, if the landowner has the simple expedient of removing the ECA by terminating the unprotected ECA-only agreement. Once the ECA has been removed, the 'ancillary land rights' agreement may itself lose code protection as the rights are, once the ECA only agreement has been terminated, no longer being used for the statutory purpose are required by paragraph 3 of the New Code.

Part 4 Other issues to consider when granting an electronic communications agreement

42.19 Introduction

42.19.1

The following sections in this part set out a number of other considerations which the drafter will wish to bear in mind in drafting an electronic communications agreement.

42.19.2

These considerations are certainly not intended to be an exhaustive list of anything and everything that should be included within a new electronic communications agreement. Rather, they form a record of some of the main terms (over and above those limited rights granted by paragraph 3 of the New Code) which should be given consideration.

3 There may be a suggestion that the ECA-only agreement is a sham, the true purpose being to use it in on or under land. The most often cited definition of 'sham' is that of Diplock LJ in *Snook v West Riding Investments Ltd* [1967] 2 QB 766 at 802C–E:

> It is I think necessary to consider what (if any) legal concept is involved in the use of this popular and pejorative word. I apprehend that if it has any meaning in law, it means acts done or documents executed by the parties to the sham which are intended by them to give to third parties or to the court the appearance of creating between the parties legal rights and obligations different from the actual legal rights and obligations (if any) which the parties intended to create. But one thing, I think is clear, in legal principle, morality and the authorities … for acts or documents to be a 'sham', with whatever legal consequences follow from this, all the parties thereto must have a common intention that the acts or documents are not to create the legal rights and obligations which they give the appearance of creating.

42.19.3

Whilst the OFCOM guidance (see section 41.10 of Chapter 41) states that legal advice will likely not be required in certain instances (e.g. in the case of small installations), landowners may feel that it would be prudent to seek professional guidance in each and every case, in order to ensure that their position is best reviewed, understood and protected, by means of an agreement which is tailored and suited to the specific circumstances. As this is a specialist area, landowners (particularly those who are allowing an operator on site as a further source of income without prior experience, or with only experience under the Old Code) should consider taking legal advice.

42.19.4

It should be noted that OFCOM has published standard terms for inclusion in electronic communications agreements (see sections 41.11 and 41.12 of Chapter 41 and Part 5 below; the terms themselves are reprinted at Appendix F to this book).

42.20 Consideration

42.20.1

The question of what consideration should be paid in lieu of rights granted under the New Code is considered elsewhere in this text (see Chapter 30).

42.20.2

The key point to bear in mind is that if the agreement grants any rights which are over and above the scope of those limited nine code rights set out at paragraph 3 of the New Code, then it is possible that the consideration provisions of paragraph 24 may not apply as straightforwardly.

42.21 Provisions for the review of the consideration that is payable for the site

42.21.1

Provision should be included within the agreement for the review of the consideration that is payable for the site, to the higher of either open market value, or the Retail Prices Index.

42.21.2

Whether the former will be within the confines of the valuation considerations at paragraph 24 of the New Code may (as discussed above) depend upon whether the agreement is a 'bare' New Code agreement, or a more extensive arrangement agreed between the parties.

42.22 Permitted use

42.22.1

Paragraph 3 of the New Code provides that the rights under the code are to be used for 'the Statutory Purposes', the definition of which is set out at paragraph 4 as follows:

> In this code 'the Statutory Purposes', in relation to an operator, means-
> (a) the purposes of providing the operator's network, or
> (b) the purposes of providing an infrastructure system.

Any permitted use under a 'bare' New Code agreement will therefore need to be expressly limited in these terms.

42.22.2

If, however, the operator requires a wider user clause (or no prohibition at all), then that is outside the remit of the New Code and a 'bare' New Code agreement.

42.22.3

Although it may be unnecessary where a permitted use clause makes clear that the use is for statutory purposes, where there is doubt as to what the primary purpose of an agreement is, it may (though see Chapter 19) assist to spell out the position in order to indicate whether the parties intended the agreement to fall within either Part 5 of the New Code, or the 1954 Act.

42.23 Alienation

42.23.1

A summary of the operator's right to share the apparatus on site and some drafting suggestions are included in section 42.11 above.

42.23.2

As discussed in Chapter 20, however, paragraph 16 of the New Code also makes void any provision in an agreement to prevent its assignment to another code operator.

42.23.3

As with the operator's new rights to upgrade sites (also considered in Chapter 20), an express grant of the assignment rights at paragraph 16 of the New Code should be included in any new agreement, i.e. that the operator can assign *the agreement* to another *code operator*. It is, of course, open to the parties to agree wider rights as a matter of contract.

42.23.4

Conversely, any alienation rights outside the strict scope of paragraph 16 can be expressly reserved to the landowner.

42.23.5

To give some measure of protection to landowners, paragraph 16(2) of the New Code provides that that the landowner may require as a term of any such assignment that the outgoing operator enter into a guarantee agreement – see paragraph 16(7):

> A 'guarantee agreement' is an agreement, in connection with the assignment of an agreement under Part 2 of this code, under which the assignor guarantees to any extent the performance by the assignee of the obligations that become binding on the assignee under sub-paragraph (4) (the 'relevant obligations').

42.23.6

Such a guarantee agreement may (see paragraph 16(9)):

(a) impose on the assignor any liability as sole or principle debtor in respect of the relevant obligations;
(b) Impose on the assignor liabilities as guarantor in respect of the assignee's performance of the relevant obligations which are no more onerous than those to which the assignor would be subject in the event that the assignor being liable as sole or principle debtor in respect of any relevant obligation;
(c) make provision incidental or supplementary to any provision within paragraph (a) or (b).

42.23.7

It plainly follows that a prudent landowner will seek provision for the operator to enter into a guarantee agreement in the event of an assignment.

42.24 Term length and break rights

42.24.1

It is important to separate out what the contractual agreement (imposed or otherwise) provides, as against the extended notice periods provided for by paragraph 31(3) of the New Code.

42.24.2

Whilst the agreement between the parties might well provide for a term of, say, ten years which can be determined by either party upon 12 months' notice at any time, paragraph

31(3) dictates that no less than 18 months' notice must be given (either during the course of the agreement pursuant to a break right or after the term of the agreement has expired).

42.24.3

In practical terms, therefore, it may be considered that any contractual notice period of less than 18 months is potentially misleading. A notice period of 18 months in respect of any right to determine the agreement should therefore be provided for, in the interests of transparency and clarity. Of course, due grounds to terminate under paragraph 31(4) of the New Code will also be required. However, note needs to be made of paragraph 33, dealing with what is essentially the renewal of a code agreement. The renewal right may be triggered by service of a notice by either the operator or site provider. The period of notice under this provisions is only six months: paragraph 33(3). Thus, it may be that a break clause which provides for a period of only six months affords the site provider more flexibility. A notice period of 18 months in circumstances where renewal is desired will, at the very least, be inconvenient.

42.24.4

In the absence of any agreement between the parties the Upper Tribunal has a discretion under paragraph 23(8) to include a break right in the agreement (and, if so, in what circumstances). Whilst there is currently no specific guidance as to what factors the Upper Tribunal would apply in exercising its discretion as to whether to include a break right in the agreement, one might sensibly assume that those types of considerations which would apply in determining this question in any 1954 Act renewal proceedings might equally apply here.[4]

42.25 Forfeiture

42.25.1

A 'bare' code agreement will probably not, as such, and without the appropriate wording, take effect as a tenancy, with the result that it would be inappropriate to include provision for forfeiture as a means of termination. Rather, reference should be made to the termination of a contractual agreement in the event of a breach. However, it may be that the tribunal is persuaded to grant a code right fleshed out as a full tenancy.

42.25.2

Subject to this, there is no reason why a forfeiture clause should not be included within an electronic communications agreement that takes effect as a tenancy, providing for termination by forfeiture in the event of operator breach. The inclusion within an imposed agreement of such a provision is within the scope of the Tribunal's discretion at paragraph 23(8) of the New Code.

4 See Reynolds and Clark, *Renewal of Business Tenancies*, 5th edition, paragraph 8.84 onwards.

42.25.3

Note that the same principles as set out at paragraph 42.24.2 above apply in terms of the overriding effect of paragraph 31(3) and the notice periods set out therein. Although the forfeiture would in effect determine the contractual term of the agreement, 18 months' notice would still be required to be given to end the operator's code rights (and Part 6 would also still need to be followed).

42.26 Decommissioning

42.26.1

Obligations for the operators to reinstate sites following the termination of any code rights in accordance with their obligations at paragraph 44 of the New Code should be expressly provided for in the agreement.

42.26.2

The costs of such removal and reinstatement can be significant, especially in the case of rooftop sites. A prudent landowner may well therefore wish to include provision in any agreement for the payment of a reinstatement bond by the operator to cover these costs, or else the purchase of an insurance policy (at the operator's cost) to cover non-compliance.

42.26.3

In the case of a performance bond, this may well need to include provision for adjustments to take account of inflation, increasing contractor costs etc.

42.26.4

It is unclear as to the extent to which it may be said that express decommissioning provisions are contrary to the terms of Part 6, given in particular the terms of paragraph 44 of the New Code, and thus ineffective having regard to the anti-avoidance terms of paragraph 100.

42.27 Lift and shift provisions

42.27.1

Landowners may from time to time need to carry out works to the land upon which electronic communications apparatus is situated – for example, roofing works in the case of a rooftop site. Paragraph 20 of the Old Code (which will carry forward to some extent under the transitional arrangements) sought to provide a mechanism to facilitate this (over and above any separate contractual arrangements made between the parties).

42.27.2

Paragraph 23(8) of the New Code envisages that there can be rights for landowners to temporarily remove apparatus to facilitate repair works to their property (colloquially known as 'lift and shift' provisions). The OFCOM guidance does not develop this, OFCOM taking the view that such provisions are a matter for negotiation between the parties (see section 41.12 of Chapter 41).

42.27.3

There is much to be said for the inclusion of a detailed lift and shift clause in the agreement, to cover the possibility of repair works being required. Such a provision potentially benefits not merely the landowner but also the operator, since each may need to relocate apparatus to another part of the site while works are carried out.

42.27.4

A lift and shift clause should ideally cover:

(a) the circumstances in which such powers can be used by either party;
(b) whether the apparatus is to be relocated, or simply removed (albeit temporarily) whilst the works are carried out;
(c) what obligations (if any) the landowner is to have in terms of undertaking the necessary works as expediently as possible; and
(d) any longstop deadlines to be complied with.

42.27.5

Note that it is not at all clear whether a lift and shift clause in fact contravenes the anti-avoidance measures at paragraph 100 of the New Code, and in particular the removal provisions set out at Part 6 of the code. Given that paragraph 23(8) of the code itself envisages that a lift and shift mechanism can form part of a New Code agreement, one would expect that this is not the case. This is considered in detail in Chapter 34.

42.28 Indemnity

42.28.1

A prudent landowner proposing to enter into an electronic communications agreement will wish to undertake a thorough examination of the circumstances of every case in order to determine exactly what risks (if any) the installation of electronic communications apparatus may present, and the measures which should best be taken in order to deal with those.

42.28.2

Some of the risks may be difficult to quantify, and it would be prudent to provide for the inclusion in the agreement of an an unlimited indemnity in favour of the landowner in respect of the use of its land. Given that it is the operators imposing the rights and also having regard to paragraph 23(5), then it is difficult to envisage why this should not in fact be the default position.

42.28.3

Conversely, however, the operator will want a cap on any such indemnity for as low an amount as it considers it can get away with.

42.29 ICNIRP

42.29.1

The agreement should provide that the operator will at all times ensure that the site complies with and is operated in accordance with the recommendations from time to time of the International Commission on Non-Ionizing Radiation Protection (INCIRP) (or any successor body) and to provide evidence of such compliance to the landowner upon request.

42.29.2

Provision should be made that, in the event of non-compliance, the apparatus must be switched off whilst compliance is effected within a specified timeframe and in default thereof, for the agreement to be terminated. It is considered that in the absence of agreement between the parties, such rights could again be sought under paragraph 23(8) of the New Code. The termination of any agreement in reliance on such a provision would not of course dispense with the requirement to comply with the terms of Part 6 to effect removal.

42.30 No contracting out

42.30.1

Paragraph 100 of the New Code seeks to provide that there can be no contracting out of the code. The Authors consider in Chapter 34 of this book whether paragraph 100 is wholly effective to achieve this aim.

42.30.2

It was not possible to contract out of the Old Code. To seek to minimise the impact of the Old Code upon landowners, however, various terms have historically been deployed in telecommunications agreements, such as a penalty rent following the expiry of the contractual agreement, contractual undertakings not to serve Old Code notices, or title transfers if apparatus remains on the site following the expiry of the contractual term.

42.30.3

The thinking behind such clauses was to impose separate and distinct contractual obligations upon the operators, as opposed to seeking to contract out of the provisions of the Old Code per se. As with so many of the vagaries of the Old Code, however, the question of whether these types of clauses were in contravention of the Old Code has never been tested and/or categorically determined by the Court. That is likewise the case under the New Code, at least for now.

42.31 Dispute resolution provisions

42.31.1

As to any agreement concerning the use of land, it is not considered sensible to limit the dispute resolution options. Whilst the OFCOM standard terms (see sections 41.11 and 41.12 of Chapter 41, and part 5 below) envisage that the parties must go through a prescribed dispute resolution process before moving on (as the last resort) to proceedings, all possible options should be left open to the parties from the outset, in order that the most appropriate dispute resolution venue can be chosen to fit the specific facts and circumstances of a dispute at any given time. This topic is discussed further in Chapter 33.

42.32 Notice clauses

42.32.1

The notice clause of any new electronic communications agreement should reflect those requirements at paragraph 91 of the New Code, namely:

(a) Notices by post must be sent by registered post service or recorded delivery.
(b) An address for service should be provided in accordance with sub-paragraph 2(a); otherwise s.394 of the 2003 Act will apply in default.
(c) The inclusion of wording which reflects sub-paragraphs (4) and (5), which cover the circumstances in which it has not been possible, following reasonable enquiries, to ascertain the name and/or address of either the occupier or the owner of an interest in the land.

42.32.2

The considerations which apply to the drafting and delivery of notices are examined in Chapter 32 of this book.

42.32.3

OFCOM has published a number of notices for use in prescribed and other circumstances – see part 6 below, and sections 41.7 and 41.8 of Chapter 41 of this book. The notices themselves are reprinted at Appendix E to this book.

42.33 Costs provisions

42.33.1

Paragraph 84(2) of the New Code provides that the Court/Tribunal can, depending upon the circumstances, order the operator to pay reasonable legal and valuation expenses incurred by the landowner.

42.33.2

It remains to be seen whether operators will agree on this basis to agree to pay any such expenses from the outset.

42.34 Site surveys/scoping visits

42.34.1

Having identified the coverage need for a new site, an operator will usually require a physical survey of the land in question in order to determine that location's suitability from an operational network perspective, bearing in mind considerations such as lines of sight.

42.34.2

In circumstances whereby the operator is seeking to impose rights under the New Code upon a landowner, the information gleaned from an initial site survey may well also be required in order that the operator can successfully demonstrate that it meets the test at paragraph 21 of the New Code.

42.34.3

In some cases, it will be possible to carry out this exercise on a desktop basis – using online facilities, maps etc. More often than not, however, the operator will need physically to visit the site in order to confirm its suitability in order to move to the next stage of seeking to agree terms with the landowner (and looking to impose those terms if no agreement can be reached).

42.34.4

It is important to note that the New Code does not include any express rights for such scoping surveys. But for the point made in paragraph 42.34.5 below, it would be arguable that, absent any implied rights, a landowner would be entitled to resist access for such purposes. If that is right, then a landowner who does not wish for electronic communications apparatus to be located on its land (for whatever reason, including if the terms/consideration to be offered in return are unlikely to be acceptable) would be entitled to refuse requests for access for this purpose and also resist any one-off payments or other such incentives which may be offered by way of consideration for the operator undertaking such an inspection.

42.34.5

However, notwithstanding the lack of any mention in the New Code of a right to carry out a survey with a view to installing electronic communications apparatus, such a right is granted by Schedule 4 to the 2003 Act, which provides that a person nominated by a code operator, and duly authorised in writing by the Secretary of State, may at any reasonable time, enter upon and survey any land in England and Wales for the purpose of ascertaining whether the land will be suitable for use by the code operator for, or in connection with, the establishment or running of the operator's network. This right is not, however, unfettered, as Section 36.5 of this Book explains. Accordingly, the omission of an express right to carry out surveys may come to cause a problem for operators in the life of the New Code.

42.34.6

Note in the case of the latter, however, that some form of legally binding agreement (not the usual heads of terms marked subject to contract) would be required in order to prevent a situation arising whereby the operator carries out its site survey, then simply seeks to impose rights under the New Code in any event.

42.35 Early access agreements

42.35.1

Where the need for access to a proposed new site is urgent, an operator may offer (sometimes in lieu of the payment of a one-off lump sum fee) for the parties to enter into an early access agreement. This will be a simple document which provides that the operator can enter the landowner's property to commence the construction and integration of the telecommunications apparatus, pending the negotiation and completion of the substantive agreement.

42.35.2

The fundamental issue with such agreements is that as soon as the operator has built and integrated the site, the real risk is that it loses all interest/incentive in completing the substantive agreement. It is therefore not uncommon to come across instances whereby early access was granted some 10–15 years previously and a substantive agreement has never been completed; sometimes no rent has been paid for the site; and more importantly there is unwelcome uncertainty as to the terms upon which the operator actually occupies the site.

42.35.3

For this reason, careful consideration has to be given to early access arrangements in order to ensure that both parties have the benefit of a satisfactory agreement governing their relationship and in conformity with the requirements of the New Code.

Part 5 The OFCOM documents

42.36 The OFCOM Code of Practice and standard terms

42.36.1

In accordance with its obligations to do so under the New Code, OFCOM has published guidance notes, a Code of Practice and also standard agreement terms which may be used by the parties. It has also published template notices, which are considered separately in section 42.32 above and part 6 below.

42.36.2

Neither the Code of Practice nor the standard terms are binding. In relation to the former, OFCOM rejected many of the proposals put forward in response to its consultation exercise, with the result that the code is a regrettably bland document (see section 41.10). In relation to the latter, OFCOM itself recognises that its standards terms (in relation to which it again rejected most of the consultation responses – see section 41.12) are merely a suggested starting point and that invariably, they will need to be tailored to individual circumstances, with the benefit of professional advice.

42.36.3

It is possible that the Tribunal will look to the Code of Practice and standard terms when considering the rights to be imposed under Part 4, although the brevity with which those documents are drafted means that the Tribunal will not glean much from such an exercise.

Part 6 Preparing and serving notices

42.37 General

42.37.1

Again in accordance with the obligations imposed upon it, OFCOM has published a full suite of notices for use under the New Code (see Chapter 41; the notices themselves are reprinted in Appendix E to this book).

42.37.2

In terms of notice requirements, a distinction is drawn in the New Code between operators on the one hand (who have slightly more onerous obligations with which to comply) and landowners on the other. The rationale for this approach is that, on the whole, operators are larger organisations, have access to greater resources and are better organised than landowners. At the end of the day, it is also the operators who are seeking to impose rights.

42.37.3

Where OFCOM has prescribed a notice, then operators are therefore required to use that precedent – paragraph 88(2). The notice is not valid if this requirement is not complied with, although the recipient landowner may choose nevertheless to treat the notice as valid (paragraph 88(4)).

42.37.4

Landowners on the other hand are only required to utilise the OFCOM precedents where serving notice for initiating:

(a) Termination of a Code agreement;
(b) Modification/renewal of such an agreement;

(c) Identifying what code apparatus is on site; or
(d) Removal of apparatus.

See paragraphs 88(1) and (2) of the New Code.

42.37.5

Any notice which does not comply with the above will be invalid, although again the recipient may opt if they wish to treat the notice as valid in any event.

42.37.6

In any other cases not involving those four situations considered in paragraph 42.37.4 above, the server will be liable for any operator costs incurred as the result of the notice not being in the OFCOM format – paragraph 88(6).

42.37.7

The practical effect of all this is that the appropriate form of notice will invariably be that issued by OFCOM. Those advising landowners or operators who wish notices to be drafted should bear this in mind.

42.37.8

Formalities in terms of the service of these notices is considered in Chapter 32.

42.37.9

Finally, paragraph 39 of the New Code provides a useful tool for the landowner, akin to a Section 40 notice under the 1954 Act. That is, the landowner may by notice require an operator to disclose whether:

(a) the operator owns electronic communications apparatus on, under or over land in which the landowner has an interest or uses such apparatus for the purposes of the operator's network; or
(b) the operator has the benefit of a code right entitling the operator to keep electronic communications apparatus on, under or over land in which the landowner has an interest.

42.37.10

Under paragraph 39(4), the operator must respond with the required information within three months. If it fails to do so and the landowner subsequently takes action to remove the apparatus, then provided that the operator does in fact own electronic communications apparatus or uses it for the purposes of the operator's network, and the operator has the benefit of a code right entitling the operator to keep electronic communications apparatus on, under or over the land, it will be liable for the landowner's costs of that removal, which could be significant.

42.37.11

In circumstances whereby it is particularly important that landowners police any installations on their land, including exactly who is utilising their property for electronic communications use, paragraph 39 should be extremely useful for these purposes.

Part 7 Venue

42.38 General

42.38.1

The venue for the determination of matters arising under the Old Code was the County Court – see paragraph 1 of that code. Often, albeit at the discretion of the Court, cases were heard in the Technology and Construction branch of the County Court, which is more likely to possess the necessary code expertise.

42.38.2

Under the New Code, the venue for determination is the Upper Tribunal (and its equivalents in other parts of the United Kingdom).

42.38.3

Whilst the jurisdiction of the County Court will continue (for a time) while Old Code cases continue (under paragraphs 20 and 21 of the Old Code: see Chapter 31), it is envisaged that this will ultimately be transferred (by the bringing into force of further regulations) to the Upper Tribunal – the rationale being the Tribunal's property background and also the opportunity to build a real bank of genuine code expertise in this forum.

42.38.4

The need for careful drafting of the proceedings that will arise under either code is obviously a matter to be taken into account by operators and landowners alike.

42.38.5

Chapter 33 of this book considers dispute resolution in further detail.

Part 8 Other practical issues

42.39 General

42.39.1

This final part of this chapter considers other practical issues to be borne in mind in drafting under the New Code.

42.40 Registration

42.40.1

New Code, as with Old Code, rights are not registrable under the Land Registration Act 2002. Although the presence of electronic communications apparatus on land or a building should in normal circumstances be apparent as the result of a physical inspection, or indeed be flagged up by the usual pre-contract enquiries, it is not uncommon for the new purchaser of a property to be caught out, particularly in the case of certain rooftop apparatus, microcells etc.

42.40.2

The provisions as to registration are considered in detail in Chapter 38 of this book.

42.41 Mortgages

42.41.1

Consider the position whereby an agreement is imposed upon a landowner, but the property is mortgaged and subject to a provision that no part of the property is to be underlet or that there is to be no parting with possession or occupation, or the sharing occupation[5] without bank consent. This is not a new problem – the same issue applied under the Old Code. Given that there are few reported cases of operators imposing agreements under the old paragraph 5[6], this was, practically speaking, a theoretical problem only.

42.41.2

It may be thought that a court imposed agreement would not give rise to a breach of the covenant to the bank. In the context of landlord and tenant involuntary *assignments* do not constitute a breach of the covenant against assignment.[7] Thus, by analogy, a court imposed agreement may be said to be a form of involuntary alienation.

42.41.3

On the other hand, it may be considered that there is merely a technical breach. Paragraph 10(2) makes it clear that, inter alia, derivative interests are bound by the

5 It has been held that roaming by operators over their networks, which involved the sending, receipt and automatic changeover on their respective frequencies and equipment used through the duration of the call and for internet access, was a breach of a covenant with respect to sharing, notwithstanding the fact that there was no sharing of frequencies; each network operator still operated its own frequencies within the radio spectrum, the customer's handset changing automatically the radio frequencies when switching between the networks: see *Arqiva v EE Ltd* [2011] EWHC 1411 (TCC).

6 See e.g. *Mercury v London and India Dock Investments Ltd* [1995] 69 P & CR 135; *Cabletel Surrey and Hampshire Ltd v Brookwood Cemetery Ltd* [2002] EWCA Civ 720; *St Leger-Davey v First Secretary of State* [2004] EWHC 512 (Admin); *Brophy v Vodafone Ltd* [2017] EWHC B9 (TCC).

7 See Woodfall, *Law of Landlord and Tenant*, vol. 1, paragraph 11.166.

code agreement. However, although a court imposed agreement takes effect as a Part 2 agreement (paragraph 22), a code agreement within Part 2 only 'binds a person with an interest in the land that is created after the right is conferred and is derived (directly or indirectly) out of the' interest of the occupier of the land (paragraph 10(2)(b)).[8] The bank's interest will have predated the court imposed agreement.

42.41.4

Time will tell whether the operators will be more proactive in terms of imposing rights under the New Code. If so, then both existing and future facility documentation may well need to be reviewed and amended to reflect the new statutory regime.

8 An agreement to oust the effect of 10(2)(b) would not appear to be caught by the anti-avoidance provisions in paragraph 100, as that only applies to Parts 3 to 6.

Part VI

The New Code – Annotated

The New Code – Annotated

Note: This part of the book reprints the text of the Electronic Communications Code introduced by the Digital Economy Act 2017 ('the New Code'), adding the Authors' commentary after each paragraph. In that commentary:

- the Communications Act 2003 is abbreviated to 'the 2003 Act';
- cross-references to the relevant part of the text of the book are referred to by paragraph or (capitalised) Section number, followed by 'of the text', to distinguish the text from paragraphs of the New Code;
- the Electronic Communications Code contained in Schedule 2 to the Telecommunications Act 1984, as amended by the 2003 Act (referred to in the New Code as 'the Existing Code') is referred to as 'the Old Code'.

Schedule 3A The Electronic Communications Code

Part 1 Key concepts

1 Introductory

(1) This Part defines some key concepts used in this code.
(2) For definitions of other terms used in this code, see –
    (a) paragraph 94 (meaning of *'the court'*).
    (b) paragraph 105 (meaning of *'occupier'*),
    (c) paragraph 108 (general interpretation),
    (d) section 32 (meaning of electronic communications networks and services), and
    (e) section 405 (general interpretation).

COMMENTARY

Paragraph 1(1) introduces paragraphs 2–7, which define some key concepts used throughout the New Code.

Paragraph 1(2) lists a number of other provisions containing definitions other than those found in Part 1. References to s.32 and s.405 are references to those sections in the 2003 Act, the New Code being Schedule 3A to that Act.

2 The operator

In this code *'operator'* means –

(a) where this code is applied in any person's case by a direction under section 106, that person, and
(b) where this code applies by virtue of section 106(3)(b), the Secretary of State or (as the case may be) the Northern Ireland department in question.

COMMENTARY

Paragraph 2 contains a definition of 'operator' by reference to s.106 of the 2003 Act.

An operator is (a) a person to whom the New Code is applied as a result of a direction given by OFCOM or (b) the Secretary of State or any Northern Ireland department providing or proposing to provide an electronic communications network.

The procedure and criteria for a direction by OFCOM applying the Code in any person's case are set out in s.107 of the 2003 Act, and are described in detail in Section 17.2 of Chapter 17 of this Book.

This paragraph is discussed at paragraphs 17.1.3–17.1.6 of the text.

3 The code rights

For the purposes of this code a *'code right'*, in relation to an operator and any land, is a right for the statutory purposes –

(a) to install electronic communications apparatus on, under or over the land,
(b) to keep installed electronic communications apparatus which is on, under or over the land,
(c) to inspect, maintain, adjust, alter, repair, upgrade or operate electronic communications apparatus which is on, under or over the land,
(d) to carry out any works on the land for or in connection with the installation of electronic communications apparatus on, under or over the land or elsewhere,
(e) to carry out any works on the land for or in connection with the maintenance, adjustment, alteration, repair, upgrading or operation of electronic communications apparatus which is on, under or over the land or elsewhere,
(f) to enter the land to inspect, maintain, adjust, alter, repair, upgrade or operate any electronic communications apparatus which is on, under or over the land or elsewhere,
(g) to connect to a power supply,
(h) to interfere with or obstruct a means of access to or from the land (whether or not any electronic communications apparatus is on, under or over the land), or
(i) to lop or cut back, or require another person to lop or cut back, any tree or other vegetation that interferes or will or may interfere with electronic communications apparatus.

COMMENTARY

Paragraph 3 contains a definition of a 'code right' in the form of a list of discrete rights from (a) to (i).

The expression 'electronic communications apparatus' is defined in paragraph 5.

The word 'land' is defined in paragraph 108.

The word 'operator' is defined in paragraph 2.

The expression 'the statutory purposes' is defined in paragraph 4.

This paragraph is discussed at Sections 15.2 to 15.6 and 18.1 of the text.

For further discussion of individual code rights, see also the following paragraphs of the text: (a) 15.3.3, 30.5.3; (b) 15.3.3, 24.5.5, 30.5.3; (c) 15.3.4, 15.6.2, 20.3.3; (d) 15.3.5; (e) 15.3.4, 20.3.3; (f) 30.5.3, 34.5.14; (g) 15.3.6; (h) 15.3.7, 15.3.8, 24.4.7, 24.4.10, 28.3.3, 30.5.3, 31.3.8; (i) 15.3.9.

4 The statutory purposes

In this code *'the statutory purposes'*, in relation to an operator, means –

(a) the purposes of providing the operator's network, or
(b) the purposes of providing an infrastructure system.

COMMENTARY

Paragraph 4 contains a definition of 'the statutory purposes' in relation to an operator, namely, the purposes of providing the operator's network or an infrastructure system.

The expression 'infrastructure system' is defined in paragraph 7.

The word 'network' is defined in paragraph 6.

The word 'operator' is defined in paragraph 2.

The word 'provide' is defined by s.405 of the 2003 Act, by reference to s.32(4) of that Act, which states that 'references to the provision of an electronic communications network include references to its establishment, maintenance or operation'.

This paragraph is discussed in Sections 15.4, and paragraphs 18.2.3 and 18.2.4 of the text.

5 Electronic communications apparatus, lines and structures

(1) In this code *'electronic communications apparatus'* means –
   (a) apparatus designed or adapted for use in connection with the provision of an electronic communications network,
   (b) apparatus designed or adapted for a use which consists of or includes the sending or receiving of communications or other signals that are transmitted by means of an electronic communications network,
   (c) lines, and
   (d) other structures or things designed or adapted for use in connection with the provision of an electronic communications network.

(2) References to the installation of electronic communications apparatus are to be construed accordingly.

(3) In this code –

'line' means any wire, cable, tube, pipe or similar thing (including its casing or coating) which is designed or adapted for use in connection with the provision of any electronic communications network or electronic communications service;

'structure' includes a building only if the sole purpose of that building is to enclose other electronic communications apparatus.

COMMENTARY

Paragraph 5 contains broad definitions of 'electronic communications apparatus', 'line' and 'structure'. The definitions in this paragraph are not dependent upon the ownership of the apparatus, nor upon the actual use to which the apparatus is put.

The word 'lines' is considered at paragraphs 28.2.5–28.2.6 of the text.

Despite being included as a general definition, the word 'structure' does not appear in the New Code, save in paragraph 5(1)(d). The word 'structure' is considered at paragraphs 18.2.27–18.2.31 of the text.

A structure may become electronic communications apparatus if it is adapted for use as such. The expression 'designed or adapted for use' is discussed at paragraphs 18.2.13–18.2.24 of the text.

The word 'apparatus' is defined in s.405 of the 2003 Act. The word 'apparatus' is considered at paragraphs 18.2.7–18.2.12 of the text.

The expression 'electronic communications network' is defined in s.32 of the 2003 Act.

The expression 'electronic communications service' is defined in s.32 of the 2003 Act.

The word 'provide' is defined by s.405 of the 2003 Act.

The word 'signal' is defined in s.405 of the 2003 Act.

This paragraph is considered in Section 18.2 of the text.

6 The operator's network

In this code *'network'* in relation to an operator means –

(a) if the operator falls within paragraph 2(a), so much of any electronic communications network or infrastructure system provided by the operator as is not excluded from the application of the code under section 106(5), and

(b) if the operator falls within paragraph 2(b), the electronic communications network which the Secretary of State or the Northern Ireland department is providing or proposing to provide.

COMMENTARY

Paragraph 6 defines an operator's 'network'.

Section 106 of the 2003 Act, to which the definition refers, is discussed in Section 17.2 of the text.

The expression 'electronic communications network' is defined in s.32 of the 2003 Act.

The expression 'infrastructure system' is defined in paragraph 7.

The word 'operator' is defined in paragraph 2.

The word 'provide' is defined by s.405 of the 2003 Act.

This paragraph is considered at paragraphs 15.4.3, 18.2.3 and 18.2.4 of the text.

7 Infrastructure system

(1) In this code *'infrastructure system'* means a system of infrastructure provided so as to be available for use by providers of electronic communications networks for the purposes of the provision by them of their networks.
(2) References in this code to provision of an infrastructure system include references to establishing or maintaining such a system.

COMMENTARY

Paragraph 7 defines the expression 'infrastructure system'.

The expression 'electronic communications network' is defined in s.32 of the 2003 Act.

The word 'network' is defined in paragraph 6.

The word 'provide' is defined by s.405 of the 2003 Act.

This paragraph is considered in Sections 15.4 and paragraphs 17.2.3–17.2.4 of the text.

Part 2 Conferral of code rights and their exercise

8 Introductory

This Part of this code makes provision about –

(a) the conferral of code rights,
(b) the persons who are bound by code rights, and
(c) the exercise of code rights.

COMMENTARY

Paragraph 8 introduces paragraphs 9 to 14. It is self-explanatory.

The expression 'code right' is defined in paragraph 3.

9 Who may confer code rights?

A code right in respect of land may only be conferred on an operator by an agreement between the occupier of the land and the operator.

COMMENTARY

Paragraph 9 concerns the conferral of code rights in respect of land on operators. A code right may only be conferred by an agreement between the occupier of the land and the operator. It is not necessary that the occupier should have a proprietary interest in order to confer code rights.

The expression 'code right' is defined in paragraph 3.

The expression 'A code right' suggests that the code rights in paragraph 3 are to be regarded as singular, rather than a composite body of rights, as to which see further Section 15.5 of the text.

The word 'land' is defined in paragraph 108.

The word 'occupier' is defined in paragraph 105.

The definition of 'occupier' is considered at Sections 16.2, 16.3 and 16.4 of the text.
The word 'operator' is defined in paragraph 2.
The procedure for designation as an 'operator' is considered at Section 17.2 of the text.

This paragraph is considered in Chapters 16 and 17 and paragraphs 19.1.4 and 34.1.6 of the text.

10 Who else is bound by code rights?

(1) This paragraph applies if, in accordance with this Part, a code right is conferred on an operator in respect of land by a person ('O') who is the occupier of the land when the code right is conferred.

(2) If O has an interest in the land when the code right is conferred, the code right also binds –
  (a) the successors in title to that interest,
  (b) a person with an interest in the land that is created after the right is conferred and is derived (directly or indirectly) out of –
    (i) O's interest, or
    (ii) the interest of a successor in title to O's interest, and
  (c) any other person at any time in occupation of the land whose right to occupation was granted by –
    (i) O, at a time when O was bound by the code right, or
    (ii) a person within paragraph (a) or (b).

(3) A successor in title who is bound by a code right by virtue of sub-paragraph (2)(a) is to be treated as a party to the agreement by which O conferred the right.

(4) The code right also binds any other person with an interest in the land who has agreed to be bound by it.

(5) If such a person ('P') agrees to be bound by the code right, the code right also binds –
  (a) the successors in title to P's interest,
  (b) a person with an interest in the land that is created after P agrees to be bound and is derived (directly or indirectly) out of –
    (i) P's interest, or
    (ii) the interest of a successor in title to P's interest, and
  (c) any other person at any time in occupation of the land whose right to occupation was granted by –
    (i) P, at a time when P was bound by the code right, or
    (ii) a person within paragraph (a) or (b).

(6) A successor in title who is bound by a code right by virtue of sub-paragraph (5)(a) is to be treated as a party to the agreement by which P agreed to be bound by the right.

COMMENTARY

Paragraph 10 identifies the persons bound by a code right once it has been conferred in accordance with paragraph 9.

If the occupier has an interest in the land when the code right is conferred, it will also automatically bind the successors in title to that interest, holders of derivative interests and occupiers as defined in paragraph 10(2)(c).

Parties with an interest in the land who would not otherwise be bound by a code right can agree to be bound by it, in which case the right will also automatically bind their successors in title to that interest, holders of derivative interests and occupiers as defined in paragraph 10(5)(c).

If successors in title are bound by a code right as a result of sub-paragraphs (3)(a) or (5)(a), they will be treated as a party to the agreement subjecting their predecessor in title to the right.

The expression 'code right' is defined in paragraph 3.

The word 'land' is defined in paragraph 108.

The word 'occupier' is defined in paragraph 105.

The word 'operator' is defined in paragraph 2.

This paragraph is considered in Sections 16.5–16.7 and paragraph 22.4.4 of the text. (See also paragraphs 31.2.10 and 31.2.11 of the text as to the interrelationship between paragraph 10 and the transitional provisions.)

11 Requirements for agreements

(1) An agreement under this Part –
  (a) must be in writing,
  (b) must be signed by or on behalf of the parties to it,
  (c) must state for how long the code right is exercisable, and
  (d) must state the period of notice (if any) required to terminate the agreement.

(2) Sub-paragraph (1)(a) and (b) also applies to the variation of an agreement under this Part.

(3) The agreement as varied must still comply with sub-paragraph (1)(c) and (d).

COMMENTARY

Paragraph 11 sets out the formal requirements for an agreement under Part 2 (i.e. an agreement between an occupier of land and an operator conferring code rights, or an agreement by a person with an interest in land who agrees to be bound by such code rights). Unlike the Old Code, which attached statutory protection to existing property agreements (which may have been subject to their own formality requirements), the New Code applies to any agreement satisfying these formal criteria. It does not appear that such an agreement need be for a fixed term.

The expression 'code right' is defined in paragraph 3.

This paragraph is considered in paragraphs 14.8.2, 19.1.5, 34.5.12, and sections 19.2–19.3 of the text. See also paragraph 31.2.6(2) of the text as to the interrelationship between paragraph 11 and the transitional provisions.

12 Exercise of code rights

(1) A code right is exercisable only in accordance with the terms subject to which it is conferred.

(2) Anything done by an operator in the exercise of a code right conferred under this Part in relation to any land is to be treated as done in the exercise of a statutory power.

(3) Sub-paragraph (2) does not apply against a person who –

(a) is the owner of the freehold estate in the land or the lessee of the land, and

(b) is not for the time being bound by the code right.

(4) Sub-paragraph (2) does not apply against a person who has the benefit of a covenant or agreement entered into as respects the land, if –

(a) the covenant or agreement was entered into under a enactment, and

(b) by virtue of the enactment, it binds or will bind persons who derive title or otherwise claim –

(i) under the covenantor, or

(ii) under a party to the agreement.

(5) In the application of sub-paragraph (3) to Scotland the reference to a person who is the owner of the freehold estate in the land is to be read as a reference to a person who is the owner of the land.

COMMENTARY

Paragraph 12(1) provides that a code right is exercisable only in accordance with the terms subject to which it is conferred (be they terms of a consensual Part 2 compliant agreement or pursuant to an agreement entered pursuant to an order of the court under Part 4). There is accordingly scope for the parties to agree (or for the court to impose) bespoke limitations on the exercise of code rights.

Paragraph 12(2) treats the operator exercising code rights as doing so in the exercise of a statutory power, save against a person with a freehold or leasehold interest in the land who is not bound by the code right (sub-paragraph (3)) or a person with the benefit of a covenant or agreement as respects the land which was entered into under an enactment which provides that it binds persons who derive title or otherwise claim under the covenantor or under a party to the agreement (sub-paragraph (4)).

The expression 'code right' is defined in paragraph 3.
The word 'enactment' is defined in paragraph 108.
The word 'land' is defined in paragraph 108.
The word 'lessee' is defined in paragraph 108.
The word 'operator' is defined in paragraph 2.

This paragraph is discussed in paragraphs 34.5.10 and 34.5.12 of the text. See also paragraph 31.3.5 of the text as to the interrelationship between paragraph 12 and the transitional provisions. See further paragraph 40.2.2 of the text as to the position in Scotland.

13 Access to land

(1) This paragraph applies to an operator by whom any of the following rights is exercisable in relation to land –

(a) a code right within paragraph (a) to (g) or (i) of paragraph 3;
(b) a right under Part 8 (street works rights);
(c) a right under Part 9 (tidal water rights);
(d) a right under paragraph 74 (power to fly lines).

(2) The operator may not exercise the right so as to interfere with or obstruct any means of access to or from any other land unless, in accordance with this code, the occupier of the other land has conferred or is otherwise bound by a code right within paragraph (h) of paragraph 3.

(3) A reference in this code to a means of access to or from land includes a means of access to or from land that is provided for use in emergencies.

(4) This paragraph does not require a person to whom sub-paragraph (5) applies to agree to the exercise of any code right on land other than the land mentioned in that sub-paragraph.

(5) This sub-paragraph applies to a person who is the occupier of, or owns an interest in, land which is –
(a) a street in England and Wales or Northern Ireland,
(b) a road in Scotland, or
(c) tidal water or lands within the meaning of Part 9 of this code.

COMMENTARY

Paragraph 13(1) lists particular rights which may not be exercised so as to interfere with or obstruct any means of access (including emergency access – sub-paragraph (3)) to or from any other land unless the occupier of the other land has conferred or is otherwise bound by a code right within paragraph 3(h) (being a right 'to interfere with or obstruct a means of access to or from the land (whether or not any electronic communications apparatus is on, under or over the land)') (sub-paragraph (2)). This paragraph does not extend code rights over streets in England and Wales, roads in Scotland or tidal water or lands so as to permit the exercise of any code right on other land (sub-paragraphs (4) and (5)).

Operators will need to be careful to agree (or seek under Part 4) code rights within paragraph 3(h) where the exercise of other code rights to be conferred could result in the interference or obstruction of a means of access to or from any other land. If they do not do this, they could find themselves unable to exercise those rights.

The expression 'code right' is defined in paragraph 3.

The drafting of paragraph 13 is consistent only with code rights being several and capable of being conferred individually, as to which see paragraph 15.5.2 of the text.

The word 'land' is defined in paragraph 108.

The word 'occupier' is defined in paragraph 105.

The word 'operator' is defined in paragraph 2.

The word 'road' is defined in paragraph 108.

The word 'street' is defined in paragraph 108.

The expression 'tidal water or lands' is defined in paragraph 61.

This paragraph is discussed in paragraph 28.3.3 of the text in relation to the right under paragraph 74 to fly lines. In respect of Scotland, see also paragraph 40.2.3 of the text.

14 Code rights and land registration

Where an enactment requires interests, charges or other obligations affecting land to be registered, the provisions of this code about who is bound by a code right have effect whether or not that right is registered.

COMMENTARY

Paragraph 14 provides that the provisions of the New Code about who is bound by a code right have effect whether or not the code right is registered. The effect of this is to treat code rights analogously to overriding interests (although strictly code rights are not a new category of overriding interest, since a code right need not be an interest in land).

This may make it difficult for a landowner or neighbour to ascertain the ownership of electronic communications apparatus or of the code rights. Paragraph 39 contains a notice procedure for landowners or occupiers to find out whether apparatus is on land pursuant to a code right.

Paragraph 14 only permits 'code rights' to bind without registration. Other rights contained in agreements conferring code rights will not bind all relevant third parties (although successors in title will be bound pursuant to sub-paragraphs 10(3) and (6)) and paragraph 14 will not save these non-code rights from any relevant failure to register.

Although there may be no need for code rights to be registered, code rights can still be registered (assuming the right in question is a registrable right or interest within the meaning of the relevant registration legislation). This may minimise disputes with third parties. However, since it is unclear whether code rights are themselves to be treated as interests in land, operators wishing to do so should attempt to register both a notice and a restriction (as to which see paragraph 38.5.11 of the text).

The expression 'code right' is defined in paragraph 3.

The word 'enactment' is defined in paragraph 108.

The word 'land' is defined in paragraph 108.

This paragraph is discussed in Sections 38.5 and 38.7 of the text and is also considered in paragraph 24.10.5 of the text. See also paragraph 31.2.6(4)(b) and Section 38.6 of the text as to the interrelationship between paragraph 14 and the transitional provisions.

Part 3 Assignment of code rights, and upgrading and sharing of apparatus

15 Introductory

This Part of this code makes provision for –

(a) operators to assign agreements under Part 2,
(b) operators to upgrade electronic communications apparatus to which such an agreement relates, and
(c) operators to share the use of any such electronic communications apparatus.

COMMENTARY

Paragraph 15 introduces paragraphs 16 to 18. It is self-explanatory.

The expression 'electronic communications apparatus' is defined in paragraph 5.

The word 'operator' is defined in paragraph 2.

As to the interrelationship between Part 3 and the transitional provisions, see Section 31.5 of the text.

16 Assignment of code rights

(1) Any agreement under Part 2 of this code is void to the extent that –
  (a) it prevents or limits assignment of the agreement to another operator, or
  (b) it makes assignment of the agreement to another operator subject to conditions (including a condition requiring the payment of money).

(2) Sub-paragraph (1) does not apply to a term that requires the assignor to enter into a guarantee agreement (see sub-paragraph (7)).

(3) In this paragraph references to *'the assignor'* or *'the assignee'* are to the operator by whom or to whom an agreement under Part 2 of this code is assigned or proposed to be assigned.

(4) From the time when the assignment of an agreement under Part 2 of this code takes effect, the assignee is bound by the terms of the agreement.

(5) The assignor is not liable for any breach of a term of the agreement that occurs after the assignment if (and only if), before the breach took place, the assignor or the assignee gave a notice in writing to the other party to the agreement which –
  (a) identified the assignee, and
  (b) provided an address for service (for the purposes of paragraph 91(2)(a)) for the assignee.

(6) Sub-paragraph (5) is subject to the terms of any guarantee agreement.

(7) A *'guarantee agreement'* is an agreement, in connection with the assignment of an agreement under Part 2 of this code, under which the assignor guarantees to any extent the performance by the assignee of the obligations that become binding on the assignee under sub-paragraph (4) (the 'relevant obligations').

(8) An agreement is not a guarantee agreement to the extent that it purports –
  (a) to impose on the assignor a requirement to guarantee in any way the performance of the relevant obligations by a person other than the assignee, or
  (b) to impose on the assignor any liability, restriction or other requirement of any kind in relation to a time after the relevant obligations cease to be binding on the assignee.

(9) Subject to sub-paragraph (8), a guarantee agreement may –
  (a) impose on the assignor any liability as sole or principal debtor in respect of the relevant obligations;
  (b) impose on the assignor liabilities as guarantor in respect of the assignee's performance of the relevant obligations which are no more onerous than those to which the assignor would be subject in the event of the

assignor being liable as sole or principal debtor in respect of any relevant obligation;
(c) make provision incidental or supplementary to any provision within paragraph (a) or (b).

(10) In the application of this paragraph to Scotland references to assignment of an agreement are to be read as references to assignation of an agreement.

(11) Nothing in the Landlord and Tenant Amendment (Ireland) Act 1860 applies in relation to an agreement under Part 2 of this code so as to –
(a) prevent or limit assignment of the agreement to another operator, or
(b) relieve the assignor from liability for any breach of a term of the agreement that occurs after the assignment.

COMMENTARY

Paragraph 16 deals with the assignment of code agreements (not code rights, which cannot be assigned individually, notwithstanding the heading).

Sub-paragraph (1) prohibits any clause which prevents, limits or imposes conditions upon the assignment of an agreement under Part 2 'to the extent' that it has that effect. Insofar as possible, the offensive part of a clause will be excised, leaving intact the remainder of the clause so far as possible. A clause to the effect that assignment may not be effected without consent, common in leasehold tenure, will be rendered void.

The prohibition on restricting assignment only applies where the proposed assignment is to another operator (sub-paragraph (1)), and does not apply to a term that requires the assignor to enter into a guarantee agreement (sub-paragraph (2)).

Paragraph 16 does not expressly prohibit clauses restricting subletting or sub-licensing.

Sub-paragraph (4) provides for the assignee to become bound by the terms of a code agreement on assignment. This is novel as a matter of English law, in which the burden of a contract cannot usually be assigned.

Subject to the terms of any guarantee agreement (sub-paragraph (6)), an assignor of a code agreement is not liable for breaches of its terms after the assignment if before the breach the assignor or the assignee gave a notice in writing in accordance with the requirements of sub-paragraph (5).

If no such notice is given, the assignee will become liable under sub-paragraph (4) from the time the assignment takes effect, but the assignor will remain liable to the occupier under the terms of the code agreement until the appropriate notice is given (as may his guarantor depending upon the terms of any guarantee).

Sub-paragraphs (7) to (9) deal with guarantee agreements, which are not prohibited by paragraph 16 (sub-paragraph (2)).

Sub-paragraphs (10) and (11) deal with the position in Scotland and Northern Ireland.

The word 'operator' is defined in paragraph 2.

This paragraph is discussed in Section 20.2 of the text. The position in Scotland is considered at paragraph 40.2.4 of the text. See also Section 31.5 of the text as to the interrelationship between Part 3 and the transitional provisions.

17 Power for operator to upgrade or share apparatus

(1) An operator ('the main operator') who has entered into an agreement under Part 2 of this code may, if the conditions in sub-paragraphs (2) and (3) are met –
  (a) upgrade the electronic communications apparatus to which the agreement relates, or
  (b) share the use of such electronic communications apparatus with another operator.
(2) The first condition is that any changes as a result of the upgrading or sharing to the electronic communications apparatus to which the agreement relates have no adverse impact, or no more than a minimal adverse impact, on its appearance.
(3) The second condition is that the upgrading or sharing imposes no additional burden on the other party to the agreement.
(4) For the purposes of sub-paragraph (3) an additional burden includes anything that –
  (a) has an additional adverse effect on the other party's enjoyment of the land, or
  (b) causes additional loss, damage or expense to that party.
(5) Any agreement under Part 2 of this code is void to the extent that –
  (a) it prevents or limits the upgrading or sharing, in a case where the conditions in sub-paragraphs (2) and (3) are met, of the electronic communications apparatus to which the agreement relates, or
  (b) it makes upgrading or sharing of such apparatus subject to conditions to be met by the operator (including a condition requiring the payment of money).
(6) References in this paragraph to sharing electronic communications apparatus include carrying out works to the apparatus to enable such sharing to take place.

COMMENTARY

Paragraph 17(1) permits operators to upgrade or share the use of electronic communications apparatus if the two conditions in sub-paragraphs (2) and (3) are met.

There is no definition of the term 'upgrading'. It is an open question whether or not 'upgrading' includes the replacement of the whole apparatus, as to which see paragraphs 20.3.3 to 20.3.4 of the text.

This paragraph is irrelevant as regards upgrading if a code right to upgrade under paragraph 3(c) has been granted, as to which see further paragraph 20.3.5 of the text.

The first condition in sub-paragraph (2) requires that there be no adverse impact (subject to a de minimis exception) on the appearance of the apparatus.

The second condition in sub-paragraph (3) requires that there be no additional burden on another party to the code agreement. By contrast with the first condition, there is no express de minimis exception, as to which see further paragraph 20.3.6 of the text.

Sub-paragraph (5) is an anti-avoidance provision which makes void agreements to the extent that they prevent, limit or subject to conditions the rights to upgrade or share use of electronic communications apparatus.

There is nothing within paragraph 17 to require an operator who is a party to a code agreement to notify the landowner of the identity of any sharer or of the apparatus installed to facilitate sharing.

The expression 'electronic communications apparatus' is defined in paragraph 5.

The word 'land' is defined in paragraph 108.

The word 'operator' is defined in paragraph 2.

This paragraph is discussed in Sections 20.3, 20.4 and 20.5, and paragraph 24.5.5(1) of the text. See also Section 31.5 of the text as to the interrelationship between Part 3 and the transitional provisions.

18 Effect of agreements enabling sharing between operators and others

(1) This paragraph applies where –
   (a) this code has been applied by a direction under section 106 in a person's case,
   (b) this code expressly or impliedly imposes a limitation on the use to which electronic communications apparatus installed by that person may be put or on the purposes for which it may be used, and
   (c) that person is a party to a relevant agreement or becomes a party to an agreement which (after the person has become a party to it) is a relevant agreement.

(2) The limitation does not preclude –
   (a) the doing of anything in relation to that apparatus, or
   (b) its use for particular purposes,

   to the extent that the doing of that thing, or the use of the apparatus for those purposes, is in pursuance of the relevant agreement.

(3) This paragraph is not to be construed, in relation to a person who is entitled or authorised by or under a relevant agreement to share the use of apparatus installed by another party to the agreement, as affecting any consent requirement imposed (whether by an agreement, an enactment or otherwise) on that person.

(4) In this paragraph –

   'consent requirement', in relation to a person, means a requirement for the person to obtain consent or permission to or in connection with –
   (a) the installation by the person of apparatus, or
   (b) the doing by the person of any other thing in relation to apparatus the use of which the person is entitled or authorised to share;

   'relevant agreement' means an agreement in relation to electronic communications apparatus which –
   (a) relates to the sharing by different parties to the agreement of the use of that apparatus, and
   (b) is an agreement that satisfies the requirements of sub-paragraph (5).

(5) An agreement satisfies the requirements of this sub-paragraph if –

(a) every party to the agreement is a person in whose case this code applies by virtue of a direction under section 106, or
(b) one or more of the parties to the agreement is a person in whose case this code so applies and every other party to the agreement is a qualifying person.

(6) A person is a qualifying person for the purposes of sub-paragraph (5) if the person is either –
(a) a person who provides an electronic communications network without being a person in whose case this code applies, or
(b) a designated provider of an electronic communications service consisting in the distribution of a programme service by means of an electronic communications network.

(7) In sub-paragraph (6) –
'*designated*' means designated by regulations made by the Secretary of State;
'*programme service*' has the same meaning as in the Broadcasting Act 1990.

COMMENTARY

Paragraph 18 deals with the effect of sharing agreements between operators and others.

The drafting of this paragraph is elaborate and obscure. It is considered in detail in Section 20.5 of the text, to which the reader should refer.

The expression 'electronic communications apparatus' is defined in paragraph 5.

The expression 'electronic communications network' is defined in s.32 of the 2003 Act.

The expression 'electronic communications service' is defined in s.32 of the 2003 Act.

The word 'enactment' is defined in paragraph 108.

The word 'provide' is defined by s.405 of the 2003 Act.

This paragraph is discussed in paragraph 20.4.11 and Section 20.5 of the text. See also Section 31.5 of the text as to the interrelationship between Part 3 and the transitional provisions.

Part 4 Power of court to impose agreement

19 Introductory

This Part of this code makes provision about –
(a) the circumstances in which the court can impose an agreement on a person by which the person confers or is otherwise bound by a code right,
(b) the test to be applied by the court in deciding whether to impose such an agreement,
(c) the effect of such an agreement and its terms,
(d) the imposition of an agreement on a person on an interim or temporary basis.

COMMENTARY

Paragraph 19 introduces paragraphs 20 to 27. It is self-explanatory.

The expression 'code right' is defined in paragraph 3.

The expression 'the court' is defined in paragraph 94. This is the county court in England, a county court in Northern Ireland, or the sheriff court in Scotland.

The Electronic Communications Code (Jurisdiction) Regulations 2017 came into force on 28 December 2017. The functions conferred on the court are to be conferred on and exercisable by certain of the Tribunals as well as by the court. Regulation 4 provides that applications under Part 4 of the Code may be commenced only in the Upper Tribunal, in relation to England and Wales, or the Lands Tribunal for Scotland, in relation to Scotland.

This paragraph is discussed generally in Chapter 33 of the text.

20 When can the court impose an agreement?

(1) This paragraph applies where the operator requires a person (a 'relevant person') to agree –
   (a) to confer a code right on the operator, or
   (b) to be otherwise bound by a code right which is exercisable by the operator.

(2) The operator may give the relevant person a notice in writing –
   (a) setting out the code right, and all of the other terms of the agreement that the operator seeks, and
   (b) stating that the operator seeks the person's agreement to those terms.

(3) The operator may apply to the court for an order under this paragraph if –
   (a) the relevant person does not, before the end of 28 days beginning with the day on which the notice is given, agree to confer or be otherwise bound by the code right, or
   (b) at any time after the notice is given, the relevant person gives notice in writing to the operator that the person does not agree to confer or be otherwise bound by the code right.

(4) An order under this paragraph is one which imposes on the operator and the relevant person an agreement between them which –
   (a) confers the code right on the operator, or
   (b) provides for the code right to bind the relevant person.

COMMENTARY

Paragraph 20 contains a mechanism by which operators who cannot obtain agreement to confer or be bound by code rights can obtain an order from the court compelling such agreement.

Paragraph 20 provides for the service of a notice in writing by an operator on a person requiring that person to confer or by bound by a code right. The notice must set out the code right and all the terms of the agreement sought by the operator and state that the operator seeks the person's agreement to those terms (sub-paragraph (2)). In default of such agreement, after 28 days from the date the notice is given, or following receipt of a notice under paragraph 20(3)(b), an operator may apply to the court for an order imposing an agreement between the operator and the relevant person.

In addition to satisfying the formal requirements of paragraph 20, any notices must also comply with the requirements of Part 15. For more detail about the service of notices generally under the New Code see Sections 32.5 to 32.9 of the text.

The expression 'code right' is defined in paragraph 3.

The word 'operator' is defined in paragraph 2.

The expression 'the court' is defined in paragraph 94. This is the county court in England, a county court in Northern Ireland, or the sheriff court in Scotland.

The Electronic Communications Code (Jurisdiction) Regulations 2017 came into force on 28 December 2017. The functions conferred on the court are to be conferred on and exercisable by certain of the Tribunals as well as by the court. Regulation 4 provides that applications under Part 4 of the Code may be commenced only in the Upper Tribunal, in relation to England and Wales, or the Lands Tribunal for Scotland, in relation to Scotland.

This paragraph is considered in Section 21.2 of the text. In respect of human rights compliance, see further Section 21.4 of the text. For cases in which a notice under the Old Code is treated as if given under paragraph 20(2) of the New Code, see Section 31.9 of the text.

21 What is the test to be applied by the court?

(1) Subject to sub-paragraph (5), the court may make an order under paragraph 20 if (and only if) the court thinks that both of the following conditions are met.
(2) The first condition is that the prejudice caused to the relevant person by the order is capable of being adequately compensated by money.
(3) The second condition is that the public benefit likely to result from the making of the order outweighs the prejudice to the relevant person.
(4) In deciding whether the second condition is met, the court must have regard to the public interest in access to a choice of high quality electronic communications services.
(5) The court may not make an order under paragraph 20 if it thinks that the relevant person intends to redevelop all or part of the land to which the code right would relate, or any neighbouring land, and could not reasonably do so if the order were made.

COMMENTARY

Paragraph 21 sets out the test to be applied by the court when dealing with an application under paragraph 20. The court may make an order if (and only if) both of the conditions in sub-paragraphs (2) and (3) are met, and may not make such an order if it thinks that the relevant person intends to redevelop all or part of the land to which the code right would relate, or any neighbouring land, and could not reasonably do so if the order were made (sub-paragraph (5)).

A number of difficulties in interpreting and applying the test are considered at paragraphs 21.3.3 to 21.3.6 of the text.

The word 'access' is to be interpreted in accordance with s.405 of the 2003 Act.

The expression 'code right' is defined in paragraph 3.

The expression 'electronic communications service' is defined in s.32 of the 2003 Act.

The word 'land' is defined in paragraph 108.

The expression 'relevant person' is defined in paragraph 20.

The expression 'the court' is defined in paragraph 94. This is the county court in England, a county court in Northern Ireland, or the sheriff court in Scotland.

The Electronic Communications Code (Jurisdiction) Regulations 2017 came into force on 28 December 2017. The functions conferred on the court are to be conferred on and exercisable by certain of the Tribunals as well as by the court. Regulation 4 provides that applications under Part 4 of the Code may be commenced only in the Upper Tribunal, in relation to England and Wales, or the Lands Tribunal for Scotland, in relation to Scotland.

This paragraph is considered in Sections 21.3 and 21.4 and paragraph 25.7.6 of the text.

22 What is the effect of an agreement imposed under paragraph 20?

An agreement imposed by an order under paragraph 20 takes effect for all purposes of this code as an agreement under Part 2 of this code between the operator and the relevant person.

COMMENTARY

Paragraph 22 provides that court-ordered agreements have the same effect for all purposes of the New Code as an agreement under Part 2 between the operator and the relevant person.

The word 'operator' is defined in paragraph 2.

The expression 'relevant person' is defined in paragraph 20.

This paragraph is considered in Section 21.5 and paragraph 34.5.13 of the text.

23 What are the terms of an agreement imposed under paragraph 20?

(1) An order under paragraph 20 may impose an agreement which gives effect to the code right sought by the operator with such modifications as the court thinks appropriate.

(2) An order under paragraph 20 must require the agreement to contain such terms as the court thinks appropriate, subject to sub-paragraphs (3) to (8).

(3) The terms of the agreement must include terms as to the payment of consideration by the operator to the relevant person for the relevant person's agreement to confer or be bound by the code right (as the case may be).

(4) Paragraph 24 makes provision about the determination of consideration under sub-paragraph (3).

(5) The terms of the agreement must include the terms the court thinks appropriate for ensuring that the least possible loss and damage is caused by the exercise of the code right to persons who –

- (a) occupy the land in question,
- (b) own interests in that land, or
- (c) are from time to time on that land.

(6) Sub-paragraph (5) applies in relation to a person regardless of whether the person is a party to the agreement.

(7) The terms of the agreement must include terms specifying for how long the code right conferred by the agreement is exercisable.

(8) The court must determine whether the terms of the agreement should include a term –

(a) permitting termination of the agreement (and, if so, in what circumstances);
(b) enabling the relevant person to require the operator to reposition or temporarily to remove the electronic communications equipment to which the agreement relates (and, if so, in what circumstances).

COMMENTARY

Paragraph 23 determines the terms to be imposed under an agreement entered into pursuant to a paragraph 20 order. The court may impose an agreement which gives effect to the code right sought with such modifications as the court thinks appropriate (sub-paragraph (1)). The order must, subject to the remainder of paragraph 23, require the agreement to contain such terms as the court considers appropriate (sub-paragraph (2)).

The terms included in the imposed agreement will therefore include: (i) the mandatory terms, which are dealt with in sub-paragraphs (3) to (8) and the provisions in Part 3 concerning assigning, upgrading and sharing; and (ii) such discretionary terms as the court may consider appropriate. For a detailed discussion of the mandatory terms, see paragraphs 21.6.6–21.6.12 of the text.

The expression 'code right' is defined in paragraph 3.

The word 'land' is defined in paragraph 108.

The word 'modification' is defined in s.405 of the 2003 Act.

The word 'operator' is defined in paragraph 2.

The expression 'relevant person' is defined in paragraph 20.

The expression 'the court' is defined in paragraph 94. This is the county court in England, a county court in Northern Ireland, or the sheriff court in Scotland.

The Electronic Communications Code (Jurisdiction) Regulations 2017 came into force on 28 December 2017. The functions conferred on the court are to be conferred on and exercisable by certain of the Tribunals as well as by the court. Regulation 4 provides that applications under Part 4 of the Code may be commenced only in the Upper Tribunal, in relation to England and Wales, or the Lands Tribunal for Scotland, in relation to Scotland.

This paragraph is discussed in Section 21.6 of the text. For more detail on 'consideration', see Chapter 30 of the text. In relation to 'lift and shift' provisions as contemplated by sub-paragraph (8)(b) see also paragraphs 34.5.5–34.5.7 of the text.

24 How is consideration to be determined under paragraph 23?

(1) The amount of consideration payable by an operator to a relevant person under an agreement imposed by an order under paragraph 20 must be an amount or amounts representing the market value of the relevant person's agreement to confer or be bound by the code right (as the case may be).
(2) For this purpose the market value of a person's agreement to confer or be bound by a code right is, subject to sub-paragraph (3), the amount that, at the date the market value is assessed, a willing buyer would pay a willing seller for the agreement –
(a) in a transaction at arm's length,
(b) on the basis that the buyer and seller were acting prudently and with full knowledge of the transaction, and

(c) on the basis that the transaction was subject to the other provisions of the agreement imposed by the order under paragraph 20.

(3) The market value must be assessed on these assumptions –

(a) that the right that the transaction relates to does not relate to the provision or use of an electronic communications network;

(b) that paragraphs 16 and 17 (assignment, and upgrading and sharing) do not apply to the right or any apparatus to which it could apply;

(c) that the right in all other respects corresponds to the code right;

(d) that there is more than one site which the buyer could use for the purpose for which the buyer seeks the right.

(4) The terms of the agreement may provide for consideration to be payable –

(a) as a lump sum or periodically,

(b) on the occurrence of a specified event or events, or

(c) in such other form or at such other time or times as the court may direct.

COMMENTARY

Paragraph 24 provides the test for the determination of the amount of consideration payable by an operator to a relevant person under an imposed agreement. The consideration must be an amount or amounts representing the market value of the relevant person's agreement to confer or be bound by the code right (sub-paragraph (1)). Sub-paragraph (2) defines 'the market value of a person's agreement to confer or be bound by a code right'. Sub-paragraph (3) provides for a number of assumptions which must be made when assessing market value.

Sub-paragraph (4) provides that the terms of the agreement may provide for consideration in whatever form or at whatever time the court may direct. Payments may be conditional, and may be made as a lump sum or be periodic.

The basis upon which consideration is to be determined under paragraph 24 gives rise to numerous questions of construction, which are considered in detail at Section 30.2 of the text.

The word 'apparatus' is defined in s.405 of the 2003 Act.

The expression 'code right' is defined in paragraph 3.

The expression 'electronic communications network' is defined in s.32 of the 2003 Act.

The word 'operator' is defined in paragraph 2.

The expression 'relevant person' is defined in paragraph 20.

The expression 'the court' is defined in paragraph 94. This is the county court in England, a county court in Northern Ireland, or the sheriff court in Scotland.

The Electronic Communications Code (Jurisdiction) Regulations 2017 came into force on 28 December 2017. The functions conferred on the court are to be conferred on and exercisable by certain of the Tribunals as well as by the court. Regulation 4 provides that applications under Part 4 of the Code may be commenced only in the Upper Tribunal, in relation to England and Wales, or the Lands Tribunal for Scotland, in relation to Scotland.

This paragraph is considered in Section 21.7, Chapter 30 (particularly Section 30.2) and paragraphs 34.4.2 to 34.4.4 of the text.

25 What rights to the payment of compensation are there?

(1) If the court makes an order under paragraph 20 the court may also order the operator to pay compensation to the relevant person for any loss or damage that has been sustained or will be sustained by that person as a result of the exercise of the code right to which the order relates.

(2) An order under sub-paragraph (1) may be made –
 (a) at the time the court makes an order under paragraph 20, or
 (b) at any time afterwards, on the application of the relevant person.

(3) An order under sub-paragraph (1) may –
 (a) specify the amount of compensation to be paid by the operator, or
 (b) give directions for the determination of any such amount.

(4) Directions under sub-paragraph (3)(b) may provide –
 (a) for the amount of compensation to be agreed between the operator and the relevant person;
 (b) for any dispute about that amount to be determined by arbitration.

(5) An order under this paragraph may provide for the operator –
 (a) to make a lump sum payment,
 (b) to make periodical payments,
 (c) to make a payment or payments on the occurrence of an event or events, or
 (d) to make a payment or payments in such other form or at such other time or times as the court may direct.

(6) Paragraph 84 makes further provision about compensation in the case of an order under paragraph 20.

COMMENTARY

Paragraph 25 (together with paragraph 84) concerns compensation for the imposition of an agreement by the court pursuant to an order under paragraph 20.

The court may order the operator to pay compensation to the relevant person for any loss or damage sustained or which will be sustained caused by the exercise of the code right in question (sub-paragraph (1)). Such an order may be made at the same time as the order under paragraph 20 or subsequently on application by the relevant person (sub-paragraph (2)).

The court may specify the amount of compensation payable (sub-paragraph (3)(a)) or may make an order giving directions for the determination of the amount of compensation to be determined by arbitration (sub-paragraphs (3)(b) and (4)(b)) (including a direction that the amount should be agreed if possible – sub-paragraph (4)).

The court has a similar degree of flexibility in determining the terms of payment of compensation under paragraph 25 as it does in determining the terms of payment of consideration under paragraph 24. An order may provide for the operator to make payments in the form of a lump sum payment, periodical payments, payments conditional upon the occurrence of an event or events or in such other form and at such time as the court may direct (sub-paragraph (5)).

The expression 'code right' is defined in paragraph 3.

The word 'operator' is defined in paragraph 2.

The expression 'relevant person' is defined in paragraph 20.

The expression 'the court' is defined in paragraph 94. This is the county court in England, a county court in Northern Ireland, or the sheriff court in Scotland.

The Electronic Communications Code (Jurisdiction) Regulations 2017 came into force on 28 December 2017. The functions conferred on the court are to be conferred on and exercisable by certain of the Tribunals as well as by the court. Regulation 4 provides that applications under Part 4 of the Code may be commenced only in the Upper Tribunal, in relation to England and Wales, or the Lands Tribunal for Scotland, in relation to Scotland.

This paragraph is considered in Section 21.8 and Chapter 30 (particularly Sections 30.3 and 30.4) of the text.

26 Interim code rights

(1) An operator may apply to the court for an order which imposes on the operator and a person, on an interim basis, an agreement between them which –
  (a) confers a code right on the operator, or
  (b) provides for a code right to bind that person.

(2) An order under this paragraph imposes an agreement on the operator and a person on an interim basis if it provides for them to be bound by the agreement –
  (a) for the period specified in the order, or
  (b) until the occurrence of an event specified in the order.

(3) The court may make an order under this paragraph if (and only if) the operator has given the person mentioned in sub-paragraph (1) a notice which complies with paragraph 20(2) stating that an agreement is sought on an interim basis and –
  (a) the operator and that person have agreed to the making of the order and the terms of the agreement imposed by it, or
  (b) the court thinks that there is a good arguable case that the test in paragraph 21 for the making of an order under paragraph 20 is met.

(4) Subject to sub-paragraphs (5) and (6), the following provisions apply in relation to an order under this paragraph and an agreement imposed by it as they apply in relation to an order under paragraph 20 and an agreement imposed by it –
  (a) paragraph 20(3) (time at which operator may apply for agreement to be imposed);
  (b) paragraph 22 (effect of agreement imposed under paragraph 20);
  (c) paragraph 23 (terms of agreement imposed under paragraph 20);
  (d) paragraph 24 (payment of consideration);
  (e) paragraph 25 (payment of compensation);
  (f) paragraph 84 (compensation where agreement imposed).

(5) The court may make an order under this paragraph even though the period mentioned in paragraph 20(3)(a) has not elapsed (and paragraph 20(3)(b) does not apply) if the court thinks that the order should be made as a matter of urgency.

(6) Paragraphs 23, 24 and 25 apply by virtue of sub-paragraph (4) as if –

(a) references to the relevant person were to the person mentioned in sub-paragraph (1) of this paragraph, and
(b) the duty in paragraph 23 to include terms as to the payment of consideration to that person in an agreement were a power to do so.

(7) Sub-paragraph (8) applies if –
(a) an order has been made under this paragraph imposing an agreement relating to a code right on an operator and a person in respect of any land, and
(b) the period specified under sub-paragraph (2)(a) has expired or, as the case may be, the event specified under sub-paragraph (2)(b) has occurred without (in either case) an agreement relating to the code right having been imposed on the person by order under paragraph 20.

(8) From the time when the period expires or the event occurs, that person has the right, subject to and in accordance with Part 6 of this code, to require the operator to remove any electronic communications apparatus placed on the land under the agreement imposed under this paragraph.

COMMENTARY

Paragraph 26 provides for a separate procedure by which an operator may apply for and obtain an order conferring code rights or providing for a code right to bind a person on an interim basis pending a full paragraph 20 determination (or for such other period as the court may specify under sub-paragraph (2)).

Such an order will only be made if a paragraph 20(2) notice has been given to the person and either (a) the operator and that person have agreed to the making of the order and the terms of the agreement imposed by it, or (b) the court thinks that there is a good arguable case that the test in paragraph 21 for the making of an order under paragraph 20 is met (sub-paragraph (3)).

The paragraphs of the New Code listed in sub-paragraph (4) apply in relation to an interim order and agreement imposed by it as they apply to orders and agreements imposed under paragraph 20 (subject to sub-paragraph (6)).

Interim orders may be made within the 28-day response period in paragraph 20(3)(a) and before a notice refusing to agree is given if the court thinks it should be made as a matter of urgency.

If an interim order expires and no paragraph 20 order has been made, the person on whom the interim agreement was imposed may require the operator to remove any electronic communications apparatus placed on the land under the imposed interim agreement (sub-paragraphs (7) and (8)).

The expression 'code right' is defined in paragraph 3.

The expression 'electronic communications apparatus' is defined in paragraph 5.

The word 'land' is defined in paragraph 108.

The word 'operator' is defined in paragraph 2.

The expression 'the court' is defined in paragraph 94. This is the county court in England, a county court in Northern Ireland, or the sheriff court in Scotland.

The Electronic Communications Code (Jurisdiction) Regulations 2017 came into force on 28 December 2017. The functions conferred on the court are to be conferred on and exercisable by certain of the Tribunals as well as by the court. Regulation 4 provides that applications under Part 4 of the Code may be commenced only in the

Upper Tribunal, in relation to England and Wales, or the Lands Tribunal for Scotland, in relation to Scotland.

This paragraph is considered at Section 21.9 of the text.

27 Temporary code rights

(1) This paragraph applies where –

(a) an operator gives a notice under paragraph 20(2) to a person in respect of any land,

(b) the notice also requires that person's agreement on a temporary basis in respect of a right which is to be exercisable (in whole or in part) in relation to electronic communications apparatus which is already installed on, under or over the land, and

(c) the person has the right to require the removal of the apparatus in accordance with paragraph 37 or as mentioned in paragraph 40(1) but the operator is not for the time being required to remove the apparatus.

(2) The court may, on the application of the operator, impose on the operator and the person an agreement between them which confers on the operator, or provides for the person to be bound by, such temporary code rights as appear to the court reasonably necessary for securing the objective in sub-paragraph (3).

(3) That objective is that, until the proceedings under paragraph 20 and any proceedings under paragraph 40 are determined, the service provided by the operator's network is maintained and the apparatus is properly adjusted and kept in repair.

(4) Subject to sub-paragraphs (5) and (6), the following provisions apply in relation to an order under this paragraph and an agreement imposed by it as they apply in relation to an order under paragraph 20 and an agreement imposed by it –

(a) paragraph 20(3) (time at which operator may apply for agreement to be imposed);

(b) paragraph 22 (effect of agreement imposed under paragraph 20);

(c) paragraph 23 (terms of agreement imposed under paragraph 20);

(d) paragraph 24 (payment of consideration);

(e) paragraph 25 (payment of compensation);

(f) paragraph 84 (compensation where agreement imposed).

(5) The court may make an order under this paragraph even though the period mentioned in paragraph 20(3)(a) has not elapsed (and paragraph 20(3)(b) does not apply) if the court thinks that the order should be made as a matter of urgency.

(6) Paragraphs 23, 24 and 25 apply by virtue of sub-paragraph (4) as if –

(a) references to the relevant person were to the person mentioned in sub-paragraph (1) of this paragraph, and

(b) the duty in paragraph 23 to include terms as to the payment of consideration to that person in an agreement were a power to do so.

(7) Sub-paragraph (8) applies where, in the course of the proceedings under paragraph 20, it is shown that a person with an interest in the land was entitled to require the removal of the apparatus immediately after it was installed.

(8) The court must, in determining for the purposes of paragraph 20 whether the apparatus should continue to be kept on, under or over the land, disregard the fact that the apparatus has already been installed there.

COMMENTARY

Paragraph 27 provides a temporary solution to operators who have electronic communications apparatus installed pursuant to a previous code agreement in the circumstances discussed in paragraph 21.10.3 of the text.

The operator may apply for temporary code rights, which the court may grant insofar as they appear reasonably necessary to secure the objective in sub-paragraph (3). The objective (that until the proceedings under paragraph 20 and any proceedings under paragraph 40 are determined, the service provided by the operator's network is maintained and the apparatus is properly adjusted and kept in repair) is entirely operator-focused and does not take into account, say, the landowner's plans for development.

The paragraphs of the New Code listed in sub-paragraph (4) apply in relation to a paragraph 27 order and agreement imposed by it as they apply to orders and agreements imposed under paragraph 20 (subject to sub-paragraph (6)).

A paragraph 27 order may be made within the 28-day response period in paragraph 20(3)(a) and before a notice refusing to agree is given if the court thinks it should be made as a matter of urgency.

The presence of the electronic communications apparatus must be disregarded for the purposes of a paragraph 20 determination if a person with an interest in the land was entitled to require the removal of the apparatus immediately after it was installed (sub-paragraphs (7) and (8)).

The expression 'code right' is defined in paragraph 3.

The expression 'electronic communications apparatus' is defined in paragraph 5.

The word 'land' is defined in paragraph 108.

The word 'network' is defined in paragraph 6.

The word 'operator' is defined in paragraph 2.

The word 'provide' is defined by s.405 of the 2003 Act.

The expression 'the court' is defined in paragraph 94. This is the county court in England, a county court in Northern Ireland, or the sheriff court in Scotland.

The Electronic Communications Code (Jurisdiction) Regulations 2017 came into force on 28 December 2017. The functions conferred on the court are to be conferred on and exercisable by certain of the Tribunals as well as by the court. Regulation 4 provides that applications under Part 4 of the Code may be commenced only in the Upper Tribunal, in relation to England and Wales, or the Lands Tribunal for Scotland, in relation to Scotland.

This paragraph is considered in Section 21.10 of the text.

Part 5 Termination and modification of agreements

28 Introductory

This Part of this code makes provision about –

(a) the continuation of code rights after the time at which they cease to be exercisable under an agreement,
(b) the procedure for bringing an agreement to an end,
(c) the procedure for changing an agreement relating to code rights, and
(d) the arrangements for the making of payments under an agreement whilst disputes under this Part are resolved.

COMMENTARY

Paragraph 28 introduces paragraphs 29 to 35. It is self-explanatory.

The expression 'code right' is defined in paragraph 3.

The Electronic Communications Code (Jurisdiction) Regulations 2017 came into force on 28 December 2017. The functions conferred on the court are to be conferred on and exercisable by certain of the Tribunals as well as by the court. Regulation 4 provides that applications under Part 5 of the Code may be commenced only in the Upper Tribunal, in relation to England and Wales, or the Lands Tribunal for Scotland, in relation to Scotland.

29 Application of this Part

(1) This Part of this code applies to an agreement under Part 2 of this code, subject to sub-paragraphs (2) to (4).
(2) This Part of this code does not apply to a lease of land in England and Wales if –
  (a) its primary purpose is not to grant code rights, and
  (b) it is a lease to which Part 2 of the Landlord and Tenant Act 1954 (security of tenure for business, professional and other tenants) applies.
(3) In determining whether a lease is one to which Part 2 of the Landlord and Tenant Act 1954 applies, any agreement under section 38A (agreements to exclude provisions of Part 2) of that Act is to be disregarded.
(4) This Part of this code does not apply to a lease of land in Northern Ireland if –
  (a) its primary purpose is not to grant code rights, and
  (b) it is a lease to which the Business Tenancies (Northern Ireland) Order 1996 (SI 1996/725 (NI 5)) applies.
(5) An agreement to which this Part of this code applies is referred to in this code as a *'code agreement'*.

COMMENTARY

Part 5 of the New Code, which provides a special regime for the termination and modification of code agreements, applies to all agreements under Part 2 of the New Code, save as set out in sub-paragraphs (2) to (4). If Part 5 applies to an agreement, it is referred to in the New Code as a code agreement.

Part 5 does not apply to a lease whose primary purpose is not to grant code rights and which is either a lease of land in England and Wales which is a business tenancy to which Part 2 of the Landlord and Tenant Act 1954 applies (or would apply, disregarding any s.38A agreement) or a lease of land in Northern Ireland to which the Business Tenancies (Northern Ireland) Order 1996 applies.

The expression 'code right' is defined in paragraph 3.

The word 'land' is defined in paragraph 108.

The word 'lease' is defined in paragraph 108 and considered further at paragraph 22.2.4 of the text.

The expression 'primary purpose' is considered at paragraphs 22.2.5 to 22.2.9 of the text.

This paragraph is discussed in Section 22.2 of the text. For the relationship between the New Code and the Landlord and Tenant Act 1954, see further Chapter 25 of the text. For the interrelationship between the transitional provisions, the New Code and the 1954 Act, see further Section 31.4 of the text.

30 Continuation of code rights

(1) Sub-paragraph (2) applies if –
  (a) a code right is conferred by, or is otherwise binding on, a person (the 'site provider') as the result of a code agreement, and
  (b) under the terms of the agreement –
    (i) the right ceases to be exercisable or the site provider ceases to be bound by it, or
    (ii) the site provider may bring the code agreement to an end so far as it relates to that right.
  (2) Where this sub-paragraph applies the code agreement continues so that –
  (a) the operator may continue to exercise that right, and
  (b) the site provider continues to be bound by the right.

(3) Sub-paragraph (2) does not apply to a code right which is conferred by, or is otherwise binding on, a person by virtue of an order under paragraph 26 (interim code rights) or 27 (temporary code rights).

(4) Sub-paragraph (2) is subject to the following provisions of this Part of this code.

COMMENTARY

This paragraph concerns situations where code rights cease to be exercisable or binding under the terms of a code agreement, or where a site provider may bring a code agreement to an end insofar as it relates to particular code rights. Save for interim or temporary code rights, and subject to the remaining provisions of Part 5, a code agreement will be continued automatically by paragraph 30(2), irrespective of the terms of the code agreement, so that the operator may continue to exercise rights, and the site provider will continue to be bound by rights, which are conferred by, or are otherwise binding on the site provider, as a result of the code agreement. So, mere expiry by effluxion of time does not bring about the cessation of a code right.

The expression 'code agreement' is defined in paragraph 29.
The expression 'code right' is defined in paragraph 3.
The word 'operator' is defined in paragraph 2.

This paragraph is discussed in Section 23.3 of the text. As to who is a 'site provider' for the purposes of the transitional provisions, see further Section 31.7 of the text.

31 How may a person bring a code agreement to an end?

(1) A site provider who is a party to a code agreement may bring the agreement to an end by giving a notice in accordance with this paragraph to the operator who is a party to the agreement.
(2) The notice must –
  (a) comply with paragraph 89 (notices given by persons other than operators),
  (b) specify the date on which the site provider proposes the code agreement should come to an end, and
  (c) state the ground on which the site provider proposes to bring the code agreement to an end.
(3) The date specified under sub-paragraph (2)(b) must fall –
  (a) after the end of the period of 18 months beginning with the day on which the notice is given, and
  (b) after the time at which, apart from paragraph 30, the code right to which the agreement relates would have ceased to be exercisable or to bind the site provider or at a time when, apart from that paragraph, the code agreement could have been brought to an end by the site provider.
(4) The ground stated under sub-paragraph (2)(c) must be one of the following –
  (a) that the code agreement ought to come to an end as a result of substantial breaches by the operator of its obligations under the agreement;
  (b) that the code agreement ought to come to an end because of persistent delays by the operator in making payments to the site provider under the agreement;
  (c) that the site provider intends to redevelop all or part of the land to which the code agreement relates, or any neighbouring land, and could not reasonably do so unless the code agreement comes to an end;
  (d) that the operator is not entitled to the code agreement because the test under paragraph 21 for the imposition of the agreement on the site provider is not met.

COMMENTARY

Paragraph 31 confers on a site provider who is a party to a code agreement the right of termination. The right to terminate is not conferred by this paragraph upon either the operator, or other parties who are bound by code rights without being a party to the code agreement. To terminate a code agreement, the procedure in this paragraph must be followed by the site provider, who must give notice to the operator party. A notice given in accordance with this paragraph will have the effect specified in paragraph 32.

The notice must fulfil the requirements of sub-paragraph (2) to be valid.

For the requirement in sub-paragraph (2)(a), see further paragraph 89.

The date to be specified under sub-paragraph (2)(b), on which the site provider proposes the code agreement should come to an end, must be after 18 months from the date of the giving of the notice, and after the date on which the code right would have ceased to bind or be exercisable, or could have been brought to an end, but for continuation under paragraph 30.

The ground for termination stated under sub-paragraph (2)(c) must be one of the four grounds listed in sub-paragraph (4). For discussion of the grounds of termination, see in particular paragraphs 22.4.9 to 22.4.17 of the text.

The expression 'code agreement' is defined in paragraph 29.

The expression 'code right' is defined in paragraph 3.

The word 'land' is defined in paragraph 108.

The word 'operator' is defined in paragraph 2.

The expression 'site provider' is defined in paragraph 30.

This paragraph is discussed in Section 22.4 of the text. For a comparison between the termination procedure under this paragraph and the modification procedure under paragraph 33, see further Section 23.4 of the text. For the interrelationship between this paragraph and the transitional provisions, see also Sections 31.6 and 31.7 of the text.

32 What is the effect of a notice under paragraph 31?

(1) Where a site provider gives a notice under paragraph 31, the code agreement to which it relates comes to an end in accordance with the notice unless –
  (a) within the period of three months beginning with the day on which the notice is given, the operator gives the site provider a counter-notice in accordance with sub-paragraph (3), and
  (b) within the period of three months beginning with the day on which the counter-notice is given, the operator applies to the court for an order under paragraph 34.

(2) Sub-paragraph (1) does not apply if the operator and the site provider agree to the continuation of the code agreement.

(3) The counter-notice must state –
  (a) that the operator does not want the existing code agreement to come to an end,
  (b) that the operator wants the site provider to agree to confer or be otherwise bound by the existing code right on new terms, or
  (c) that the operator wants the site provider to agree to confer or be otherwise bound by a new code right in place of the existing code right.

(4) If, on an application under sub-paragraph (1)(b), the court decides that the site provider has established any of the grounds stated in the site provider's notice under paragraph 31, the court must order that the code agreement comes to an end in accordance with the order.

(5) Otherwise the court must make one of the orders specified in paragraph 34.

COMMENTARY

Absent agreement (sub-paragraph (2)), a code agreement to which a paragraph 31 notice relates will come to an end in accordance with the terms of the notice unless the operator gives a counter-notice within three months of the date on which the notice was given, and then within three months of the date of the counter-notice applies to the court for an order under paragraph 34 (sub-paragraph (1)).

The counter-notice must state one of the three options listed in sub-paragraph (3).

If an application is made for an order under paragraph 34, but the site provider establishes a ground for termination stated in his paragraph 31 notice, the court must order that the code agreement comes to an end in accordance with the order (sub-paragraph (4)). If not, the court must make a paragraph 34 order (sub-paragraph (5)).

The expression 'code agreement' is defined in paragraph 29.

The expression 'code right' is defined in paragraph 3.

The word 'operator' is defined in paragraph 2.

The expression 'site provider' is defined in paragraph 30.

The expression 'the court' is defined in paragraph 94. This is the county court in England, a county court in Northern Ireland, or the sheriff court in Scotland.

The Electronic Communications Code (Jurisdiction) Regulations 2017 came into force on 28 December 2017. The functions conferred on the court are to be conferred on and exercisable by certain of the Tribunals as well as by the court. Regulation 4 provides that applications under Part 5 of the Code may be commenced only in the Upper Tribunal, in relation to England and Wales, or the Lands Tribunal for Scotland, in relation to Scotland.

This paragraph is discussed in Section 22.5 of the text. As to transitional provisions, see further Section 31.8 of the text.

33 How may a party to a code agreement require a change to the terms of an agreement which has expired?

(1) An operator or site provider who is a party to a code agreement by which a code right is conferred by or otherwise binds the site provider may, by notice in accordance with this paragraph, require the other party to the agreement to agree that –
  (a) the code agreement should have effect with modified terms,
  (b) where under the code agreement more than one code right is conferred by or otherwise binds the site provider, that the agreement should no longer provide for an existing code right to be conferred by or otherwise bind the site provider,
  (c) the code agreement should –
    (i) confer an additional code right on the operator, or
    (ii) provide that the site provider is otherwise bound by an additional code right, or
  (d) the existing code agreement should be terminated and a new agreement should have effect between the parties which –
    (i) confers a code right on the operator, or
    (ii) provides for a code right to bind the site provider.
(2) The notice must –
  (a) comply with paragraph 88 or 89, according to whether the notice is given by an operator or a site provider,
  (b) specify –

(i) the day from which it is proposed that the modified terms should have effect,
(ii) the day from which the agreement should no longer provide for the code right to be conferred by or otherwise bind the site provider,
(iii) the day from which it is proposed that the additional code right should be conferred by or otherwise bind the site provider, or
(iv) the day on which it is proposed the existing code agreement should be terminated and from which a new agreement should have effect,

(as the case may be), and

(c) set out details of –
(i) the proposed modified terms,
(ii) the code right it is proposed should no longer be conferred by or otherwise bind the site provider,
(iii) the proposed additional code right, or
(iv) the proposed terms of the new agreement,

(as the case may be).

(3) The day specified under sub-paragraph (2)(b) must fall –
(a) after the end of the period of 6 months beginning with the day on which the notice is given, and
(b) after the time at which, apart from paragraph 30, the code right to which the existing code agreement relates would have ceased to be exercisable or to bind the site provider or at a time when, apart from that paragraph, the code agreement could have been brought to an end by the site provider.

(4) Sub-paragraph (5) applies if, after the end of the period of 6 months beginning with the day on which the notice is given, the operator and the site provider have not reached agreement on the proposals in the notice.

(5) Where this paragraph applies, the operator or the site provider may apply to the court for the court to make an order under paragraph 34.

COMMENTARY

Paragraph 33 deals with the procedure for modifying, rather than terminating, a code agreement. This procedure may be exercised by a party who is either a site provider or an operator. The heading is rather misleading, since the code agreement will be continued under paragraph 30 rather than expire by effluxion of time.

By notice, the party seeking a modification must require the other party to agree that the modification specified in accordance with sub-paragraph (1) should have effect.

The notice must comply with the formal requirements in sub-paragraph (2).

Sub-paragraph (2)(b) requires a date to be specified for the relevant modification to take effect, which must fall after the expiry of a six-month period from the date on which the notice was given, and after the date on which the code right would have ceased to bind or be exercisable, or could have been brought to an end by the site provider, but for continuation under paragraph 30 (sub-paragraph (3)).

In default of agreement as to the modification proposals within six months of the giving of the notice, either party may apply for an order under paragraph 34 (sub-paragraphs (4) and (5)).

The expression 'code agreement' is defined in paragraph 29.

The expression 'code right' is defined in paragraph 3.

The word 'operator' is defined in paragraph 2.

The expression 'site provider' is defined in paragraph 30.

The expression 'the court' is defined in paragraph 94. This is the county court in England, a county court in Northern Ireland, or the sheriff court in Scotland.

The Electronic Communications Code (Jurisdiction) Regulations 2017 came into force on 28 December 2017. The functions conferred on the court are to be conferred on and exercisable by certain of the Tribunals as well as by the court. Regulation 4 provides that applications under Part 5 of the Code may be commenced only in the Upper Tribunal, in relation to England and Wales, or the Lands Tribunal for Scotland, in relation to Scotland.

This paragraph is discussed in Chapter 23 of the text. For a comparison between the modification procedure under this paragraph and the termination procedure under paragraph 31, see in particular Section 23.4 of the text. As to transitional provisions, see further Section 31.8 of the text.

34 What orders may a court make on an application under paragraph 32 or 33?

(1) This paragraph sets out the orders that the court may make on an application under paragraph 32(1)(b) or 33(5).

(2) The court may order that the operator may continue to exercise the existing code right in accordance with the existing code agreement for such period as may be specified in the order (so that the code agreement has effect accordingly).

(3) The court may order the modification of the terms of the code agreement relating to the existing code right.

(4) Where under the code agreement more than one code right is conferred by or otherwise binds the site provider, the court may order the modification of the terms of the code agreement so that it no longer provides for an existing code right to be conferred by or otherwise bind the site provider.

(5) The court may order the terms of the code agreement relating to the existing code right to be modified so that –

(a) it confers an additional code right on the operator, or

(b) it provides that the site provider is otherwise bound by an additional code right.

(6) The court may order the termination of the code agreement relating to the existing code right and order the operator and the site provider to enter into a new agreement which –

(a) confers a code right on the operator, or

(b) provides for a code right to bind the site provider.

(7) The existing code agreement continues until the new agreement takes effect.

(8) This code applies to the new agreement as if it were an agreement under Part 2 of this code.

(9) The terms conferring or providing for an additional code right under sub-paragraph (5), and the terms of a new agreement under sub-paragraph (6), are to be such as are agreed between the operator and the site provider.

(10) If the operator and the site provider are unable to agree on the terms, the court must on an application by either party make an order specifying those terms.

(11) Paragraphs 23(2) to (8), 24, 25 and 84 apply –
  (a) to an order under sub-paragraph (3), (4) or (5), so far as it modifies or specifies the terms of the agreement, and
  (b) to an order under sub-paragraph (10)

  as they apply to an order under paragraph 20.

(12) In the case of an order under sub-paragraph (10) the court must also have regard to the terms of the existing code agreement.

(13) In determining which order to make under this paragraph, the court must have regard to all the circumstances of the case, and in particular to –
  (a) the operator's business and technical needs,
  (b) the use that the site provider is making of the land to which the existing code agreement relates,
  (c) any duties imposed on the site provider by an enactment, and
  (d) the amount of consideration payable by the operator to the site provider under the existing code agreement.

(14) Where the court makes an order under this paragraph, it may also order the operator to pay the site provider the amount (if any) by which A exceeds B, where –
  (a) A is the amount of consideration that would have been payable by the operator to the site provider for the relevant period if that amount had been assessed on the same basis as the consideration payable as the result of the order, and
  (b) B is the amount of consideration payable by the operator to the site provider for the relevant period.

(15) In sub-paragraph (14) the relevant period is the period (if any) that –
  (a) begins on the date on which, apart from the operation of paragraph 30, the code right to which the existing code agreement relates would have ceased to be exercisable or to bind the site provider or from which, apart from that paragraph, the code agreement could have been brought to an end by the site provider, and
  (b) ends on the date on which the order is made.

COMMENTARY

Paragraph 34 sets out the possible orders which a court might make on an application under paragraphs 32(1)(b) or 33(5). The court is provided with an extensive menu of options in sub-paragraphs (2) to (6).

For discussion of the ancillary matters in sub-paragraphs (7) to (15), see further paragraphs 22.6.5–22.6.12 of the text.

The word 'business' is defined in s.405 of the 2003 Act.

The expression 'code agreement' is defined in paragraph 29.

The expression 'code right' is defined in paragraph 3.

The word 'enactment' is defined in paragraph 108.

The word 'land' is defined in paragraph 108.

The word 'modification' is defined in s.405 of the 2003 Act.

The word 'operator' is defined in paragraph 2.

The expression 'site provider' is defined in paragraph 30.

The expression 'the court' is defined in paragraph 94. This is the county court in England, a county court in Northern Ireland, or the sheriff court in Scotland.

The Electronic Communications Code (Jurisdiction) Regulations 2017 came into force on 28 December 2017. The functions conferred on the court are to be conferred on and exercisable by certain of the Tribunals as well as by the court. Regulation 4 provides that applications under Part 5 of the Code may be commenced only in the Upper Tribunal, in relation to England and Wales, or the Lands Tribunal for Scotland, in relation to Scotland.

This paragraph is discussed in Section 22.6 of the text. For discussion of related transitional provisions, see further Section 31.8 of the text.

35 What arrangements for payment can be made pending determination of the application?

(1) This paragraph applies where –
 (a) a code right continues to be exercisable under paragraph 30 after the time at which, apart from the operation of that paragraph, the code right would have ceased to be exercisable or to bind the site provider or from which, apart from that paragraph, the code agreement relating to the right could have been brought to an end by the site provider, and
 (b) the operator or the site provider has applied to the court for an order under paragraph 32(1)(b) or 33(5).

(2) The site provider may –
 (a) agree with the operator that, until the application has been finally determined, the site provider will continue to receive the payments of consideration from the operator to which the site provider is entitled under the agreement relating to the existing code right,
 (b) agree with the operator that, until that time, the site provider will receive different payments of consideration under that agreement, or
 (c) apply to the court for the court to determine the payments of consideration to be made by the operator to the site provider under that agreement until that time.

(3) The court must determine the payments under sub-paragraph (2)(c) on the basis set out in paragraph 24 (calculation of consideration).

COMMENTARY

Paragraph 35 is similar to a much simplified version of the interim rent procedure under the Landlord and Tenant Act 1954.

Where an application is made under paragraph 32(1)(b) or 33(5) and a code right continues to be exercisable under paragraph 30 after the time at which it would otherwise have ceased to be exercisable or to bind the site provider, or after the time from which the code agreement relating to the right could have been brought to an end by the site provider, the site provider may reach an agreement with the operator under sub-paragraphs (2)(a) or (b), or apply to the court under sub-paragraph (2)(c).

The agreements which could be reached are that, pending final determination of the application, the site provider will receive either (a) the same or (b) different payments of consideration under the code agreement. Alternatively, the site provider may apply for the court to determine the consideration to be paid pending determination of the application, in which case the court must determine the payments on the basis set out in paragraph 24.

The expression 'code agreement' is defined in paragraph 29.
The expression 'code right' is defined in paragraph 3.
The word 'operator' is defined in paragraph 2.
The expression 'site provider' is defined in paragraph 30.
The expression 'the court' is defined in paragraph 94. This is the county court in England, a county court in Northern Ireland, or the sheriff court in Scotland.

The Electronic Communications Code (Jurisdiction) Regulations 2017 came into force on 28 December 2017. The functions conferred on the court are to be conferred on and exercisable by certain of the Tribunals as well as by the court. Regulation 4 provides that applications under Part 5 of the Code may be commenced only in the Upper Tribunal, in relation to England and Wales, or the Lands Tribunal for Scotland, in relation to Scotland.

This paragraph is discussed in Section 22.7 of the text. For the basis of determining applications under sub-paragraph (2)(c) and paragraph 24, see further the commentary to paragraph 24 above.

Part 6 Rights to require removal of electronic communications apparatus

36 Introductory

This Part of this code makes provision about –

(a) the cases in which a person has the right to require the removal of electronic communications apparatus or the restoration of land,
(b) the means by which a person can discover whether apparatus is on land pursuant to a code right, and
(c) the means by which a right to require removal of apparatus or restoration of land can be enforced.

COMMENTARY

Paragraph 36 introduces paragraphs 37 to 44. It is self-explanatory.
The word 'apparatus' is defined in s.405 of the 2003 Act.
The expression 'code right' is defined in paragraph 3.
The expression 'electronic communications apparatus' is defined in paragraph 5.
The word 'land' is defined in paragraph 108.

The Electronic Communications Code (Jurisdiction) Regulations 2017 came into force on 28 December 2017. The functions conferred on the court are to be conferred on and exercisable by certain of the Tribunals as well as by the court. Regulation 4 provides that applications under Part 6 of the Code may be commenced only in the Upper Tribunal, in relation to England and Wales, or the Lands Tribunal for Scotland, in relation to Scotland.

37 When does a landowner have the right to require removal of electronic communications apparatus?

(1) A person with an interest in land (a 'landowner') has the right to require the removal of electronic communications apparatus on, under or over the land if (and only if) one or more of the following conditions are met.

(2) The first condition is that the landowner has never since the coming into force of this code been bound by a code right entitling an operator to keep the apparatus on, under or over the land.
This is subject to sub-paragraph (4).

(3) The second condition is that a code right entitling an operator to keep the apparatus on, under or over the land has come to an end or has ceased to bind the landowner –
(a) as mentioned in paragraph 26(7) and (8),
(b) as the result of paragraph 32(1), or
(c) as the result of an order under paragraph 32(4) or 34(4) or (6), or
(d) where the right was granted by a lease to which Part 5 of this code does not apply.

This is subject to sub-paragraph (4).

(4) The landowner does not meet the first or second condition if –
(a) the land is occupied by a person who –
(i) conferred a code right (which is in force) entitling an operator to keep the apparatus on, under or over the land, or
(ii) is otherwise bound by such a right, and
(b) that code right was not conferred in breach of a covenant enforceable by the landowner.

(5) In the application of sub-paragraph (4)(b) to Scotland the reference to a covenant enforceable by the landowner is to be read as a reference to a contractual term which is so enforceable.

(6) The third condition is that –
(a) an operator has the benefit of a code right entitling the operator to keep the apparatus on, under or over the land, but
(b) the apparatus is not, or is no longer, used for the purposes of the operator's network, and
(c) there is no reasonable likelihood that the apparatus will be used for that purpose.

(7) The fourth condition is that –
(a) this code has ceased to apply to a person so that the person is no longer entitled under this code to keep the apparatus on, under or over the land,
(b) the retention of the apparatus on, under or over the land is not authorised by a scheme contained in an order under section 117, and
(c) there is no other person with a right conferred by or under this code to keep the apparatus on, under or over the land.

(8) The fifth condition is that –
(a) the apparatus was kept on, under or over the land pursuant to –
(i) a transport land right (see Part 7), or
(ii) a street work right (see Part 8),
(b) that right has ceased to be exercisable in relation to the land by virtue of paragraph 54(9), and
(c) there is no other person with a right conferred by or under this code to keep the apparatus on, under or over the land.

(9) This paragraph does not affect rights to require the removal of apparatus under another enactment (see paragraph 41).

COMMENTARY

Paragraph 37 confers on a landowner the right to require removal of electronic communications apparatus on, under or over its land. It should be noted that it is only a person with an interest in the land who may require the removal of the apparatus, not merely an occupier, notwithstanding that occupiers may grant code rights. For further discussion of the meaning of 'the landowner', see Section 24.2 of the text.

The landowner will have such a right if one of the five conditions set out in sub-paragraphs (2), (3), (6), (7) and (8) is satisfied (or if there is a right under another enactment to require its removal – sub-paragraph (9)).

For discussion of the first condition, see paragraphs 24.3.3–24.3.16 of the text.

For discussion of the second condition, see paragraphs 24.3.17–24.3.22 of the text.

For discussion of the third condition, see Sections 12.5–12.8 and paragraphs 24.3.23–24.3.26 of the text.

For discussion of the fourth condition, see paragraphs 24.3.27–24.3.28 of the text.

For discussion of the fifth condition, see paragraphs 24.3.29–24.3.40 of the text.

The word 'apparatus' is defined in s.405 of the 2003 Act.

The expression 'code right' is defined in paragraph 3.

The expression 'electronic communications apparatus' is defined in paragraph 5.

The word 'enactment' is defined in paragraph 108.

The word 'land' is defined in paragraph 108.

The word 'lease' is defined in paragraph 108.

The word 'network' is defined in paragraph 6.

The word 'operator' is defined in paragraph 2.

The expression 'street work right' is defined in paragraph 59.

The expression 'transport land right' is defined in paragraph 48.

This paragraph is discussed in Sections 12.5–12.8 and 24.2–24.3 of the text. For the effect of the Landlord and Tenant Act 1954 on the right to require removal of apparatus, see further Section 25.5 and paragraph 25.7.5 of the text. For the position in Scotland, see also paragraph 40.2.5 of the text.

38 When does a landowner or occupier of neighbouring land have the right to require removal of electronic communications apparatus?

(1) A landowner or occupier of any land ('neighbouring land') has the right to require the removal of electronic communications apparatus kept on, under or over other land in exercise of a right mentioned in paragraph 13(1), if both of the following conditions are met.

(2) The first condition is that the apparatus interferes with or obstructs a means of access to or from the neighbouring land.

(3) The second condition is that the landowner or occupier of the neighbouring land is not bound by a code right within paragraph 3(h) entitling an operator to cause the interference or obstruction.

(4) A landowner of neighbouring land who is not the occupier of the land does not meet the second condition if –

(a) the land is occupied by a person who –

(i) conferred a code right (which is in force) entitling an operator to cause the interference or obstruction, or
(ii) is otherwise bound by such a right, and
(b) that code right was not conferred in breach of a covenant enforceable by the landowner.
(5) In the application of sub-paragraph (4)(b) to Scotland the reference to a covenant enforceable by the landowner is to be read as a reference to a contractual term which is so enforceable.

COMMENTARY

Paragraph 38 confers on landowners or occupiers of neighbouring land the right to require removal of electronic communications apparatus if both of the conditions in sub-paragraphs (2) and (3) are met.

For the first condition, see further paragraph 24.4.6 of the text.

For the second condition, see further paragraphs 24.4.7–24.4.9 of the text.

The expression 'code right' is defined in paragraph 3.

The expression 'electronic communications apparatus' is defined in paragraph 5.

The word 'land' is defined in paragraph 108.

The word 'landowner' is defined in paragraph 37.

The word 'occupier' is defined in paragraph 105.

The word 'operator' is defined in paragraph 2.

This paragraph is discussed in Section 24.4 of the text. For the position in Scotland, see further paragraph 40.2.5 of the text.

39 How does a landowner or occupier find out whether apparatus is on land pursuant to a code right?

(1) A landowner may by notice require an operator to disclose whether –
(a) the operator owns electronic communications apparatus on, under or over land in which the landowner has an interest or uses such apparatus for the purposes of the operator's network, or
(b) the operator has the benefit of a code right entitling the operator to keep electronic communications apparatus on, under or over land in which the landowner has an interest.
(2) A landowner or occupier of neighbouring land may by notice require an operator to disclose whether –
(a) the operator owns electronic communications apparatus on, under or over land that forms (or, but for the apparatus, would form) a means of access to the neighbouring land, or uses such apparatus for the purposes of the operator's network, or
(b) the operator has the benefit of a code right entitling the operator to keep electronic communications apparatus on, under or over land that forms (or, but for the apparatus, would form) a means of access to the neighbouring land.
(3) The notice must comply with paragraph 89 (notices given by persons other than operators).

(4) Sub-paragraph (5) applies if –
 (a) the operator does not, before the end of the period of three months beginning with the date on which the notice under sub-paragraph (1) or (2) was given, give a notice to the landowner or occupier that –
  (i) complies with paragraph 88 (notices given by operators), and
  (ii) discloses the information sought by the landowner or occupier,
 (b) the landowner or occupier takes action under paragraph 40 to enforce the removal of the apparatus, and
 (c) it is subsequently established that –
  (i) the operator owns the apparatus or uses it for the purposes of the operator's network, and
  (ii) the operator has the benefit of a code right entitling the operator to keep the apparatus on, under or over the land.

(5) The operator must nevertheless bear the costs of any action taken by the landowner or occupier under paragraph 40 to enforce the removal of the apparatus.

COMMENTARY

Paragraph 39 provides a notice procedure by which landowners (sub-paragraph (1)) and landowners and occupiers of neighbouring land (sub-paragraph (2)) may require an operator to disclose whether it owns, uses or has a code right in respect of certain electronic communications equipment by serving a paragraph 89 compliant (sub-paragraph (3)) notice upon them.

If a person gives a paragraph 39 notice and the operator does not give a paragraph 88 compliant counter-notice disclosing the information sought within three months, the operator will have to bear any costs incurred by the landowner or occupier under paragraph 40 to attempt to enforce the removal of the apparatus (sub-paragraphs (4) and (5)).

The expression 'code right' is defined in paragraph 3.
The expression 'electronic communications apparatus' is defined in paragraph 5.
The word 'information' is defined in s.405 of the 2003 Act.
The word 'land' is defined in paragraph 108.
The word 'landowner' is defined in paragraph 37.
The expression 'neighbouring land' is defined in paragraph 38.
The word 'network' is defined in paragraph 6.
The word 'occupier' is defined in paragraph 105.
The word 'operator' is defined in paragraph 2.

This paragraph is discussed in Sections 12.6 and 24.5 of the text.

40 How does a landowner or occupier enforce removal of apparatus?

(1) The right of a landowner or occupier to require the removal of electronic communications apparatus on, under or over land, under paragraph 37 or 38, is exercisable only in accordance with this paragraph.

(2) The landowner or occupier may give a notice to the operator whose apparatus it is requiring the operator –

(a) to remove the apparatus, and
(b) to restore the land to its condition before the apparatus was placed on, under or over the land.

(3) The notice must –
(a) comply with paragraph 89 (notices given by persons other than operators), and
(b) specify the period within which the operator must complete the works.

(4) The period specified under sub-paragraph (3) must be a reasonable one.

(5) Sub-paragraph (6) applies if, within the period of 28 days beginning with the day on which the notice was given, the landowner or occupier and the operator do not reach agreement on any of the following matters –
(a) that the operator will remove the apparatus;
(b) that the operator will restore the land to its condition before the apparatus was placed on, under or over the land;
(c) the time at which or period within which the apparatus will be removed;
(d) the time at which or period within which the land will be restored.

(6) The landowner or occupier may make an application to the court for –
(a) an order under paragraph 44(1) (order requiring operator to remove apparatus etc.), or
(b) an order under paragraph 44(3) (order enabling landowner to sell apparatus etc.).

(7) If the court makes an order under paragraph 44(1), but the operator does not comply with the agreement imposed on the operator and the landowner or occupier by virtue of paragraph 44(7), the landowner or occupier may make an application to the court for an order under paragraph 44(3).

(8) On an application under sub-paragraph (6) or (7) the court may not make an order in relation to apparatus if an application under paragraph 20(3) has been made in relation to the apparatus and has not been determined.

COMMENTARY

A person with a right to require the removal of apparatus under paragraphs 37 or 38 may exercise that right in accordance with paragraph 40 by giving a notice to the operator whose apparatus it is (i.e. the owner of the apparatus) requiring removal of the apparatus and restoration of the land to its previous condition (sub-paragraph (2)).

The notice must comply with paragraph 89 and must specify a reasonable (sub-paragraph (4)) period within which the works must be completed (sub-paragraph (3)).

The landowner or occupier may apply to the court under sub-paragraph (6) for an order under paragraph 44(1) or 44(3) if the parties have not reached agreement as to the matters set out in sub-paragraph (5) within 28 days of the date on which the notice was given.

The landowner or occupier may also apply for an order under paragraph 44(3) if the court makes an order under paragraph 44(1) but the operator fails to comply with the agreement imposed by that order.

The court may not make an order under paragraphs 44(1) or 44(3) on an application under sub-paragraphs (6) or (7) if a pending application under paragraph 20(3) has been made. For further discussion of this see paragraph 24.4.10 of the text.

The word 'apparatus' is defined in s.405 of the 2003 Act.

The expression 'electronic communications apparatus' is defined in paragraph 5.
The word 'land' is defined in paragraph 108.
The word 'landowner' is defined in paragraph 37.
The word 'occupier' is defined in paragraph 105.
The word 'operator' is defined in paragraph 2.
The expression 'the court' is defined in paragraph 94. This is the county court in England, a county court in Northern Ireland or the sheriff court in Scotland.

The Electronic Communications Code (Jurisdiction) Regulations 2017 came into force on 28 December 2017. The functions conferred on the court are to be conferred on and exercisable by certain of the Tribunals as well as by the court. Regulation 4 provides that applications under Part 6 of the Code may be commenced only in the Upper Tribunal, in relation to England and Wales, or the Lands Tribunal for Scotland, in relation to Scotland.

This paragraph is discussed in Section 24.6 and paragraph 24.4.10 of the text. For the interrelationship between this paragraph and the Landlord and Tenant Act 1954, see further paragraphs 25.7.5 et seq of the text.

41 How are other rights to require removal of apparatus enforced?

(1) The right of a person (a 'third party') under an enactment other than this code, or otherwise than under an enactment, to require the removal of electronic communications apparatus on, under or over land is exercisable only in accordance with this paragraph.
(2) The third party may give a notice to the operator whose apparatus it is, requiring the operator –
   (a) to remove the apparatus, and
   (b) to restore the land to its condition before the apparatus was placed on, under or over the land.
(3) The notice must –
   (a) comply with paragraph 89 (notices given by persons other than operators), and
   (b) specify the period within which the operator must complete the works.
(4) The period specified under sub-paragraph (3) must be a reasonable one.
(5) Within the period of 28 days beginning with the day on which notice under sub-paragraph (2) is given, the operator may give the third party notice ('counter-notice') –
   (a) stating that the third party is not entitled to require the removal of the apparatus, or
   (b) specifying the steps which the operator proposes to take for the purpose of securing a right as against the third party to keep the apparatus on the land.
(6) If the operator does not give counter-notice within that period, the third party is entitled to enforce the removal of the apparatus.
(7) If the operator gives the third party counter-notice within that period, the third party may enforce the removal of the apparatus only in pursuance of an order of the court that the third party is entitled to enforce the removal of the apparatus.

(8) If the counter-notice specifies steps under paragraph (5)(b), the court may make an order under sub-paragraph (7) only if it is satisfied –
   (a) that the operator is not intending to take those steps or is being unreasonably dilatory in taking them; or
   (b) that taking those steps has not secured, or will not secure, for the operator as against the third party any right to keep the apparatus installed on, under or over the land or to reinstall it if it is removed.

(9) Where the third party is entitled to enforce the removal of the apparatus, under sub-paragraph (6) or under an order under sub-paragraph (7), the third party may make an application to the court for –
   (a) an order under paragraph 44(1) (order requiring operator to remove apparatus etc.), or
   (b) an order under paragraph 44(3) (order enabling third party to sell apparatus etc.).

(10) If the court makes an order under paragraph 44(1), but the operator does not comply with the agreement imposed on the operator and the third party by virtue of paragraph 44(7), the third party may make an application to the court for an order under paragraph 44(3).

(11) An order made on an application under this paragraph need not include provision within paragraph 44(1)(b) or (3)(d) unless the court thinks it appropriate.

(12) Sub-paragraph (9) is without prejudice to any other method available to the third party for enforcing the removal of the apparatus.

COMMENTARY

Any right to require the removal of electronic communications apparatus on, under or over land not derived from the New Code (including those under other enactments) may only be exercised in accordance with paragraph 41 by giving a notice to the operator whose apparatus it is (i.e. the owner of the apparatus) requiring removal of the apparatus and restoration of the land to its previous condition (sub-paragraph (2)).

The notice must comply with paragraph 89 and must specify a reasonable (sub-paragraph (4)) period within which the works must be completed (sub-paragraph (3)).

The third party may enforce it if the operator does not give a counter-notice fulfilling the requirements of sub-paragraph (5) within 28 days of the giving of the notice (sub-paragraph (6)), either via a court order under sub-paragraph (9) or by such other lawful means for enforcing the removal of the apparatus as are appropriate (sub-paragraph (12)).

However, if a counter-notice is given within 28 days, the third party must apply to the court to be able to enforce removal of the apparatus (sub-paragraph (7)). For the procedure and tests to be applied once a valid counter-notice is given and for discussion of the further applications which may be made under sub-paragraphs (9) and (10) see the discussion at paragraphs 24.7.11–24.7.14 of the text.

The expression 'electronic communications apparatus' is defined in paragraph 5.
The word 'enactment' is defined in paragraph 108.
The word 'land' is defined in paragraph 108.
The word 'operator' is defined in paragraph 2.

The expression 'the court' is defined in paragraph 94. This is the county court in England, a county court in Northern Ireland, or the sheriff court in Scotland.

The Electronic Communications Code (Jurisdiction) Regulations 2017 came into force on 28 December 2017. The functions conferred on the court are to be conferred on and exercisable by certain of the Tribunals as well as by the court. Regulation 4 provides that applications under Part 6 of the Code may be commenced only in the Upper Tribunal, in relation to England and Wales, or the Lands Tribunal for Scotland, in relation to Scotland.

The expression 'third party' is considered further in paragraph 24.7.3 of the text.

This paragraph is discussed in Sections 24.7 and paragraphs 24.2.32 and 24.6.20 of the text.

42 How does paragraph 40 apply if a person is entitled to require apparatus to be altered in consequence of street works?

(1) This paragraph applies where the third party's right in relation to which paragraph 41 applies is a right to require the alteration of the apparatus in consequence of the stopping up, closure, change or diversion of a street or road or the extinguishment or alteration of a public right of way.

(2) The removal of the apparatus in pursuance of paragraph 41 constitutes compliance with a requirement to make any other alteration.

(3) A counter-notice under paragraph 41(5) may state (in addition to, or instead of, any of the matters mentioned in paragraph 41(5)(b)) that the operator requires the third party to reimburse the operator in respect of any expenses incurred by the operator in or in connection with the making of any alteration in compliance with the requirements of the third party.

(4) An order made under paragraph 41 on an application by the third party in respect of a counter-notice containing a statement under sub-paragraph (3) must, unless the court otherwise thinks fit, require the third party to reimburse the operator in respect of the expenses referred to in the statement.

(5) Paragraph 44(3)(b) to (e) do not apply.

(6) In this paragraph –
'*road*' means a road in Scotland;
'*street*' means a street in England and Wales or Northern Ireland.'

COMMENTARY

Paragraph 42 amends the paragraph 41 procedure in the ways specified insofar as concerns rights to require the alteration of apparatus in consequence of the stopping up, closure, change or diversion of a street or road or the extinguishment or alteration of a public right of way.

There appears to be an error in the heading, in that the terms of paragraph 42 plainly relate to paragraph 41, and not paragraph 40.

The expression 'alteration' is defined in paragraph 108(2) to include 'moving, removal or replacement', as to which see further paragraph 24.8.3 of the text.

The expression 'counter-notice' is defined in paragraph 41.

The word 'operator' is defined in paragraph 2.

The word 'road' is defined in paragraph 108.

The word 'street' is defined in paragraph 108.

The expression 'the court' is defined in paragraph 94. This is the county court in England, a county court in Northern Ireland, or the sheriff court in Scotland.

The Electronic Communications Code (Jurisdiction) Regulations 2017 came into force on 28 December 2017. The functions conferred on the court are to be conferred on and exercisable by certain of the Tribunals as well as by the court. Regulation 4 provides that applications under Part 6 of the Code may be commenced only in the Upper Tribunal, in relation to England and Wales, or the Lands Tribunal for Scotland, in relation to Scotland.

The expression 'third party' is defined in paragraph 41.

This paragraph is discussed in Section 24.8 of the text. For the labelling differences in Scotland, see further paragraph 40.2.3 of the text.

43 When can a separate application for restoration of land be made?

(1) This paragraph applies if –
   (a) the condition of the land has been affected by the exercise of a code right, and
   (b) restoration of the land to its condition before the code right was exercised does not involve the removal of electronic communications apparatus from any land.

(2) The occupier of the land, the owner of the freehold estate in the land or the lessee of the land ('the relevant person') has the right to require the operator to restore the land if the relevant person is not for the time being bound by the code right. This is subject to sub-paragraph (3).

(3) The relevant person does not have that right if –
   (a) the land is occupied by a person who –
      (i) conferred a code right (which is in force) entitling the operator to affect the condition of the land in the same way as the right mentioned in sub-paragraph (1), or
      (ii) is otherwise bound by such a right, and
   (b) that code right was not conferred in breach of a covenant enforceable by the relevant person.

(4) In the application of sub-paragraph (3)(b) to Scotland the reference to a covenant enforceable by the relevant person is to be read as a reference to a contractual term which is so enforceable.

(5) A person who has the right conferred by this paragraph may give a notice to the operator requiring the operator to restore the land to its condition before the code right was exercised.

(6) The notice must –
   (a) comply with paragraph 89 (notices given by persons other than operators), and
   (b) specify the period within which the operator must complete the works.

(7) The period specified under sub-paragraph (6) must be a reasonable one.

(8) Sub-paragraph (9) applies if, within the period of 28 days beginning with the day on which the notice was given, the landowner and the operator do not reach agreement on any of the following matters –

(a) that the operator will restore the land to its condition before the code right was exercised;
(b) the time at which or period within which the land will be restored.

(9) The landowner may make an application to the court for –
(a) an order under paragraph 44(2) (order requiring operator to restore land), or
(b) an order under paragraph 44(4) (order enabling landowner to recover cost of restoring land).

(10) If the court makes an order under paragraph 44(2), but the operator does not comply with the agreement imposed on the operator and the landowner by virtue of paragraph 44(7), the landowner may make an application to the court for an order under paragraph 44(4).

(11) In the application of sub-paragraph (2) to Scotland the reference to a person who is the owner of the freehold estate in the land is to be read as a reference to a person who is the owner of the land.

COMMENTARY

Paragraph 43, broadly, assists a relevant person (as defined in sub-paragraph (2)) who is not bound by a code right and who wishes the land to be restored to its condition prior to the installation of the electronic communications apparatus, but without wishing to have the apparatus removed. This paragraph, like paragraphs 40 to 42 before it, imposes a notice procedure and permits applications to be made for orders under the provisions of paragraph 44. The procedure in this paragraph is examined in detail in paragraphs 24.9.2–24.9.7 of the text.

The expression 'code right' is defined in paragraph 3.

The expression 'electronic communications apparatus' is defined in paragraph 5.

The word 'land' is defined in paragraph 108.

The word 'landowner' is defined in paragraph 37.

The word 'lessee' is defined in paragraph 108.

The word 'occupier' is defined in paragraph 105.

The word 'operator' is defined in paragraph 2.

The expression 'the court' is defined in paragraph 94. This is the county court in England, a county court in Northern Ireland, or the sheriff court in Scotland.

The Electronic Communications Code (Jurisdiction) Regulations 2017 came into force on 28 December 2017. The functions conferred on the court are to be conferred on and exercisable by certain of the Tribunals as well as by the court. Regulation 4 provides that applications under Part 6 of the Code may be commenced only in the Upper Tribunal, in relation to England and Wales, or the Lands Tribunal for Scotland, in relation to Scotland.

This paragraph is discussed in Section 24.9 of the text. For labelling differences in Scotland, see further paragraph 40.2.5 of the text.

44 What orders may the court make on an application under paragraphs 40 to 43?

(1) An order under this sub-paragraph is an order that the operator must, within the period specified in the order –

(a) remove the electronic communications apparatus, and
(b) restore the land to its condition before the apparatus was placed on, under or over the land.

(2) An order under this sub-paragraph is an order that the operator must, within the period specified in the order, restore the land to its condition before the code right was exercised.

(3) An order under this sub-paragraph is an order that the landowner, occupier or third party may do any of the following –
(a) remove or arrange the removal of the electronic communications apparatus;
(b) sell any apparatus so removed;
(c) recover the costs of any action under paragraph (a) or (b) from the operator;
(d) recover from the operator the costs of restoring the land to its condition before the apparatus was placed on, under or over the land;
(e) retain the proceeds of sale of the apparatus to the extent that these do not exceed the costs incurred by the landowner, occupier or third party as mentioned in paragraph (c) or (d).

(4) An order under this sub-paragraph is an order that the landowner may recover from the operator the costs of restoring the land to its condition before the code right was exercised.

(5) An order under this paragraph on an application under paragraph 40 may require the operator to pay compensation to the landowner for any loss or damage suffered by the landowner as a result of the presence of the apparatus on the land during the period when the landowner had the right to require the removal of the apparatus from the land but was not able to exercise that right.

(6) Paragraph 84 makes further provision about compensation under sub-paragraph (5).

(7) An order under sub-paragraph (1) or (2) takes effect as an agreement between the operator and the landowner, occupier or third party that –
(a) requires the operator to take the steps specified in the order, and
(b) otherwise contains such terms as the court may so specify.

COMMENTARY

Paragraph 44 details the court's order making powers under paragraphs 40 to 43, by listing the four orders which might be made in sub-paragraphs (1) to (4). The substantive effects of those orders are fairly self-explanatory. Paragraph 44 also provides for various self-explanatory ancillary aspects to those orders, which are considered further at paragraphs 24.10.3–24.10.5 of the text.

It should be borne in mind that, if the court makes an order under sub-paragraph (1), and the operator does not comply with the terms of the agreement imposed by sub-paragraph (7), a further application might be made for an order under sub-paragraph (3) pursuant to paragraphs 40(7) or 41(10). Similarly, if the court makes an order under sub-paragraph (2), and the operator does not comply with the terms of the agreement imposed by sub-paragraph (7), a further application might be made for an order under sub-paragraph (4) pursuant to paragraph 43(10).

The expression 'code right' is defined in paragraph 3.

The expression 'electronic communications apparatus' is defined in paragraph 5.
The word 'land' is defined in paragraph 108.
The word 'landowner' is defined in paragraph 37.
The word 'occupier' is defined in paragraph 105.
The word 'operator' is defined in paragraph 2.
The expression 'the court' is defined in paragraph 94. This is the county court in England, a county court in Northern Ireland, or the sheriff court in Scotland.

The Electronic Communications Code (Jurisdiction) Regulations 2017 came into force on 28 December 2017. The functions conferred on the court are to be conferred on and exercisable by certain of the Tribunals as well as by the court. Regulation 4 provides that applications under Part 6 of the Code may be commenced only in the Upper Tribunal, in relation to England and Wales, or the Lands Tribunal for Scotland, in relation to Scotland.

The expression 'third party' is defined in paragraph 41.

This paragraph is discussed in Section 24.10 of the text. For the interaction between the Part 6 procedure for removal of apparatus and the Landlord and Tenant Act 1954, see further paragraph 25.7.5 of the text. As to compensation, see further Chapter 30 of the text, in particular paragraphs 30.5.5 and 30.5.6.

Part 7 Conferral of transport land rights and their exercise

45 Introductory

This Part of this code makes provision about –

(a) the conferral of transport land rights, and
(b) the exercise of transport land rights.

COMMENTARY

Paragraph 45 introduces paragraphs 46 to 55. It is self-explanatory.

The expression 'transport land right' is defined in paragraph 48.

46 Transport land and transport undertakers

In this Part of this code –

'*transport land*' means land which is used wholly or mainly –

(a) as a railway, canal or tramway, or
(b) in connection with a railway, canal or tramway on the land;

'*transport undertaker*', in relation to transport land, means the person carrying on the railway, canal or tramway undertaking.

COMMENTARY

Paragraph 46 defines 'transport land' and 'transport undertaker' for the purposes of Part 7. It is self-explanatory.

There is no statutory language limiting the vertical extent of the 'transport land', as to which see further the discussion at paragraph 26.1.6 of the text.

The word 'land' is defined in paragraph 108.

This paragraph is referred to in paragraphs 26.1.2 and 26.1.3 of the text.

47 Conferral of transport land rights

(1) An operator may exercise a transport land right for the statutory purposes.
(2) But that is subject to the following provisions of this Part of this code.

COMMENTARY

Paragraph 47 confers transport land rights, which may be exercised by an operator for the statutory purposes, subject to the remaining provisions of Part 7.

The word 'operator' is defined in paragraph 2.

The expression 'the statutory purposes' is defined in paragraph 4.

The expression 'transport land' is defined in paragraph 46.

The expression 'transport land right' is defined in paragraph 48.

This paragraph is referred to in paragraph 26.1.4 of the text.

48 The transport land rights

(1) For the purposes of this code a *'transport land right'*, in relation to an operator, is –
   (a) a right to cross any transport land with a line;
   (b) a right, for the purposes of crossing any transport land with a line –
      (i) to install and keep the line and any other electronic communications apparatus on, under or over the transport land;
      (ii) to inspect, maintain, adjust, alter, repair, upgrade or operate electronic communications apparatus on, under or over the transport land;
      (iii) a right to carry out any works on the transport land for or in connection with the exercise of a right under sub-paragraph (i) or (ii);
      (iv) a right to enter the transport land to inspect, maintain, adjust, alter, repair, upgrade or operate the line or other electronic communications apparatus.
(2) A line installed in the exercise of a transport land right need not cross the transport land in question by a direct route or the shortest route from the point at which the line enters the transport land.
(3) But the line must not cross the transport land by any route which, in the horizontal plane, exceeds that shortest route by more than 400 metres.
(4) The transport land rights do not authorise an operator to install a line or other electronic communications apparatus in any position on transport land in which the line or other apparatus would interfere with traffic on the railway, canal or tramway.

COMMENTARY

Paragraph 48 defines the expression 'transport land right'.

The word 'apparatus' is defined in s.405 of the 2003 Act.

The expression 'electronic communications apparatus' is defined in paragraph 5.

The word 'line' is defined in paragraph 5.
The word 'operator' is defined in paragraph 2.
The expression 'transport land' is defined in paragraph 46.

This paragraph is considered in paragraphss 26.1.4–26.1.5 of the text.

49 Non-emergency works: when can an operator exercise the transport land rights?

(1) Before exercising a transport land right in order to carry out non-emergency works, the operator must give the transport undertaker notice of the intention to carry out the works ('notice of proposed works').
(2) Notice of proposed works must contain a plan and section of the works; but, if the transport undertaker agrees, the notice may instead contain a description of the works (whether or not in the form of a diagram).
(3) The operator must not begin the proposed works until the notice period has ended.
(4) But the operator's power to carry out the proposed works is subject to paragraph 50.
(5) In this paragraph –
'non-emergency works' means any works which are not emergency works under paragraph 51;
'notice period' means the period of 28 days beginning with the day on which notice of proposed works is given.

COMMENTARY

Paragraph 49 deals with the procedure for exercising transport land rights to carry out non-emergency works (as defined in sub-paragraph (5)).

Sub-paragraph (1) provides that the operator must give notice of proposed works to the transport undertaker (complying with the formal requirements of sub-paragraph (2)) before exercising such a right.

The operator must not begin the proposed works until after 28 days from the day on which the notice of proposed works is given (sub-paragraph (3)) and the power to begin them is subject to paragraph 50 (sub-paragraph (4)).

The expression 'emergency works' is defined in paragraph 51.
The word 'operator' is defined in paragraph 2.
The expression 'transport land right' is defined in paragraph 48.
The expression 'transport undertaker' is defined in paragraph 46.

This paragraph is discussed in Section 26.2 of the text.

50 What is the effect of the transport undertaker giving notice of objection to the operator?

(1) This paragraph applies if an operator gives a transport undertaker notice of proposed works under paragraph 49.
(2) The transport undertaker may, within the notice period, give the operator notice objecting to the proposed works ('notice of objection').

(3) If notice of objection is given, the operator or the transport undertaker may, within the arbitration notice period, give the other notice that the objection is to be referred to arbitration under paragraph 52 ('arbitration notice').

(4) In a case where notice of objection is given, the operator may exercise a transport land right in order to carry out the proposed works only if they are permitted under sub-paragraph (5) or (6).

(5) Works are permitted in a case where –

(a) the arbitration notice period has ended, and

(b) no arbitration notice has been given.

(6) In a case where arbitration notice has been given, works are permitted in accordance with an award made on the arbitration.

(7) In this paragraph –

(a) *'arbitration notice period'* means the period of 28 days beginning with the day on which objection notice is given;

(b) expressions defined in paragraph 49 have the same meanings as in that paragraph.

COMMENTARY

Paragraph 50 permits a transport undertaker served with a notice of proposed works to give a notice of objection within 28 days. If it does so, then within 28 days of the notice of objection, either party may give notice to the other referring the dispute to arbitration. If neither party does so, the operator may exercise the transport land right in order to carry out the proposed works after the expiry of the 28-day period (sub-paragraph (5)). If the objection is referred to arbitration, the operator may exercise the transport land right in order to carry out the proposed works in accordance with an award made on the arbitration (sub-paragraph (6)).

Paragraph 52 makes provision about arbitration.

The expression 'notice of proposed works' is defined in paragraph 49.

The expression 'notice period' is defined in paragraph 49.

The word 'operator' is defined in paragraph 2.

The expression 'transport land right' is defined in paragraph 48.

The expression 'transport undertaker' is defined in paragraph 46.

This paragraph is discussed in paragraphs 26.2.4 and 26.2.5 of the text.

51 Emergency works: when can an operator exercise the transport land rights?

(1) An operator may exercise a transport land right in order to carry out emergency works.

(2) If the operator exercises a transport land right to carry out emergency works, the operator must give the transport undertaker an emergency works notice as soon as reasonably practicable after starting the works.

(3) An *'emergency works notice'* is a notice which –

(a) identifies the emergency works;

(b) contains a statement of the reason why the works are emergency works; and

(c) contains either –

(i) the matters which would be included in a notice of proposed works (if one were given in relation to the works), or
(ii) a reference to a notice of proposed works which relates to the works that are emergency works (if one has been given).

(4) A transport undertaker may, within the compensation notice period, give the operator notice which requires the operator to pay compensation for loss or damage sustained in consequence of the carrying out of emergency works ('compensation notice').

(5) The operator must pay the transport undertaker any compensation which is required by a compensation notice (if given within the compensation notice period).

(6) The amount of compensation payable under sub-paragraph (5) is to be agreed between the operator and the transport undertaker.

(7) But if –
(a) the compensation agreement period has ended, and
(b) the operator and the transport undertaker have not agreed the amount of compensation payable under sub-paragraph (6),

the operator or the transport undertaker may give the other notice that the disagreement is to be referred to arbitration under paragraph 52.

(8) A reference in this paragraph to emergency works includes a reference to any works which are included in a notice of proposed works but become emergency works before the operator is authorised by paragraph 50 or 51 to carry them out.

(9) In this paragraph –
'compensation agreement period' means the period of 28 days beginning with the day on which a compensation notice is given;
'compensation notice period' means the period of 28 days beginning with the day on which an emergency works notice is given;
'emergency works' means works carried out in order to stop anything already occurring, or to prevent anything imminent from occurring, which is likely to cause –
(a) danger to persons or property,
(b) the interruption of any service provided by the operator's network, or
(c) substantial loss to the operator,

and any other works which it is reasonable (in all the circumstances) to carry out with those works;
'notice of proposed works' means such notice given under paragraph 49.

COMMENTARY

Paragraph 51 contains in sub-paragraph (9) the definition of the expression 'emergency works'.

In an emergency there will be no time to comply with the paragraph 50 notice procedure. Accordingly, paragraph 51 makes different provision for transport land rights to be exercisable by the operator to enable it to undertake emergency works. The operator must give an emergency works notice (as defined in sub-paragraph (3)) to the transport undertaker as soon as practicable after starting the works. Within 28 days of the emergency works notice, the transport undertaker may give a compensation notice to the operator, requiring compensation for loss or damage sustained in consequence of the carrying out of the emergency works. Absent agreement as to the amount of compensation within a further 28-day period from the date of giving the compensation notice, either the operator or the transport operator may give notice to the other referring the dispute to arbitration.

Paragraph 52 makes provision about arbitration.

The word 'network' is defined in paragraph 6.

The word 'operator' is defined in paragraph 2.

The word 'provide' is defined by s.405 of the 2003 Act.

The expression 'transport land right' is defined in paragraph 48.

The expression 'transport undertaker' is defined in paragraph 46.

This paragraph is discussed in Section 26.3 of the text.

52 What happens if a dispute about the transport land rights is referred to arbitration?

(1) This paragraph applies if notice is given under paragraph 50(3) or 51(7) that the following matter (the 'matter in dispute') is to be referred to arbitration –
   (a) an objection to proposed works;
   (b) a disagreement about an amount of compensation.

(2) The matter in dispute is to be referred to the arbitration of a single arbitrator appointed –
   (a) by agreement between the parties, or
   (b) in the absence of such agreement, by the President of the Institution of Civil Engineers.

(3) If the matter in dispute is an objection to proposed works, the arbitrator has the following powers –
   (a) power to require the operator to give the arbitrator a plan and section in such form as the arbitrator thinks appropriate;
   (b) power to require the transport undertaker to give the arbitrator any observations on such a plan or section in such form as the arbitrator thinks appropriate;
   (c) power to impose on either party any other requirements which the arbitrator thinks appropriate (including a requirement to provide information in such form as the arbitrator thinks appropriate);
   (d) power to make an award –
      (i) requiring modifications to the proposed works, and
      (ii) specifying the terms on which, and the conditions subject to which, the proposed works may be carried out;
   (e) power to award one or both of the following, payable to the transport undertaker –
      (i) compensation for loss or damage sustained by that person in consequence of the carrying out of the works;
      (ii) consideration for the right to carry out the works.

(4) If the matter in dispute is a disagreement about an amount of compensation, the arbitrator has the following powers –
   (a) power to impose on either party any requirements which the arbitrator thinks appropriate (including a requirement to provide information in such form as the arbitrator thinks appropriate);

(b) power to award compensation, payable to the transport undertaker, for loss or damage sustained by that person in consequence of the carrying out of the emergency works.

(5) The arbitrator may make an award conditional upon a party complying with a requirement imposed under sub-paragraph (3)(a), (b) or (c) or (4)(a).

(6) In determining what award to make, the matters to which the arbitrator must have regard include the public interest in there being access to a choice of high quality electronic communications services.

(7) The arbitrator's power under sub-paragraph (3) or (4) to award compensation for loss includes power to award compensation for any increase in the expenses incurred by the transport undertaker in carrying on its railway, canal or tramway undertaking.

(8) An award of consideration under sub-paragraph (3)(e)(ii) must be determined on the basis of what would have been fair and reasonable if the transport undertaker had willingly given authority for the works to be carried out on the same terms, and subject to the same conditions (if any), as are contained in the award.

(9) In this paragraph *'party'* means –
(a) the operator, or
(b) the transport undertaker.

COMMENTARY

Paragraph 52 makes provision in respect of disputes about transport land rights which are referred to arbitration by a notice given under either paragraph 50(3) or 51(7).

The word 'access' is to be interpreted in accordance with s.405 of the 2003 Act.

The expression 'electronic communications service' is defined in s.32 of the 2003 Act.

The word 'information' is defined in s.405 of the 2003 Act.

The word 'modification' is defined in s.405 of the 2003 Act.

The word 'operator' is defined in paragraph 2.

The expression 'transport undertaker' is defined in paragraph 46.

This paragraph and the arbitrator's powers including those concerning compensation are discussed in Section 26.4 and paragraph 26.3.5 of the text. As to arbitration under the codes generally, see further Section 33.11 of the text.

53 When can a transport undertaker require an operator to alter communications apparatus?

(1) A transport undertaker may give an operator notice which requires the operator to alter a line or other electronic communications apparatus specified in the notice ('notice requiring alterations') on the ground that keeping the apparatus on, under or over transport land interferes with, or is likely to interfere with –
(a) the carrying on of the transport undertaker's railway, canal or tramway undertaking, or
(b) anything done or to be done for the purposes of its railway, canal or tramway undertaking.

(2) The operator may, within the notice period, give the transport undertaker notice ('counter-notice') specifying the respects in which the operator is not prepared to comply with the notice requiring alterations.
(3) The operator must comply with the notice requiring alterations, within a reasonable time and to the reasonable satisfaction of the transport undertaker, if –
(a) the notice period has ended, and
(b) no counter-notice has been given.
(4) If counter-notice has been given (within the notice period), the transport undertaker may apply to the court for an order requiring the operator to alter any of the specified apparatus.
(5) The court must not make an order unless it is satisfied that the order is necessary on one of the grounds mentioned in sub-paragraph (1).
(6) In determining whether to make an order, the matters to which the court must also have regard include the public interest in there being access to a choice of high quality electronic communications services.
(7) An order under this paragraph may take such form and be on such terms as the court thinks fit.
(8) In particular, the order –
(a) may impose such conditions, and
(b) may contain such directions to the operator or the transport undertaker, as the court thinks necessary for resolving any difference between the operator and the transport undertaker and for protecting their respective interests.
(9) In this paragraph –
'*notice period*' means the period of 28 days beginning with the day on which notice requiring alterations is given;
'*specified apparatus*' means the line or other electronic communications apparatus specified in notice requiring alterations.

COMMENTARY

Paragraph 53 provides a procedure by which a transport undertaker can require an operator to make alterations to alter a line or other electronic communications apparatus specified in a notice on the basis that the apparatus interferes or is likely to interfere with the carrying on of, or anything done or to be done for the purposes of the transport undertaker's undertaking.

The word 'access' is to be interpreted in accordance with s.405 of the 2003 Act.

The expression 'electronic communications apparatus' is defined in paragraph 5.

The expression 'electronic communications service' is defined in s.32 of the 2003 Act.

The word 'line' is defined in paragraph 5.

The word 'operator' is defined in paragraph 2.

The expression 'the court' is defined in paragraph 94. This is the county court in England, a county court in Northern Ireland, or the sheriff court in Scotland.

However, proceedings under this paragraph may be commenced only in the Upper Tribunal, in relation to England and Wales, or the Lands Tribunal for Scotland, in relation to Scotland: regulation 2(1)(b) and 4 of the Electronic Communications Code (Jurisdiction) Regulations 2017.

The expression 'transport land' is defined in paragraph 46.

The expression 'transport undertaker' is defined in paragraph 46.

This paragraph is discussed in Section 26.5 of the text.

54 What happens to the transport land rights if land ceases to be transport land?

(1) This paragraph applies if an operator is exercising a transport land right in relation to land immediately before a time when it ceases to be transport land.
(2) After that time, this Part of this code – except for paragraph 53 – continues to apply to the land as if it were still transport land (and, accordingly, the operator may continue to exercise any transport land right in relation to the land as if it were still transport land).
(3) But sub-paragraph (2) is subject to sub-paragraphs (4) to (9).
(4) In the application of this Part of this code to land in accordance with sub-paragraph (2), references to the transport undertaker have effect as references to the occupier of the land.
(5) The application of this Part of this code to land in accordance with sub-paragraph (2) does not authorise the operator –
  (a) to cross the land with any line that is not in place at the time when the land ceases to be transport land, or
  (b) to install and keep any line or other electronic communications apparatus that is not in place at the time when the land ceases to be transport land.
(6) But sub-paragraph (5) does not affect the power of the operator to replace an existing line or other apparatus (whether in place at the time when the land ceased to be transport land or a replacement itself authorised by this sub-paragraph) with a new line or apparatus which –
  (a) is not substantially different from the existing line or apparatus, and
  (b) is not in a significantly different position.
(7) The occupier of the land may, at any time after the land ceases to be transport land, give the operator notice specifying a date on which this Part of this code is to cease to apply to the land in accordance with this paragraph ('notice of termination').
(8) That date specified in the notice of termination must fall after the end of the period of 12 months beginning with the day on which the notice of termination is given.
(9) On the date specified in notice of termination in accordance with sub-paragraph (8), the transport land rights cease to be exercisable in relation to the land in accordance with this paragraph.

COMMENTARY

By paragraph 54, Part 7 of the New Code (except paragraph 53) continues to apply to land which ceases to be transport land as if it were still transport land, as modified by sub-paragraphs (4) to (6). The occupier of the land may give the operator a notice specifying a date on which Part 7 will cease to apply, which must fall after the end of a 12-month period from the date of the notice.

The word 'apparatus' is defined in s.405 of the 2003 Act.

The expression 'electronic communications apparatus' is defined in paragraph 5.

The word 'land' is defined in paragraph 108.

The word 'line' is defined in paragraph 5.

The word 'occupier' is defined in paragraph 105.
The word 'operator' is defined in paragraph 2.
The expression 'transport land' is defined in paragraph 46.
The expression 'transport land right' is defined in paragraph 48.

This paragraph is discussed in Sections 26.6 and 26.7 of the text.

55 Offence: operators who do not comply with this Part of this code

(1) An operator is guilty of an offence if the operator starts any works in contravention of any provision of paragraph 49, paragraph 50 or paragraph 51.
(2) An operator guilty of an offence under this paragraph is liable on summary conviction to a fine not exceeding level 3 on the standard scale.
(3) In a case where this Part of this code applies in accordance with paragraph 54, the reference in this paragraph to paragraph 49, paragraph 50 or paragraph 51 is a reference to that paragraph as it applies in accordance with paragraph 54.

COMMENTARY

Paragraph 55 creates criminal offences where an operator starts works in contravention of paragraphs 49 to 51. Operators must accordingly be careful to check whether works are 'emergency works' within the definition in paragraph 51(9).
The word 'contravention' is defined in s.405 of the 2003 Act.
The word 'operator' is defined in paragraph 2.

This paragraph is discussed in Section 26.8 of the text.

Part 8 Conferral of street work rights and their exercise

56 Introductory

This Part of this code makes provision about –
(a) the conferral of street work rights, and
(b) the exercise of street work rights.

COMMENTARY

Paragraph 56 introduces paragraphs 57 to 59. It is self-explanatory.
The expression 'street work right' is defined in paragraph 59.

57 Streets and roads

In this Part of this code –
'road' means –
(a) a road in Scotland which is a public road;
(b) a road in Northern Ireland;

'street' means a street in England and Wales which is a maintainable highway (within the meaning of Part 3 of New Roads and Street Works Act 1991), other than one which is a footpath, bridleway or restricted byway that crosses, and forms part of, any agricultural land or any land which is being brought into use for agriculture.

COMMENTARY

Paragraph 57 defines the words 'street' and 'road' for the purposes of Part 8.
The word 'agricultural' is defined in paragraph 108.
The word 'agriculture' is defined in paragraph 108.
The word 'bridleway' is defined in paragraph 108.
The word 'footpath' is defined in paragraph 108.
The word 'land' is defined in paragraph 108.
The expression 'restricted byway' is defined in paragraph 108.
The word 'road' is defined generally in paragraph 108.
The word 'street' is defined generally in paragraph 108.

This paragraph is discussed in paragraphs 27.1.2–27.1.3 of the text. For labelling differences in Scotland, see further paragraph 40.2.3 of the text.

58 Conferral of street work rights

(1) An operator may exercise a street work right for the statutory purposes.
(2) But that is subject to the following provisions of this Part of this code.

COMMENTARY

Paragraph 58 permits an operator to exercise a street work right for the statutory purposes, subject to paragraph 59.
The word 'operator' is defined in paragraph 2.
The expression 'street work right' is defined in paragraph 59.
The expression 'the statutory purposes' is defined in paragraph 4.

This paragraph is referred to in paragraph 27.1.4 of the text.

59 The street work rights

(1) For the purposes of this code a *'street work right'*, in relation to an operator, is –
  (a) a right to install and keep electronic communications apparatus in, on, under, over, along or across a street or a road;
  (b) a right to inspect, maintain, adjust, alter, repair, upgrade or operate electronic communications apparatus which is installed or kept by the exercise of the right under paragraph (a);
  (c) a right to carry out any works in, on, under, over, along or across a street or road for or in connection with the exercise of a right under paragraph (a) or (b);
  (d) a right to enter any street or road to inspect, maintain, adjust, alter, repair, upgrade or operate electronic communications apparatus which is installed or kept by the exercise of the right under paragraph (a).

(2) The works that may be carried out under sub-paragraph (1)(c) include –
- (a) breaking up or opening a street or a road;
- (b) tunnelling or boring under a street or a road;
- (c) breaking up or opening a sewer, drain or tunnel.

COMMENTARY

Paragraph 59 defines the expression 'street work right'.

The expression 'electronic communications apparatus' is defined in paragraph 5.
The word 'operator' is defined in paragraph 2.
The word 'road' is defined in paragraph 57.
The word 'street' is defined in paragraph 57.

This paragraph is considered in paragraphs 27.1.4–27.1.5 of the text.

Part 9 Conferral of tidal water rights and their exercise

60 Introductory

This Part of this code makes provision about –
- (a) the conferral of tidal water rights, and
- (b) the exercise of tidal water rights.

COMMENTARY

Paragraph 60 introduces paragraphs 61 to 64. It is self-explanatory.

The expression 'tidal water right' is defined in paragraph 63.

61 Tidal water or lands

In this Part of this code '*tidal water or lands*' includes –
- (a) any estuary or branch of the sea,
- (b) the shore below mean high water springs, and
- (c) the bed of any tidal water.

COMMENTARY

Paragraph 61 defines 'tidal water or lands' for the purposes of Part 9.

This paragraph is referred to in paragraph 27.2.2 of the text.

62 Conferral of tidal water rights

(1) An operator may exercise a tidal water right for the statutory purposes.
(2) But that is subject to the following provisions of this Part of this code.

COMMENTARY

Paragraph 62 permits an operator to exercise a street work right for the statutory purposes, subject to paragraphs 63 and 64.

The word 'operator' is defined in paragraph 2.

The expression 'the statutory purposes' is defined in paragraph 4.

The expression 'tidal water right' is defined in paragraph 63.

This paragraph is referred to in paragraph 27.2.3 of the text.

63 The tidal water rights

(1) For the purposes of this code a *'tidal water right'*, in relation to an operator, is –
 (a) a right to install and keep electronic communications apparatus on, under or over tidal water or lands;
 (b) a right to inspect, maintain, adjust, alter, repair, upgrade or operate electronic communications apparatus on, under or over the tidal water or lands;
 (c) a right to carry out any works on, under or over any tidal water or lands for or in connection with the exercise of a right under paragraph (a) or (b);
 (d) a right to enter any tidal water or lands to inspect, maintain, adjust, alter, repair, upgrade or operate electronic communications apparatus which is installed or kept by the exercise of the right under paragraph (a).

(2) The works that may be carried out under sub-paragraph (1)(c) include placing a buoy or seamark.

COMMENTARY

Paragraph 63 defines the expression 'tidal water right'.

The expression 'electronic communications apparatus' is defined in paragraph 5.

The word 'operator' is defined in paragraph 2.

The expression 'tidal water or lands' is defined in paragraph 61.

This paragraph is considered in paragraph 27.2.4 of the text.

64 Exercise of tidal water right: Crown land

(1) An operator may not exercise a tidal water right in relation to land in which a Crown interest subsists unless agreement to the exercise of the right in relation to the land has been given in respect of that interest by the appropriate authority in accordance with paragraph 104.

(2) Where, in connection with an agreement between the operator and the appropriate authority for the exercise of such a right, the operator and the appropriate authority cannot agree the consideration to be paid by the operator, the operator or the appropriate authority may apply to the appointed valuer for a determination of the market value of the right.

(3) An application under sub-paragraph (2) must be made in writing and must include –
   (a) the proposed terms of the agreement, and
   (b) the reasoned evidence of the operator and of the appropriate authority as to the market value of the right.

(4) As soon as reasonably practicable after receiving such an application, the appointed valuer must –
   (a) determine the market value of the tidal water right; and
   (b) notify the operator and the appropriate authority in writing of its determination and the reasons for it.

(5) If the agreement mentioned in sub-paragraph (2) or an agreement in substantially the same terms is concluded following a determination under sub-paragraph (4), the consideration payable by the operator must not be more than the market value notified under sub-paragraph (4)(b).

(6) For this purpose the market value of a tidal water right is, subject to sub-paragraph (7), the amount that, at the date the market value is assessed, a willing buyer would pay a willing seller for the right –
   (a) in a transaction at arm's length,
   (b) on the basis that the buyer and seller were acting prudently and with full knowledge of the transaction, and
   (c) on the basis that the transaction was subject to the proposed terms set out in the application.

(7) The market value must be assessed on these assumptions –
   (a) that the right that the transaction relates to does not relate to the provision or use of an electronic communications network;
   (b) that the right in all other respects corresponds to the tidal water right;
   (c) that there is more than one site which the buyer could use for the purpose for which the buyer seeks the right.

(8) The appointed valuer may charge a fee in respect of the consideration of an application under sub-paragraph (4) and may apportion the fee between the operator and the appropriate authority as the appointed valuer considers appropriate.

(9) In this paragraph *'the appointed valuer'* means –
   (a) such person as the operator and the appropriate authority may agree;
   (b) if no person is agreed, such person as may be nominated, on the application of the operator or the appropriate authority, by the President of the Royal Institution of Chartered Surveyors.

COMMENTARY

Paragraph 64 deals with the exercise of tidal water rights in respect of Crown land.

Sub-paragraph (1) prevents an operator from exercising a tidal water right in relation to land in which a Crown interest subsists unless agreement to the exercise of the right in relation to the land has been given in respect of that interest by the appropriate authority in accordance with paragraph 104.

The remainder of this paragraph deals with the procedure and tests for resolution of disputes as to the appropriate consideration to be paid for the exercise of the right, by application to the appointed valuer.

The expression 'Crown interest' is defined in paragraph 104.
The expression 'electronic communications network' is defined in s.32 of the 2003 Act.
The word 'land' is defined in paragraph 108.
The word 'operator' is defined in paragraph 2.
The expression 'tidal water right' is defined in paragraph 63.

This paragraph is discussed in paragraphs 27.2.5–27.2.8 of the text.

Part 10 Undertaker's works affecting electronic communications apparatus

65 Introductory

This Part of this code makes provision about the carrying out of undertaker's works by undertakers or operators.

COMMENTARY

Paragraph 65 introduces paragraphs 66 to 72. It is self-explanatory.
The word 'operator' is defined in paragraph 2.
The word 'undertaker' is defined in paragraph 66.
The expression 'undertaker's works' is defined in paragraph 66.

66 Key definitions

(1) In this Part of this code –
'undertaker' means a person (including a local authority) of a description set out in any of the entries in the first column of the following table;
'undertaker's works', in relation to an undertaker of a description set out in a particular entry in the first column of the table, means works of the description set out in the corresponding entry in the second column of the table.

'undertaker'	*'undertaker's works'*
A person authorised by any enactment (whether public general or local) or by any order or scheme made under or confirmed by any enactment to carry on any railway, tramway, road transport, water transport, canal, inland navigation, dock, harbour, pier or lighthouse undertaking	Works that the undertaker is authorised to carry out for the purposes of, or in connection with, the undertaking which it carries on
A person (apart from the operator) to whom this code is applied by a direction under section 106 of the 2003 Act	Works that the undertaker is authorised to carry out by or in accordance with any provision of this code
Any person to whom this Part of this code is applied by any enactment (whenever passed or made)	Works for the purposes of which this paragraph is applied to the undertaker

(2) In this Part of this code –

(a) a reference to undertaker's works which interfere with a network is a reference to any undertaker's works which involve, or are likely to involve, an alteration of any electronic communications apparatus kept on, under or over any land for the purposes of an operator's network;

(b) a reference to an alteration of any electronic communications apparatus is a reference to a temporary or permanent alteration of the apparatus.

COMMENTARY

Paragraph 66 defines the expressions 'undertaker', 'undertaker's works', 'undertaker's works which interfere with a network' and 'an alteration of any electronic communications apparatus' for the purposes of Part 10.

The expression 'electronic communications apparatus' is defined in paragraph 5.

The word 'enactment' is defined in paragraph 108.

The word 'land' is defined in paragraph 108.

The word 'network' is defined in paragraph 6.

The word 'operator' is defined in paragraph 2.

The word 'road' is defined in paragraph 108.

This paragraph is referred to in paragraphs 27.3.2–27.3.3 of the text.

67 When can an undertaker carry out non-emergency undertaker's works?

(1) Before carrying out non-emergency undertaker's works which interfere with a network, an undertaker must give the operator notice of the intention to carry out the works ('notice of proposed works').

(2) Notice of proposed works must specify –

(a) the nature of the proposed undertaker's works,

(b) the alteration of the electronic communications apparatus which the works involve or are likely to involve, and

(c) the time and place at which the works will begin.

(3) The undertaker must not begin the proposed undertaker's works (including the proposed alteration of electronic communications apparatus) until the notice period has ended.

(4) But the undertaker's power to alter electronic communications apparatus (in carrying out the proposed undertaker's works) is subject to paragraph 68.

(5) In this paragraph –

'non-emergency undertaker's works' means any undertaker's works which are not emergency works under paragraph 71;

'notice period' means the period of 10 days beginning with the day on which notice of proposed works is given.

COMMENTARY

Paragraph 67 provides that any non-emergency undertaker's works which interfere with a network must not be begun until a notice of proposed works is given specifying the matters set out in sub-paragraph (2) and a 10-day notice period has ended.

However, insofar as the proposed undertaker's works involve altering electronic communications apparatus, the power to carry out the works is subject to paragraph 68.

The expression 'electronic communications apparatus' is defined in paragraph 5.

The word 'network' is defined in paragraph 6.

The word 'operator' is defined in paragraph 2.

The word 'undertaker' is defined in paragraph 66.

The expression 'undertaker's works' is defined in paragraph 66.

This paragraph is discussed in paragraphs 27.3.4–27.3.8 of the text.

68 What is the effect of the operator giving counter-notice to the undertaker?

(1) This paragraph applies if an undertaker gives an operator notice of proposed works under paragraph 67.

(2) The operator may, within the notice period, give the undertaker notice ('counter-notice') stating either –

(a) that the operator requires the undertaker to make any alteration of the electronic communications apparatus that is necessary or expedient because of the proposed undertaker's works –

(i) under the supervision of the operator, and

(ii) to the satisfaction of the operator; or

(b) that the operator intends to make any alteration of the electronic communications apparatus that is necessary or expedient because of the proposed undertaker's works.

(3) In a case where counter-notice contains a statement under sub-paragraph (2)(a), the undertaker must act in accordance with the counter-notice when altering electronic communications apparatus (in carrying out the proposed undertaker's works).

(4) But, if the operator unreasonably fails to provide the required supervision, the undertaker must act in accordance with the counter-notice only insofar as it requires alterations to be made to the satisfaction of the operator.

(5) In a case where counter-notice contains a statement under sub-paragraph (2)(b) (operator intends to make alteration), the undertaker must not alter electronic communications apparatus (in carrying out the proposed undertaker's works).

(6) But that does not prevent the undertaker from making any alteration of electronic communications apparatus which the operator fails to make within a reasonable time.

(7) Expressions defined in paragraph 67 have the same meanings in this paragraph.

COMMENTARY

Paragraph 68 permits an operator who is given a notice of proposed works under paragraph 67 to give a counter-notice in accordance with the requirements of sub-paragraph (2) within the 10 day notice period. The effect of the counter-notice is set out in sub-paragraphs (3) to (6), which are self-explanatory (although there may be some scope for argument in individual cases as to what constitutes an unreasonable failure to provide the required supervision under sub-paragraph (4) or a reasonable time for the purposes of sub-paragraph (6)).

The expression 'electronic communications apparatus' is defined in paragraph 5.
The expression 'notice of proposed works' is defined in paragraph 67.
The expression 'notice period' is defined in paragraph 67.
The word 'operator' is defined in paragraph 2.
The word 'undertaker' is defined in paragraph 66.
The expression 'undertaker's works' is defined in paragraph 66.

This paragraph is discussed in paragraph 27.3.9 of the text.

69 What expenses must the undertaker pay?

(1) This paragraph applies if an undertaker carries out any non-emergency undertaker's works in accordance with paragraph 67 (including in a case where counter-notice is given under paragraph 68).
(2) The undertaker must pay the operator the amount of any loss or damage sustained by the operator in consequence of any alteration being made to electronic communications apparatus (in carrying out the works).
(3) The undertaker must pay the operator any expenses incurred by the operator in, or in connection with, supervising the undertaker when altering electronic communications apparatus (in carrying out the works).
(4) Any amount which is not paid in accordance with this paragraph is to be recoverable by the operator from the undertaker in any court of competent jurisdiction.

COMMENTARY

Paragraph 69(2) makes the undertaker liable to the operator for the amount of loss or damage sustained by the operator in consequence of any alteration being made to electronic communications apparatus in carrying out any non-emergency undertaker's works under paragraph 67.

Paragraph 69(3) makes the undertaker liable to the operator for any expenses incurred by the operator in, or in connection with, supervising the undertaker when altering electronic communications apparatus in carrying out any non-emergency undertaker's works under paragraph 67.

The expression 'counter-notice' is defined in paragraph 68.
The expression 'electronic communications apparatus' is defined in paragraph 5.
The expression 'non-emergency undertaker's works' is defined in paragraph 67.
The word 'operator' is defined in paragraph 2.
The word 'undertaker' is defined in paragraph 66.
The expression 'undertaker's works' is defined in paragraph 66.

This paragraph is discussed in paragraph 27.3.10 of the text.

70 When can the operator alter apparatus in connection with non-emergency undertaker's works?

(1) An operator may make an alteration of electronic communications apparatus if –

(a) notice of proposed works has been given,
(b) the notice period has ended, and
(c) counter-notice has been given which states (in accordance with paragraph 68(2)(b)) that the operator intends to make the alteration.

(2) If the operator makes any alteration in accordance with this paragraph, the undertaker must pay the operator –
(a) any expenses incurred by the operator in, or in connection with, making the alteration; and
(b) the amount of any loss or damage sustained by the operator in consequence of the alteration being made.

(3) Any amount which is not paid in accordance with sub-paragraph (2) is to be recoverable by the operator from the undertaker in any court of competent jurisdiction.

(4) Expressions defined in paragraph 67 have the same meanings in this paragraph.

COMMENTARY

Paragraph 70 deals with the operator's power to alter electronic communications apparatus having given a counter-notice under paragraph 68(2)(b) after the end of the notice period.

Sub-paragraph (2) requires the undertaker to pay the operator for any expenses incurred in, or in connection with, making the alteration, and the amount of any loss or damage sustained by the operator in consequence of the alteration being made.

The expression 'counter-notice' is defined in paragraph 68.
The expression 'electronic communications apparatus' is defined in paragraph 5.
The expression 'notice of proposed works' is defined in paragraph 67.
The expression 'notice period' is defined in paragraph 67.
The word 'operator' is defined in paragraph 2.
The word 'undertaker' is defined in paragraph 66.

This paragraph is discussed in paragraph 27.3.11 of the text.

71 When can an undertaker carry out emergency undertaker's works?

(1) An undertaker may, in carrying out emergency undertaker's works, make an alteration of any electronic communications apparatus kept on, under or over any land for the purposes of an operator's network.

(2) The undertaker must give the operator notice of the emergency undertaker's works as soon as practicable after beginning them.

(3) This paragraph does not authorise the undertaker to make an alteration of apparatus after any failure by the undertaker to give notice in accordance with sub-paragraph (2).

(4) The undertaker must make the alteration to the satisfaction of the operator.

(5) If the undertaker makes any alteration in accordance with this paragraph, the undertaker must pay the operator –
(a) any expenses incurred by the operator in, or in connection with, supervising the undertaker when making the alteration; and

(b) the amount of any loss or damage sustained by the operator in consequence of the alteration being made.

(6) Any amount which is not paid in accordance with sub-paragraph (5) is to be recoverable by the operator from the undertaker in any court of competent jurisdiction.

(7) In this paragraph *'emergency undertaker's works'* means undertaker's works carried out in order to stop anything already occurring, or to prevent anything imminent from occurring, which is likely to cause –

(a) danger to persons or property,

(b) interference with the exercise of any functions conferred or imposed on the undertaker by or under any enactment, or

(c) substantial loss to the undertaker,

and any other works which it is reasonable (in all the circumstances) to carry out with those works.

COMMENTARY

Paragraph 71 deals with the carrying out of emergency undertaker's works, as defined in sub-paragraph (7), including provision for compensation under sub-paragraphs (5) and (6). It is self-explanatory, although there may be some scope for dispute under sub-paragraph (7) as to whether other works are reasonable in all the circumstances of a given case so as to amount to 'emergency undertaker's works'.

The expression 'electronic communications apparatus' is defined in paragraph 5.
The word 'enactment' is defined in paragraph 108.
The word 'land' is defined in paragraph 108.
The word 'network' is defined in paragraph 6.
The word 'operator' is defined in paragraph 2.
The word 'undertaker' is defined in paragraph 66.
The expression 'undertaker's works' is defined in paragraph 66.

This paragraph is discussed in paragraph 27.3.12 of the text.

72 Offence: undertakers who do not comply with this Part of this code

(1) An undertaker, or an agent of an undertaker, is guilty of an offence if that person –

(a) makes an alteration of electronic communications apparatus in carrying out non-emergency undertaker's works, and

(b) does so –

(i) without notice of proposed works having been given in accordance with paragraph 67, or

(ii) (in a case where such notice is given) before the end of the notice period under paragraph 67.

(2) An undertaker, or an agent of an undertaker, is guilty of an offence if that person –

(a) makes an alteration of electronic communications apparatus in carrying out non-emergency undertaker's works, and

(b) unreasonably fails to comply with any reasonable requirement of the operator under this Part of this code when doing so.

(3) An undertaker, or an agent of an undertaker, is guilty of an offence if that person –
   (a) makes an alteration of electronic communications apparatus in carrying out emergency undertaker's works, and
   (b) does so without notice of emergency undertaker's works having been given in accordance with paragraph 71.

(4) A person guilty of an offence under this paragraph is liable on summary conviction to –
   (a) a fine not exceeding level 4 on the standard scale, if the service provided by the operator's network is interrupted by the works or failure, or
   (b) a fine not exceeding level 3 on the standard scale, if that service is not interrupted.

(5) This paragraph does not apply to a Northern Ireland department.

COMMENTARY

Paragraph 72 creates a criminal offence for undertakers or their agents who do not comply with Part 10.

The expression 'electronic communications apparatus' is defined in paragraph 5.
The word 'network' is defined in paragraph 6.
The expression 'non-emergency undertaker's works' is defined in paragraph 67.
The expression 'notice of proposed works' is defined in paragraph 67.
The expression 'notice period' is defined in paragraph 67.
The word 'operator' is defined in paragraph 2.
The word 'provide' is defined by s.405 of the 2003 Act.
The word 'undertaker' is defined in paragraph 66.
The expression 'undertaker's works' is defined in paragraph 66.

This paragraph is discussed in paragraph 27.3.13 of the text.

Part 11 Overhead apparatus

73 Introductory

This Part of this code –
   (a) confers a power on operators to install and keep certain overhead apparatus, and
   (b) imposes a duty on operators to affix notices to certain overhead apparatus.

COMMENTARY

Paragraph 73 introduces paragraphs 74 to 75. It is self-explanatory.

For the right to object to the powers conferred by Part 11 see further Part 12 of the New Code.

The word 'apparatus' is defined in s.405 of the 2003 Act.
The word 'operator' is defined in paragraph 2.

74 Power to fly lines

(1) This paragraph applies where any electronic communications apparatus is kept on or over any land for the purposes of an operator's network.

(2) The operator has the right, for the statutory purposes, to install and keep lines which –

(a) pass over other land adjacent to, or in the vicinity of, the land on or over which the apparatus is kept,

(b) are connected to that apparatus, and

(c) are not, at any point where they pass over the other land, less than three metres above the ground or within two metres of any building over which they pass.

(3) Sub-paragraph (2) does not authorise the installation or keeping on or over any land of –

(a) any electronic communications apparatus used to support, carry or suspend a line installed under sub-paragraph (2), or

(b) any line which, as a result of its position, interferes with the carrying on of any business carried on on that land.

(4) In this paragraph *'business'* includes a trade, profession or employment and includes any activity carried on by a body of persons (whether corporate or unincorporate).

COMMENTARY

Paragraph 74 makes provision for conferring upon an operator the power to install and keep particular overhead apparatus on or over land for the purposes of the operator's network.

The word 'body' is defined in s.405 of the 2003 Act.

The expression 'electronic communications apparatus' is defined in paragraph 5.

The word 'land' is defined in paragraph 108.

The word 'line' is defined in paragraph 5.

The word 'network' is defined in paragraph 6.

The word 'operator' is defined in paragraph 2.

The expression 'the statutory purposes' is defined in paragraph 4.

This paragraph is discussed in Sections 28.2 and 28.3 of the text. For transitional provisions, see further Section 28.7 of the text. For the right to object to the powers conferred by this paragraph see further Part 12 of the New Code and Sections 28.4 and 28.5 of the text.

75 Duty to attach notices to overhead apparatus

(1) This paragraph applies where –

(a) an operator has, for the purposes of the operator's network, installed any electronic communications apparatus, and

(b) the whole or part of the apparatus is at a height of three metres or more above the ground.

(2) The operator must, before the end of the period of three days beginning with the day after that on which the installation is completed, in a secure and durable manner attach a notice –

(a) to every major item of apparatus installed, or
(b) if no major item of apparatus is installed, to the nearest major item of electronic communications apparatus to which the apparatus that is installed is directly or indirectly connected.

(3) A notice attached under sub-paragraph (2) above –
(a) must be attached in a position where it is reasonably legible, and
(b) must give the name of the operator and an address in the United Kingdom at which any notice of objection may be given under paragraph 77(5) in respect of the apparatus in question.

(4) Any person giving such a notice at that address in respect of that apparatus is to be treated as having given that address for the purposes of paragraph 91(2).

(5) An operator who breaches the requirements of this paragraph is guilty of an offence and liable on summary conviction to a fine not exceeding level 2 on the standard scale.

(6) In any proceedings for an offence under this paragraph it is a defence for the person charged to prove that the person took all reasonable steps and exercised all due diligence to avoid committing the offence.

COMMENTARY

Paragraph 75, which replicates paragraph 18 of the Old Code, makes provision for the operator to affix a notice to certain overhead apparatus.

It is an offence under sub-paragraph (5) for the operator to fail to provide the notice in accordance with the provisions of this paragraph, save that sub-paragraph (6) provides a defence if the person charged can prove that they took all reasonable steps and exercised all due diligence to avoid committing the offence.

The word 'apparatus' is defined in s.405 of the 2003 Act.
The word 'line' is defined in paragraph 5.
The word 'network' is defined in paragraph 6.
The word 'operator' is defined in paragraph 2.

This paragraph is discussed in Section 28.6 of the text.

Part 12 Rights to object to certain apparatus

76 Introductory

This Part of this code makes provision conferring rights to object to certain kinds of apparatus, and makes provision about –
(a) the cases in which and persons by whom a right can be exercised, and
(b) the power and procedures of the court if an objection is made.

COMMENTARY

Paragraph 76 introduces paragraphs 77 to 81. It is self-explanatory.

The word 'apparatus' is defined in s.405 of the 2003 Act.

The expression 'the court' is defined in paragraph 94. This is the county court in England, a county court in Northern Ireland or the sheriff court in Scotland.

The Electronic Communications Code (Jurisdiction) Regulations 2017 came into force on 28 December 2017. The functions conferred on the court are to be conferred on and exercisable by certain of the Tribunals as well as by the court. Regulation 4 provides that applications under Part 12 of the Code may be commenced only in the Upper Tribunal, in relation to England and Wales, or the Lands Tribunal for Scotland, in relation to Scotland.

77 When and by whom can a right to object be exercised?

(1) A right to object under this Part of this code is available where, pursuant to the right in paragraph 62, an operator keeps electronic communications apparatus installed on, under or over tidal water or lands within the meaning of Part 9 of this code.

(2) In that case a person has a right to object under this Part of this code if the person –

(a) is an occupier of, or has an interest in, the tidal water or lands,

(b) is not bound by a code right enabling the operator to keep the apparatus installed on, under or over the tidal water or lands, and

(c) is not a person with the benefit of a Crown interest in the tidal water or lands.

(3) A right to object under this Part of this code is available where an operator keeps a line installed over land pursuant to the right in paragraph 74.

(4) In that case a person has a right to object under this Part of this code if the person –

(a) is an occupier of, or has an interest in, the land, and

(b) is not bound by a code right enabling the operator to keep the apparatus installed over the land.

(5) A right to object under this Part of this code is available where –

(a) electronic communications apparatus is kept on or over land for the purposes of an operator's network, and

(b) the whole or any part of that apparatus is at a height of three metres or more above the ground.

(6) In that case a person has a right to object under this Part of this code if –

(a) the person is an occupier of, or has an interest in, any neighbouring land, and

(b) because of the nearness of the neighbouring land to the land on or over which the apparatus is kept –

(i) the enjoyment of the neighbouring land is capable of being prejudiced by the apparatus, or

(ii) any interest in that land is capable of being prejudiced by the apparatus.

(7) There is no right to object under this Part of this code in respect of electronic communications apparatus if the apparatus –

(a) replaces any electronic communications apparatus which is not substantially different from the new apparatus, and

(b) is not in a significantly different position.

COMMENTARY

Paragraph 77 confers miscellaneous rights to object, save that there will be no right to object under this paragraph in respect of electronic communications apparatus which replaces existing apparatus which is not substantially different from the new apparatus and is not in a significantly different position (sub-paragraph (7)).

There is a right to object, in accordance with sub-paragraph (1), where, pursuant to the right in paragraph 62, an operator keeps electronic communications apparatus installed on, under or over tidal water or lands within the meaning of Part 9. There is also a right to object, in accordance with sub-paragraph (3), where lines have been installed over land pursuant to paragraph 74. There is a further right to object, in accordance with sub-paragraph (5), where an operator keeps installed on or over land electronic communications apparatus for the purposes of the operator's network (this apparatus need not be confined to lines or have been erected pursuant to paragraph 74), and at least part of it is at a height of 3 metres or more above the ground.

For discussion of the right in sub-paragraph (3), see further paragraphs 28.4.3–28.4.5 of the text.

For discussion of the right in sub-paragraph (5), see further paragraphs 28.4.6–28.4.8 of the text.

For each right to object, the person who has the right to object is identified in paragraphs (2), (4) and (6) respectively.

The expression 'code right' is defined in paragraph 3.
The expression 'Crown interest' is defined in paragraph 104.
The expression 'electronic communications apparatus' is defined in paragraph 5.
The word 'land' is defined in paragraph 108.
The word 'line' is defined in paragraph 5.
The expression 'neighbouring land' is defined in paragraph 38.
The word 'network' is defined in paragraph 6.
The word 'occupier' is defined in paragraph 105.
The word 'operator' is defined in paragraph 2.
The expression 'tidal water or lands' is defined in paragraph 61.

This paragraph is discussed in Section 28.4 of the text.

78 How may a right to object be exercised?

(1) A person with a right to object under this Part ('the objector') may exercise the right by giving a notice to the operator.
(2) The right to object that the person has, and the procedure that applies to that right, depends on whether –
  (a) the notice is given before the end of the period of 12 months beginning with the date on which installation of the apparatus was completed (see paragraph 79), or
  (b) the notice is given after the end of that period (see paragraph 80).

COMMENTARY

The right to object under paragraph 77 is exercised by giving notice to the operator under sub-paragraph (1). A different procedure will apply to the right depending upon whether or not the notice is given before the end of the period of 12 months beginning with the date on which installation was completed of the apparatus to which the objection relates, as to which see further commentary to paragraphs 79 and 80 below.

The word 'apparatus' is defined in s.405 of the 2003 Act.

The word 'operator' is defined in paragraph 2.

This paragraph is discussed in Section 28.5 of the text.

79 What is the procedure if the objection is made within 12 months of installation?

(1) This paragraph applies if the notice is given before the end of the period of 12 months beginning with the date on which installation of the apparatus was completed.
(2) At any time after the end of the period of two months beginning with the date on which the notice is given, but before the end of the period of four months beginning with that date, the objector may apply to the court to have the objection upheld.
(3) The court must uphold the objection if the following conditions are met.
(4) The first condition is that the apparatus appears materially to prejudice the objector's enjoyment of, or interest in, the land by reference to which the objection is made.
(5) The second condition is that the court is not satisfied that the only possible alterations of the apparatus will –
 (a) substantially increase the cost or diminish the quality of the service provided by the operator's network to persons who have, or may in future have, access to it,
 (b) involve the operator in substantial additional expenditure (disregarding any expenditure caused solely by the fact that any proposed alteration was not adopted originally or, as the case may be, that the apparatus has been unnecessarily installed), or
 (c) give to any person a case at least as good as the objector has to have an objection under this paragraph upheld.
(6) If the court upholds an objection under this paragraph it may by order do any of the following –
 (a) direct the alteration of the apparatus to which the objection relates;
 (b) authorise the installation (instead of the apparatus to which the objection relates), in a manner and position specified in the order, of any apparatus specified in the order;
 (c) direct that no objection may be made under this paragraph in respect of any apparatus the installation of which is authorised by the court.
(7) Where an objector has both given a notice under paragraph 78 and applied for compensation under any of the other provisions of this code –
 (a) the court may give such directions as it thinks fit for ensuring that no compensation is paid until any proceedings under this paragraph have been disposed of, and
 (b) if the court makes an order under this paragraph, it may provide in that order for some or all of the compensation otherwise payable under this code to the objector not to be so payable, or, if the case so requires, for

some or all of any compensation paid under this code to the objector to be repaid to the operator.

(8) For the purposes of sub-paragraph (5)(c), the court has the power on an application under this paragraph to give the objector directions for bringing the application to the notice of such other interested persons as it thinks fit.

(9) This paragraph is subject to paragraph 81.

COMMENTARY

Paragraph 79 deals with the procedure for objections where the right to object has been exercised by giving notice within 12 months of installation of the relevant apparatus.

The court's powers may be limited, as to which see further paragraph 81.

The word 'access' is to be interpreted in accordance with s.405 of the 2003 Act.

The word 'apparatus' is defined in s.405 of the 2003 Act.

The word 'land' is defined in paragraph 108.

The word 'network' is defined in paragraph 6.

The word 'operator' is defined in paragraph 2.

The word 'provide' is defined by s.405 of the 2003 Act.

The expression 'the court' is defined in paragraph 94. This is the county court in England, a county court in Northern Ireland or the sheriff court in Scotland.

The Electronic Communications Code (Jurisdiction) Regulations 2017 came into force on 28 December 2017. The functions conferred on the court are to be conferred on and exercisable by certain of the Tribunals as well as by the court. Regulation 4 provides that applications under Part 12 of the Code may be commenced only in the Upper Tribunal, in relation to England and Wales, or the Lands Tribunal for Scotland, in relation to Scotland.

The expression 'the objector' is defined in paragraph 78.

This paragraph is discussed in paragraphs 28.5.3–28.5.8 of the text.

80 What is the procedure if the objection is made later than 12 months after installation?

(1) This paragraph applies if the notice is given after the end of the period of 12 months beginning with the date on which installation of the apparatus was completed.

(2) At any time after the end of the period of two months beginning with the date on which the notice is given, but before the end of the period of four months beginning with that date, the objector may apply to the court to have the objection upheld.

(3) The court may uphold the objection only if it is satisfied that –

(a) the alteration is necessary to enable the objector to carry out a proposed improvement of the land by reference to which the objection is made, and

(b) the alteration will not substantially interfere with any service which is or is likely to be provided using the operator's network.

(4) If the court upholds an objection under this paragraph it may by order direct the alteration of the apparatus to which the objection relates.

(5) An order under this paragraph may provide for the alteration to be carried out with such modifications, on such terms and subject to such conditions as the court thinks fit.

(6) But the court must not include any such modifications, terms or conditions in its order without the consent of the objector, and if such consent is not given may refuse to make an order under this paragraph.

(7) An order made under this paragraph must, unless the court otherwise thinks fit, require the objector to reimburse the operator in respect of any expenses which the operator incurs in or in connection with the execution of any works in compliance with the order.

(8) This paragraph is subject to paragraph 81.

(9) In this paragraph *'improvement'* includes development and change of use.

COMMENTARY

Paragraph 80 deals with the procedure for objections where the right to object has been exercised by giving notice after the 12-month period from installation of the relevant apparatus.

The court's powers may be limited, as to which see further paragraph 81.

The word 'land' is defined in paragraph 108.

The word 'modification' is defined in s.405 of the 2003 Act.

The word 'network' is defined in paragraph 6.

The word 'operator' is defined in paragraph 2.

The word 'provide' is defined by s.405 of the 2003 Act.

The expression 'the court' is defined in paragraph 94. This is the county court in England, a county court in Northern Ireland or the sheriff court in Scotland.

The Electronic Communications Code (Jurisdiction) Regulations 2017 came into force on 28 December 2017. The functions conferred on the court are to be conferred on and exercisable by certain of the Tribunals as well as by the court. Regulation 4 provides that applications under Part 12 of the Code may be commenced only in the Upper Tribunal, in relation to England and Wales, or the Lands Tribunal for Scotland, in relation to Scotland.

The expression 'the objector' is defined in paragraph 78.

This paragraph is discussed in paragraphs 28.5.9–28.5.11 of the text.

81 What limitations are there on the court's powers under paragraph 79 or 80?

(1) This paragraph applies where the court is considering making –
  (a) an order under paragraph 79 directing the alteration of any apparatus or authorising the installation of any apparatus, or
  (b) an order under paragraph 80 directing the alteration of any apparatus.

(2) The court must not make the order unless it is satisfied –
  (a) that the operator has all such rights as it appears to the court appropriate that the operator should have for the purpose of making the alteration or, as the case may be, installing the apparatus, or
  (b) that –

(i) the operator would have all those rights if the court, on an application under paragraph 20, imposed an agreement on the operator and another person, and
(ii) it would be appropriate for the court, on such an application, to impose such an agreement.

(3) For the purposes of avoiding the need for the agreement of any person to the alteration or installation of any apparatus, the court has the same powers as it would have if an application had been duly made under paragraph 20 above for an order imposing such an agreement.

(4) For the purposes of this paragraph, the court has the power on an application under paragraph 79 or 80 to give the objector directions for bringing the application to the notice of such other interested persons as it thinks fit.

COMMENTARY

Sub-paragraph (2) prevents the court from making an order under paragraphs 79 or 80 directing the alteration of any apparatus, or an order under paragraph 79 authorising the installation of any apparatus unless it is satisfied that the criteria set out in that sub-paragraph are made out. Those criteria ensure that the operator has or will have suitable rights to undertake the relevant alteration or installation.

The word 'apparatus' is defined in s.405 of the 2003 Act.

The word 'operator' is defined in paragraph 2.

The expression 'the court' is defined in paragraph 94. This is the county court in England, a county court in Northern Ireland or the sheriff court in Scotland.

The Electronic Communications Code (Jurisdiction) Regulations 2017 came into force on 28 December 2017. The functions conferred on the court are to be conferred on and exercisable by certain of the Tribunals as well as by the court. Regulation 4 provides that applications under Part 12 of the Code may be commenced only in the Upper Tribunal, in relation to England and Wales, or the Lands Tribunal for Scotland, in relation to Scotland.

The expression 'the objector' is defined in paragraph 78.

This paragraph is discussed in paragraphs 28.5.12 and 28.5.13 of the text.

Part 13 Rights to lop trees

82 Rights to lop trees

(1) This paragraph applies where –
(a) a tree or other vegetation overhangs a street in England and Wales or Northern Ireland or a road in Scotland, and
(b) the tree or vegetation –
(i) obstructs, or will or may obstruct, relevant electronic communications apparatus, or
(ii) interferes with, or will or may interfere with, such apparatus.

(2) In sub-paragraph (1) *'relevant electronic communications apparatus'* means electronic communications apparatus which –
(a) is installed, or about to be installed, on land, and
(b) is used, or to be used, for the purposes of an operator's network.

(3) The operator may, by notice to the occupier of the land on which the tree or vegetation is growing, require the tree to be lopped or the vegetation to be cut back to prevent the obstruction or interference.

(4) If, within the period of 28 days beginning with the day on which the notice is given, the occupier gives the operator a counter-notice objecting to the lopping of the tree or cutting back of the vegetation, the notice has effect only if confirmed by an order of the court.

(5) Sub-paragraph (6) applies if at any time a notice under sub-paragraph (3) has not been complied with and –

(a) the period of 28 days beginning with the day on which the notice was given has expired without a counter-notice having been given, or

(b) an order of the court confirming the notice has come into force.

(6) The operator may cause the tree to be lopped or the vegetation to be cut back.

(7) Where the operator lops a tree or cuts back vegetation in exercise of the power in sub-paragraph (6) the operator must do so in a husband-like manner and in such a way as to cause the minimum damage to the tree or vegetation.

(8) Sub-paragraph (9) applies where –

(a) a notice under sub-paragraph (3) is complied with (either without a counter-notice having been given or after the notice has been confirmed), or

(b) the operator exercises the power in sub-paragraph (6).

(9) The court must, on an application made by a person who has sustained loss or damage in consequence of the lopping of the tree or cutting back of the vegetation or who has incurred expenses in complying with the notice, order the operator to pay that person such compensation in respect of the loss or damage as it thinks fit.

COMMENTARY

Paragraph 82 provides a procedure by which an operator may deal with trees or vegetation overhanging a street or road, which will or may obstruct electronic communications apparatus which is or is about to be installed on land and is or is to be used for the purposes of the operator's network.

Sub-paragraph (3) provides that the operator may, by notice to the occupier of the land on which the tree or vegetation is growing, require the tree to be lopped or the vegetation to be cut back to prevent the obstruction or interference.

The occupier then has 28 days to give a counter-notice objecting to this. The operator may cause the tree to be lopped or the vegetation to be cut back in a husband-like manner if no counter-notice is given within that period, or if the court makes an order confirming that the notice has come into force.

Sub-paragraph (9) provides that a person suffering loss or damage in consequence of the lopping of the tree or cutting back of the vegetation or who has incurred expenses in complying with the notice may apply for compensation. The court may order the operator to pay such compensation as it thinks fit.

The expression 'electronic communications apparatus' is defined in paragraph 5.

The word 'land' is defined in paragraph 108.

The word 'network' is defined in paragraph 6.

The word 'occupier' is defined in paragraph 105.

The word 'operator' is defined in paragraph 2.

The word 'road' is defined in paragraph 108.
The word 'street' is defined in paragraph 108.
The expression 'the court' is defined in paragraph 94. This is the county court in England, a county court in Northern Ireland or the sheriff court in Scotland.
The Electronic Communications Code (Jurisdiction) Regulations 2017 came into force on 28 December 2017. The functions conferred on the court are to be conferred on and exercisable by certain of the Tribunals as well as by the court. Regulation 4 provides that applications under Part 13 of the Code may be commenced only in the Upper Tribunal, in relation to England and Wales, or the Lands Tribunal for Scotland, in relation to Scotland.

This paragraph is discussed in Chapter 29 of the text. In particular, for procedural matters see Section 29.3, for the nature of the work see Section 29.4 and for compensation see Section 29.5. For consideration of the transitional provisions see Sections 29.6 and 31.13 of the text.

Part 14 Compensation under the code

83 Introductory

This Part of this code makes provision about compensation under this code.

COMMENTARY

Paragraph 83 introduces paragraphs 84 to 86. It is self-explanatory.

84 Compensation where agreement imposed or apparatus removed

(1) This paragraph applies to the following powers of the court to order an operator to pay compensation to a person –
  (a) the power in paragraph 25(1) (compensation where order made imposing agreement on person);
  (b) the power in paragraph 44(5) (compensation in relation to removal of the apparatus from the land).
(2) Depending on the circumstances, the power of the court to order the payment of compensation for loss or damage includes power to order payment for –
  (a) expenses (including reasonable legal and valuation expenses, subject to the provisions of any enactment about the powers of the court by whom the order for compensation is made to award costs or, in Scotland, expenses),
  (b) diminution in the value of the land, and
  (c) costs of reinstatement.
(3) For the purposes of assessing such compensation for diminution in the value of land, the following provisions apply with any necessary modifications as they apply for the purposes of assessing compensation for the compulsory purchase of any interest in land –
  (a) in relation to England and Wales, rules (2) to (4) set out in section 5 of the Land Compensation Act 1961;
  (b) in relation to Scotland, rules (2) to (4) set out in section 12 of the Land Compensation (Scotland) Act 1963;

(c) in relation to Northern Ireland, rules (2) to (4) set out in Article 6(1) of the Land Compensation (Northern Ireland) Order 1982 (SI 1982/712 (NI 9)).

(4) In the application of this paragraph to England and Wales, section 10(1) to (3) of the Land Compensation Act 1973 (compensation in respect of mortgages, trusts of land and settled land) applies in relation to such compensation for diminution in the value of land as it applies in relation to compensation under Part 1 of that Act.

(5) In the application of this paragraph to Scotland, section 10(1) and (2) of the Land Compensation (Scotland) Act 1973 (compensation in respect of restricted interests in land) applies in relation to such compensation for diminution in the value of land as it applies in relation to compensation under Part 1 of that Act.

(6) In the application of this paragraph to Northern Ireland, Article 13(1) to (3) of the Land Acquisition and Compensation (Northern Ireland) Order 1973 (SI 1973/1896 (NI 21)) (compensation in respect of mortgages, trusts for sale and settlements) applies in relation to such compensation for diminution in the value of land as it applies in relation to compensation under Part II of that Order.

(7) Where a person has a claim for compensation to which this paragraph applies and a claim for compensation under any other provision of this code in respect of the same loss, the compensation payable to that person must not exceed the amount of that person's loss.

COMMENTARY

Paragraph 84 deals with the assessment of compensation.

Where compensation is payable under paragraphs 25(1) or 44(5), the power of the court to order the payment of compensation for loss or damage includes power to order payment for expenses, diminution in the value of the land and costs of reinstatement (sub-paragraph (1) and (2)).

The legislation specified in sub-paragraphs (3) to (6) is applied, with any necessary modifications, in determining the compensation for diminution in value of the land. In respect of sub-paragraph (3) it applies as it applies for the purposes of assessing compensation for the compulsory purchase of any interest in land. In respect of sub-paragraphs (4) to (6), it applies as it apples in relation to compensation under the Parts specified in those sub-paragraphs.

Sub-paragraph (7) prevents double compensation by prohibiting a person from being compensated under multiple provisions of the New Code in excess of that person's loss. For the possibility of multiple heads of claim for the diminution in the value of the land, see further paragraphs 30.2.21–30.2.22 of the text.

Disputes in relation to compensation under Part 14 are not referred to in regulation 2 of the Electronic Communications Code (Jurisdiction) Regulations 2017, but disputes under paragraph 84 (being disputes under the specified legislation) are already resolved by specialist tribunals, as to which see further paragraph 33.5.7 of the text.

The word 'enactment' is defined in paragraph 108.

The word 'land' is defined in paragraph 108.

The word 'modification' is defined in s.405 of the 2003 Act.

The word 'operator' is defined in paragraph 2.

The expression 'the court' is defined in paragraph 94. This is the county court in England, a county court in Northern Ireland or the sheriff court in Scotland.

This paragraph is discussed in Sections 30.3 to 30.5 of the text. For the different position in Scotland, see further Section 40.4 of the text.

85 Compensation for injurious affection to neighbouring land etc.

(1) This paragraph applies where a right conferred by or in accordance with any provision of Parts 2 to 9 of this code is exercised by an operator.

(2) In the application of this paragraph to England and Wales, compensation is payable by the operator under section 10 of the Compulsory Purchase Act 1965 (compensation for injurious affection to neighbouring land) as if that section applied in relation to injury caused by the exercise of such a right as it applies in relation to injury caused by the execution of works on land that has been compulsorily acquired.

(3) In the application of this paragraph to Scotland, compensation is payable by the operator under section 6 of the Railway Clauses Consolidation (Scotland) Act 1845 as if that section applied in relation to injury caused by the exercise of such a right as it applies in relation to injury caused by the execution of works on land that has been taken or used for the purpose of a railway.

(4) Any question as to a person's entitlement to compensation by virtue of sub-paragraph (3), or as to the amount of that compensation, is, in default of agreement, to be determined by the Lands Tribunal for Scotland.

(5) In the application of this paragraph to Northern Ireland, compensation is payable by the operator under Article 18 of the Land Compensation (Northern Ireland) Order 1982 (SI 1982/712 (NI 9)) as if that section applied in relation to injury caused by the exercise of such a right as it applies in relation to injury caused by the execution of works on land that has been compulsorily acquired.

(6) Any question as to a person's entitlement to compensation by virtue of sub-paragraph (5), or as to the amount of that compensation, is, in default of agreement, to be determined by the Lands Tribunal for Northern Ireland.

(7) Compensation is payable on a claim for compensation under this paragraph only if the amount of the compensation exceeds £50.

(8) Compensation is payable to a person under this paragraph irrespective of whether the person claiming the compensation has any interest in the land in relation to which the right referred to in sub-paragraph (1) is exercised.

(9) Compensation under this paragraph may include reasonable legal and valuation expenses, subject to the provisions of any enactment about the powers of the court or tribunal by whom an order for compensation is made to award costs or, in Scotland, expenses.

COMMENTARY

Paragraph 85 provides for the payment of compensation for injurious affection to neighbouring land where a right conferred by or in accordance with any provision of

Parts 2 to 9 of the New Code is exercised by an operator (including reasonable legal and valuation expenses, subject to the provisions of any enactment concerning powers to award costs or expenses- sub-paragraph (9)). The sections specified in sub-paragraphs (2), (3) and (5) apply as if those sections applied in relation to injury caused by the exercise of such a right as they apply to their own subject matter. Compensation is only payable if the amount exceeds £50 (sub-paragraph (7)). The party to whom compensation is payable need not have an interest in the land in relation to which the right under the New Code is exercised (sub-paragraph (8)).

Disputes in relation to compensation under Part 14 are not referred to in regulation 2 of the Electronic Communications Code (Jurisdiction) Regulations 2017, but disputes under paragraph 85 (being disputes under the specified legislation) are already resolved by specialist tribunals, as to which see further paragraph 33.5.7 of the text.

The word 'enactment' is defined in paragraph 108.

The word 'land' is defined in paragraph 108.

The word 'operator' is defined in paragraph 2.

The expression 'the court' is defined in paragraph 94. This is the county court in England, a county court in Northern Ireland or the sheriff court in Scotland.

This paragraph is discussed in Section 30.6 of the text. For the position in Scotland see further Section 40.4 of the text.

86 No other compensation available

Except as provided by any provision of Parts 2 to 13 of this code or this Part, an operator is not liable to compensate any person for, and is not subject to any other liability in respect of, any loss or damage caused by the lawful exercise of any right conferred by or in accordance with any provision of those Parts.

COMMENTARY

Paragraph 86 provides that, if an operator lawfully exercises any right conferred by or in accordance with any provision of Parts 2 to 14 of the New Code, it will not be liable to compensate any person in respect of loss or damage caused thereby, save as provided in those Parts of the New Code.

The word 'operator' is defined in paragraph 2.

This paragraph is discussed in paragraph 30.3.11 of the text.

Part 15 Notices under the code

87 Introductory

This Part makes provision –

(a) about requirements for the form of notices given under this code by operators,
(b) about requirements for the form of notices given under this code by persons other than operators, and
(c) about procedures for giving notices.

COMMENTARY

Paragraph 87 introduces paragraphs 88 to 91. It is self-explanatory.

The word 'operator' is defined in paragraph 2.

88 Notices given by operators

(1) A notice given under this code by an operator must –
   (a) explain the effect of the notice,
   (b) explain which provisions of this code are relevant to the notice, and
   (c) explain the steps that may be taken by the recipient in respect of the notice.

(2) If OFCOM have prescribed the form of a notice which may or must be given by an operator under a provision of this code, a notice given by an operator under that provision must be in that form.

(3) A notice which does not comply with this paragraph is not a valid notice for the purposes of this code.

(4) Sub-paragraph (3) does not prevent the person to whom the notice is given from relying on the notice if the person chooses to do so.

(5) In any proceedings under this code a certificate issued by OFCOM stating that a particular form of notice has been prescribed by them as mentioned in this paragraph is conclusive evidence of that fact.

COMMENTARY

Sub-paragraphs (1) and (2) stipulate formality requirements which must be satisfied in order for a notice given by an operator under the New Code to be valid. The notice must explain (a) the effect of the notice, (b) which provisions of the New Code are relevant to the notice and (c) the steps which may be taken by the recipient in respect of the notice. It must also be in the prescribed form if applicable.

Even if the notice is not valid due to non-compliance with those formality requirements, its recipient may rely upon it as though it were a valid notice (sub-paragraph (4)).

The acronym 'OFCOM' is defined in s.405 of the 2003 Act.

The word 'operator' is defined in paragraph 2.

This paragraph is discussed in paragraphs 32.5.3–32.5.5 of the text. In relation to OFCOM's prescribed forms of notices see further Sections 41.7 and 41.8 of the text.

89 Notices given by others

(1) Sub-paragraph (2) applies to a notice given under paragraph 31(1), 33(1), 39(1) or 40(2) by a person other than an operator.

(2) If OFCOM have prescribed the form of a notice given under the provision in question by a person other than an operator, the notice must be in that form.

(3) A notice which does not comply with sub-paragraph (2) is not a valid notice for the purposes of this code.

(4) Sub-paragraph (3) does not prevent the operator to whom the notice is given from relying on the notice if the operator chooses to do so.

(5) Sub-paragraph (6) applies to a notice given under any other provision of this code by a person other than an operator if –
   (a) OFCOM have prescribed the form of a notice given under that provision by a person other than an operator,
   (b) the notice is given in response to a notice given by an operator, and
   (c) the operator has, in giving the notice, drawn the person's attention to the form prescribed by OFCOM.

(6) The notice is a valid notice for the purposes of this code, but the person giving the notice must bear any costs incurred by the operator as a result of the notice not being in that form.

(7) In any proceedings under this code a certificate issued by OFCOM stating that a particular form of notice has been prescribed by them as mentioned in this paragraph is conclusive evidence of that fact.

COMMENTARY

Paragraph 89(2) requires notices given under paragraphs 31(1), 33(1), 39(1) or 40(2) by a person other than an operator to be in the prescribed form, if any. If it is not in the prescribed form, it will not be valid, although the operator may rely upon the notice as if it were valid should it choose to do so (sub-paragraph (3) and (4)).

As regards other notices given by non-operators under the New Code in response to a notice given by an operator, where the operator, in giving the notice, has drawn the person's attention to a prescribed form, if the notice is not in the applicable prescribed form it remains valid, but the non-operator must bear any costs incurred by the operator as a result of the notice not being in that form (sub-paragraph (6)).

The acronym 'OFCOM' is defined in s.405 of the 2003 Act.

The word 'operator' is defined in paragraph 2.

This paragraph is discussed in paragraph 21.2.14 and Section 32.6 of the text. In relation to OFCOM's prescribed forms of notices see further Sections 41.7 and 41.8 of the text.

90 Prescription of notices by OFCOM

(1) OFCOM must prescribe the form of a notice to be given under each provision of this code that requires a notice to be given.

(2) OFCOM may from time to time amend or replace a form prescribed under sub-paragraph (1).

(3) Before prescribing a form for the purposes of this code, OFCOM must consult operators and such other persons as OFCOM think appropriate.

(4) Sub-paragraph (3) does not apply to the amendment or replacement of a form prescribed under sub-paragraph (1).

COMMENTARY

Paragraph 90 deals with the prescription of forms for notices under the New Code.

The acronym 'OFCOM' is defined in s.405 of the 2003 Act.

The word 'operator' is defined in paragraph 2.

This paragraph is discussed in Section 32.9 and Chapter 41 of the text.

91 Procedures for giving notice

(1) A notice given under this code must not be sent by post unless it is sent by a registered post service or by recorded delivery.
(2) For the purposes, in the case of a notice under this code, of section 394 of this Act (service of notifications and other documents) and section 7 of the Interpretation Act 1978 (references to service by post), the proper address of a person ('P') is –
   (a) if P has given the person giving the notice an address for service under this code, that address, and
   (b) otherwise, the address given by section 394.
(3) Sub-paragraph (4) applies if it is not practicable, for the purposes of giving a notice under this code, to find out after reasonable enquiries the name and address of a person who is the occupier of land for the purposes of this code.
(4) A notice may be given under this code to the occupier –
   (a) by addressing it to a person by the description of 'occupier' of the land (and describing the land), and
   (b) by delivering it to a person who is on the land or, if there is no person on the land to whom it can be delivered, by affixing it, or a copy of it, to a conspicuous object on the land.
(5) Sub-paragraph (6) applies if it is not practicable, for the purposes of giving a notice under this code, to find out after reasonable enquiries the name and address of the owner of an interest in land.
(6) A notice may be given under this code to the owner –
   (a) by addressing it to a person by the description of 'owner' of the interest (and describing the interest and the land), and
   (b) by delivering it to a person who is on the land or, if there is no person on the land to whom it can be delivered, by affixing it, or a copy of it, to a conspicuous object on the land.

COMMENTARY

Paragraph 91 deals with the service of notices under the New Code.

For a discussion of the general principles applicable to the service of notices see Section 32.3 of the text.

A person giving an address in respect of apparatus under paragraph 75 is treated as having given that address for the purposes of sub-paragraph (2) (paragraph 75(4)).

The word 'land' is defined in paragraph 108.

The word 'occupier' is defined in paragraph 105.

This paragraph is discussed in Sections 32.5 and 32.7, and Chapter 41 of the text.

Part 16 Enforcement and dispute resolution

92 Introductory

This Part of this code makes provision about –
   (a) the court or tribunal by which agreements and rights under this code may be enforced,

(b) the meaning of references to *'the court'* in this code, and
(c) the power of the Secretary of State by regulations to confer jurisdiction under this code on other tribunals.

COMMENTARY

Paragraph 92 introduces paragraphs 93 to 98. It is self-explanatory.

93 Enforcement of agreements and rights

An agreement under this code, and any right conferred by this code, may be enforced –
(a) in the case of an agreement imposed by a court or tribunal, by the court or tribunal which imposed the agreement,
(b) in the case of any agreement or right, by any court or tribunal which for the time being has the power to impose an agreement under this code, or
(c) in the case of any agreement or right, by any court of competent jurisdiction.

COMMENTARY

Paragraph 93 is self-explanatory.

94 Meaning of 'the court'

(1) In this code *'the court'* means –
(a) in relation to England and Wales, the county court,
(b) in relation to Scotland, the sheriff court, and
(c) in relation to Northern Ireland, a county court.
(2) Sub-paragraph (1) is subject to provision made by regulations under paragraph 95.

COMMENTARY

Paragraph 94 defines 'the court' for the purposes of the New Code. Subject to regulations, the court means in relation to England and Wales, the county court, in relation to Scotland, the sheriff court, and in relation to Northern Ireland, a county court.

In practice, disputes under the New Code will be commenced in a specialist tribunal rather than the court, as to which see Commentary to paragraph 95 and Sections 33.3 to 33.9 of the text.

This paragraph is discussed in Section 33.3 of the text.

95 Power to confer jurisdiction on other tribunals

(1) The Secretary of State may by regulations provide for a function conferred by this code on the court to be exercisable by any of the following –
(a) in relation to England, the First-tier Tribunal;
(b) in relation to England and Wales, the Upper Tribunal;

(c) in relation to Scotland, the Lands Tribunal for Scotland;
(d) in relation to Northern Ireland, the Lands Tribunal for Northern Ireland.

(2) Regulations under sub-paragraph (1) may make provision for the function to be exercisable by a tribunal to which the regulations apply –
(a) instead of by the court, or
(b) as well as by the court.

(3) The Secretary of State may by regulations make provision –
(a) requiring proceedings to which regulations under sub-paragraph (1) apply to be commenced in the court or in a tribunal to which the regulations apply;
(b) enabling the court or such a tribunal to transfer such proceedings to a tribunal which has jurisdiction in relation to them by virtue of such regulations or to the court.

(4) The power in section 402(3)(c) for regulations under sub-paragraph (1) or (3) to make consequential provision includes power to make provision which amends, repeals or revokes or otherwise modifies the application of any enactment.

(5) Before making regulations under sub-paragraph (1) or (3) the Secretary of State must –
(a) so far as the regulations relate to Scotland, consult the Scottish Ministers;
(b) so far as the regulations relate to Northern Ireland, consult the Department of Justice in Northern Ireland.

COMMENTARY

Paragraph 95 provides for regulations to be made permitting specified specialist Tribunals to exercise the functions of the court under the New Code.

The Electronic Communications Code (Jurisdiction) Regulations 2017 came into force on 28 December 2017. The functions conferred on the court are to be conferred on and exercisable by certain of the Tribunals as well as by the court. Regulation 4 provides that applications under parts 4, 5, 6, 12 and 13 and paragraph 53 of the Code may be commenced only in the Upper Tribunal, in relation to England and Wales, or the Lands Tribunal for Scotland, in relation to Scotland.

The word 'enactment' is defined in paragraph 108.

The expression 'the court' is defined in paragraph 94. This is the county court in England, a county court in Northern Ireland, or the sheriff court in Scotland.

This paragraph is discussed in Sections 33.3–33.9 of the text.

96 Award of costs by tribunal

(1) Where in any proceedings a tribunal exercises functions by virtue of regulations under paragraph 95(1), it may make such order as it thinks fit as to costs, or, in Scotland, expenses.

(2) The matters a tribunal must have regard to in making such an order include in particular the extent to which any party is successful in the proceedings.

COMMENTARY

The awarding of costs (or, in Scotland, expenses) by a tribunal is considered at Section 33.8 of the text. When exercising its discretion as to costs, the tribunal must have regard in particular to the extent to which any party is successful in the proceedings.

97 Applications to the court

Regulation 3 of the Electronic Communications and Wireless Telegraphy Regulations 2011 (SI 2011/1210) makes provision about the time within which certain applications to the court under this code must be determined.

COMMENTARY

The expression 'the court' is defined in paragraph 94. This is the county court in England, a county court in Northern Ireland or the sheriff court in Scotland.

Regulation 3 applies where: (a) a person authorised to provide public electronic communications networks applies to a competent authority for the granting of rights to install facilities on, over or under public or private property for the purposes of such a network, (b) a person authorised to provide electronic communications networks other than to the public applies to a competent authority for the granting of rights to install facilities on, over or under public property for the purposes of such a network, or (c) a person applies to OFCOM for a direction applying the electronic communications code in the person's case.

Regulation 3 provides a six-month time limit from receipt of a completed application by a competent authority for that authority to make its decision, except in cases of expropriation.

This paragraph is discussed in Sections 33.2 and 33.3 of the text, and more generally in Chapter 4.

98 Appeals in Northern Ireland

Article 60 of the County Courts (Northern Ireland) Order 1980 (ordinary appeals from the county court in civil cases) is to apply in relation to any determination of the court in Northern Ireland under this code in the same manner as it applies in relation to any decree of the court made in the exercise of the jurisdiction conferred by Part 3 of that Order.

COMMENTARY

The expression 'the court' is defined in paragraph 94. This is a county court in Northern Ireland.

PART 17 SUPPLEMENTARY PROVISIONS

99 Relationship between this code and existing law

(1) This code does not authorise the contravention of any provision of an enactment passed or made before the coming into force of this code.

(2) Sub-paragraph (1) does not apply if and to the extent that an enactment makes provision to the contrary.

COMMENTARY

Paragraph 99 is a saving provision designed to ensure that the potentially expropriatory measures contained in the New Code do not conflict with other legislation.

It is not possible for the parties to 'contract out' of paragraph 99, as to which see paragraph 100(2).

The word 'contravention' is defined in s.405 of the 2003 Act.

The word 'enactment' is defined in paragraph 108.

This paragraph is discussed in Section 18.4 of the text.

100 Relationship between this code and agreements with operators

(1) This code does not affect any rights or liabilities arising under an agreement to which an operator is a party.
(2) Sub-paragraph (1) does not apply in relation to paragraph 99 or Parts 3 to 6 of this code.

COMMENTARY

Parties may contract out of provisions of the New Code save for paragraph 99 and Parts 3 to 6. The provisions identified in sub-paragraph (2) can affect rights and liabilities arising under an agreement to which an operator is a party. The provisions of the agreement which are inconsistent with those provisions will simply have no effect (although they are not void).

The word 'operator' is defined in paragraph 2.

This paragraph and the ability of the parties to contract out of the New Code are discussed in Sections 34.1, 34.3, 34.4 and 34.5 of the text. For the interrelationship between this paragraph and the transitional provisions see also paragraph 31.2.6(4)(c) of the text.

101 Ownership of property

The ownership of property does not change merely because the property is installed on or under, or affixed to, any land by any person in exercise of a right conferred by or in accordance with this code.

COMMENTARY

Paragraph 101 appears to disapply the normal law of fixtures insofar as concerns property installed on or under, or affixed to, any land by any person in exercise of a right conferred by or in accordance with the New Code. An operator's apparatus remains the property of its owner (which need not be the operator) absent agreement to the contrary.

The word 'land' is defined in paragraph 108.

This paragraph is discussed in Section 18.5 of the text.

102 Conduits

(1) This code does not authorise an operator to do anything inside a relevant conduit without the agreement of the authority with control of the conduit.
(2) The agreement of the authority with control of a public sewer is sufficient in all cases to authorise an operator to exercise any of the rights under this code in order to do anything wholly inside that sewer.
(3) In this paragraph the following expressions have the same meanings as in section 98 of the Telecommunications Act 1984 –
(a) *'public sewer'* and *'relevant conduit'*;
(b) references to the authority with control of a relevant conduit.

COMMENTARY

The consent of the authority with control of a relevant conduit or public sewer is required before apparatus can be placed inside that conduit or other rights granted by or pursuant to the New Code can be exercised inside that conduit.

The word 'operator' is defined in paragraph 2.

The expression 'public sewer' is defined in s.98(9) of the Telecommunications Act 1984.

The expression 'relevant conduit' is defined in s.98(6) of the Telecommunications Act 1984.

References to the authority with control of a relevant conduit are dealt with in s.98(7) and (8) of the Telecommunications Act 1984.

This paragraph is discussed in paragraph 18.4.6 of the text. As to transitional provisions see also the footnote to paragraph 31.10.2 of the text.

103 Duties for OFCOM to prepare codes of practice

(1) OFCOM must prepare and publish a code of practice dealing with –
(a) the provision of information for the purposes of this code by operators to persons who occupy or have an interest in land;
(b) the conduct of negotiations for the purposes of this code between operators and such persons;
(c) the conduct of operators in relation to persons who occupy or have an interest in land adjoining land on, under or over which electronic communications apparatus is installed;
(d) such other matters relating to the operation of this code as OFCOM think appropriate.
(2) OFCOM must prepare and publish standard terms which may (but need not) be used in agreements under this code.
(3) OFCOM may from time to time –
(a) amend or replace a code of practice or standard terms published under this paragraph;
(b) publish the code or terms as amended or (as the case may be) the replacement code or terms.

(4) Before publishing a code of practice or standard terms under this paragraph, OFCOM must consult operators and such other persons as OFCOM think appropriate.
(5) Sub-paragraph (4) does not apply to –
  (a) the publication of amendments to a code of practice or standard terms, or
  (b) the publication of a replacement code or replacement terms.

COMMENTARY

The drafting of paragraph 103, concerning OFCOM's duties, is self-explanatory.
The expression 'electronic communications apparatus' is defined in paragraph 5.
The word 'information' is defined in s.405 of the 2003 Act.
The word 'land' is defined in paragraph 108.
The acronym 'OFCOM' is defined in s.405 of the 2003 Act.
The word 'operator' is defined in paragraph 2.

This paragraph is discussed in paragraphs 41.5.3–41.5.4 and Sections 41.9–41.10 of the text.

104 Application of this code to the Crown

(1) This code applies in relation to land in which there subsists, or at any material time subsisted, a Crown interest as it applies in relation to land in which no such interest subsists.
(2) In this code *'Crown interest'* means –
  (a) an interest which belongs to Her Majesty in right of the Crown,
  (b) an interest which belongs to Her Majesty in right of the Duchy of Lancaster,
  (c) an interest which belongs to the Duchy of Cornwall,
  (d) an interest which belongs to a government department or which is held in trust for Her Majesty for the purposes of a government department, or
  (e) an interest which belongs to an office-holder in the Scottish Administration or which is held in trust for Her Majesty for the purposes of the Scottish Administration by such an office-holder.
(3) This includes, in particular –
  (a) an interest which belongs to Her Majesty in right of Her Majesty's Government in Northern Ireland, and
  (b) an interest which belongs to a Northern Ireland department or which is held in trust for Her Majesty for the purposes of a Northern Ireland department.
(4) Where an agreement is required by this code to be given in respect of any Crown interest subsisting in any land, the agreement must be given by the appropriate authority.
(5) Where a notice under this code is required to be given in relation to land in which a Crown interest subsists, the notice must be given by or to the appropriate authority (as the case may require).
(6) In this paragraph *'the appropriate authority'* means –

(a) in the case of land belonging to Her Majesty in right of the Crown, the Crown Estate Commissioners or the relevant person or, as the case may be, the government department or office-holder in the Scottish Administration having the management of the land in question;
(b) in the case of land belonging to Her Majesty in right of the Duchy of Lancaster, the Chancellor of the Duchy of Lancaster;
(c) in the case of land belonging to the Duchy of Cornwall, such person as the Duke of Cornwall, or the possessor for the time being of the Duchy of Cornwall, appoints;
(d) in the case of land belonging to an office-holder in the Scottish Administration or held in trust for Her Majesty by such an office-holder for the purposes of the Scottish Administration, the office-holder;
(e) in the case of land belonging to Her Majesty in right of Her Majesty's Government in Northern Ireland, the Northern Ireland department having the management of the land in question;
(f) in the case of land belonging to a government department or a Northern Ireland department or held in trust for Her Majesty for the purposes of a government department or a Northern Ireland department, that department.

(7) In sub-paragraph (6)(a) *'relevant person'*, in relation to land to which section 90B(5) of the Scotland Act 1998 applies, means the person having the management of that land.
(8) Any question as to the authority that is the appropriate authority in relation to any land is to be referred to the Treasury, whose decision is final.
(9) Paragraphs 55 (offence in relation to transport land rights) and 75(5) (offence in relation to notices on overhead apparatus) do not apply where this code applies in the case of the Secretary of State or a Northern Ireland department by virtue of section 106(3)(b).
(10) References in this paragraph to an office-holder in the Scottish Administration are to be construed in accordance with section 126(7) of the Scotland Act 1998.

COMMENTARY

Paragraph 104 provides, for the avoidance of doubt, that the New Code affects Crown land just as it affects other land (although this does not apply to tidal waters, as to which see paragraph 64(1)). For the purposes of the New Code, agreements and notices in respect of land in which a Crown interest (as defined by sub-paragraph (2)) subsists are to be given by or to the appropriate authority (as defined by sub-paragraph (6); where a question as to the appropriate authority arises the Treasury will determine the question on a reference under sub-paragraph (8)).

Section 106(3)(b) of the 2003 Act, referred to in sub-paragraph (9), applies the New Code to a Secretary of State or Northern Ireland department providing or proposing to provide an electronic communications network. Such a Secretary of State or Northern Ireland department do not commit the specified offences.

The word 'land' is defined in paragraph 108.

This paragraph is discussed in paragraph 18.4.5 of the text. For the position in Scotland see also paragraph 40.2.8 of the text.

105 Meaning of 'occupier'

(1) References in this code to an occupier of land are to the occupier of the land for the time being.
(2) References in this code to an occupier of land, in relation to a footpath or bridleway that crosses and forms part of agricultural land, are to the occupier of that agricultural land.
(3) Sub-paragraph (4) applies in relation to land which is –
 (a) a street in England and Wales or Northern Ireland, other than a footpath or bridleway within sub-paragraph (2), or
 (b) a road in Scotland, other than such a footpath or bridleway.
(4) References in this code to an occupier of land –
 (a) in relation to such a street in England and Wales, are to the street managers within the meaning of Part 3 of the New Roads and Street Works Act 1991,
 (b) in relation to such a street in Northern Ireland, are to the street managers within the meaning of the Street Works (Northern Ireland) Order 1995 (SI 1995/3210 (NI 19)), and
 (c) in relation to such a road in Scotland, are to the road managers within the meaning of Part 4 of the New Roads and Street Works Act 1991.
(5) Sub-paragraph (6) applies in relation to land which –
 (a) is unoccupied, and
 (b) is not a street in England and Wales or Northern Ireland or a road in Scotland.
(6) References in this code to an occupier of land, in relation to land within sub-paragraph (5), are to –
 (a) the person (if any) who for the time being exercises powers of management or control over the land, or
 (b) if there is no person within paragraph (a), to every person whose interest in the land would be prejudicially affected by the exercise of a code right in relation to the land.
(7) In this paragraph –
 (a) *'agricultural land'* includes land which is being brought into use for agriculture, and
 (b) references in relation to England and Wales to a footpath or bridleway include a restricted byway.

COMMENTARY

Paragraph 105, headed 'meaning of "occupier"' does not actually define the term 'occupier' (this concept is considered at Section 16.2 of the text) but sub-paragraphs (2) to (4) identify the occupier of particular types of land. The general sub-paragraph (1) simply provides that a reference to an occupier of land is a reference to the occupier of that land for the time being. References to the occupier of unoccupied land which is not a street in England and Wales or Northern Ireland or a road in Scotland are references to the person exercising powers of management or control over the land or if there is no such person, every person whose interest in the land would be prejudicially affected by the exercise of a code right (sub-paragraphs (5) and (6)).

The word 'agricultural' is defined in paragraph 108.
The word 'agriculture' is defined in paragraph 108.
The word 'bridleway' is defined in paragraph 108.
The expression 'code right' is defined in paragraph 3.
The word 'footpath' is defined in paragraph 108.
The word 'land' is defined in paragraph 108.
The expression 'restricted byway' is defined in paragraph 108.
The word 'road' is defined in paragraph 108.
The word 'street' is defined in paragraph 108.

This paragraph is discussed in Sections 16.2 and 16.3 of the text. For the position in Scotland see also paragraph 40.2.3 of the text.

106 Lands Tribunal for Scotland procedure rules

The power to make rules under section 3(6) of the Lands Tribunal Act 1949 (Lands Tribunal for Scotland procedure rules) for the purposes of this code or regulations made under it is exercisable by the Scottish Ministers instead of by the Secretary of State (and any reference there to the approval of the Treasury does not apply).

COMMENTARY

Paragraph 106 is self-explanatory.

This paragraph is set out at paragraph 33.4.5 of the text.

107 Arbitrations in Scotland

Until the Arbitration (Scotland) Act 2010 is in force in relation to any arbitrations carried out under or by virtue of this code, that Act applies as if it were in force in relation to those arbitrations.

COMMENTARY

Paragraph 107 is self-explanatory.

As to the commencement of the Arbitration (Scotland) Act 2010, see Arbitration (Scotland) Act 2010 (Commencement No. 1 and Transitional Provisions) Order 2010/195.

This paragraph is discussed in Section 33.12 of the text.

108 General interpretation

(1) In this code –
'agriculture' and *'agricultural'* –
(a) in relation to England and Wales, have the same meanings as in the Highways Act 1980,
(b) in relation to Scotland, have the same meanings as in the Town and Country Planning (Scotland) Act 1997, and

(c) in relation to Northern Ireland, have the same meanings as in the Agriculture Act (Northern Ireland) 1949;

'bridleway' and *'footpath'* –

(a) in relation to England and Wales, have the same meanings as in the Highways Act 1980,

(b) in relation to Scotland, have the same meanings as Part 3 of the Countryside (Scotland) Act 1967, and

(c) in relation to Northern Ireland, mean a way over which the public have, by virtue of the Access to the Countryside (Northern Ireland) Order 1983 (SI 1983/1895 (NI 18)), a right of way (respectively) on horseback and on foot;

'code agreement' has the meaning given by paragraph 29(5);

'Crown interest' has the meaning given by paragraph 104(2) and (3);

'enactment' includes –

(a) an enactment comprised in subordinate legislation within the meaning of the Interpretation Act 1978,

(b) an enactment comprised in, or in an instrument made under, a Measure or Act of the National Assembly for Wales,

(c) an enactment comprised in, or in an instrument made under, an Act of the Scottish Parliament, and

(d) an enactment comprised in, or in an instrument made under, Northern Ireland legislation;

'land' does not include electronic communications apparatus;

'landowner' has the meaning given by paragraph 37(1);

'lease' includes –

(a) in relation to England and Wales and Northern Ireland, any leasehold tenancy (whether in the nature of a head lease, sub-lease or underlease) and any agreement to grant such a tenancy but not a mortgage by demise or sub-demise, and

(b) in relation to Scotland, any sub-lease and any agreement to grant a sub-lease,

and *'lessee'* is to be construed accordingly;

'relevant person' has the meaning given by paragraph 20(1);

'restricted byway' has the same meaning as in Part 2 of the Countryside and Rights of Way Act 2000;

'road' –

(a) in relation to Scotland, has the same meaning as in Part 4 of the New Roads and Street Works Act 1991;

(b) in relation to Northern Ireland, has the same meaning as in the Roads (Northern Ireland) Order 1993 (SI 1993/3160 (NI 15));

'site provider' has the meaning given by paragraph 30(1);

'street' –

(a) in relation to England and Wales, has the same meaning as in Part 3 of the New Roads and Street Works Act 1991, and

(b) in relation to Northern Ireland, has the same meaning as in the Street Works (Northern Ireland) Order 1995 (SI 1995/3210 (NI 19)).

(2) In this code, references to the alteration of any apparatus include references to the moving, removal or replacement of the apparatus.'

COMMENTARY

Paragraph 108 provides definitions of various expressions, most of which are self-explanatory.

In respect of the definition of 'land', see further paragraph 30.5.3 of the text. 'Land' is defined only negatively, so as not to include electronic communications apparatus.

In relation to sub-paragraph (2) and 'lift and shift' provisions, see further paragraph 34.5.5 of the text. As to how sub-paragraph (2) affects the interpretation of paragraph 42, see further paragraph 24.8.3 of the text.

The word 'apparatus' is defined in s.405 of the 2003 Act.

The expression 'electronic communications apparatus' is defined in paragraph 5.

As to the position in respect of labelling differences in Scotland see also Section 40.2 of the text.

Part VII

Appendices

Appendix A
Extracts from the Telecommunications Act 1984 and the Communications Act 2003

PART I

Telecommunications Act 1984

Section 10

The telecommunications code.

10(1) Subject to the following provisions of this section, the code (to be known as 'the telecommunications code') which is contained in Schedule 2 to this Act shall have effect –

(a) where it is applied to a particular person by a licence granted by the Secretary of State under section 7 above authorising that person to run a telecommunication system; and

(b) where the Secretary of State or a Northern Ireland department is running or is proposing to run a telecommunication system.

(2) The telecommunications code shall not be applied to a person authorised by a licence under section 7 above to run a telecommunication system unless –

(a) that licence is a licence to which section 8 above applies; or

(b) it appears to the Secretary of State –

(i) that the running of the system will benefit the public; and

(ii) that it is not practicable for the system to be run without the application of that code to that person.

(3) Where the telecommunications code is applied to any person by a licence under section 7 above it shall have effect subject to such exceptions and conditions as may be included in the licence for the purpose of qualifying the rights exercisable by that person by virtue of the code.

(3A) Where –

(a) the telecommunications code expressly or impliedly imposes any limitation on the use to which any telecommunication apparatus installed by a person authorised by a licence under section 7 above may be put, and

(b) that person is a party to a relevant agreement,

that limitation shall not have effect so as to preclude the doing of anything which is done in relation to that apparatus in pursuance of that agreement; and anything which is so done shall be disregarded in determining, for the purposes of the telecommunications code as it applies in relation to that person, the purposes for which the apparatus is used.

(3B) Subsection (3A) above shall not be construed, in relation to a person who is authorised by a relevant agreement to share the use of any apparatus installed by another party to the agreement, as affecting any requirement on him (whether imposed by any statutory provision or otherwise) to obtain any consent or permission in connection with the installation by him of any apparatus, or the doing by him of any other thing, in pursuance of the agreement.

(3C) In subsections (3A) and (3B) above 'relevant agreement', in relation to any telecommunication apparatus, means an agreement in writing –

(a) to which the parties are two or more persons to whom the telecommunications code has been applied by a licence granted under section 7 above; and

(b) which relates to the sharing by those persons of the use of that apparatus,

and in subsection (3B) above 'statutory provision' means any provision of an enactment or of an instrument having effect under an enactment.

(4) Without prejudice to the generality of subsection (3) above, the exceptions and conditions there mentioned shall include such exceptions and conditions as appear to the Secretary of State to be requisite or expedient for the purpose of securing –

(a) that the physical environment is protected and, in particular, that the natural beauty and amenity of the countryside is conserved;

(b) that there is no greater damage to streets or interference with traffic than is reasonably necessary;

(c) that funds are available for meeting any liabilities which may arise from the exercise of rights conferred by or in accordance with the code; and any condition falling within this subsection may impose on the person to whom the code is applied a requirement to comply with directions given in a manner specified in the condition and by a person so specified or of a description so specified.

(5) A licence under section 7 above which applies the telecommunications code to any person in relation to any part or locality of the United Kingdom shall include a condition requiring that person to cause copies of –

(a) the exceptions and conditions subject to which the telecommunications code has effect as so applied; and

(b) every direction given in a manner specified in any such condition by a person so specified or of a description so specified,

to be open for inspection by members of the public free of charge at such premises in that part or locality as are specified in the licence or are of a description so specified.

(6) Before granting under section 7 above a licence which applies the telecommunications code to a particular person in relation to any part or locality of the United Kingdom, the Secretary of State shall publish a notice –

(a) stating that he proposes to apply the code to that person in relation to that part or locality and setting out the effect of the exceptions and conditions subject to which he proposes that the code should have effect as so applied;
(b) stating the reasons why he proposes to apply the code to that person in relation to that part or locality and why he proposes that the code as so applied should have effect subject to those exceptions and conditions; and
(c) specifying the time (not being less than 28 days from the date of publication of the notice) within which representations or objections with respect to the proposed application of the code to that person in relation to that part or locality and with respect to the proposed exceptions and conditions may be made,

and shall reconsider his proposals in the light of any representations or objections which are duly made and not withdrawn.

(7) If the Secretary of State, on reconsidering in pursuance of subsection (6) above any proposals specified in a notice under that subsection, grants a licence under section 7 above applying the telecommunications code to any person in relation to any part or locality of the United Kingdom, he shall on granting that licence publish a further notice –
(a) stating that the code has been applied to that person in relation to that part or locality and setting out the effect of the exceptions and conditions subject to which the code has effect as so applied; and
(b) stating the reasons why the code has been applied to that person in relation to that part or locality and why the code as so applied has effect subject to those exceptions and conditions.
(8) Where the Secretary of State has granted a licence under section 7 above which applies the telecommunications code to a particular person in relation to any part or locality of the United Kingdom, he may –
(a) with the consent of that person; or
(b) if it appears to him requisite or expedient to do so for the purpose mentioned in subsection (4) above, modify the exceptions and conditions subject to which the code has effect as so applied.
(9) Before modifying the exceptions and conditions subject to which the telecommunications code has effect as applied to any person in relation to any part or locality of the United Kingdom by a licence granted under section 7 above, the Secretary of State shall publish a notice –
(a) stating that he proposes to make the modifications and setting out their effect;
(b) stating the reasons why he proposes to make the modifications; and
(c) specifying the time (not being less than 28 days from the date of publication of this notice) within which representations or objections with respect to the proposed modifications may be made, and shall reconsider his proposals in the light of any representations or objections which are duly made and not withdrawn.
(10) If the Secretary of State, on reconsidering in pursuance of subsection (9) above any proposals specified in a notice under that subsection, modifies the exceptions and conditions subject to which the telecommunications code has effect as applied to any person in relation to any part or locality of the United Kingdom

by a licence granted under section 7 above, he shall on making the modifications publish a further notice –

(a) stating that the modifications have been made and setting out their effect; and

(b) stating the reasons why the modifications have been made.

(11) A notice under this section shall be published in such manner as the Secretary of State considers appropriate for bringing the matters to which the notice relates to the attention of persons likely to be affected by them.

Section 11

Provisions supplementary to section 10.

(1) ...

(2) ....

(3) The Secretary of State may from time to time by order provide that the telecommunications code shall have effect for all purposes as if an amount specified in the order were substituted for the amount specified, or for the time being having effect as if specified, in sub-paragraph (3) of paragraph 16 of the code as the minimum amount of compensation payable under that paragraph; and an order under this subsection may contain such transitional provisions as the Secretary of State considers appropriate.

(4) In any case where it appears to the Secretary of State that it is expedient for transitional provision to be made in connection with the telecommunications code ceasing to apply to any person by reason of the expiry or revocation of a person's licence under section 7 above, the Secretary of State may make a scheme giving effect to such transitional provision as the Secretary of State thinks fit.

(5) Without prejudice to the generality of subsection (4) above, a scheme under that subsection may –

(a) impose obligations on a person to whom the telecommunications code has ceased to apply as mentioned in subsection (4) above to remove anything installed in pursuance of any right conferred by or in accordance with the telecommunications code, to restore land to its condition before anything was done in pursuance of any such right or to pay the expenses of any such removal or restoration;

(b) provide for those obligations to be enforceable in such manner (otherwise than by criminal penalties) and by such persons as may be specified in the scheme;

(c) authorise the retention of apparatus on any land pending the grant of a licence under section 7 above authorising the running by any person of a telecommunication system for the purposes of which that apparatus may be used;

(d) provide for the purposes of any provision contained in the scheme by virtue of paragraph (a), (b) or (c) above for such questions arising under the scheme as are specified in the scheme, or are of a description so specified, to be referred to, and determined by, the Director.

Schedule 2: The Telecommunications Code

1. Interpretation of code (England and Wales)

(1) In this code, except in so far as the context otherwise requires –

'agriculture' and *'agricultural'* –

(a) in England and Wales, have the same meanings as in the Highways Act 1980;
(b) in Scotland, have the same meanings as in the Town and Country Planning (Scotland) Act 1972; and
(c) in Northern Ireland, have the same meanings as in the Agriculture Act (Northern Ireland) 1949;

'alter', *'alteration'* and *'altered'* shall be construed in accordance with sub-paragraph (2) below;

'bridleway' and *'footpath'* –

(a) in England and Wales, have the same meanings as in the Highways Act 1980;
(b) in Scotland, have the same meanings as in Part III of the Countryside (Scotland) Act 1967; and
(c) in Northern Ireland, mean a way over which the public have, by virtue of the Access to the Countryside (Northern Ireland) Order 1983, a right of way on horseback and on foot, respectively;

'conduit' includes a tunnel, subway, tube or pipe; *'conduit system'* means a system of conduits provided so as to be available for use by providers of electronic communications networks for the purposes of the provision by them of their networks;
'the court' means, without prejudice to any right of appeal conferred by virtue of paragraph 25 below or otherwise –

(a) in relation to England and Wales and Northern Ireland, the county court: and
(b) in relation to Scotland, the sheriff;

'electronic communications apparatus' means –

(a) any apparatus (within the meaning of the Communications Act 2003) which is designed or adapted for use in connection with the provision of an electronic communications network;
(b) any apparatus (within the meaning of that Act) that is designed or adapted for a use which consists of or includes the sending or receiving of communications or other signals that are transmitted by means of an electronic communications network;
(c) any line;
(d) any conduit, structure, pole or other thing in, on, by or from which any electronic communications apparatus is or may be installed, supported, carried or suspended;

and references to the installation of electronic communications apparatus are to be construed accordingly; *'electronic communications network'* has the same meaning as in the Communications Act 2003, and references to the provision of such a network are to be construed in accordance with the provisions of that Act; *'electronic communications service'* has the same meaning as in the Communications Act 2003, and references to the provision of such a service are to be construed in accordance with the provisions of that Act;

'emergency works', in relation to the operator or a relevant undertaker for the purposes of paragraph 23 below, means works the execution of which at the time it is

proposed to execute them is requisite in order to put an end to, or prevent, the arising of circumstances then existing or imminent which are likely to cause –

(a) danger to persons or property,
(b) the interruption of any service provided by the operator's network or, as the case may be, interference with the exercise of any functions conferred or imposed on the undertaker by or under any enactment; or
(c) substantial loss to the operator or, as the case may be, the undertaker,

and such other works as in all the circumstances it is reasonable to execute with those works;

'line' means any wire, cable, tube, pipe or similar thing (including its casing or coating) which is designed or adapted for use in connection with the provision of any electronic communications network or electronic communications service;

'maintainable highway' –

(a) in England and Wales, means a maintainable highway within the meaning of Part III of the New Roads and Street Works Act 1991 other than one which is a footpath, bridleway or restricted byway that crosses, and forms part of, any agricultural land or any land which is being brought into use for agriculture; and
(b) in Northern Ireland, means a highway maintainable by the Department of the Environment for Northern Ireland;

'the operator' means –

(a) where the code is applied in any person's case by a direction under section 106 of the Communications Act 2003, that person; and
(b) where it applies by virtue of section 106(3)(b) of that Act, the Secretary of State or (as the case may be) the Northern Ireland department in question;

'the operator's network' means –

(a) in relation to an operator falling within paragraph (a) of the definition of 'operator', so much of any electronic communications network or conduit system provided by that operator as is not excluded from the application of the code under section 106(5) of the Communications Act 2003; and
(b) in relation to an operator falling within paragraph (b) of that definition, the electronic communications network which the Secretary of State or the Northern Ireland department is providing or proposing to provide;

'railway' includes a light railway;

'restricted byway' has the same meaning as in Part 2 of the Countryside and Rights of Way Act 2000;

'signal' has the same meaning as in section 32 of the Communications Act 2003;

'the statutory purposes' means the purposes of [the provision of the operator's network;

'street' has the same meaning as in Part III of the New Roads and Street Works Act 1991;

'structure' does not include a building

(2) In this code, references to the alteration of any apparatus include references to the moving, removal or replacement of the apparatus.

(3) In relation to any land which, otherwise than in connection with a street on that land, is divided horizontally into different parcels, the references in this code to a place over or under the land shall have effect in relation to each parcel as not including references to any place in a different parcel.

(3A) References in this code to the provision of a conduit system include references to establishing or maintaining such a system.

[...]

(5) For the purposes of the definition in this paragraph of 'street' Part III of the New Roads and Street Works Act 1991 shall be deemed to extend to Northern Ireland.

1. Interpretation of the Code (Scotland)

(1) In this code, except in so far as the context otherwise requires –

'agriculture' and *'agricultural'* –

(a) in England and Wales, have the same meanings as in the Highways Act 1980;
(b) in Scotland, have the same meanings as in the Town and Country Planning (Scotland) Act 1972; and
(c) in Northern Ireland, have the same meanings as in the Agriculture Act (Northern Ireland) 1949;

'alter', *'alteration'* and *'altered'* shall be construed in accordance with sub-paragraph (2) below;

'bridleway' and *'footpath'* –

(a) in England and Wales, have the same meanings as in the Highways Act 1980;
(b) in Scotland, have the same meanings as in Part III of the Countryside (Scotland) Act 1967; and
(c) in Northern Ireland, mean a way over which the public have, by virtue of the Access to the Countryside (Northern Ireland) Order 1983, a right of way on horseback and on foot, respectively;

'conduit' includes a tunnel, subway, tube or pipe;] [*'conduit system'* means a system of conduits provided so as to be available for use by providers of electronic communications networks for the purposes of the provision by them of their networks;

'the court' means, without prejudice to any right of appeal conferred by virtue of paragraph 25 below or otherwise –

(a) in relation to England and Wales and Northern Ireland, the county court: and
(b) in relation to Scotland, the sheriff;

'electronic communications apparatus' means –

(a) any apparatus (within the meaning of the Communications Act 2003) which is designed or adapted for use in connection with the provision of an electronic communications network;
(b) any apparatus (within the meaning of that Act) that is designed or adapted for a use which consists of or includes the sending or receiving of communications or other signals that are transmitted by means of an electronic communications network;
(c) any line;
(d) any conduit, structure, pole or other thing in, on, by or from which any electronic communications apparatus is or may be installed, supported, carried or suspended;

and references to the installation of electronic communications apparatus are to be construed accordingly;] [*'electronic communications network'* has the same meaning as in the Communications Act 2003, and references to the provision of such a network are to be construed in accordance with the provisions of that Act; *'electronic communications service'* has the same meaning as in the Communications Act 2003, and references to the provision of such a service are to be construed in accordance with the provisions of that Act;

'emergency works', in relation to the operator or a relevant undertaker for the purposes of paragraph 23 below, means works the execution of which at the time it is proposed to execute them is requisite in order to put an end to, or prevent, the arising of circumstances then existing or imminent which are likely to cause –

(a) danger to persons or property,
(b) the interruption of any service provided by [the operator's network] [3] or, as the case may be, interference with the exercise of any functions conferred or imposed on the undertaker by or under any enactment; or
(c) substantial loss to the operator or, as the case may be, the undertaker,

and such other works as in all the circumstances it is reasonable to execute with those works;

'line' means any wire, cable, tube, pipe or similar thing (including its casing or coating) which is designed or adapted for use in connection with the provision of any electronic communications network or electronic communications service;

…

'the operator' means –

(a) where the code is applied in any person's case by a direction under section 106 of the Communications Act 2003, that person; and
(b) where it applies by virtue of section 106(3)(b) of that Act, the Secretary of State or (as the case may be) the Northern Ireland department in question;

'the operator's network' means –

(a) in relation to an operator falling within paragraph (a) of the definition of 'operator', so much of any electronic communications network or conduit system provided by that operator as is not excluded from the application of the code under section 106(5) of the Communications Act 2003; and

(b) in relation to an operator falling within paragraph (b) of that definition, the electronic communications network which the Secretary of State or the Northern Ireland department is providing or proposing to provide;

'public road' means a public road within the meaning of Part IV of the New Roads and Street Works Act 1991 other than one which is a footpath or a bridleway that crosses, and forms part of, any agricultural land or any land which is being brought into use for agriculture;

'railway' includes a light railway;

'road' has [the same meaning as in Part IV of the New Roads and Street Works Act 1991;

'signal' has the same meaning as in section 32 of the Communications Act 2003;

'the statutory purposes' means the purposes of [the provision of the operator's network;

[...]

'structure' does not include a building

(2) In this code, references to the alteration of any apparatus include references to the moving, removal or replacement of the apparatus.

(3) In relation to any land which, otherwise than in connection with a road on that land, is divided horizontally into different parcels, the references in this code to a place over or under the land shall have effect in relation to each parcel as not including references to any place in a different parcel.

(3A) References in this code to the provision of a conduit system include references to establishing or maintaining such a system.

2. *Agreement required to confer right to execute works etc.*

(1) The agreement in writing of the occupier for the time being of any land shall be required for conferring on the operator a right for the statutory purposes –

(a) to execute any works on that land for or in connection with the installation, maintenance, adjustment, repair or alteration of electronic communications apparatus; or

(b) to keep electronic communications apparatus installed on, under or over that land; or

(c) to enter that land to inspect any apparatus kept installed (whether on, under or over that land or elsewhere) for the purposes of the operator's network.

(2) A person who is the owner of the freehold estate in any land or is a lessee of any land shall not be bound by a right conferred in accordance with sub-paragraph (1) above by the occupier of that land unless –

(a) he conferred the right himself as occupier of the land; or

(b) he has agreed in writing to be bound by the right; or

(c) he is for the time being treated by virtue of sub-paragraph (3) below as having so agreed; or

(d) he is bound by the right by virtue of sub-paragraph (4) below.

(3) If a right falling within sub-paragraph (1) above has been conferred by the occupier of any land for purposes connected with the provision, to the occupier from time to time of that land, of any electronic communications services and –
(a) the person conferring the right is also the owner of the freehold estate in that land or is a lessee of the land under a lease for a term of a year or more, or
(b) in a case not falling within paragraph (a) above, a person owning the freehold estate in the land or a lessee of the land under a lease for a term of a year or more has agreed in writing that his interest in the land should be bound by the right,

then, subject to paragraph 4 below, that right shall (as well as binding the person who conferred it) have effect, at any time when the person who conferred it or a person bound by it under sub-paragraph (2)(b) or (4) of this paragraph is the occupier of the land, as if every person for the time being owning an interest in that land had agreed in writing to the right being conferred for the said purposes and, subject to its being exercised solely for those purposes, to be bound by it.

(4) In any case where a person owning an interest in land agrees in writing (whether when agreeing to the right as occupier or for the purposes of sub-paragraph (3)(b) above or otherwise) that his interest should be bound by a right falling within sub-paragraph (1) above, that right shall (except in so far as the contrary intention appears) bind the owner from time to time of that interest and also –
(a) the owner from time to time of any other interest in the land, being an interest created after the right is conferred and not having priority over the interest to which the agreement relates; and
(b) any other person who is at any time in occupation of the land and whose right to occupation of the land derives (by contract or otherwise) from a person who at the time the right to occupation was granted was bound by virtue of this sub-paragraph.

(5) A right falling within sub-paragraph (1) above shall not be exercisable except in accordance with the terms (whether as to payment or otherwise) subject to which it is conferred; and, accordingly, every person for the time being bound by such a right shall have the benefit of those terms.

(6) A variation of a right falling within sub-paragraph (1) above or of the terms on which such a right is exercisable shall be capable of binding persons who are not parties to the variation in the same way as, under sub-paragraphs (2), (3) and (4) above, such a right is capable of binding persons who are not parties to the conferring of the right.

(7) It is hereby declared that a right falling within sub-paragraph (1) above is not subject to the provisions of any enactment requiring the registration of interests in, charges on or other obligations affecting land.

(8) In this paragraph and paragraphs 3 and 4 below –
(a) references to the occupier of any land shall have effect –
(i) in relation to any footpath, bridleway or restricted byway that crosses and forms part of any agricultural land or any land which is being brought into use for agriculture, as references to the occupier of that land;
(ii) in relation to any street or, in Scotland, road (not being such a footpath, bridleway or restricted byway), as references –

in England and Wales or Northern Ireland, to the street managers within the meaning of Part III of the New Roads and Street Works Act 1991 (which for this purpose shall be deemed to extend to Northern Ireland), and in Scotland, to the road managers within the meaning of Part IV of that Act; and

(iii) in relation to any land (not being a street or, in Scotland, road) which is unoccupied, as references to the person (if any) who for the time being exercises powers of management or control over the land or, if there is no such person, to every person whose interest in the land would be prejudicially affected by the exercise of the right in question;

(b) *'lease'* includes any leasehold tenancy (whether in the nature of a head lease, sub-lease or underlease) and any agreement to grant such a tenancy but not a mortgage by demise or sub-demise and *'lessee'* shall be construed accordingly; and

(c) references to the owner of a freehold estate shall, in relation to land in Scotland, have effect as references to the person –
   (i) who is infeft proprietor of the land; or
   (ii) who has right to the land but whose title thereto is not complete; or
   (iii) in the case of land subject to a heritable security constituted by *ex facie* absolute disposition, who is the debtor in the security, except where the creditor is in possession of the land,

other than a person having a right as a superior only.

(9) Subject to paragraphs 9(2) and 11(2) below, this paragraph shall not require any person to give his agreement to the exercise of any right conferred by any of paragraphs 9 to 12 below.

3. Agreement required for obstructing access etc.

(1) A right conferred in accordance with paragraph 2 above or by paragraph 9, 10 or 11 below to execute any works on any land, to keep electronic communications apparatus installed on, under or over any land or to enter any land shall not be exercisable so as to interfere with or obstruct any means of entering or leaving any other land unless the occupier for the time being of the other land conferred, or is otherwise bound by, a right to interfere with or obstruct that means of entering or leaving the other land.

(2) The agreement in writing of the occupier for the time being of the other land shall be required for conferring any right for the purposes of sub-paragraph (1) above on the operator.

(3) The references in sub-paragraph (1) above to a means of entering or leaving any land include references to any means of entering or leaving the land provided for use in emergencies.

(4) Sub-paragraphs (2) to (7) of paragraph 2 above except subparagraph (3) shall apply (subject to the following provisions of this code) in relation to a right falling within sub-paragraph (1) above as they apply in relation to a right falling within paragraph 2(1) above.

(5) Nothing in this paragraph shall require the person who is the occupier of, or owns any interest in, any land which is a street or to which paragraph 11 below applies to agree to the exercise of any right on any other land.

4. Effect of rights and compensation

(1) Anything done by the operator in exercise of a right conferred in relation to any land in accordance with paragraph 2 or 3 above shall be deemed to be done in exercise of a statutory power except as against –
  (a) a person who, being the owner of the freehold estate in that land or a lessee of the land, is not for the time being bound by the right; or
  (b) a person having the benefit of any covenant or agreement which has been entered into as respects the land under any enactment and which, by virtue of that enactment, binds or will bind persons deriving title or otherwise claiming under the covenantor or, as the case may be, a person who was a party to the agreement.

(2) Where a right has been conferred in relation to any land in accordance with paragraph 2 or 3 above and anything has been done in exercise of that right, any person who, being the occupier of the land, the owner of the freehold estate in the land or a lessee of the land, is not for the time being bound by the right shall have the right to require the operator to restore the land to its condition before that thing was done.

(3) Any duty imposed by virtue of sub-paragraph (2) above shall, to the extent that its performance involves the removal of any electronic communications apparatus from any land, be enforceable only in accordance with paragraph 21 below.

(4) Where –
  (a) on a right in relation to any land being conferred or varied in accordance with paragraph 2 above, there is a depreciation in the value of any relevant interest in the land, and
  (b) that depreciation is attributable to the fact that paragraph 21 below will apply to the removal from the land, when the owner for the time being of that interest becomes the occupier of the land, of any [electronic communications apparatus] [1] installed in pursuance of that right,

the operator shall pay compensation to the person who, at the time the right is conferred or, as the case may be, varied, is the owner of that relevant interest; and the amount of that compensation shall be equal (subject to sub-paragraph (9) below) to the amount of the depreciation.

(5) In sub-paragraph (4) above *'relevant interest'*, in relation to land subject to a right conferred or varied in accordance with paragraph 2 above, means any interest in respect of which the following two conditions are satisfied at the time the right is conferred or varied, namely –
  (a) the owner of the interest is not the occupier of the land but may become the occupier of the land by virtue of that interest; and
  (b) the owner of the interest becomes bound by the right or variation by virtue only of paragraph 2(3) above.

(6) Any question as to a person's entitlement to compensation under sub-paragraph (4) above, or as to the amount of any compensation under that sub-paragraph, shall, in default of agreement, be referred to and determined by the appropriate tribunal; and section 4 of the Land Compensation Act 1961 (costs) shall apply, with the necessary modifications, in relation to any such determination.

(7) A claim to compensation under sub-paragraph (4) above shall be made by giving the operator notice of the claim and specifying in that notice particulars of –
  (a) the land in respect of which the claim is made;
  (b) the claimant's interest in the land and, so far as known to the claimant, any other interests in the land;
  (c) the right or variation in respect of which the claim is made; and
  (d) the amount of the compensation claimed;

and such a claim shall be capable of being made at any time before the claimant becomes the occupier of the land in question, or at any time in the period of three years beginning with that time.

(8) For the purposes of assessing any compensation under sub-paragraph (4) above, rules (2) to (4) set out in section 5 of the Land Compensation Act 1961 shall, subject to any necessary modifications, have effect as they have effect for the purposes of assessing compensation for the compulsory acquisition of any interest in land.

(9) Without prejudice to the powers of the appropriate tribunal in respect of the costs of any proceedings before the Tribunal by virtue of this paragraph, where compensation is payable under sub-paragraph (4) above there shall also be payable, by the operator to the claimant, any reasonable valuation or legal expenses incurred by the claimant for the purposes of the preparation and prosecution of his claim for that compensation.

(10) Subsections (1) to (3) of section 10 of the Land Compensation Act 1973 (compensation in respect of mortgages, trusts of land and settled land) shall apply in relation to compensation under sub-paragraph (4) above as they apply in relation to compensation under Part I of that Act.

(10A) In this paragraph *'the appropriate tribunal'* means –
  (a) in the application of this Act to England and Wales, the Upper Tribunal;
  (b) in the application of this Act to Scotland, the Lands Tribunal for Scotland;
  (c) in the application of this Act to Northern Ireland, the Lands Tribunal for Northern Ireland.

(11) In the application of this paragraph to Scotland –
  (a) for any reference to costs there is substituted a reference to expenses;
  (b) for the reference in sub-paragraph (6) above to section 4 of the Land Compensation Act 1961 there is substituted a reference to sections 9 and 11 of the Land Compensation (Scotland) Act 1963;
  (c) for the reference in sub-paragraph (8) above to section 5 of the Land Compensation Act 1961 there is substituted a reference to section 12 of the Land Compensation (Scotland) Act 1963;
  (d) for the reference in sub-paragraph (10) above to subsections (1) to (3) of section 10 of the Land Compensation Act 1973 there is substituted a reference to subsections (1) and (2) of section 10 of the Land Compensation (Scotland) Act 1973.

(12) In the application of this paragraph to Northern Ireland –
  [...]
  (b) for the references in sub-paragraphs (6) and (8) above to sections 4 and 5 of the Land Compensation Act 1961 there are substituted references to

Articles 4, 5 and 6 of the Land Compensation (Northern Ireland) Order 1982, respectively;

(c) for the references in sub-paragraph (10) above to subsections (1) to (3) of section 10 of the Land Compensation Act 1973 and to Part I of that Act there are substituted references to paragraphs (1) to (3) of Article 13 of the Land Acquisition and Compensation (Northern Ireland) Order 1973 and to Part II of that Order, respectively.

5. Power to dispense with the need for required agreement

(1) Where the operator requires any person to agree for the purposes of paragraph 2 or 3 above that any right should be conferred on the operator, or that any right should bind that person or any interest in land, the operator may give a notice to that person of the right and of the agreement that he requires.

(2) Where the period of 28 days beginning with the giving of a notice under sub-paragraph (1) above has expired without the giving of the required agreement, the operator may apply to the court for an order conferring the proposed right or providing for it to bind any person or any interest in land, and (in either case) dispensing with the need for the agreement of the person to whom the notice was given.

(3) The court shall make an order under this paragraph if, but only if, it is satisfied that any prejudice caused by the order –

(a) is capable of being adequately compensated for by money; or

(b) is outweighed by the benefit accruing from the order to the persons whose access to an electronic communications network or to electronic communications services will be secured by the order;

and in determining the extent of the prejudice, and the weight of that benefit, the court shall have regard to all the circumstances and to the principle that no person should unreasonably be denied access to an electronic communications network or to electronic communications services.

(4) An order under this paragraph made in respect of a proposed right may, in conferring that right or providing for it to bind any person or any interest in land and in dispensing with the need for any person's agreement, direct that the right shall have effect with such modifications, be exercisable on such terms and be subject to such conditions as may be specified in the order.

(5) The terms and conditions specified by virtue of sub-paragraph (4) above in an order under this paragraph, shall include such terms and conditions as appear to the court appropriate for ensuring that the least possible loss and damage is caused by the exercise of the right in respect of which the order is made to persons who occupy, own interests in or are from time to time on the land in question.

(6) For the purposes of proceedings under this paragraph in a county court in England and Wales or Northern Ireland, section 63(1) of the County Courts Act 1984 and Article 33(1) of the County Courts (Northern Ireland) Order 1980 (assessors) shall have effect as if the words 'on the application of any party' were omitted; and where an assessor is summoned, or, in Northern Ireland, appointed, by virtue of this sub-paragraph –

(a) he may, if so directed by the judge, inspect the land to which the proceedings relate without the judge and report on the land to the judge in writing; and
(b) the judge may take the report into account in determining whether to make an order under this paragraph and what order to make.

In relation to any time before 1st August 1984, the reference in this sub-paragraph to section 63(1) of the County Courts Act 1984 shall have effect as a reference to section 91(1) of the County Courts Act 1959.

(7) Where an order under this paragraph, for the purpose of conferring any right or making provision for a right to bind any person or any interest in land, dispenses with the need for the agreement of any person, the order shall have the same effect and incidents as the agreement of the person the need for whose agreement is dispensed with and accordingly (without prejudice to the foregoing) shall be capable of variation or release by a subsequent agreement.

6. Acquisition of rights in respect of apparatus already installed

(1) The following provisions of this paragraph apply where the operator gives notice under paragraph 5(1) above to any person and –
  (a) that notice requires that person's agreement in respect of a right which is to be exercisable (in whole or in part) in relation to electronic communications apparatus already kept installed on, under or over the land in question, and
  (b) that person is entitled to require the removal of that apparatus but, by virtue of paragraph 21 below, is not entitled to enforce its removal.
(2) The court may, on the application of the operator, confer on the operator such temporary rights as appear to the court reasonably necessary for securing that, pending the determination of any proceedings under paragraph 5 above or paragraph 21 below, the service provided by the operator's network is maintained and the apparatus properly adjusted and kept in repair.
(3) In any case where it is shown that a person with an interest in the land was entitled to require the removal of the apparatus immediately after it was installed, the court shall, in determining for the purposes of paragraph 5 above whether the apparatus should continue to be kept installed on, under or over the land, disregard the fact that the apparatus has already been installed there.

7. Court to fix financial terms where agreement dispensed with

(1) The terms and conditions specified by virtue of sub-paragraph (4) of paragraph 5 above in an order under that paragraph dispensing with the need for a person's agreement, shall include –
  (a) such terms with respect to the payment of consideration in respect of the giving of the agreement, or the exercise of the rights to which the order relates, as it appears to the court would have been fair and reasonable if the agreement had been given willingly and subject to the other provisions of the order; and
  (b) such terms as appear to the court appropriate for ensuring that that person and persons from time to time bound by virtue of paragraph 2(4) above by the rights to which the order relates are adequately compensated (whether by the

payment of such consideration or otherwise) for any loss or damage sustained by them in consequence of the exercise of those rights.

(2) In determining what terms should be specified in an order under paragraph 5 above for requiring an amount to be paid to any person in respect of –
  (a) the provisions of that order conferring any right or providing for any right to bind any person or any interest in land, or
  (b) the exercise of any right to which the order relates,

the court shall take into account the prejudicial effect (if any) of the order or, as the case may be, of the exercise of the right on that person's enjoyment of, or on any interest of his in, land other than the land in relation to which the right is conferred.

(3) In determining what terms should be specified in an order under paragraph 5 above for requiring an amount to be paid to any person, the court shall, in a case where the order is made in consequence of an application made in connection with proceedings under paragraph 21 below, take into account, to such extent as it thinks fit, any period during which that person –
  (a) was entitled to require the removal of any electronic communications apparatus from the land in question, but
  (b) by virtue of paragraph 21 below, was not entitled to enforce its removal;

but where the court takes any such period into account, it may also take into account any compensation paid under paragraph 4(4) above.

(4) The terms specified by virtue of sub-paragraph (1) above in an order under paragraph 5 above may provide –
  (a) for the making of payments from time to time to such persons as may be determined under those terms; and
  (b) for questions arising in consequence of those terms (whether as to the amount of any loss or damage caused by the exercise of a right or otherwise) to be referred to arbitration or to be determined in such other manner as may be specified in the order.

(5) The court may, if it thinks fit –
  (a) where the amount of any sum required to be paid by virtue of terms specified in an order under paragraph 5 above has been determined, require the whole or any part of any such sum to be paid into court;
  (b) pending the determination of the amount of any such sum, order the payment into court of such amount on account as the court thinks fit.

(6) Where terms specified in an order under paragraph 5 above require the payment of any sum to a person who cannot be found or ascertained, that sum shall be paid into court.

8. Notices and applications by potential subscribers

(1) Where –
  (a) it is reasonably necessary for the agreement of any person to the conferring of any right, or to any right's binding any person or any interest in land, to

be obtained by the operator before another person ('the potential subscriber') may be afforded access to the operator's network, and

(b) the operator has not given a notice or (if he has given a notice) has not made an application in respect of that right under paragraph 5 above,

the potential subscriber may at any time give a notice to the operator requiring him to give a notice or make an application under paragraph 5 above in respect of that right.

(2) At any time after notice has been given to the operator under sub-paragraph (1) above, the operator may apply to the court to have the notice set aside on the ground that the conditions mentioned in that, sub-paragraph are not satisfied or on the ground that, even if the agreement were obtained, the operator would not afford the potential subscriber access to the operator's network and could not be required to afford him access to that network.

(3) Subject to any order of the court made in or pending any proceedings under sub-paragraph (2) above, if at any time after the expiration of the period of 28 days beginning with the giving to the operator of a notice under sub-paragraph (1) above the operator has not complied with the notice, the potential subscriber may himself, on the operator's behalf, give the required notice and (if necessary) make an application under paragraph 5 above or, as the case may be, make the required application.

(4) The court may, on an application made by virtue of subparagraph (3) above, give such directions as it thinks fit –
   (a) with respect to the separate participation of the operator in the proceedings to which the application gives rise, and
   (b) requiring the operator to provide information to the court.

(5) A covenant, condition or agreement which would have the effect of preventing or restricting the taking by any person as a potential subscriber of any step under this paragraph shall be void to the extent that it would have that effect.

(6) Nothing in this paragraph shall be construed as requiring the operator to reimburse the potential subscriber for any costs incurred by the potential subscriber in or in connection with the taking of any step under this paragraph on the operator's behalf.

9. Street works

(1) The operator shall, for the statutory purposes, have the right to do any of the following things, that is to say –
   (a) install electronic communications apparatus, or keep electronic communications apparatus installed, under, over, in, on, along or across a street or, in Scotland, a road;
   (b) inspect, maintain, adjust, repair or alter any electronic communications apparatus so installed; and
   (c) execute any works requisite for or incidental to the purposes of any works falling within paragraph (a) or (b) above, including for those purposes the following kinds of works, that is to say –
      (i) breaking up or opening a street or, in Scotland, a road;
      (ii) tunnelling or boring under a street or, in Scotland, a road; and

(iii) breaking up or opening a sewer, drain or tunnel;

[...]

(2) This paragraph has effect subject to paragraph 3 above and the following provisions of this code, and the rights conferred by this paragraph shall not be exercisable in a street which is not a maintainable highway or, in Scotland, a road which is not a public road without either the agreement required by paragraph 2 above or an order of the court under paragraph 5 above dispensing with the need for that agreement.

(3) The rights conferred by this paragraph shall not be exercisable on any land comprised in the route of a special road within the meaning of the Roads (Northern Ireland) Order 1980.

10. Power to fly lines

(1) Subject to paragraph 3 above and the following provisions of this code, where any electronic communications apparatus is kept installed on or over any land for the purposes of the operator's network, the operator shall, for the statutory purposes, have the right to install and keep installed lines which –

(a) pass over other land adjacent to or in the vicinity of the land on or over which that apparatus is so kept;

(b) are connected to that apparatus; and

(c) are not at any point in the course of passing over the other land less than 3 metres above the ground or within 2 metres of any building over which they pass.

(2) Nothing in sub-paragraph (1) above shall authorise the installation or keeping on or over any land of –

(a) any electronic communications apparatus used to support, carry or suspend a line installed in pursuance of that sub-paragraph: or

(b) any line which by reason of its position interferes with the carrying on of any business carried on on that land.

(3) In this paragraph *'business'* includes a trade, profession or employment and includes any activity carried on by a body of persons (whether corporate or unincorporate).

11. Tidal waters etc.

(1) Subject to paragraph 3 above and the following provisions of this code, the operator shall have the right for the statutory purposes –

(a) to execute any works (including placing any buoy or sea-mark) on any tidal water or lands for or in connection with the installation, maintenance, adjustment, repair or alteration of electronic communications apparatus;

(b) to keep electronic communications apparatus installed on, under or over tidal water or lands; and

(c) to enter any tidal water or lands to inspect any electronic communications apparatus so installed.

(2) A right conferred by this paragraph shall not be exercised in relation to any land in which a Crown interest, within the meaning of paragraph 26 below, subsists unless agreement to the exercise of the right in relation to that land has been given, in accordance with sub-paragraph (3) of that paragraph, in respect of that interest.

[...]

(11) In this paragraph –

[...]

'tidal water or lands' includes any estuary or branch of the sea, the shore below mean high water springs and the bed of any tidal water.

12. Linear obstacles

(1) Subject to the following provisions of this code, the operator shall, for the statutory purposes, have the right in order to cross any relevant land with a line, to install and keep the line and other electronic communications apparatus on, under or over that land and –
  (a) to execute any works on that land for or in connection with the installation, maintenance, adjustment, repair or alteration of that line or the other electronic communications apparatus; and
  (b) to enter on that land to inspect the line or the other apparatus.
(2) A line installed in pursuance of any right conferred by this paragraph need not cross the relevant land in question by a direct route or by the shortest route from the point at which the line enters that land, but it shall not cross that land by any route which, in the horizontal plane, exceeds the said shortest route by more than 400 metres.
(3) Electronic communications apparatus shall not be installed in pursuance of any right conferred by this paragraph in any position on the relevant land in which it interferes with traffic on the railway, canal or tramway on that land.
(4) The operator shall not execute any works on any land in pursuance of any right conferred by this paragraph unless –
  (a) he has given the person with control of the land 28 days' notice of his intention to do so; or
  (b) the works are emergency works.
(5) A notice under sub-paragraph (4) above shall contain a plan and section of the proposed works or (in lieu of a plan and section) any description of the proposed works (whether or not in the form of a diagram) which the person with control of the land has agreed to accept for the purposes of this sub-paragraph.
(6) If, at any time before a notice under sub-paragraph (4) above expires, the person with control of the land gives the operator notice of objection to the works, the operator shall be entitled to execute the works only –
  (a) if, within the period of 28 days beginning with the giving of the notice of objection, neither the operator not that person has given notice to the other requiring him to agree to an arbitrator to whom the objection may be referred under paragraph 13 below; or
  (b) in accordance with an award made on such a reference; or
  (c) to the extent that the works have at any time become emergency works.
(7) If the operator exercises any power conferred by this paragraph to execute emergency works on any land, he shall, as soon as reasonably practicable after commencing those works, give the person with control of the land a notice identifying the works and containing –
  (a) a statement of the reason why the works are emergency works; and
  (b) either the matters which would be required to be contained in a notice under sub-paragraph (4) above with respect to those works or, as the case may

require, a reference to an earlier notice under that sub-paragraph with respect to those works.

(8) If within the period of 28 days beginning with the giving of a notice under sub-paragraph (7) above the person to whom that notice was given gives a notice to the operator requiring him to pay compensation, the operator shall be liable to pay that person compensation in respect of loss or damage sustained in consequence of the carrying out of the emergency works in question; and any question as to the amount of that compensation shall, in default of agreement, be referred to arbitration under paragraph 13 below.

(9) If the operator commences the execution of any works in contravention of any provision of this paragraph, he shall be guilty of an offence and liable on summary conviction to a fine not exceeding level 3 on the standard scale.

(10) In this paragraph *'relevant land'* means land which is used wholly or mainly either as a railway, canal or tramway or in connection with a railway, canal or tramway on that land, and a reference to the person with control of any such land is a reference to the person carrying on the railway, canal or tramway undertaking in question.

13. Arbitration in relation to linear obstacles

(1) Any objection or question which, in accordance with paragraph 12 above, is referred to arbitration under this paragraph shall be referred to the arbitration of a single arbitrator appointed by agreement between the parties concerned or, in default of agreement, by the President of the Institution of Civil Engineers.

(2) Where an objection under paragraph 12 above is referred to arbitration under this paragraph the arbitrator shall have the power –

- (a) to require the operator to submit to the arbitrator a plan and section in such form as the arbitrator may think requisite for the purposes of the arbitration;
- (b) to require the observations on any such plan or section of the person who objects to the works to be submitted to the arbitrator in such form as the arbitrator may think requisite for those purposes;
- (c) to direct the operator or that person to furnish him with such information and to comply with such other requirements as the arbitrator may think requisite for those purposes;
- (d) to make an award requiring modifications to the proposed works and specifying the terms on which and the conditions subject to which the works may be executed; and
- (e) to award such sum as the arbitrator may determine in respect of one or both of the following matters, that is to say –
  - (i) compensation to the person who objects to the works in respect of loss or damage sustained by that person in consequence of the carrying out of the works, and
  - (ii) consideration payable to that person for the right to carry out the works.

(3) Where a question as to compensation in respect of emergency works is referred to arbitration under this paragraph, the arbitrator –

- (a) shall have the power to direct the operator or the person who requires the payment of compensation to furnish him with such information and to comply with such other requirements as the arbitrator may think requisite for the purposes of the arbitration; and

(b) shall award to the person requiring the payment of compensation such sum (if any) as the arbitrator may determine in respect of the loss or damage sustained by that person in consequence of the carrying out of the emergency works in question.

(4) The arbitrator may treat compliance with any requirement made in pursuance of sub-paragraph (2)(a) to (c) or (3)(a) above as a condition of his making an award.

(5) In determining what award to make on a reference under this paragraph, the arbitrator shall have regard to all the circumstances and to the principle that no person should unreasonably be denied access to an electronic communications network or to electronic communications services.

(6) For the purposes of the making of an award under this paragraph –

(a) the references in sub-paragraphs (2)(e) and (3)(b) above to loss shall, in relation to a person carrying on a railway, canal or tramway undertaking, include references to any increase in the expenses of carrying on that undertaking; and

(b) the consideration mentioned in sub-paragraph (2)(e) above shall be determined on the basis of what would have been fair and reasonable if the person who objects to the works had given his authority willingly for the works to be executed on the same terms and subject to the same conditions (if any) as are contained in the award.

(7) In the application of this paragraph to Scotland, the reference to an arbitrator shall have effect as a reference to an arbiter and the arbiter may and, if so directed by the Court of Session, shall state a case for the decision of that Court on any question of law arising in the arbitration.

[...]

14. Alteration of apparatus crossing a linear obstacle

(1) Without prejudice to the following provisions of this code, the person with control of any relevant land may, on the ground that any electronic communications apparatus kept installed on, under or over that land for the purposes of the operator's network interferes, or is likely to interfere, with –

(a) the carrying on of the railway, canal or tramway undertaking carried on by that person, or

(b) anything done or to be done for the purposes of that undertaking,

give notice to the operator requiring him to alter that apparatus.

(2) The operator shall within a reasonable time and to the reasonable satisfaction of the person giving the notice comply with a notice under sub-paragraph (1) above unless before the expiration of the period of 28 days beginning with the giving of the notice he gives a counter-notice to the person with control of the land in question specifying the respects in which he is not prepared to comply with the original notice.

(3) Where a counter-notice has been given under sub-paragraph (2) above the operator shall not be required to comply with the original notice but the person with control of the relevant land may apply to the court for an order requiring the alteration of any electronic communications apparatus to which the notice relates.

(4) The court shall not make an order under this paragraph unless it is satisfied that the order is necessary on one of the grounds mentioned in sub-paragraph (1) above and in determining whether to make such an order the court shall also have regard to all the circumstances and to the principle that no person should unreasonably be denied access to an electronic communications network or to electronic communications services.

(5) An order under this paragraph may take such form and be on such terms as the court thinks fit and may impose such conditions and may contain such directions to the operator or the person with control of the land in question as the court thinks necessary for resolving any difference between the operator and that person and for protecting their respective interests.

(6) In this paragraph references to relevant land and to the person with control of such land have the same meaning as in paragraph 12 above.

15. Use of certain conduits

(1) Nothing in the preceding provision of this code shall authorise the doing of anything inside a relevant conduit without the agreement of the authority with control of that conduit.

(2) The agreement of the authority with control of a public sewer shall be sufficient in all cases to confer a right falling within any of the preceding provisions of this code where the right is to be exercised wholly inside that sewer.

(3) In this paragraph –

(a) *'relevant conduit'* and *'public sewer'* have the same meanings as in section 98 of this Act; and

(b) a reference to the authority with control of a relevant conduit shall be construed in accordance with subsections (7) and (8) of that section.

16. Compensation for injurious affection to neighbouring land etc.

(1) Where a right conferred by or in accordance with any of the preceding provisions of this code is exercised, compensation shall be payable by the operator under section 10 of the Compulsory Purchase Act 1965 (compensation for injurious affection to neighbouring land etc.) as if that section had effect in relation to injury caused by the exercise of such a right as it has effect in relation to injury caused by the execution of works on land that has been compulsorily purchased.

(2) Sub-paragraph (1) above shall not confer any entitlement to compensation on any person in respect of the exercise of a right conferred in accordance with paragraph 2 or 3 above, if that person conferred the right or is bound by it by virtue of paragraph 2(2)(b) or (d) above, but, save as aforesaid, the entitlement of any person to compensation under this paragraph shall be determined irrespective of his ownership of any interest in the land where the right is exercised.

(3) Compensation shall not be payable on any claim for compensation under this paragraph unless the amount of the compensation exceeds £50.

(4) In the application of this paragraph to Scotland –

(a) for any reference in sub-paragraph (1) to section 10 of the Compulsory Purchase Act 1965 there is substituted a reference to section 6 of the Railway Clauses Consolidation (Scotland) Act 1845;

(b) for the reference in that sub-paragraph to land that has been compulsorily purchased there is substituted a reference to land that has been taken or used for the purpose of a railway;
(c) any question as to a person's entitlement to compensation by virtue of that sub-paragraph, or as to the amount of that compensation, shall, in default of agreement, be determined by the Lands Tribunal for Scotland.

(5) In the application of this paragraph to Northern Ireland –
(a) for any reference in sub-paragraph (1) to section 10 of the Compulsory Purchase Act 1965 there is substituted a reference to Article 18 of the Land Compensation (Northern Ireland) Order 1982;
(b) any question as to a person's entitlement to compensation by virtue of sub-paragraph (1) above, or as to the amount of that compensation, shall, in default of agreement, be determined by the Lands Tribunal for Northern Ireland.

17. Objections to overhead apparatus

(1) This paragraph applies where the operator has completed the installation for the purposes of the operator's network of any electronic communications apparatus the whole or part of which is at a height of 3 metres or more above the ground.

(2) At any time before the expiration of the period of 3 months beginning with the completion of the installation of the apparatus a person who is the occupier of or owns an interest in –
(a) any land over or on which the apparatus has been installed, or
(b) any land the enjoyment of which, or any interest in which, is, because of the nearness of the land to the land on or over which the apparatus has been installed, capable of being prejudiced by the apparatus,

may give the operator notice of objection in respect of that apparatus.

(3) No notice of objection may be given in respect of any apparatus if the apparatus –
(a) replaces any electronic communications apparatus which is not substantially different from the new apparatus; and
(b) is not in a significantly different position.

(4) Where a person has both given a notice under this paragraph and applied for compensation under any of the preceding provisions of this code, the court –
(a) may give such directions as it thinks fit for ensuring that no compensation is paid until any proceedings under this paragraph have been disposed of, and
(b) if the court makes an order under this paragraph, may provide in that order for some or all of the compensation otherwise payable under this code to that person not to be so payable, or, if the case so requires, for some or all of any compensation paid under this code to that person to be repaid to the operator.

(5) At any time after the expiration of the period of 2 months beginning with the giving of a notice of objection but before the expiration of the period of 4 months beginning with the giving of that notice, the person who gave the notice may apply to the court to have the objection upheld.

(6) Subject to sub-paragraph (7) below, the court shall uphold the objection if the apparatus appears materially to prejudice the applicant's enjoyment of, or interest

in, the land in right of which the objection is made and the court is not satisfied that the only possible alterations of the apparatus will –

(a) substantially increase the cost or diminish the quality of the service provided by the operator's network to persons who have, or may in future have, access to it, or
(b) involve the operator in substantial additional expenditure (disregarding any expenditure occasioned solely by the fact that any proposed alteration was not adopted originally or, as the case may be, that the apparatus has been unnecessarily installed), or
(c) give to any person a case at least as good as the applicant has to have an objection under this paragraph upheld.

(7) The court shall not uphold the objection if the applicant is bound by a right of the operator falling within paragraph 2 or 3(1) above to install the apparatus and it appears to the court unreasonable, having regard to the fact that the applicant is so bound and the circumstances in which he became so bound, for the applicant to have given notice of objection.

(8) In considering the matters specified in sub-paragraph (6) above the court shall have regard to all the circumstances and to the principle that no person should unreasonably be denied access to an electronic communications network or to electronic communications services.

(9) If it upholds an objection under this paragraph the court may by order –

(a) direct the alteration of the apparatus to which the objection relates;
(b) authorise the installation (instead of the apparatus to which the objection relates), in a manner and position specified in the order, of any apparatus so specified;
(c) direct that no objection may be made under this paragraph in respect of any apparatus the installation of which is authorised by the court.

(10) The court shall not make any order under this paragraph directing the alteration of any apparatus or authorising the installation of any apparatus unless it is satisfied either –

(a) that the operator has all such rights as it appears to the court appropriate that he should have for the purpose of making the alteration or, as the case may be, installing the apparatus, or
(b) that –
  (i) he would have all those rights if the court, on an application under paragraph 5 above, dispensed with the need for the agreement of any person, and
  (ii) it would be appropriate for the court, on such an application, to dispense with the need for that agreement;

and, accordingly, for the purposes of dispensing with the need for the agreement of any person to the alteration or installation of any apparatus, the court shall have the same powers as it would have if an application had been duly made under paragraph 5 above for an order dispensing with the need for that person's agreement.

(11) For the purposes of sub-paragraphs (6)(c) and (10) above, the court shall have power on an application under this paragraph to give the applicant directions for bringing the application to the notice of such other interested persons as it thinks fit.

18. Obligation to affix notices to overhead apparatus

(1) Where the operator has for the purposes of the operator's network installed any electronic communications apparatus the whole or part of which is at a height of 3 metres or more above the ground, the operator shall, before the expiration of the period of 3 days beginning with the completion of the installation, in a secure and durable manner affix a notice –
   (a) to every major item of apparatus installed; or
   (b) if no major item of apparatus is installed, to the nearest major item of electronic communications apparatus to which the apparatus that is installed is directly or indirectly connected.

(2) A notice affixed under sub-paragraph (1) above shall be affixed in a position where it is reasonably legible and shall give the name of the operator and an address in the United Kingdom at which any notice of objection may be given under paragraph 17 above in respect of the apparatus in question; and any person giving such a notice at that address in respect of that apparatus shall be deemed to have been furnished with that address for the purposes of paragraph 24(2A)(a) below.

(3) If the operator contravenes the requirements of this paragraph he shall be guilty of an offence and liable on summary conviction to a fine not exceeding level 2 on the standard scale.

(4) In any proceedings for an offence under this paragraph it shall be a defence for the person charged to prove that he took all reasonable steps and exercised all due diligence to avoid committing the offence

19. Tree lopping

(1) Where any tree overhangs any street and, in doing so, either –
   (a) obstructs or interferes with the working of any electronic communications apparatus used for the purposes of the operator's network, or
   (b) will obstruct or interfere with the working of any electronic communications apparatus which is about to be installed for those purposes.

the operator may by notice to the occupier of the land on which the tree is growing require the tree to be lopped so as to prevent the obstruction or interference.

(2) If within the period of 28 days beginning with the giving of the notice by the operator, the occupier of the land on which the tree is growing gives the operator a counter-notice objecting to the lopping of the tree, the notice shall have effect only if confirmed by an order of the court.

(3) If at any time a notice under sub-paragraph (1) above has not been complied with and either –
   (a) a period of 28 days beginning with the giving of the notice has expired without a counter-notice having been given, or
   (b) an order of the court confirming the notice has come into force,

the operator may himself cause the tree to be lopped as mentioned in sub-paragraph (1) above.

(4) Where the operator lops a tree in exercise of the power conferred by sub-paragraph (3) above he shall do so in a husband-like manner and in such a way as to cause the minimum damage to the tree.

(5) Where –

(a) a notice under sub-paragraph (1) above is complied with either without a counter-notice having been given or after the notice has been confirmed, or

(b) the operator exercises the power conferred by sub-paragraph (3) above,

the court shall, on an application made by a person who has sustained loss or damage in consequence of the lopping of the tree or who has incurred expenses in complying with the notice, order the operator to pay that person such compensation in respect of the loss, damage or expenses as it thinks fit.

20. Power to require alteration of apparatus

(1) Where any electronic communications apparatus is kept installed on, under or over any land for the purposes of the operator's network, any person with an interest in that land or adjacent land may (notwithstanding the terms of any agreement binding that person) by notice given to the operator require the alteration of the apparatus on the ground that the alteration is necessary to enable that person to carry out a proposed improvement of the land in which he has an interest.

(2) Where a notice is given under sub-paragraph (1) above by any person to the operator, the operator shall comply with it unless he gives a counter-notice under this sub-paragraph within the period of 28 days beginning with the giving of the notice.

(3) Where a counter-notice is given under sub-paragraph (2) above to any person, the operator shall make the required alteration only if the court on an application by that person makes an order requiring the alteration to be made.

(4) The court shall make an order under this paragraph for an alteration to be made only if, having regard to all the circumstances and the principle that no person should unreasonably be denied access to an electronic communications network or to electronic communications services, it is satisfied –

(a) that the alteration is necessary as mentioned in sub-paragraph (1) above; and

(b) that the alteration will not substantially interfere with any service which is or is likely to be provided using the operator's network.

(5) The court shall not make an order under this paragraph for the alteration of any apparatus unless it is satisfied either –

(a) that the operator has all such rights as it appears to the court appropriate that he should have for the purpose of making the alteration, or

(b) that –

(i) he would have all those rights if the court, on an application under paragraph 5 above, dispensed with the need for the agreement of any person, and

(ii) it would be appropriate for the court, on such an application, to dispense with the need for that agreement;

and, accordingly, for the purposes of dispensing with the need for the agreement of any person to the alteration of any apparatus, the court shall have the same powers as it would have if an application had been duly made under paragraph 5 above for an order dispensing with the need for that person's agreement.

(6) For the purposes of sub-paragraph (5) above, the court shall have power on an application under this paragraph to give the applicant directions for bringing the application to the notice of such other interested persons as it thinks fit.

(7) An order under this paragraph may provide for the alteration to be carried out with such modifications, on such terms and subject to such conditions as the court thinks fit, but the court shall not include any such modifications, terms or conditions in its order without the consent of the applicant, and if such consent is not given may refuse to make an order under this paragraph.

(8) An order made under this paragraph on the application of any person shall, unless the court otherwise thinks fit, require that person to reimburse the operator in respect of any expenses which the operator incurs in or in connection with the execution of any works in compliance with the order.

(9) In sub-paragraph (1) above *'improvement'* includes development and change of use.

21. Restriction on right to require the removal of apparatus

(1) Where any person is for the time being entitled to require the removal of any of the operator's electronic communications apparatus from any land (whether under any enactment or because that apparatus is kept on, under or over that land otherwise than in pursuance of a right binding that person or for any other reason) that person shall not be entitled to enforce the removal of the apparatus except, subject to sub-paragraph (12) below, in accordance with the following provisions of this paragraph.

(2) The person entitled to require the removal of any of the operator's electronic communications apparatus shall give a notice to the operator requiring the removal of the apparatus.

(3) Where a person gives a notice under sub-paragraph (2) above and the operator does not give that person a counter-notice within the period of 28 days beginning with the giving of the notice, that person shall be entitled to enforce the removal of the apparatus.

(4) A counter-notice given under sub-paragraph (3) above to any person by the operator shall do one or both of the following, that is to say –
    (a) state that that person is not entitled to require the removal of the apparatus;
    (b) specify the steps which the operator proposes to take for the purpose of securing a right as against that person to keep the apparatus on the land.

(5) Those steps may include any steps which the operator could take for the purpose of enabling him, if the apparatus is removed, to re-install the apparatus; and the fact that by reason of the following provisions of this paragraph any proposed re-installation is only hypothetical shall not prevent the operator from taking those steps or any court or person from exercising any function in consequence of those steps having been taken.

(6) Where a counter-notice is given under sub-paragraph (3) above to any person, that person may only enforce the removal of the apparatus in pursuance of an order of the court; and, where the counter-notice specifies steps which the operator is proposing to take to secure a right to keep the apparatus on the land, the court shall not make such an order unless it is satisfied –

(a) that the operator is not intending to take those steps or is being unreasonably dilatory in the taking of those steps; or

(b) that the taking of those steps has not secured, or will not secure, for the operator as against that person any right to keep the apparatus installed on, under or over the land or, as the case may be, to re-install it if it is removed.

(7) Where any person is entitled to enforce the removal of any apparatus under this paragraph (whether by virtue of sub-paragraph (3) above or an order of the court under sub-paragraph (6) above), that person may, without prejudice to any method available to him apart from this sub-paragraph for enforcing the removal of that apparatus, apply to the court for authority to remove it himself; and, on such an application, the court may, if it thinks fit, give that authority.

(8) Where any apparatus is removed by any person under an authority given by the court under sub-paragraph (7) above, any expenses incurred by him in or in connection with the removal of the apparatus shall be recoverable by him from the operator in any court of competent jurisdiction; and in so giving an authority to any person the court may also authorise him, in accordance with the directions of the court, to sell any apparatus removed under the authority and to retain the whole or a part of the proceeds of sale on account of those expenses.

(9) Any electronic communications apparatus kept installed on, under or over any land shall (except for the purposes of this paragraph and without prejudice to paragraphs 6(3) and 7(3) above) be deemed, as against any person who was at any time entitled to require the removal of the apparatus, but by virtue of this paragraph not entitled to enforce its removal, to have been lawfully so kept at that time.

(10) Where this paragraph applies (whether in pursuance of an enactment amended by Schedule 4 to this Act or otherwise) in relation to electronic communications apparatus the alteration of which some person ('the relevant person') is entitled to require in consequence of the stopping up, closure, change or diversion of any street (or, in Scotland, any road) or the extinguishment or alteration of any public right of way –

(a) the removal of the apparatus shall constitute compliance with a requirement to make any other alteration;

(b) a counter-notice under sub-paragraph (3) above may state (in addition to, or instead of, any of the matters mentioned in sub-paragraph (4) above) that the operator requires the relevant person to reimburse him in respect of any expenses which he incurs in or in connection with the making of any alteration in compliance with the requirements of the relevant person;

(c) an order made under this paragraph on an application by the relevant person in respect of a counter-notice containing such a statement shall, unless the court otherwise thinks fit, require the relevant person to reimburse the operator in respect of any expenses which he so incurs; and

(d) sub-paragraph (8) above shall not apply.

(11) References in this paragraph to the operator's electronic communications apparatus include references to electronic communications apparatus which (whether or not vested in the operator) is being, is to be or has been used for the purposes of the operator's network.

(12) A person shall not, under this paragraph, be entitled to enforce the removal of any apparatus on the ground only that he is entitled to give a notice under paragraph 11, 14, 17 or 20 above; and this paragraph is without prejudice to paragraph 23 below and to the power to enforce an order of the court under the said paragraph 11, 14, 17 or 20.

22. Abandonment of apparatus

Without prejudice to the preceding provisions of this code, where the operator has a right conferred by or in accordance with this code for the statutory purposes to keep electronic communications apparatus installed on, under or over any land, he is not entitled to keep that apparatus so installed if, at a time when the apparatus is not, or is no longer, used for the purposes of the operator's network, there is no reasonable likelihood that it will be so used.

23. Undertakers' Works

(1) The following provisions of this paragraph apply where a relevant undertaker is proposing to execute any undertaker's works which involve or are likely to involve a temporary or permanent alteration of any electronic communications apparatus kept installed on, under or over any land for the purposes of the operator's network.
(2) The relevant undertaker shall, not less than 10 days before the works are commenced, give the operator a notice specifying the nature of the undertaker's works, the alteration or likely alteration involved and the time and place at which the works will be commenced.
(3) Sub-paragraph (2) above shall not apply in relation to any emergency works of which the relevant undertaker gives the operator notice as soon as practicable after commencing the works.
(4) Where a notice has been given under sub-paragraph (2) above by a relevant undertaker to the operator, the operator may within the period of 10 days beginning with the giving of the notice give the relevant undertaker a counter-notice which may state either –
    (a) that the operator intends himself to make any alteration made necessary or expedient by the proposed undertaker's works; or
    (b) that he requires the undertaker in making any such alteration to do so under the supervision and to the satisfaction of the operator.
(5) Where a counter-notice given under sub-paragraph (4) above states that the operator intends himself to make any alteration –
    (a) the operator shall (subject to sub-paragraph (7) below) have the right, instead of the relevant undertaker, to execute any works for the purpose of making that alteration; and
    (b) any expenses incurred by the operator in or in connection with the execution of those works and the amount of any loss or damage sustained by the operator in consequence of the alteration shall be recoverable by the operator from the undertaker in any court of competent jurisdiction.
(6) Where a counter-notice given under sub-paragraph (4) above states that any alteration is to be made under the supervision and to the satisfaction of the operator –
    (a) the relevant undertaker shall not make the alteration except as required by the notice or under sub-paragraph (7) below; and
    (b) any expenses incurred by the operator in or in connection with the provision of that supervision and the amount of any loss or damage sustained by the operator in consequence of the alteration shall be recoverable by the operator from the undertaker in any court of competent jurisdiction.
(7) Where –
    (a) no counter-notice is given under sub-paragraph (4) above, or

(b) the operator, having given a counter-notice falling within that sub-paragraph, fails within a reasonable time to make any alteration made necessary or expedient by the proposed undertaker's works or, as the case may be, unreasonably fails to provide the required supervision,

the relevant undertaker may himself execute works for the purpose of making the alteration or, as the case may be, may execute such works without the supervision of the operator; but in either case the undertaker shall execute the works to the satisfaction of the operator.

(8) If the relevant undertaker or any of his agents –

(a) executes any works without the notice required by sub-paragraph (2) above having been given, or

(b) unreasonably fails to comply with any reasonable requirement of the operator under this paragraph,

he shall, subject to sub-paragraph (9) below, be guilty of an offence and liable on summary conviction to a fine which –

(i) if the service provided by the operator's network is interrupted by the works or failure, shall not exceed level 4 on the standard scale; and

(ii) if that service is not so interrupted, shall not exceed level 3 on the standard scale.

(9) Sub-paragraph (8) above does not apply to a Northern Ireland department.

(10) In this paragraph –

'relevant undertaker' means –

(a) any person (including a local authority) authorised by any Act (whether public general or local) or by any order or scheme made under or confirmed by any Act to carry on –

(i) any railway, tramway, road transport, water transport, canal, inland navigation, dock, harbour, pier or lighthouse undertaking; or

[...]

(b) any person (apart from the operator) to whom this code is applied by a direction under section 106 of the Communications Act 2003; and

(c) any person to whom this paragraph is applied by any Act amended by or under or passed after this Act;

'undertaker's works' means –

(a) in relation to a relevant undertaker falling within paragraph (a) of the preceding definition, any works which that undertaker is authorised to execute for the purposes of, or in connection with, the carrying on by him of the undertaking mentioned in that paragraph;

(b) in relation to a relevant undertaker falling within paragraph (b) of that definition, any works which that undertaker is authorised to execute by or in accordance with any provision of this code; and

(c) in relation to a relevant undertaker falling within paragraph (c) of that definition, the works for the purposes of which this paragraph is applied to that undertaker.

(11) The application of this paragraph by virtue of paragraph (c) of each of the definitions in sub-paragraph (10) above to any person for the purposes of any works shall be without prejudice to its application by virtue of paragraph (a) of each of those definitions to that person for the purposes of any other works.

24. Notices under code

(1) Any notice required to be given by the operator to any person for the purposes of any provision of this code must be in a form approved by OFCOM as adequate for indicating to that person the effect of the notice and of so much of this code as is relevant to the notice and to the steps that may be taken by that person under this code in respect of that notice.

(2) A notice required to be given to any person for the purposes of any provision of this code is not to be sent to him by post unless it is sent by a registered post service or by recorded delivery.

(2A) For the purposes, in the case of such a notice, of section 394 of the Communications Act 2003 and the application of section 7 of the Interpretation Act 1978 in relation to that section, the proper address of a person is –

- (a) if the person to whom the notice is to be given has furnished the person giving the notice with an address for service under this code, that address; and
- (b) only if he has not, the address given by that section of the Act of 2003.

[...]

(5) If it is not practicable, for the purposes of giving any notice under this code, after reasonable inquiries to ascertain the name and address –

- (a) of the person who is for the purposes of any provision of this code the occupier of any land, or
- (b) of the owner of any interest in any land,

a notice may be given under this code by addressing it to a person by the description of 'occupier' of the land (describing it) or, as the case may be, 'owner' of the interest (describing both the interest and the land) and by delivering it to some person on the land or, if there is no person on the land to whom it can be delivered, by affixing it, or a copy of it, to some conspicuous object on the land.

(6) In any proceedings under this code a certificate issued by OFCOM and stating that a particular form of notice has been approved by them as mentioned in sub-paragraph (1) above shall be conclusive evidence of the matter certified.

24A. Electronic communications networks: determination of applications to install facilities

Regulation 3 of the Electronic Communications and Wireless Telegraphy Regulation 2011 makes provision about the time within which certain applications under this code for the granting of rights to install facilities must be determined.

25. Appeals in Northern Ireland

Article 60 of the County Courts (Northern Ireland) Order 1980 (ordinary appeals from the country court in civil cases) shall apply in relation to any determination of the court in Northern Ireland under this code in like manner as it applies in relation to any decree of the court made in the exercise of the jurisdiction conferred by Part III of that Order

26. Application to the Crown

(1) This code shall apply in relation to land in which there subsists, or at any material time subsisted, a Crown interest as it applies in relation to land in which no such interest subsists.

(2) In this paragraph *'Crown interest'* means an interest which belongs to Her Majesty in right of the Crown or of the Duchy of Lancaster or to the Duchy of Cornwall or to a Government department or which is held in trust for Her Majesty for the purposes of a Government department and, without prejudice to the foregoing, includes any interest which belongs to Her Majesty in right of Her Majesty's Government in Northern Ireland or to a Northern Ireland department or which is held in trust for Her Majesty for the purposes of a Northern Ireland department.

(3) An agreement required by this code to be given in respect of any Crown interest subsisting in any land shall be given by the appropriate authority, that is to say –
   (a) in the case of land belonging to Her Majesty in right of the Crown, the Crown Estate Commissioners or, as the case may require, the government department having the management of the land in question or the relevant person;
   (b) in the case of land belonging to Her Majesty in right of the Duchy of Lancaster, the Chancellor of that Duchy;
   (c) in the case of land belonging to the Duchy of Cornwall, such person as the Duke of Cornwall, or the possessor for the time being of the Duchy of Cornwall, appoints;
   (d) in the case of land belonging to Her Majesty in right of Her Majesty's Government in Northern Ireland, the Northern Ireland department having the management of the land in question;
   (e) in the case of land belonging to a government department or a Northern Ireland department or held in trust for Her Majesty for the purposes of a government department or a Northern Ireland department, that department;

and if any question arises as to what authority is the appropriate authority in relation to any land that question shall be referred to the Treasury, whose decision shall be final.

(3A) In sub-paragraph (3), *'relevant person'*, in relation to any land to which section 90B(5) of the Scotland Act 1998 applies, means the person who manages that land.

(4) Paragraphs 12(9) and 18(3) above shall not apply where this code applies in the case of the Secretary of State or a Northern Ireland department by virtue of section 106(3)(b) of the Communications Act 2003.

27. Savings for and exclusion of certain remedies etc.

(1) Except in so far as provision is otherwise made by virtue of Schedule 4 to this Act, this code shall not authorise the contravention of any provision made by or under any enactment passed before this Act.

(2) The provisions of this code, except paragraphs 8(5) and 21 and sub-paragraph (1) above, shall be without prejudice to any rights or liabilities arising under any agreement to which the operator is a party.

(3) Except as provided under the preceding provisions of this code, the operator shall not be liable to compensate any person for, or be subject to any other liability in respect of, any loss or damage caused by the lawful exercise of any right conferred by or in accordance with this code.

28. Application of code to existing systems

(1) Subject to the following provisions of this paragraph, references in this code to electronic communications apparatus installed on, under or over any land include references to [electronic communications apparatus so installed before this code comes into force.

(2) Without prejudice to sub-paragraph (1) above, any line or other apparatus lawfully installed before this code comes into force which if this code had come into force could have been installed under paragraph 12 of this code shall (subject to sub-paragraph (6) below) be treated for the purposes of this code as if it had been so installed.

(3) Any consent given (or deemed to have been given) for the purposes of any provision of the Telegraph Acts 1863 to 1916 before this code comes into force shall –

  (a) have effect after this code comes into force as an agreement given for the purposes of this code, and

  (b) so have effect, to any extent that is necessary for ensuring that the same persons are bound under this code as were bound by the consent, as if it were an agreement to confer a right or, as the case may require, to bind any interest in land of the person who gave (or is deemed to have given) the consent.

(4) Where by virtue of sub-paragraph (3) above any person is bound by any right, that right shall not be exercisable except on the same terms and subject to the same conditions as the right which, by virtue of the giving of the consent, was exercisable before this code comes into force; and where under any enactment repealed by this Act those terms or conditions included a requirement for the payment of compensation or required the determination of any matter by any court or person, the amount of the compensation or, as the case may be, that matter shall be determined after the coming into force of this code in like manner as if this Act had not been passed.

(5) A person shall not be entitled to compensation under any provision of this code if he is entitled to compensation in respect of the same matter by virtue of sub-paragraph (4) above.

(6) Neither this code nor the repeal by this Act of any provision of the Telegraph Acts 1863 to 1916 (which contain provisions confirming or continuing in force certain agreements) shall prejudice any rights or liabilities (including any rights or liabilities transferred by virtue of section 60 of this Act) which arise at any time under any agreement which was entered into before this code comes into force and relates to the installation, maintenance, adjustment, repair, alteration or inspection of any electronic communications apparatus or to keeping any such apparatus installed on, under or over any land.

(7) Any person who before the coming into force of this code has –
   (a) given a notice ('the Telegraph Acts notice') under or for the purposes of any provision of the Telegraph Acts 1863 to 1916 to any person, or
   (b) made an application under or for the purposes of any such provision (including, in particular, an application for any matter to be referred to any court or person),

may give a notice to the person to whom the Telegraph Acts notice was given or, as the case may be, to every person who is or may be a party to the proceedings resulting from the application stating that a specified step required to be taken under or for the purposes of this code, being a step equivalent to the giving of the Telegraph Acts notice or the making of the application, and any steps required to be so taken before the taking of that step should be treated as having been so taken.

(8) A notice may be given under sub-paragraph (7) above with respect to an application notwithstanding that proceedings resulting from the application have been commenced.

(9) Where a notice has been given to any person under sub-paragraph (7) above, that person may apply to the court for an order setting aside the notice on the ground that it is unreasonable in all the circumstances to treat the giving of the Telegraph Acts notice or the making of the application in question as equivalent to the taking of the steps specified in the notice under that sub-paragraph; but unless the court sets aside the notice under that sub-paragraph, the steps specified in the notice shall be treated as having been taken and any proceedings already commenced shall be continued accordingly.

(10) Where before this code comes into force anything has, in connection with the exercise by the operator of any power conferred on him by the Telegraph Acts 1863 to 1916, been done under or for the purposes of the street works code contained in the Public Utilities Street Works Act 1950, that thing shall, in so far as it could have been done in connection with the exercise of any power conferred by this code, have effect after this code comes into force, without any notice being given under sub-paragraph (7) above, as if it had been done in connection with the power conferred by this code.

(11) In relation to anything done under section 5 of Schedule 3 to the Water Act 1945 or section 5 of Schedule 4 to the Water (Scotland) Act 1980 before the coming into force of this code, the preceding provisions of this paragraph shall have effect, so far as the context permits, as if references to the Telegraph Acts 1863 to 1916 included references to that section.

(12) References in this paragraph to the coming into force of this code shall have effect as references to the time at which the code comes into force in relation to the operator.

29. Effect of agreements concerning sharing of apparatus

(1) This paragraph applies where –
   (a) this code has been applied by a direction under section 106 of the Communications Act 2003 in a person's case;

(b) this code expressly or impliedly imposes a limitation on the use to which electronic communications apparatus installed by that person may be put or on the purposes for which it may be used; and
(c) that person is a party to a relevant agreement or becomes a party to an agreement which (after he has become a party to it) is a relevant agreement.

(2) The limitation is not to preclude –
(a) the doing of anything in relation to that apparatus, or
(b) its use for particular purposes,

to the extent that the doing of that thing, or the use of the apparatus for those purposes, is in pursuance of the agreement.

(3) This paragraph is not to be construed, in relation to a person who is entitled or authorised by or under a relevant agreement to share the use of apparatus installed by another party to the agreement, as affecting any consent requirement imposed (whether by a statutory provision or otherwise) on that person.
(4) In this paragraph –

'consent requirement', in relation to a person, means a requirement for him to obtain consent or permission to or in connection with –

(a) the installation by him of apparatus; or
(b) the doing by him of any other thing in relation to apparatus the use of which he is entitled or authorised to share;

'relevant agreement' means an agreement in relation to electronic communications apparatus which –

(a) relates to the sharing by different parties to the agreement of the use of that apparatus; and
(b) is an agreement that satisfies the requirements of sub-paragraph (5);

'statutory provision' means a provision of an enactment or of an instrument having effect under an enactment.

(5) An agreement satisfies the requirements of this sub-paragraph if –
(a) every party to the agreement is a person in whose case this code applies by virtue of a direction under section 106 of the Communications Act 2003; or
(b) one or more of the parties to the agreement is a person in whose case this code so applies and every other party to the agreement is a qualifying person.

(6) A person is a qualifying person for the purposes of sub-paragraph (5) if he is either –
(a) a person who provides an electronic communications network without being a person in whose case this code applies; or
(b) a designated provider of an electronic communications service consisting in the distribution of a programme service by means of an electronic communications network.

(7) In sub-paragraph (6) –

'designated' means designated by an order made by the Secretary of State;
'programme service' has the same meaning as in the Broadcasting Act 1990.

PART II

Communications Act 2003

Section 32 Meaning of electronic communications networks and services

(1) In this Act *'electronic communications network'* means –
   (a) a transmission system for the conveyance, by the use of electrical, magnetic or electro-magnetic energy, of signals of any description; and
   (b) such of the following as are used, by the person providing the system and in association with it, for the conveyance of the signals –
      (i) apparatus comprised in the system;
      (ii) apparatus used for the switching or routing of the signals;
      (iii) software and stored data; and
      (iv) (except for the purposes of sections 125 to 127) other resources, including network elements which are not active.

(2) In this Act *'electronic communications service'* means a service consisting in, or having as its principal feature, the conveyance by means of an electronic communications network of signals, except in so far as it is a content service.

(3) In this Act *'associated facility'* means a facility, element or service which is available for use, or has the potential to be used, in association with the use of an electronic communications network or electronic communications service (whether or not one provided by the person making the facility, element or service available) for the purpose of –
   (a) making the provision of that network or service possible;
   (b) making possible the provision of other services provided by means of that network or service; or
   (c) supporting the provision of such other services.

(4) In this Act –
   (a) references to the provision of an electronic communications network include references to its establishment, maintenance or operation;
   (b) references, where one or more persons are employed or engaged to provide the network or service under the direction or control of another person, to the person by whom an electronic communications network or electronic communications service is provided are confined to references to that other person; and
   (c) references, where one or more persons are employed or engaged to make facilities available under the direction or control of another person, to the person by whom any associated facilities are made available are confined to references to that other person.

(5) Paragraphs (a) and (b) of subsection (4) apply in relation to references in subsection (1) to the provision of a transmission system as they apply in relation to references in this Act to the provision of an electronic communications network.

(6) The reference in subsection (1) to a transmission system includes a reference to a transmission system consisting of no more than a transmitter used for the conveyance of signals.

(7) In subsection (2) *'a content service'* means so much of any service as consists in one or both of the following –

(a) the provision of material with a view to its being comprised in signals conveyed by means of an electronic communications network;
(b) the exercise of editorial control over the contents of signals conveyed by means of a such a network.

(8) In this section references to the conveyance of signals include references to the transmission or routing of signals or of parts of signals and to the broadcasting of signals for general reception.
(9) For the purposes of this section the cases in which software and stored data are to be taken as being used for a particular purpose include cases in which they –
(a) have been installed or stored in order to be used for that purpose; and
(b) are available to be so used.
(10) In this section *'signal'* includes –
(a) anything comprising speech, music, sounds, visual images or communications or data of any description; and
(b) signals serving for the impartation of anything between persons, between a person and a thing or between things, or for the actuation or control of apparatus.

Section 106

APPLICATION OF THE ELECTRONIC COMMUNICATIONS CODE

(1) In this Chapter 'the electronic communications code' means the code set out in Schedule 2 to the Telecommunications Act 1984 (c. 12).
(2) Schedule 3 (which amends Schedule 2 to the Telecommunications Act 1984 (c. 12) for the purpose of translating the telecommunications code into a code applicable in the context of the new regulatory regime established by this Act) shall have effect.
(3) The electronic communications code shall have effect –
(a) in the case of a person to whom it is applied by a direction given by OFCOM; and
(b) in the case of the Secretary of State or any Northern Ireland department where the Secretary of State or that department is providing or proposing to provide an electronic communications network.
(4) The only purposes for which the electronic communications code may be applied in a person's case by a direction under this section are –
(a) the purposes of the provision by him of an electronic communications network; or
(b) the purposes of the provision by him of a system of conduits which he is making available, or proposing to make available, for use by providers of electronic communications networks for the purposes of the provision by them of their networks.
(5) A direction applying the electronic communications code in any person's case may provide for that code to have effect in his case –
(a) in relation only to such places or localities as may be specified or described in the direction;
(b) for the purposes only of the provision of such electronic communications network, or part of an electronic communications network, as may be so specified or described; or
(c) for the purposes only of the provision of such conduit system, or part of a conduit system, as may be so specified or described.
(6) The Secretary of State may by order provide for the electronic communications code to have effect for all purposes with a different amount substituted for the

amount for the time being specified in paragraph 16(3) of the code (minimum compensation).

(7) In this section 'conduit' includes a tunnel, subway, tube or pipe.

Section 107

PROCEDURE FOR DIRECTIONS APPLYING CODE

(1) OFCOM are not to give a direction applying the electronic communications code in any person's case except on an application made for the purpose by that person.

(2) If OFCOM publish a notification setting out their requirements with respect to –
  (a) the content of an application for a direction applying the electronic communications code, and
  (b) the manner in which such an application is to be made,

such an application must be made in accordance with the requirements for the time being in force.

(3) OFCOM may –
  (a) from time to time review the requirements for the time being in force for the purposes of subsection (2); and
  (b) on any such review, modify them in such manner as they think fit by giving a notification of the revised requirements.

(4) In considering whether to apply the electronic communications code in any person's case, OFCOM must have regard, in particular, to each of the following matters –
  (a) the benefit to the public of the electronic communications network or conduit system by reference to which the code is to be applied to that person;
  (b) the practicability of the provision of that network or system without the application of the code;
  (c) the need to encourage the sharing of the use of electronic communications apparatus;
  (d) whether the person in whose case it is proposed to apply the code will be able to meet liabilities arising as a consequence of –
    (i) the application of the code in his case; and
    (ii) any conduct of his in relation to the matters with which the code deals.

(5) For the purposes of subsections (6) and (7) of section 3 OFCOM's duty under subsection (4) ranks equally with their duties under that section.

(6) Before giving a direction under section 106, OFCOM must –
  (a) publish a notification of their proposal to give the direction; and
  (b) consider any representations about that proposal that are made to them within the period specified in the notification.

(7) A notification for the purposes of subsection (6)(a) must contain the following –
  (a) a statement of OFCOM's proposal;
  (b) a statement of their reasons for that proposal;
  (c) a statement of the period within which representations may be made to them about the proposal.

(8) The statement of OFCOM's proposal must –
  (a) contain a statement that they propose to apply the code in the case of the person in question;

(b) set out any proposals of theirs to impose terms under section 106(5);

but this subsection is subject to sections 113(7) and 115(5).

(9) The period specified as the period within which representations may be made must end no less than one month after the day of the publication of the notification.
(10) The publication by OFCOM of a notification for any of the purposes of this section must be a publication in such manner as OFCOM consider appropriate for bringing the notification to the attention of the persons who, in their opinion, are likely to be affected by it.

Section 108

REGISTER OF PERSONS IN WHOSE CASE CODE APPLIES

(1) It shall be the duty of OFCOM to establish and maintain a register of persons in whose case the electronic communications code applies by virtue of a direction under section 106.
(2) OFCOM must record in the register every direction given under that section.
(3) Information recorded in the register must be recorded in such manner as OFCOM consider appropriate.
(4) It shall be the duty of OFCOM to publish a notification setting out –
    (a) the times at which the register is for the time being available for public inspection; and
    (b) the fees that must be paid for, or in connection with, an inspection of the register.
(5) The publication of a notification under subsection (4) must be a publication in such manner as OFCOM consider appropriate for bringing it to the attention of the persons who, in their opinion, are likely to be affected by it.
(6) OFCOM must make the register available for public inspection –
    (a) during such hours, and
    (b) on payment of such fees,

as are set out in the notification for the time being in force under subsection (4).

Section 109

RESTRIC TIONS AND CONDITIONS SUBJECT TO WHICH CODE APPLIES

(1) Where the electronic communications code is applied in any person's case by a direction given by OFCOM, that code is to have effect in that person's case subject to such restrictions and conditions as may be contained in regulations made by the Secretary of State.
(2) In exercising his power to make regulations under this section it shall be the duty of the Secretary of State to have regard to each of the following –
    (a) the duties imposed on OFCOM by sections 3 and 4;
    (b) the need to protect the environment and, in particular, to conserve the natural beauty and amenity of the countryside;
    (c) the need to ensure that highways are not damaged or obstructed, and traffic not interfered with, to any greater extent than is reasonably necessary;

(d) the need to encourage the sharing of the use of electronic communications apparatus;
(e) the need to secure that a person in whose case the code is applied will be able to meet liabilities arising as a consequence of –
(i) the application of the code in his case; and
(ii) any conduct of his in relation to the matters with which the code deals.

(3) The power of the Secretary of State to provide by regulations for the restrictions and conditions subject to which the electronic communications code has effect includes power to provide for restrictions and conditions which are framed by reference to any one or more of the following –
(a) the making of a determination in accordance with the regulations by a person specified in the regulations;
(b) the giving of an approval or consent by a person so specified; or
(c) the opinion of any person.

(4) Before making any regulations under this section, the Secretary of State must consult –
(a) OFCOM; and
(b) such other persons as he considers appropriate.

Section 110

ENFORCEMENT OF RESTRICTIONS AND CONDITIONS

(1) Where OFCOM determine that there are reasonable grounds for believing that a person in whose case the electronic communications code applies is contravening, or has contravened, a requirement imposed by virtue of any restrictions or conditions under section 109, they may give him a notification under this section.

(2) A notification under this section is one which –
(a) sets out the determination made by OFCOM;
(b) specifies the requirement and the contravention in respect of which that determination has been made; and
(c) specifies the period during which the person notified has an opportunity of doing the things specified in subsection (3).

(3) Those things are –
(a) making representations about the matters notified;
(b) complying with any notified requirement of which he remains in contravention; and
(c) remedying the consequences of notified contraventions.

(4) Subject to subsections (5) to (7), the period for doing those things must be the period of one month beginning with the day after the one on which the notification was given.

(5) OFCOM may, if they think fit, allow a longer period for doing those things either –
(a) by specifying a longer period in the notification; or
(b) by subsequently, on one or more occasions, extending the specified period.

(6) The person notified shall have a shorter period for doing those things if a shorter period is agreed between OFCOM and the person notified.

(7) The person notified shall also have a shorter period if –
(a) OFCOM have reasonable grounds for believing that the contravention is a repeated contravention;

(b) they have determined that, in those circumstances, a shorter period would be appropriate; and

(c) the shorter period has been specified in the notification.

(8) A notification under this section –

(a) may be given in respect of more than one contravention; and

(b) if it is given in respect of a continuing contravention, may be given in respect of any period during which the contravention has continued.

(9) Where a notification under this section has been given to a person in respect of a contravention of a requirement, OFCOM may give a further notification in respect of the same contravention of that requirement if, and only if –

(a) the contravention is one occurring after the time of the giving of the earlier notification;

(b) the contravention is a continuing contravention and the subsequent notification is in respect of so much of a period as falls after a period to which the earlier notification relates; or

(c) the earlier notification has been withdrawn without a penalty having been imposed in respect of the notified contravention.

(10) For the purposes of this section a contravention is a repeated contravention, in relation to a notification with respect to that contravention, if –

(a) a previous notification under this section has been given in respect of the same contravention or in respect of another contravention of the same requirement; and

(b) the subsequent notification is given no more than twelve months after the day of the making by OFCOM of a determination for the purposes of section 111(2) or 112(2) that the contravention to which the previous notification related did occur.

Section 111

ENFORCEMENT NOTIFICATION FOR CONTRAVENTION OF CODE RESTRICTIONS

(1) This section applies where –

(a) a person ('the notified provider') has been given a notification under section 110;

(b) OFCOM have allowed the notified provider an opportunity of making representations about the matters notified; and

(c) the period allowed for the making of the representations has expired.

(2) OFCOM may give the notified provider an enforcement notification if they are satisfied –

(a) that he has been in contravention, in one or more of the respects notified, of a requirement specified in the notification under section 110; and

(b) that he has not, during the period allowed under section 110, taken all such steps as they consider appropriate –

(i) for complying with that requirement; and

(ii) for remedying the consequences of the notified contravention of that requirement.

(3) An enforcement notification is a notification which imposes one or both of the following requirements on the notified provider –

(a) a requirement to take such steps for complying with the notified requirement as may be specified in the notification;

(b) a requirement to take such steps for remedying the consequences of the notified contravention as may be so specified.

(4) A decision of OFCOM to give an enforcement notification to a person –

(a) must be notified by them to that person, together with the reasons for the decision, no later than one week after the day on which it is taken; and

(b) must fix a reasonable period for the taking of the steps required by the notification.

(5) It shall be the duty of a person to whom an enforcement notification has been given to comply with it.

(6) That duty shall be enforceable in civil proceedings by OFCOM –

(a) for an injunction;

(b) for specific performance of a statutory duty under section 45 of the Court of Session Act 1988 (c. 36); or

(c) for any other appropriate remedy or relief.

Section 112

PENALTIES FOR CONTRAVENTION OF CODE RESTRICTIONS

(1) This section applies (in addition to section 111) where –

(a) a person ('the notified provider') has been given a notification under section 110;

(b) OFCOM have allowed the notified provider an opportunity of making representations about the matters notified; and

(c) the period allowed for the making of the representations has expired.

(2) OFCOM may impose a penalty on the notified provider if he –

(a) has been in contravention, in any of the respects notified, of a requirement specified in the notification under section 110; and

(b) has not, during the period allowed under that section, taken all such steps as they consider appropriate –

(i) for complying with the notified requirement; and

(ii) for remedying the consequences of the notified contravention of that requirement.

(3) Where a notification under section 110 relates to more than one contravention, a separate penalty may be imposed in respect of each contravention.

(4) Where such a notification relates to a continuing contravention, no more than one penalty may be imposed under this section in respect of the period of contravention specified in the notification.

(5) OFCOM may also impose a penalty on the notified provider if he has contravened, or is contravening, a requirement of an enforcement notification.

(6) The amount of a penalty imposed under this section is to be such amount not exceeding £10,000 as OFCOM determine to be –

(a) appropriate; and

(b) proportionate to the contravention in respect of which it is imposed.

(7) In making that determination OFCOM must have regard to –

(a) any representations made to them by the notified provider;

(b) any steps taken by him towards complying with the requirements contraventions of which have been notified to him under section 110; and

(c) any steps taken by him for remedying the consequences of those contraventions.

(8) Where OFCOM impose a penalty on a person under this section, they shall –
  (a) within one week of making their decision to impose the penalty, notify that person of that decision and of their reasons for that decision; and
  (b) in that notification, fix a reasonable period after it is given as the period within which the penalty is to be paid.
(9) A penalty imposed under this section –
  (a) must be paid to OFCOM; and
  (b) if not paid within the period fixed by them, is to be recoverable by them accordingly.
(10) The Secretary of State may by order amend this section so as to substitute a different maximum penalty for the maximum penalty for the time being specified in subsection (6).
(11) No order is to be made containing provision authorised by subsection (10) unless a draft of the order has been laid before Parliament and approved by a resolution of each House.

Section 113

SUSPENSION OF APPLICATION OF CODE

(1) OFCOM may suspend the application of the electronic communications code in any person's case if they are satisfied –
  (a) that he is or has been in serious and repeated contravention of requirements to pay administrative charges fixed under section 38 (whether in respect of the whole or a part of the charges);
  (b) that the bringing of proceedings for the recovery of the amounts outstanding has failed to secure complete compliance by the contravening provider with the requirements to pay the charges fixed in his case, or has no reasonable prospect of securing such compliance;
  (c) that an attempt, by the imposition of penalties under section 41, to secure such compliance has failed; and
  (d) that the suspension of the application of the code is appropriate and proportionate to the seriousness (when repeated as they have been) of the contraventions.
(2) OFCOM may, to the extent specified in subsection (3), suspend the application in that person's case of the electronic communications code if –
  (a) the electronic communications code has been applied by a direction under section 106 in any person's case; and
  (b) OFCOM give a direction under section 42, 100, 132 or 140 for the suspension or restriction of that person's entitlement to provide an electronic communications network, or a part of such a network.
(3) The extent, in any person's case, of a suspension under subsection (2) must not go beyond the application of the code for the purposes of so much of an electronic communications network as that person is prohibited from providing by virtue of the suspension or restriction of his entitlement to provide such a network, or part of a network.
(4) OFCOM may, to the extent specified in subsection (5), suspend the application in that person's case of the electronic communications code if –

(a) the electronic communications code has been applied by a direction under section 106 in any person's case; and
(b) that person is a person in whose case there have been repeated and serious contraventions of requirements imposed by virtue of any restrictions or conditions under section 109.

(5) The extent, in any person's case, of a suspension under subsection (4) must not go beyond the following applications of the code in his case –
(a) its application for the purposes of electronic communications networks, or parts of such a network, which are not yet in existence at the time of the suspension;
(b) its application for the purposes of conduit systems, or parts of such systems, which are not yet in existence or not yet used for the purposes of electronic communications networks; and
(c) its application for other purposes in circumstances in which the provision of an electronic communications network, or part of such a network, would not have to cease if its application for those purposes were suspended.

(6) A suspension under this section of the application of the code in any person's case must be by a further direction given to that person by OFCOM under section 106.

(7) The statement required by section 107(8) to be included, in the case of a direction for the purposes of this section, in the statement of OFCOM's proposal is a statement of their proposal to suspend the application of the code.

(8) A suspension of the application of the electronic communications code in any person's case –
(a) shall cease to have effect if the suspension is under subsection (2) and the network suspension or restriction ceases to have effect; but
(b) subject to that shall continue in force until such time (if any) as it is withdrawn by OFCOM.

(9) In subsection (8) the reference to the network suspension or restriction, in relation to a suspension of the application of the electronic communications code, is a reference to the suspension or restriction of an entitlement to provide an electronic communications network, or part of such a network, which is the suspension or restriction by reference to which the application of the code was suspended under subsection (2).

(10) Subject to subsection (11), where the application of the electronic communications code is suspended in a person's case, he shall not, while it is so suspended, be entitled to exercise any right conferred on him by or by virtue of the code.

(11) The suspension, in a person's case, of the application of the electronic communications code does not, except so far as otherwise provided by a scheme contained in an order under section 117 –
(a) affect (as between the original parties to it) any agreement entered into for the purposes of the code or any agreement having effect in accordance with it;
(b) affect anything done under the code before the suspension of its application; or
(c) require the removal of, or prohibit the use of, any apparatus lawfully installed on, in or over any premises before that suspension.

(12) Subsection (9) of section 42 applies for the purposes of subsection (1) as it applies for the purposes of that section.

Section 114

PROCEDURE FOR DIRECTIONS UNDER S. 113

(1) Except in an urgent case, OFCOM are not to give a direction under section 113(4) suspending the application of the electronic communications code in the case of any person ('the operator') unless they have –
  (a) notified the operator of the proposed suspension and of the steps (if any) that they are proposing to take under section 117;
  (b) provided him with an opportunity of making representations about the proposals and of proposing steps for remedying the situation that has given rise to the proposed suspension; and
  (c) considered every representation and proposal made to them during the period allowed by them for the operator to take advantage of that opportunity.
(2) That period must be one ending not less than one month after the day of the giving of the notification.
(3) As soon as practicable after giving a direction under section 113 in an urgent case, OFCOM must provide the operator with an opportunity of –
  (a) making representations about the effect of the direction and of any steps taken under section 117 in connection with the suspension; and
  (b) proposing steps for remedying the situation that has given rise to the situation.
(4) A case is an urgent case for the purposes of this section if OFCOM –
  (a) consider that it would be inappropriate, because the circumstances appearing to OFCOM to require the suspension fall within subsection (5), to allow time, before giving a direction under section 113, for the making and consideration of representations; and
  (b) decide for that reason to act in accordance with subsection (3), instead of subsection (1).
(5) Circumstances fall within this subsection if they have resulted in, or create an immediate risk of –
  (a) a serious threat to the safety of the public, to public health or to national security;
  (b) serious economic or operational problems for persons (apart from the operator) who are communications providers or persons who make associated facilities available; or
  (c) serious economic or operational problems for persons who make use of electronic communications networks, electronic communications services or associated facilities.

Section 115

MODIFICATION AND REVOCATION OF APPLICATION OF CODE

(1) OFCOM may at any time modify the terms on which, by virtue of section 106(5), the code is applied in a person's case.
(2) OFCOM may revoke a direction applying the electronic communications code in a person's case if an application for the revocation has been made by that person.
(3) If at any time it appears to OFCOM that a person in whose case the electronic communications code has been applied is not the provider of an electronic communications network or conduit system for the purposes of which the code applies, OFCOM may revoke the direction applying the code in his case.

(4) A modification or revocation under this section shall be by a further direction under section 106 to the person in whose case the electronic communications code has been applied by the direction being modified or revoked.
(5) The matters required by section 107(8) to be included, in the case of a direction for the purposes of this section, in the statement of OFCOM's proposal are whichever of the following is applicable –
(a) a statement of their proposal to modify terms imposed under section 106(5);
(b) a statement of their proposal to revoke the direction applying the code.

Section 116

NOTIFICATION OF CESSATION BY PERSON TO WHOM CODE APPLIES

(1) This section applies where, by virtue of a direction under section 106, the electronic communications code applies in any person's case for the purposes of the provision by him of –
(a) an electronic communications network which is not of a description designated for the purposes of section 33; or
(b) such a system of conduits as is mentioned in section 106(4)(b).
(2) If that person ceases to provide that network or conduit system, he must notify OFCOM of that fact.
(3) A notification under this section must be given within such period and in such manner as may be required by OFCOM.
(4) OFCOM may impose a penalty on a person who fails to comply with a requirement imposed by or under this section.
(5) The amount of a penalty imposed on a person under this section is to be such amount not exceeding £1,000 as OFCOM may determine to be both –
(a) appropriate; and
(b) proportionate to the matter in respect of which it is imposed.
(6) Where OFCOM impose a penalty on a person under this section, they shall –
(a) within one week of making their decision to impose the penalty, notify that person of that decision and of their reasons for that decision; and
(b) in that notification, fix a reasonable period after it is given as the period within which the penalty is to be paid.
(7) A penalty imposed under this section –
(a) must be paid to OFCOM; and
(b) if not paid within the period fixed by them, is to be recoverable by them accordingly.
(8) The Secretary of State may by order amend this section so as to substitute a different maximum penalty for the maximum penalty for the time being specified in subsection (5).
(9) No order is to be made containing provision authorised by subsection (8) unless a draft of the order has been laid before Parliament and approved by a resolution of each House.

Section 117

TRANSITIONAL SCHEMES ON CESSATION OF APPLICATION OF CODE

(1) Where it appears to OFCOM –

(a) that the electronic communications code has ceased or is to cease to apply, to any extent, in the case of any person ('the former operator'),
(b) that it has ceased or will cease so to apply for either of the reasons specified in subsection (2), and
(c) that it is appropriate for transitional provision to be made in connection with it ceasing to apply in the case of the former operator,

they may by order make a scheme containing any such transitional provision as they think fit in that case.

(2) Those reasons are –
(a) the suspension under section 113 of the application of the code in the former operator's case;
(b) the revocation or modification under section 115 of the direction applying the code in his case.
(3) A scheme contained in an order under this section may, in particular –
(a) impose any one or more obligations falling within subsection (4) on the former operator;
(b) provide for those obligations to be enforceable in such manner (otherwise than by criminal penalties) and by such persons as may be specified in the scheme;
(c) authorise the retention of apparatus on any land pending its subsequent use for the purposes of an electronic communications network, electronic communications service or conduit system to be provided by any person;
(d) provide for the transfer to such persons as may be specified in, or determined in accordance with, the scheme of any rights or liabilities arising out of any agreement or other obligation entered into or incurred in pursuance of the code by the former operator;
(e) provide, for the purposes of any provision contained in the scheme by virtue of any of the preceding paragraphs, for such questions arising under the scheme as are specified in the scheme, or are of a description so specified, to be referred to, and determined by, OFCOM.
(4) The obligations referred to in subsection (3)(a) are –
(a) an obligation to remove anything installed in pursuance of any right conferred by or in accordance with the code;
(b) an obligation to restore land to its condition before anything was done in pursuance of any such right; or
(c) an obligation to pay the expenses of any such removal or restoration.
(5) Sections 110 to 112 apply in relation to the requirements imposed by virtue of a scheme contained in an order under this section as they apply in relation to a requirement imposed by virtue of restrictions or conditions under section 109.
(6) Section 403 applies to the power of OFCOM to make an order under this section.

Section 118

COMPULSORY ACQUISITION OF LAND ETC.

Schedule 4 (which provides for compulsory acquisition of land by the provider of an electronic communications network in whose case the electronic communications code applies and for entry on land by persons nominated by such a provider) shall have effect.

Section 119

POWER TO GIVE ASSISTANCE IN RELATION TO CERTAIN PROCEEDINGS

(1) This section applies where any actual or prospective party to any proceedings falling within subsection (2) (other than the operator, within the meaning of the electronic communications code) applies to OFCOM for assistance under this section in relation to those proceedings.

(2) The proceedings falling within this subsection are any actual or prospective proceedings in which there falls to be determined any question arising under, or in connection with –

(a) the electronic communications code as applied in any person's case by a direction under section 106; or

(b) any restriction or condition subject to which that code applies.

(3) OFCOM may grant the application if, on any one or more of the following grounds, they think fit to do so –

(a) on the ground that the case raises a question of principle;

(b) on the ground that it is unreasonable, having regard to the complexity of the case or to any other matter, to expect the applicant to deal with the case without assistance under this section;

(c) by reason of any other special consideration.

(4) Assistance by OFCOM under this section may include –

(a) giving advice or arranging for the giving of advice by a solicitor or counsel;

(b) procuring or attempting to procure the settlement of the matter in dispute;

(c) arranging for the giving of any assistance usually given by a solicitor or counsel –

(i) in the steps preliminary or incidental to proceedings; or

(ii) in arriving at, or giving effect to, a compromise to avoid proceedings or to bring them to an end;

(d) arranging for representation by a solicitor or counsel;

(e) arranging for the giving of any other assistance by a solicitor or counsel;

(f) any other form of assistance which OFCOM consider appropriate.

(5) Nothing in subsection (4)(d) shall be taken to affect the law and practice regulating the descriptions of persons who may appear in, conduct or defend any proceedings, or who may address the court in any proceedings.

(6) In so far as expenses are incurred by OFCOM in providing the applicant with assistance under this section, the recovery of those expenses (as taxed or assessed in such manner as may be prescribed by rules of court) shall constitute a first charge for the benefit of OFCOM –

(a) on any costs or expenses which (whether by virtue of a judgment or order of a court, or an agreement or otherwise) are payable to the applicant by any other person in respect of the matter in connection with which the assistance is given; and

(b) so far as relates to costs or expenses, on the applicant's rights under a compromise or settlement arrived at in connection with that matter to avoid proceedings, or to bring them to an end.

(7) A charge conferred by subsection (6) is subject to –

(a) any charge imposed by section 10(7) of the Access to Justice Act 1999 (c. 22) and any provision made by or under Part 1 of that Act for the payment of any sum to the Legal Services Commission;

(b) any charge or obligation for payment in priority to other debts under the Legal Aid (Scotland) Act 1986 (c. 47); or
(c) any charge under the Legal Aid, Advice and Assistance (Northern Ireland) Order 1981

151 Interpretation of Chapter 1

(1) In this Chapter –

'the Access Directive' means Directive 2002/19/EC of the European Parliament and of the Council on access to, and interconnection of, electronic communications networks and associated facilities, as amended by Directive 2009/140/EC of the European Parliament and of the Council;

'access-related condition' means a condition set as an access-related condition under section 45;

'allocation' and *'adoption'*, in relation to telephone numbers, and cognate expressions, are to be construed in accordance with section 56;

'apparatus market', in relation to a market power determination, is to be construed in accordance with section 46(9)(b);

'designated universal service provider' means a person who is for the time being designated in accordance with regulations under section 66 as a person to whom universal service conditions are applicable;

'electronic communications apparatus' –

(a) in relation to SMP apparatus conditions and in section 141, means apparatus that is designed or adapted for a use which consists of or includes the sending or receiving of communications or other signals (within the meaning of section 32) that are transmitted by means of an electronic communications network; and
(b) in all other contexts, has the same meaning as in the electronic communications code;

'the electronic communications code' has the meaning given by section 106(1);

'end-user', in relation to a public electronic communications service, means –

(a) a person who, otherwise than as a communications provider, is a customer of the provider of that service;
(b) a person who makes use of the service otherwise than as a communications provider; or
(c) a person who may be authorised, by a person falling within paragraph (a), so to make use of the service;

'the Framework Directive' means Directive 2002/21/EC of the European Parliament and of the Council on a common regulatory framework for electronic communications networks and services, as amended by Directive 2009/140/EC of the European Parliament and of the Council;

'general condition' means a condition set as a general condition under section 45;

'interconnection' is to be construed in accordance with subsection (2);

'market power determination' means –

(a) a determination, for the purposes of provisions of this Chapter, that a person has significant market power in an identified services market or an identified apparatus market, or
(b) a confirmation for such purposes of a market power determination reviewed on a further analysis under section 84 or 85;

'misuse', in relation to an electronic communications network or electronic communications service, is to be construed in accordance with section 128(5) and (8), and cognate expressions are to be construed accordingly;

'network access' is to be construed in accordance with subsection (3);

'persistent' and *'persistently'*, in relation to misuse of an electronic communications network or electronic communications service, are to be construed in accordance with section 128(6) and (7);

'premium rate service' is to be construed in accordance with section 120(7);

'privileged supplier condition' means a condition set as a privileged supplier condition under section 45;

'provider', in relation to a premium rate service, is to be construed in accordance with section 120(9) to (12), and cognate expressions are to be construed accordingly;

'public communications provider' means –

(a) a provider of a public electronic communications network;
(b) a provider of a public electronic communications service; or
(c) a person who makes available facilities that are associated facilities by reference to a public electronic communications network or a public electronic communications service;

'public electronic communications network' means an electronic communications network provided wholly or mainly for the purpose of making electronic communications services available to members of the public;

'public electronic communications service' means any electronic communications service that is provided so as to be available for use by members of the public;

'regulatory authorities' is to be construed in accordance with subsection (5);

'relevant international standards' means –

(a) any standards or specifications from time to time drawn up and published in accordance with Article 17 of the Framework Directive;
(b) the standards and specifications from time to time adopted by –
   (i) the European Committee for Standardisation,
   (ii) the European Committee for Electrotechnical Standardisation; or
   (iii) the European Telecommunications Standards Institute; and
(c) the international standards and recommendations from time to time adopted by –
   (i) the International Telecommunication Union;
   (ii) the International Organisation for Standardisation; or
   (iii) the International Electrotechnical Committee;

'service interoperability' means interoperability between different electronic communications services;

'services market', in relation to a market power determination or market identification, is to be construed in accordance with section 46(8)(a);

'significant market power' is to be construed in accordance with section 78;

'SMP condition' means a condition set as an SMP condition under section 45, and *'SMP services condition'* and *'SMP apparatus condition'* are to be construed in accordance with subsections (8) and (9) of that section respectively;

'telephone number' has the meaning given by section 56(5);

'the Universal Service Directive' means Directive 2002/22/EC of the European Parliament and of the Council on universal service and users' rights relating to electronic communications networks and services, as amended by Directive 2009/136/EC of the European Parliament and of the Council;

'universal service condition' means a condition set as a universal service condition under section 45;

'the universal service order' means the order for the time being in force under section 65.

(2) In this Chapter references to interconnection are references to the linking (whether directly or indirectly by physical or logical means, or by a combination of physical and logical means) of one public electronic communications network to another for the purpose of enabling the persons using one of them to be able –
   (a) to communicate with users of the other one; or
   (b) to make use of services provided by means of the other one (whether by the provider of that network or by another person).

(3) In this Chapter references to network access are references to –
   (a) interconnection of public electronic communications networks; or
   (b) any services, facilities or arrangements which –
      (i) are not comprised in interconnection; but
      (ii) are services, facilities or arrangements by means of which [a person] [4] is able, for the purposes of the provision of an electronic communications service (whether by him or by another), to make use of anything mentioned in subsection (4);

and references to providing network access include references to providing any such services, making available any such facilities or entering into any such arrangements.

(4) The things referred to in subsection (3)(b) are –
   (a) any electronic communications network or electronic communications service provided by another communications provider;
   (b) any apparatus comprised in such a network or used for the purposes of such a network or service;
   (ba) any electronic communications apparatus;
   (c) any facilities made available by another that are associated facilities by reference to any network or service (whether one provided by that provider or by another);
   (d) any other services or facilities which are provided or made available by another person and are capable of being used for the provision of an electronic communications service

(4A) In subsections (3)(b)(ii) and (4)(d), the references to an electronic communications service include the conveyance by means of an electronic communications network of signals, including an information society service or content service so conveyed.

(4B) In subsection (4A) –

'*content service*' has the meaning given by section 32(7), and

'*information society service*' has the meaning given by Article 2(a) of Directive 2000/31/EC of the European Parliament and of the Council of 8 June 2000 on certain legal aspects of information society services, in particular electronic commerce, in the Internal Market

(5) References in this Chapter to the regulatory authorities of member States are references to such of the authorities of the member States as have been notified to the European Commission as the regulatory authorities of those States for the purposes of the Framework Directive.

(6) For the purposes of this Chapter, where there is a contravention of an obligation that requires a person to do anything within a particular period or before a particular time, that contravention shall be taken to continue after the end of that period, or after that time, until that thing is done.

(7) References in this Chapter to remedying the consequences of a contravention include references to paying an amount to a person –
   (a) by way of compensation for loss or damage suffered by that person; or
   (b) in respect of annoyance, inconvenience or anxiety to which he has been put.

(8) In determining for the purposes of provisions of this Chapter whether a contravention is a repeated contravention for any purposes, a notification of a contravention under that provision shall be disregarded if it has been withdrawn before the imposition of a penalty in respect of the matters notified.

(9) For the purposes of this section a service is made available to members of the public if members of the public are customers, in respect of that service, of the provider of that service.

2003 Code

As contained in schedule 2 to the Telecommunications Act 1984, above

Appendix B
2017 Code and extracts from the Digital Economy Act 2017

PART III

Digital Economy Act 2017

4 The electronic communications code

(1) In the Telecommunications Act 1984 omit Schedule 2 (the telecommunications code).
(2) Before Schedule 4 to the Communications Act 2003 insert Schedule 3A set out in Schedule 1 to this Act.
(3) Section 106 of the Communications Act 2003 (application of the electronic communications code) is amended as follows.
(4) In subsection (1) for 'the code set out in Schedule 2 to the Telecommunications Act 1984 (c 12)' substitute 'the code set out in Schedule 3A'.
(5) Omit subsection (2).
(6) In subsection (4)(b) for 'conduits' substitute 'infrastructure'.
(7) In subsection (5)(c) for 'conduit system' in each place substitute 'system of infrastructure'.
(8) In subsection (6) for '16(3)' substitute '85(7)'.
(9) Omit subsection (7).
(10) Schedules 2 (transitional provisions) and 3 (consequential amendments) have effect.

5 Power to make transitional provision in connection with the code

(1) The Secretary of State may by regulations made by statutory instrument make transitional, transitory or saving provision in connection with the coming into force of section 4 and Schedule 1.
(2) Regulations under this section may amend Schedule 2.
(3) A statutory instrument containing regulations under this section –
    (a) if it includes provision made by virtue of subsection (2), may not be made unless a draft of the instrument has been laid before and approved by a resolution of each House of Parliament;
    (b) otherwise, is subject to annulment in pursuance of a resolution of either House of Parliament

6 Power to make consequential provision etc. in connection with the code

(1) The Secretary of State may by regulations make consequential provision in connection with any provision made by or under section 4 or this section or Schedule 1 or 3.

(2) Regulations under subsection (1) may amend, repeal, revoke or otherwise modify the application of any enactment (but, in the case of primary legislation, only if the primary legislation was passed or made before the end of the Session in which this Act is passed).

(3) Regulations under this section –
  (a) are to be made by statutory instrument;
  (b) may make different provision for different purposes;
  (c) may include incidental, supplementary, consequential, transitional, transitory or saving provision.

(4) A statutory instrument containing regulations under this section (whether alone or with other provisions) which amend, repeal or modify the application of primary legislation may not be made unless a draft of the instrument has been laid before and approved by a resolution of each House of Parliament.

(5) Any other statutory instrument containing regulations under this section is subject to annulment in pursuance of a resolution of either House of Parliament.

(6) In this section –
'enactment' includes –
  (a) an enactment comprised in subordinate legislation within the meaning of the Interpretation Act 1978,
  (b) an enactment comprised in, or in an instrument made under, a Measure or Act of the National Assembly for Wales,
  (c) an enactment comprised in, or in an instrument made under, an Act of the Scottish Parliament, and
  (d) an enactment comprised in, or in an instrument made under, Northern Ireland legislation;

'primary legislation' means –
  (a) an Act of Parliament,
  (b) a Measure or Act of the National Assembly for Wales,
  (c) an Act of the Scottish Parliament, or
  (d) Northern Ireland legislation.

7 Application of the code: protection of the environment

For section 109(2A) of the Communications Act 2003 (under which regulations that set restrictions and conditions to the application of the electronic communications code are deemed by subsection (2B) to comply with duties under National Parks and other legislation if they comply with the duty to have regard to the need to protect the environment, but only if they expire before 6 April 2018) substitute –

'(2A) Subsection (2B) applies if the Secretary of State has complied with subsection (2)(b) in connection with any particular exercise of the power to make regulations under this section.'

Schedule One: The Electronic Communications Code

Please see annotated version of the Code, after Chapter 42.

Schedule 2 THE ELECTRONIC COMMUNICATIONS CODE: TRANSITIONAL PROVISION

Interpretation

1

(1) This paragraph has effect for the purposes of this Schedule.
(2) The *'existing code'* means Schedule 2 to the Telecommunications Act 1984.
(3) The *'new code'* means Schedule 3A to the Communications Act 2003.
(4) A *'subsisting agreement'* means –
  (a) an agreement for the purposes of paragraph 2 or 3 of the existing code, or
  (b) an order under paragraph 5 of the existing code,

which is in force, as between an operator and any person, at the time the new code comes into force (and whose terms do not provide for it to cease to have effect at that time).

(5) Expressions used in this Schedule and in the new code have the same meaning as in the new code, subject to any modification made by this Schedule.

Effect of subsisting agreement

2

(1) A subsisting agreement has effect after the new code comes into force as an agreement under Part 2 of the new code between the same parties, subject to the modifications made by this Schedule.
(2) A person who is bound by a right by virtue of paragraph 2(4) of the existing code in consequence of a subsisting agreement is, after the new code comes into force, treated as bound pursuant to Part 2 of the new code.

Limitation of code rights

3

In relation to a subsisting agreement, references in the new code to a code right are –
(a) in relation to the operator and the land to which an agreement for the purposes of paragraph 2 of the existing code relates, references to a right for the statutory purposes to do the things listed in paragraph 2(1)(a) to (c) of the existing code;
(b) in relation to land to which an agreement for the purposes of paragraph 3 of the existing code relates, a right to do the things mentioned in that paragraph.

Limitation of persons bound

4

(1) A person bound by a code right by virtue only of paragraph 2(3) of the existing code continues to be bound by it so long as they would be bound if paragraph 2(3) of the existing code continued to have effect.
(2) In relation to such a person, paragraph 4(4) to (12) of the existing code continue to have effect, but as if in paragraph 4(4)(b) the reference to paragraph 21 of the existing code were a reference to Part 6 of the new code.

Exclusion of assignment, upgrading and sharing provisions

5

(1) Part 3 of the new code (assignment of code rights, and upgrading and sharing of apparatus) does not apply in relation to a subsisting agreement.
(2) Part 3 of the new code does not apply in relation to a code right conferred under the new code if, at the time when it is conferred, the exercise of the right depends on a right that has effect under a subsisting agreement.

Termination and modification of agreements

6

(1) This paragraph applies in relation to a subsisting agreement, in place of paragraph 29(2) to (4) of the new code.
(2) Part 5 of the new code (termination and modification of agreements) does not apply to a subsisting agreement that is a lease of land in England and Wales, if –
  (a) it is a lease to which Part 2 of the Landlord and Tenant Act 1954 applies, and
  (b) there is no agreement under section 38A of that Act (agreements to exclude provisions of Part 2) in relation the tenancy.
(3) Part 5 of the new code does not apply to a subsisting agreement that is a lease of land in England and Wales, if –
  (a) the primary purpose of the lease is not to grant code rights (the rights referred to in paragraph 3 of this Schedule), and
  (b) there is an agreement under section 38A of the 1954 Act in relation the tenancy.
(4) Part 5 of the new code does not apply to a subsisting agreement that is a lease of land in Northern Ireland, if it is a lease to which the Business Tenancies (Northern Ireland) Order 1996 (SI 1996/725 (NI 5)) applies.

Termination and modification of agreements

7

(1) Subject to paragraph 6, Part 5 of the new code applies to a subsisting agreement with the following modifications.
(2) The *'site provider'* (see paragraph 30(1)(a) of the new code) does not include a person who was under the existing code bound by the agreement only by virtue of paragraph 2(2)(c) of that code.
(3) Where the unexpired term of the subsisting agreement at the coming into force of the new code is less than 18 months, paragraph 31 applies (with necessary modification) as if for the period of 18 months referred to in sub-paragraph (3)(a) there were substituted a period equal to the unexpired term or 3 months, whichever is greater.
(4) Paragraph 34 applies with the omission of sub-paragraph (13)(d).

Apparatus, works etc.

8

(1) Paragraphs 9 to 14 of the existing code (rights in relation to street works, flying lines, tidal waters, linear obstacles) continue to apply in relation to anything in the process of being done when the new code comes into force.

(2) Apparatus lawfully installed under any of those provisions (before or after the time when the new code comes into force) is to be treated as installed under the corresponding provision of the new code if it could have been installed under that provision if the provision had been in force or applied to its installation.
(3) The corresponding provisions are –
  (a) Part 7 (transport land rights), in relation to paragraph 12 of the existing code;
  (b) Part 8 (street work rights), in relation to paragraph 9 of the existing code;
  (c) Part 9 (tidal water rights), in relation to paragraph 11 of the existing code;
  (d) paragraph 74 (power to fly lines), in relation to paragraph 10 of the existing code.

Apparatus, works etc.

9

Any agreement given in accordance with paragraph 26(3) of the existing code for the purposes of paragraph 11(2) of that code has effect for the purposes of paragraph 64 of the new code as if given in accordance with paragraph 104 of that code.

Apparatus, works etc.

10

Any agreement that has effect under paragraph 15 of the existing code and that would be sufficient for the purpose of doing anything wholly inside a sewer if that paragraph continued in force is sufficient for that purpose under paragraph 102(2) of the new code.

Court applications for required rights etc.

11

(1) This paragraph applies where –
  (a) before the time when the new code comes into force, a notice has been given under paragraph 5(1) of the existing code, and
  (b) at that time no application has been made to the court in relation to the notice.
(2) The notice has effect as if given under paragraph 20(2) of the new code.

Court applications for required rights etc.

12

(1) This paragraph applies where before the time when the new code comes into force –
  (a) a notice has been given under paragraph 5(1) of the existing code, and
  (b) an application has been made to the court in relation to the notice.
(2) Subject to sub-paragraph (3), the existing code continues to apply in relation to the application.
(3) An order made under the existing code by virtue of sub-paragraph (2) has effect as an order under paragraph 20 of the new code.

Temporary code rights

13

The coming into force of the new code does not affect any application or order made under paragraph 6 of the existing code.

Compensation

14

The repeal of the existing code does not affect paragraph 16 of that code, or any other right to compensation, as it applies in relation to the exercise of a right before the new code comes into force.

Objections in relation to apparatus

15

The repeal of the existing code does not affect paragraphs 17 and 18 of that code as they apply in relation to anything whose installation was completed before the repeal comes into force.

Objections in relation to apparatus

16

(1) Subject to the following provisions of this paragraph, the repeal of the existing code does not affect paragraph 20 of that code as it applies in relation to anything whose installation was completed before the repeal comes into force.
(2) A right under paragraph 20 is not by virtue of sub-paragraph (1) exercisable in relation to any apparatus by a person who is a party to, or is bound by, an agreement under the new code in relation to the apparatus.
(3) A subsisting agreement is not an agreement under the new code for the purposes of sub-paragraph (2).

Objections in relation to apparatus

17

Part 12 of the new code does not apply in relation to apparatus whose installation was completed before the new code came into force.

Tree lopping

18

(1) This paragraph applies where –
  (a) before the time when the new code comes into force, a notice has been given under paragraph 19 of the existing code, and
  (b) at that time no application has been made to the court in relation to the notice.
(2) The notice and any counter-notice under that paragraph have effect as if given under paragraph 82 of the new code.

Tree lopping

19

(1) This paragraph applies where before the time when the new code comes into force –
  (a) a notice has been given under paragraph 19 of the existing code, and
  (b) an application has been made to the court in relation to the notice.
(2) The existing code continues to apply in relation to the application.

Right to require removal of apparatus

20

(1) This paragraph applies where before the repeal of the existing code comes into force a person has given notice under paragraph 21(2) of that code requiring the removal of apparatus.
(2) The repeal does not affect the operation of paragraph 21 in relation to anything done or that may be done under that paragraph following the giving of the notice.
(3) For the purposes of applying that paragraph after the repeal comes into force, steps specified in a counter-notice under sub-paragraph (4)(b) of that paragraph as steps which the operator proposes to take under the existing code are to be read as including any corresponding steps that the operator could take under the new code or by virtue of this Schedule.

Undertaker's works

21

The repeal of the existing code does not affect the operation of paragraph 23 of that code in relation to works –
(a) in relation to which a notice has been given under that paragraph before the time when that repeal comes into force, or
(b) which have otherwise been commenced before that time.

Digital Economy Act 2017 c. 30

Supplementary

22

Any agreement which, immediately before the repeal of the existing code, is a relevant agreement for the purposes of paragraph 29 of that code is to be treated in relation to times after the coming into force of that repeal as a relevant agreement for the purposes of paragraph 18 of the new code.

Supplementary

23

Part 15 of the new code applies in relation to notices under this Schedule as it applies in relation to notices under that code.

Supplementary

24

Paragraphs 24 to 27 of the existing code continue to have effect in relation to any provision of that code so far as the provision has effect by virtue of this Schedule.

Supplementary

25

A person entitled to compensation by virtue of this Schedule is not entitled to compensation in respect of the same matter under any provision of the new code.

Appendix C
Statutory Instruments

STATUTORY INSTRUMENTS

2017 No. 765 (C. 60)

BROADCASTING

CONSUMER PROTECTION

DATA PROTECTION

DISCLOSURE OF INFORMATION

ELECTRONIC COMMUNICATIONS

INTELLECTUAL PROPERTY

The Digital Economy Act 2017 (Commencement No. 1)

Regulations 2017

Made - - - - 12th July 2017

The Secretary of State makes the following Regulations in exercise of the powers conferred by section 118(4), (6) and (7) of the Digital Economy Act 2017([1]).

Citation and interpretation

1. – (1) These Regulations may be cited as the Digital Economy Act 2017 (Commencement No.

(2) Regulations 2017.

In these Regulations 'the 2017 Act' means the Digital Economy Act 2017.

Provisions coming into force on 31st July 2017

2. The following provisions of the 2017 Act come into force on 31st July 2017 –

(a) section 5 (power to make transitional provision in connection with the code);

(b) section 6 (power to make consequential provision etc. in connection with the code);

(c) section 8 (regulation of dynamic spectrum access services);

(d) section 14 (internet pornography: requirement to prevent access by persons under 18) but only for the purpose of making regulations under subsection (2);

(e) section 15 (meaning of 'pornographic material') so far as it relates to the purpose specified in paragraph (d) and to the provisions specified in paragraphs (h), (j) and (l);

(f) section 16 (the age-verification regulator: designation and funding);

(g) section 17 (parliamentary procedure for designation of age-verification regulator);

(h) section 21(5) (meaning of 'ancillary service provider') so far as it relates to the provision specified in paragraph (l);

1 2017 c. 30.

(i) section 22 (meaning of 'extreme pornographic material') so far as it relates to the provisions specified in paragraphs (e), (h) and (j);
(j) section 25 (guidance to be published by regulator);
(k) section 26(2) (exercise of functions by regulator);
(l) section 27 (guidance by Secretary of State to regulator);
(m) section 30(1) and (2) (interpretation and general provisions relating to this Part) so far as it relates to the provisions specified in paragraphs (d) to (l);
(n) section 34 (copyright etc. where broadcast retransmitted by cable);
(o) section 46 (disclosure of information by civil registration officials), but only for the purpose of issuing the code of practice under section 19AC of the Registration Service Act 1953([2]);
(p) section 47 (consequential provision: civil registration) so far as it relates to the purpose specified in paragraph (o);
(q) section 74 (disclosure of non-identifying information by the Revenue and Customs);
(r) section 76 (disclosure of non-identifying information by Revenue Scotland);
(s) section 77 (disclosure of employer reference information by the Revenue and Customs);
(t) section 78 (disclosure of information by the Revenue and Customs to the Statistics Board);
(u) section 79 (disclosure of information by public authorities to the Statistics Board), except for subsection (3), and in relation to England and Wales and Scotland only;
(v) section 81 (disclosure by the Statistics Board to devolved administrations), in relation to England and Wales and Scotland only;
(w) section 87 (appeals from decisions of OFCOM and others: standard of review);
(x) section 92 (digital additional services: seriously harmful extrinsic material);
(y) section 93 (on-demand programme services: accessibility for people with disabilities), except subsection (3);
(z) section 95 (electronic programme guides and public service channels);
(aa) section 98 (strategic priorities and provision of information);
(bb) section 100 (retention by OFCOM of amounts paid under Wireless Telegraphy Act 2006);
(cc) section 101 (international recognition of satellite frequency assignments: power of OFCOM to charge fees);
(dd) section 104 (internet filters);
(ee) section 106 (power to create offence of breaching limits on internet and other ticket sales), in relation to England and Wales and Scotland only;
(ff) section 108 (regulations about charges payable to the Information Commissioner);
(gg) section 109 (functions relating to regulations under section 108);
(hh) section 110 (supplementary provision relating to section 108);
(ii) Schedule 1 (the electronic communications code), but only for the purpose of making regulations under paragraph 95 (power to confer jurisdiction on other tribunals) of Schedule 3A to the Communications Act

2 1953 c. 37.

2003[3], and section 4 (the electronic communications code) so far as is necessary for that purpose;

(jj) paragraph 47 of Schedule 3 (electronic communications code: consequential amendments), and section 4 (the electronic communications code) so far as it relates to that paragraph.

Provisions coming into force on 1st October 2017

3. The following provisions of the 2017 Act come into force on 1st October 2017 –

(a) section 32 (offences: infringing copyright and making available right);

(b) section 33 (registered designs: infringement: marking product with internet link);

(c) section 35 (disclosure of information to improve public service delivery), but only for the purpose of making regulations, and in relation to England and Wales (except so far as it relates to the disclosure of information to or by a water or sewerage undertaker for an area which is wholly or mainly in Wales) and Scotland only;

(d) section 36 (disclosure of information to gas and electricity suppliers etc.), but only for the purpose of making regulations;

(e) section 43 (code of practice: public service delivery), except so far as it relates to the disclosure of information to or by a water or sewerage undertaker for an area which is wholly or mainly in Wales, and in relation to England and Wales and Scotland only;

(f) section 44 (regulations under this Chapter) so far as it relates to the purposes specified in paragraphs (c) and (d);

(g) section 48 (disclosure of information to reduce debt owed to the public sector), but only for the purpose of making regulations, and in relation to England and Wales and Scotland only;

(h) section 52 (code of practice: debt owed to the public sector), in relation to England and Wales and Scotland only;

(i) section 54 (regulations under this Chapter) so far as it relates to the purpose specified in paragraph (g);

(j) section 56 (disclosure of information to combat fraud against the public sector), but only for the purpose of making regulations, and in relation to England and Wales and Scotland only;

(k) section 60 (code of practice: fraud against the public sector), in relation to England and Wales and Scotland only;

(l) section 62 (regulations under this Chapter) so far as it relates to the purpose specified in paragraph (j);

(m) section 70 (code of practice: sharing for research purposes), except so far as it relates to the disclosure of information by the Welsh Revenue Authority, and in relation to England and Wales and Scotland only;

(n) section 80 (access to information by the Statistics Board), but only for the purposes of preparing and publishing the statement under section 45E, and the code of practice under section 45G, of the Statistics and Registration Service Act 2007[4], and in relation to England and Wales and Scotland only.

---

3 2003 c. 21.
4 2007 c. 18.

Provision coming into force on 1st October 2018

4. Section 102 (billing limits for mobile phones) of the 2017 Act comes into force on 1st October 2018.

Matthew Hancock
Minister of State
12th July 2017 Department for Culture, Media and Sport

EXPLANATORY NOTE

(This note is not part of the Regulations)

These Regulations bring into force specified provisions of the Digital Economy Act 2017 (c.30) ('the 2017 Act'). The dates of commencement of certain other provisions are set out in section 118 of the 2017 Act.

Regulation 2 lists provisions which come into force on 31st July 2017 and regulation 3 lists provisions which come into force on 1st October 2017. Regulation 4 brings into force a provision on 1st October 2018.

Printed and published in the UK by The Stationery Office Limited under the authority and superintendence of Jeff James, Controller of Her Majesty's Stationery Office and Queen's Printer of Acts of Parliament.

DRAFT STATUTORY INSTRUMENTS

2017 No. 1284

ELECTRONIC COMMUNICATIONS

The Electronic Communications Code (Jurisdiction) Regulations 2017

Made – – – – *14th December 2017*

Coming into force in accordance with regulation 1(1)

The Secretary of State, i n exercise of the powers conferred by section 402(3) of, and paragraphs 95(1) to (4) of Schedule 3A to, the Communications Act 2003([1]), makes the following Regulations.

The Secretary of State has consulted the Scottish Ministers in accordance with paragraph 95(5)(a) of Schedule 3A to that Act.

A draft of these Regulations was laid before Parliament and approved by a resolution of each House of Parliament in accordance with section 402(2A) of that Act([2]).

Citation, commencement and extent

1. – (1) These Regulations may be cited as the Electronic Communications Code (Jurisdiction) Regulations 2017 and come into force on the day that section 4 of, and Schedule 1 to, the Digital Economy Act 2017 come fully into force.

(2) These Regulations extend to England and Wales and to Scotland.

Interpretation

2. – (1) In these Regulations –

'the code' means the electronic communications code set out in Schedule 3A to the Communications Act 2003;

'relevant proceedings' means proceedings under any of the following provisions of the code –

(a) Parts 4, 5, 6, 12 or 13, or
(b) paragraph 53.

1 2003 c. 21. Schedule 3A was inserted by Schedule 1 to the Digital Economy Act 2017 (c. 30).
2 Section 402(2A) was inserted by paragraph 47(3) of Schedule 3 to the Digital Economy Act 2017.

Conferral of jurisdiction on tribunals

3. Subject to regulation 4, the functions conferred by the code on the court([3]) are also exercisable by the following tribunals –

(a) in relation to England, the First-tier Tribunal (in a case where relevant proceedings are transferred to it by the Upper Tribunal),

(b) in relation to England and Wales, the Upper Tribunal, and

(c) in relation to Scotland, the Lands Tribunal for Scotland,

and any provision of the code which confers a function on the court is to be read as if the reference to the court included references to these tribunals.

Restriction on jurisdiction for commencement of relevant proceedings

4. Relevant proceedings must be commenced –

(a) in relation to England and Wales, in the Upper Tribunal, or

(b) in relation to Scotland, the Lands Tribunal for Scotland.

Transfer of relevant proceedings to the court

5.–(1) A tribunal referred to in regulation 3 may transfer relevant proceedings to –

(a) in relation to England or Wales, the county court, or

(b) in relation to Scotland the sheriff court,

if that tribunal considers the court to be a more appropriate forum for the determination of those proceedings.

(2) A tribunal may transfer proceedings in accordance with paragraph (1) of its own motion or on the application of a party to those proceedings.

Name
Minister of State

Date Department for Digital, Culture, Media and Sport

EXPLANATORY NOTE

(This note is not part of the Regulations)

The Electronic Communications Code ('the code') is set out in Schedule 3A to the Communications Act 2003. Schedule 3A was inserted by Part 2 of the Digital Economy Act 2017. The code replaces the previous code set out in Schedule 2 to the Telecommunications Act 1984.

The code sets out the basis on which electronic communications operators authorised by Ofcom under section 106 of the Communications Act 2003 may exercise rights to deploy and maintain their electronic communications apparatus on, over and under land. Electronic communications apparatus is defined in paragraph 5 of the code. Under the provisions of the code, 'the court' has jurisdiction for most disputes.

3 See paragraph 94(1) of Schedule 3A to the Communications Act 2003 for the definition of 'the court'.

Regulation 3 permits functions conferred on the court by the code to be exercised by certain tribunals, and modifies the code accordingly. These regulations do not extend to Northern Ireland. Accordingly, all functions conferred by the code on a county court in Northern Ireland are exercisable in Northern Ireland only by a county court.

Regulation 4 provides that 'relevant proceedings' (defined in regulation 2) under the code must be commenced only in the Upper Tribunal or the Lands Tribunal for Scotland.

The First-tier Tribunal can hear relevant proceedings only if the Upper Tribunal transfers a case to it in accordance with rule 5(3)(k)(ii) of the Tribunal Procedure (Upper Tribunal) (Lands Chamber) Rules 2010 (S.I. 2010/2600).

Regulation 5 gives each of the tribunals listed in regulation 3 power to transfer relevant proceedings to the respective court, if the court would be a more appropriate forum.

The county court (in England and Wales) and the sheriff court (in Scotland) can hear relevant proceedings, brought after the date on which these Regulations come into force, only if a tribunal transfers those proceedings in accordance with the provisions of regulation 5.

An Impact Assessment has not been produced for this instrument as no impact on the private or voluntary sectors is foreseen. An Explanatory Memorandum is published alongside this instrument on www.legislation.gov.uk.

Printed and published in the UK by The Stationery Office Limited under the authority and superintendence of Jeff James, Controller of Her Majesty's Stationery Office and Queen's Printer of Acts of Parliament.

STATUTORY INSTRUMENTS

2017 No. 1136 (C. 106)

ELECTRONIC COMMUNICATIONS

The Digital Economy Act 2017 (Commencement No. 2)

Regulations 2017

Made – – – – *21st November 2017*

The Secretary of State makes the following Regulations in exercise of the powers conferred by section 118(6) and (7) of the Digital Economy Act 2017[1].

Citation

1. These Regulations may be cited as the Digital Economy Act 2017 (Commencement No. 2) Regulations 2017.

Provisions coming into force on 22nd November 2017

2. Schedule 1 (the electronic communications code) to the Digital Economy Act 2017, but only in relation to paragraph 106 (Lands Tribunal for Scotland procedure rules) of Schedule 3A to the Communications Act 2003[2], and section 4 (the electronic communications code) so far as it relates to that paragraph, come into force on 22nd November 2017.

Matthew Hancock
Minister of State
21st November 2017 Department for Digital, Culture, Media and Sport

1 2017 c. 30.
2 2003 c. 21.

EXPLANATORY NOTE

(This note is not part of the Regulations)

These Regulations bring into force Schedule 1 to the Digital Economy Act 2017 (c. 30) in relation only to paragraph 106 of Schedule 3A to the Communications Act 2003 (c. 21), which makes provision for the Scottish Ministers to make rules for the Lands Tribunal for Scotland in relation to the electronic communications code.

NOTE AS TO EARLIER COMMENCEMENT REGULATIONS

(This note is not part of the Regulations)

The following provisions of the Digital Economy Act 2017 (c. 30) have been or are to be brought into force by commencement regulations made before the date of these Regulations.

Provision	*Date of Commencement*	*S.I. No.*
Section 4 (partially)	31.07.17	2017/765
Section 5	31.07.17	2017/765
Section 6	31.07.17	2017/765
Section 8	31.07.17	2017/765
Section 14 (partially)	31.07.17	2017/765
Section 15 (partially)	31.07.17	2017/765
Section 16	31.07.17	2017/765
Section 17	31.07.17	2017/765
Section 21(5) (partially)	31.07.17	2017/765
Section 22 (partially)	31.07.17	2017/765
Section 25	31.07.17	2017/765
Section 26(2)	31.07.17	2017/765
Section 27	31.07.17	2017/765
Section 30(1) and (2) (partially)	31.07.17	2017/765
Section 32	01.10.17	2017/765
Section 33	01.10.17	2017/765
Section 34	31.07.17	2017/765
Section 35 (partially)	01.10.17	2017/765
Section 36 (partially)	01.10.17	2017/765
Section 43 (partially)	01.10.17	2017/765
Section 44 (partially)	01.10.17	2017/765
Section 46 (partially)	31.07.17	2017/765
Section 47 (partially)	31.07.17	2017/765
Section 48 (partially)	01.10.17	2017/765
Section 52 (partially)	01.10.17	2017/765
Section 54 (partially)	01.10.17	2017/765
Section 56 (partially)	01.10.17	2017/765
Section 60 (partially)	01.10.17	2017/765
Section 62 (partially)	01.10.17	2017/765
Section 70 (partially)	01.10.17	2017/765
Section 74	31.07.17	2017/765
Section 76	31.07.17	2017/765
Section 77	31.07.17	2017/765
Section 78	31.07.17	2017/765
Section 79, except subsection (3) (partially)	31.07.17	2017/765
Section 80 (partially)	01.10.17	2017/765

Section 81 (partially)	31.07.17	2017/765
Section 87	31.07.17	2017/765
Section 92	31.07.17	2017/765
Section 93, except subsection (3)	31.07.17	2017/765
Section 95	31.07.17	2017/765
Section 98	31.07.17	2017/765
Section 100	31.07.17	2017/765
Section 101	31.07.17	2017/765
Section 102	01.10.18	2017/765
Section 104	31.07.17	2017/765
Section 106 (partially)	31.07.17	2017/765
Section 108	31.07.17	2017/765
Section 109	31.07.17	2017/765
Section 110	31.07.17	2017/765
Schedule 1 (partially)	31.07.17	2017/765
Schedule 3, paragraph 47	31.07.17	2017/765

Printed and published in the UK by The Stationery Office Limited under the authority and superintendence of Jeff James, Controller of Her Majesty's Stationery Office and Queen's Printer of Acts of Parliament.

STATUTORY INSTRUMENTS

2017 No. 1286 (C. 119)

ELECTRONIC COMMUNICATIONS

The Digital Economy Act 2017 (Commencement No. 3)

Regulations 2017

Made – – – – *14th December 2017*

The Secretary of State makes the following Regulations in exercise of the powers conferred by section 118(6) and (7) of the Digital Economy Act 2017[1].

Citation

1. These Regulations may be cited as the Digital Economy Act 2017 (Commencement No. 3) Regulations 2017.

Provisions coming into force on 28th December 2017

2. The following provisions of the Digital Economy Act 2017 come into force on 28th December 2017 –

(a) section 4 (the electronic communications code) in so far as not already in force;
(b) Schedule 1 (the electronic communications code), in so far as not already in force;
(c) Schedule 2 (the electronic communications code: transitional provision);
(d) Schedule 3 (electronic communications code: consequential amendments), in so far as not already in force.

Matthew Hancock
Minister of State
14th December 2017 Department for Digital, Culture, Media and Sport

1 2017 c. 30.

EXPLANATORY NOTE

(This note is not part of the Regulations)

These Regulations commence provisions of the Digital Economy Act 2017 (c. 30) ('the 2017 Act') relating to the electronic communications code insofar as those provisions are not already in force. They are the third commencement regulations made under the 2017 Act.

The Explanatory Notes to the 2017 Act provide an explanation of the reforms to the electronic communications code introduced by the 2017 Act and are available online (www.legislation.gov.uk/ukpga/2017/30/pdfs/ukpgaen_20170030_en.pdf). A full impact assessment for the reforms is available from the Department for Digital, Culture, Media and Sport at 100 Parliament Street, London SW1A 2BQ. A copy may be obtained online (www.gov.uk/government/uploads/system/uploads/attachment_data/file/524895/ECC_Imp act_Assessment.pdf).

NOTE AS TO EARLIER COMMENCEMENT REGULATIONS

(This note is not part of the Regulations)

The following provisions of the Digital Economy Act 2017 (c. 30) have been or are to be brought into force by commencement regulations made before the date of these Regulations.

Provision	*Date of Commencement*	*S.I. No.*
Section 4 (partially)	31.07.17	2017/765
Section 4 (further purposes)	22.11.17	2017/1136
Section 5	31.07.17	2017/765
Section 6	31.07.17	2017/765
Section 8	31.07.17	2017/765
Section 14 (partially)	31.07.17	2017/765
Section 15 (partially)	31.07.17	2017/765
Section 16	31.07.17	2017/765
Section 17	31.07.17	2017/765
Section 21(5) (partially)	31.07.17	2017/765
Section 22 (partially)	31.07.17	2017/765
Section 25	31.07.17	2017/765
Section 26(2)	31.07.17	2017/765
Section 27	31.07.17	2017/765
Section 30(1) and (2) (partially)	31.07.17	2017/765
Section 32	01.10.17	2017/765
Section 33	01.10.17	2017/765
Section 34	31.07.17	2017/765
Section 35 (partially)	01.10.17	2017/765
Section 36 (partially)	01.10.17	2017/765
Section 43 (partially)	01.10.17	2017/765
Section 44 (partially)	01.10.17	2017/765
Section 46 (partially)	31.07.17	2017/765
Section 47 (partially)	31.07.17	2017/765
Section 48 (partially)	01.10.17	2017/765
Section 52 (partially)	01.10.17	2017/765
Section 54 (partially)	01.10.17	2017/765
Section 56 (partially)	01.10.17	2017/765

Section 60 (partially)	01.10.17	2017/765
Section 62 (partially)	01.10.17	2017/765
Section 70 (partially)	01.10.17	2017/765
Section 74	31.07.17	2017/765
Section 76	31.07.17	2017/765
Section 77	31.07.17	2017/765
Section 78	31.07.17	2017/765
Section 79, except subsection (3) (partially)	31.07.17	2017/765
Section 80 (partially)	01.10.17	2017/765
Section 81 (partially)	31.07.17	2017/765
Section 87	31.07.17	2017/765
Section 92	31.07.17	2017/765
Section 93, except subsection (3)	31.07.17	2017/765
Section 95	31.07.17	2017/765
Section 98	31.07.17	2017/765
Section 100	31.07.17	2017/765
Section 101	31.07.17	2017/765
Section 102	01.10.18	2017/765
Section 104	31.07.17	2017/765
Section 106 (partially)	31.07.17	2017/765
Section 108	31.07.17	2017/765
Section 109	31.07.17	2017/765
Section 110	31.07.17	2017/765
Schedule 1 (partially)	31.07.17	2017/765
Schedule 1 (further purposes)	22.11.17	2017/1136
Schedule 3, paragraph 47	31.07.17	2017/765

Printed and published in the UK by The Stationery Office Limited under the authority and superintendence of Jeff James, Controller of Her Majesty's Stationery Office and Queen's Printer of Acts of Parliament.

STATUTORY INSTRUMENTS

2017 No. 1169

TRIBUNALS AND INQUIRIES

The First-tier Tribunal and Upper Tribunal (Chambers)

(Amendment No. 2) Order 2017

Made - - - -	*28th November 2017*
Laid before Parliament	*30th November 2017*
Coming into force in accordance with article 1	

The Lord Chancellor, with the concurrence of the Senior President of Tribunals, makes the following Order, in exercise of the power conferred by section 7(9) of the Tribunals, Courts and Enforcement Act 2007[1].

Citation and commencement

1. This Order may be cited as the First-tier Tribunal and Upper Tribunal (Chambers) (Amendment No. 2) Order 2017 and subject to article 2 comes into force 21 days after the day on which it is laid.

2. Insofar as it relates to proceedings under Schedule 3A to the Communications Act 2003[2], article 5 comes into force on the day on which, and immediately after, section 4 of and Schedule 1 to the Digital Economy Act 2017[3] come into force.

Amendment to the First-tier Tribunal and Upper Tribunal (Chambers) Order 2010

3. The First-tier Tribunal and Upper Tribunal (Chambers) Order 2010[4] is amended as follows.

4. In article 7 (functions of the Tax Chamber)[5], after paragraph (d) insert – ';
(e) a function of the Welsh Revenue Authority.'.

5. In article 12(a) (functions of the Lands Chamber)[6], after sub-paragraph (iv) insert – ';
(v) proceedings under Schedule 3A to the Communications Act 2003;
(vi) proceedings under the Riot Compensation Act 2016[7].'.

1 2007 c. 15.
2 2003 c. 21; Schedule 3A was inserted by section 4 of, and Schedule 1 to, the Digital Economy Act 2017 (c. 30). (**c**) 2017 c. 30.
3 2017 c. 30.
4 S.I. 2010/2655.
5 Article 7 was amended by section 15(3) of, and paragraph 190 of Schedule 8 to, the Crime and Courts Act 2013 (c. 22). (**f**) Article 12 was amended by S.I. 2013/1187.
6 Article 12 was amended by S.I. 2013/1187.
7 2016 c. 8.

6. In article 13(1) (functions of the Tax and Chancery Chamber)[8], after paragraph (h) insert – ';

(i) an application under section 151, 181E or 181F of the Tax Collection and Management (Wales) Act 2016[9].'.

David Lidington
28th November 2017 Lord Chancellor

EXPLANATORY NOTE

(This note is not part of the Order)

This Order amends the First-tier Tribunal and Upper Tribunal (Chambers) Order 2010 (S.I. 2010/2655) to reflect the conferral of further jurisdiction on the First-tier Tribunal and Upper Tribunal relating to proceedings in respect of functions of the Welsh Revenue Authority and the conferral of further jurisdiction on the Upper Tribunal relating to proceedings under Schedule 3A to the Communications Act 2003 (c. 21) and the Riot Compensation Act 2016 (c. 8).

A full impact assessment has not been produced for this instrument as no, or no significant, impact on the private, voluntary or public sectors is foreseen.

Printed and published in the UK by The Stationery Office Limited under the authority and superintendence of Jeff James, Controller of Her Majesty's Stationery Office and Queen's Printer of Acts of Parliament.

8 Article 13 was amended by section 15(3) of, and paragraph 190 of Schedule 8 to, the Crime and Courts Act 2013 (c. 22) and S.I. 2013/1187, 2014/1901 and 2017/722.

9 2016 (anaw 6). Sections 181E and 181F were inserted by section 76 of, and paragraph 63 of Schedule 23 to, the Land Transaction Tax and Anti-avoidance of Devolved Taxes (Wales) Act 2017 (anaw 1). Sections 181E and 181F are modified by section 181H of the Tax Collection and Management (Wales) Act 2016, which was also inserted by section 76 of, and paragraph 63 of Schedule 23 to, the Land Transaction Tax and Anti-avoidance of Devolved Taxes (Wales) Act 2017.

TRIBUNAL JUDICIARY

UPPER TRIBUNAL (LANDS CHAMBER)

P RACTICE NOTE[1]

ELECTRONIC COMMUNICATIONS CODE

1. The Electronic Communications Code ('the Code') is found in Schedule 3A of the Communications Act 2003, into which it was inserted by Part 2 of the Digital Economy Act 2017. The Code sets out the basis on which electronic communications operators may exercise rights to deploy and maintain their electronic communications apparatus on, over and under land.

2. The Electronic Communications Code (Jurisdiction) Regulations 2017 provides for dispute resolution functions conferred by the Code on the court to be exercisable in relation to England and Wales by the Upper Tribunal and by the First-tier Tribunal. The same Regulations provide that certain proceedings under the Code may be commenced in England and Wales only in the Upper Tribunal.

3. Proceedings which may be commenced in England and Wales only in the Upper Tribunal are applications under the following provisions of the Code:

Part 4	(Power of the court to impose agreement)
Part 5	(Termination and modification of agreements)
Part 6	(Removal of Electronic Communications Apparatus)
Para. 53	(Alteration of apparatus at request of transport undertakers) Part 12 (Rights to object to certain apparatus)
Part 12	(Rights to object to certain apparatus)
Part 13	(Rights to lop trees)

1 This is not a formal practice direction, but is issued as guidance to parties at the commencement of this new jurisdiction.

4. Within the Upper Tribunal proceedings under the Code are assigned to the Lands Chamber by the First-tier Tribunal and Upper Tribunal (Chambers) (Amendment No. 2) Order 2017.

5. Code disputes should be referred to the Tribunal using the forms and procedures applicable to references under Part 5 of the Tribunal Procedure (Upper Tribunal) (Lands Chamber) Rules 2010 ('the Rules'). A copy of the Rules and the forms for use in references is available on the Tribunal's website at www.gov.uk/courts-tribunals/upper-tribunal-lands-chamber.

6. The Electronic Communications and Wireless Telegraphy Regulations 2011 require that in certain Code disputes the Tribunal must make its decision within 6 months of receiving a completed reference. In view of this requirement pre-action engagement by parties is strongly encouraged.

7. On receipt of a notice of reference in a Code dispute the Tribunal will fix an appointment for a case management hearing at which directions will be given to enable a final hearing to take place within 5 months of receiving the reference. Parties commencing a reference in a Code dispute should seek to agree in advance what directions will be required.

8. The first case management hearing in a Code dispute is likely to take place within 2 or 3 weeks of receipt of the notice of reference and will usually be held on a Friday. If it is more convenient to one or both of the parties, the case management hearing may be conducted over the telephone.

9. By rule 10(1)(e) of the Rules (as amended) the Tribunal has power to award costs in Code disputes.

10. Although the First-tier Tribunal has jurisdiction to determine Code disputes, such disputes may not be commenced in the First-tier Tribunal. By rule 5(k) of the Rules the Upper Tribunal may transfer proceedings to another court or tribunal if that other court or tribunal has jurisdiction in relation to the proceedings. For the time being it is not anticipated that Code disputes would normally be transferred by the Tribunal to the First-tier Tribunal, but parties who agree that their dispute should be determined in the First-tier Tribunal may apply for transfer to be considered.

The Honourable Mr Justice Holgate
Chamber President

26 January 2018

Appendix D
OFCOM Code of Practice[1]

STATEMENT:
Publication Date: 15 December 2017

About this document

This document contains the Code of Practice. The Code of Practice deals with

a) the provision of information for the purposes of the new Code by operators to persons who occupy or have an interest in land;
b) the conduct of negotiations for the purposes of the new Code between operators and such persons;
c) the conduct of operators in relation to persons who occupy or have an interest in land adjoining land on, under or over which electronic communications apparatus is installed; and
d) such other matters relating to the operation of the new Code as Ofcom think appropriate.

The Code of Practice does not represent a guide to the new Electronic Communications Code nor does it replace or supplement its provisions by imposing any new rights or obligations on the respective parties. Instead it is designed to complement the new Code by suggesting best practice to facilitate positive and productive engagement between all parties across a range of issues, roles and responsibilities. Whilst the Code of Practice provides some examples of best practice these are not intended to be exhaustive.

Contents

1. Code of Practice

Introduction

1.1 Electronic communications services (such as landlines, mobile phones and internet services) are now regarded as essential services. In order that these services can be

1 See Sections 41.9 and 41.10 for a commentary on the Code of Practice.

provided where they are needed, The Electronic Communications Code ('Code') provides a statutory basis whereby communications providers (known in this context as 'Operators[2]') can place their Apparatus[3] on land or buildings owned by another person or organisation.

1.2 In view of the ever increasing and critical needs of local communities (and the UK economy as a whole) to have access to 21st century communications networks, such as high speed broadband connection or a 4G mobile connection (and 5G in due course), the Code has been reformed under the Digital Economy Act 2017 so as to make it more straightforward for Operators to gain access to the locations they need, to improve coverage, capability and capacity.

Purpose of the Code of Practice

1.3 The purpose of this Code of Practice, which has also been established under the Digital Economy Act, is to set out expectations for the conduct of the parties to any agreement made under the Code. It is not a guide to the Code or the Code regulations, but it is intended to complement them and to make it simple for Operators, Landowners and Occupiers[4] to come to agreement over a range of issues relating to the occupation of a site. References to landowners should also be taken, where appropriate, to encompass Occupiers as defined in the Code. Agreements under the Code are binding and so Landowners may wish to consider seeking independent professional advice before entering into such an agreement (see below).

1.4 'Site' in this Code of Practice is used in a broad sense[5] as any place to install Apparatus, such as under or on top of open land, the rooftop of a building, a tunnel or a lamp-post.

1.5 All parties to whom this Code of Practice applies should treat each other professionally and with respect, remembering always that the goal is to improve and maintain essential communications services for all. Operators should take adequate steps to satisfy themselves that they are negotiating with a party who has a lawful right to grant the necessary agreement if not negotiating with the Landowner. Landowners and Operators must respect the needs and legitimate concerns of Occupiers of land when rights under the Code are exercised. Operators ought to be responsible for the behaviour and conduct of any contractors that they instruct to carry out work on their behalf.

Scope

1.6 This Code of Practice:

- Provides a reference framework to support Landowners and Operators to establish, develop and maintain effective working relationships, to the benefit of users of all communications services;

2 An Operator is an organisation which has been granted Code Powers by Ofcom, for example, a communications provider that is providing a landline, broadband, cable or mobile network, or a person who provides infrastructure which supports such a network. A list of those with Code Powers is maintained by Ofcom.

3 'Apparatus' is a broad term and refers to what is defined in the Code as electronic communications apparatus; it includes such items as antennae for mobile signals, masts, cabinets, cables, ducts and telegraph poles.

4 The meaning of 'Landowner', 'Operator', and 'Occupier' is as defined in the Code.

5 'Site' is equivalent to the term 'Land' in the Code, as set out in paragraph 108.

- Sets out what Landowners and Operators should expect from each other in the context of:
    - Establishing new agreements for the installation of apparatus;
    - The ongoing access to and operation, maintenance and upgrading of existing sites and apparatus;
    - The decommissioning of sites that are no longer required;
    - The redevelopment of sites;
- Provides a framework for site provision, whereby the commercial process of coming to an agreement, and of maintaining an agreement, can take account of all the practical requirements of both parties;
- Sets out clear lines of communication through which disputed matters can be escalated;
- Does **not** address the financial aspects of the relationship between the Landowner and the Operator

1.7 While the Code of Practice sets out some clear principles and expectations about how Landowners and Operators should behave towards each other, it should be noted that there are some special regimes in place (e.g. transport land, public maintainable highway and tidal waters), where different specific considerations may apply.

1.8 The Code of Practice covers a wide range of scenarios, from the construction of a full mobile mast to the installation of just one telegraph pole or a very small length of cable and it should be noted that not all the procedural elements should be required in each and every case.

Communication and contact information

1.9 Central to the purpose of this Code of Practice is the maintenance of good communications between the parties in order to facilitate good working relationships.

Keeping contact information up to date

1.10 The Operator should ensure that the Landowner and any relevant Occupier of the site or of access routes to the site have **up-to-date site and contact information** available to them, so that the Landowner can easily assess which point of contact to use in all the circumstances which may arise, such as:

- In the event of an emergency
- For routine estate or management issues
- To change or confirm access arrangements
- For escalation of redevelopment/decommissioning issues

1.11 In turn, the Landowner and Occupier should provide email address/contact details in writing directly to the registered office of the Operator, and ensure the Operator is notified of any changes so that the Operator knows which point of contact to use in all the circumstances which may arise.

Professional advice

1.12 Landowners and Operators may choose to negotiate directly with each other. Alternatively, the parties may wish to seek professional advice from a suitably qualified and experienced person such as a surveyor or valuer. This could also include taking legal advice before concluding an agreement[6].

1.13 In all cases, both Operators and Landowners should act in a consistent, fair and open manner with each other in relation to any proposed works.

New agreements for the installation of Apparatus

1.14 Additional Apparatus can be required for a number of reasons, such as:

- Customer demand
- To provide coverage to new areas
- To provide additional network capacity
- To provide new services
- To replace obsolete sites or sites that are being redeveloped

1.15 Where new apparatus needs to be deployed on a new site, the Operators will follow a sequence of steps, depending on the nature of the apparatus to be installed. For minor installations of apparatus (for example, the placement of a telegraph pole), it may be possible to reach an agreement on standard terms and conditions and without the need for a site visit. For more complex situations (such as a new mobile mast), a site visit may be required to assess the suitability of the location and to find out other background information.

Stage 1: Site Survey

1.16 Once it has been determined that new Apparatus is required in a given area, the Operator should identify various options for new sites and survey possible solutions based on technical and planning considerations.

1.17 Although access to maps, satellite imagery, building plans etc. can enable much of the site feasibility to be conducted remotely, direct access to a potential site and the ability to discuss practical matters with Landowners may be required.

1.18 Where access is necessary, the Operator should request such access in writing, covering the matters set out in Annex A, where relevant. The Operator should generally request that access is given within a reasonable period (e.g. this may be a period of around 7 days). The access request should set out the nature of the visit and a basic outline of the proposed installation/s.

1.19 To ensure the site survey is productive, the parties may choose to meet on site. At the appropriate moment in the assessment process, the Landowner, on the Operator's request, should seek to provide relevant information such as:

- Who owns/occupies the site;
- The current use of the site;
- Whether there are any multiple occupancy management arrangements in place
- Any planned change or intended change in ownership, occupation or use;
- Any proposals there may be to change the use of or develop the land, including whether there are any existing planning permissions in place;
- Details of known pipes, drains, cables or structures…etc.;

6 A list of such advisers can be provided by professional bodies such as Central Association of Agricultural Valuers, Law Society, Law Society for Scotland, Royal Institution of Chartered Surveyors and Scottish Agricultural Arbiters and Valuers Association.

- Whether there is/are any harmful materials, liquids, vegetation, sites of special scientific interest, protected flora, fauna, listed buildings, archaeological considerations or public rights of way on or adjacent to the site
- Any other rights of public access on the site or adjacent to the site

Stage 2: Consultation and agreement

1.20 The type of apparatus that can be deployed on, over or under a site can vary enormously. It could include, for example:
- A telegraph pole being placed in a field;
- A cable being laid in an existing duct in a shopping centre;
- An antenna system for mobile coverage being installed on the roof of an office block;
- A lattice tower being erected in a wood

1.21 Each of these examples could require different consultation processes.

1.22 When a suitable location has been identified for the installation of apparatus, the Operator should proceed to secure any necessary consents for the site, in accordance with relevant regulations, consulting with the Local Planning Authority, and other parties, where required, and any applicable guidelines or codes of practice[7].

1.23 Where a proposal is straightforward, with standard apparatus, such as a single cabinet or pole, it may be appropriate for the Operator to send the Landowner a simple written agreement with a request to sign it and return. Where the proposal is less simple, it may be appropriate for the Operator to send a summary of the proposed terms of an agreement for the Landowner to consider and review. In such cases the documentation might include, for example, a plan showing the proposed design, access routes and cable routes; loading calculations for rooftop sites; and proposals for electricity provision.

1.24 Bfore concluding an agreement, the Landowner and Operator should agree access arrangements for construction, installation, subsequent planned maintenance, upgrades and emergency maintenance to repair service affecting faults. The key points for access arrangements are covered in **Annex B.**

1.25 Although the Code provides a mechanism for the court to impose terms of occupation on the Landowner and the Operator, the parties should make every effort to reach voluntary agreement first.

1.26 Whilst some agreements should be expected to be completed within a matter of weeks, and some simple cases might potentially be signed on site during the survey stage, agreements for larger or more complex arrangements may generally take longer, but in all cases the parties should endeavour to respond promptly to correspondence from the other side and aim to complete the process as swiftly as possible.

1.27 In the absence of terms being agreed between the parties in the circumstances described in paragraph 20(3) of the Code, the Code provides for a process whereby a court can impose the terms of occupation and/or the conferring of code rights pursuant to paragraph 19 of the Code. It must be emphasised, though, that one of

7 For example: Code of Best Practice on Mobile Network Development in England, www.mobileuk.org/codes-of-practice.html.

the principal purposes of this Code of Practice is to establish a voluntary process, which avoids recourse to the courts.

Stage 3: Deployment stage

1.28 When the Operator is carrying out works on a Landowner's property it should endeavour to cause minimal disruption and inconvenience. The Operator should notify the Landowner of the following:

- Contact details for the Operator, the name and contact details of the contractor managing the scheme and also the person to whom the Landowner can escalate any matters of concern
- Drawings detailing the apparatus to be deployed with an accompanying written description of the works
- Any requirement to be able to have access across other land (whether belonging to the
- Landowner or a third party)
- Timing of the work, including the estimated start date and duration of the works
- Working times
- Procedures for safeguarding the Landowner's property (e.g. livestock)

1.29 Where applicable, the Operator should retain a dated photographic record of the condition of the site prior to the commencement of works and on completion of the works.

Neighbours and other occupiers

1.30 Persons with an interest in land adjoining a proposed site may need to be consulted in accordance with national regulations, guidelines and any applicable Codes of Practice[8].

1.31 Operators should also negotiate access arrangements with the owner and/or occupier of land adjoining a site, where use of that land is required for either constructing and/or maintaining the site (using Code powers, if no agreement can be reached).

1.32 Any requirement for access by the Operator with respect to such adjoining land ought to cover the matters set out in Annex B (i.e. the same considerations as for the Landowner, where applicable).

The ongoing access to and operation, maintenance and upgrading of existing sites and apparatus

1.33 All electronic communications sites are an integral part of a wider network. Individual sites variously provide coverage, capacity and functionality to that wider network and Operators require access to their apparatus in order to be able

8 For example: Cabinet and pole siting Code of Practice, and The Code of Best Practice on Mobile Network Development in England, www.mobileuk.org/codes-of-practice.html.

to maintain a quality of service to their customers. In the case of service affecting faults, access should be required as soon as possible.

1.34 As set out in Stage 2 Consultation Phase, any agreements between the Operator and the Landowner should set out how to access sites for operational needs. Annex B sets out key points for access arrangements. Where necessary, Operators and Landowners should meet, prior to entering into a contract, to discuss preferred access routes and processes and agree clear expectations as to what should happen when access is required.

1.35 In the case of emergencies, such as where there is a service-affecting fault or the Apparatus is malfunctioning, Operators need to access the Apparatus without delay, in order to resolve the issue and maintain service for customers, including the ability to make calls to the emergency services. Wherever possible, Operators should contact the Landowner to explain when and why access is required and Landowners should seek to cooperate with the restoration of service.

1.36 Access for routine maintenance should be organised so that Operators can give sufficient notice in accordance with the access arrangements agreed with the Landowner.

1.37 Where Operators are physically sharing a site or using any apparatus on a site, and no additional consents are required under the Code, the Operators should nevertheless notify Landowners of the name and contact details of other sharers and users, so that the Landowner, for security purposes, can know who is in lawful occupation of the site.

1.38 Where access may be required to other parts of the land owned by the Landowner, such as where an area of land is required to use a crane or cherry picker, the access arrangements should cover such scenarios and provide that the Operator should return the land to the condition it was in prior to the land being used or accessed.

1.39 Operators should seek to ensure that anyone accessing a site on their behalf:

- Carries photographic identification
- Can explain why they are there and for whom they are working
- Can advise Landowners who to contact within the Operator for more information or to comment on any visit

1.40 Operators should, upon reasonable request, provide verification of which contractor was on site at any given point in time and confirmation of why they were there – e.g. To inspect, maintain and effect an emergency repair or physical upgrade etc.

1.41 Operators should adhere to any legal or regulatory requirements for managing location specific risks. This might include notifiable diseases (such as Foot and Mouth, Avian Flu etc.). For sites at sensitive locations, it might include arranging accompanied access to secure areas. Operators should comply with any reasonable procedures implemented by Landowners for these purposes. Landowners should, so far as is possible, preserve the ability for Operators to access their apparatus, particularly in the case of operational emergency.

Decommissioning sites that are no longer required

1.42 The Code makes provision for Landowners to request the removal of apparatus, if it is not being used and there is no prospect of it being so.

1.43 As a general principle, Operators should ensure that redundant sites are decommissioned within a reasonable period after use ceases. However, in the case of apparatus below ground (such as ducts for cables), it may be preferable to the parties for the Apparatus to be made safe and left in place. Operators should discuss decommissioning proposals with Landowners in order to agree the way to proceed.

1.44 When requested to remove redundant apparatus by a Landowner, the Operator should, within a reasonable time, respond, either by explaining that the apparatus will still be needed or by agreeing a date by when the apparatus will be made safe or removed, and the site reinstated, if relevant.

Other

Renewal of existing sites and the Code

1.45 When an existing site agreement is due to expire, the parties should seek to agree terms for the continued use of the site before the existing agreement comes to an end.

1.46 Parties should commence negotiations sufficiently far in advance of the expiry of an existing agreement to allow adequate time for terms to be agreed.

Repairs to a Landowner's property

1.47 From time to time, Landowners/Occupiers will have to carry out essential repairs to their property and where possible it may be necessary for apparatus to be moved temporarily to effect such repairs. In such circumstances, the parties should negotiate in good faith so as to allow the works to be completed, and to avoid, so far as possible, any resultant interruption to public communications services and to allow continuity of services. In relation to repairs to the Landowner's property, as part of the good faith negotiations, the parties should discuss the detail of the timings, duration and extent of the works.

Redevelopment by the Landowner

1.48 The Code makes provision for Landowners to redevelop their property (Paragraphs 30–31), requiring that the Landowner should give 18 months' notice of the intention to redevelop. Paragraphs 30–31 of the Code are intended for use by Landowners who genuinely intend to redevelop their property. Landowners are encouraged to give Operators as much prior notice as possible, in order that adequate time can be afforded to allow the Operator to identify alternative suitable sites.

1.49 Operators may request to see evidence of the Landowner's intention to redevelop but they should act reasonably at all times, so as not to hinder the Landowner's progress where there is a genuine intention to redevelop. For example, Operators should act in a timely manner to locate suitable new sites with the principal aim that communications services in a locality can be maintained, with the minimum of disruption to the users.

1.50 Where a Landowner is progressing a redevelopment opportunity, consideration should be given to the possibility of incorporating the communications apparatus within the Landowner's property if this is a reasonable and practicable option.

Escalation procedures

1.51 The Code sets out formal dispute resolution procedures.

1.52 Nevertheless, where disputes arise, the parties should seek to resolve them informally (i.e. without recourse to litigation) in the first instance. There may be occasions, though, where one party or the other may need to serve legal notices, while still continuing to pursue an informal resolution.

1.53 To facilitate this process, Operators and Landowners should make available to each other, and, where applicable, those with an interest in adjoining land, contact details for the relevant person, through whom matters of dispute can be raised. Those matters may include failure to abide by the Code of Practice.

Schedules to the Code of Practice

Schedule A – Requesting access for a survey

1.54 An Operator wishing to access land for the purpose of surveying its suitability for siting electronic communications apparatus should contact the Landowner of a potential site and provide the following information:

- Identity of operator, points of contact for operator and any agent
- Areas of search for possible installation of apparatus
- Requirements for initial survey:
  - What access is desired?
  - With what apparatus?
  - Over what timescale?
- Description of likely apparatus and any ancillary links required, for example power connections
- Confirmation of whether planning consent would be required
- Likely impact of apparatus on the site and/or adjoining land, for example line of sight requirements, possible interference with existing equipment etc.
- Type of agreement sort (e.g. temporary or long-term)
- Proposed timescale for construction/installation
- The letter may also include information about what action an Operator might take, in the event that the Landowner fails to respond

1.55 In some instances, though, when an Operator is surveying at a neighbouring property, and it becomes apparent that the Apparatus would be better suited on an adjoining property, it may possible to agree with the Landowner to complete a survey immediately and then follow-up in writing once the survey has been completed.

Schedule B – Key points for access arrangements

1.56 Access arrangements should cover the following points, where appropriate[9]:

- Contact details (including in emergencies) for:
  - The Operator
  - The Landowner

9 Note: for many fixed line installations, this will be covered by an Operators standard wayleave.

  - Any Occupier of the land, if different from the Landowner

- Description of access arrangements (including any out of hours or weekend factors (e.g. for business premises that are closed at the weekend)
- Recovery of reasonable costs (e.g. if a supervisor is necessary at sensitive locations)
- An undertaking from the Operator to make good any damage to the Landowner's property
- Notifying the Operator of any site-specific considerations, for example:
  - Requirements for supervision at sensitive or hazardous sites
  - Bio-security and any other appropriate security arrangements
  - Any relevant environmental schemes (where care has been taken not to contravene the rules of the scheme)

- Parking and access routes across land or through buildings for construction and maintenance personnel, vehicles, equipment and apparatus
- Adherence to the Countryside Code, or the Scottish Outdoor Access Code where relevant

Appendix E
OFCOM Template Notices[1]

STATUTORY NOTICE
OF THE ASSIGNMENT OF AN AGREEMENT UNDER THE ELECTRONIC COMMUNICATIONS CODE
Paragraph 16(5) of Part 3 of Schedule 3A of the Communications Act 2003

1. This is a statutory notice pursuant to paragraph 16(5) of the electronic communications code, set out in Schedule 3A to the Communications Act 2003 (the '**Code**').[2]
2. We [*Insert name of assignee Code operator*] ('**we**' or '**us**') understand that you, [*Insert name of site provider*], are currently party to an agreement under Part 2 of the Code with [*Insert name of assigning Code operator*] (the '**Operator**' and the '**Agreement**'). Under the Agreement, you agreed to [*confer / be bound by*] a number of Code rights in order to facilitate the deployment by the Operator of its [*electronic communications network / infrastructure system*]. These Code rights relate to land occupied by you at [*Insert address*].
3. The purpose of this notice is to inform you that, on [*Insert date*], the Operator assigned the Agreement to us. This means that the Operator has transferred the benefit of the Code rights [*conferred by / binding on*] you under the Agreement to us, and that we are (from the date of the assignment) bound by the terms of the Agreement.
4. Please note that, from the date on which this notice has been given to you, the Operator will not be liable for any breach of a term of the Agreement (unless that breach took place before the date on which this notice was given to you). As a result, should you have any concerns in the future about the exercise of Code rights on your land, please contact us.
5. Our contact address is [*Insert*].
6. If you have any questions about this notice, please do not hesitate to contact us via telephone (*Insert number*) or e-mail (*insert email address*).

[*Insert date of Notice*]

1 See Sections 41.7 and 41.8 for a commentary.

2 A copy of the Communications Act 2003 is available online at www.legislation.gov.uk.

STATUTORY NOTICE
OF THE ASSIGNMENT OF AN AGREEMENT UNDER THE ELECTRONIC COMMUNICATIONS CODE
Paragraph 16(5) of Part 3 of Schedule 3A of the Communications Act 2003

1. This is a statutory notice pursuant to paragraph 16(5) of the electronic communications code, set out in Schedule 3A to the Communications Act 2003 (the '**Code**').[1]
2. [*Insert name of Code operator*] ('**we**' or '**us**') are currently party to an agreement under Part 2 of the Code with you, [*Insert name of site provider*] (the '**Agreement**'). Under the Agreement, you agreed to [*confer / be bound by*] a number of Code rights in order to facilitate the deployment by us of our [*electronic communications network / infrastructure system*]. These Code rights relate to land occupied by you at [*Insert address*].
3. The purpose of this notice is to inform you that, on [*Insert date*], we assigned the Agreement to [*Insert name of assignee*] (the '**Assignee**'). This means that we have transferred the benefit of the Code rights [*conferred by / binding on*] you under the Agreement to the Assignee, and the Assignee is (from the date of the assignment) bound by the terms of the Agreement.
4. Please note that, from the date on which this notice has been given to you, we will not be liable for any breach of a term of the Agreement (unless that breach took place before the date on which this notice was given to you). As a result, should you have any concerns in the future about the exercise of Code rights on your land, please contact the Assignee.
5. The contact address for the Assignee is [*Insert*].
6. If you have any questions about this notice, please do not hesitate to contact us via telephone (*Insert number*) or e-mail (*insert email address*).

[*Insert date of Notice*]

1 A copy of the Communications Act 2003 is available online at www.legislation.gov.uk.

STATUTORY NOTICE
SEEKING AGREEMENT TO THE CONFERRAL OF RIGHTS UNDER THE ELECTRONIC COMMUNICATIONS CODE
Paragraph 20(2) [and Paragraph 27(1)] of Part 4 of Schedule 3A of the Communications Act 2003

IMPORTANT NOTICE

If you are willing to enter into a Code Agreement, you should respond within 28 days

1. This is a statutory notice pursuant to paragraph 20(2) [*and paragraph 27(1)*] of the electronic communications code, set out in Schedule 3A to the Communications Act 2003 (the '**Code**').[1]
2. This notice has been issued by [*Name of Code operator*] ('**we**' or '**us**') to you, [*Insert name*], because we would like to [*insert brief description of rights sought, e.g. to install apparatus and carry out related works*] on land occupied by you for the purposes of our [*electronic communications network and/or infrastructure system*]. We are seeking your agreement to confer these rights on us.

   [OR – delete appropriate version of paragraph 2]
2. This notice has been issued by [*Name of Code operator*] ('**we**' or '**us**') to you, [*Insert name*], because we have certain rights to [*insert brief description of rights already exercisable by operator in relation to the land, e.g. keep apparatus installed on land in relation to which you have an interest]* for the purpose of our [*electronic communications network and/or infrastructure system*]. We are seeking your agreement to be bound by these rights.
3. [*We also require your agreement on a temporary basis in relation to electronic communications apparatus that is already installed on, under or over your land. This is in order to secure that the service provided by our [electronic communications network and/or infrastructure system] is maintained, and the apparatus is properly adjusted and kept in repair.*]

BACKGROUND

4. We provide an [*electronic communications network and/or infrastructure system*] in the United Kingdom. This is used in order to provide consumers with [*insert a brief description of the retail services which are dependent on this network and/or infrastructure system (e.g. fixed voice and broadband services)*].
5. For this purpose, the Office of Communications (OFCOM) has given a direction applying the Code to us. The Code regulates the relationships between us and occupiers of land, thereby facilitating the deployment of electronic communications apparatus.

1 A copy of the Communications Act 2003 is available online at www.legislation.gov.uk.

INTERPRETATION

6. In this notice:
   a. '**Apparatus**' means the electronic communications apparatus described in Annex 1;
   b. '**Land**' means the [*land*] at [*Insert address / description of land, etc.*]; and
   c. words used but not defined in this Notice shall have the meaning ascribed to them in the Code.

DETAILS OF THE AGREEMENT WE ARE SEEKING

7. In this notice, we are seeking your agreement [*to confer on us / to be bound by*] the following rights:
   a. [*the right to install the Apparatus on, under or over the Land*];
   b. [*the right to keep installed the Apparatus which is on, under or over the Land*];
   c. [*the right to inspect, maintain, adjust, alter, repair, upgrade or operate the Apparatus which is on, under or over the Land*];
   d. [*the right to carry out any works on the Land for or in connection with the installation of the Apparatus on, under or over the Land [or the installation of electronic communications apparatus elsewhere]*];
   e. [*the right to carry out any works on the Land for or in connection with the maintenance, adjustment, alteration, repair, upgrading or operation of the Apparatus which is on, under or over the Land [or of electronic communications apparatus elsewhere]*];
   f. [*the right to enter the Land to inspect, maintain, adjust, alter, repair, upgrade or operate the Apparatus which is on, under or over the Land [or any electronic communications apparatus elsewhere]*];
   g. [*the right to connect the Apparatus to a power supply*];
   h. [*the right to interfere with or obstruct a means of access to or from the Land (whether or not the Apparatus is on, under or over the Land)*]; and
   i. [*the right to lop or cut back, or require another person to lop or cut back, any tree or other vegetation that interferes or will or may interfere with the Apparatus*].

(together, the '**Code Rights**').

8. In addition to the Code Rights, we are also seeking in this notice your agreement to the additional terms set out in Annex 2.
9. [*As the electronic communications apparatus described in Annex 3 (the '**Existing Apparatus**') is already installed on, under or over the Land, we are also seeking your agreement on a temporary basis to [confer/be bound by] the Code Rights set out at paragraph 7 above in respect of the Existing Apparatus (the '**Temporary Code Rights**'.) [And, in addition to the Temporary Code Rights we are also seeking your agreement on a temporary basis to the additional terms set out in Annex 2].*

CONSEQUENCES OF NOT REACHING AGREEMENT ON THE CODE RIGHTS

10. If either:
    a. you do not, before the end of 28 days beginning with the day on which this notice is given, agree [*to confer / to be bound by*] the Code Rights; or

  b. at any time after this notice is given, you give notice in writing to us that you do not agree [*to confer / to be bound by*] the Code Rights,

we will be entitled to apply to the court for an order under paragraph 20(4) of the Code.

11. For more information on the circumstances in which a court may impose such an order, and on the type of agreement that the court may impose, please see the supplementary information at the back of this notice.

[CONSEQUENCES OF NOT REACHING AGREEMENT ON THE TEMPORARY CODE RIGHTS]

12. If:
  a. you have the right to require the removal of the Existing Apparatus under paragraph 37 or 41(1) of the Code but we are not for the time being required to remove it; and
  b. either:
    i. you do not, before the end of 28 days beginning with the day on which this notice is given, agree [*to confer / to be bound by*] the Temporary Code Rights; or
    ii. at any time after this notice is given, you give notice in writing to us that you do not agree [*to confer / to be bound by*] the Temporary Code Rights,

we will have the right to apply to the courts for an order under paragraph 27(2) of the Code. Further detail on these orders is provided in the supplementary information at the back of this notice.

13. We consider that the agreement sought in this notice in relation to Temporary Code Rights is [*not*] a matter of urgency and therefore [*do not*] intend to apply for such an order prior to the end of the 28-day period referred to above.[2]

YOUR OPTIONS

14. In response to this notice, you may:
  a. agree [*to confer the Code Rights on us / to be bound by the Code Rights*] [and/or *to confer the Temporary Code Rights on us / to be bound by the Temporary Code Rights*];
  b. give notice to us that you do not agree [*to confer / to be bound by*] the Code Rights [*and/or the Temporary Code Rights*]; or
  c. do nothing.
15. In deciding how to respond to this notice, you may wish to seek independent legal advice.
16. If you agree [*to confer the Code Rights on us / to be bound by the Code Rights*], [we will send you an agreement reflecting the terms set out in this notice and ask you to sign it] [we ask you to sign the agreement attached at Annex 2]. Similarly, if you agree [*to confer the Temporary Code Rights on us / to be bound by the Temporary Code Rights*], we will also send you an agreement reflecting the terms set out in

2 In limited circumstances, where the court agrees that it is a matter of urgency for an order to be made under paragraph 27(5) of the Code, it may make such an order even though the 28-day period referred to at paragraph [12]a. above has not elapsed (and paragraph [12]b. does not apply).

this notice and ask you to sign it. You would be entitled to seek independent legal advice in relation to [this/these] agreement[s].

17. Alternatively, and as explained at paragraph[s] 10 [and 12] above, if you do nothing or give notice to us that you do not agree [*to confer / to be bound by*] the Code rights [*and Temporary Code Rights*], we will be entitled to apply to the court for an order under paragraph 20(4) [*and an order under paragraph 27(2)*] of the Code.
18. Please submit any notification pursuant to paragraph 14a. or b. to us in writing as soon as possible and, in any event, before the end of 28 days beginning with the day on which this notice is given.
19. To be effective, such notification must be **delivered by hand** or sent by **registered post** or **recorded delivery** to the following address:

[*Insert address details*]

20. If you have any questions about this notice, please do not hesitate to contact us via telephone (*Insert number*) or e-mail (*insert email address*).

[*INSERT DATE OF NOTICE*]

ANNEX 1

THE APPARATUS

[*Insert a description of the electronic communications apparatus to which the notice relates*]

ANNEX 2

ADDITIONAL TERMS OF AGREEMENT SOUGHT

[*Insert description of the additional contractual terms sought or attach a draft agreement*]

ANNEX 3

THE EXISTING APPARATUS

[*Insert a description of the electronic communications apparatus already installed on, under or over the Land and in respect of which you are seeking the Temporary Code Rights*]

STATUTORY NOTICE
SEEKING AGREEMENT TO THE CONFERRAL OF INTERIM RIGHTS UNDER THE ELECTRONIC COMMUNICATIONS CODE
Paragraph 26(3) of Part 4 of Schedule 3A of The Communications Act 2003

IMPORTANT NOTICE

If you are willing to enter into a Code Agreement, you should respond within 28 days

1. This is a statutory notice pursuant to paragraph 26(3) of the electronic communications code, set out in Schedule 3A to the Communications Act 2003 (the '**Code**').[1]
2. This notice has been issued by [*Name of Code operator*] ('**we**' or '**us**') to you, [*Insert name*], because we would like to [*insert brief description of rights sought, e.g. to install apparatus and carry out related works*] on land occupied by you for the purposes of our [*electronic communications network and/or infrastructure system*]. We are seeking your agreement, on an interim basis, to confer these rights on us.
   [OR – delete appropriate version of paragraph 2]
2. This notice has been issued by [*Name of Code operator*] ('**we**' or '**us**') to you, [*Insert name*], because we have certain rights to [*insert brief description of rights already exercisable by operator in relation to the land, e.g. keep apparatus installed on land in relation to which you have an interest]* for the purpose of our [*electronic communications network and/or infrastructure system*]. We are seeking your agreement, on an interim basis, to be bound by these rights.

BACKGROUND

3. We provide an [*electronic communications network and/or infrastructure system*] in the United Kingdom. This is used in order to provide consumers with [*insert a brief description of the retail services which are dependent on this network and/or infrastructure system* (*e.g. fixed voice and broadband services)*].
4. For this purpose, the Office of Communications (OFCOM) has given a direction applying the Code to us. The Code regulates the relationships between us and occupiers of land, thereby facilitating the deployment of electronic communications apparatus.

INTERPRETATION

5. In this notice:
   a. '**Apparatus**' means the electronic communications apparatus described in Annex 1;
   b. '**Land**' means the land at [*Insert address / description of land, etc.*].

1 A copy of the Communications Act 2003 is available online at www.legislation.gov.uk.

DETAILS OF THE AGREEMENT WE ARE SEEKING

6. In this notice, we are seeking your agreement to [*confer on us / be bound by*] the following rights, on the interim basis specified at paragraph 8 below:
   a. [*the right to install the Apparatus on, under or over the Land*];
   b. [*the right to keep installed the Apparatus which is on, under or over the Land*];
   c. [*the right to inspect, maintain, adjust, alter, repair, upgrade or operate the Apparatus which is on, under or over the Land*];
   d. [*the right to carry out any works on the Land for or in connection with the installation of the Apparatus on, under or over the Land [or the installation of electronic communications apparatus elsewhere]*];
   e. [*the right to carry out any works on the Land for or in connection with the maintenance, adjustment, alteration, repair, upgrading or operation of the Apparatus which is on, under or over the Land [or of electronic communications apparatus elsewhere]*];
   f. [*the right to enter the Land to inspect, maintain, adjust, alter, repair, upgrade or operate the Apparatus which is on, under or over the Land [or any electronic communications apparatus elsewhere]*];
   g. [*the right to connect the Apparatus to a power supply*];
   h. [*the right to interfere with or obstruct a means of access to or from the Land (whether or not the Apparatus is on, under or over the Land)*]; and
   i. [*the right to lop or cut back, or require another person to lop or cut back, any tree or other vegetation that interferes or will or may interfere with the Apparatus*].

(together, the '**Code Rights**').

7. In addition to the Code Rights, we are also seeking in this notice your agreement to the additional terms set out in Annex 2.
8. We would like the Code Rights, and additional terms set out in Annex 2, to be exercisable [*for a period of [Insert period (e.g. 3 months)]* / until the *occurrence of [Insert details of a particular event (e.g. until redevelopment of alternative property)]*].

CONSEQUENCES OF NOT REACHING AGREEMENT

9. If either:
   a. you do not, before the end of 28 days beginning with the day on which this notice is given, agree [*to confer / to be bound by*] the Code Rights; or
   b. at any time after this notice is given, you give notice in writing to us that you do not agree [*to confer / to be bound by*] the Code Rights,

we will be entitled to apply to the court for an order under paragraph 26 of the Code. Further detail on these orders is provided in the supplementary information at the back of this notice.

10. We consider that the agreement sought in this notice is [*not*] a matter of urgency and therefore [*do not*] intend to apply for such an order prior to the end of the 28-day period referred to above.[2]

2 In limited circumstances, where the court agrees that it is a matter of urgency for an order to be made under paragraph 26(5) of the Code, it may make such an order even though the 28-day period referred to at paragraph 9a. above has not elapsed (and paragraph 9b. does not apply).

YOUR OPTIONS

11. In response to this notice, you may:
    a. agree [*to confer the Code Rights on us / to be bound by the Code Rights*] on the interim basis requested in this notice;
    b. give notice to us that you do not agree [*to confer / to be bound by*] the Code Rights on the interim basis requested in this notice; or
    c. do nothing.
12. In deciding how to respond to this notice, you may wish to seek independent legal advice.
13. If you agree [*to confer the Code Rights on us / to be bound by the Code Rights*] on the interim basis requested in this notice, [we will send you an agreement reflecting the terms set out in this notice and ask you to sign it.] [we ask you to sign the agreement attached at Annex 2.] You would be entitled to seek independent legal advice in relation to this agreement.
14. Alternatively, and as explained at paragraph 9 above, if you do nothing or give notice to us that you do not agree [*to confer / to be bound by*] the Code rights on the interim basis requested in this notice, we will be entitled to apply to the court for an order under paragraph 26 of the Code.
15. Please submit any notification pursuant to paragraph 11a. or b. to us in writing as soon as possible and, in any event, before the end of 28 days beginning with the day on which this notice is given.
16. To be effective, such notification must be **delivered by hand** or sent by **registered post** or **recorded delivery** to the following address:

[*Insert address details*]

17. If you have any questions about this notice, please do not hesitate to contact us via telephone (*Insert number*) or e-mail (*insert email address*).

[*INSERT DATE OF NOTICE*]

ANNEX 1
THE APPARATUS

[*Insert a description of the electronic communications apparatus to which the notice relates*]

ANNEX 2
ADDITIONAL TERMS OF AGREEMENT SOUGHT

[*Insert description of the additional contractual terms sought or attach a draft agreement*]

SUPPLEMENTARY INFORMATION FOR THE RECIPIENT OF THIS NOTICE

Orders under paragraph 25(3) of the Code

1. An order under paragraph 25 of the Code is an order which imposes on us and you an agreement. The effect of such an agreement would be [*to confer the Code Rights on us / provide for the Code Rights to bind you*], on an *interim* basis.

2. Paragraphs 23 and 24 of the Code contain further detail about the terms of the agreement that the court may impose. And paragraph 22 of the Code states that such an agreement takes effect for all purposes of the Code as an agreement under Part 2 of the Code between the operator and the relevant person.
3. The court may only make an order if:
   a. you have agreed with us to the making of the order and the terms of the agreement imposed by it; **or**
   b. it thinks that there is a good arguable case that **both** of the following conditions are met:
      i. the prejudice caused to you by the order is capable of being adequately compensated by money; and
      ii. the public benefit likely to result from the making of the order (having regard to the public interest in access to a choice of high quality electronic communications services) outweighs the prejudice to you.
4. The court may **not** make such an order if it thinks that you intend to redevelop all or part of the land to which the Code Rights would relate, or any neighbouring land, and could not reasonably do so if the order were made.
5. The court also has the power to order us to pay you compensation for any loss or damage that you have sustained or will sustain as a result of the exercise of the Code Rights. Paragraphs 25 and Part 14 of the Code contain further detail about this.

STATUTORY NOTICE
BRINGING AN AGREEMENT UNDER THE ELECTRONIC COMMUNICATIONS CODE TO AN END
Paragraph 31(1) of Part 5 of Schedule 3A of the Communications Act 2003

1. This is a statutory notice pursuant to paragraph 31(1) of the electronic communications code, set out in Schedule 3A to the Communications Act 2003 (the '**Code**').[1] **[see note (a)]**
2. [*I / We*] [*Insert name*] [*am/are*] currently party to an agreement under Part 2 of the Code with you, [*Name of Code operator*] (the '**Agreement**'). Under the Agreement, [*I/we*] agreed to confer or be otherwise bound by a number of Code rights in order to facilitate the deployment by you of your electronic communications network and/or infrastructure system. These code rights relate to land occupied by [*me/us*] at [*Insert address*].
3. The purpose of this notice is to inform you that [*I/we*] would like to bring the Agreement to an end.
4. [*I/we*] propose that the Agreement be brought to an end on [*Insert date*]. **[see note (b)]**
5. [*I/we*] propose that the Agreement be brought to an end: **[see note (c)]**
   a. [*as a result of substantial breaches by you of your obligations under the Agreement;*]
   b. [*because of persistent delays by you in making payments to [me/us] under the Agreement;*]
   c. [*because* [*I/we*] *intend to redevelop all or part of the land to which the Agreement relates, or any neighbouring land, and could not reasonably do so unless the Agreement comes to an end;*]
   d. [*because you are not entitled to the Agreement because the test under paragraph 21 of the Code for the imposition of the Agreement on [me/us] is not met.*]
6. In accordance with paragraph 32 of the Code, the Agreement should come to an end on the terms set out in this notice unless:
   a. within the period of three months beginning with the day on which the notice is given, you give [*me/us*] a counter-notice in accordance with paragraph 32(3) of the Code; and
   b. within the period of three months beginning with the day on which the counter-notice is given, you apply to the court for an order under paragraph 34 of the Code.

[*Insert date of Notice*]

NOTES FOR COMPLETING THIS NOTICE

You may wish to obtain independent legal advice before completing this notice.

a) *This notice should be* ***delivered by hand*** *or sent by* ***registered post*** *or* ***recorded delivery*** *to the operator at:*
   - *the address for service that the operator has given to you for the purposes of the Code; or*

1 A copy of the Communications Act 2003 is available online at www.legislation.gov.uk.

- *if no such address has been given to you, at the address given by section 394 of the Communications Act 2003 (available online at* www.legislation.gov.uk).

b) *Paragraph 31(3) of the Code provides that the date on which the Agreement is proposed to come to an end must fall:*
- *after the end of the period of 18 months beginning with the day on which this notice is given to the operator;* ***and***
- *after the time at which, apart from paragraph 30 of the Code, the code right(s) to which the Agreement relates would have ceased to be exercisable or to bind you or at a time when, apart from that paragraph, the Agreement could have been brought to an end by you.*

For the purposes of the first bullet point, the period of 18 months should not be calculated from the date of this notice, but from the date on which it is **given** *to the operator. If the notice is sent by post, it will be deemed to have been given to the operator at the time at which it would be delivered in the ordinary course of post.*

In accordance with the second bullet point, one of the following conditions must apply on the date on which you propose the Agreement should come to an end: (i) the code right(s) to which the Agreement relates are no longer exercisable or no longer bind you (e.g. because the term of the Agreement has expired) **or** *(ii) you are able to bring the Agreement to an end (e.g. by giving notice to terminate under the Agreement). The effect of paragraph 30 of the Code should be disregarded in considering whether condition (i) or (ii) applies (paragraph 30 provides for the continuation of the Agreement notwithstanding that, under its terms, the code right(s) are no longer exercisable or no longer bind you, or that you may bring the Agreement to an end).*

c) *The effect of paragraph 31(4) of the Code is that the Agreement may only be brought to an end if one of the grounds set out in a. to d. applies. Please delete as appropriate to state the ground on which you propose to end the Agreement.*

STATUTORY NOTICE
REQUIRING A CHANGE TO THE TERMS OF AN AGREEMENT UNDER THE ELECTRONIC COMMUNICATIONS CODE
Paragraph 33(1) of Part 5 of Schedule 3A of the Communications Act 2003

IMPORTANT NOTICE

If you agree to the changes we are requesting, you should respond within six months

1. This is a statutory notice pursuant to paragraph 33(1) of the electronic communications code, set out in Schedule 3A to the Communications Act 2003 (the '**Code**').[1]
2. The purpose of this notice is to require a change to the terms of an agreement between you, [*Insert name of site provider*] and [*us*], [*Insert name of Code operator*] under Part 2 of the Code. We are seeking your agreement to this change.

BACKGROUND

3. We have entered into an agreement under Part 2 of the Code (the '**Agreement**'). Under the Agreement, you [*have conferred on us / become bound by*] certain rights under the Code. The purpose of these rights is to facilitate the deployment by us of our [*electronic communications network and/or system of infrastructure*] at [*Insert address*].
4. Paragraph 33 of the Code explains how a party to a Code agreement may require a change to the terms of an agreement which has expired. It provides that, in the first instance, the party seeking the change should provide notice to the other party of the change that it is seeking and the date on which that change would take place.

THE CHANGE WE ARE REQUESTING

5. We are asking you to agree, from the date set out in paragraph 6 below, that:
   a. [the Agreement should have effect subject to the modified terms set out in Annex [X];]
   b. [our existing Code right to [*insert details of Code right*] should no longer [*be conferred by / bind*] you;]
   c. [the Agreement should also [*confer on us / bind you to*] [*Insert details of additional Code right sought*];]
   d. [the Agreement should be terminated and a new agreement should have effect between us on the terms set out in Annex [X].]
6. The day from which we propose that:
   a. [the modified terms should have effect;]
   b. [the Code right referred to in paragraph 5 above should no longer [*be conferred by / bind*] you;]
   c. [the additional Code right referred to in paragraph 5 above should [*be conferred by / bind*] you;]

1 A copy of the Communications Act 2003 is available online at www.legislation.gov.uk.

d. [the Agreement should be terminated, and from which the new agreement set out in Annex [X] should have effect]
is [*Insert Date*].[2]

CONSEQUENCES OF NOT REACHING AGREEMENT

7. If, after the end of six months beginning with the day on which this notice is given, we have not reached agreement with you on the proposals in this notice, we may apply to the court for an order under paragraph 34 of the Code.
8. Further detail on these orders is provided in the supplementary information at the back of this notice.

YOUR OPTIONS

9. In response to this notice, you may:
   a. agree to the change requested above;
   b. give notice to us that you do not agree to the change requested above; or
   c. do nothing.
10. In deciding how to respond to this notice, you may wish to seek independent legal advice.
11. If you agree to the change requested above, we will send you [*a modified version of the Agreement reflecting the terms set out in this notice / a new agreement reflecting the terms set out in Annex [X] together with a notice of confirmation that you agree to termination of the Agreement*]. We will ask you to sign [*this/these documents*]. You would be entitled to seek independent legal advice in relation to [*this/these*] document[s].
12. Alternatively, and as explained at paragraph 7 above, if you do nothing or give notice to us that you do not agree the change requested above, we will be entitled to apply to the court for an order under paragraph 34 of the Code after the end of six months beginning with the day on which this notice is given.
13. Please submit any notification pursuant to paragraph 9a. or b. to us in writing as soon as possible and, in any event, before the end of six months beginning with the day on which this notice is given.
14. To be effective, such notification must be **delivered by hand** or sent by **registered post** or **recorded delivery** to the following address:

[*Insert address details*]

15. If you have any questions about this notice, please do not hesitate to contact us via telephone (*Insert number*) or e-mail (*insert email address*).

[*INSERT DATE OF NOTICE*]

ANNEX [X]

2 Regulation 33(3) of the Code requires that the date must fall: (a) after the end of the period of six months beginning with the day on which this notice is given; and (b) after the time at which, apart from paragraph 30, the Code right to which the existing Code agreement relates would have ceased to be exerciseable or to bind the site provider or at a time when, apart from that paragraph, the Code agreement could have been brought to an end by the site provider.

SUPPLEMENTARY INFORMATION FOR THE RECIPIENT OF THIS NOTICE

Orders under paragraph 34 of the Code

1. The types of orders which the court may make under paragraph 34 include an order which has the effect of:
   a. [modifying the terms of the Agreement;]
   b. [modifying the terms of the Agreement so that one of the Code rights set out therein is no longer [*conferred by / binding on*] you;]
   c. [modifying the terms of the Agreement so that it [*confers an additional Code right on you / provides that you are bound by an additional Code right*];]
   d. [terminating the Agreement and ordering you to enter into a new agreement which [*confers a Code right on us / provides for a Code right to bind you*];]
2. In determining whether to make an order under paragraph 34, the court must have regard to all the circumstances of the case, and in particular to:
   a. the operator's business and technical needs;
   b. the use that the site provider is making of the land to which the existing code agreement relates;
   c. any duties imposed on the site provider by an enactment; and
   d. the amount of consideration payable by the operator to the site provider under the existing code agreement.
3. If the court makes an order under paragraph 34, it may also order the operator to pay the site provider consideration. See paragraph 34(14) for details of how the consideration should be calculated by the court in this case.

STATUTORY NOTICE
REQUESTING DISCLOSURE OF WHETHER APPARATUS IS ON LAND PURSUANT TO THE ELECTRONIC COMMUNICATIONS CODE
Paragraph 39(1) of Part 6 of Schedule 3A of the Communications Act 2003

1. This is a statutory notice pursuant to paragraph 39(1) of the electronic communications code, set out in Schedule 3A to the Communications Act 2003 (the '**Code**').[1] **[see note (a)]**
2. [*I/We*], [*Insert name of landowner*], have issued this notice to you, [*Name of Code operator*], in order to find out if you: **[see note (b)]**
   a. [own electronic communications apparatus on, under or over land in which [I/we] have an interest or use such apparatus for the purpose of your network; and/or]
   b. [have the benefit of a Code right (as defined in paragraph 3 of the Code) entitling you to keep electronic communications apparatus on, under or over land in which [I/we] have an interest.
3. The land in which [*I/we*] have an interest, and to which this notice relates, is [*Insert address / description of land, etc.*] (the '**Land**').
4. Please provide the information requested above as soon as possible and, in any event, before the end of **three months** beginning with the date on which this notice is given.
5. If you do not provide the information requested by this date[2] and BOTH:
   a. [*I/We*] take action under paragraph 40 of the Code to enforce the removal of the apparatus at the Land; and
   b. it is subsequently established that:
      i. you own the apparatus or use it for the purposes of your network; and
      ii. you have the benefit of a Code right entitling you to keep the apparatus on, under or over the Land,

you must bear the costs of any such action taken by [*me/us*] under paragraph 40 of the Code.

[*INSERT DATE OF NOTICE*]

NOTES FOR COMPLETING THIS NOTICE

You may wish to obtain independent legal advice before completing this notice.

(a) *This notice should be* ***delivered by hand*** *or sent by* ***registered post*** *or* ***recorded delivery*** *to the operator at:*
- *the address for service that the operator has given to you for the purposes of the Code; or*
- *if no such address has been given to you, at the address given by section 394 of the Communications Act 2003 (available online at* www.legislation.gov.uk).

1 A copy of the Communications Act 2003 is available online at www.legislation.gov.uk.

2 This includes if you do not provide the information requested in a manner which complies with paragraph 88 of the Code.

(b) *In a notice under paragraph 39(1) of the Code, you can only require disclosure of the information set out at paragraphs a. and b., and you must have an interest in the land to which the notice relates. Please delete a. or b. as appropriate, based on the information that you are seeking.*

STATUTORY NOTICE
REQUIRING THE REMOVAL OF APPARATUS INSTALLED UNDER THE ELECTRONIC COMMUNICATIONS CODE
Paragraph 40(2) of Part 6 of Schedule 3A of the Communications Act 2003

1. This is a statutory notice pursuant to paragraph 40(2) of the electronic communications code, set out in Schedule 3A to the Communications Act 2003 (the '**Code**').[1] **[see note (a)]**
2. You, [*Insert name of Code operator*], have installed electronic communications apparatus [*on / under / over*] land at [*Insert address*] (the '**Land**').
3. [*I/we*], [*Insert name*], [currently have an interest in the Land / [*am/are*] currently the owner or occupier of land that neighbours the Land]. **[see note (b)]**
4. The purpose of this notice is to inform you that [*I/we*] would like you to remove that apparatus on the basis of [state the conditions relied on under paragraph 37 or 38 of the Code] and to restore the Land to its condition before the apparatus was placed [*on / under / over*] it.
5. [*I/ We*] would like you to complete these works on or before [*Insert Date*]. **[see note (c)]**
6. Please confirm, within the period of 28 days beginning with the day on which this notice is given, if you agree to complete these works by the date(s) specified in paragraph 5 above.
7. Alternatively, if you consider that the date(s) specified by [*me/us*] at paragraph 5 above is not reasonable, please indicate as soon as possible (and at least within the period of 28 days beginning with the day on which this notice is given) what date(s) you consider to be reasonable for completion of the works.
8. [*I/we*] will be entitled to make an application to the court if we do not reach agreement on any of the following matters within the period of 28 days beginning with the day on which this notice is given:
   a. that you will remove the apparatus;
   b. that you will restore the Land to its condition before the apparatus was placed on, under or over the Land;
   c. the time at which or period within which the apparatus is removed;
   d. the time at which or period within which the Land will be restored.
9. The application to the court may be for:
   a. an order under paragraph 44(1) of the Code requiring you to remove the apparatus; or
   b. an order under paragraph 44(3) of the Code enabling [*me/us*] to sell the apparatus,

NOTES FOR COMPLETING THIS NOTICE

You may wish to obtain independent legal advice before completing this notice.

(a) *This notice should be* ***delivered by hand*** *or sent by* ***registered post*** *or* ***recorded delivery*** *to the operator at:*

1 A copy of the Communications Act 2003 is available online at www.legislation.gov.uk.

- *the address for service that the operator has given to you for the purposes of the Code; or*
- *if no such address has been given to you, at the address given by section 394 of the Communications Act 2003 (available online at* www.legislation.gov.uk).

(b) *You are only entitled to send this notice if:*

- *you have an interest in the Land and have the right, under paragraph 37 of the Code, to require the removal of electronic communications apparatus on, under or over the Land. To have this right, one or more of the five conditions set out at paragraph 37 must be met; or*
- *you are the owner or occupier of neighbouring land and have the right, under paragraph 38 of the Code, to require the removal of electronic communications apparatus on, under or over the Land. To have this right, both of the conditions set out at paragraph 38 must be met.*

(c) *The date specified by you as the deadline for completion of the works must be a reasonable one. What is reasonable will depend on the individual circumstances of your case, including the complexity of the works to be undertaken.*

Please note that you are entitled to specify a date for removal of the apparatus which is different to the date specified by you for restoring the Land to its previous condition.

[*Insert date of Notice*]

STATUTORY NOTICE
COUNTER-NOTICE REGARDING THE REMOVAL OF APPARATUS INSTALLED UNDER THE ELECTRONIC COMMUNICATIONS CODE
Paragraph 41(5) of Part 6 of Schedule 3A of the Communications Act 2003

1. This is a statutory notice pursuant to paragraph 41(5) of the electronic communications code, set out in Schedule 3A to the Communications Act 2003 (the '**Code**').[1]
2. On [*Insert date*], we [*Insert name of Code operator*], were given a notice by you, [*Insert name of third party*], under paragraph 41(2) of the Code. In that notice, you required us to remove electronic communications apparatus installed [*on / under / over*] land at [*Insert address*] (the '**Apparatus**'), and to restore that land to its previous condition.
3. The purpose of this counter-notice is to inform you that you are not entitled to require the removal of the Apparatus. This is because [*Insert details*].
   [*OR – delete appropriate version of paragraph 3*]
3. The purpose of this counter-notice is to inform you of the steps which we propose to take for the purpose of securing a right as against you to keep the Apparatus on the land. In particular, we propose to [*Insert details of proposed steps*].[2]
4. As a result of this notice, you may only enforce the removal of the Apparatus in pursuance of an order of the court under paragraph 41(7) of the Code.

[*Insert date of Notice*]

1 A copy of the Communications Act 2003 is available online at www.legislation.gov.uk.

2 Paragraph 41(8) of the Code provides that the court may only make an order under paragraph 41(7) if it is satisfied that either: (a) we are not intending to take those steps or are being unreasonably dilatory in taking them; or (b) taking those steps has not secured, or will not secure, for us as against you any right to keep the Apparatus installed on, under or over the land or to reinstall it if it is removed.

STATUTORY NOTICE
REQUIRING THE RESTORATION OF LAND TO ITS CONDITION BEFORE THE EXERCISE OF RIGHTS UNDER THE ELECTRONIC COMMUNICATIONS CODE
Paragraph 43(5) of Part 6 of Schedule 3A of the Communications Act 2003

1. This is a statutory notice pursuant to paragraph 43(5) of the electronic communications code, set out in Schedule 3A to the Communications Act 2003 (the '**Code**').[1] [see note (a)]
2. The purpose of this notice is to require you, [*Insert name of Code operator*], to restore land in which [*I / we*], [*Insert name of relevant person*], have an interest, to its condition before the exercise of a Code right by you.
3. The land to which this notice relates is at [*Insert address and any other relevant details*] (the '**Land**'). [*I am / We are*] the [*occupier of / owner of the freehold estate in*[2] */ lessee of*] the Land and have a right, under paragraph 43(2) of the Code to require you to restore the Land to its condition before the exercise of a Code right by you. [see note (b)]
4. In order to restore the Land to its previous condition, [*I / we*] ask that you complete the works on or before [*Insert date*]. [see note (c)]
5. Please confirm, within the period of 28 days beginning with the day on which this notice is given, if you agree to complete these works by the date specified in paragraph 4 above.
6. Alternatively, if you consider that the date specified by [*me/us*] at paragraph 4 above is not reasonable, please indicate as soon as possible (and at least within the period of 28 days beginning with the day on which this notice is given) what date you consider to be reasonable for completion of the works.
7. If we do not reach agreement on any of the above matters within 28 days beginning with the day on which this notice is given, [*I/we*] will be entitled to make an application to the court for:
    a. an order under paragraph 44(2) of the Code requiring you to restore the Land; or
    b. an order under paragraph 44(4) of the Code enabling [*me/us*] to recover the cost of restoring the Land.

NOTES FOR COMPLETING THIS NOTICE

You may wish to obtain independent legal advice before completing this notice.

(a) *This notice should be* ***delivered by hand*** *or sent by* ***registered post*** *or* ***recorded delivery*** *to the operator at:*
- *the address for service that the operator has given to you for the purposes of the Code; or*
- *if no such address has been given to you, at the address given by section 394 of the Communications Act 2003 (available online at* www.legislation.gov.uk).

1 A copy of the Communications Act 2003 is available online at www.legislation.gov.uk.
2 When applied in Scotland, this should be read as a reference to the owner of the Land.

(b) *You only have a right to require the restoration of the Land if a number of conditions are satisfied. These are as follows:*

i. The condition of the Land has been affected by the exercise of a Code right and restoration of the Land to its condition before the Code right was exercised does not involve the removal of electronic communications apparatus from any land;

ii. You are the occupier of the Land, the owner of the freehold estate in the Land (or, in Scotland, the owner of the Land) or the lessee of the Land;

iii. You are not for the time being bound by the Code right (i.e. you have not entered into an agreement with the Code operator regarding the conferral of the Code right, or had such an agreement imposed on you by the courts).

Further, even if you satisfy the conditions set out above, you will not have a right to require the restoration of the Land if:

i. *it is occupied by a person who either:*

a. *conferred a Code right (which is in force) entitling the Code operator to affect the condition of the Land in the same way as the Code right mentioned at paragraph (b) i. above; or*

b. *is otherwise bound by such a right; and*

ii. *that Code right was not conferred in breach of a covenant enforceable by you (or, in Scotland, in breach of a contractual term enforceable by you).*

(c) *The date specified by you as the deadline for completion of the works must be a reasonable one. What is reasonable will depend on the individual circumstances of your case, including the complexity of the works to be undertaken.*

[*Insert date of Notice*]

STATUTORY NOTICE
REGARDING THE EXERCISE OF A TRANSPORT LAND RIGHT UNDER THE ELECTRONIC COMMUNICATIONS CODE TO CARRY OUT NON-EMERGENCY WORKS
Paragraph 49(1) of Part 7 of Schedule 3A of the Communications Act 2003

IMPORTANT NOTICE

If you object to the works proposed in this notice, you should give a notice of objection within 28 days uested in this notice, you should respond within six months

1. This is a statutory notice pursuant to paragraph 49(1) of the electronic communications code, set out in Schedule 3A to the Communications Act 2003 (the '**Code**').[1]
2. We, [*Insert name of Code operator*], are writing to you in your capacity as a transport undertaker (as defined in paragraph 46 of the Code) and, in particular, in respect of land at [*Insert address / description of land, etc.*] (the '**Transport Land**').
3. The purpose of this notice is to inform you that we intend to exercise a transport land right (as defined at paragraph 48(1) of the Code) in order to carry out non-emergency works (as defined in paragraph 49(5) of the Code) relating to a line crossing the Transport Land or any other electronic communications apparatus installed [*on / under / over*] the Transport Land.

BACKGROUND

4. We provide an [*electronic communications network and/or infrastructure system*] in the United Kingdom. For this purpose, the Office of Communications (OFCOM) has given a direction applying the Code to us.
5. The Code regulates the relationships between us and occupiers of land, thereby facilitating the deployment of electronic communications apparatus. Amongst other things, the Code entitles us to exercise transport land rights subject to the provisions of paragraphs 48 to 55 of the Code.
6. Paragraph 48(1) of the Code requires us to provide transport undertakers with a notice of proposed works before exercising a transport land right in order to carry out non-emergency works. This is a notice under paragraph 49(1) of the Code.

THE PROPOSED WORKS

7. We intend to carry out non-emergency works on the Transport Land for or in connection with the exercise of a transport land right.
8. A detailed plan and section of the works is set out in the Annex to this notice. **[see note (a)]**
9. We intend to commence the proposed works on [*Insert date*] [*and expect that they will be completed by [Insert date]*]. **[see note (b)]**

1 A copy of the Communications Act 2003 is available online at www.legislation.gov.uk.

YOUR OPTIONS

10. If you do not object to the proposed works set out at paragraphs 7 and 8 above, you do not have to do anything in response to this notice. We will then proceed to carry out the proposed works on the date(s) specified above.
11. However, if you object to the proposed works, you are entitled to give us a notice informing us of your objection (a '**notice of objection**') within 28 days of the day on which this notice was given to you.[2] Please give any notice of objection to us at the following address:

[*insert address*]

12. Where a notice of objection is given, either of us may, within the period of 28 days beginning with the day on which the notice of objection is given (the '**arbitration notice period**'), give the other notice that the objection is to be referred to arbitration under paragraph 52 of the Code (an '**arbitration notice**').
13. Where an arbitration notice has been given, works will only be permitted in accordance with an award made in the arbitration. However, if the arbitration notice period has ended and no arbitration notice has been given, we will be permitted to carry out the proposed works. This is the case notwithstanding your notice of objection.

[*Insert date of Notice*]

NOTES FOR COMPLETING THIS NOTICE

(a) If the transport undertaker agrees, this notice may instead contain a description of the proposed works (whether or not in the form of a diagram) rather than a plan and section of the works.

(b) Under paragraphs 49(3) and (5) of the Code, Code operators are not entitled to begin their proposed works until after the period of 28 days beginning with the day on which notice of proposed works is given.

2 See, in particular, paragraph 50(2) of the Code.

STATUTORY NOTICE
NOTICE OF OBJECTION TO THE EXERCISE OF A TRANSPORT LAND RIGHT UNDER THE ELECTRONIC COMMUNICATIONS CODE TO CARRY OUT NON-EMERGENCY WORKS
Paragraph 50(2) of Part 7 of Schedule 3A of the Communications Act 2003

1. This is a statutory notice pursuant to paragraph 50(2) of the electronic communications code, set out in Schedule 3A to the Communications Act 2003 (the '**Code**').[1] **[see note (a)]**
2. We, [*Insert name of Transport Undertaker*], have received a notice from you, [*Insert name of Code operator*], pursuant to paragraph 49(1) of the Code. In that notice, dated [*Insert date*], you informed us of your intention to carry out non-emergency works (as defined in paragraph 49(5) of the Code) at [*Insert address / description of land, etc.*] (the '**Proposed Works**').
3. The purpose of this notice is to inform you that we object to the Proposed Works.

CONSEQUENCES OF THIS NOTICE

4. As a result of this notice, either one of us may give the other a notice that the objection is to be referred to arbitration under paragraph 52 of the Code.
5. Any notice of arbitration must be given within the period of 28 days beginning with the day on which this notice of objection has been given (the '**arbitration notice period**').
6. Please note that you may only exercise a transport land right in order to carry out the Proposed Works if **either**:
   a. the arbitration notice period has ended and no arbitration notice has been given by either of us; **or**
   b. an arbitration notice has been given within the arbitration notice period but the works are permitted in accordance with an award made on the arbitration.

[*Insert date of Notice*]

NOTES FOR COMPLETING THIS NOTICE

(a) *This notice should be* ***delivered by hand*** *or sent by* ***registered post*** *or* ***recorded delivery*** *to the operator at:*

- *the address for service that the operator has given to you for the purposes of the Code; or*
- *if no such address has been given to you, at the address given by section 394 of the Comunications Act 2003 (available online at* www.legislation.gov.uk).

1 A copy of the Communications Act 2003 is available online at www.legislation.gov.uk.

STATUTORY NOTICE
REFERRAL TO ARBITRATION OF AN OBJECTION TO PROPOSED WORKS UNDER THE ELECTRONIC COMMUNICATIONS CODE
Paragraph 50(3) of Part 7 of Schedule 3A of the Communications Act 2003

1. This is a statutory notice pursuant to paragraph 50(3) of the electronic communications code, set out in Schedule 3A to the Communications Act 2003 (the '**Code**').[1] **[see note (a)]**
2. We, [*Insert name of Code operator*], recently gave you, [*Insert name of Transport Undertaker*] notice of our intention to carry out non-emergency works at [*Insert address / description of land, etc.*] (the '**Proposed Works**'). You subsequently gave us a notice of objection to the Proposed Works under paragraph 50(2) of the Code.
   [OR – delete appropriate version of paragraph 2]
2. You, [*Insert name of Code operator*], recently gave us, [*Insert name of Transport Undertaker*] notice of your intention to carry out non-emergency works at [*Insert address / description of land, etc.*] (the '**Proposed Works**'). We subsequently gave you a notice of objection to the Proposed Works under paragraph 50(2) of the Code.
3. The purpose of this notice is to inform you that [*our/your*] objection to the Proposed Works is to be referred to arbitration under paragraph 52 of the Code.
4. This means that it will be referred to the arbitration of a single arbitrator[2] appointed **either**:
   a. by agreement between both of us; **or**
   b. in the absence of such agreement, by the President of the Institution of Civil Engineers.
5. [We propose that [*Insert name*] be appointed as arbitrator in respect of this matter. A copy of this individual's relevant qualifications is provided at Annex 1 of this notice. If you do not agree with this individual's appointment as arbitrator, please provide the name of one or more individuals that you consider would be suitable for this role and provide details of their relevant qualifications]. **[see note (b)]**
6. Please note that, as a result of this notice, [*you/we*] will not be permitted to carry out the Proposed Works unless they are permitted in accordance with an award made on the arbitration.

[*Insert date of Notice*]
[ANNEX 1
QUALIFICATIONS OF PROPOSED ARBITRATOR]

NOTES FOR COMPLETING THIS NOTICE

You may wish to obtain independent legal advice before completing this notice.

(a) *This notice should be* ***delivered by hand*** *or sent by* ***registered post*** *or* ***recorded delivery*** *to the Code operator/transport undertaker (as applicable) at:*

1 A copy of the Communications Act 2003 is available online at www.legislation.gov.uk.

2 Paragraph 52(3) of the Code sets out in detail the powers of the arbitrator in relation to this dispute. Paragraph 52(6) of the Code also explains that, in determining what award to make, the matters to which the arbitrator must have regard include the public interest in there being access to a choice of high quality electronic communications services.

- v *address for service that the operator/undertaker has given to you for the purposes of the Code; or*
- *if no such address has been given to you, at the address given by section 394 of the Communications Act 2003 (available online at* www.legislation.gov.uk).

(b) *You are not required to put forward the name of an individual that you think would be suitable to act as the arbitrator. However, the parties should seek to reach agreement on this, where possible.*

To facilitate this, it may be advisable for more than one individual's name to be put forward as the potential arbitrator.

STATUTORY NOTICE
REQUIRING THE PAYMENT OF COMPENSATION FOR LOSS OR DAMAGE SUSTAINED IN CONSEQUENCE OF THE CARRYING OUT OF EMERGENCY WORKS UNDER THE ELECTRONIC COMMUNICATIONS CODE
Paragraph 51(4) of Part 7 of Schedule 3A of the Communications Act 2003

1. This is a statutory notice pursuant to paragraph 51(4) of the electronic communications code, set out in Schedule 3A to the Communications Act 2003 (the '**Code**').[1] **[see note (a)]**
2. On [*Insert Date*], you [*Insert Name of Code operator*], gave us [*Insert name of Transport Undertaker*], notice that you had started exercising a transport land right in order to carry out emergency works within the meaning set out in paragraph 51(9) of the Code relating to electronic communications apparatus at [*Insert address / description of land, etc.*] (the '**Emergency Works**').
3. The purpose of this notice is to require you to pay compensation for loss or damage sustained by us in consequence of the carrying out of the Emergency Works.

DETAILS OF OUR LOSS OR DAMAGE

4. As a consequence of the Emergency Works, [*Please provide a brief summary of the loss or damage sustained by you as a result of those works*].
5. To compensate us for this loss or damage, we request a compensation payment of £[*Insert amount*]. Further details on how we have calculated the amount of this compensation are set out at Annex 1.

CONSEQUENCE OF THIS NOTICE

6. You are required, by paragraph 51(5) of the Code, to pay us any compensation which is required by this notice.
7. The amount of compensation payable under paragraph 51(5) of the Code is to be agreed between us. However, if (within the period of 28 days beginning with the day on which this notice was given) we have not agreed the amount of compensation payable, either of us may give the other notice that the disagreement is to be referred to arbitration under paragraph 52 of the Code.

YOUR OPTIONS

8. We would like to agree the amount of compensation payable by you, without recourse to arbitration. Accordingly, please respond to this notice (within 28 days from the day on which it was given to you) to confirm whether you agree to pay us the compensation referred to at paragraph 5 above. If you do not agree, please explain why and explain what compensation (if any) you think should be payable.
9. Please note that, whilst you are not required to respond to this notice, we will be entitled to refer this matter to arbitration under paragraph 52 of the Code if you fail to do so within the period specified at paragraph 7 above.

[*Insert date of Notice*]

1 A copy of the Communications Act 2003 is available online at www.legislation.gov.uk.

ANNEX 1

AMOUNT OF COMPENSATION SOUGHT

[Please insert details and where relevant, any supporting materials (such as invoices)]

NOTES FOR COMPLETING THIS NOTICE

You may wish to obtain independent legal advice before completing this notice.

(a) *This notice should be* ***delivered by hand*** *or sent by* ***registered post*** *or* ***recorded delivery*** *to the operator at:*

- *the address for service that the operator has given to you for the purposes of the Code; or*
- *if no such address has been given to you, at the address given by section 394 of the Communications Act 2003 (available online at* www.legislation.gov.uk).

Also, this notice should be given within the period of 28 days beginning with the day on which the emergency works notice was given to you by the Code operator. If you give this notice to the operator after this date, it will not be required to pay you any compensation under paragraph 51(5) of the Code.

STATUTORY NOTICE
REQUIRING THE ALTERATION OF ELECTRONIC COMMUNICATIONS APPARATUS UNDER THE ELECTRONIC COMMUNICATIONS CODE
Paragraph 53(1) of Part 7 of Schedule 3A of the Communications Act 2003

1. This is a statutory notice pursuant to paragraph 53(1) of the electronic communications code, set out in Schedule 3A to the Communications Act 2003 (the '**Code**').[1] [see note (a)]
2. The purpose of this notice is to require you, [*Insert name of Code operator*], to alter the electronic communications apparatus described in Annex 1 of this Notice in the manner specified in that Annex. That apparatus is installed at [*Insert address*] (the '**Transport Land**').
3. We, [*Insert name of transport undertaker*], require this alteration on the ground that keeping the apparatus [*on / under / over*] the Transport Land [*interferes with / is likely to interfere with*]: [see note (b)]
   a. [*the carrying on of our [railway / canal / tramway] undertaking*]; or
   b. [*anything [done / to be done] for the purposes of our [railway / canal / tramway] undertaking*]

CONSEQUENCES OF THIS NOTICE

4. You may, within the period of 28 days beginning with the day on which this notice is given to you (the '**notice period**'), give us a notice specifying the respects in which you are not prepared to comply with this notice (a '**counter-notice**').
5. If the notice period has ended and you have not given any counter-notice, you must comply with this notice within a reasonable time and to our reasonable satisfaction.
6. If you decide to give a counter-notice (within the notice period), we may apply to the court for an order requiring you to alter any of the apparatus specified in Annex 1. The court must not make an order unless it is satisfied that it is necessary on one of the grounds mentioned in paragraph 53(1) of the Code.[2]

[*Insert date of Notice*]

ANNEX 1
DETAILS OF ELECTRONIC COMMUNICATIONS APPARATUS AND ALTERATION REQUIRED

1 A copy of the Communications Act 2003 is available online at www.legislation.gov.uk.
2 See paragraphs 53(6) to 53(8) for further details on the matters to which the court must have regard when deciding whether or not to make such an order, as well as the court's powers in making an order.

The electronic communications apparatus to which this notice relates

[*Insert details of the line or other electronic communications apparatus to which this notice relates*]

The alterations required to that apparatus

[*Insert details of the alterations that you require*]

NOTES FOR COMPLETING THIS NOTICE

You may wish to obtain independent legal advice before completing this notice.

(a) *This notice should be* ***delivered by hand*** *or sent by* ***registered post*** *or* ***recorded delivery*** *to the operator at:*
- *the address for service that the operator has given to you for the purposes of the Code; or*
- *if no such address has been given to you, at the address given by section 394 of the Communications Act 2003 (available online at* www.legislation.gov.uk).

(b) *You are only entitled to require the alteration of electronic communications apparatus under paragraph 53(1) of the Code on one or both of the grounds set out above. Please delete as appropriate.*

(c) *Paragraph 108(2) of the Code defines 'alteration' to include the moving, removal or replacement of apparatus.*

STATUTORY NOTICE
NOTICE OF OBJECTION TO THE ALTERATION OF ELECTRONIC COMMUNICATIONS APPARATUS UNDER THE ELECTRONIC COMMUNICATIONS CODE
Paragraph 53(2) of Part 7 of Schedule 3A of the Communications Act 2003

1. This is a statutory notice pursuant to paragraph 53(2) of the electronic communications code, set out in Schedule 3A to the Communications Act 2003 (the '**Code**').[1,2]
2. We, [*Insert name of Code operator*], have received a notice from you, [*Insert name of Transport Undertaker*], pursuant to paragraph 53(1) of the Code (the '**Notice requiring Alterations**'). In that notice, dated [*Insert date*], you required us to alter the electronic communications apparatus described in Annex 1 of that Notice in the manner specified in that Annex. That apparatus is installed at [*Insert address*].
3. The purpose of this notice is to inform you that we are not prepared to comply with the Notice requiring Alterations in the following respects:

[*Insert details of the respects in which you are not prepared to comply with the Notice requiring Alterations*].

CONSEQUENCES OF THIS NOTICE

4. As a result of this notice, you may apply to the court for an order requiring us to alter any of the specified apparatus under paragraph 53(4) of the Code.
5. The court must not make such an order unless it is satisfied that the order is necessary on the ground that keeping the apparatus on, under or over transport land interferes with, or is likely to interfere with, EITHER:
   a. the carrying on of your railway, canal or tramway undertaking; OR
   b. anything done or to be done for the purposes of your railway, canal or tramway undertaking.[3]

[*Insert date of Notice*]

1 A copy of the Communications Act 2003 is available online at www.legislation.gov.uk.
2 This notice must be given to the relevant Transport Undertaker within the period of 28 days beginning with the day on which the Notice requiring Alterations was given. If you do not give this notice before the expiry of that time period, then you are required by paragraph 53(3) of the Code to comply with the Notice requiring Alterations within a reasonable time and to the reasonable satisfaction of the Transport Undertaker.
3 Paragraphs 53(6) to 53(8) of the Code provide further details on the court's powers to impose such an order, and on the matters to which it must have regard.

STATUTORY NOTICE
REGARDING THE TERMINATION OF TRANSPORT LAND RIGHTS UNDER THE ELECTRONIC COMMUNICATIONS CODE
Paragraph 54(7) of Part 7 of Schedule 3A of the Communications Act 2003

1. This is a statutory notice pursuant to paragraph 54(7) of the electronic communications code, set out in Schedule 3A to the Communications Act 2003 (the '**Code**'). [see note (a)]
2. You, [*Insert name of Code operator*], have been exercising a transport land right (as defined in paragraph 48 of the Code) in relation to land occupied by us, [*Insert name of Transport Undertaker*]. This land is at [*Insert address / description of land, etc.*] (the '**Land**').
3. The Land has ceased to be transport land (as defined in paragraph 46 of the Code).
4. The purpose of this notice is to inform you that, from [*Insert date*], Part 7 of the Code is to cease to apply to the Land in accordance with paragraph 54 of the Code. This means that the transport land rights which are currently exercisable in relation to the Land will cease to be exercisable. [see notes (b) and (c)]

[*Insert date of Notice*]

NOTES FOR COMPLETING THIS NOTICE

You may wish to obtain independent legal advice before completing this notice.

(a) *This notice should be* ***delivered by hand*** *or sent by* ***registered post*** *or* ***recorded delivery*** *to the operator at:*

- *the address for service that the operator has given to you for the purposes of the Code; or*
- *if no such address has been given to you, at the address given by section 394 of the Communications Act 2003 (available online at* www.legislation.gov.uk).

This notice may only be given ***after*** *the land ceases to be transport land.*

Also, the notice only needs to be given if the Code operator was exercising a transport land right in relation to the land immediately before the time when it ceased to be transport land.

(b) *Part 7 of the Code makes provision about the conferral, and exercise, of transport land rights. In particular, it provides that Code operators may exercise certain rights in respect of transport land and that some of these rights will continue to be exercisable even if the land ceases to be transport land.*

In particular, if a Code operator is exercising a transport land right in relation to land immediately before a time when it ceases to be transport land, paragraph 54(2) of the Code provides that Part 7 of the Code will continue to apply to the land as if it were still transport land (and, accordingly, the Code operator may continue to exercise any transport land right in relation to the land as if it were still transport land).

However, paragraphs 54(4) to (9) of the Code set out a number of exceptions to this. One such exception is where the occupier of the land gives the Code operator a notice of termination under paragraph 54(7) of the Code. This standard form notice is intended to be such a notice of termination.

(c) *Under paragraph 54(8) of the Code, the date specified in this notice of termination must fall after the end of the period of 12 months beginning with the day on which this notice is given.*

For example, if this notice is given on 1 November 2017, the earliest date which may be specified in this notice is 2 November 2018.

STATUTORY NOTICE
REGARDING NON-EMERGENCY UNDERTAKER'S WORKS WHICH INTERFERE WITH AN ELECTRONIC COMMUNICATIONS NETWORK
Paragraph 67(1) of Part 10 of Schedule 3A of the Communications Act 2003

1. This is a statutory notice pursuant to paragraph 67(1) of the electronic communications code, set out in Schedule 3A to the Communications Act 2003 (the '**Code**').[1] **[see note (a)]**
2. The purpose of this notice is to inform you that we, [*Insert name of undertaker*] intend to carry out non-emergency works at [*Insert address, and any other relevant description*] (the '**Land**'). **[see note (b)]**
3. These works will involve, or are likely to involve, an alteration of electronic communications apparatus kept by you, [*Insert name of Code operator*] on, under or over the Land for the purposes of your network.

THE PROPOSED WORKS AND ALTERATIONS

4. We propose to [*Insert details of the proposed non-emergency works, with more detail in an annex if necessary*] (the '**Proposed Works**'). These works will begin on [*Insert date*] at [*the location specified in paragraph 2 above*].
5. The Proposed Works [*will involve / are likely to involve*] the alteration of the electronic communications apparatus specified in the Annex to this notice in the manner specified in that Annex. This alteration would be [*permanent / temporary*].

YOUR OPTIONS

6. You are entitled, within the period of 10 days beginning with the day on which this notice is given (the '**notice period**'), to give us a counter-notice under paragraph 68(2) of the Code. Such a counter-notice can state either:
   a. that you require us to make any alteration of the electronic communications apparatus that is necessary or expedient because of the Proposed Works under your supervision and to your satisfaction; or
   b. that you intend to make any alteration of the electronic communications apparatus that is necessary or expedient because of the Proposed Works.[2]
7. We must not begin the proposed works (including the proposed alteration of your electronic communications apparatus) until the notice period has ended. **[see note (c)]**
8. Further, if you give us a counter-notice within the notice period, paragraphs 68(3) to (6) of the Code will apply.
9. Even if you do not provide us with a counter-notice within the notice period, we are required by paragraph 69 of the Code to pay you the amount of any loss or

1 A copy of the Communications Act 2003 is available online at www.legislation.gov.uk.

2 Under paragraphs 69(2) and 70(2)(b) of the Code, we would be required to pay you the amount of any loss or damage sustained by you in consequence of the alteration referred to in this notice. Further, if you choose to supervise us when altering the electronic communications apparatus, we are required by paragraph 69(3) of the Code to pay you any expenses incurred by you in, or in connection with, that supervision. Similarly, if you choose to make the alterations yourself, we are required under paragraph 70(2)(a) to pay you any expenses incurred by you in, or in connection with, making the alteration.

damage sustained by you in consequence of any alteration being made to your electronic communications apparatus (in carrying out the Proposed Works).

[*Insert date of Notice*]

ANNEX 1
THE APPARATUS AND THE PROPOSED ALTERATION

The electronic communications apparatus

[*Insert details of the electronic communications apparatus to which the alteration will relate*]

The proposed alteration(s)

[*Insert details of the alterations which will, or are likely, to be made to the electronic communications apparatus described above as a result of the Proposed Works*]

NOTES FOR COMPLETING THIS NOTICE

You may wish to obtain independent legal advice before completing this notice.

(a) *This notice should be* ***delivered by hand*** *or sent by* ***registered post*** *or* ***recorded delivery*** *to the operator at:*
- *the address for service that the operator has given to you for the purposes of the Code; or*
- *if no such address has been given to you, at the address given by section 394 of the Communications Act 2003 (available online at* www.legislation.gov.uk).

Please note that you will be guilty of an offence if you make an alteration of electronic communications apparatus in carrying out non-emergency undertaker's works without giving this notice.

(b) *This notice is only suitable when providing notice of non-emergency undertaker's works. These works are defined at paragraph 67(5) of the Code as any undertaker's works (defined at paragraph 66(1) of the Code) which are not emergency works. Emergency undertaker's works are defined in paragraph 71 of the Code as follows:*

'undertaker's works carried out in order to stop anything already occurring, or to prevent anything imminent from occurring, which is likely to cause:

(a) *danger to persons or property;*

(b) *interference with the exercise of any functions conferred or imposed on the undertaker by or under any enactment; or*

(c) *substantial loss to the undertaker,*

and any other works which it is reasonable (in all the circumstances) to carry out with those works.'

(c) *Please note that you will be guilty of an offence under paragraph 72 of the Code if you make an alteration of electronic communications apparatus in carrying out non-emergency undertaker's works before the end of the notice period.*

(d) *Paragraph 108(2) of the Code defines 'alteration' to include the moving, removal or replacement of apparatus.*

STATUTORY NOTICE
REGARDING THE CARRYING OUT OF EMERGENCY UNDERTAKER'S WORKS UNDER THE ELECTRONIC COMMUNICATIONS CODE
Paragraph 71(2) of Part 10 of Schedule 3A of the Communications Act 2003

1. This is a statutory notice pursuant to paragraph 71(2) of the electronic communications code, set out in Schedule 3A to the Communications Act 2003 (the '**Code**'). [see note (a)]
2. We, [*Insert name of undertaker*], are writing to you in your capacity as a Code operator and, in particular, in respect of electronic communications apparatus kept by you [*on / under / over*] land at [*Insert address / description of land, etc.*] (the '**Apparatus**').
3. The purpose of this notice is to inform you that we have begun carrying out emergency undertaker's works (as defined in paragraph 71 of the Code) and, in carrying out those works, [*have made / have commenced making / will make*] an alteration to the Apparatus in accordance with paragraph 71(1) of the Code.
4. Further details on the alteration that we [*have made / have commenced making / will make*] are set out in the Annex to this notice.

YOUR RIGHTS

5. We are required by paragraph 71(4) of the Code to make the alteration to your satisfaction. [*If you are not satisfied with the alterations made to the Apparatus / If you would like to supervise us in making these Alterations*[1]] [see note (b)], please let us know by contacting:

   [*Insert contact details*].
6. We are also required by paragraph 71(5)(b) to pay you the amount of any loss or damage sustained by you in consequence of the alteration being made. If you sustain any loss or damage in consequence of the alteration being made, please provide details to the contact referred to in the paragraph above.

[*Insert date of Notice*]

ANNEX
THE ALTERATIONS

[*Insert details of the relevant alterations. These can be provided by descriptive text and/or diagrams*].

NOTES FOR COMPLETING THIS NOTICE

You may wish to obtain independent legal advice before completing this notice.

(a) *This notice should be* ***delivered by hand*** *or sent by* ***registered post*** *or* ***recorded delivery*** *to the operator at:*
- *the address for service that the operator has given to you for the purposes of the Code; or*

1 If you would like to supervise us in making these alterations, we are required by paragraph 71(5)(a) of the Code to pay you any expenses incurred in, or in connection with, that supervision.

- *if no such address has been given to you, at the address given by section 394 of the Communications Act 2003 (available online at* www.legislation.gov.uk).

This notice is only suitable for emergency undertaker's works and must be given to the relevant Code operator ***as soon as practicable*** *after those works are started. Emergency undertaker's works are defined in paragraph 71 of the Code as follows:*

'undertaker's works carried out in order to stop anything already occurring, or to prevent anything imminent from occurring, which is likely to cause:

(a) *danger to persons or property;*
(b) *interference with the exercise of any functions conferred or imposed on the undertaker by or under any enactment; or*
(c) *substantial loss to the undertaker,*

and any other works which it is reasonable (in all the circumstances) to carry out with those works.'

Please note that an undertaker, or an agent of an undertaker, is guilty of an offence if that person makes an alteration of electronic communications apparatus in carrying out emergency undertaker's works and does so without notice of those works having been given in accordance with paragraph 71 of the Code.

(b) *Please choose appropriate text depending on whether or not the alterations have already been completed. If they have not been completed, you should provide the operator with the opportunity to supervise them being made.*
(c) *Paragraph 108(2) of the Code defines 'alteration' to include the moving, removal or replacement of apparatus.*

STATUTORY NOTICE
REGARDING THE INSTALLATION OF ELECTRONIC COMMUNICATIONS APPARATUS UNDER THE ELECTRONIC COMMUNICATIONS CODE
Paragraph 75(2) of Part 11 of Schedule 3A of the Communications Act 2003

1. This is a statutory notice pursuant to paragraph 75(2) of the electronic communications code, set out in Schedule 3A to the Communications Act 2003 (the '**Code**'). [see note (a)]
2. We, [*Insert name of Code operator*], have installed [*this electronic communications apparatus / electronic communications apparatus that is directly or indirectly connected to this piece of apparatus*] for the purposes of our network (the '**Apparatus**'). We are required to provide this notice because the Apparatus is at a height of three or more metres above the ground.

RIGHT TO OBJECT

3. You **will** have a right to object to the Apparatus under paragraph 77 of the Code if BOTH:
   a. you are an occupier of, or have an interest in, any land neighbouring the land on or over which the Apparatus is kept; AND
   b. because of the nearness of that neighbouring land to the land on or over which the Apparatus is kept:
      i. the enjoyment of that neighbouring land is capable of being prejudiced by the Apparatus; or
      ii. any interest in that land is capable of being prejudiced by the Apparatus.
4. You **will not** however have a right to object, even if the criteria referred to at paragraph 3 above are satisfied, if the Apparatus:
   a. replaces any electronic communications apparatus which is not substantially different from the Apparatus; and
   b. is not in a significantly different position.
5. If you have a right to object to the Apparatus, you may exercise that right by giving us notice of your objection under paragraph 78 of the Code. Any such notice should be provided in writing to:

[*Insert appropriate UK postal address*]

6. Any notice of objection should be provided as soon as practicable. Please note that your right to object, and the procedure that applies to that right, will depend on whether you provide notice of your objection within 12 months of installation of the Apparatus, or after. The circumstances in which a court may uphold an objection are significantly more limited if the Apparatus has been installed for over 12 months.[1]

[*Insert date of Notice*]

NOTES FOR COMPLETING THIS NOTICE

This notice should be given before the end of the period of three days beginning with the day after that on which the relevant installation is completed.

1 See paragraphs 79 and 80 of the Code for more detail on your right to object and the procedure that will apply in respect of it.

It must be attached, in a secure and durable manner, to every major item of apparatus installed or, if no major item of apparatus is installed, to the nearest major item of electronic communications apparatus to which the apparatus that is installed is directly or indirectly connected. It must also be attached in a position where it is reasonably legible.

STATUTORY NOTICE
NOTICE OF OBJECTION TO ELECTRONIC COMMUNICATIONS APPARATUS KEPT ON, UNDER OR OVER TIDAL WATER OR LAND UNDER THE ELECTRONIC COMMUNICATIONS CODE
Paragraph 78(1) of Part 12 of Schedule 3A of the Communications Act 2003

1. This is a statutory notice pursuant to paragraph 78(1) of the electronic communications code, set out in Schedule 3A to the Communications Act 2003 (the '**Code**').[1] [see note (a)]
2. The purpose of this notice is to inform you, [*Insert name of Code operator*], that [*I/ we*], [*Insert name of objector(s)*], object to electronic communications apparatus installed by you on, under or over [*tidal water / tidal lands*]. [see note (b)]

THE APPARATUS

3. The apparatus to which this notice relates is kept [*on / under / over*] [*tidal water / tidal lands*] at [*Insert address / description of location of tidal water or lands*] (the '**Apparatus**').
4. [A map which identifies the approximate location of the Apparatus and the [*tidal water / tidal lands*] [*on / under / over*] which the Apparatus is kept is provided in the Annex to this notice.] [see note (c)]
5. [*I/We*] *[[am/are] an occupier of / have an interest in]* that tidal water or lands.

MY OBJECTION

6. [I/We] object to the Apparatus on the ground that it materially prejudices [*my/our*] enjoyment of, or interest in, the tidal water or lands. In particular, [*explain why the Apparatus materially prejudices your enjoyment of, or interest in, the tidal water or lands and in what way the Apparatus should be altered*].

   [OR – delete appropriate version of paragraph 6] [see note (d)]
6. [*I/We*] object to the Apparatus on the ground that it is necessary to alter the Apparatus to enable [*me/us*] to carry out a proposed improvement of the tidal water or lands. In particular, [*explain what proposed improvements you would like to make, why these are necessary and in what way the Apparatus should be altered*].
7. For all future correspondence on this matter, please contact me:

[*Insert relevant contact details (post, phone, email)*].

[*Insert date of Notice*]

[ANNEX
APPROXIMATE LOCATION OF THE APPARATUS
[*Insert map and mark, as closely as possible, where the Apparatus is located*].]

NOTES FOR COMPLETING THIS NOTICE

You may wish to obtain independent legal advice before completing this notice.

(a) *This notice should be* ***delivered by hand*** *or sent by* ***registered post*** *or* ***recorded delivery*** *to the operator at:*

1 A copy of the Communications Act 2003 is available online at www.legislation.gov.uk.

- *the address for service (if any) that the operator has given to you for the purposes of the Code; or*
- *if no such address has been given to you, at the address given by section 394 of the Communications Act 2003 (available online at* www.legislation.gov.uk).

Please note that you will also need to make a separate application to the court (in addition to giving this notice) if you want to have your objection upheld. You can apply to the court to have your objection upheld after the end of the period of two months beginning with the date on which this notice is given, but before the end of the period of four months beginning with that date.

For example, if you give your notice of objection to the Code operator on 1 November 2017, you will only be entitled to apply to the court to have your objection upheld between 2 January 2017 and 1 March 2018.

(b) *You will only have a right to object to the Apparatus, under paragraph 77(1) of the Code, if:*
   a. *you are an occupier, or have an interest in, the tidal water or lands;*
   b. *you are not bound by a Code right enabling the Code operator to keep the Apparatus installed on, under or over the tidal water or lands; and*
   c. *you are not a person with the benefit of a Crown interest in the tidal water or lands.*

You **will not** however have a right to object, even if the above criteria are satisfied, if the Apparatus:

a. replaces any electronic communications apparatus which is not substantially different from the Apparatus; and
b. is not in a significantly different position.

(c) *You are not required to provide a map showing the approximate location. However, you may find this helpful, particularly if you think that the Code operator may find it difficult to identify the relevant apparatus from the address / description given in paragraph 3 above.*

(d) *If your objection is made* **within 12 months** *of the Apparatus being installed, you should choose the first version of paragraph 6. This is because the court must be satisfied, amongst other things, that the Apparatus appears materially to prejudice your enjoyment of, or interest in, the tidal water or lands before it can uphold your objection. See paragraphs 79(4) and (5) for the conditions which must be met before the court will uphold your objection.*

However, if your objection is made **later than 12 months** *after the Apparatus was installed, you should choose the second version of paragraph 6. This is because the court must be satisfied, amongst other things, that the Apparatus needs to be altered to enable you to carry out a proposed improvement of the tidal water or lands before it can uphold your objection. See paragraph 80(3) for the conditions which must be met before the court may uphold your objection.*

(e) *Paragraph 108(2) of the Code defines 'alteration' to include the moving, removal or replacement of apparatus.*

(f)

STATUTORY NOTICE

NOTICE OF OBJECTION TO A LINE INSTALLED OVER LAND UNDER THE ELECTRONIC COMMUNICATIONS CODE

Paragraph 78(1) of Part 12 of Schedule 3A of the Communications Act 2003

1. This is a statutory notice pursuant to paragraph 78(1) of the electronic communications code, set out in Schedule 3A to the Communications Act 2003 (the '**Code**').[1] [see note (a)]
2. The purpose of this notice is to inform you, [*Insert name of Code operator*], that [*I/we*], [*Insert name of objector(s)*], object to a line installed over land by you pursuant to paragraph 74 of the Code. [see note (b)]

THE APPARATUS

3. The line to which this notice relates has been installed over land at [*Insert address*] (the '**Line**').
4. [A map which identifies the approximate location of the Line and the land over which it has been installed is provided in the Annex to this notice.] [see note (c)]
5. [*I/We*] [[*am/are*] *the occupier of / have an interest in*] that land.

MY OBJECTION

6. [*I/We*] object to the Line on the ground that it materially prejudices [*my/our*] enjoyment of, or interest in, the land. In particular, [*explain why the Line materially prejudices your enjoyment of, or interest in, the land and in what way the Line should be altered*].

   [OR – delete appropriate version of paragraph 6] [see note (d)]

6. [*I/We*] object to the Line on the ground that it is necessary to alter the Line to enable [*me/us*] to carry out a proposed improvement of the land. In particular, [*explain what proposed improvements you would like to make, why these are necessary and in what way the Line should be altered*].
7. For all future correspondence on this matter, please contact [*me/us*]:

[*Insert relevant contact details (post, phone, email)*].

[*Insert date of Notice*]

[ANNEX

APPROXIMATE LOCATION OF THE LINE

[*Insert map and mark, as closely as possible, where the Line is located*].]

NOTES FOR COMPLETING THIS NOTICE

You may wish to obtain independent legal advice before completing this notice.

(a) *This notice should be* ***delivered by hand*** *or sent by* ***registered post*** *or* ***recorded delivery*** *to the operator at:*

- *the address for service that the operator has given to you for the purposes of the Code; or*

1 A copy of the Communications Act 2003 is available online at www.legislation.gov.uk.

- *if no such address has been given to you, at the address given by section 394 of the Communications Act 2003 (available online at* www.legislation.gov.uk).

Please note that you will also need to make a separate application to the court (in addition to giving this notice) if you want to have your objection upheld. You can apply to the court to have your objection upheld after the end of the period of two months beginning with the date on which this notice is given, but before the end of the period of four months beginning with that date.

For example, if you give your notice of objection to the Code operator on 1 November 2017, you will only be entitled to apply to the court to have your objection upheld between 2 January 2017 and 1 March 2018.

(b) *You will only have a right to object to the Line, pursuant to paragraph 77(3) of the Code, if:*
   a. *you are an occupier, or have an interest in, the land over which it has been installed; and*
   b. *you are not bound by a Code right enabling the Code operator to keep the Line installed over the land.*

You **will not** however have a right to object, even if the above criteria are satisfied, if the Line:
   a. replaces any line which is not substantially different from the Line; and
   b. is not in a significantly different position.

(c) *You are not required to provide a map showing the approximate location. However, you may find this helpful, particularly if you think that the Code operator may find it difficult to identify the relevant apparatus from the address / description given in paragraph 3 above.*

(d) *If your objection is made* **within 12 months** *of the Line being installed, you should choose the first version of paragraph 6. This is because the court must be satisfied, amongst other things, that the Line appears materially to prejudice your enjoyment of, or interest in, the land before it can uphold your objection. See paragraphs 79(4) and (5) for the conditions which must be met before the court will uphold your objection.*

However, if your objection is made **later than 12 months** *after the Line was installed, you should choose the second version of paragraph 6. This is because the court must be satisfied, amongst other things, that the Line needs to be altered to enable you to carry out a proposed improvement of the land before it can uphold your objection. See paragraph 80(3) for the conditions which must be met before the court may uphold your objection.*

(e) *Paragraph 108(2) of the Code defines 'alteration' to include the moving, removal or replacement of apparatus.*

STATUTORY NOTICE
NOTICE OF OBJECTION TO ELECTRONIC COMMUNICATIONS APPARATUS KEPT ON OR OVER LAND UNDER THE ELECTRONIC COMMUNICATIONS CODE
Paragraph 78(1) of Part 12 of Schedule 3A of the Communications Act 2003

1. This is a statutory notice pursuant to paragraph 78(1) of the electronic communications code, set out in Schedule 3A to the Communications Act 2003 (the '**Code**').[1] **[see note (a)]**
2. The purpose of this notice is to inform you, [*Insert name of Code operator*], that [*I/we*], [*Insert name of objector(s)*], object to electronic communications apparatus kept by you on or over land for the purposes of your network. **[see note (b)]**

THE APPARATUS

3. The apparatus to which this notice relates has been installed [*on / over*] land at [*Insert address*] (the '**Apparatus**').
4. [*I/we*] [*occupy / have an interest in*] neighbouring land at [*Insert address*] (the '**neighbouring land**')
5. [A map which identifies the approximate location of the Apparatus, the land [*on / over*] which the Apparatus is kept, and the neighbouring land is provided in the Annex to this notice.] **[see note (c)]**

MY OBJECTION

6. [*I/we*] object to the Apparatus on the ground that it materially prejudices [*my/our*] enjoyment of, or interest in, the neighbouring land. In particular, [*explain why the Apparatus materially prejudices your enjoyment of, or interest in, the neighbouring land and in what way the Apparatus should be altered*].

    [OR – delete appropriate version of paragraph 6] **[see note (d)]**

6. [*I/we*] object to the Apparatus on the ground that it is necessary to alter the Apparatus to enable [*me/us*] to carry out a proposed improvement of the neighbouring land. In particular, [*explain what proposed improvements you would like to make, why these are necessary and in what way the Apparatus should be altered*].
7. For all future correspondence on this matter, please contact [*me/us*]:

[*Insert relevant contact details (post, phone, email)*].

[*Insert date of Notice*]

[ANNEX

APPROXIMATE LOCATION OF THE APPARATUS

[*Insert map and mark, as closely as possible, where the Apparatus is located*].]

1 A copy of the Communications Act 2003 is available online at www.legislation.gov.uk.

NOTES FOR COMPLETING THIS NOTICE

You may wish to obtain independent legal advice before completing this notice.

(a) *This notice should be* ***delivered by hand*** *or sent by* ***registered post*** *or* ***recorded delivery*** *to the operator at:*
- *the address for service that the operator has given to you for the purposes of the Code; or*
- *if no such address has been given to you, at the address given by section 394 of the Communications Act 2003 (available online at* www.legislation.gov.uk).

Please note that you will also need to make a separate application to the court (in addition to giving this notice) if you want to have your objection upheld. You can apply to the court to have your objection upheld after the end of the period of two months beginning with the date on which this notice is given, but before the end of the period of four months beginning with that date.

For example, if you give your notice of objection to the Code operator on 1 November 2017, you will only be entitled to apply to the court to have your objection upheld between 2 January 2017 and 1 March 2018.

(b) *You will only have a right to object to the Apparatus, under paragraph 77(5) of the Code, if:*
  a. *the whole or any part of the Apparatus is at a height of three metres or more above the ground;*
  b. *you are an occupier, or have an interest in, any land neighbouring the land on which the apparatus is kept; and*
  c. *because of the nearness of the neighbouring land to the land on or over which the Apparatus is kept:*
    i. *the enjoyment of the neighbouring land is capable of being prejudiced by the Apparatus; or*
    ii. *any interest in that land is capable of being prejudiced by the Apparatus.*

You **will not** however have a right to object, even if the above criteria are satisfied, if the Apparatus:
  a. *replaces any electronic communications apparatus which is not substantially different from the Apparatus; and*
  b. *is not in a significantly different position.*

(c) *You are not required to provide a map showing the approximate location. However, you may find this helpful, particularly if you think that the Code operator may find it difficult to identify the relevant apparatus from the address / description given in paragraph 3 above.*

PLEASE SEE OVERLEAF

NOTES FOR COMPLETING THIS NOTICE (CONTINUED)

(d) *If your objection is made* ***within 12 months*** *of the Apparatus being installed, you should choose the first version of paragraph 6. This is because the court must be satisfied, amongst other things, that the Apparatus appears materially* *to prejudice your enjoyment of, or interest in, the neighbouring land before it can uphold your objection. See paragraphs 79(4) and (5) of the Code for the conditions which must be met before the court will uphold your objection.*

However, if your objection is made ***later than 12 months*** *after the Apparatus was installed, you should choose the second version of paragraph 6. This is because the court must be satisfied, amongst other things, that the Apparatus needs to be altered to enable you to carry out a proposed improvement of the neighbouring land before it can uphold your objection. See paragraph 80(3) of the Code for the conditions which must be met before the court may uphold your objection.*

(e) *Paragraph 108(2) of the Code defines 'alteration' to include the moving, removal or replacement of apparatus.*

STATUTORY NOTICE
REQUIRING [*A TREE TO BE LOPPED / VEGETATION TO BE CUT BACK*]
PURSUANT TO THE ELECTRONIC COMMUNICATIONS CODE
Paragraph 82(3) of Part 13 of Schedule 3A of the Communications Act 2003

IMPORTANT NOTICE

If you object to the works proposed in this notice, you should send a counter notice within 28 days

1. This is a statutory notice pursuant to paragraph 82(3) of the electronic communications code, set out in Schedule 3A to the Communications Act 2003 (the '**Code**').[1]
2. We [*Insert name of Code operator*] are giving you, [*Insert name*], this notice because we understand that you are the occupier of land at [*Insert address and any other relevant details*] (the '**Land**').
3. We are concerned that [*a tree / vegetation*] growing on the Land is overhanging a [*street / road*] and [*obstructing / will or may obstruct / interferes with / will or may interfere with*] electronic communications apparatus [*installed / about to be installed*] by us and which is [*used / to be used*] for the purposes of our network. Further details on the [*tree / vegetation*] to which this notice relates are set out in the Annex to this notice.
4. The purpose of this notice is to require you to [*lop the tree / cut back the vegetation*] to prevent the [*obstruction / interference*] referred to above. We ask that you do this on or before [*Insert date*].

YOUR OPTIONS

5. In response to this notice, you may:
   a. comply with this notice and therefore [*lop the tree / cut back the vegetation*] by the deadline specified above;[2]
   b. within the period of 28 days beginning with the day on which this notice is given, give us a counter-notice under paragraph 82(4) of the Code objecting to the [*lopping of the tree / cutting back of the vegetation*]. If you do submit such a notice, we will only be entitled to [*lop the tree / cut back the vegetation*] in pursuance of an order of the court; or **[see note (a)]**
   c. do nothing. If you do nothing, we will be entitled to cause the [*tree to be lopped / vegetation to be cut back*] after the expiry of the 28-day period referred to above (i.e. without an order of the court).[3]

1 A copy of the Communications Act 2003 is available online at www.legislation.gov.uk.

2 If you do so and sustain any loss or damage or incur any expenses as a result, you will be entitled under paragraph 82(9) of the Code to apply to the court for an order requiring us to pay such compensation as the court thinks fit.

3 Paragraph 82(7) of the Code requires us to carry out any such works in a husband-like manner and in such a way as to cause the minimum damage to the tree or vegetation. This is also the case if we obtain an order of the court entitling us to carry out the works ourselves. Where we cause the tree to be lopped or vegetation to be cut back and you sustain any loss or damage as a result, you will be entitled under paragraph 82(9) of the Code to apply to the court for an order requiring us to pay such compensation as the court thinks fit.

6. Please submit any counter-notice pursuant to paragraph 5b. above to us in writing as soon as possible and, in any event, before the end of 28 days beginning with the day on which this notice is given.
7. To be effective, such notification must be **delivered by hand** or sent by **registered post** or **recorded delivery** to the following address:

   [*Insert address details*]
8. If you have any questions about this notice, please do not hesitate to contact us via telephone (*Insert number*) or e-mail (*insert email address*).

[*INSERT DATE OF NOTICE*]

ANNEX

THE [*TREE / VEGETATION*] TO WHICH THIS NOTICE RELATES

[*Insert further details, such as a map showing the precise location of the tree/vegetation and the precise works which you consider need to be carried out in order to prevent the obstruction or interference with your apparatus*].

Appendix F
OFCOM Standard Terms[1]

STATEMENT:
Publication Date: 15 December 2017

About this document

This document contains Standard Terms which may (but need not) be used by Code operators and landowners or occupiers when negotiating agreements to confer Code rights.

Contents

Section

1. Standard Terms

[Name of Grantor]
and
[Name of Operator]
CODE AGREEMENT
relating to the installation of electronic communications apparatus at [*address*]

This Agreement is made on [Insert date] between:

i) [Insert name], whose [address/registered office] is at [Insert address] (the 'Grantor'); and where applicable, whose Company Registration Number is [insert CRN]; and
ii) [Insert name], whose registered office is at [Insert address] (the 'Operator'), and whose Company Registration Number is [insert CRN].

RECITALS

a) The Code (as defined in clause 1) facilitates the deployment of electronic communications apparatus by persons in whose case it is applied.
b) The Code has been applied to the Operator by virtue of a direction under section 106 of the Communications Act 2003.
c) The Grantor is the occupier of certain land.

1 See Sections 41.11 and 41.12 for a commentary on the OFCOM Standard Terms.

d) This Agreement is an agreement pursuant to paragraph 9 of Part 2 of the Code. It sets out the contractual basis upon which the Grantor is willing to confer code rights in respect of that land on the Operator.

IT IS AGREED AS FOLLOWS:

1. DEFINITIONS AND INTERPRETATION

1.1. In this Agreement, the following words shall have the following meanings: 'Act' means the Communications Act 2003;

'**Apparatus**' means the Electronic Communications Apparatus described in Schedule 1, and shall be deemed to include any future alterations to or upgrades of the Apparatus that are made in accordance with this Agreement and/or pursuant to the Code;

'**Code**' means the electronic communications code set out at Schedule 3A to the Act;

'**Code Rights**' means the rights set out at clause [2.1];

'**Electronic Communications Apparatus**' shall have the meaning ascribed to that term in paragraph 5

of the Code;

'**Land**' means the land at [insert address] [and marked on the plan attached at Schedule 2];

'**Term**' means the period of time during which this Agreement is in force, which shall be a period starting on [*insert date*] and ending on [*insert date*].

1.2. In this Agreement, unless expressly stated otherwise:

a) a reference to either party includes that party's employees, agents, contractors and sub- contractors;
b) a reference to any statute or statutory provision includes that statute or statutory provision as amended, re-enacted, consolidated or replaced;
c) a reference to a clause or schedule is to the relevant clause or schedule of this Agreement;
d) words importing the singular shall include the plural, and vice versa.

2. RIGHTS OF THE OPERATOR

2.1 In consideration of the covenants set out at clause [4] of this Agreement [and payment of the sum set out at clause [3]], the Grantor hereby agrees that the Operator shall have the right for the Term:

a) to install the Apparatus on, under or over the Land;
b) to keep installed the Apparatus which is on, under or over the Land;
c) to inspect, maintain, adjust, alter, repair, operate or (subject to clause [2.3]) upgrade the Apparatus which is on, under or over the Land;
d) to carry out any works on the Land for or in connection with the installation of the Apparatus on, under or over the Land [or the installation of Electronic Communications Apparatus elsewhere];
e) to carry out any works on the Land for or in connection with the maintenance, adjustment, alteration, repair, operation or (subject to clause [2.3], the upgrading of the Apparatus which is on, under or over the Land [or any Electronic Communications Apparatus elsewhere];

f) to enter the Land to inspect, maintain, adjust, alter, repair, operate or (subject to clause [2.3]) upgrade the Apparatus which is on, under or over the Land [or any Electronic Communications Apparatus elsewhere];
g) to connect the Apparatus to a power supply;
h) to interfere with or obstruct a means of access to or from the Land (whether or not the Apparatus is on, under or over the Land);
i) to lop or cut back, or require another person to lop or cut back, any tree or vegetation that interferes or will or may interfere with [any Electronic Communications Apparatus/the Apparatus],

(together, the '**Code Rights**').

2.2. Subject to clause [2.3], the Operator may also share the use of the Apparatus with another Operator, and carry out any works to the Apparatus to enable that sharing to take place.

2.3. The Operator may only upgrade or share the Apparatus (and exercise the associated rights set out in clauses [2.1(e), 2.1(f) and 2.2]) if:
a) any changes to the Apparatus as a result of the upgrading or sharing have no adverse impact, or no more than a minimal adverse impact, on its appearance;
b) the upgrading or sharing does not impose any additional burden on the Grantor, including:
   i) anything that has an additional adverse effect on the Grantor's enjoyment of the Land; or
   ii) anything that causes additional loss, damage or expense to the Grantor.

2.4. The right of entry set out in clause [2.1(f)] may be exercised by the Operator with or without workmen, vehicles (where appropriate), plant equipment or machinery.

3. PAYMENT

The Operator agrees that it will pay to the Grantor, in respect of this Agreement, the sum of [Insert amount] pounds (£[Insert amount]) [per annum / for the Term].]

4. OPERATOR'S OBLIGATIONS

4.1. The Operator covenants with the Grantor that it will:
a) save in the event of an [emergency[2]], give the Grantor not less than seven days' prior written notice of its intention to enter the Land;
b) in the event of an [emergency], seek to contact the Grantor (which may be by electronic or verbal communication) as soon as reasonably practicable to inform him:
   i) that the Operator intends to enter the Land, or has entered the Land;
   ii) why entry is or was required; and
   iii) when entry took place or is intended to take place.
c) otherwise exercise its right to enter the land in accordance with the access arrangements set out in Schedule [3] to this Agreement;
d) exercise its Code Rights in a proper and workmanlike manner taking all reasonable precautions to avoid obstructions to, or interference with, the use of the Land or any adjoining land and so as to cause as little damage, nuisance

2 The parties may wish to define what constitutes an 'emergency'.

and inconvenience as possible to the Grantor and any occupiers of any adjoining land;

e) do as little physical damage as is reasonably practicable in exercising its Code Rights and, as soon as reasonably practicable, make good to the reasonable satisfaction of the Grantor all resulting damage caused to the Land or any adjoining land;

f) exercise its Code Rights and use and operate the Apparatus in accordance with all applicable legislation;

g) obtain and maintain in force any necessary consents for the installation and retention of the Apparatus, provide evidence of any such consents upon demand to the Grantor and carry out all works in accordance with such consents;

h) maintain and keep the Apparatus in good repair and condition and so as not to be a danger to the Grantor, its employees or property, or the occupiers or any adjoining land;

i) pay all rates or other charges that may be levied in respect of the Apparatus or the exercise of its Code Rights; and

j) maintain insurance with a reputable insurance company against any liability to the public or other third party liability in connection with any injury, death, loss or damage to any persons or property belonging to any third party arising out of the exercise by the Operator of the rights granted under this Agreement, and provide details of such insurance and evidence that it is in force to the Grantor upon reasonable request.

5. THE GRANTOR'S OBLIGATIONS

5.1. The Grantor agrees that it will:

a) not build or place anything on the Land, or permit any third party to do the same, that makes it more difficult for the Operator to access the Apparatus, or which might interfere with the Apparatus, without the Operator's express written consent (which should not be unreasonably withheld);

b) not cause damage to or interfere with the Apparatus or its operation and not permit any third party to do the same; and

c) give reasonable prior written notice to the Operator of any action that it intends to take that would or might affect the continuous operation of the Apparatus, including (but not limited to) causing an interruption to any power supply to which the Apparatus is connected.

6. OWNERSHIP OF THE APPARATUS

The Apparatus shall remain the absolute property of the Operator at all times.

7. GENERAL

7.1. It is agreed that no relationship of landlord and tenant is created by this Agreement between the Grantor and the Operator.

7.2. This Agreement will not apply to any part of the Land which is or (from the date of such adoption) becomes adopted as highway maintainable at the public expense.

7.3. In the event of any inconsistency between this Agreement and any provision of the Code, the Code will prevail.

8. INDEMNITY FOR THIRD PARTY CLAIMS

8.1. The Operator will indemnify the Grantor up to a maximum of [£1 million / £3 million / £5 million] [per annum / in aggregate in respect of a claim or series of claims arising from the same incident] against any third party actions, claims, costs, proceedings or demands ('Third Party Claim') arising as a result of any act or omission by the Operator in exercising its rights under this Agreement, except to the extent that the Grantor's acts or omissions have caused or contributed to any such Third Party Claim and provided that:

a) the Grantor shall as soon as reasonably practicable give notice in writing to the Operator of any Third Party Claim brought, made or threatened against the Grantor;
b) the Grantor shall not compromise or settle such Third Party Claim without the express written consent of the Operator (which shall not be unreasonably withheld or delayed);
c) the Grantor shall permit the Operator to defend any Third Party Claim in the name of the Grantor at the expense of the Operator.

9. LIMITATION OF LIABILITY

9.1. Nothing in this Agreement limits or excludes the liability of either party:

a) for death or personal injury resulting from its negligence;
b) for any damage or liability incurred as a result of fraud or fraudulent misrepresentation by that party; or
c) where or to the extent that it is otherwise prohibited by law.

9.2. Subject to clause [9.1], the Operator's liability under this Agreement to the Grantor shall be limited to the sum of [Insert amount] pounds (£[Insert amount]) [per annum/in aggregate]. This limitation of liability shall not apply to the indemnity granted under clause [8.1].

9.3. Subject to clause [9.1], the Grantor's liability under this Agreement to the Operator shall be limited to the sum of [Insert amount] pounds (£[Insert amount]) [per annum/in aggregate in respect of a claim or series of claims arising from the same incident].

10. TERMINATION

The Grantor may terminate this Agreement[3] by giving the Operator [thirty (30)] days' notice in writing if:

a) the Operator is in [material/substantial] breach of any of its obligations under this Agreement and:
   i) the breach is incapable of remedy; or
   ii) the Operator has failed to remedy the breach within [thirty (30)] days] after the Grantor notifies the Operator of the breach;
b) the Operator has persistently delayed making payments due to the Grantor under the terms of this Agreement;
c) the Grantor intends to redevelop all or part of the Land or any neighbouring land, and could not reasonably do so unless this Agreement comes to an end; or

3 Please note that, if the Grantor exercises any of these termination rights, the agreement will nevertheless continue under paragraph 29 of the Code, unless the Grantor also gives 18 months' notice to terminate under paragraph 30 of the Code.

d) both:
   i) the prejudice caused to the Grantor by the continuation of the Agreement is incapable of being adequately compensated by money; and
   ii) the public benefit likely to result from the continuation of the Agreement does not outweigh the prejudice to the Grantor.

10.3. The Operator may terminate this Agreement by giving the Grantor [x days / months] notice in writing.

10.4. On termination of this Agreement (except where the Agreement continues in accordance with paragraph 29(2) of the Code), the Operator will as soon as reasonably practicable remove the Apparatus from the Land and make good any damage to the Land caused by its removal to the reasonable satisfaction of the Grantor.

11. ASSIGNMENT/ASSIGNATION

11.1. The Operator may assign this Agreement to another Operator who will be bound by its terms with effect from the date of the assignment.

11.2. The Operator will not be liable for any breach of this Agreement occurring on or after the date of the assignment if:
   a) the Grantor is given written notice of the name of the Operator assignee and its address for the purposes of clause [15.2]; and
   b) that notice was given prior to the breach occurring.

12. CONTRACTS (RIGHTS OF THIRD PARTIES) ACT

Unless expressly stated, nothing in this Agreement will create any rights in favour of any person pursuant to the Contracts (Rights of Third Parties) Act 1999.

13. ENTIRE AGREEMENT

This Agreement is the entire agreement between the Grantor and the Operator relating to the Apparatus at the Land.

14. SEVERANCE

Each provision of this Agreement will be construed as a separate provision and if one or more of them is considered illegal, invalid or unenforceable then that provision will be deemed deleted but the enforceability of the remainder of this Agreement will not be affected.

15. NOTICES

15.1. Any notice given under this Agreement must be in writing and signed by or on behalf of the person giving it.[4]

15.2. Any such notice will be deemed to have been given if it is personally delivered or sent by registered, recorded or first class post, and (in each case) addressed:
   a) to the Grantor at b) [insert address]
   c) [marked for the attention of [insert name]]
   d) to the Operator at:

4 In addition, please also have regard to paragraphs 88 and 89 of the Code which apply to notices given by operators and by persons other than operators under paragraphs of the Code.

e) [insert address]
f) [marked for the attention of [insert name]]

15.3. Following the execution of this Agreement, either party may amend its address for the purposes of clause [15.2] by notice to the other party.

15.4. Each party agrees that the address set out in clause [15.2] (as it may be subsequently amended under clause [15.3]) will also constitute their address for service for the purposes of paragraph 87(2)(a) of the Code.

16. MEDIATION

16.1. If any dispute arises in connection with this agreement, the parties may agree to enter into mediation in good faith to settle such a dispute and will do so in accordance with the [Centre for Effective Dispute Resolution ('CEDR') Model Mediation Procedure]. To initiate the mediation a party must give notice in writing ('ADR notice') to the other party to the dispute, referring the dispute to mediation. A copy of the referral should be sent to [CEDR]. Unless otherwise agreed between the parties, the mediator will be nominated by [CEDR] within [14] days of notice of the dispute.

16.2. Unless otherwise agreed, the mediation will start not later than [28] days after the date of the ADR notice. [The commencement of a mediation will not prevent the parties commencing or continuing court proceedings]

16.3. [No party may commence any court proceedings in relation to any dispute arising out of this agreement until it has attempted to settle the dispute by mediation and either the mediation has terminated or the other party has failed to participate in the mediation, provided that the right to issue proceedings is not prejudiced by a delay].

17. GOVERNING LAW AND JURISDICTION

17.1. This agreement is governed by, and shall be construed in accordance with, the laws of England and Wales/Scotland/Northern Ireland.

17.2. [Subject to clause 16] the parties agree to submit to the exclusive jurisdiction of the [courts of England/sheriff in Scotland/county courts of Northern Ireland] as regards any disputes or claims arising out of this Agreement.[5]

Signed for and on behalf of [Grantor] by:

Signature
Name
Date

Signed for an on behalf of [Operator] by:

Signature
Name
Date

5 Please note paragraph *95 of the Code which provides the Secretary of State with the power to confer jurisdiction on tribunals other than the county court in England and Wales, the sheriff court in Scotland and a county court in Northern Ireland.*

SCHEDULE 1
THE APPARATUS
[Insert description of the electronic communications apparatus to be installed]

[SCHEDULE 2
THE LAND]
[Insert plan showing location of the Land]
SCHEDULE [3]
ACCESS ARRANGEMENTS
[To be agreed between the parties – see Schedule B of the ECC Code of Practice]

Index